Complete Solutions Guide to Accompany

CALCULUS

ALTERNATE SIXTH EDITION

Larson / Hostetler / Edwards

Volume I
Chapters 1–7

Dianna L. Zook
Indiana University
Purdue University at Fort Wayne, Indiana

Houghton Mifflin Company Boston New York

Editor in Chief, Mathematics: Charles Hartford
Managing Editor: Cathy Cantin
Senior Associate Editor: Maureen Brooks
Associate Editor: Michael Richards
Assistant Editor: Carolyn Johnson
Supervising Editor: Karen Carter
Art Supervisor: Gary Crespo
Marketing Manager: Sara Whittern
Associate Marketing Manager: Ros Kane
Marketing Assistant: Carrie Lipscomb
Design: Henry Rachlin
Composition and Art: Meridian Creative Group

Printed in the U.S.A.

ISBN: 0-395-88841-7

123456789-MA 01 00 99 98

Preface

This solutions guide is a supplement to *Calculus with Analytic Geometry, Alternate Sixth Edition,* by Roland E. Larson, Robert P. Hostetler, and Bruce H. Edwards. Solutions to every exercise in the text are given with all essential algebraic steps included. There are three volumes in the complete set of solutions guides. Volume I contains Chapters 1–7, Volume II contains Chapters 8–14, and Volume III contains Chapters 15–18. Also available is a one-volume *Study and Solutions Guide to Accompany Calculus, Alternate Sixth Edition,* written by David E. Heyd, which contains worked-out solutions to *selected* representative exercises form the text.

I have made every effort to see that the solutions are correct. However, I would appreciate hearing about any errors or other suggestions for improvement.

I would like to thank the staff at Larson Texts, Inc. for their help in the preparation of this guide. I would also like to thank the students in my mathematics classes. Finally, I would like to thank my husband, Ed Schlindwein, for his support during the many months I have worked on this project.

Dianna L. Zook
Indiana University
Purdue University at Fort Wayne, Indiana 46805

CONTENTS

CHAPTER 1
The Cartesian Plane and Functions

Section 1.1 Real Numbers and the Real Line

1. $0.7 = \frac{7}{10}$
Rational

2. -3678
Rational

3. $\dfrac{3\pi}{2}$
Irrational (since π is irrational)

4. $3\sqrt{2} - 1$
Irrational

5. $4.3451\overline{451}$
Rational

6. $\frac{22}{7}$
Rational

7. $\sqrt[3]{64} = 4$
Rational

8. $0.8177\overline{8177}$
Rational

9. $4\frac{5}{8} = \frac{45}{8}$
Rational

10. $(\sqrt{2})^3 = 2\sqrt{2}$
Irrational

11. Let $x = 0.36\overline{36}$.

$$\begin{aligned} 100x &= 36.36\overline{36} \\ -x &= -0.36\overline{36} \\ \hline 99x &= 36 \\ x &= \tfrac{36}{99} = \tfrac{4}{11} \end{aligned}$$

12. Let $x = 0.3\overline{18}$.

$$\begin{aligned} 1000x &= 318.18\overline{18} \\ -10x &= -3.18\overline{18} \\ \hline 990x &= 315 \\ x &= \tfrac{315}{990} = \tfrac{7}{22} \end{aligned}$$

13. Let $x = 0.297\overline{297}$.

$$\begin{aligned} 1000x &= 297.297\overline{297} \\ -x &= -0.297\overline{297} \\ \hline 999x &= 297 \\ x &= \tfrac{297}{999} = \tfrac{11}{37} \end{aligned}$$

14. Let $x = 0.9900\overline{9900}$.

$$\begin{aligned} 10{,}000x &= 9900.9900\overline{9900} \\ -x &= -0.9900\overline{9900} \\ \hline 9999x &= 9900 \\ x &= \tfrac{9900}{9999} = \tfrac{100}{101} \end{aligned}$$

15. Given $a < b$:

(a) $a + 2 < b + 2$; True

(b) $5b < 5a$; False

(c) $5 - a > 5 - b$; True

(d) $\dfrac{1}{a} < \dfrac{1}{b}$; False

(e) $(a - b)(b - a) > 0$; False

(f) $a^2 < b^2$; False

16. $A = \{x\colon 0 < x\}$, $B = \{x\colon -2 \le x \le 2\}$, $C = \{x\colon x < 1\}$

(a) $A \cup B = \{x\colon -2 \le x\}$ $[-2, \infty)$

(b) $A \cap B = \{x\colon 0 < x \le 2\}$ $(0, 2]$

(c) $B \cap C = \{x\colon -2 \le x < 1\}$ $[-2, 1)$

(d) $A \cup C = \{x\colon x \text{ is a real number}\}$ $(-\infty, \infty)$

(e) $A \cap B \cap C = \{x\colon 0 < x < 1\}$ $(0, 1)$

17.

Interval Notation	Set Notation	Graph
$[-2,\ 0)$	$\{x:\ -2 \le x < 0\}$	
$(-\infty,\ -4]$	$\{x:\ x \le -4\}$	
$[3,\ \frac{11}{2}]$	$\{x:\ 3 \le x \le \frac{11}{2}\}$	
$(-1,\ 7)$	$\{x:\ -1 < x < 7\}$	

18.

Interval Notation	Set Notation	Graph
$[100,\ \infty)$	$\{x:\ x \ge 100\}$	
$(10,\ \infty)$	$\{x:\ 10 < x\}$	
$(\sqrt{2},\ 8]$	$\{x:\ \sqrt{2} < x \le 8\}$	
$(\frac{1}{3},\ \frac{22}{7}]$	$\{x:\ \frac{1}{3} < x \le \frac{22}{7}\}$	

19. $x - 5 \ge 7$

$x \ge 12$

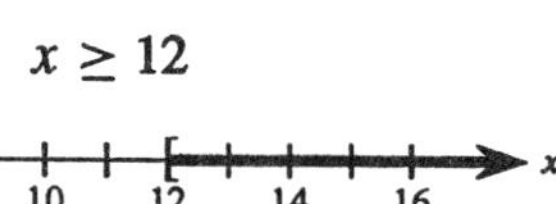

20. $2x > 3$

$x > \frac{3}{2}$

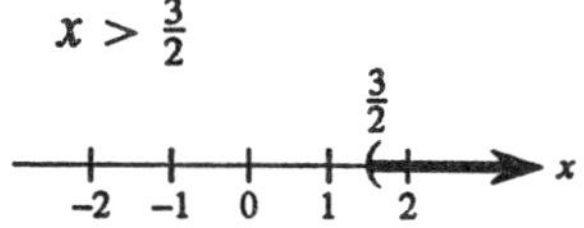

21. $4x + 1 < 2x$

$4x < 2x - 1$

$2x < -1$

$x < -\frac{1}{2}$

22. $2x + 7 < 3$

$2x < -4$

$x < -2$

23. $2x - 1 \ge 0$

$2x \ge 1$

$x \ge \frac{1}{2}$

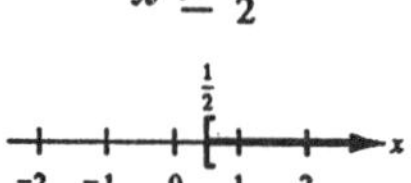

24. $3x + 1 \ge 2x + 2$

$3x \ge 2x + 1$

$x \ge 1$

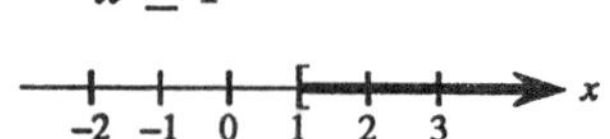

25. $-4 < 2x - 3 < 4$

$-1 < 2x < 7$

$-\frac{1}{2} < x < \frac{7}{2}$

26. $0 \le x + 3 < 5$

$-3 \le x < 2$

27. $\frac{3x}{4} > x + 1$

$-\frac{x}{4} > 1$

$x < -4$

28. $-1 < -x/3 < 1$

$-3 < -x < 3$

$3 > x > -3$

$-3 < x < 3$

29. $\frac{x}{2} + \frac{x}{3} > 5$

$3x + 2x > 30$

$5x > 30$

$x > 6$

30. $x > \frac{1}{x}$

If $x > 0$: $x^2 > 1 \Rightarrow x > 1$

If $x < 0$: $x^2 < 1 \Rightarrow -1 < x < 0$

31. $|x| < 1 \Rightarrow -1 < x < 1$

32. $\frac{x}{2} - \frac{x}{3} > 5$

$3x - 2x > 30$

$x > 30$

33. $\left|\frac{x-3}{2}\right| \ge 5$

$x - 3 \ge 10$ or $x - 3 \le -10$

$x \ge 13$ $\quad$ $x \le -7$

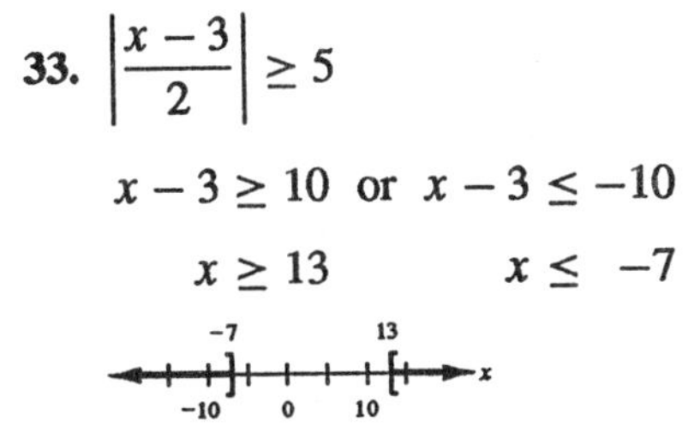

34. $\left|\frac{x}{2}\right| > 3 \Rightarrow x > 6$ or $x < -6$

35. $|x - a| < b$

$-b < x - a < b$

$a - b < x < a + b$

36. $|x + 2| < 5$

$-5 < x + 2 < 5$

$-7 < x < 3$

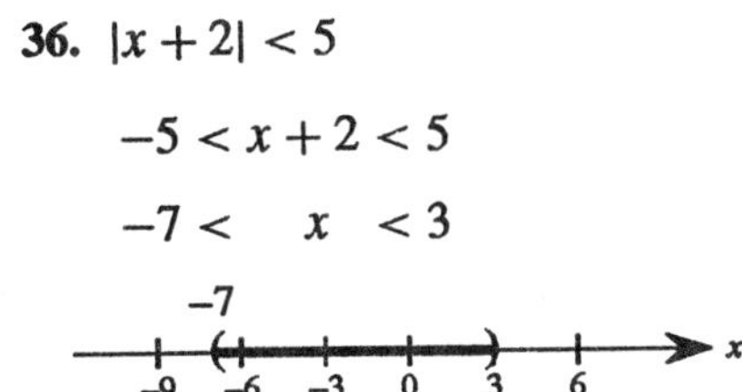

37. $|2x + 1| < 5$

$-5 < 2x + 1 < 5$

$-6 < 2x < 4$

$-3 < x < 2$

38. $|3x + 1| \ge 4$

$3x + 1 \ge 4$ or $3x + 1 \le -4$

$3x \ge 3$ $\quad$ $3x \le -5$

$x \ge 1$ $\quad$ $x \le -\frac{5}{3}$

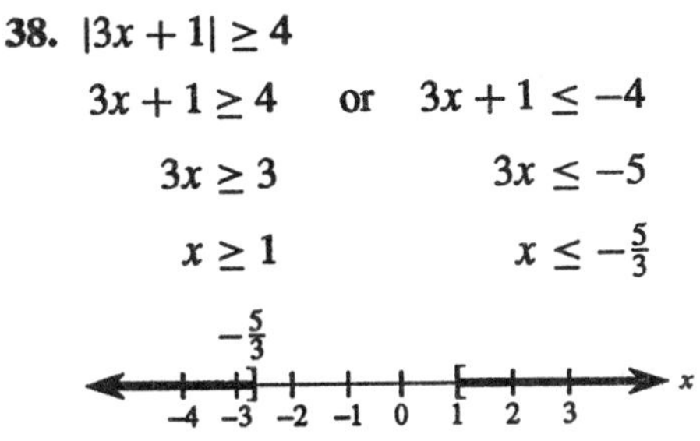

39. $\left|1 - \frac{2x}{3}\right| < 1$

$-1 < 1 - \frac{2x}{3} < 1$

$-2 < -\frac{2x}{3} < 0$

$3 > x > 0$

40. $|9 - 2x| < 1$

$-1 < 9 - 2x < 1$

$-10 < -2x < -8$

$5 > x > 4$

41. $x^2 \le 3 - 2x$

$x^2 + 2x - 3 \le 0$

$(x+3)(x-1) \le 0$

$x = -3$

$x = 1$

Solution: $-3 \le x \le 1$

Test Intervals

$(-)(-) > 0$	$(+)(-) < 0$	$(+)(+) > 0$
$x < -3$	$-3 < x < 1$	$x > 1$

42. $x^4 - x \le 0$

$x(x^3 - 1) \le 0$

$x = 0$

$x = 1$

Solution: $0 \le x \le 1$

Test Intervals

$(-)(-) > 0$	$(+)(-) < 0$	$(+)(+) > 0$
$x < 0$	$0 < x < 1$	$x > 1$

43. $x^2 + x - 1 \le 5$

$x^2 + x - 6 \le 0$

$(x+3)(x-2) \le 0$

$x = -3$

$x = 2$

Solution: $-3 \le x \le 2$

Test Intervals

$(-)(-) > 0$	$(+)(-) < 0$	$(+)(+) > 0$
$x < -3$	$-3 < x < 2$	$x > 2$

44. $2x^2 + 1 < 9x - 3$

$2x^2 - 9x + 4 < 0$

$(2x-1)(x-4) < 0$

$x = \frac{1}{2}$

$x = 4$

Solution: $\frac{1}{2} < x < 4$

Test Intervals

$(-)(-) > 0$	$(+)(-) < 0$	$(+)(+) > 0$
$x < \frac{1}{2}$	$\frac{1}{2} < x < 4$	$x > 4$

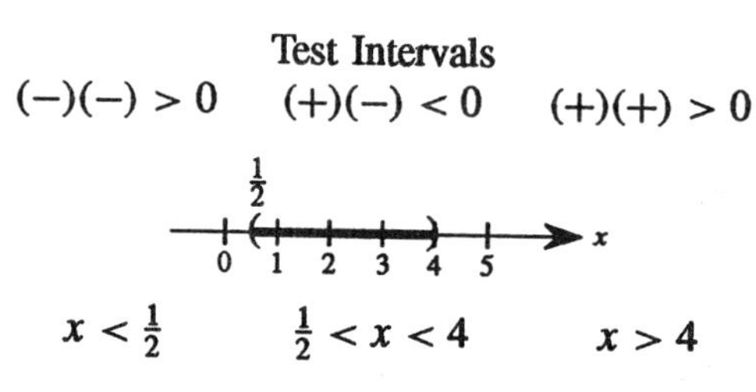

45. $a = -1, \quad b = 3$

Directed distance from a to b: 4

Directed distance from b to a: -4

Distance between a and b: 4

46. $a = -\frac{5}{2}, \quad b = \frac{13}{4}$

Directed distance from a to b: $\frac{23}{4}$

Directed distance from b to a: $-\frac{23}{4}$

Distance between a and b: $\frac{23}{4}$

47. (a) $a = 126, \quad b = 75$

Directed distance from a to b: -51

Directed distance from b to a: 51

Distance between a and b: 51

(b) $a = -126, \quad b = -75$

Directed distance from a to b: 51

Directed distance from b to a: -51

Distance between a and b: 51

48. (a) $a = 9.34, \quad b = -5.65$

Directed distance from a to b: -14.99

Directed distance from b to a: 14.99

Distance between a and b: 14.99

(b) $a = \frac{16}{5}, \quad b = \frac{112}{75}$

Directed distance from a to b: $-\frac{128}{75}$

Directed distance from b to a: $\frac{128}{75}$

Distance between a and b: $\frac{128}{75}$

49. $a = -1, \quad b = 3$

Midpoint: $\dfrac{-1+3}{2} = 1$

50. $a = -5, \quad b = -\dfrac{3}{2}$

Midpoint: $\dfrac{-5 + (-3/2)}{2} = -\dfrac{13}{4}$

51. (a) $[7, 21]$

Midpoint: 14

(b) $[8.6, 11.4]$

Midpoint: 10

52. (a) $[-6.85, 9.35]$

Midpoint: 1.25

(b) $[-4.6, -1.3]$

Midpoint: -2.95

53. $a = -2, \quad b = 2$

Midpoint: 0

Distance between midpoint and each endpoint: 2

$$|x - 0| \le 2$$

$$|x| \le 2$$

54. $a = -3, \quad b = 3$

Midpoint: 0

Distance between midpoint and each endpoint: 3

$$|x - 0| \ge 3$$

$$|x| \ge 3$$

55. $a = 0, \quad b = 4$

Midpoint: 2

Distance between midpoint and each endpoint: 2

$$|x - 2| > 2$$

56. $a = 20, \quad b = 24$

Midpoint: 22

Distance between midpoint and each endpoint: 2

$$|x - 22| \ge 2$$

57. (a) All numbers that are at most 10 units from 12

$$|x - 12| \le 10$$

(b) All numbers that are at least 10 units from 12

$$|x - 12| \ge 10$$

58. (a) y is at most 2 units from a: $|y - a| \le 2$

(b) y is less than δ units from c: $|y - c| < \delta$

59. $A = P + Prt$

$P = 1000, \quad A > 1250, \quad t = 2$

$$1000 + 1000r(2) > 1250$$

$$2000r > 250$$

$$r > 0.1250$$

$$r > 12.5\%$$

60. $R = 115.95x, \quad C = 95x + 750, \quad R > C$

$$115.95x > 95x + 750$$

$$20.95x > 750$$

$$x > 35.7995$$

$$x \ge 36 \text{ units}$$

61. $C = 0.32m + 2300, \quad C < 10{,}000$

$$0.32m + 2300 < 10{,}000$$

$$0.32m < 7700$$

$$m < 24{,}062.5 \text{ miles}$$

62. $\left|\dfrac{h - 68.5}{2.7}\right| \le 1$

$$-1 \le \frac{h - 68.5}{2.7} \le 1$$

$$-2.7 \le h - 68.5 \le 2.7$$

$$65.8'' \le h \le 71.2''$$

63. $\left|\dfrac{x-50}{5}\right| \ge 1.645$

$$\frac{x-50}{5} \le -1.645 \quad \text{or} \quad \frac{x-50}{5} \ge 1.645$$

$$x - 50 \le -8.225 \qquad x - 50 \ge 8.225$$

$$x \le 41.775 \qquad x \ge 58.225$$

$$x \le 41 \qquad x \ge 59$$

64. $|p - 2{,}250{,}000| < 125{,}000$

$$-125{,}000 < p - 2{,}250{,}000 < 125{,}000$$

$$2{,}125{,}000 < p < 2{,}375{,}000$$

High = 2,375,000 barrels

Low = 2,125,000 barrels

65. (a) $\pi \approx 3.1415926535$

$\frac{355}{113} = 3.141592920$

$\frac{355}{113} > \pi$

(b) $\pi \approx 3.1415926535$

$\frac{22}{7} \approx 3.142857143$

$\frac{22}{7} > \pi$

66. (a) $\frac{224}{151} \approx 1.483443709$

$\frac{144}{97} \approx 1.484536082$

$\frac{144}{97} > \frac{224}{151}$

(b) $\frac{73}{81} \approx 0.901234568$

$\frac{6427}{7132} \approx 0.901149748$

$\frac{73}{81} > \frac{6427}{7132}$

67. If $a \ge 0$ and $b \ge 0$, then
$|ab| = ab = |a|\,|b|$.
If $a < 0$ and $b < 0$, then
$|ab| = ab = (-a)(-b) = |a|\,|b|$.
If $a \ge 0$ and $b < 0$, then
$|ab| = -ab = a(-b) = |a|\,|b|$.
If $a < 0$ and $b \ge 0$, then
$|ab| = -ab = (-a)b = |a|\,|b|$.

68. $|a - b| = |(-1)(b - a)|$

$= |-1|\,|b - a| = (1)|b - a| = |b - a|$

69. $\left|\dfrac{a}{b}\right| = \left|a\left(\dfrac{1}{b}\right)\right| = |a|\left|\dfrac{1}{b}\right| = |a| \cdot \dfrac{1}{|b|} = \dfrac{|a|}{|b|}, \quad b \ne 0$

70. If $a \ge 0$, then $|a| = a = \sqrt{a^2}$.
If $a < 0$, then $|a| = -a = \sqrt{(-a)^2} = \sqrt{a^2}$.

71. $n = 1, \quad |a| = |a|$

$n = 2, \quad |a^2| = |a \cdot a| = |a|\,|a| = |a|^2$

$n = 3, \quad |a^3| = |a^2 \cdot a| = |a^2|\,|a| = |a|^2|a| = |a|^3$

$\vdots$

$|a^n| = |a^{n-1}a| = |a^{n-1}|\,|a| = |a|^{n-1}|a| = |a|^n$

72. If $a \ge 0$, then $a = |a|$. Thus, $-|a| \le a \le |a|$.
If $a < 0$, then $a = -|a|$. Thus, $-|a| \le a \le |a|$.

73. $|a| \le k \Leftrightarrow \sqrt{a^2} \le k \Leftrightarrow a^2 \le k^2 \Leftrightarrow a^2 - k^2 \le 0 \Leftrightarrow (a+k)(a-k) \le 0 \Leftrightarrow -k \le a \le k$

74. $k \le |a| \Leftrightarrow k \le \sqrt{a^2} \Leftrightarrow k^2 \le a^2 \Leftrightarrow 0 \le a^2 - k^2 \Leftrightarrow 0 \le (a+k)(a-k) \Leftrightarrow k \le a$ or $a \le -k$

Section 1.2 The Cartesian Plane

1. $d = \sqrt{(4-2)^2 + (5-1)^2}$

$= \sqrt{4+16} = \sqrt{20} = 2\sqrt{5}$

$x = \dfrac{4+2}{2} = 3$

$y = \dfrac{5+1}{2} = 3$

Midpoint: (3, 3)

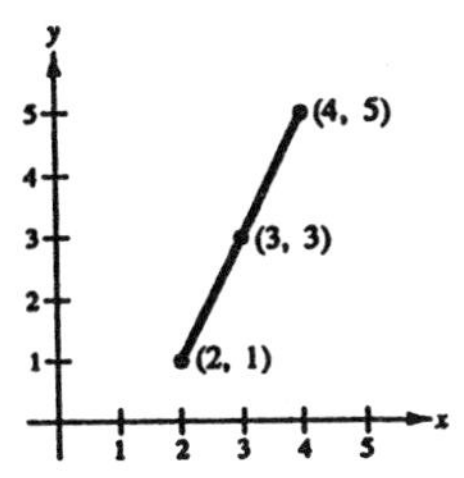

2. $d = \sqrt{(3+3)^2 + (-2-2)^2}$

$= \sqrt{36+16} = 2\sqrt{13}$

$x = \dfrac{-3+3}{2} = 0$

$y = \dfrac{2+(-2)}{2} = 0$

Midpoint: (0, 0)

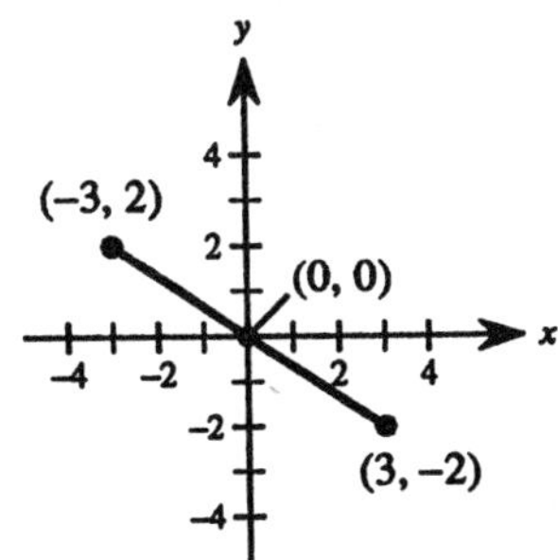

3. $d = \sqrt{\left(\dfrac{1}{2}+\dfrac{3}{2}\right)^2 + (1+5)^2}$

$= \sqrt{4+36} = \sqrt{40} = 2\sqrt{10}$

$x = \dfrac{(-3/2)+(1/2)}{2} = -\dfrac{1}{2}$

$y = \dfrac{-5+1}{2} = -2$

Midpoint: $\left(-\dfrac{1}{2}, -2\right)$

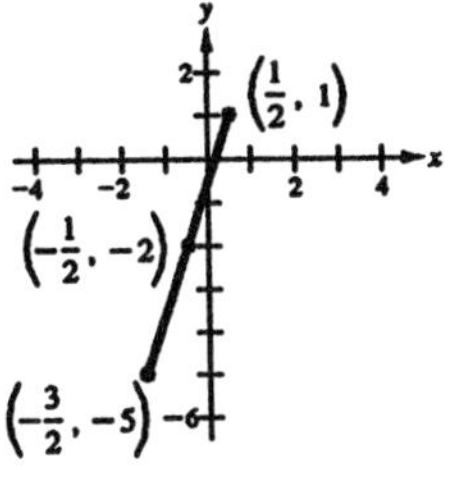

4. $d = \sqrt{\left(\dfrac{5}{6}-\dfrac{4}{6}\right)^2 + \left(\dfrac{3}{3}+\dfrac{1}{3}\right)^2}$

$= \sqrt{\dfrac{1}{36}+\dfrac{64}{36}} = \dfrac{\sqrt{65}}{6}$

$x = \dfrac{(2/3)+(5/6)}{2} = \dfrac{9}{12} = \dfrac{3}{4}$

$y = \dfrac{(-1/3)+1}{2} = \dfrac{1}{3}$

Midpoint: $\left(\dfrac{3}{4}, \dfrac{1}{3}\right)$

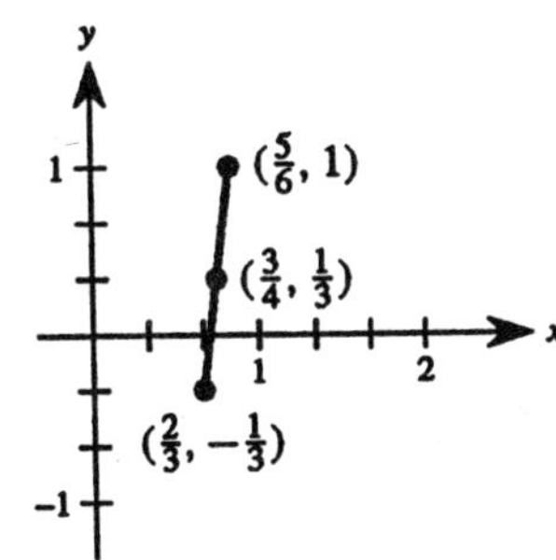

5. $d = \sqrt{(-1 + 1)^2 + (1 - \sqrt{3})^2} = \sqrt{4 + 1 - 2\sqrt{3} + 3} = \sqrt{8 - 2\sqrt{3}}$

$x = \dfrac{-1+1}{2} = 0$

$y = \dfrac{1+\sqrt{3}}{2}$

Midpoint: $\left(0, \dfrac{1+\sqrt{3}}{2}\right)$

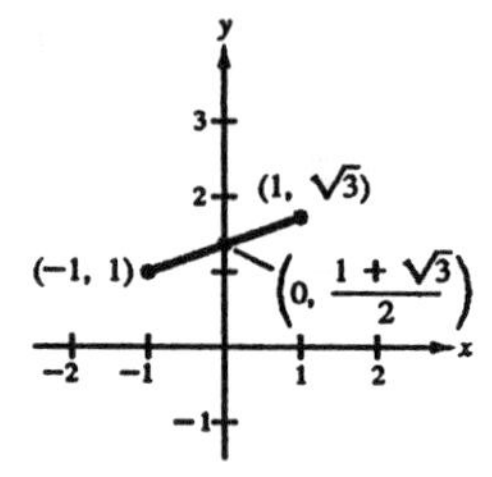

6. $d = \sqrt{(-2+0)^2 + (0 - \sqrt{2})^2} = \sqrt{4+2} = \sqrt{6}$

$x = \dfrac{-2+0}{2} = -1$

$y = \dfrac{0+\sqrt{2}}{2} = \dfrac{\sqrt{2}}{2}$

Midpoint: $\left(-1, \dfrac{\sqrt{2}}{2}\right)$

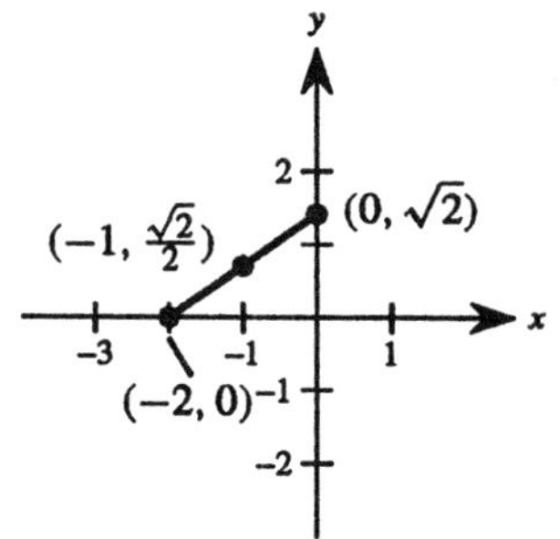

7. $d_1 = \sqrt{9+36} = \sqrt{45}$

$d_2 = \sqrt{4+1} = \sqrt{5}$

$d_3 = \sqrt{25+25} = \sqrt{50}$

$(d_1)^2 + (d_2)^2 = (d_3)^2$

Right triangle

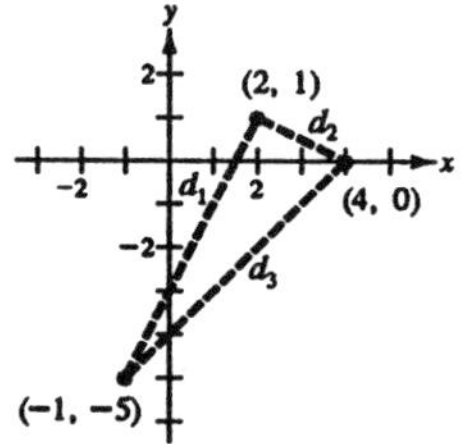

8. $d_1 = \sqrt{9+49} = \sqrt{58}$

$d_2 = \sqrt{25+4} = \sqrt{29}$

$d_3 = \sqrt{4+25} = \sqrt{29}$

$d_2 = d_3$

Isosceles triangle

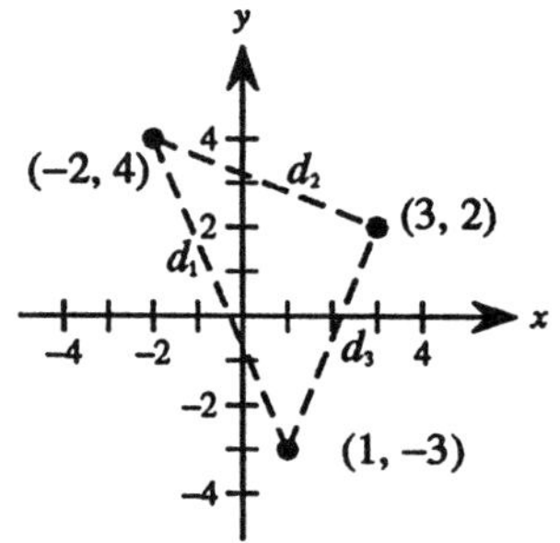

9. $d_1 = d_2 = d_3 = d_4 = \sqrt{5}$

Rhombus

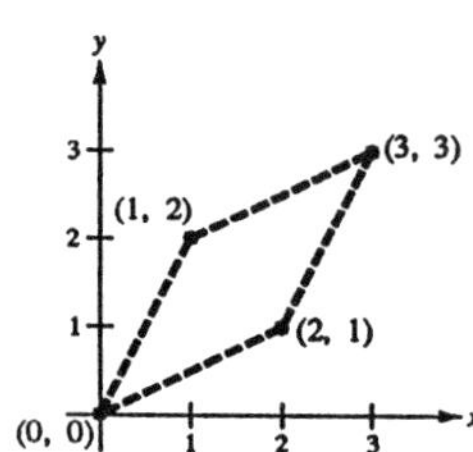

10. $d_1 = \sqrt{9+36} = \sqrt{45} = d_3$

$d_2 = \sqrt{1+9} = \sqrt{10} = d_4$

Parallelogram

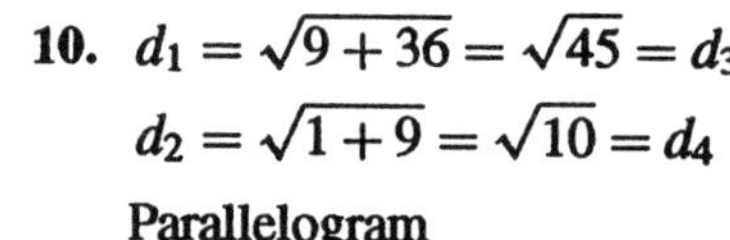

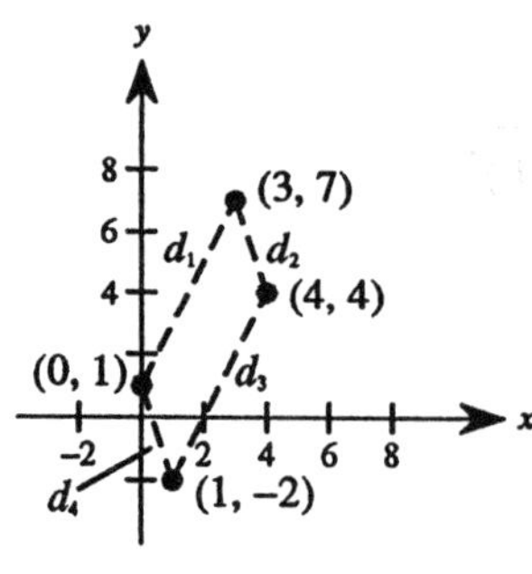

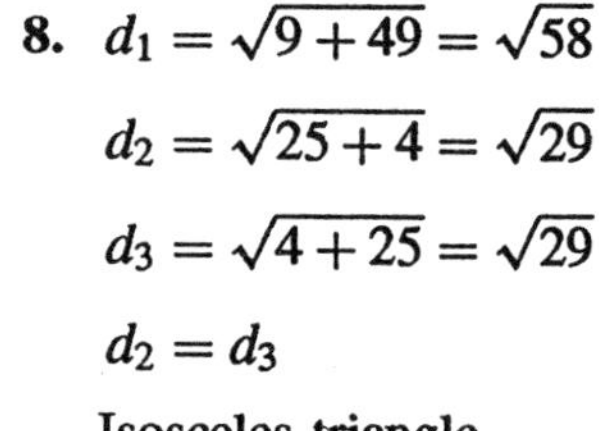

11. $d_1 = \sqrt{4+16} = \sqrt{20} = 2\sqrt{5}$

$d_2 = \sqrt{1+4} = \sqrt{5}$

$d_3 = \sqrt{9+36} = 3\sqrt{5}$

$d_1 + d_2 = d_3$

Collinear

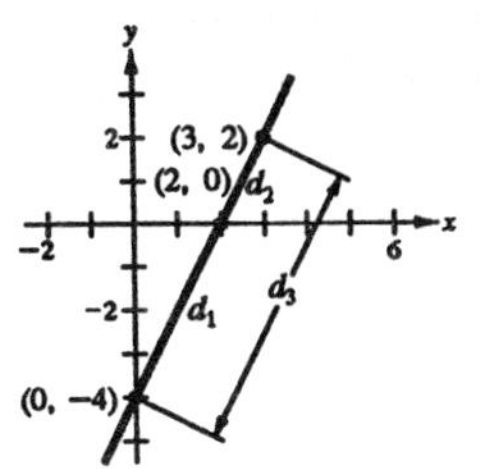

12. $d_1 = \sqrt{49+100} = \sqrt{149} = 12.2066$

$d_2 = \sqrt{25+49} = \sqrt{74} = 8.6023$

$d_3 = \sqrt{144+289} = \sqrt{433} = 20.8087$

$d_1 + d_2 \neq d_3$

Not collinear

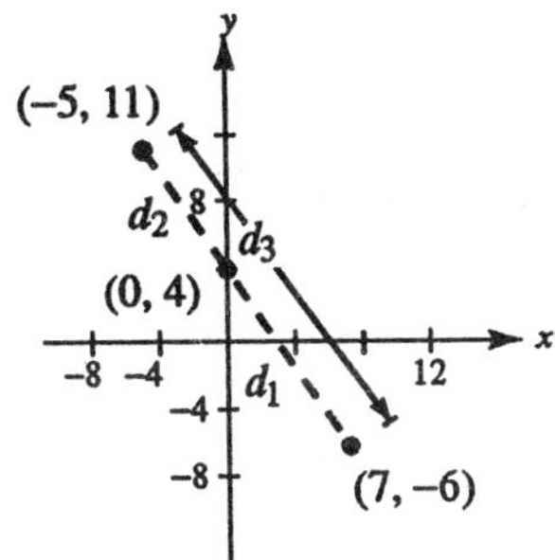

13. $d_1 = \sqrt{1+1} = \sqrt{2}$

$d_2 = \sqrt{9+4} = \sqrt{13}$

$d_3 = \sqrt{16+9} = 5$

$d_1 + d_2 \neq d_3$

Not collinear

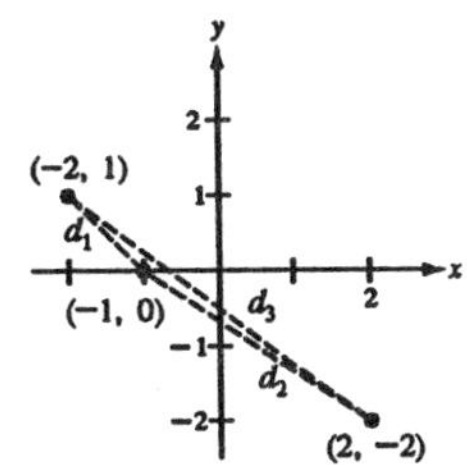

14. $d_1 = \sqrt{16+4} = \sqrt{20} = 2\sqrt{5}$

$d_2 = \sqrt{4+4} = \sqrt{8} = 2\sqrt{2}$

$d_3 = \sqrt{36+16} = \sqrt{52} = 2\sqrt{13}$

$d_1 + d_2 \neq d_3$

Not collinear

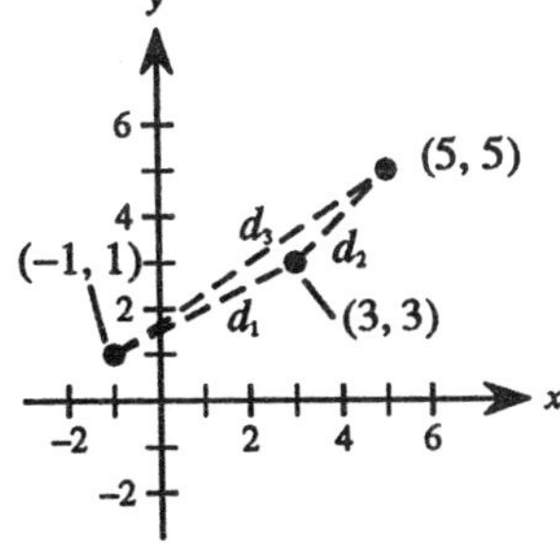

15. $5 = \sqrt{x^2 + 16}$

$25 = x^2 + 16$

$9 = x^2$

$x = \pm 3$

16. $5 = \sqrt{(x-2)^2 + 9}$

$25 = (x-2)^2 + 9$

$16 = (x-2)^2$

$\pm 4 = x - 2$

$x = 2 \pm 4 = -2,\ 6$

17. $8 = \sqrt{(3-0)^2 + (y-0)^2}$

$8 = \sqrt{9 + y^2}$

$64 = 9 + y^2$

$55 = y^2$

$y = \pm\sqrt{55}$

18. $8 = \sqrt{(5-5)^2 + (y-1)^2}$

$8 = \sqrt{(y-1)^2}$

$8 = |y-1|$

$y - 1 = 8$ or $y - 1 = -8$

$y = 9$ $\qquad$ $y = -7$

19.
$$\begin{aligned}\sqrt{(x+2)^2+(y-3)^2} &= \sqrt{(x-4)^2+(y+1)^2}\\ x^2+4x+4+y^2-6y+9 &= x^2-8x+16+y^2+2y+1\\ 12x-8y-4 &= 0\\ 3x-2y-1 &= 0\end{aligned}$$

20.
$$\begin{aligned}\sqrt{(x+7)^2+(y+1)^2} &= \sqrt{(x-3)^2+\left(y-\tfrac{5}{2}\right)^2}\\ x^2+14x+49+y^2+2y+1 &= x^2-6x+9+y^2-5y+\tfrac{25}{4}\\ 20x+7y+50-\tfrac{61}{4} &= 0\\ 80x+28y+200-61 &= 0\\ 80x+28y+139 &= 0\end{aligned}$$

21. The midpoint of the given line segment is $\left(\dfrac{x_1+x_2}{2}, \dfrac{y_1+y_2}{2}\right)$.

The midpoint between (x_1, y_1) and $\left(\dfrac{x_1+x_2}{2}, \dfrac{y_1+y_2}{2}\right)$ is

$$\left(\frac{x_1+(x_1+x_2)/2}{2}, \frac{y_1+(y_1+y_2)/2}{2}\right) = \left(\frac{3x_1+x_2}{4}, \frac{3y_1+y_2}{4}\right).$$

The midpoint between $\left(\dfrac{x_1+x_2}{2}, \dfrac{y_1+y_2}{2}\right)$ and (x_2, y_2) is

$$\left(\frac{(x_1+x_2)/2+x_2}{2}, \frac{(y_1+y_2)/2+y_2}{2}\right) = \left(\frac{x_1+3x_2}{4}, \frac{y_1+3y^2}{4}\right).$$

Thus, the three points are

$$\left(\frac{3x_1+x_2}{4}, \frac{3y_1+y_2}{4}\right), \left(\frac{x_1+x_2}{2}, \frac{y_1+y_2}{2}\right), \left(\frac{x_1+3x_2}{4}, \frac{y_1+3y_2}{4}\right).$$

22. (a)
$$\left(\frac{3(1)+4}{4}, \frac{3(-2)+(-1)}{4}\right) = \left(\frac{7}{4}, -\frac{7}{4}\right)$$
$$\left(\frac{1+4}{2}, \frac{-2+(-1)}{2}\right) = \left(\frac{5}{2}, -\frac{3}{2}\right)$$
$$\left(\frac{1+3(4)}{4}, \frac{-2+3(-1)}{4}\right) = \left(\frac{13}{4}, -\frac{5}{4}\right)$$

(b)
$$\left(\frac{3(-2)+0}{4}, \frac{3(-3)+0}{4}\right) = \left(-\frac{3}{2}, -\frac{9}{4}\right)$$
$$\left(\frac{-2+0}{2}, \frac{-3+0}{2}\right) = \left(-1, -\frac{3}{2}\right)$$
$$\left(\frac{-2+3(0)}{4}, \frac{-3+3(0)}{4}\right) = \left(-\frac{1}{2}, -\frac{3}{4}\right)$$

23. (a)
$$\begin{aligned}x^2+5x &= x^2+5x+\left(\tfrac{5}{2}\right)^2-\left(\tfrac{5}{2}\right)^2\\ &= x^2+5x+\tfrac{25}{4}-\tfrac{25}{4}\\ &= \left(x+\tfrac{5}{2}\right)^2-\tfrac{25}{4}\end{aligned}$$

(b)
$$\begin{aligned}x^2+8x+7 &= x^2+8x+\left(\tfrac{8}{2}\right)^2-\left(\tfrac{8}{2}\right)^2+7\\ &= x^2+8x+16-16+7\\ &= (x+4)^2-9\end{aligned}$$

24. (a)
$$\begin{aligned}4x^2-4x-39 &= 4\left[x^2-x-\tfrac{39}{4}\right]\\ &= 4\left[x^2-x+\tfrac{1}{4}-\tfrac{1}{4}-\tfrac{39}{4}\right]\\ &= 4\left[\left(x-\tfrac{1}{2}\right)^2-10\right]\\ &= 4\left(x-\tfrac{1}{2}\right)^2-40\end{aligned}$$

(b)
$$\begin{aligned}5x^2+x &= 5\left[x^2+\tfrac{1}{5}x\right]\\ &= 5\left[x^2+\tfrac{1}{5}x+\tfrac{1}{100}-\tfrac{1}{100}\right]\\ &= 5\left[\left(x+\tfrac{1}{10}\right)^2-\tfrac{1}{100}\right]\\ &= 5\left(x+\tfrac{1}{10}\right)^2-\tfrac{1}{20}\end{aligned}$$

25. Center: (0, 0)
Radius: 1
Matches graph (d)

26. Center: (1, 3)
Radius: 2
Matches graph (c)

27. Center: (1, 0)
Radius: 0
Matches graph (b)

28. Center: $\left(-\frac{1}{2}, \frac{3}{4}\right)$
Radius: $\frac{1}{2}$
Matches graph (e)

29. Center: (−3, 1)
Radius: 4
Matches graph (a)

30. Center: (0, 1)
Radius: 1
Matches graph (f)

31. $$(x-0)^2+(y-0)^2=(3)^2$$
$$x^2+y^2-9=0$$

32. $$(x-0)^2+(y-0)^2=(5)^2$$
$$x^2+y^2-25=0$$

33. $$(x-2)^2+(y+1)^2=(4)^2$$
$$x^2+y^2-4x+2y-11=0$$

34. $$(x+4)^2+(y-3)^2=\left(\tfrac{5}{8}\right)^2$$
$$64(x+4)^2+64(y-3)^2=25$$
$$64x^2+64y^2+512x-384y+1575=0$$

35. $$\text{Radius}=\sqrt{(-1-0)^2+(2-0)^2}=\sqrt{5}$$
$$(x+1)^2+(y-2)^2=5$$
$$x^2+2x+1+y^2-4y+4=5$$
$$x^2+y^2+2x-4y=0$$

36. $$\text{Radius}=\sqrt{[3-(-1)]^2+(-2-1)^2}=5$$
$$(x-3)^2+(y+2)^2=25$$
$$x^2-6x+9+y^2+4y+4=25$$
$$x^2+y^2-6x+4y-12=0$$

37. Center: (3, 2)
Radius: $\sqrt{10}$
$$(x-3)^2+(y-2)^2=10$$
$$x^2-6x+9+y^2-4y+4=10$$
$$x^2+y^2-6x-4y+3=0$$

38. Center: (0, 0)
Radius: $\sqrt{2}$
$$(x-0)^2+(y-0)^2=(\sqrt{2})^2$$
$$x^2+y^2-2=0$$

39. (0, 0), (0, 8), (6, 0)

$(0-h)^2+(0-k)^2=r^2, \quad h^2+k^2=r^2$ Equation 1
$(0-h)^2+(8-k)^2=r^2, \quad h^2+(8-k)^2=r^2$ Equation 2
$(6-h)^2+(0-k)^2=r^2, \quad (6-h)^2+k^2=r^2$ Equation 3

$(8-k)^2-k^2=0, \quad 64-16k=0, \quad k=4$ Equation 2 − Equation 1
$(6-h)^2-h^2=0, \quad 36-12h=0, \quad h=3$ Equation 3 − Equation 1
$(3)^2+(4)^2=r^2, \quad 25=r^2$ Equation 1

$$(x-3)^2+(y-4)^2=25$$
$$x^2+y^2-6x-8y=0$$

Alternate Solution

Note that the given points form the vertices of a right triangle. Thus, the hypotenuse of the triangle must lie on a diameter of the circle and the center must be the midpoint of the line segment joining (0, 8) and (6, 0).

$$(h, k)=(3, 4)$$
$$r=\sqrt{(3-6)^2+(4-0)^2}=\sqrt{9+16}=5$$
$$(x-3)^2+(y-4)^2=25$$
$$x^2-6x+9+y^2-8y+16=25$$
$$x^2+y^2-6x-8y=0$$

40. $(1,\ -1),\ (2,\ -2),\ (0,\ -2)$

$(1-h)^2+(-1-k)^2=r^2$ Equation 1
$(2-h)^2+(-2-k)^2=r^2$ Equation 2
$h^2+(-2-k)^2=r^2$ Equation 3

$(2-h)^2-h^2=0,\quad 4-4h=0,\quad h=1$ Equation 2 − Equation 3
$(-1-k)^2=r^2$ Equation 1
$(2-1)^2+(-2-k)^2=r^2$ Equation 2

$(-1-k)^2=1+(-2-k)^2$

$1+2k+k^2=1+4+4k+k^2,\quad -4=2k,\quad k=-2$

$(1-1)^2+(-1+2)^2=r^2,\quad 1=r^2$

$(x-1)^2+(y+2)^2=1,\quad x^2+y^2-2x+4y+4=0$

41.

$$x^2+y^2-2x+6y=-6$$
$$(x^2-2x+1)+(y^2+6y+9)=-6+1+9$$
$$(x-1)^2+(y+3)^2=4$$

Center: $(1,\ -3)$
Radius: 2

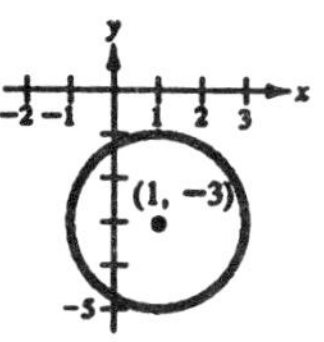

42.

$$x^2+y^2-2x+6y=15$$
$$(x^2-2x+1)+(y^2+6y+9)=15+1+9$$
$$(x-1)^2+(y+3)^2=25$$

Center: $(1,\ -3)$
Radius: 5

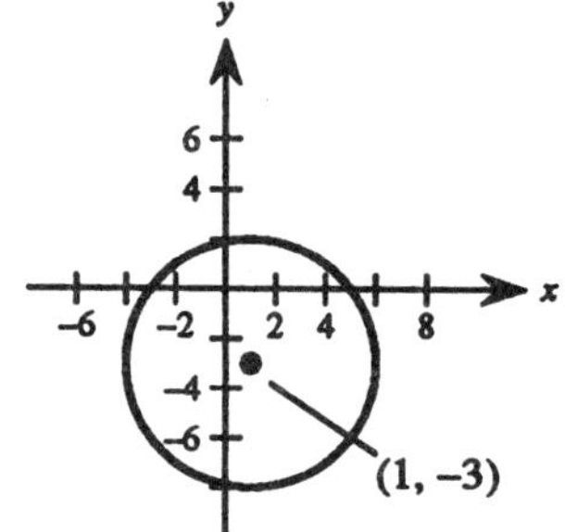

43.

$$x^2+y^2-2x+6y=-10$$
$$(x^2-2x+1)+(y^2+6y+9)=-10+1+9$$
$$(x-1)^2+(y+3)^2=0$$

Only a point $(1,\ -3)$

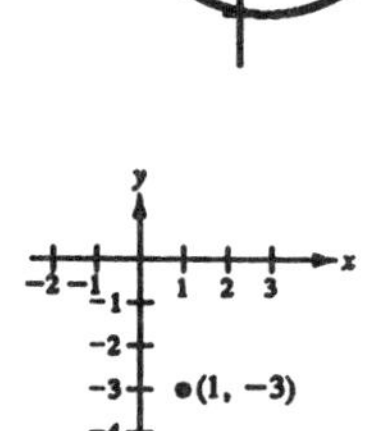

44.

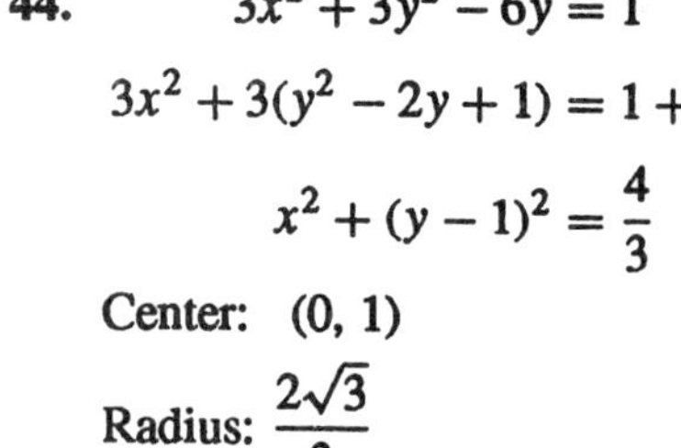

$$3x^2+3y^2-6y=1$$
$$3x^2+3(y^2-2y+1)=1+3$$
$$x^2+(y-1)^2=\frac{4}{3}$$

Center: $(0,\ 1)$

Radius: $\dfrac{2\sqrt{3}}{3}$

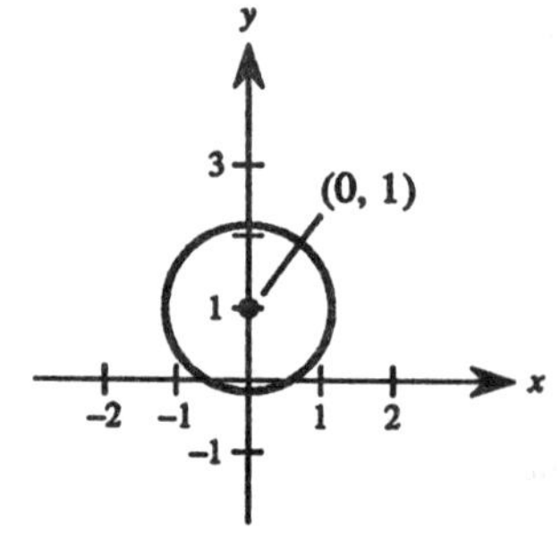

45.

$$2x^2 + 2y^2 - 2x - 2y - 3 = 0$$

$$2(x^2 - x + \tfrac{1}{4}) + 2(y^2 - y + \tfrac{1}{4}) = 3 + \tfrac{1}{2} + \tfrac{1}{2}$$

$$\left(x - \tfrac{1}{2}\right)^2 + \left(y - \tfrac{1}{2}\right)^2 = 2$$

Center: $\left(\frac{1}{2}, \frac{1}{2}\right)$

Radius: $\sqrt{2}$

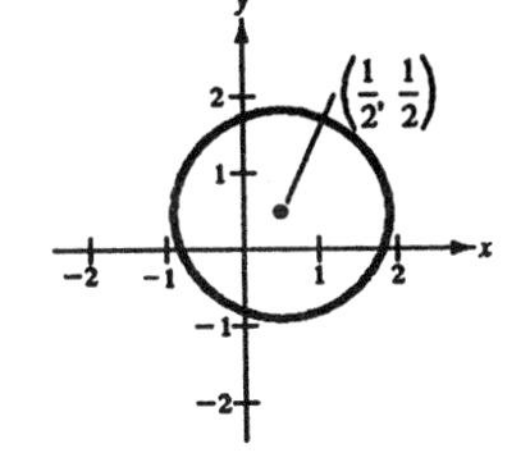

46.

$$4x^2 + 4y^2 - 4x + 2y - 1 = 0$$

$$4\left(x^2 - x + \frac{1}{4}\right) + 4\left(y^2 + \frac{y}{2} + \frac{1}{16}\right) = 1 + 1 + \frac{1}{4}$$

$$4\left(x - \frac{1}{2}\right)^2 + 4\left(y + \frac{1}{4}\right)^2 = \frac{9}{4}$$

$$\left(x - \frac{1}{2}\right)^2 + \left(y + \frac{1}{4}\right)^2 = \frac{9}{16}$$

Center: $\left(\frac{1}{2}, -\frac{1}{4}\right)$

Radius: $\frac{3}{4}$

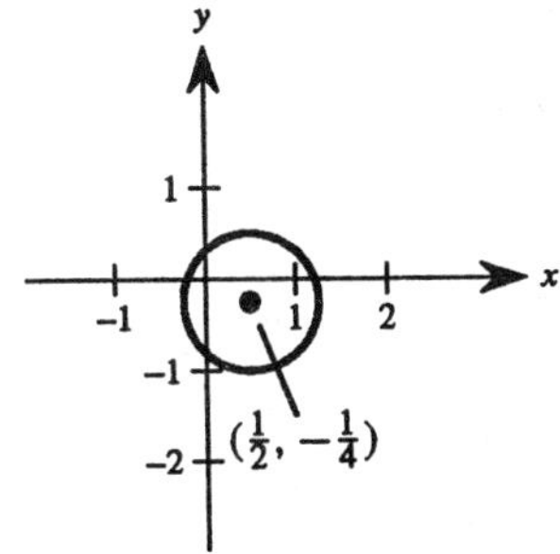

47.

$$16x^2 + 16y^2 + 16x + 40y - 7 = 0$$

$$16\left(x^2 + x + \frac{1}{4}\right) + 16\left(y^2 + \frac{5y}{2} + \frac{25}{16}\right) = 7 + 4 + 25$$

$$16\left(x + \frac{1}{2}\right)^2 + 16\left(y + \frac{5}{4}\right)^2 = 36$$

$$\left(x + \frac{1}{2}\right)^2 + \left(y + \frac{5}{4}\right)^2 = \frac{9}{4}$$

Center: $\left(-\frac{1}{2}, -\frac{5}{4}\right)$

Radius: $\frac{3}{2}$

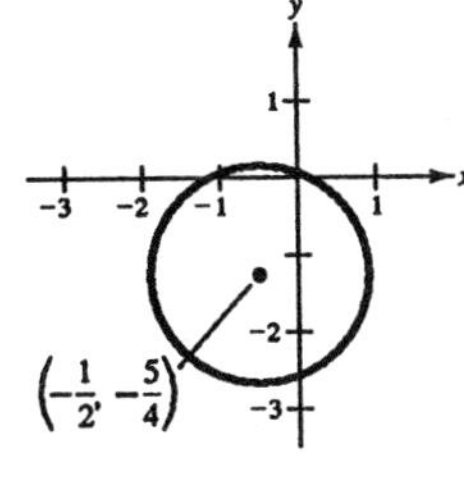

48.

$$x^2 + y^2 - 4x + 2y + 3 = 0$$

$$(x^2 - 4x + 4) + (y^2 + 2y + 1) = -3 + 4 + 1$$

$$(x - 2)^2 + (y + 1)^2 = 2$$

Center: $(2, -1)$

Radius: $\sqrt{2}$

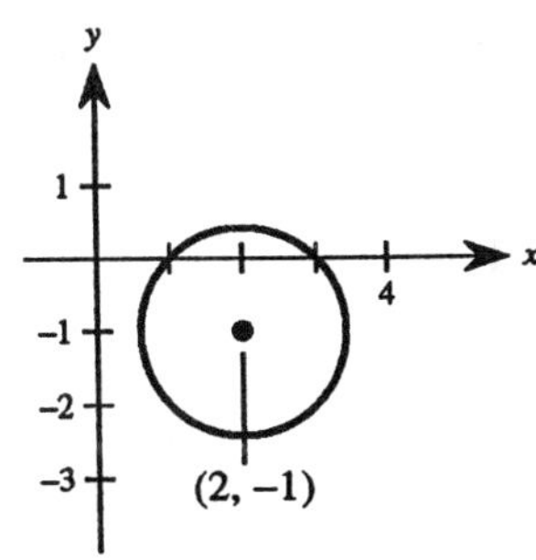

49. Assume the center of the earth is $(0, 0)$.

$$x^2 + y^2 = (22{,}000 + 4000)^2$$

$$x^2 + y^2 = 26{,}000^2$$

50. Substitution of the points (1, 2), (−1, 2), and (2, 1) into the equation $x^2 + y^2 + Cx + Dy + E = 0$ yields the following system of equations.

(Let $x = 1$ and $y = 2$.) $\quad C + 2D + E = -5$
(Let $x = -1$ and $y = 2$.) $\quad -C + 2D + E = -5$
(Let $x = 2$ and $y = 1$.) $\quad 2C + D + E = -5$

The simultaneous solution is $C = 0$, $D = 0$, and $E = -5$. Thus, the equation of the circle is $x^2 + y^2 = 5$.

51. The given points form the vertices of a right triangle. From geometry, the hypotenuse of the triangle must lie on a diameter of the circle. Thus, the center must be the midpoint of the line segment joining (4, 3) and (−2, −5).

$$(h,\ k) = \left[\frac{4-2}{2},\ \frac{3-5}{2}\right] = (1,\ -1)$$

$$r = \sqrt{(4-1)^2 + [3-(-1)]^2} = 5$$

$$(x-1)^2 + [y-(-1)]^2 = 5^2$$

$$(x-1)^2 + (y+1)^2 = 25$$

52. Substitution of the points (4, 1) and (6, 3) into the equation $(x-h)^2 + (y-k)^2 = 10$ yields the following system of equations.

(Let $x = 4$ and $y = 1$.) $\quad (4-h)^2 + (1-k)^2 = 10$
(Let $x = 6$ and $y = 3$.) $\quad (6-h)^2 + (3-k)^2 = 10$

Solving these two equations for h and k yields $h = 7$ and $k = 0$ or $h = 3$ and $k = 4$. Therefore, there are two possible solutions.

$$(x-7)^2 + (y-0)^2 = 10 \quad \text{and} \quad (x-3)^2 + (y-4)^2 = 10$$

$$x^2 + y^2 - 14x + 39 = 0 \qquad x^2 + y^2 - 6x - 8y + 15 = 0$$

53.

$$x^2 + y^2 - 4x + 2y + 1 \le 0$$

$$(x^2 - 4x + 4) + (y^2 + 2y + 1) \le -1 + 4 + 1$$

$$(x-2)^2 + (y+1)^2 \le 4$$

Center: (2, −1)
Radius: 2

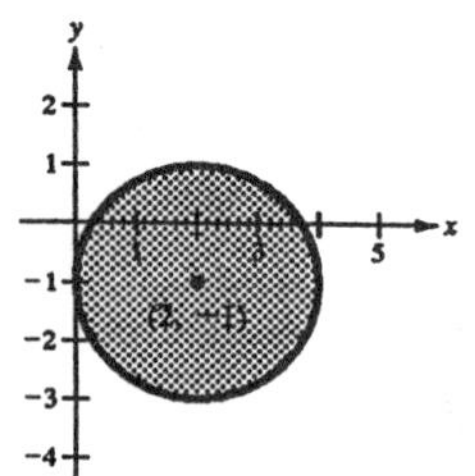

54.

$$x^2 + y^2 - 4x + 2y + 1 > 0$$

$$(x^2 - 4x + 4) + (y^2 + 2y + 1) > -1 + 4 + 1$$

$$(x-2)^2 + (y+1)^2 > 4$$

Center: (2, −1)
Radius: 2

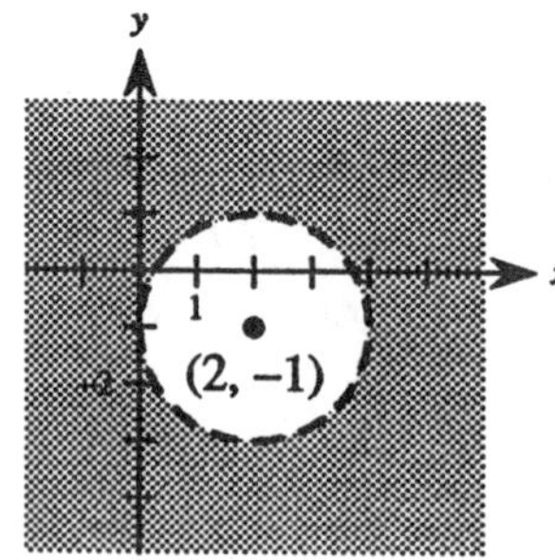

55. $(x+3)^2+(y-1)^2<9$
Center: $(-3, 1)$
Radius: 3

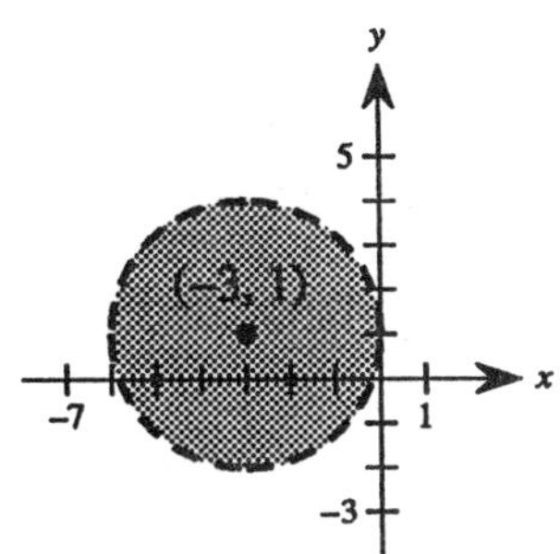

56. $(x-1)^2+\left(y-\frac{1}{2}\right)^2>1$
Center: $\left(1, \frac{1}{2}\right)$
Radius: 1

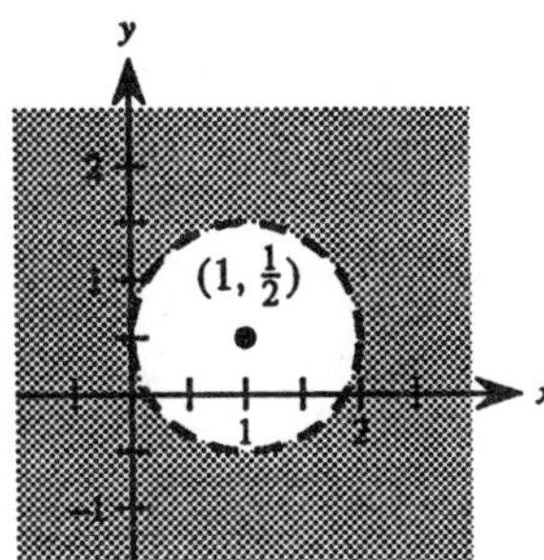

57. The distance between (x_1, y_1) and $\left(\dfrac{2x_1+x_2}{3}, \dfrac{2y_1+y_2}{3}\right)$ is

$$d=\sqrt{\left(x_1-\frac{2x_1+x_2}{3}\right)^2+\left(y_1-\frac{2y_1+y_2}{3}\right)^2}$$

$$=\sqrt{\left(\frac{x_1-x_2}{3}\right)^2+\left(\frac{y_1-y_2}{3}\right)^2}=\sqrt{\frac{1}{9}[(x_1-x_2)^2+(y_1-y_2)^2}=\frac{1}{3}\sqrt{(x_1-x_2)^2+(y_1-y_2)^2}$$

which is $\frac{1}{3}$ of the distance between (x_1, y_1) and (x_2, y_2).

$$\left(\frac{\left(\frac{2x_1+x_2}{3}\right)+x_2}{2}, \frac{\left(\frac{2y_1+y_2}{3}\right)+y_2}{2}\right)=\left(\frac{x_1+2x_2}{3}, \frac{y_1+2y_2}{3}\right)$$

is the second point of trisection.

58. (a) $\left(\dfrac{2(1)+4}{3}, \dfrac{2(-2)+1}{3}\right)=(2, -1)$

$\left(\dfrac{1+2(4)}{3}, \dfrac{-2+2(1)}{3}\right)=(3, 0)$

(b) $\left(\dfrac{2(-2)+0}{3}, \dfrac{2(-3)+0}{3}\right)=\left(-\dfrac{4}{3}, -2\right)$

$\left(\dfrac{-2+2(0)}{3}, \dfrac{-3+2(0)}{3}\right)=\left(-\dfrac{2}{3}, -1\right)$

59. Let one vertex be at $(0, 0)$ and another at $(a, 0)$.

Midpoint of $(0, 0)$ and (d, e) is $\left(\dfrac{d}{2}, \dfrac{e}{2}\right)$.

Midpoint of (b, c) and $(a, 0)$ is $\left(\dfrac{a+b}{2}, \dfrac{c}{2}\right)$.

Midpoint of $(0, 0)$ and $(a, 0)$ is $\left(\dfrac{a}{2}, 0\right)$.

Midpoint of (b, c) and (d, e) is $\left(\dfrac{b+d}{2}, \dfrac{c+e}{2}\right)$.

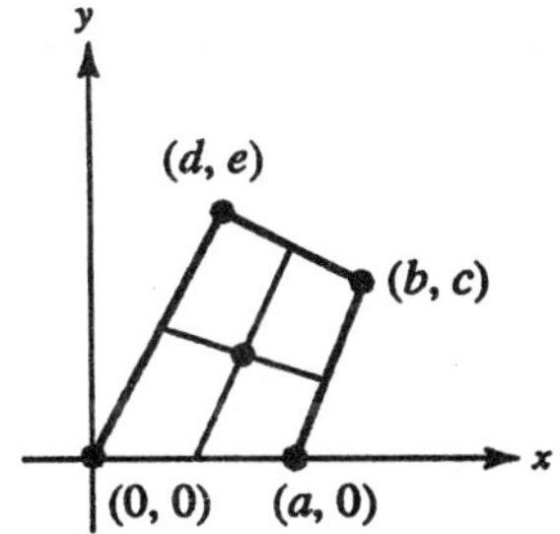

Midpoint of line segment joining $\left(\dfrac{d}{2}, \dfrac{e}{2}\right)$ and $\left(\dfrac{a+b}{2}, \dfrac{c}{2}\right)$ is $\left(\dfrac{a+b+d}{4}, \dfrac{c+e}{4}\right)$.

Midpoint of line segment joining $\left(\dfrac{a}{2}, 0\right)$ and $\left(\dfrac{b+d}{2}, \dfrac{c+e}{2}\right)$ is $\left(\dfrac{a+b+d}{4}, \dfrac{c+e}{4}\right)$.

Therefore the line segments intersect at their midpoints.

60. Let the three vertices be $(0, 0)$, $(a, 0)$, and $(0, b)$.

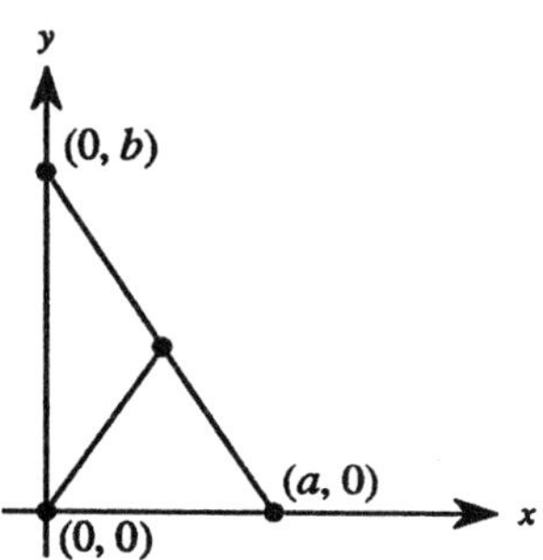

Then the midpoint of the hypotenuse is $\left(\frac{a}{2}, \frac{b}{2}\right)$.

The distance between $\left(\frac{a}{2}, \frac{b}{2}\right)$ and each of the vertices is

$$d = \frac{1}{2}\sqrt{a^2 + b^2}.$$

61. For simplicity, we assume the semicircle is centered at the origin with radius of r. Since (x, y) lies on the semicircle, it must satisfy the equation

$$x^2 + y^2 = r^2$$
$$y^2 = r^2 - x^2$$
$$y = \sqrt{r^2 - x^2}.$$

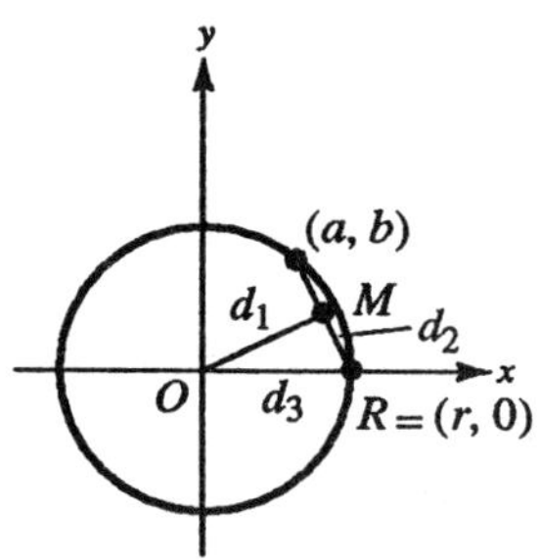

The slope of the line containing (x, y) and $(r, 0)$ is

$$m_1 = \frac{0 - y}{r - x} = \frac{-\sqrt{r^2 - x^2}}{r - x} = \frac{-\sqrt{(r - x)(r + x)}}{r - x} = \frac{-\sqrt{r + x}}{\sqrt{r - x}}.$$

The slope of the line containing (x, y) and $(-r, 0)$ is

$$m_2 = \frac{0 - y}{-r - x} = \frac{-\sqrt{r^2 - x^2}}{-(r + x)} = \frac{\sqrt{(r - x)(r + x)}}{r + x} = \frac{\sqrt{r - x}}{\sqrt{r + x}}.$$

Since $m_1 = -1/m_2$, the lines joining $(r, 0)$ and $(-r, 0)$ to (x, y) must be perpendicular. Hence, the angle θ is a right angle.

62. Assume that the center of the circle is $(0, 0)$ and that one end of the chord is $(r, 0)$ and the other end is (a, b). Then, the midpoint of the chord is $[(a + r)/2, b/2]$. The slope of the line segment joining $(0, 0)$ and $[(a + r)/2, b/2]$ is

$$m_1 = \frac{b}{a + r} = \frac{\sqrt{r^2 - a^2}}{r + a} = \frac{\sqrt{(r - a)(r + a)}}{r + a} = \sqrt{\frac{r - a}{r + a}}.$$

The slope of the line segment joining $(r, 0)$ and (a, b) is

$$m_2 = \frac{b}{a - r} = \frac{\sqrt{r^2 - a^2}}{-(r - a)} = \frac{\sqrt{(r + a)(r - a)}}{-(r - a)} = -\sqrt{\frac{r + a}{r - a}}.$$

Since $m_1 = -\dfrac{1}{m_2}$, the lines are perpendicular.

Therefore, the line joining $(0, 0)$ and $[(a + r)/2, b/2]$ is the perpendicular bisection of the chord and passes through the center of the circle.

63. To show that $[(x_1 + x_2)/2,\ (y_1 + y_2)/2]$ is the midpoint of the line segment joining $(x_1,\ y_1)$ and $(x_2,\ y_2)$ we must show that $d_1 = d_2$ and $d_1 + d_2 = d_3$ (see graph).

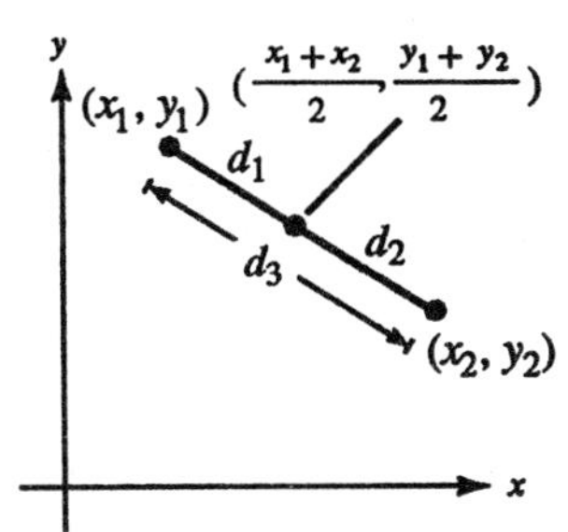

$$d_1 = \sqrt{\left(\frac{x_1 + x_2}{2} - x_1\right)^2 + \left(\frac{y_1 + y_2}{2} - y_1\right)^2}$$

$$= \sqrt{\left(\frac{x_2 - x_1}{2}\right)^2 + \left(\frac{y_2 - y_1}{2}\right)^2} = \frac{1}{2}\sqrt{(x_2 - x_1)^2(y_2 - y_1)^2}$$

$$d_2 = \sqrt{\left(x_2 - \frac{x_1 + x_2}{2}\right)^2 + \left(y_2 - \frac{y_1 - y_2}{2}\right)^2}$$

$$= \sqrt{\left(\frac{x_2 - x_1}{2}\right)^2 + \left(\frac{y_2 - y_1}{2}\right)^2} = \frac{1}{2}\sqrt{(x_2 - x_1)^2 + (y_2 - y_1)^2}$$

$$d_3 = \sqrt{(x_2 - x_1)^2 + (y^2 - y_1)^2}$$

Therefore, $d_1 = d_2$ and $d_1 + d_2 = d_3$.

Section 1.3 Graphs of Equations

1. $y = x - 2$

   x-intercept: $(2, 0)$

   y-intercept: $(0,\ -2)$

   Matches graph (c)

2. $y = -\frac{1}{2}x + 2$

   x-intercept: $(4, 0)$

   y-intercept: $(0, 2)$

   Matches graph (d)

3. $y = x^2 + 2x$

   x-intercepts: $(0,\ 0),\ (-2,\ 0)$

   y-intercept: $(0, 0)$

   Matches graph (b)

4. $y = \sqrt{9 - x^2}$

   x-intercepts: $(-3,\ 0),\ (3,\ 0)$

   y-intercept: $(0, 3)$

   Matches graph (f)

5. $y = 4 - x^2$

   x-intercepts: $(2,\ 0),\ (-2,\ 0)$

   y-intercept: $(0, 4)$

   Matches graph (a)

6. $y = x^3 - x$

   x-intercepts: $(0,\ 0),\ (-1,\ 0),\ (1,\ 0)$

   y-intercept: $(0, 0)$

   Matches graph (e)

7. $y = 2x - 3$

   y-intercept: $y = 2(0) - 3 = -3;\ (0,\ -3)$

   x-intercept: $0 = 2x - 3$

   $x = \frac{3}{2};\ \left(\frac{3}{2},\ 0\right)$

8. $y = (x - 1)(x - 3)$

   y-intercept: $y = (0 - 1)(0 - 3)$

   $y = 3;\ (0,\ 3)$

   x-intercepts: $0 = (x - 1)(x - 3)$

   $x = 1,\ 3;\ (1,\ 0),\ (3,\ 0)$

9. $y = x^2 + x - 2$

y-intercept: $y = 0^2 + 0 - 2$

$y = -2$; $(0, -2)$

x-intercepts: $0 = x^2 + x - 2$

$0 = (x + 2)(x - 1)$

$x = -2, 1$; $(-2, 0)$, $(1, 0)$

10. $y^2 = x^3 - 4x$

y-intercept: $y^2 = 0^3 - 4(0)$

$y = 0$; $(0, 0)$

x-intercepts: $0 = x^3 - 4x$

$0 = x(x - 2)(x + 2)$

$x = 0, \pm 2$; $(0, 0)$, $(\pm 2, 0)$

11. $y = x^2\sqrt{9 - x^2}$

y-intercept: $y = 0^2\sqrt{9 - 0^2}$

$y = 0$; $(0, 0)$

x-intercepts: $0 = x^2\sqrt{9 - x^2}$

$0 = x^2\sqrt{(3 - x)(3 + x)}$

$x = 0, \pm 3$; $(0, 0)$, $(\pm 3, 0)$

12. $xy = 4$

No intercepts

13. $y = \dfrac{x - 1}{x - 2}$

y-intercept: $y = \dfrac{0 - 1}{0 - 2}$

$y = \dfrac{1}{2}$; $\left(0, \dfrac{1}{2}\right)$

x-intercept: $0 = \dfrac{x - 1}{x - 2}$

$x = 1$; $(1, 0)$

14. $y = \dfrac{x^2 + 3x}{(3x + 1)^2}$

y-intercept: $y = \dfrac{0^2 + 3(0)}{[3(0) + 1]^2}$

$y = 0$; $(0, 0)$

x-intercepts: $0 = \dfrac{x^2 + 3x}{(3x + 1)^2}$

$0 = \dfrac{x(x + 3)}{(3x + 1)^2}$

$x = 0, -3$; $(0, 0)$, $(-3, 0)$

15. $x^2y - x^2 + 4y = 0$

y-intercept:

$0^2(y) - 0^2 + 4y = 0$

$y = 0$; $(0, 0)$

x-intercept:

$x^2(0) - x^2 + 4(0) = 0$

$x = 0$; $(0, 0)$

16. $y = 2x - \sqrt{x^2 + 1}$

y-intercept: $y = 2(0) - \sqrt{0^2 + 1}$

$y = -1$; $(0, -1)$

x-intercepts: $0 = 2x - \sqrt{x^2 + 1}$

$2x = \sqrt{x^2 + 1}$

$4x^2 = x^2 + 1$

$3x^2 = 1$

$x^2 = \dfrac{1}{3}$

$x = \pm\dfrac{\sqrt{3}}{3}$

$x = \dfrac{\sqrt{3}}{3}$; $\left(\dfrac{\sqrt{3}}{3}, 0\right)$

Note: $x = -\sqrt{3}/3$ is an extraneous solution.

17. Symmetric with respect to the y-axis since

$$y = (-x)^2 - 2 = x^2 - 2.$$

18. Symmetric with respect to the y-axis since

$$\begin{aligned} y &= (-x)^4 - (-x)^2 + 3 \\ &= x^4 - x^2 + 3. \end{aligned}$$

19. Symmetric with respect to the y-axis since

$$\begin{aligned} (-x)^2y - (-x)^2 + 4y &= 0 \\ x^2y - x^2 + 4y &= 0. \end{aligned}$$

20. Symmetric with respect to the origin since

$$\begin{aligned} (-x)(-y) - \sqrt{4-(-x)^2} &= 0 \\ xy - \sqrt{4-x^2} &= 0. \end{aligned}$$

21. Symmetric with respect to the x-axis since

$$(-y)^2 = y^2 = x^3 - 4x.$$

22. Symmetric with respect to the x-axis since

$$x(-y)^2 = xy^2 = -10.$$

23. Symmetric with respect to the origin since

$$\begin{aligned} (-y) &= (-x)^3 + (-x) \\ -y &= -x^3 - x \\ y &= x^3 + x. \end{aligned}$$

24. Symmetric with respect to the origin since

$$(-x)(-y) = xy = 1.$$

25. Symmetric with respect to the origin since

$$\begin{aligned} -y &= \frac{-x}{(-x)^2+1} \\ y &= \frac{x}{x^2+1}. \end{aligned}$$

26. No symmetry with respect to either axis or the origin.

27. $y = 2x - 3$

$(1, 2)$:	$2 \neq 2(1) - 3$	Not on the graph
$(1, -1)$:	$-1 = 2(1) - 3$	On the graph
$(4, 5)$:	$5 = 2(4) - 3$	On the graph

28. $x^2 + y^2 = 4$

$(1, -\sqrt{3})$:	$1^2 + (-\sqrt{3})^2 = 4$	On the graph
$\left(\frac{1}{2}, -1\right)$:	$\left(\frac{1}{2}\right)^2 + (-1)^2 \neq 4$	Not on the graph
$\left(\frac{3}{2}, \frac{7}{2}\right)$:	$\left(\frac{3}{2}\right)^2 + \left(\frac{7}{2}\right)^2 \neq 4$	Not on the graph

29. $x^2y - x^2 + 4y = 0$

$$y = \frac{x^2}{x^2+4}$$

$\left(1, \frac{1}{5}\right)$:	$\frac{1}{5} = \frac{1}{1^2+4}$	On the graph
$\left(2, \frac{1}{2}\right)$:	$\frac{1}{2} = \frac{4}{2^2+4}$	On the graph
$(-1, -2)$:	$-2 \neq \frac{1}{(-1)^2+4}$	Not on the graph

30. $x^2 - xy + 4y = 3$

$y = \dfrac{x^2 - 3}{x - 4}$

$(0, 2)$: $2 \neq \dfrac{0-3}{0-4}$ Not on the graph

$\left(-2, -\dfrac{1}{6}\right)$: $-\dfrac{1}{6} = \dfrac{(-2)^2 - 3}{(-2) - 4}$ On the graph

$(3, -6)$: $-6 = \dfrac{3^2 - 3}{3 - 4}$ On the graph

31. $y = x$

Intercept: (0, 0)

Symmetry: origin

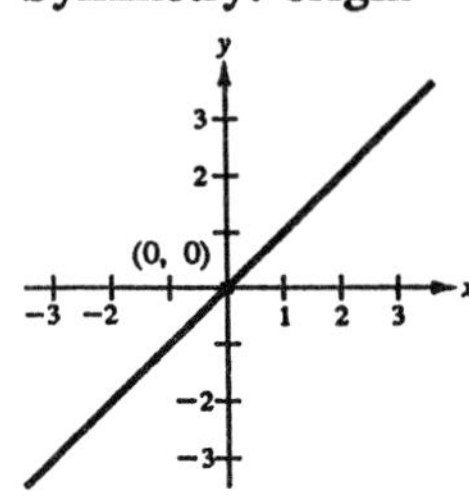

32. $y = x - 2$

Intercepts:

(0, −2), (2, 0)

Symmetry: none

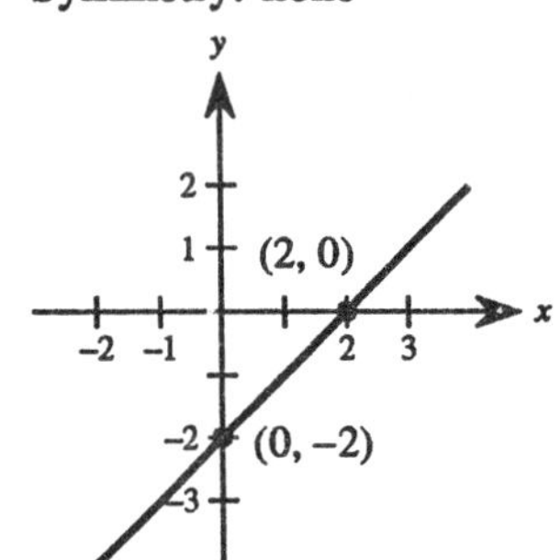

33. $y = x + 3$

Intercepts:

(−3, 0), (0, 3)

Symmetry: none

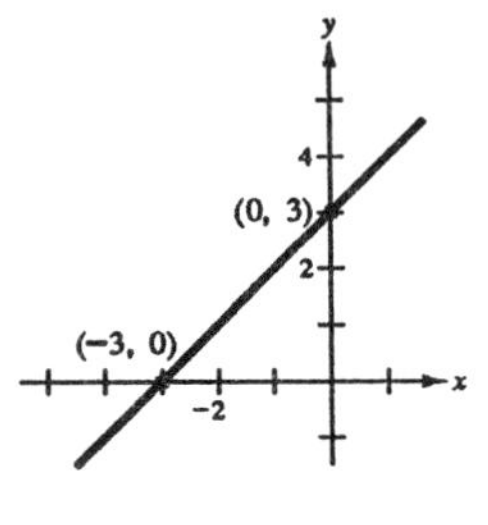

34. $y = 2x - 3$

Intercepts:

$\left(\frac{3}{2}, 0\right)$, (0, −3)

Symmetry: none

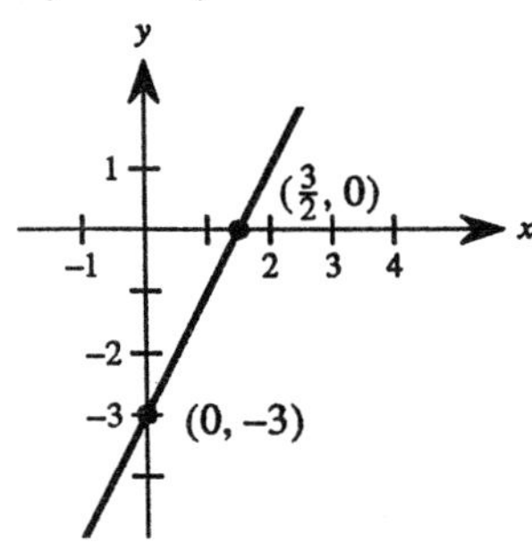

35. $y = -3x + 2$

Intercepts:

$\left(\frac{2}{3}, 0\right)$, (0, 2)

Symmetry: none

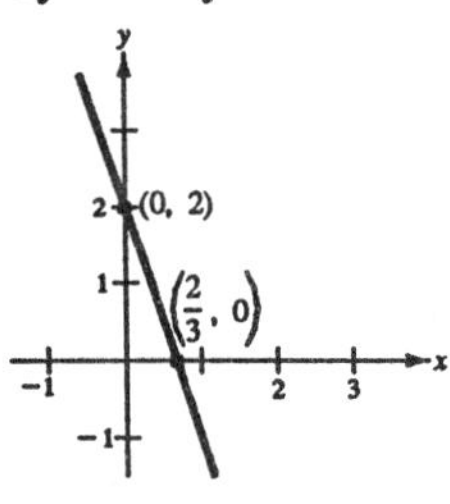

36. $y = -\dfrac{x}{2} + 2$

Intercepts:

(4, 0), (0, 2)

Symmetry: none

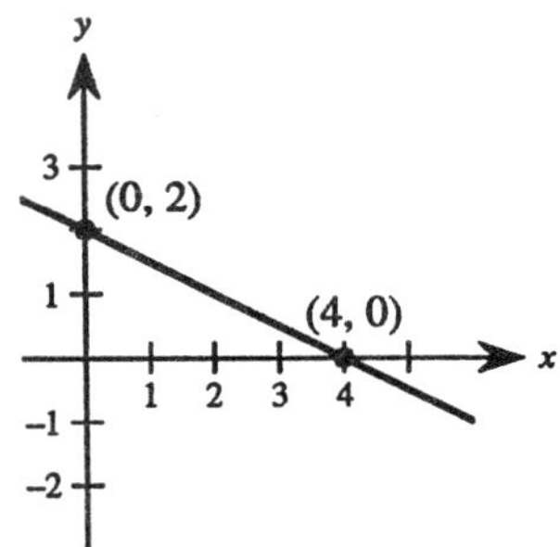

37. $y = \dfrac{x}{2} - 4$

Intercepts:

(8, 0), (0, −4)

Symmetry: none

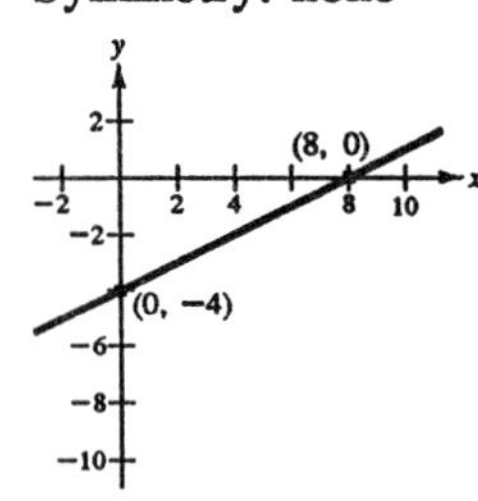

38. $y = x^2 + 3$

Intercept: (0, 3)

Symmetry: y-axis

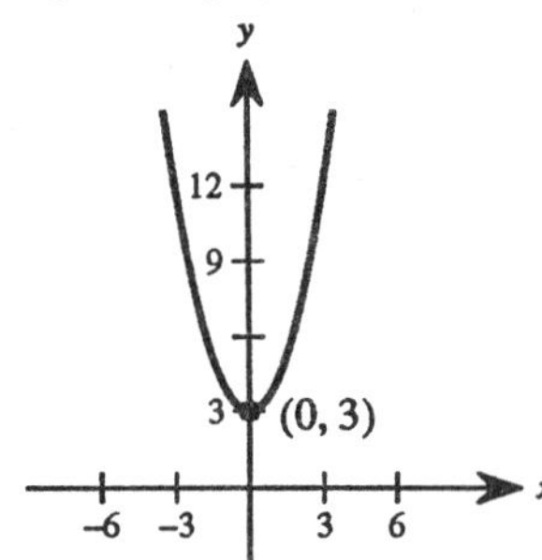

39. $y = 1 - x^2$

Intercepts:

(1, 0), (−1, 0), (0, 1)

Symmetry: y-axis

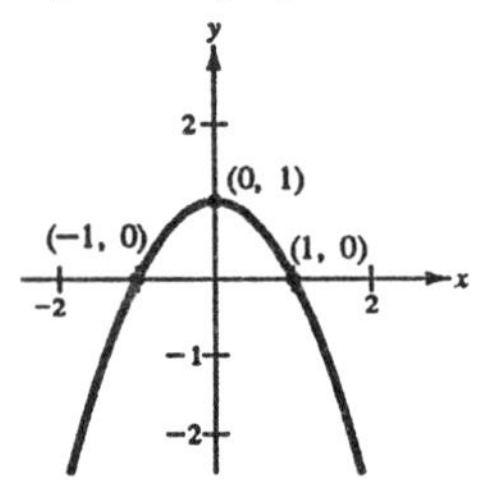

40. $y = 2x^2 + x = x(2x + 1)$

Intercepts:

$(0, 0)$, $\left(-\frac{1}{2}, 0\right)$

Symmetry: none

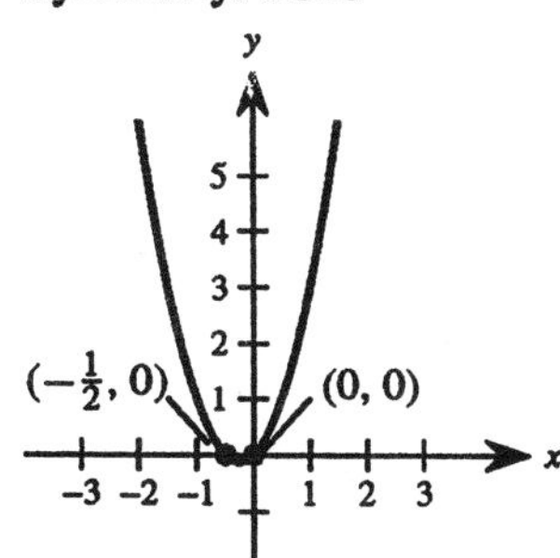

41. $y = x^3 + 2$

Intercepts:

$(-\sqrt[3]{2}, 0)$, $(0, 2)$

Symmetry: none

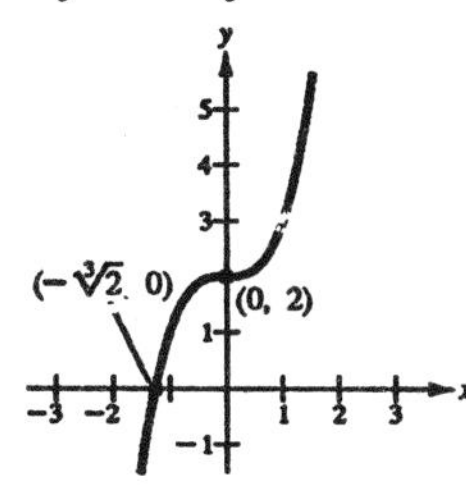

42. $y = \sqrt{9 - x^2}$

Intercepts:

$(-3, 0)$, $(3, 0)$, $(0, 3)$

Symmetry: y-axis

Domain: $[-3, 3]$

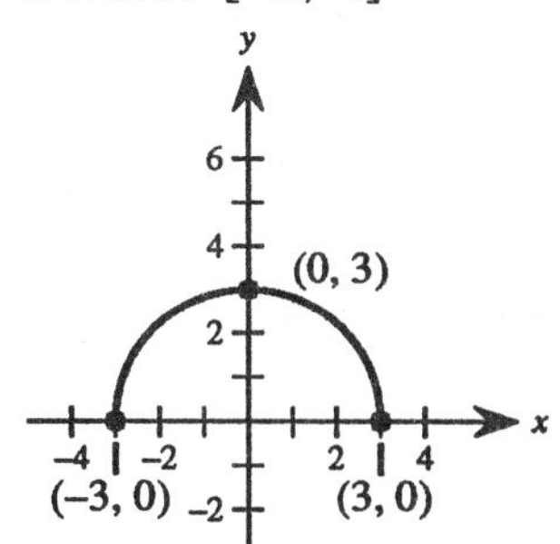

43. $y = (x + 2)^2$

Intercepts:

$(-2, 0)$, $(0, 4)$

Symmetry: none

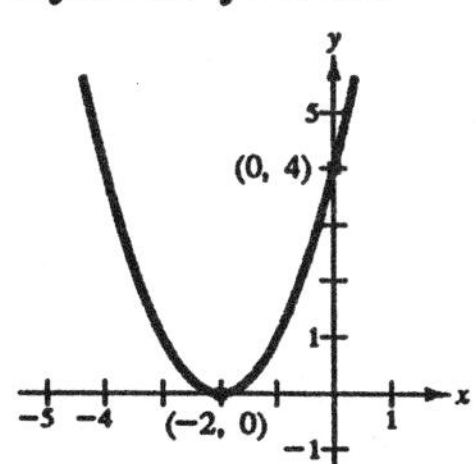

44. $x = y^2 - 4$

Intercepts:

$(0, 2)$, $(0, -2)$, $(-4, 0)$

Symmetry: x-axis

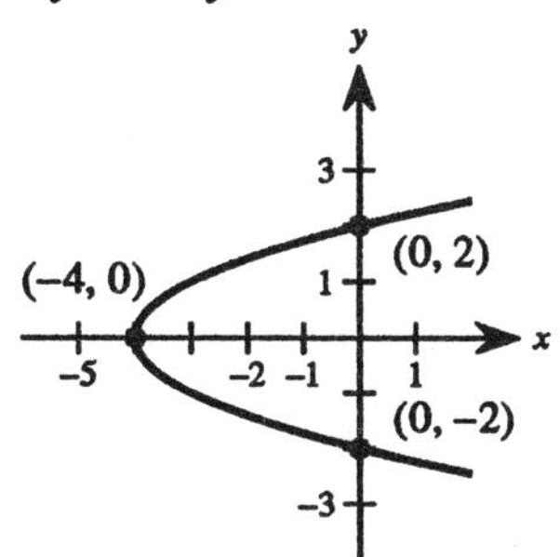

45. $y = \dfrac{1}{x}$

Intercepts: none

Symmetry: origin

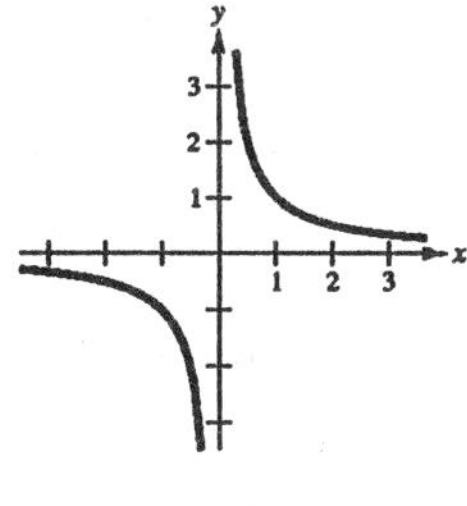

46. $y = 2x^4$

Intercept: $(0, 0)$

Symmetry: y-axis

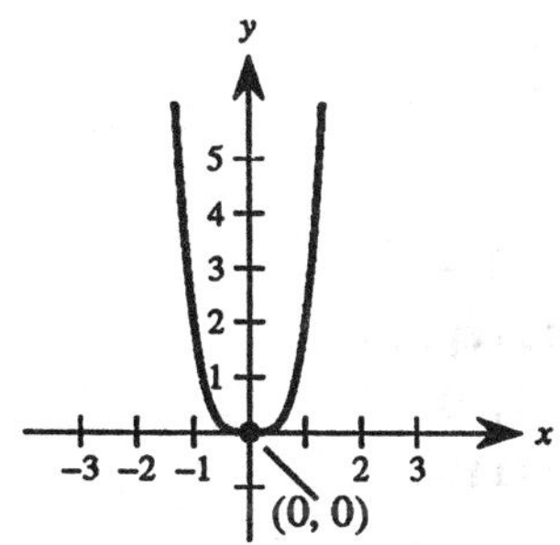

47. $y = -2x^2 + x + 1$

$= (2x + 1)(-x + 1)$

Intercepts:

$\left(-\frac{1}{2}, 0\right)$, $(1, 0)$, $(0, 1)$

Symmetry: none

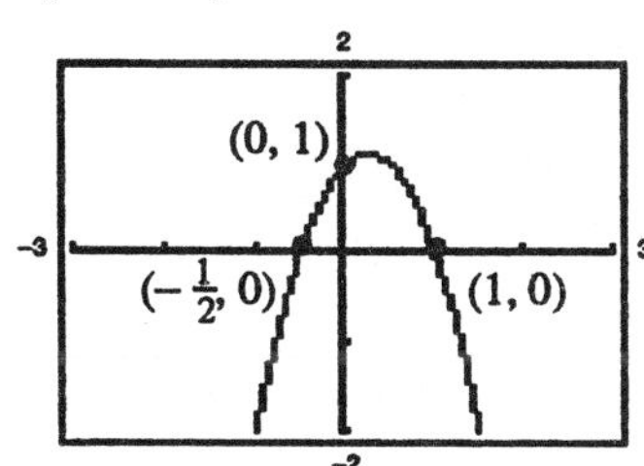

48. $y = x^3 - 1$

Intercepts: $(1, 0)$, $(0, -1)$

Symmetry: none

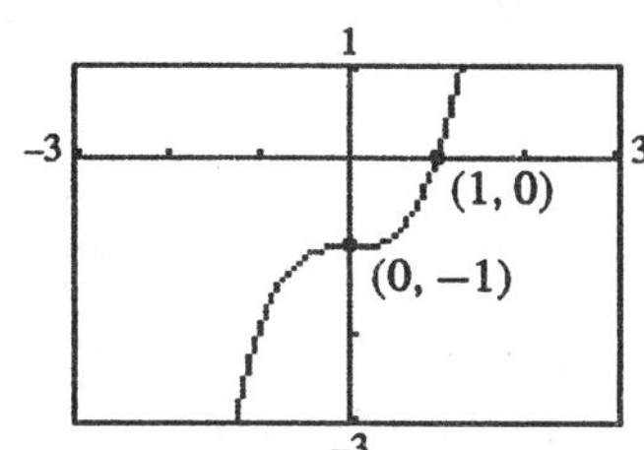

49. $y = x\sqrt{25 - x^2}$

Intercepts: (0, 0), (−5, 0), (5, 0)

Symmetry: origin

Domain: [−5, 5]

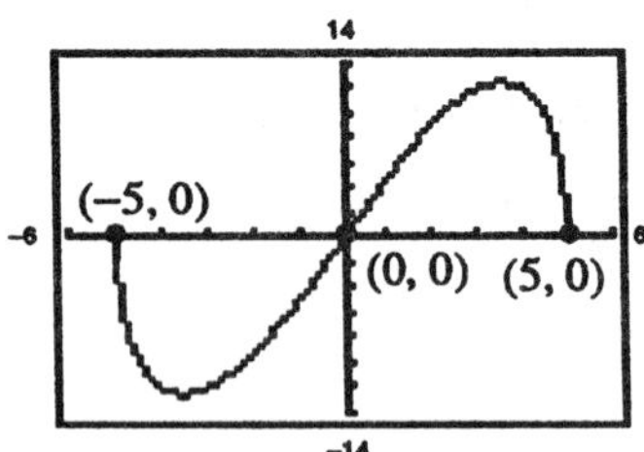

50. $y = \dfrac{5}{x^2 + 1} - 1$

Intercepts: (0, 4), (−2, 0), (2, 0)

Symmetry: y-axis

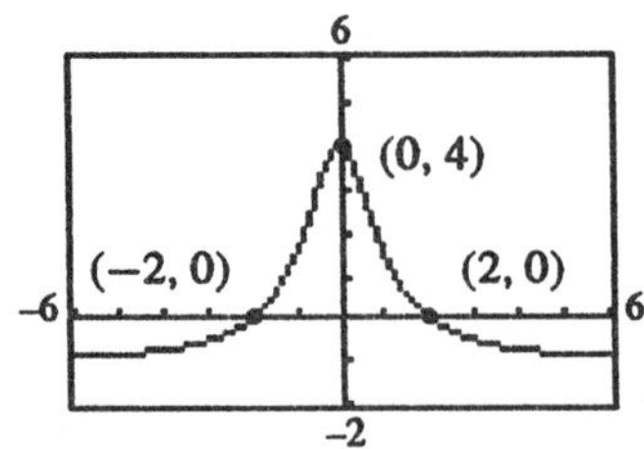

51. $x^2 + 4y^2 = 4 \Rightarrow y = \pm\dfrac{\sqrt{4 - x^2}}{2}$

Intercepts: (−2, 0), (2, 0), (0, −1), (0, 1)

Symmetry: origin and both axes

Domain: [−2, 2]

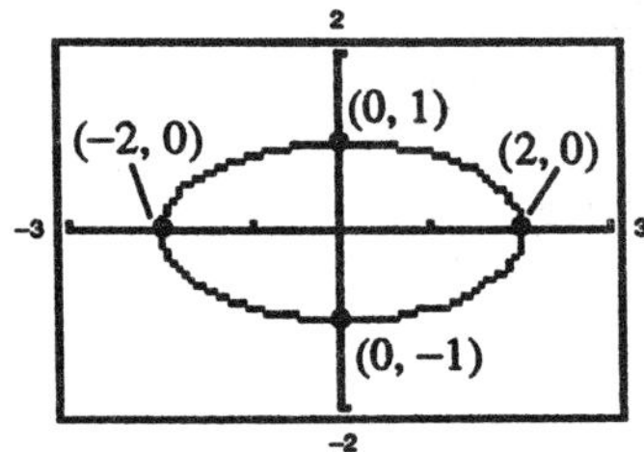

52. $9x^2 + y^2 = 9 \Rightarrow y = \pm 3\sqrt{1 - x^2}$

Intercepts: (1, 0), (−1, 0), (0, 3), (0, −3)

Symmetry: origin and both axes

Domain: [−1, 1]

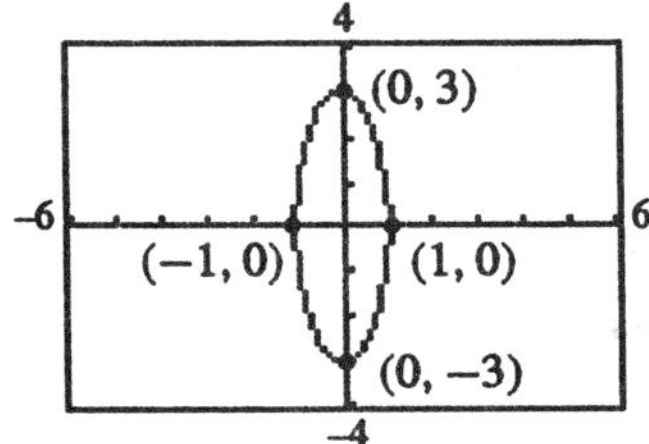

53. $y = (x + 2)(x - 4)(x - 6)$

54. $y = \left(x + \frac{5}{2}\right)(x - 2)\left(x - \frac{3}{2}\right)$

55. Some possible equations:

$$y = x$$
$$y = x^3$$
$$y = 3x^3 - x$$
$$y = \sqrt[3]{x}$$

56. Some possible equations:

$$x = y^2$$
$$x = |y|$$
$$x = y^4 + 1$$
$$x^2 + y^2 = 25$$

57. $x + y = 2 \Rightarrow y = 2 - x$

$2x - y = 1 \Rightarrow y = 2x - 1$

$2 - x = 2x - 1$

$3 = 3x$

$1 = x$

The corresponding y-value is $y = 1$.

Point of intersection: (1, 1)

58. $2x - 3y = 13 \Rightarrow y = \dfrac{2x - 13}{3}$

$5x + 3y = 1 \Rightarrow y = \dfrac{1 - 5x}{3}$

$\dfrac{2x - 13}{3} = \dfrac{1 - 5x}{3}$

$2x - 13 = 1 - 5x$

$7x = 14$

$x = 2$

The corresponding y-value is $y = -3$.

Point of intersection: (2, −3)

59. $x + y = 7 \Rightarrow y = 7 - x$

$3x - 2y = 11 \Rightarrow y = \dfrac{3x - 11}{2}$

$7 - x = \dfrac{3x - 11}{2}$

$14 - 2x = 3x - 11$

$-5x = -25$

$x = 5$

The corresponding y-value is $y = 2$.
Point of intersection: $(5, 2)$

60. $x^2 + y^2 = 25 \Rightarrow y^2 = 25 - x^2$

$2x + y = 10 \Rightarrow y = 10 - 2x$

$25 - x^2 = (10 - 2x)^2$

$25 - x^2 = 100 - 40x + 4x^2$

$0 = 5x^2 - 40x + 75 = 5(x - 3)(x - 5)$

$x = 3$ or $x = 5$

The corresponding y-values are $y = 4$ and $y = 0$.
Points of intersection: $(3, 4)$, $(5, 0)$

61. $x^2 + y^2 = 5 \Rightarrow y^2 = 5 - x^2$

$x - y = 1 \Rightarrow y = x - 1$

$5 - x^2 = (x - 1)^2$

$5 - x^2 = x^2 - 2x + 1$

$0 = 2x^2 - 2x - 4 = 2(x + 1)(x - 2)$

$x = -1$ or $x = 2$

The corresponding y-values are $y = -2$ and $y = 1$.
Points of intersection: $(-1, -2)$, $(2, 1)$

62. $x^2 + y = 4 \Rightarrow y = 4 - x^2$

$2x - y = 1 \Rightarrow y = 2x - 1$

$4 - x^2 = 2x - 1$

$0 = x^2 + 2x - 5$

$x = \dfrac{-2 \pm \sqrt{24}}{2} = -1 \pm 6$

The corresponding y-values are $y = -3 \pm 2\sqrt{6}$.
Points of intersection: $\left(-1 - \sqrt{6}, -3 - 2\sqrt{6}\right)$, $\left(-1 + \sqrt{6}, -3 + 2\sqrt{6}\right)$

63. $y = x^3$

$y = x$

$x^3 = x$

$x^3 - x = 0$

$x(x + 1)(x - 1) = 0$

$x = 0$, $x = -1$, or $x = 1$

The corresponding y-values are $y = 0$, $y = -1$, and $y = 1$.
Points of intersection: $(0, 0)$, $(-1, -1)$, $(1, 1)$

64. $x = 3 - y^2 \Rightarrow y^2 = 3 - x$

$y = x - 1$

$3 - x = (x - 1)^2$

$3 - x = x^2 - 2x + 1$

$0 = x^2 - x - 2 = (x + 1)(x - 2)$

$x = -1$ or $x = 2$

The corresponding y-values are $y = -2$ and $y = 1$.
Points of intersection: $(-1, -2)$, $(2, 1)$

65. $y = x^3 - 2x^2 + x - 1$

$y = -x^2 + 3x - 1$

$x^3 - 2x^2 + x - 1 = -x^2 + 3x - 1$

$x^3 - x^2 - 2x = 0$

$x(x - 2)(x + 1) = 0$

$x = -1, 0, 2$

$(-1, -5)$, $(0, -1)$, $(2, 1)$

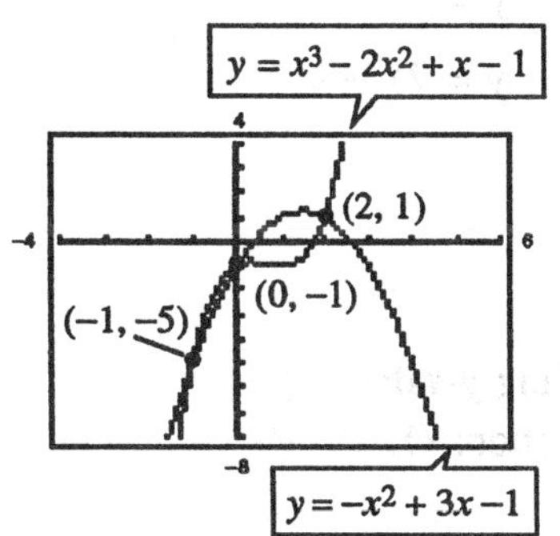

66.
$$y = x^4 - 2x^2 + 1$$
$$y = 1 - x^2$$
$$1 - x^2 = x^4 - 2x^2 + 1$$
$$0 = x^4 - x^2$$
$$0 = x^2(x+1)(x-1)$$
$$x = -1,\ 0,\ 1$$
$(-1,\ 0),\ (0,\ 1),\ (1,\ 0)$

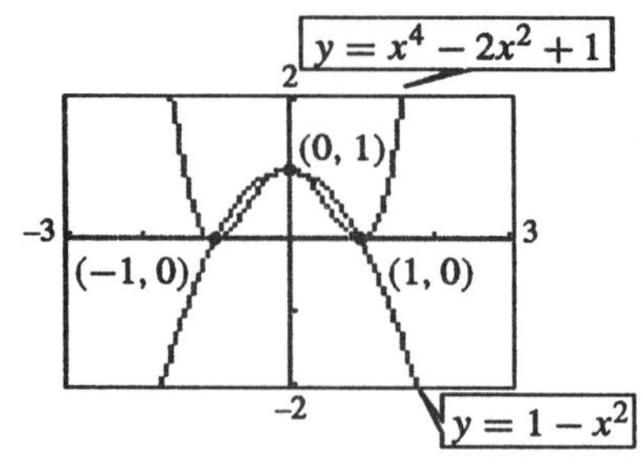

67. $8650x + 250{,}000 = 9950x$
$$250{,}000 = 1300x$$
$$x \approx 192 \text{ units}$$

68. $5.5\sqrt{x} + 10,000 = 3.29x$
$$\left(5.5\sqrt{x}\right)^2 = (3.29x - 10{,}000)^2$$
$$30.25x = 10.8241x^2 - 65{,}800x + 100{,}000{,}000$$
$$0 = 10.8241x^2 - 65830.25x + 100{,}000{,}000 \quad \text{Use the Quadratic Formula}$$
$$x \approx 3133 \text{ units}$$
The other root, $x \approx 2949$, does not satisfy the equation $R = C$.

69. (a) $4 = k(1)^3$
$k = 4$

(b) $1 = k(-2)^3$
$k = -\frac{1}{8}$

(c) $0 = k(0)^3$
$k =$ any real number

(d) $-1 = k(-1)^3$
$k = 1$

70. $y = kx + 5$ matches (b).
Use $(1,\ 7)$: $7 = k(1) + 5 \Rightarrow k = 2$, thus, $y = 2x + 5$.
$y = x^2 + k$ matches (d).
Use $(1,\ -9)$: $-9 = (1)^2 + k \Rightarrow k = -10$, thus, $y = x^2 - 10$.
$y = kx^{3/2}$ matches (a).
Use $(1,\ 3)$: $3 = k(1)^{3/2} \Rightarrow k = 3$, thus, $y = 3x^{3/2}$.
$xy = k$ matches (c).
Use $(1,\ 36)$: $(1)(36) = k \Rightarrow k = 36$, thus, $xy = 36$.

71. (a)

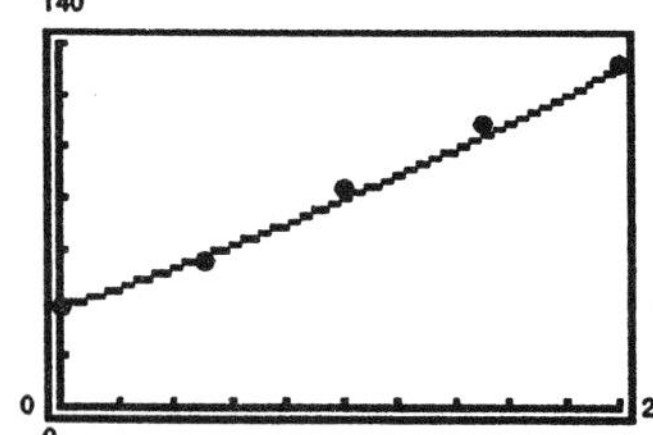

(b) $\dfrac{130.7}{38.8} \approx 3.369$

(c) $y = 0.05(30)^2 + 3.63(30) + 37.68$
≈ 191.58 CPI for the year 2000

72. (a)

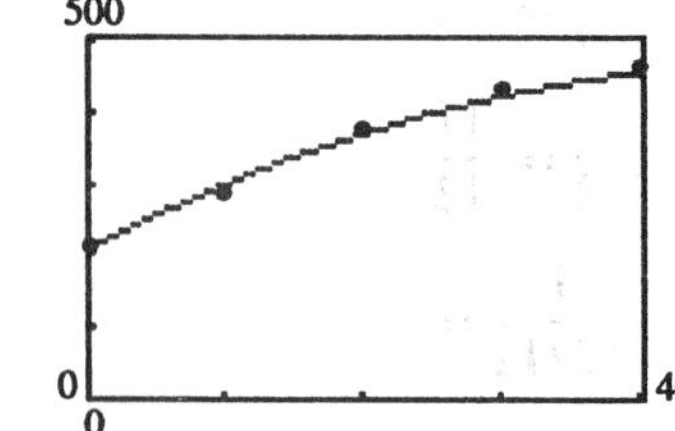

(b) $y = -0.09(50)^2 + 9.80(50) + 211.49$
≈ 476.49 acres per farm in the year 2000

73. (a)

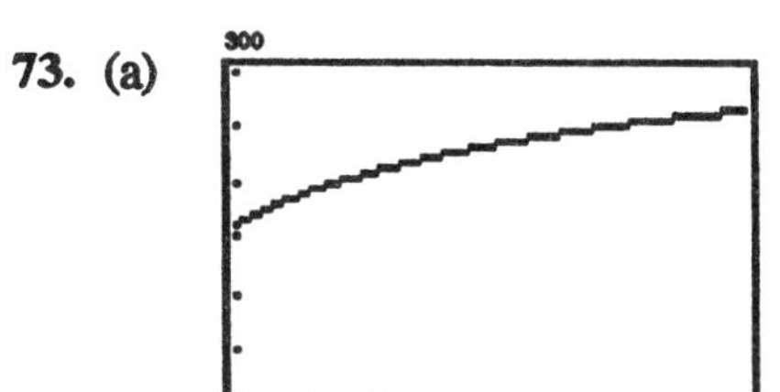

(b) $y = 75.82 - 2.11(14.696) + 43.51\sqrt{14.696}$

$\approx 211.61°$

(c) When $x \approx 24.725$ lb/in.^{2}, $y \approx 240°$.

74. 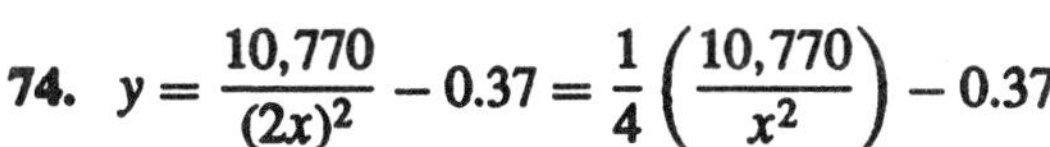

$y = \dfrac{10{,}770}{(2x)^2} - 0.37 = \dfrac{1}{4}\left(\dfrac{10{,}770}{x^2}\right) - 0.37$

The resistance is changed by approximately a factor of $\frac{1}{4}$.

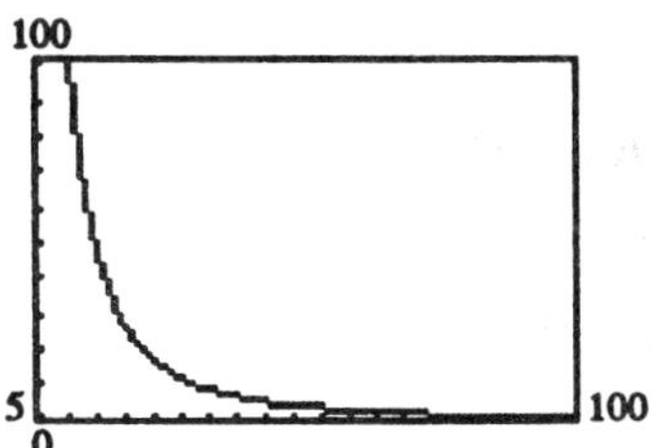

75. If (x, y) is on the graph, then so is $(-x, y)$ by y-axis symmetry. Since $(-x, y)$ is on the graph, then so is $(-x, -y)$ by x-axis symmetry. Hence, the graph is symmetric with respect to the origin. The converse is not true. For example, $y = x^3$ has origin symmetry but is not symmetric with respect to either the x-axis or the y-axis.

76. Assume that the graph has x-axis and origin symmetry. If (x, y) is on the graph, so is $(x, -y)$ by x-axis symmetry. Since $(x, -y)$ is on the graph, then so is $(-x, -(-y)) = (-x, y)$ by origin symmetry. Therefore, the graph is symmetric with respect to the y-axis. The argument is the same for y-axis and origin symmetry.

77. False; x-axis symmetry means that if $(1, -2)$ is on the graph, then $(1, 2)$ is also on the graph.

78. True

79. True; the x-intercepts are

$$\left(\frac{-b \pm \sqrt{b^2 - 4ac}}{2a}, 0\right).$$

80. True; the x-intercept is

$$\left(-\frac{b}{2a}, 0\right).$$

81. Distance to the origin = $K \times$ Distance to (2, 0)

$$\sqrt{x^2 + y^2} = K\sqrt{(x-2)^2 + y^2}, \quad K \neq 1$$

$$x^2 + y^2 = K^2(x^2 - 4x + 4 + y^2)$$

$$(1 - K^2)x^2 + (1 - K^2)y^2 + 4K^2x - 4K^2 = 0$$

Note: This is the equation of a circle!

82. $x_n = \dfrac{1}{2}x_{n-1} + \dfrac{1}{x_{n-1}}, \quad n = 1, 2, \ldots,$ and $x_0 = 1$

$$x_1 = \frac{1}{2}(1) + \frac{1}{1} = \frac{3}{2} = 1.5$$

$$x_2 = \frac{1}{2}\left(\frac{3}{2}\right) + \frac{1}{3/2} = \frac{17}{12} \approx 1.41667$$

$$x_3 = \frac{1}{2}\left(\frac{17}{12}\right) + \frac{1}{17/12} = \frac{577}{408} \approx 1.41422$$

$$x_4 = \frac{1}{2}\left(\frac{577}{408}\right) + \frac{1}{577/408} = \frac{665{,}857}{470{,}832} \approx 1.41421$$

As n gets larger, $x_n \to \sqrt{2}$.

Section 1.4 Lines in the Plane

1. $m = 1$ **2.** $m = 2$ **3.** $m = 0$ **4.** $m = -1$ **5.** $m = -3$ **6.** $m = \frac{3}{2}$

7. $m = \dfrac{2-(-4)}{5-3} = \dfrac{6}{2} = 3$

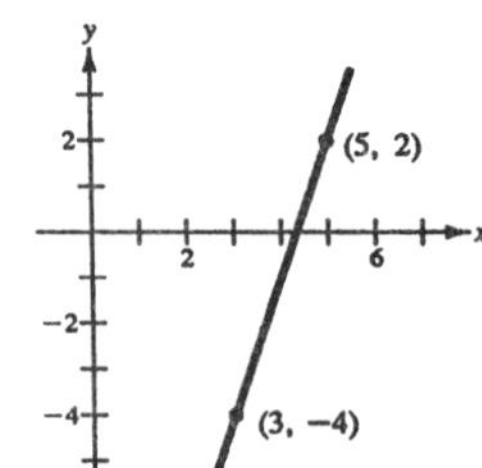

8. $m = \dfrac{1-(-3)}{-2-4} = \dfrac{4}{-6} = -\dfrac{2}{3}$

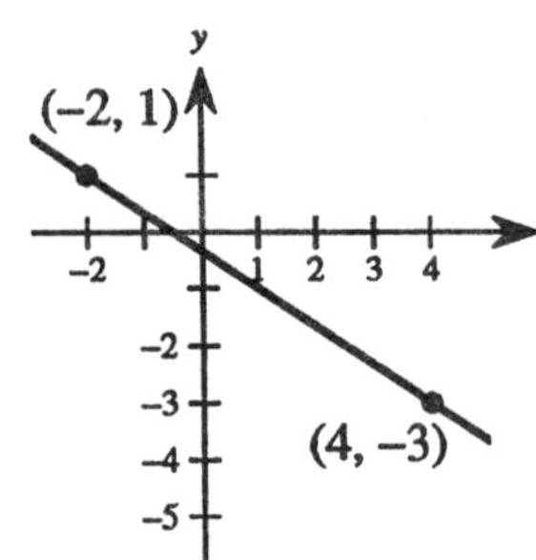

9. $m = \dfrac{2-2}{6-(1/2)} = 0$

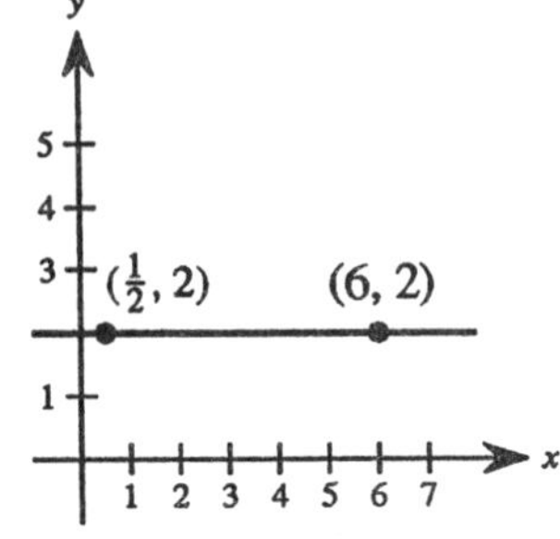

10. $m = \dfrac{5-1}{2-2} = \dfrac{4}{0}$ undefined

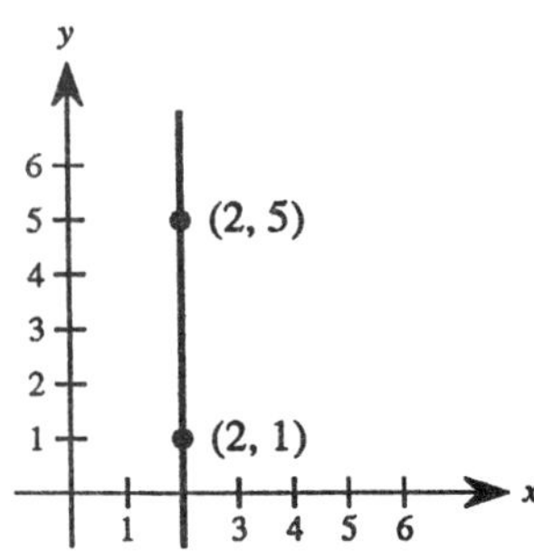

11. $m = \dfrac{4-2}{-2-1} = -\dfrac{2}{3}$

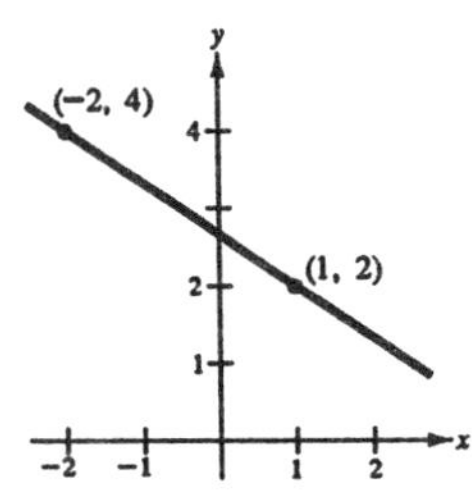

12. $m = \dfrac{(3/4)-(-1/4)}{(7/8)-(5/4)}$

$= \dfrac{1}{-3/8} = -\dfrac{8}{3}$

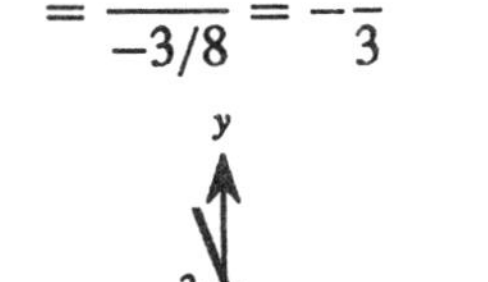

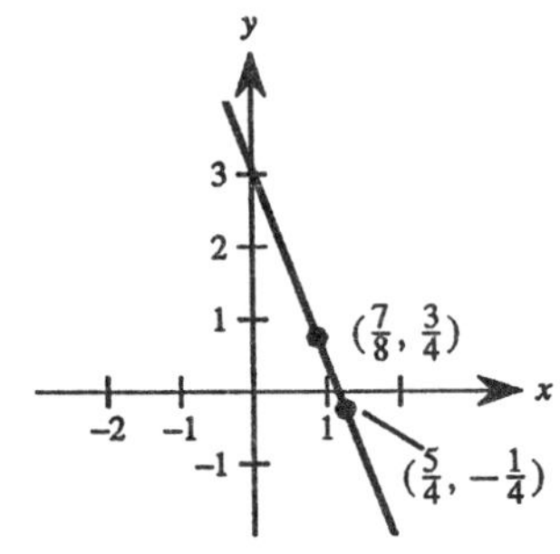

13. Since the slope is 0, the line is horizontal and its equation is $y = 1$. Therefore, three additional points are (0, 1), (1, 1), and (3, 1).

14. Since the slope is undefined, the line is vertical and its equation is $x = -3$. Therefore, three additional points are (−3, 2), (−3, 3), and (−3, 5).

15. The equation of this line is

$$y - 7 = -3(x - 1)$$
$$y = -3x + 10$$

Therefore, three additional points are (0, 10), (2, 4), and (3, 1).

16. The equation of this line is

$$y + 2 = 2(x + 2)$$
$$y = 2x + 2$$

Therefore, three additional points are (−3, −4), (−1, 0), and (0, 2).

17. $x + 5y = 20$

$$y = -\tfrac{1}{5}x + 4$$

Therefore, the slope is $m = -\frac{1}{5}$ and the y-intercept is (0, 4).

18. $6x - 5y = 15$

$$y = \tfrac{6}{5}x - 3$$

Therefore, the slope is $m = \frac{6}{5}$ and the y-intercept is (0, −3).

19. $x = 4$

The line is vertical. Therefore, the slope is undefined and there is no y-intercept.

20. $y = -1$

The line is horizontal. Therefore, the slope is $m = 0$ and the y-intercept is (0, −1).

21. $m = \dfrac{1-(-3)}{2-0} = 2$

$y - 1 = 2(x - 2)$

$y - 1 = 2x - 4$

$0 = 2x - y - 3$

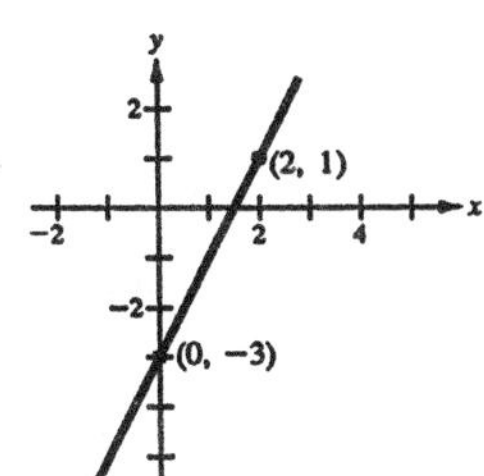

22. $m = \dfrac{4-(-4)}{1-(-3)} = \dfrac{8}{4} = 2$

$y - 4 = 2(x - 1)$

$y - 4 = 2x - 2$

$0 = 2x - y + 2$

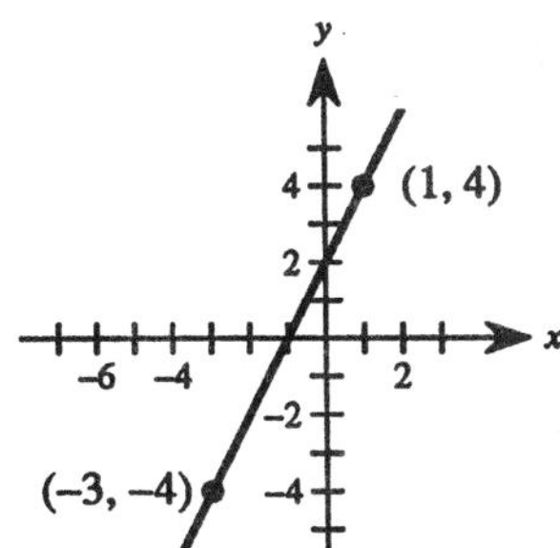

23. $m = \dfrac{3-0}{-1-0} = -3$

$y - 0 = -3(x - 0)$

$y = -3x$

$3x + y = 0$

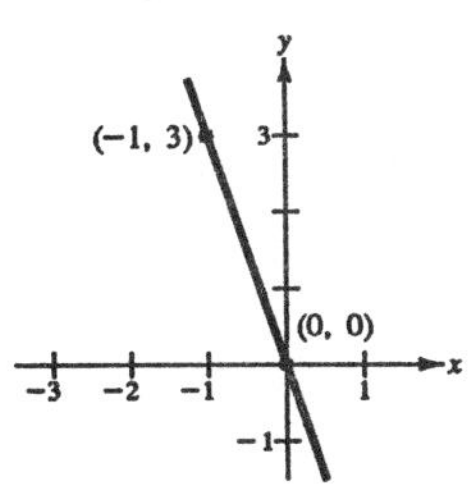

24. $m = \dfrac{6-2}{-3-1} = \dfrac{4}{-4} = -1$

$y - 2 = -1(x - 1)$

$y - 2 = -x + 1$

$x + y - 3 = 0$

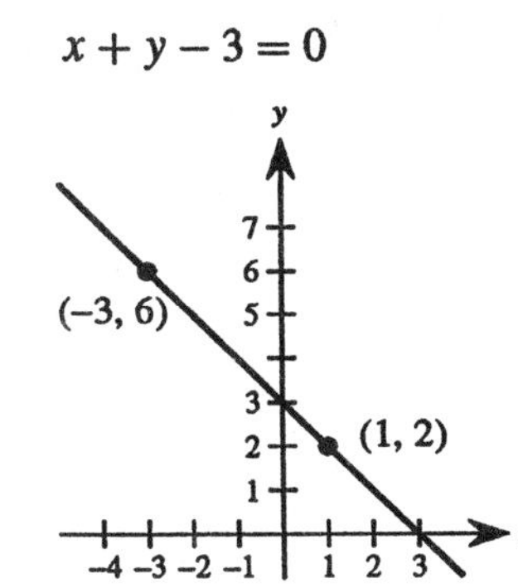

25. $m = 0$

$y = -2$

$y + 2 = 0$

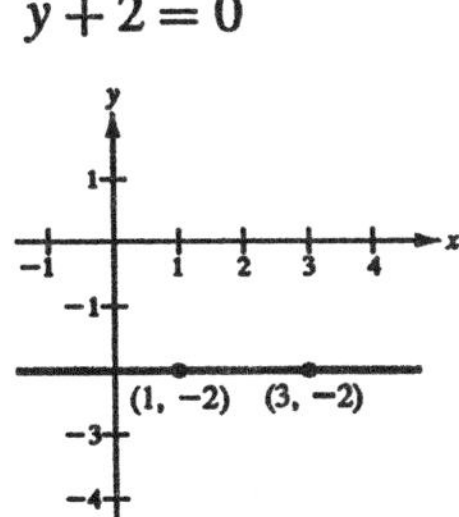
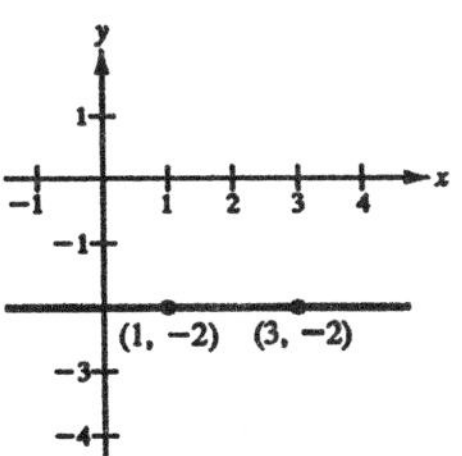

26. $m = \dfrac{(3/4) - (-1/4)}{(7/8) - (5/4)}$

$= \dfrac{1}{-3/8} = -\dfrac{8}{3}$

$y + \dfrac{1}{4} = \dfrac{-8}{3}\left(x - \dfrac{5}{4}\right)$

$12y + 3 = -32x + 40$

$32x + 12y - 37 = 0$

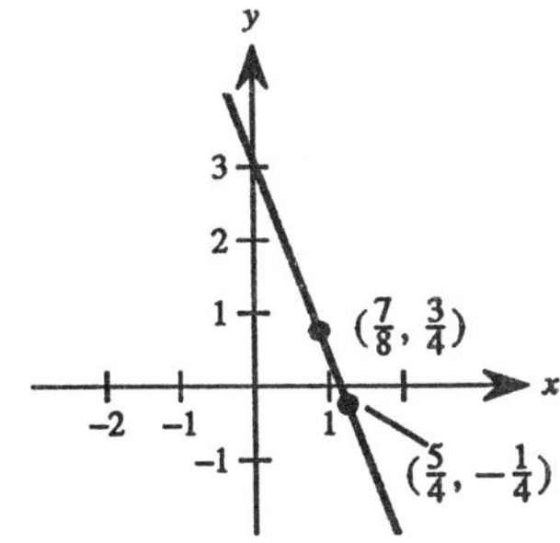

27. $y = \frac{3}{4}x + 3$

$4y = 3x + 12$

$0 = 3x - 4y + 12$

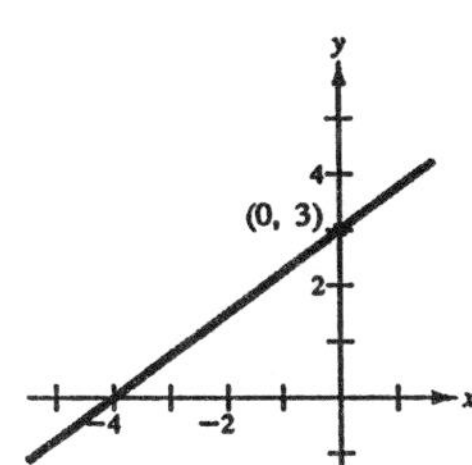

28. $x = -1$

$x + 1 = 0$

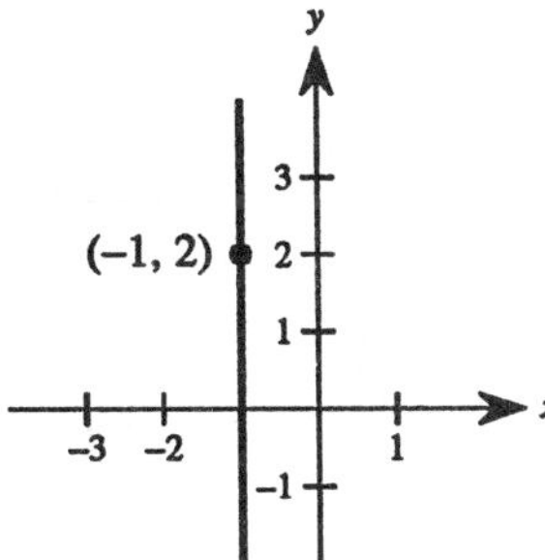

29. $y = \frac{2}{3}x$

$3y = 2x$

$2x - 3y = 0$

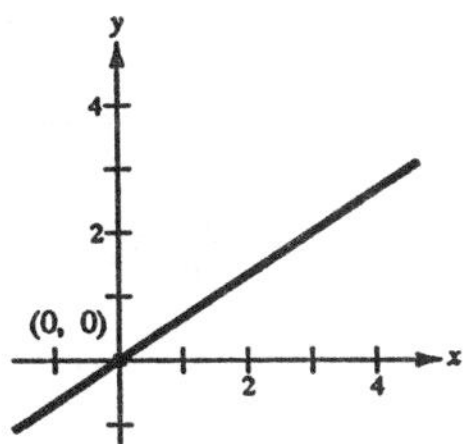

30. $y - 4 = -\frac{3}{5}(x + 2)$

$5y - 20 = -3x - 6$

$3x + 5y - 14 = 0$

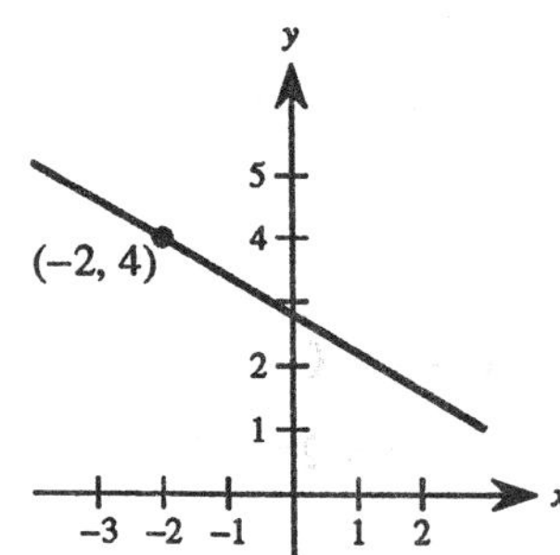

31. $y - 2 = 4(x - 0)$

$y = 4x + 2$

$0 = 4x - y + 2$

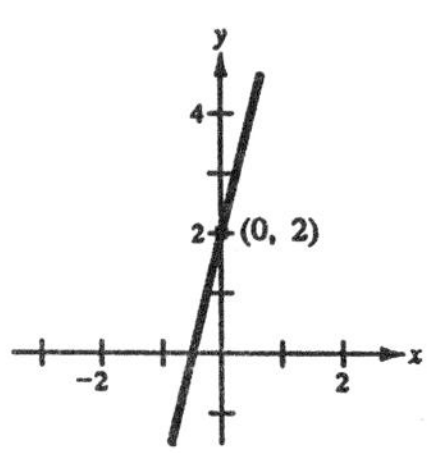

32. $y = 4$

$y - 4 = 0$

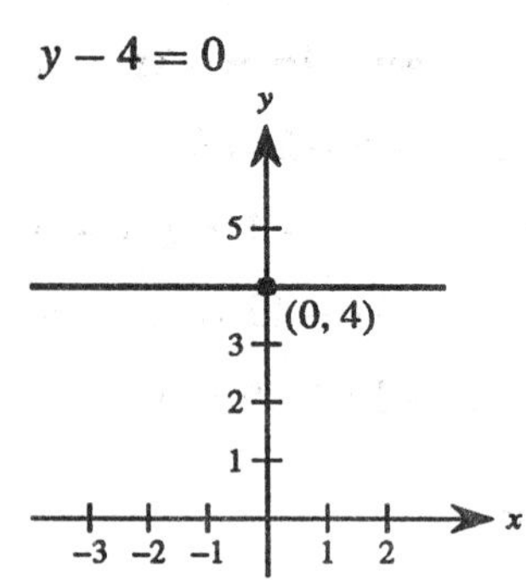

33. $x = 3$

$x - 3 = 0$

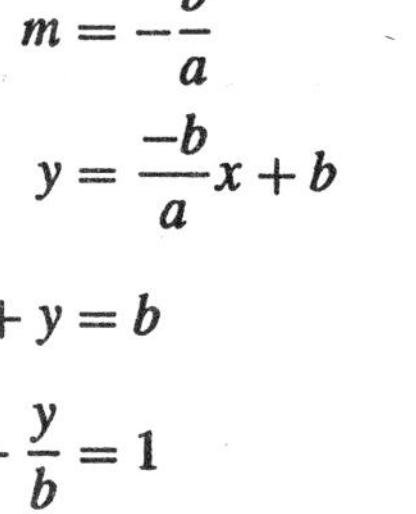

34. $m = -\frac{b}{a}$

$y = \frac{-b}{a}x + b$

$\frac{b}{a}x + y = b$

$\frac{x}{a} + \frac{y}{b} = 1$

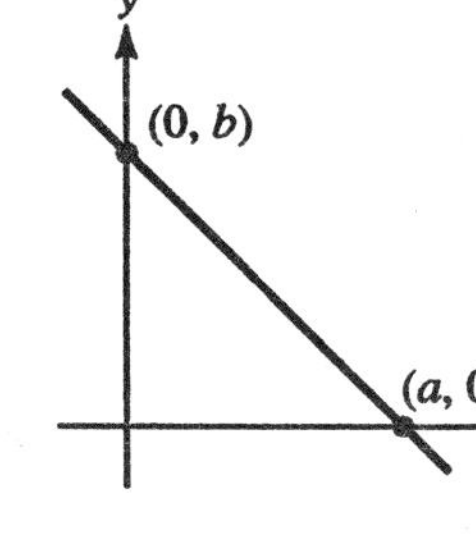

35. $\frac{x}{2} + \frac{y}{3} = 1$

$3x + 2y - 6 = 0$

36. $\frac{x}{-3} + \frac{y}{4} = 1$

$4x - 3y + 12 = 0$

37. $\frac{x}{-1/6} + \frac{y}{-2/3} = 1$

$-6x - \frac{3}{2}y = 1$

$12x + 3y + 2 = 0$

38. $\frac{x}{-2/3} + \frac{y}{-2} = 1$

$\frac{-3x}{2} - \frac{y}{2} = 1$

$3x + y + 2 = 0$

39. $\frac{x}{a} + \frac{y}{a} = 1$

$\frac{1}{a} + \frac{2}{a} = 1$

$\frac{3}{a} = 1$

$a = 3 \Rightarrow \quad x + y = 3$

$x + y - 3 = 0$

40. $\frac{x}{a} + \frac{y}{a} = 1$

$\frac{-3}{a} + \frac{4}{a} = 1$

$\frac{1}{a} = 1$

$a = 1 \Rightarrow \quad x + y = 1$

$x + y - 1 = 0$

41. (a) $y - 1 = 2(x - 2)$

$y - 1 = 2x - 4$

$2x - y - 3 = 0$

(b) $y - 1 = -\frac{1}{2}(x - 2)$

$2y - 2 = -x + 2$

$x + 2y - 4 = 0$

42. (a) $y - 2 = -1(x + 3)$

$y - 2 = -x - 3$

$x + y + 1 = 0$

(b) $y - 2 = 1(x + 3)$

$y - 2 = x + 3$

$x - y + 5 = 0$

43. (a) $y - \frac{3}{4} = -\frac{5}{3}\left(x - \frac{7}{8}\right)$

$24y - 18 = -40x + 35$

$40x + 24y - 53 = 0$

(b) $y - \frac{3}{4} = \frac{3}{5}\left(x - \frac{7}{8}\right)$

$40y - 30 = 24x - 21$

$24x - 40y + 9 = 0$

44. (a) $y - 4 = -\frac{3}{4}(x + 6)$

$4y - 16 = -3x - 18$

$3x + 4y + 2 = 0$

(b) $y - 4 = \frac{4}{3}(x + 6)$

$3y - 12 = 4x + 24$

$4x - 3y + 36 = 0$

45. (a) $x - 2 = 0$

(b) $y - 5 = 0$

46. (a) $y = 0$

(b) $x + 1 = 0$

47. $y = -3$

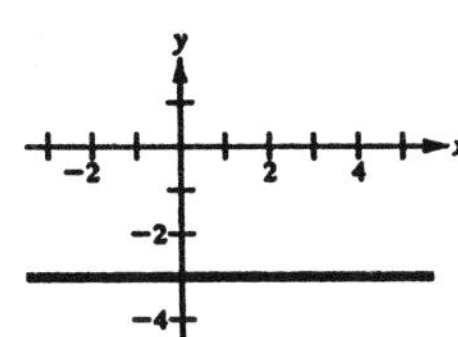

48. $x = 4$

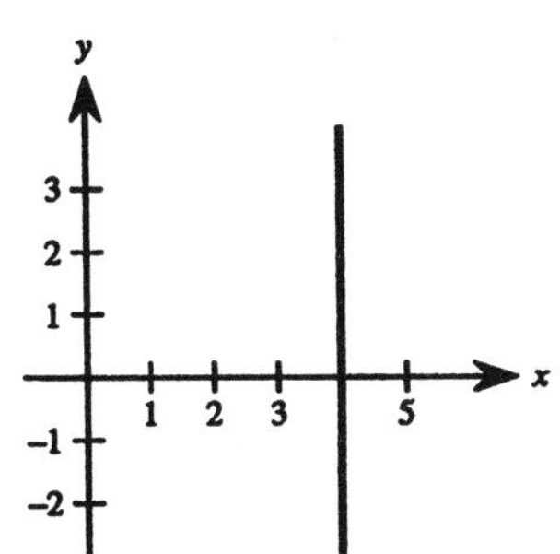

49. $2x - y - 3 = 0$

$y = 2x - 3$

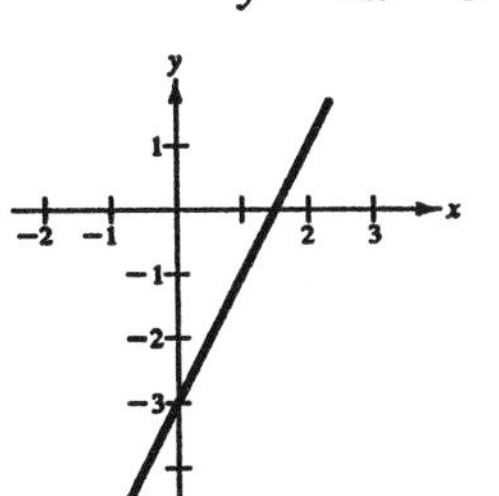

50. $x + 2y + 6 = 0$

$y = -\frac{1}{2}x - 3$

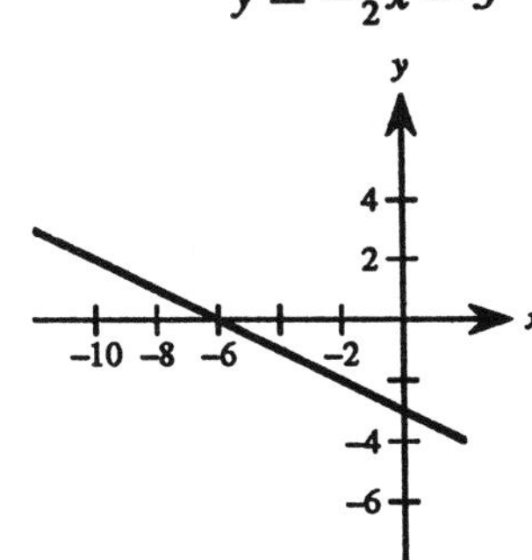

51. $y = -2x + 1$

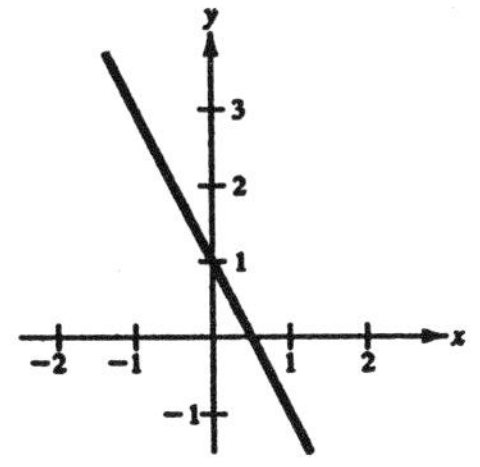

52. $y - 1 = 3(x + 4)$

$y = 3x + 13$

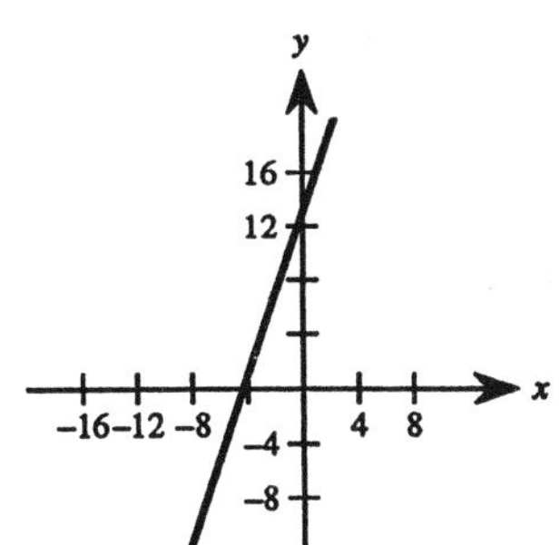

53. $x^2 = 4x - x^2$

$2x^2 - 4x = 0$

$2x(x - 2) = 0$

$x = 0 \qquad x = 2$

$y = 0 \qquad y = 4$

Points of intersection: (0, 0), (2, 4)

$m = 2$

$y = 2x$

$2x - y = 0$

54. $x^2 - 4x + 3 = -x^2 + 2x + 3$

$2x^2 - 6x = 0$

$2x(x - 3) = 0$

$x = 0 \qquad x = 3$

$y = 3 \qquad y = 0$

Points of intersection: (0, 3), (3, 0)

$m = -1$

$y - 3 = -(x - 0)$

$y = -x + 3$

$x + y - 3 = 0$

55. $m_1 = \dfrac{1-0}{-2-(-1)} = -1$

$m_2 = \dfrac{-2-0}{2-(-1)} = -\dfrac{2}{3}$

$m_1 \neq m_2$

The lines are not collinear.

56. $m_1 = \dfrac{-6-4}{7-0} = -\dfrac{10}{7}$

$m_2 = \dfrac{11-4}{-5-0} = -\dfrac{7}{5}$

$m_1 \neq m_2$

The lines are not collinear.

57. Equations of perpendicular bisectors:

$$x = 0$$

$$y - \frac{c}{2} = \frac{a-b}{c}\left(x - \frac{a+b}{2}\right)$$

$$y - \frac{c}{2} = \frac{a+b}{-c}\left(x - \frac{b-a}{2}\right)$$

Solving simultaneously, the point of intersection is

$$\left(0, \frac{-a^2+b^2+c^2}{2c}\right).$$

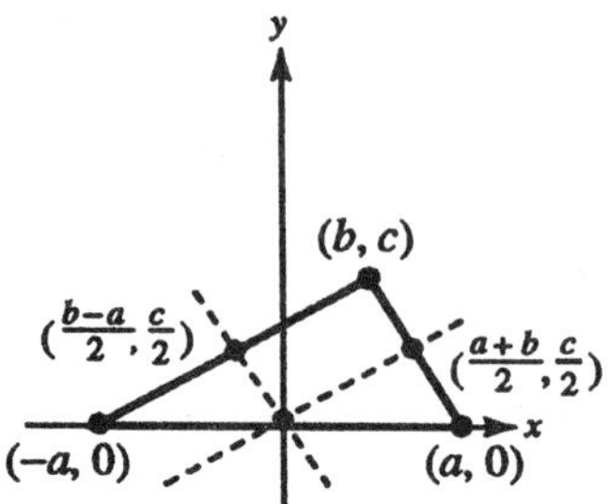

58. Equations of the medians:

$$y = \frac{c}{b}x$$

$$y = \frac{c}{3a+b}(x+a)$$

$$y = \frac{c}{-3a+b}(x-a)$$

Solving simultaneously, the point of intersection is

$$\left(\frac{b}{3}, \frac{c}{3}\right).$$

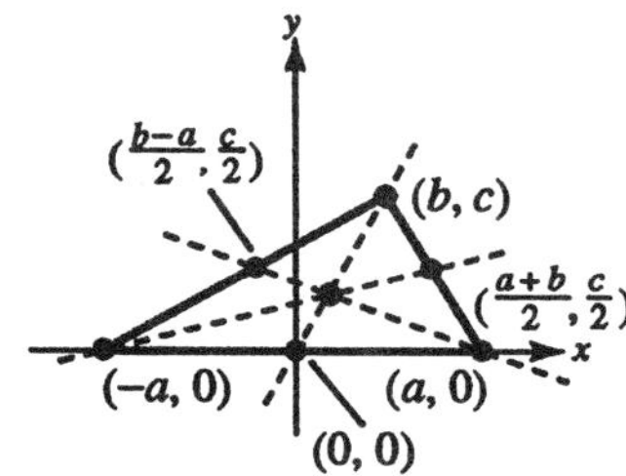

59. Equations of altitudes:

$$y = \frac{a-b}{c}(x+a)$$

$$x = b$$

$$y = -\frac{a+b}{c}(x-a)$$

Solving simultaneously, the point of intersection is

$$\left(b, \frac{a^2-b^2}{c}\right).$$

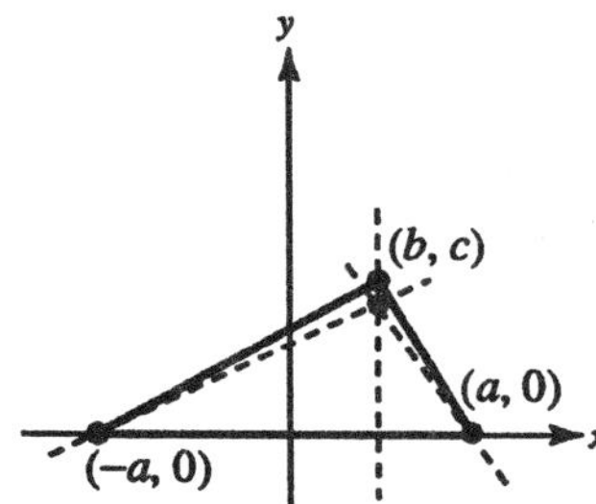

60. The slope of the line segment from $\left(\frac{b}{3}, \frac{c}{3}\right)$ to $\left(b, \frac{a^2-b^2}{c}\right)$ is:

$$m_1 = \frac{[(a^2-b^2)/c] - (c/3)}{b-(b/3)} = \frac{(3a^2-3b^2-c^2)/(3c)}{(2b)/3} = \frac{3a^2-3b^2-c^2}{2bc}.$$

The slope of the line segment from $\left(\frac{b}{3}, \frac{c}{3}\right)$ to $\left(0, \frac{-a^2+b^2+c^2}{2c}\right)$ is:

$$m_2 = \frac{[(-a^2+b^2+c^2)/(2c)] - (c/3)}{0-(b/3)} = \frac{(-3a^2+3b^2+3c^2-2c^2)/(6c)}{-b/3} = \frac{3a^2-3b^2-c^2}{2bc}$$

$m_1 = m_2$

Therefore, the points are collinear.

61. Find the equation of the line through the points (0, 32) and (100, 212).

$$m = \tfrac{180}{100} = \tfrac{9}{5}$$

$$F = \tfrac{9}{5}C + 32$$

$$5F - 9C - 160 = 0$$

62.

C	−17.8	−10	10	20	32.2	177
F	0	14	50	68	90	350.6

63. $C = 0.25x + 95$

64. $W = 0.75x + 9.50$

65. Depreciation per year:

$$\frac{875}{5} = \$175$$

$$y = 875 - 175t$$

where $0 \le t \le 5$.

66. Depreciation per year:

$$\frac{825{,}000 - 75{,}000}{25} = \$30{,}000$$

$$y = 825{,}000 - 30{,}000t$$

where $0 \le t \le 25$.

67. (a) (50, 380), (47, 425)

$$m = \frac{425-380}{47-50} = -15$$

$$p - 380 = -15(x-50)$$

$$p = -15x + 1130 \text{ or}$$

$$x = \frac{1}{15}(1130 - p)$$

(b) $x = \frac{1}{15}(1130 - 455) = 45$ units

(c) $x = \frac{1}{15}(1130 - 395) = 49$ units

68. (a) (0, 16), (6, 37.5)

$$m = \frac{37.5-16}{6} = \frac{21.5}{6} = \frac{43}{12}$$

$$y - 16 = \frac{43}{12}(x - 0)$$

$$y = \frac{43}{12}x + 16$$

(b) $y = \frac{43}{12}(10) + 16 \approx 51.83$ million subscribers

(c) $y = \frac{43}{12}(5) + 16 \approx 33.92$ million subscribers

(d) Every twelve years the number of subscribers will increase by 43 million.

69. $4x + 3y - 10 = 0 \Rightarrow d = \dfrac{|4(0) + 3(0) - 10|}{\sqrt{4^2+3^2}} = \dfrac{10}{5} = 2$

70. $4x + 3y - 10 = 0 \Rightarrow d = \dfrac{|4(2) + 3(3) - 10|}{\sqrt{4^2+3^2}} = \dfrac{7}{5}$

71. $x - y - 2 = 0 \Rightarrow \dfrac{|1(-2) + (-1)(1) - 2|}{\sqrt{1^2+1^2}} = \dfrac{5}{\sqrt{2}} = \dfrac{5\sqrt{2}}{2}$

72. $x + 1 = 0 \Rightarrow d = \dfrac{|1(6) + (0)(2) + 1|}{\sqrt{1^2 + 0^2}} = 7$

73. A point on the line $x + y = 1$ is $(0, 1)$. The distance from the point $(0, 1)$ to $x + y = 5$ is

$$d = \frac{|1 - 5|}{\sqrt{2}} = \frac{4}{\sqrt{2}} = 2\sqrt{2}.$$

74. A point on the line $3x - 4y = 1$ is $(-1, -1)$. The distance from the point $(-1, -1)$ to $3x - 4y - 10 = 0$ is

$$d = \frac{|-3 + 4 - 10|}{5} = \frac{9}{5}.$$

75. For simplicity, let the vertices of the rhombus be $(0, 0)$, $(a, 0)$, (b, c), and $(a + b, c)$, as shown in the figure. The slopes of the diagonals are then $m_1 = c/(a + b)$ and $m_2 = c/(b - a)$. Since $a^2 = b^2 + c^2$, we have

$$m_1 m_2 = \frac{c}{a + b} \cdot \frac{c}{b - a} = \frac{c^2}{b^2 - a^2} = \frac{c^2}{-c^2} = -1.$$

Therefore, the diagonals are perpendicular.

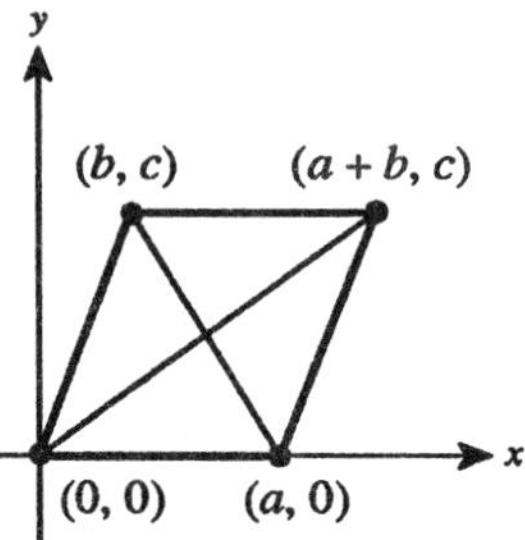

76. For simplicity, let the vertices of the quadrilateral be $(0, 0)$, $(a, 0)$, (b, c), and (d, e), as shown in the figure. The midpoints of the sides are

$$\left(\frac{a}{2}, 0\right), \left(\frac{a + b}{c}, \frac{c}{2}\right), \left(\frac{b + d}{2}, \frac{c + e}{2}\right), \text{ and } \left(\frac{d}{2}, \frac{e}{2}\right).$$

The slopes of the opposite sides are equal.

$$m_1 = \frac{\frac{c}{2} - 0}{\frac{a + b}{2} - \frac{a}{2}} = \frac{\frac{c + e}{2} - \frac{e}{2}}{\frac{b + d}{2} - \frac{d}{2}} = \frac{c}{b}$$

$$m_2 = \frac{0 - \frac{e}{2}}{\frac{a}{2} - \frac{d}{2}} = \frac{\frac{c}{2} - \frac{c + e}{2}}{\frac{a + b}{2} - \frac{b + d}{2}} = -\frac{e}{a - d}$$

Therefore, the figure is a parallelogram.

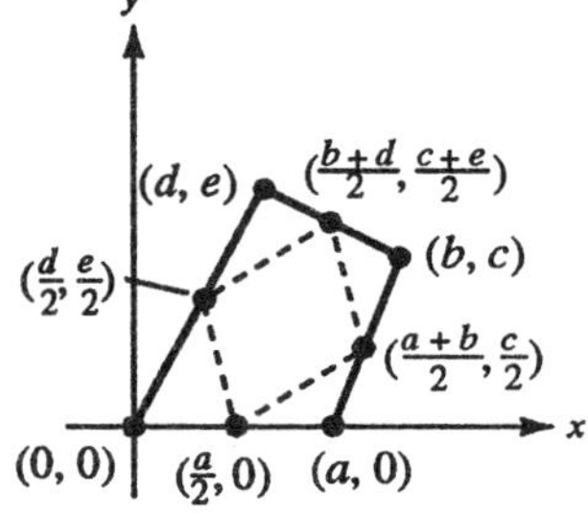

77. Since the triangles are similar, the result immediately follows.

$$\frac{y_2{}^* - y_1{}^*}{x_2{}^* - x_1{}^*} = \frac{y_2 - y_1}{x_2 - x_1}$$

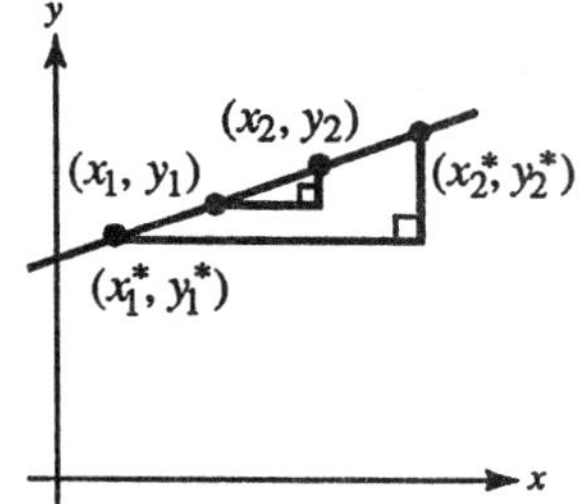

78. If $m_1 = -1/m_2$, then $m_1 m_2 = -1$. Let L_3 be a line with slope m_2 that is perpendicular to L_1. Then $m_1 m_3 = -1$. Hence, $m_2 = m_3 \Rightarrow L_2$ and L_3 are parallel. Therefore, L_2 and L_1 are also perpendicular.

Section 1.5 Functions

1. (a) $f(0) = 2(0) - 3 = -3$

(b) $f(-3) = 2(-3) - 3 = -9$

(c) $f(b) = 2b - 3$

(d) $f(x-1) = 2(x-1) - 3 = 2x - 5$

2. (a) $f\left(\frac{1}{2}\right) = \left(\frac{1}{2}\right)^2 - 2\left(\frac{1}{2}\right) + 2 = \frac{5}{4}$

(b) $f(-1) = (-1)^2 - 2(-1) + 2 = 5$

(c) $f(c) = c^2 - 2c + 2$

(d) $f(x+\Delta x) = (x+\Delta x)^2 - 2(x+\Delta x) + 2$

$= x^2 + 2x\Delta x + (\Delta x)^2 - 2x - 2\Delta x + 2$

3. (a) $f(-2) = \sqrt{-2+3} = \sqrt{1} = 1$

(b) $f(6) = \sqrt{6+3} = \sqrt{9} = 3$

(c) $f(c) = \sqrt{c+3}$

(d) $f(x+\Delta x) = \sqrt{x+\Delta x+3}$

4. (a) $f(2) = \dfrac{1}{\sqrt{2}} = \dfrac{\sqrt{2}}{2}$

(b) $f\left(\dfrac{1}{4}\right) = \dfrac{1}{\sqrt{1/4}} = 2$

(c) $f(x+\Delta x) = \dfrac{1}{\sqrt{x+\Delta x}}$

(d) $f(x+\Delta x) - f(x) = \dfrac{1}{\sqrt{x+\Delta x}} - \dfrac{1}{\sqrt{x}} = \dfrac{\sqrt{x} - \sqrt{x+\Delta x}}{\sqrt{x}\sqrt{x+\Delta x}} = \dfrac{\sqrt{x} - \sqrt{x-\Delta x}}{\sqrt{x^2 + x(\Delta x)}}$

5. (a) $f(2) = \dfrac{|2|}{2} = 1$

(b) $f(-2) = \dfrac{|-2|}{-2} = -1$

(c) $f(x^2) = \dfrac{|x^2|}{x^2} = 1$

(d) $f(x-1) = \dfrac{|x-1|}{x-1}$

6. (a) $f(2) = 6$

(b) $f(-2) = 6$

(c) $f(x^2) = x^2 + 4$

(d) $f(x+\Delta x) - f(x) = |x+\Delta x| + 4 - (|x| + 4)$

$= |x+\Delta x| - |x|$

7. $\dfrac{f(2+\Delta x) - f(2)}{\Delta x} = \dfrac{(2+\Delta x)^2 - (2+\Delta x) + 1 - [(2)^2 - 2 + 1]}{\Delta x}$

$= \dfrac{4 + 4\Delta x + (\Delta x)^2 - 2 - \Delta x + 1 - 4 + 2 - 1}{\Delta x} = 4 - 1 + \Delta x = 3 + \Delta x$

8. $\dfrac{f(1+\Delta x) - f(1)}{\Delta x} = \dfrac{1/(1+\Delta x) - 1}{\Delta x} = \dfrac{1 - (1+\Delta x)}{\Delta x(1+\Delta x)} = \dfrac{-1}{1+\Delta x}$

9. $\dfrac{f(x+\Delta x) - f(x)}{\Delta x} = \dfrac{(x+\Delta x)^3 - x^3}{\Delta x} = \dfrac{x^3 + 3x^2\Delta x + 3x(\Delta x)^2 + (\Delta x)^3 - x^3}{\Delta x} = 3x^2 + 3x\Delta x + (\Delta x)^2$

10. $\dfrac{f(x) - f(1)}{x-1} = \dfrac{3x - 1 - (3-1)}{x-1} = \dfrac{3(x-1)}{x-1} = 3$

11. $\dfrac{f(x) - f(2)}{x-2} = \dfrac{\left(1/\sqrt{x-1} - 1\right)}{x-2} = \dfrac{1 - \sqrt{x-1}}{(x-2)\sqrt{x-1}} \cdot \dfrac{1+\sqrt{x-1}}{1+\sqrt{x-1}} = \dfrac{2-x}{(x-2)\sqrt{x-1}\left(1+\sqrt{x-1}\right)}$

$= \dfrac{-1}{\sqrt{x-1}\left(1+\sqrt{x-1}\right)}$

12. $\dfrac{f(x)-f(1)}{x-1} = \dfrac{x^3-x-0}{x-1} = \dfrac{x(x+1)(x-1)}{x-1} = x(x+1)$

13. $f(x) = 4 - x$
Domain: $(-\infty, \ \infty)$
Range: $(-\infty, \ \infty)$

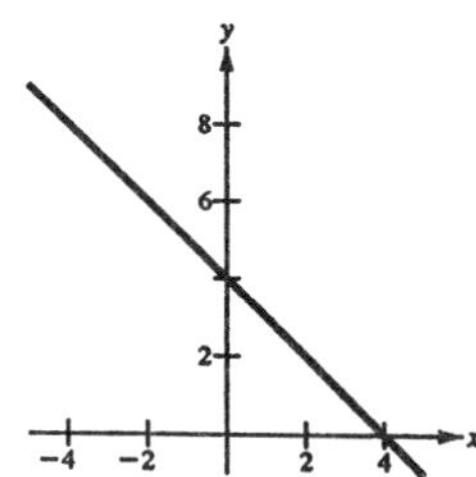

14. $f(x) = \frac{1}{3}x$
Domain: $(-\infty, \ \infty)$
Range: $(-\infty, \ \infty)$

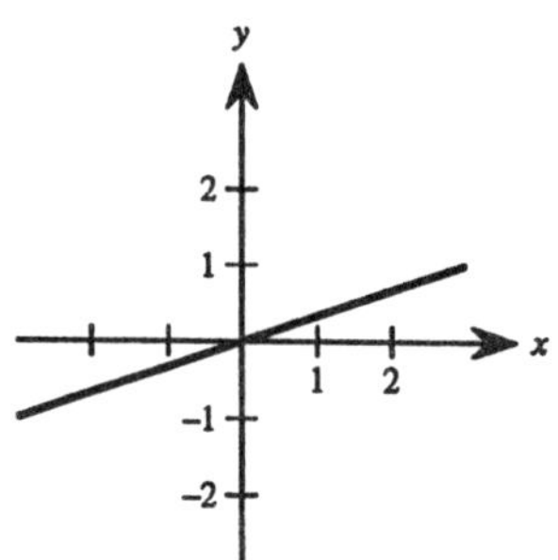

15. $f(x) = 4 - x^2$
Domain: $(-\infty, \ \infty)$
Range: $(-\infty, \ 4]$

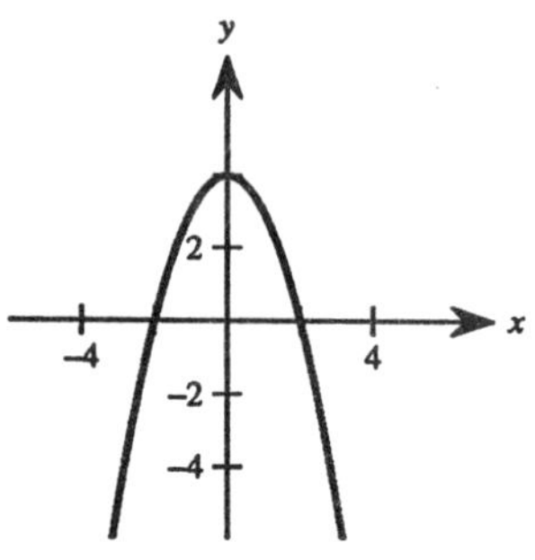

16. $g(x) = \dfrac{4}{x}$
Domain: $(-\infty, \ 0), \ (0, \ \infty)$
Range: $(-\infty, \ 0), \ (0, \ \infty)$

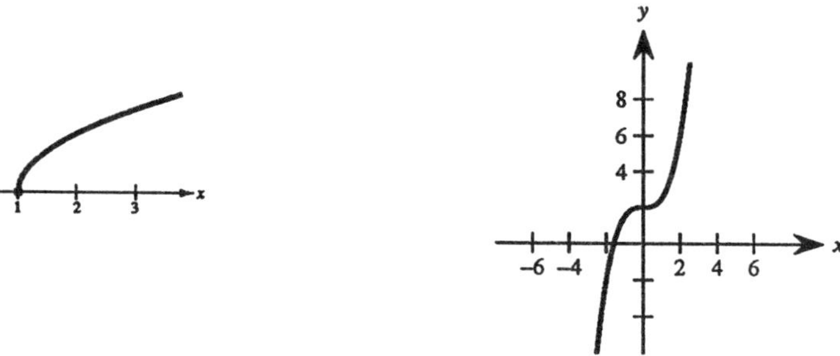

17. $h(x) = \sqrt{x-1}$
Domain: $[1, \ \infty)$
Range: $[0, \ \infty)$

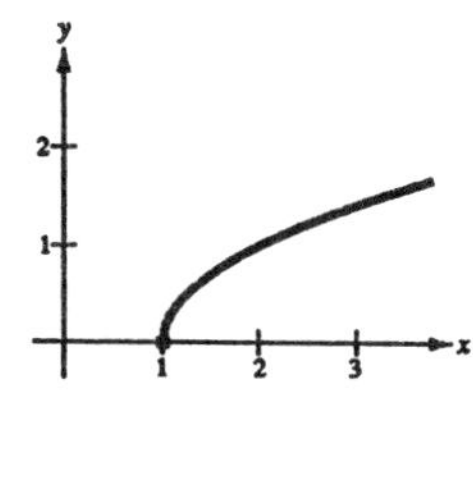

18. $f(x) = \frac{1}{2}x^3 + 2$
Domain: $(-\infty, \ \infty)$
Range: $(-\infty, \ \infty)$

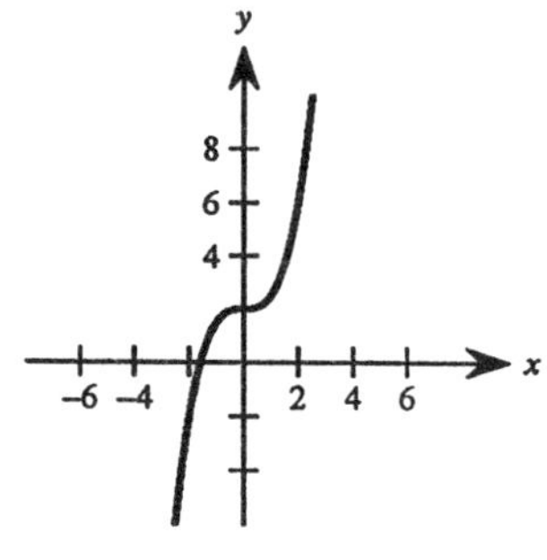

19. $f(x) = \sqrt{9 - x^2}$
Domain: $[-3, \ 3]$
Range: $[0, \ 3]$

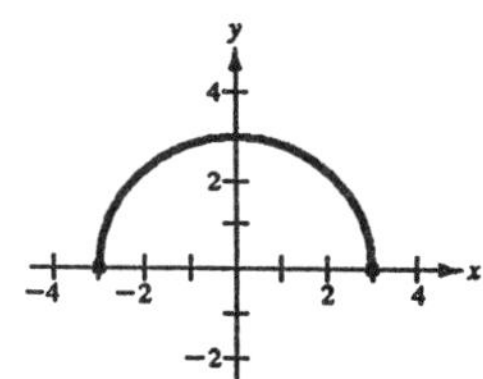

20. $h(x) = \sqrt{25 - x^2}$
Domain: $[-5, \ 5]$
Range: $[0, 5]$

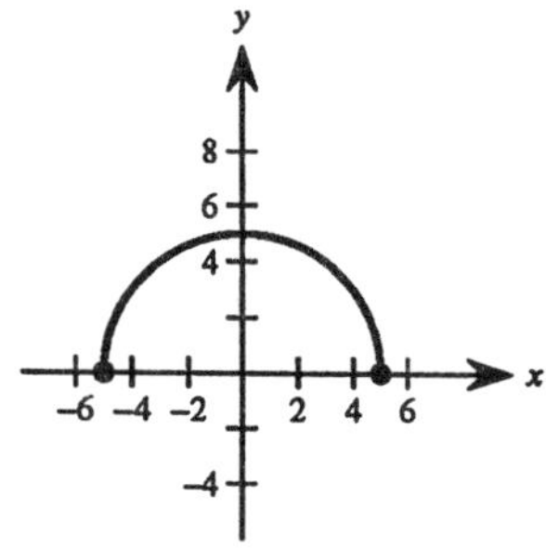

21. $f(x) = |x - 2|$
Domain: $(-\infty, \ \infty)$
Range: $[0, \ \infty)$

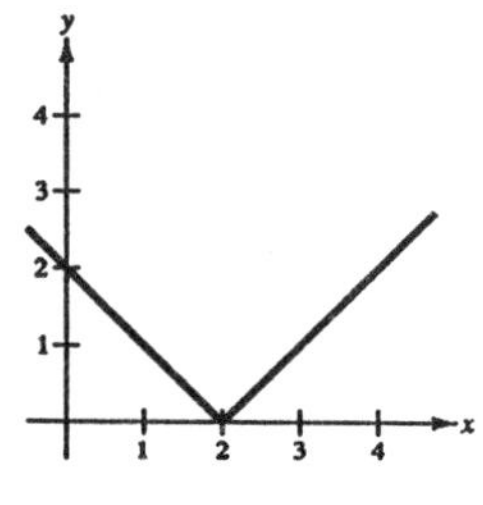

22. $f(x) = \frac{|x|}{x}$

Domain: $(-\infty, 0), (0, \infty)$

Range: $-1, 1$

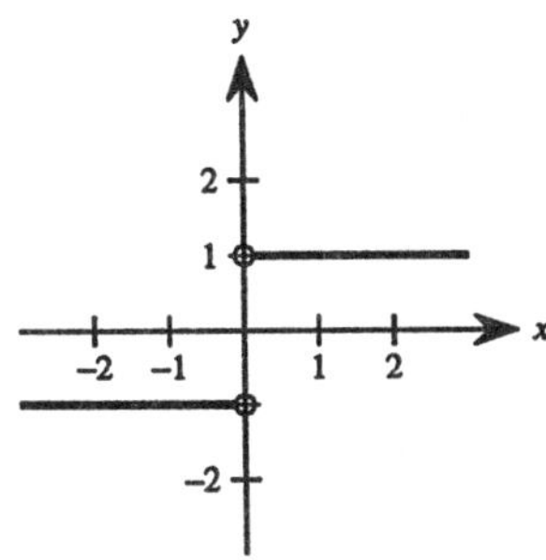

23. $y = x^2$

y is a function of x.

24. $y = x^3 - 1$

y is a function of x.

25. $x - y^2 = 0$

y is not a function of x.

26. $x^2 + y^2 = 9$

y is not a function of x.

27. $\sqrt{x^2 - 4} - y = 0$

y is a function of x.

28. $x - xy + y + 1 = 0$

y is a function of x.

29. $x^2 + y^2 = 4 \Rightarrow y = \pm\sqrt{4 - x^2}$

y is not a function of x since there are two values of y for some x.

30. $x = y^2 \Rightarrow y = \pm\sqrt{x}$

y is not a function of x since there are two values of y for some x.

31. $x^2 + y = 4 \Rightarrow y = 4 - x^2$

y is a function of x since there is one value of y for each x.

32. $x + y^2 = 4 \Rightarrow y = \pm\sqrt{4 - x}$

y is not a function of x since there are two values of y for some x.

33. $2x + 3y = 4 \Rightarrow y = \frac{4 - 2x}{3}$

y is a function of x since there is one value of y for each x.

34. $x^2 + y^2 - 4y = 0 \Rightarrow y^2 - 4y + 4 = -x^2 + 4$

$$(y - 2)^2 = 4 - x^2$$

$$y - 2 = \pm\sqrt{4 - x^2}$$

$$y = 2 \pm \sqrt{4 - x^2}$$

y is not a function of x since there are two values of y for some x.

35. $y^2 = x^2 - 1 \Rightarrow y = \pm\sqrt{x^2 - 1}$

y is not a function of x since there are two values of y for some x.

36. $x^2y - x^2 + 4y = 0 \Rightarrow y = \frac{x^2}{x^2 + 4}$

y is a function of x since there is one value of y for each x.

37. (a) $y = \sqrt{x} + 2$

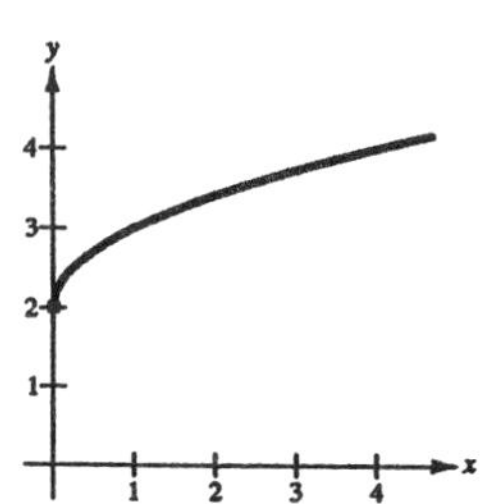

(b) $y = -\sqrt{x}$

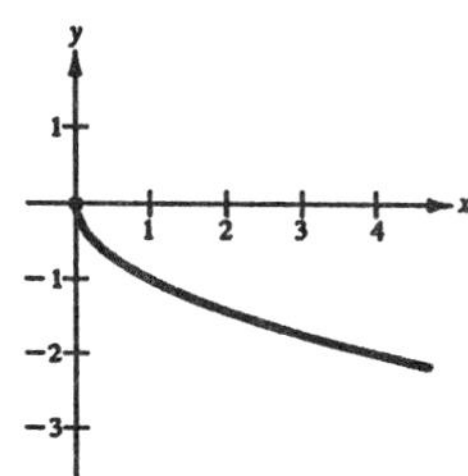

(c) $y = \sqrt{x - 2}$

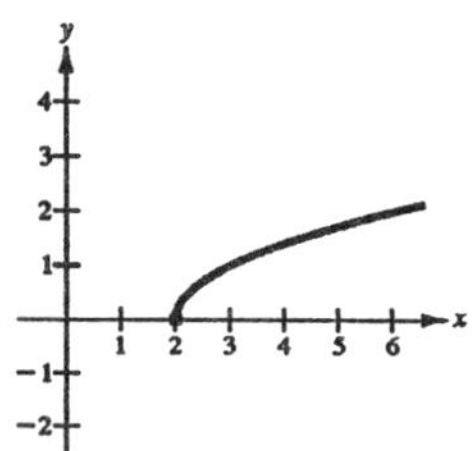

(d) $y = \sqrt{x + 3}$

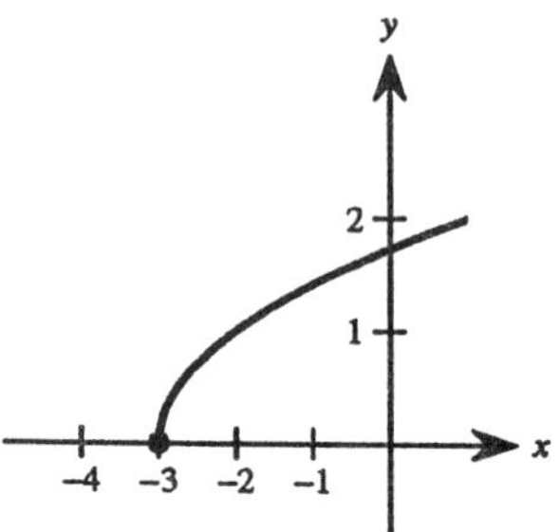

(e) $y = \sqrt{x - 4}$

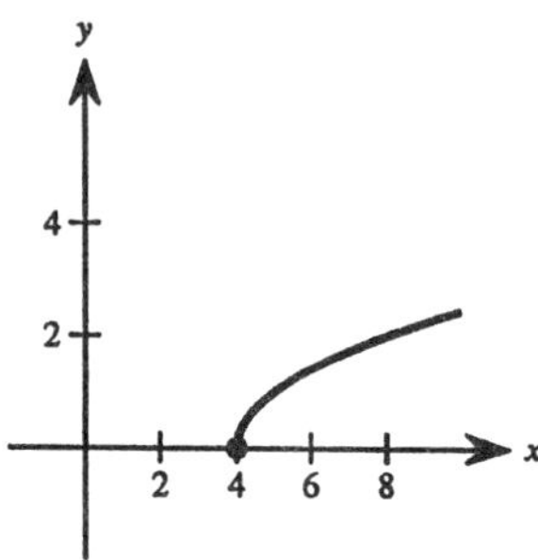

(f) $y = 2\sqrt{x}$

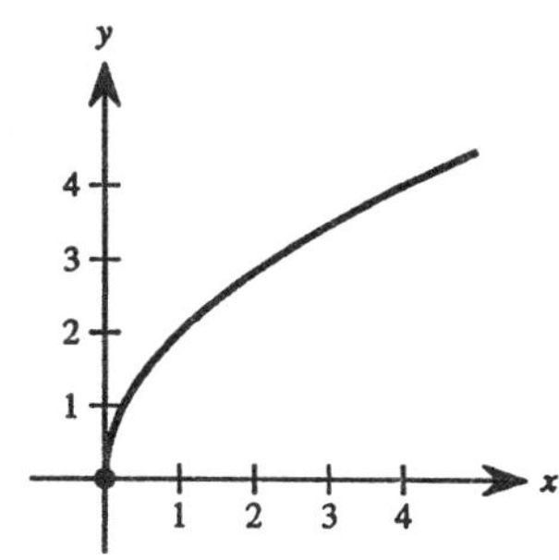

38. (a) $y = \dfrac{1}{x} - 1$

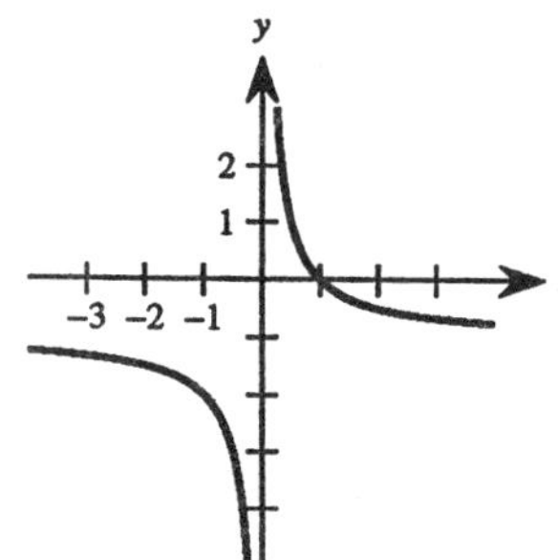

(b) $y = \dfrac{1}{x + 1}$

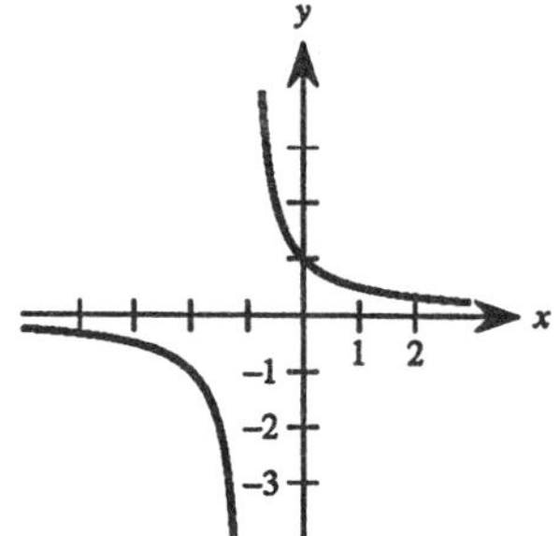

(c) $y = \dfrac{1}{x - 1}$

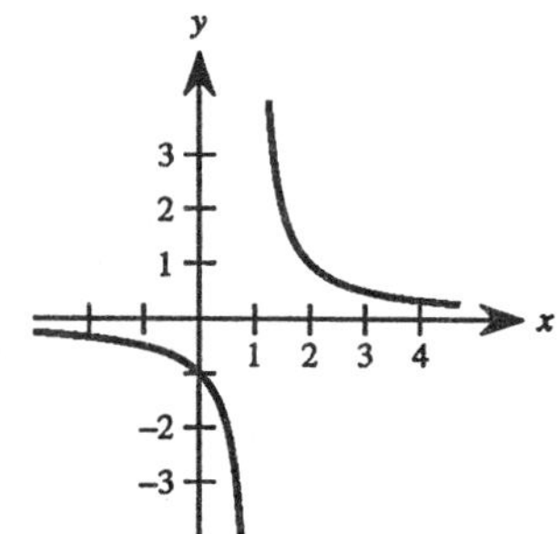

(d) $y = -\dfrac{1}{x}$

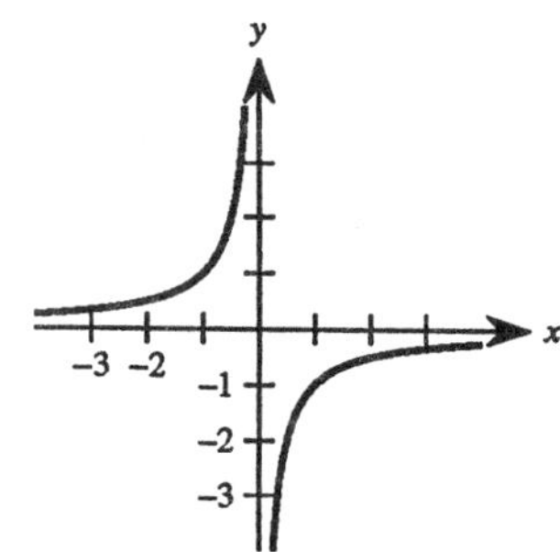

(e) $y = \dfrac{4}{x}$

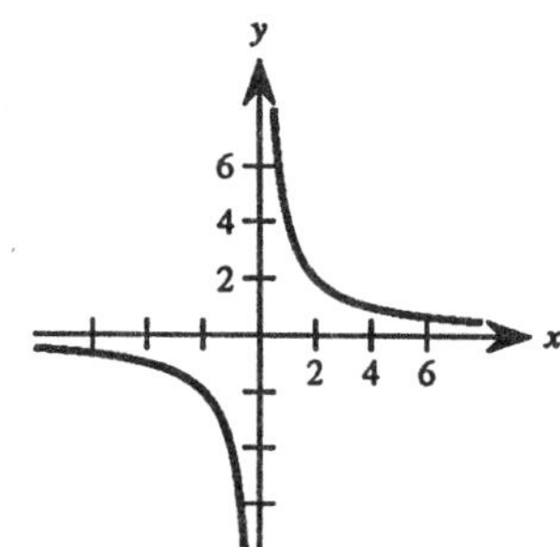

(f) $y = -\dfrac{1}{x} + 2$

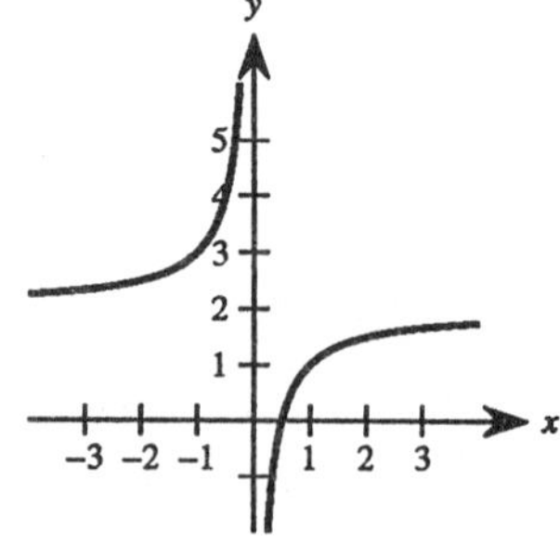

39. (a) $y = (x + 3)^2$

(b) $y = x^2 + 3$

40. (a) $f(x) = 3x^4 + 4x^3$

$f \to \infty$ as $x \to -\infty$

$f \to \infty$ as $x \to \infty$

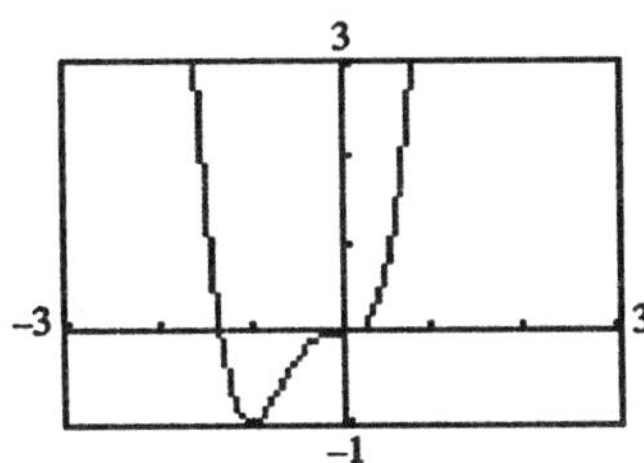

(b) $f(x) = 2x^5 - 3x^2 + 1$

$f \to -\infty$ as $x \to -\infty$

$f \to \infty$ as $x \to \infty$

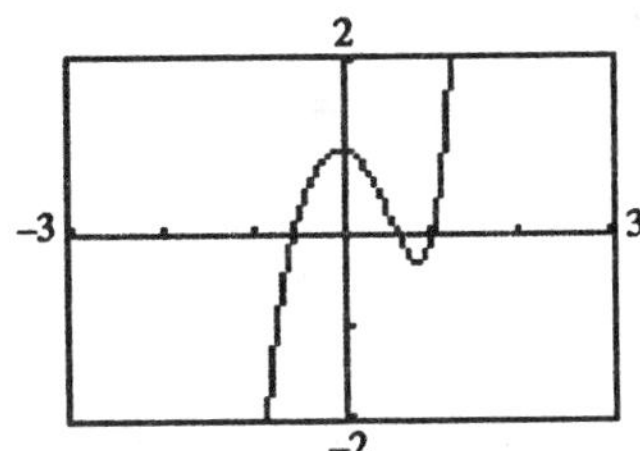

(c) $f(x) = -2x^3 + 3x - 4$

$f \to \infty$ as $x \to -\infty$

$f \to -\infty$ as $x \to \infty$

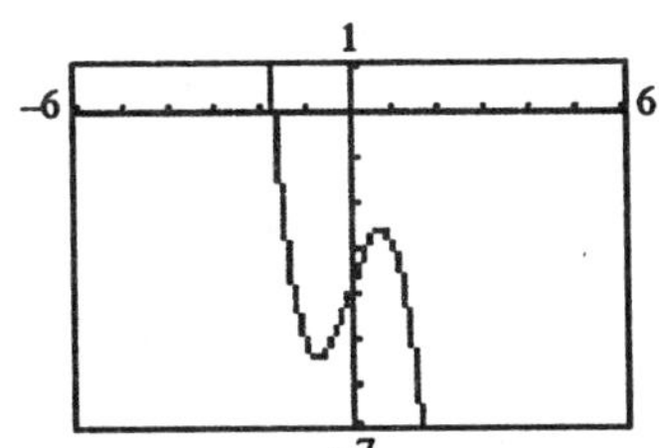

(d) $f(x) = -10x^6 - 3x^5 + 2x$

$f \to -\infty$ as $x \to -\infty$

$f \to -\infty$ as $x \to \infty$

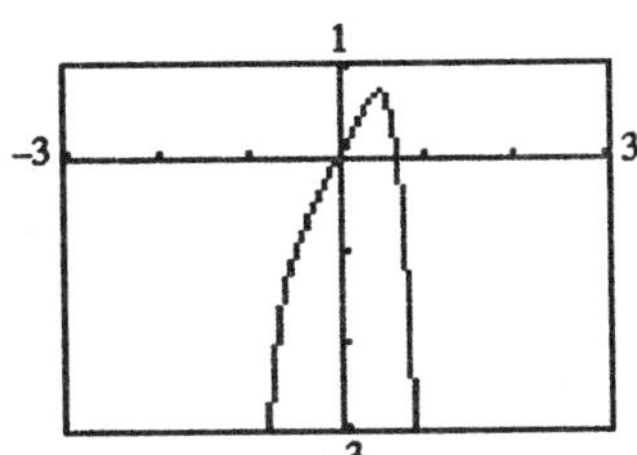

If the degree is even and the leading coefficient is positive, then the graph rises to the left and to the right. If the degree is even and the leading coefficient is negative, then the graph falls to the left and to the right. If the degree is odd and the leading coefficient is positive, then the graph rises to the right and falls to the left. If the degree is odd and the leading coefficient is negative, then the graph falls to the right and rises to the left.

41. (a) $f(g(1)) = f(0) = 0$

(b) $g(f(1)) = g(1) = 0$

(c) $g(f(0)) = g(0) = -1$

(d) $f(g(-4)) = f(15) = \sqrt{15}$

(e) $f(g(x)) = f(x^2 - 1) = \sqrt{x^2 - 1}$

(f) $g(f(x)) = g(\sqrt{x}) = (\sqrt{x})^2 - 1 = x - 1$

42. (a) $f(g(2)) = f(3) = \dfrac{1}{3}$

(b) $g(f(2)) = g\left(\dfrac{1}{2}\right) = -\dfrac{3}{4}$

(c) $f\left(g\left(\dfrac{1}{\sqrt{2}}\right)\right) = f\left(-\dfrac{1}{2}\right) = -2$

(d) $g\left(f\left(\dfrac{1}{\sqrt{2}}\right)\right) = g(\sqrt{2}) = 1$

(e) $g(f(x)) = g\left(\dfrac{1}{x}\right) = \left(\dfrac{1}{x}\right)^2 - 1 = \dfrac{1 - x^2}{x^2}$

(f) $f(g(x)) = f(x^2 - 1) = \dfrac{1}{x^2 - 1}$

43. $f(x) = x^2$, $g(x) = \sqrt{x}$

$(f \circ g)(x) = f(g(x)) = f(\sqrt{x}) = (\sqrt{x})^2 = x;$ Domain: $[0, \infty)$

$(g \circ f)(x) = g(f(x)) = g(x^2) = \sqrt{x^2} = |x|;$ Domain: $(-\infty, \infty)$

Yes, $(f \circ g) = (g \circ f)$ for $x \geq 0$.

44. $f(x) = x^3$, $g(x) = \sqrt[3]{x}$

$(f \circ g)(x) = f(g(x)) = f(\sqrt[3]{x}) = (\sqrt[3]{x})^3 = x;$ Domain: $(-\infty, \infty)$

$(g \circ f)(x) = g(f(x)) = g(x^3) = \sqrt[3]{x^3} = x;$ Domain: $(-\infty, \infty)$

Yes, $(f \circ g) = (g \circ f)$.

45. $f(x) = x + 1,\ g(x) = \dfrac{1}{x}$

$(f \circ g)(x) = f(g(x)) = f\left(\dfrac{1}{x}\right) = \dfrac{1}{x} + 1 = \dfrac{1+x}{x}$; Domain: $(-\infty,\ 0),\ (0,\ \infty)$

$(g \circ f)(x) = g(f(x)) = g(x+1) = \dfrac{1}{x+1}$; Domain: $(-\infty,\ -1),\ (-1,\ \infty)$

No, $(f \circ g) \neq (g \circ f)$.

46. $f(x) = x^2 - 1,\ g(x) = x$

$(f \circ g)(x) = f(g(x)) = f(x) = x^2 - 1$; Domain: $(-\infty,\ \infty)$

$(g \circ f)(x) = g(f(x)) = g(x^2 - 1) = x^2 - 1$; Domain: $(-\infty,\ \infty)$

Yes, $(f \circ g) = (g \circ f)$.

47.
$$x^2 - 9 = 0$$
$$x^2 = 9$$
$$x = \pm 3$$

48.
$$x^3 - x = 0$$
$$x(x^2 - 1) = 0$$
$$x(x+1)(x-1) = 0$$
$$x = 0,\ x = -1,\ x = 1$$

49.
$$\frac{3}{x-1} + \frac{4}{x-2} = 0$$
$$\frac{3}{x-1} = -\frac{4}{x-2}$$
$$3(x-2) = -4(x-1)$$
$$3x - 6 = -4x + 4$$
$$7x = 10$$
$$x = \frac{10}{7}$$

50.
$$a + \frac{b}{x} = 0$$
$$\frac{b}{x} = -a$$
$$b = -ax$$
$$\frac{b}{-a} = x$$
$$x = -\frac{b}{a}$$

51.
$$f(-x) = 4 - (-x)^2$$
$$= 4 - x^2 = f(x)$$

Even

52.
$$f(-x) = \sqrt[3]{-x}$$
$$= -\sqrt[3]{x} = -f(x)$$

Odd

53.
$$f(-x) = (-x)[4 - (-x)^2]$$
$$= -x(4 - x^2) = -f(x)$$

Odd

54. $f(-x) = 4(-x) - (-x)^2 = -4x - x^2$

Neither odd nor even

55.
$$f(-x) = a_{2n+1}(-x)^{2n+1} + \cdots + a_3(-x)^3 + a_1(-x)$$
$$= -[a_{2n+1}x^{2n+1} + \cdots + a_3x^3 + a_1x] = -f(x)$$

Odd

56.
$$f(-x) = a_{2n}(-x)^{2n} + a_{2n-2}(-x)^{2n-2} + \cdots + a_2(-x)^2 + a_0$$
$$= a_{2n}x^{2n} + a_{2n-2}x^{2n-2} + \cdots + a_2x^2 + a_0 = f(x)$$

Even

57. Let $F(x) = f(x)g(x)$ where f and g are even. Then

$$F(-x) = f(-x)g(-x) = f(x)g(x) = F(x).$$

Thus, $F(x)$ is even. Let $F(x) = f(x)g(x)$ where f and g are odd. Then

$$F(-x) = f(-x)g(-x) = [-f(x)][-g(x)] = f(x)g(x) = F(x).$$

Thus, $F(x)$ is even.

58. Let $F(x) = f(x)g(x)$ where f is even and g is odd. Then

$$F(-x) = f(-x)g(-x) = f(x)[-g(x)] = -f(x)g(x) = -F(x).$$

Thus, $F(x)$ is odd.

59. $R = 4 - \dfrac{x^2}{2}$

$r = 2$

60. $R = x^2$

$r = x^3$

61. $h = x^2$

$p = x$

62. $h = \sqrt{4 - x^2}$

$p = x$

63. $2x + 2y = 100 \Rightarrow A = xy = \dfrac{x(100 - 2x)}{2} = x(50 - x)$

64. $4x + 3y = 200 \Rightarrow A = 2xy = 2x\left[\dfrac{200 - 4x}{3}\right] = \dfrac{8}{3}(50x - x^2)$

65. $V = lwh = x(12 - 2x)^2$

66. $y = \sqrt{25 - x^2}$

$A = 2xy = 2x\sqrt{25 - x^2}$

67. $4x + y = 108$

$V = x^2y = x^2(108 - 4x) = 108x^2 - 4x^3$

68. $2x^2 + 4xy = 100$

$V = x^2y = x^2\left(\dfrac{100 - 2x^2}{4x}\right) = 25x - \dfrac{x^3}{2}$

69. $T = \dfrac{D}{R}$

$$\begin{aligned} T &= T_{\text{row}} + T_{\text{walk}} \\ &= \frac{\sqrt{x^2 + 4}}{2} + \frac{\sqrt{1 + (3 - x)^2}}{4} \\ &= \frac{\sqrt{x^2 + 4}}{2} + \frac{\sqrt{x^2 - 6x + 10}}{4} \end{aligned}$$

70. $A = \pi r^2 = \pi y^2 = \pi\left(\sqrt{x}\right)^2 = \pi x$

71. $(A \circ r)(t) = A(r(t)) = A(0.6t) = \pi(0.6t)^2 = 0.36\pi t^2$

$(A \circ r)(t)$ represents the area of the circle at time t.

72. (a)

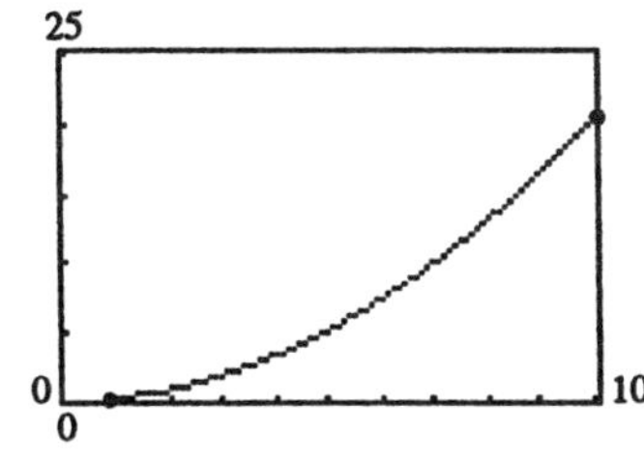

(b) $H(1.6x) = 0.002(1.6x)^2 + 0.005(1.6x) - 0.029$

$= 0.00512x^2 + 0.008x - 0.029,$

$\dfrac{10}{1.6} < x < \dfrac{100}{1.6} \Rightarrow 6.25 < x < 62.5$

Chapter 1 Review Exercises

1. $$|x-2| \le 3$$
$$-3 \le x-2 \le 3$$
$$-1 \le x \le 5$$

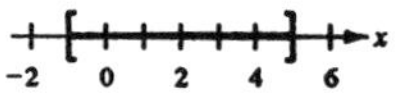

2. $|3x-2| \le 0 \Rightarrow 3x-2=0$
$$x = \frac{2}{3}$$

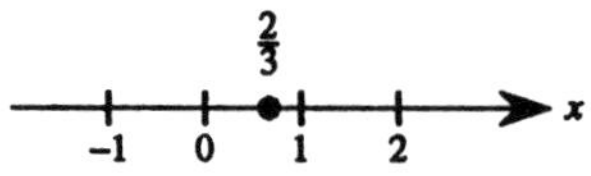

3. $4 < (x+3)^2$
$$4 < x^2+6x+9$$
$$0 < x^2+6x+5$$
$$0 < (x+1)(x+5)$$
$$x < -5 \text{ or } x > -1$$

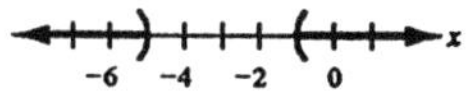

4. $\dfrac{1}{|x|} < 1$
$$|x| > 1$$
$$x > 1 \text{ or } x < -1$$

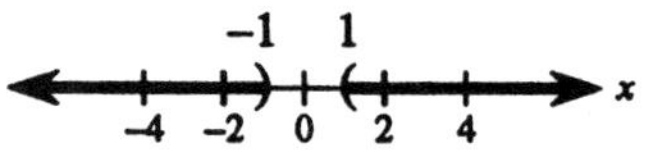

5. $\dfrac{(7/8)+(10/4)}{2} = \dfrac{27/8}{2} = \dfrac{27}{16}$

6. $\dfrac{-1+(3/2)}{2} = \dfrac{1}{4}$

7. $\left(\dfrac{1-3}{2}, \dfrac{4+2}{2}\right) = (-1, 3)$
$$\left(\frac{1+5}{2}, \frac{4+0}{2}\right) = (3, 2)$$
$$\left(\frac{5-3}{2}, \frac{0+2}{2}\right) = (1, 1)$$

8. Let the coordinates of the vertices be (x_1, y_2), (x_2, y_2), and (x_3, y_3).
$$\left(\frac{x_1+x_2}{2}, \frac{y_1+y_2}{2}\right) = (0, 2) \Rightarrow \frac{x_1+x_2}{2} = 0, \quad \frac{y_1+y_2}{2} = 2$$
$$\left(\frac{x_1+x_3}{2}, \frac{y_1+y_3}{2}\right) = (2, 1) \Rightarrow \frac{x_1+x_3}{2} = 2, \quad \frac{y_1+y_3}{2} = 1$$
$$\left(\frac{x_2+x_3}{2}, \frac{y_2+y_3}{2}\right) = (1, -1) \Rightarrow \frac{x_2+x_3}{2} = 1, \quad \frac{y_2+y_3}{2} = -1$$
From the first set of equations we have $x_2 = -x_1$, $y_2 = 4 - y_1$.
From the second set of equations we have $x_3 = 4 - x_1$, $y_3 = 2 - y_1$.
From the third set of equations we have
$$\frac{-x_1+(4-x_1)}{2} = 1 \Rightarrow x_1 = 1$$
$$\frac{(4-y_1)+(2-y_1)}{2} = -1 \Rightarrow y_1 = 4.$$
Using these values we have $x_2 = -1$, $y_2 = 0$, $x_3 = 3$, and $y_3 = -2$. Therefore, the vertices are $(1, 4)$, $(-1, 0)$, and $(3, -2)$.

9. $(x^2+6x+9)+(y^2-2y+1)=-1+9+1$

$$(x+3)^2+(y-1)^2=9$$

Center: $(-3,\ 1)$
Radius: 3

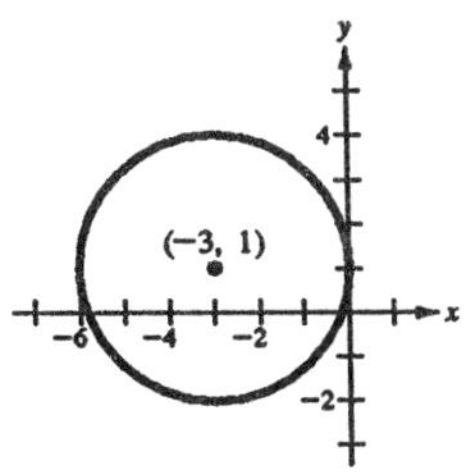

10. $4\left(x^2-x+\frac{1}{4}\right)+4(y^2+2y+1)=11+1+4$

$$\left(x-\tfrac{1}{2}\right)^2+(y+1)^2=4$$

Center: $\left(\frac{1}{2},\ -1\right)$
Radius: 2

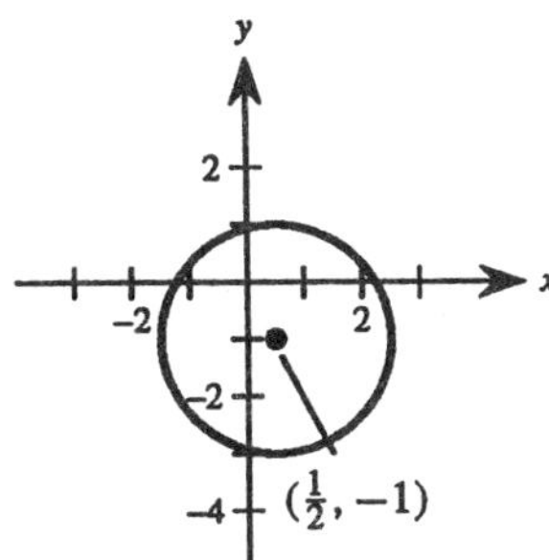

11. $(x^2+6x+9)+(y^2-2y+1)=-10+9+1$

$$(x+3)^2+(y-1)^2=0$$

Point: $(-3,\ 1)$

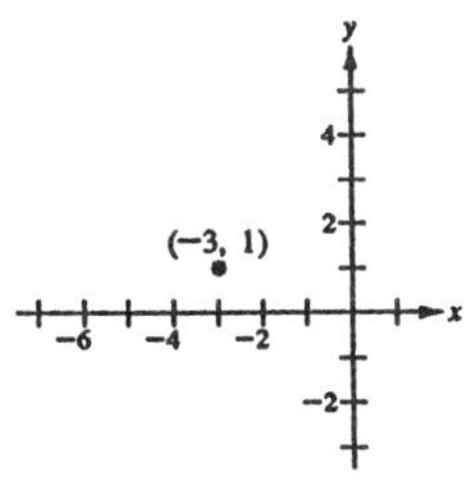

12. $(x^2-6x+9)+(y^2+8y+16)=9+16$

$$(x-3)^2+(y+4)^2=25$$

Center: $(3,\ -4)$
Radius: 5

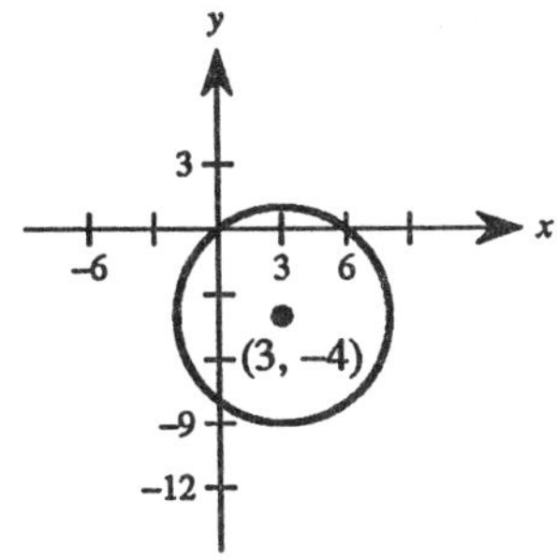

13. $(x^2-6x+9)+(y^2+8y+16)=c+9+16$

$$(x-3)^2+(y+4)^2=c+25$$

If the radius is 2, then

$$c+25=4$$

$$c=-21.$$

14. $\sqrt{(x+2)^2+(y-0)^2}=2\sqrt{(x-3)^2+(y-1)^2}$

$$(x+2)^2+y^2=4[(x-3)^2+(y-1)^2]$$

$$x^2+4x+4+y^2=4x^2-24x+4y^2-8y+40$$

$$0=3x^2+3y^2-28x-8y+36$$

$$\left(x-\tfrac{14}{3}\right)^2+\left(y-\tfrac{4}{3}\right)^2=\tfrac{104}{9}$$

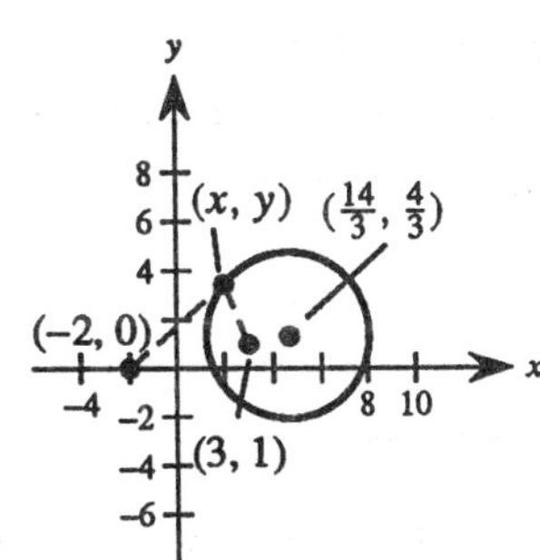

15. $(x-1)^2+(y-2)^2=9$

$x^2-2x+1+y^2-4y+4=9$

$x^2+y^2-2x-4y=4$

(a) $(1, 5)$ on the circle since $1^2+5^2-2(1)-4(5)=4$

(b) $(0, 0)$ inside the circle since $0^2+0^2-2(0)-4(0)<4$

(c) $(-2, 1)$ outside the circle since $(-2)^2+1^2-2(-2)-4(1)>4$

(d) $(0, 4)$ inside the circle since $0^2+4^2-2(0)-4(4)<4$

16. $(x-2)^2+(y-1)^2=4$

$x^2+y^2-4x-2y=-1$

(a) $(1, 1)$ inside the circle since $(1)^2+(1)^2-4(1)-2(1)<-1$

(b) $(4, 2)$ outside the circle since $(4)^2+(2)^2-4(4)-2(2)>-1$

(c) $(0, 1)$ on the circle since $(0)^2+(1)^2-4(0)-2(1)=-1$

(d) $(3, 1)$ inside the circle since $(3)^2+(1)^2-4(3)-2(1)<-1$

17. $y=-\frac{1}{2}x+\frac{3}{2}$

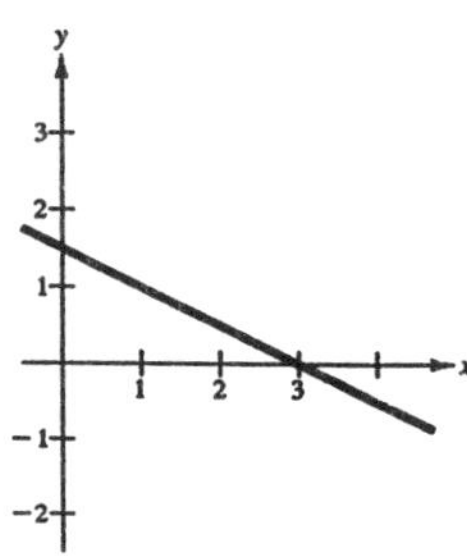

18. $y=1+\dfrac{1}{x}$

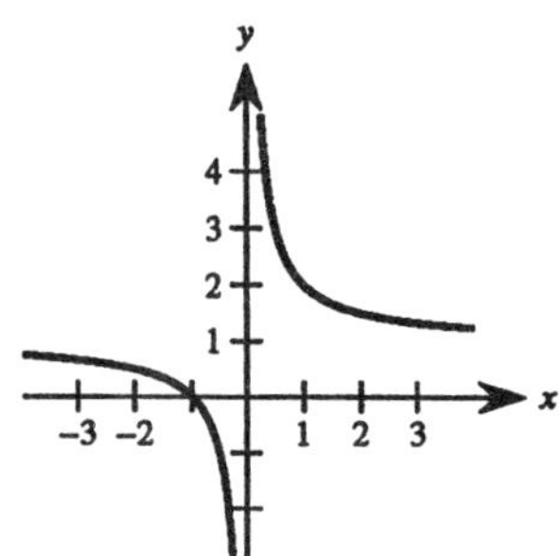

19. $y=7-6x-x^2$

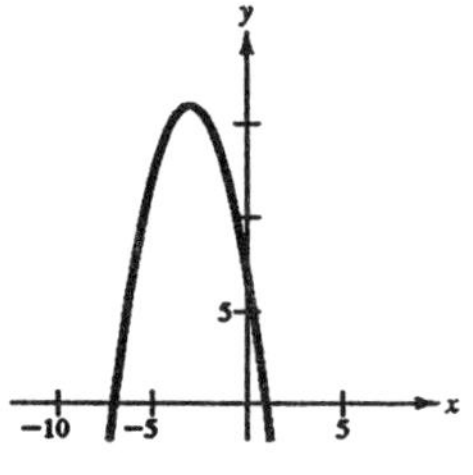

20. $y=x(6-x)$

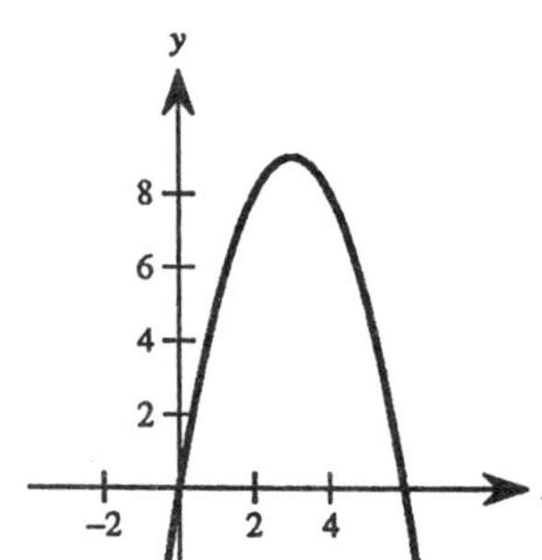

21. The slope of the line segment from $(-1, 3)$ to $(2, 9)$ is $(9-3)/(2+1)=2$. The slope of the line segment from $(2, 9)$ to $(3, 1)$ is $(1-9)/(3-2)=-8$. The points do not lie on the same line.

22. The slope of the line segment from $(2, 5)$ to $(4, 10)$ is $(10-5)/(4-2)=5/2$. The slope of the line segment from $(4, 10)$ to $(6, 20)$ is $(20-10)/(6-4)=5$. The points do not lie on the same line.

23. $4x - 2y = 6$

$$y = 2x - 3$$

Slope: 2
y-intercept: -3

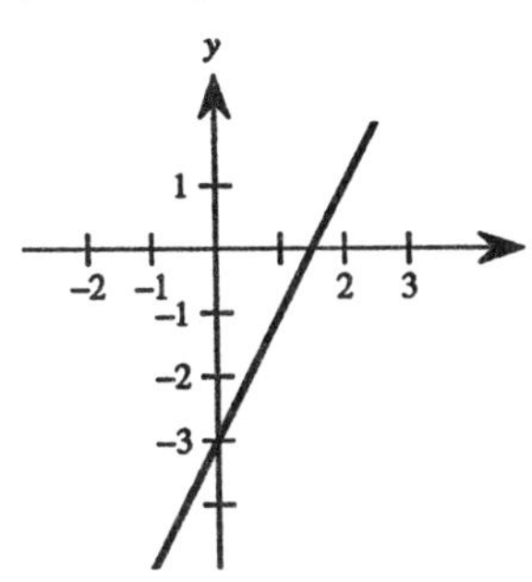

24. $0.02x + 0.15y = 0.25$

$$2x + 15y = 25$$

$$y = -\tfrac{2}{15}x + \tfrac{5}{3}$$

Slope: $-\frac{2}{15}$
y-intercept: $\frac{5}{3}$

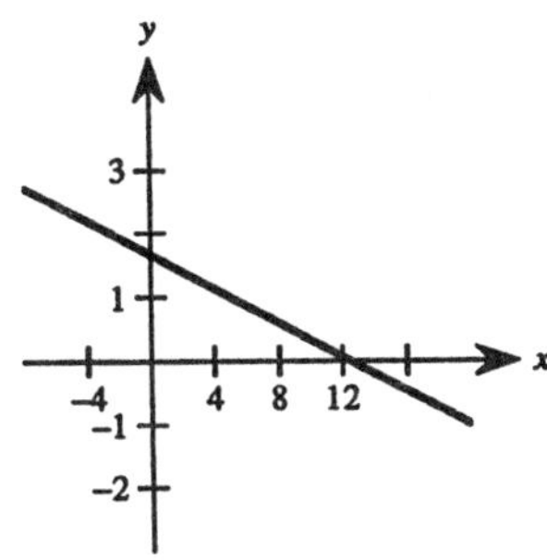

25. $-\frac{1}{3}x + \frac{5}{6}y = 1$

$$-\tfrac{2}{5}x + y = \tfrac{6}{5}$$

$$y = \tfrac{2}{5}x + \tfrac{6}{5}$$

Slope: $\frac{2}{5}$
y-intercept: $\frac{6}{5}$

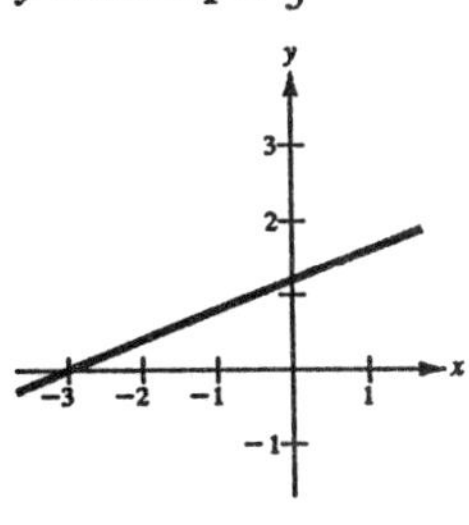

26. $51x + 17y = 102$

$$y = -3x + 6$$

Slope: -3
y-intercept: 6

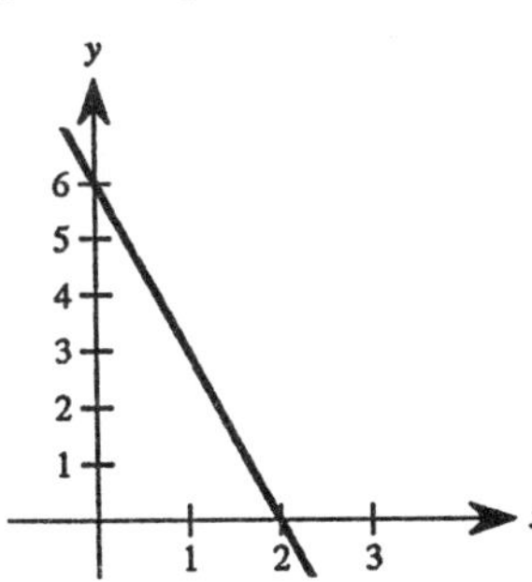

27. (a) $y - 4 = \dfrac{7}{16}(x + 2)$

$$y = \frac{7}{16}x + \frac{39}{8}$$

$$0 = 7x - 16y + 78$$

(b) $y - 4 = \dfrac{5}{3}(x + 2)$

$$y = \frac{5}{3}x + \frac{22}{3}$$

$$0 = 5x - 3y + 22$$

(c) $m = \dfrac{4 - 0}{-2 - 0} = -2$

$$y = -2x$$

$$2x + y = 0$$

(d) $x = -2$

$$x + 2 = 0$$

28. (a) $y - 3 = -\dfrac{2}{3}(x - 1)$

$$y = -\frac{2}{3}x + \frac{11}{3}$$

$$2x + 3y - 11 = 0$$

(b) $y - 3 = 1(x - 1)$

$$y = x + 2$$

$$0 = x - y + 2$$

(c) $m = \dfrac{4 - 3}{2 - 1} = 1$

$$y - 3 = 1(x - 1)$$

$$y = x + 2$$

$$0 = x - y + 2$$

(d) $y = 3$

$$y - 3 = 0$$

29. $\left(\dfrac{x + 2}{2}, \dfrac{y + 3}{2}\right) = (-1, 4)$

$$x = -4$$

$$y = 5$$

Other endpoint: $(-4, 5)$

30. $\sqrt{(x-0)^2+(y-0)^2} = \sqrt{(x-2)^2+(y-3)^2} = \sqrt{(x-3)^2+(y+2)^2}$

$x^2+y^2 = x^2+y^2-4x-6y+13 = x^2+y^2-6x+4y+13$

$4x+6y=13 \qquad 2x-10y=0$

$$\begin{aligned} 4x+6y &= 13 \\ -4x+20y &= 0 \\ \hline 26y &= 13 \\ y &= \tfrac{1}{2} \end{aligned} \qquad \begin{aligned} 2x-10\left(\tfrac{1}{2}\right) &= 0 \\ x &= \tfrac{5}{2} \end{aligned}$$

Point: $\left(\frac{5}{2}, \frac{1}{2}\right)$

31.
$$\begin{aligned} 3x-4y &= 8 \\ 4x+4y &= 20 \\ \hline 7x \quad &= 28 \\ x &= 4 \\ y &= 1 \end{aligned}$$

Point: (4, 1)

32. $y = x+1$

$(x+1)-x^2 = 7$

$0 = x^2-x+6$

No real solution
No points of intersection

33. $v = 850a + 300{,}000$
Domain: $\{a\colon a \ge 0\}$

34. $v = 3.25b$
Domain: $\{b\colon b \ge 0\}$

35. $s = 6x^2$
Domain: $\{x\colon x \ge 0\}$

36. $s = 4\pi r^2$
Domain: $\{r\colon r \ge 0\}$

37. $d = 45t$
Domain: $\{t\colon t \ge 0\}$

38. $a = \dfrac{\sqrt{3}x^2}{4}$
Domain: $\{x\colon x \ge 0\}$

39. $x + y = 500$

$P = xy = x(500-x) = 500x - x^2$

40. $xy = 120$

$S = x + y = x + \dfrac{120}{x} = \dfrac{x^2+120}{x}$

41. $x^2 - y = 0$

Function since there is one value of y for each x

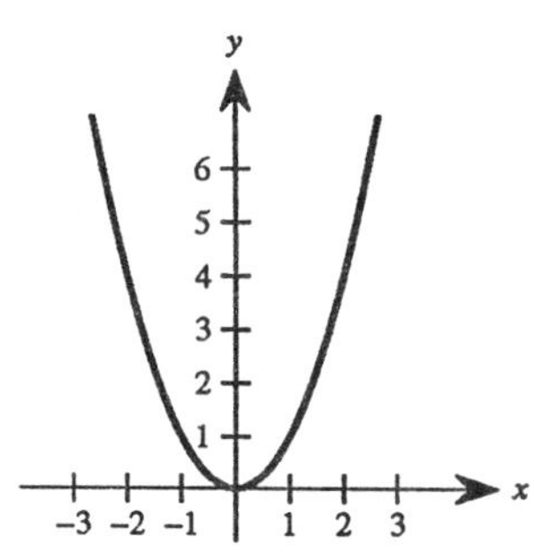

42. $x^2 + 4y^2 = 16$

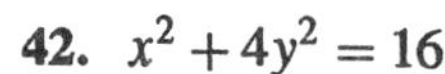

$y = \pm\dfrac{\sqrt{16-x^2}}{2}$

Not a function since there are two values of y for some x

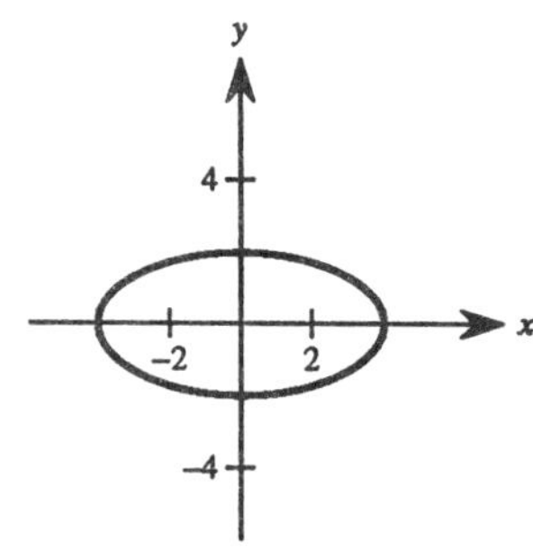

43. $x - y^2 = 0$

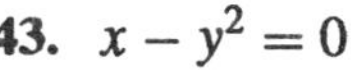

$y = \pm\sqrt{x}$

Not a function since there are two values of y for some x

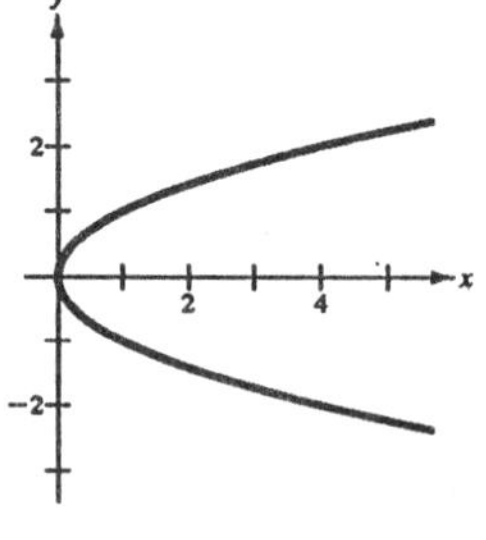

44. $x^3 - y^2 + 1 = 0$

$$y = \pm\sqrt{x^3 + 1}$$

Not a function since there are two values of y for some x

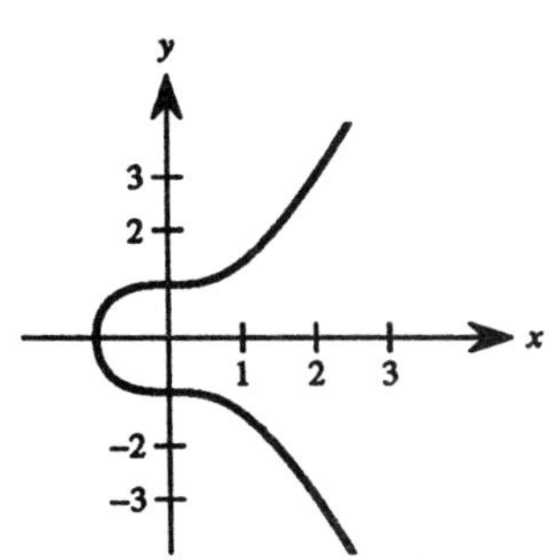

45. $y = x^2 - 2x$

Function since there is one value of y for each x

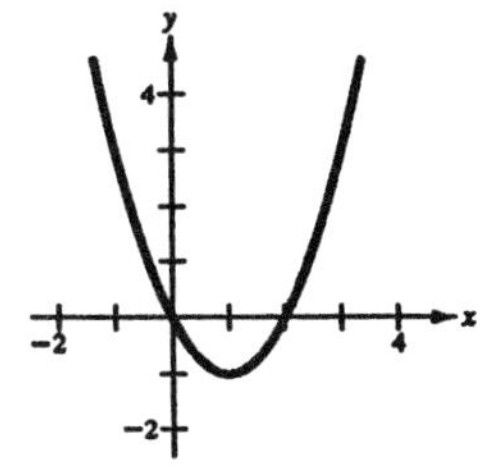

46. $y = 36 - x^2$

Function since there is one value of y for each x

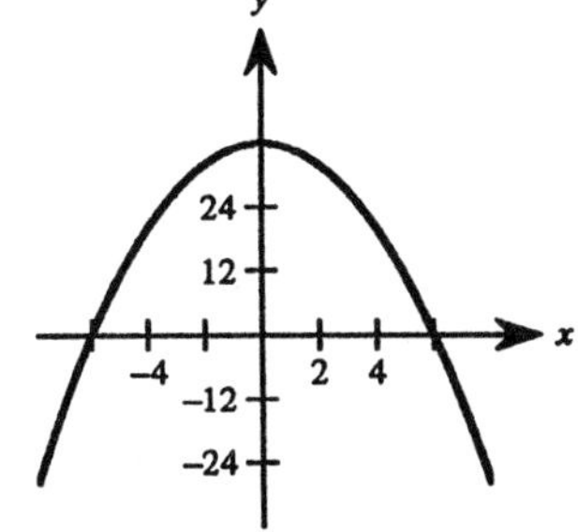

47. $f(x) = 1 - x^2, \quad g(x) = 2x + 1$

(a) $f(x) + g(x) = -x^2 + 2x + 2$

(b) $f(x) - g(x) = -x^2 - 2x$

(c) $f(x)g(x) = (1 - x^2)(2x + 1)$

$$= -2x^3 - x^2 + 2x + 1$$

(d) $\dfrac{f(x)}{g(x)} = \dfrac{1 - x^2}{2x + 1}$

(e) $f[g(x)] = 1 - (2x + 1)^2$

$$= 1 - (4x^2 + 4x + 1) = -4x^2 - 4x$$

(f) $g[f(x)] = 2(1 - x^2) + 1 = 3 - 2x^2$

48. $f(x) = 2x - 3, \quad g(x) = \sqrt{x + 1}$

(a) $f(x) + g(x) = 2x - 3 + \sqrt{x + 1}$

(b) $f(x) - g(x) = 2x - 3 - \sqrt{x + 1}$

(c) $f(x)g(x) = (2x - 3)\sqrt{x + 1}$

(d) $\dfrac{f(x)}{g(x)} = \dfrac{2x - 3}{\sqrt{x + 1}}$

(e) $f(g(x)) = 2\sqrt{x + 1} - 3$

(f) $g(f(x)) = \sqrt{(2x - 3) + 1}$

$$= \sqrt{2x - 2} = \sqrt{2(x - 1)}$$

49. (a)

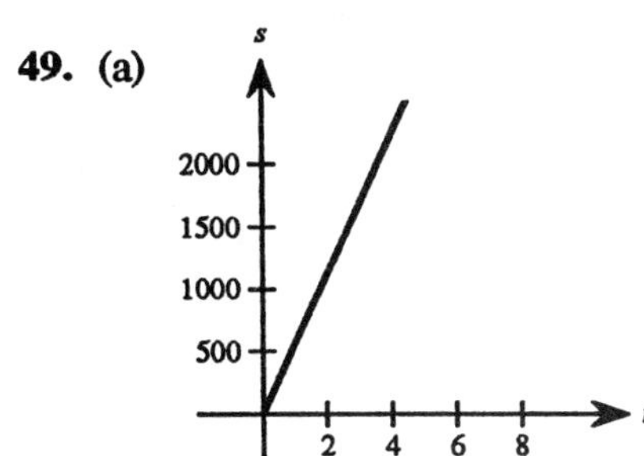

(b) The speed of the plane is 560 miles per hour.

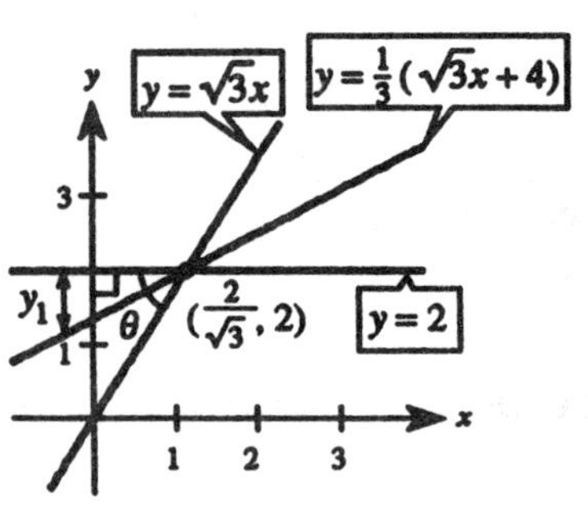

50. $\sqrt{3}x = 2 \Rightarrow x = \dfrac{2}{\sqrt{3}}$

$$\tan\theta = \frac{2}{2/\sqrt{3}} = \sqrt{3} \Rightarrow \theta = 60^\circ$$

$$\frac{1}{2}\theta = 30^\circ$$

$$\tan 30^\circ = \frac{y_1}{2/\sqrt{3}} = \frac{\sqrt{3}\,y_1}{2}$$

$$\frac{1}{\sqrt{3}} = \frac{\sqrt{3}\,y_1}{2}$$

$$2 = 3y_1 \Rightarrow y_1 = \frac{2}{3}$$

$$y = 2 - \frac{2}{3} = \frac{4}{3}$$

Two points on the line are $(2/\sqrt{3},\ 2)$ and $(0,\ 4/3)$.

$$m = \frac{(4/3) - 2}{0 - (2/\sqrt{3})} = \frac{\sqrt{3}}{3}$$

$$y - \frac{4}{3} = \frac{\sqrt{3}}{3}(x - 0)$$

$$y = \frac{\sqrt{3}}{3}x + \frac{4}{3} = \frac{1}{3}(\sqrt{3}x + 4)$$

51. $C = 0.30x + 150$

52. (a) $C = 9.25t + 13.50t + 36{,}500$

$= 22.75t + 36{,}500$

(b) $R = 30t$

(c) $30t = 22.75t + 36{,}00$

$7.25t = 36{,}500$

$t \approx 5034.48$ hrs to break even

CHAPTER 2
Limits and Their Properties

Section 2.1 An Introduction to Limits

1.

x	1.9	1.99	1.999	2.001	2.01	2.1
$f(x)$	0.3448	0.3344	0.3334	0.3332	0.3322	0.3226

$\lim_{x \to 2} \frac{x-2}{x^2-x-2} \approx 0.3333$ (Actual limit is $\frac{1}{3}$.)

2.

x	1.9	1.99	1.999	2.001	2.01	2.1
$f(x)$	0.2564	0.2506	0.2501	0.2499	0.2494	0.2439

$\lim_{x \to 2} \frac{x-2}{x^2-4} \approx 0.25$ (Actual limit is $\frac{1}{4}$.)

3.

x	−0.1	−0.01	−0.001	0.001	0.01	0.1
$f(x)$	0.2911	0.2889	0.2887	0.2887	0.2884	0.2863

$\lim_{x \to 0} \frac{\sqrt{x+3}-\sqrt{3}}{x} \approx 0.2887.$ $\left(\text{Actual limit is } \frac{1}{2\sqrt{3}}\right).$

4.

x	−3.1	−3.01	−3.001	−2.999	−2.99	−2.9
$f(x)$	−0.2485	−0.2498	−0.2500	−0.2500	−0.2502	−0.2516

$\lim_{x \to -3} \frac{\sqrt{1-x}-2}{x+3} \approx -0.25$ (Actual limit is $-\frac{1}{4}$.)

5.

x	2.9	2.99	2.999	3.001	3.01	3.1
$f(x)$	−0.0641	−0.0627	−0.0625	−0.0625	−0.0623	−0.0610

$\lim_{x \to 3} \frac{[1/(x+1)]-(1/4)}{x-3} \approx -0.0625$ (Actual limit is $-\frac{1}{16}$.)

6.

x	3.9	3.99	3.999	4.001	4.01	4.1
$f(x)$	0.0408	0.0401	0.0400	0.0400	0.0399	0.0392

$\lim_{x \to 4} \frac{[x/(x+1)]-(4/5)}{x-4} \approx 0.04$ (Actual limit is $\frac{1}{25}$.)

7. 1

8. 3

9. 2

10. 3

11. Limit does not exist

12. Limit does not exist

13. $\lim_{x\to 2} x^2 = 2^2 = 4$

14. $\lim_{x\to -3} (3x+2) = 3(-3)+2 = -7$

15. $\lim_{x\to 0} (2x-1) = 2(0)-1 = -1$

16. $\lim_{x\to 1} (-x^2+1) = -(1)^2+1 = 0$

17. $\lim_{x\to 2} (-x^2+x-2) = -(2)^2+(2)-2 = -4$

18. $\lim_{x\to 1} (3x^3-2x^2+4) = 3(1)^3-2(1)^2+4 = 5$

19. $\lim_{x\to 3} \sqrt{x+1} = \sqrt{3+1} = 2$

20. $\lim_{x\to 4} \sqrt[3]{x+4} = \sqrt[3]{4+4} = 2$

21. $\lim_{x\to -4} (x+3)^2 = (-4+3)^2 = 1$

22. $\lim_{x\to 0} (2x-1)^3 = [2(0)-1]^3 = -1$

23. $\lim_{x\to 2} \frac{1}{x} = \frac{1}{2}$

24. $\lim_{x\to -3} \frac{2}{x+2} = \frac{2}{-3+2} = -2$

25. $\lim_{x\to -1} \frac{x^2+1}{x} = \frac{(-1)^2+1}{-1} = -2$

26. $\lim_{x\to 3} \frac{\sqrt{x+1}}{x-4} = \frac{\sqrt{3+1}}{3-4} = -2$

27. (a) $\lim_{x\to c} [5g(x)] = 5\lim_{x\to c} g(x) = 5(3) = 15$

(b) $\lim_{x\to c} [f(x)+g(x)] = \lim_{x\to c} f(x) + \lim_{x\to c} g(x) = 2+3 = 5$

(c) $\lim_{x\to c} [f(x)g(x)] = \left[\lim_{x\to c} f(x)\right]\left[\lim_{x\to c} g(x)\right] = (2)(3) = 6$

(d) $\lim_{x\to c} \frac{f(x)}{g(x)} = \frac{\lim_{x\to c} f(x)}{\lim_{x\to c} g(x)} = \frac{2}{3}$

28. (a) $\lim_{x\to c} [4f(x)] = 4\lim_{x\to c} f(x) = 4\left(\frac{3}{2}\right) = 6$

(b) $\lim_{x\to c} [f(x)+g(x)] = \lim_{x\to c} f(x) + \lim_{x\to c} g(x) = \frac{3}{2}+\frac{1}{2} = 2$

(c) $\lim_{x\to c} [f(x)g(x)] = \left[\lim_{x\to c} f(x)\right]\left[\lim_{x\to c} g(x)\right] = \left(\frac{3}{2}\right)\left(\frac{1}{2}\right) = \frac{3}{4}$

(d) $\lim_{x\to c} \frac{f(x)}{g(x)} = \frac{\lim_{x\to c} f(x)}{\lim_{x\to c} g(x)} = \frac{3/2}{1/2} = 3$

29. (a) $\lim_{x\to c} [f(x)]^3 = \left[\lim_{x\to c} f(x)\right]^3 = (4)^3 = 64$

(b) $\lim_{x\to c} \sqrt{f(x)} = \sqrt{\lim_{x\to c} f(x)} = \sqrt{4} = 2$

(c) $\lim_{x\to c} [3f(x)] = 3\lim_{x\to c} f(x) = 3(4) = 12$

(d) $\lim_{x\to c} [f(x)]^{3/2} = \left[\lim_{x\to c} f(x)\right]^{3/2} = (4)^{3/2} = 8$

30. (a) $\lim_{x\to c} \sqrt[3]{f(x)} = \sqrt[3]{\lim_{x\to c} f(x)} = \sqrt[3]{27} = 3$

(b) $\lim_{x\to c} \frac{f(x)}{18} = \frac{\lim_{x\to c} f(x)}{\lim_{x\to c} 18} = \frac{27}{18} = \frac{3}{2}$

(c) $\lim_{x\to c} [f(x)]^2 = \left[\lim_{x\to c} f(x)\right]^2 = (27)^2 = 729$

(d) $\lim_{x\to c} [f(x)]^{2/3} = \left[\lim_{x\to c} f(x)\right]^{2/3} = (27)^{2/3} = 9$

31. $\lim_{x \to 4} f(x) = \frac{1}{6}$

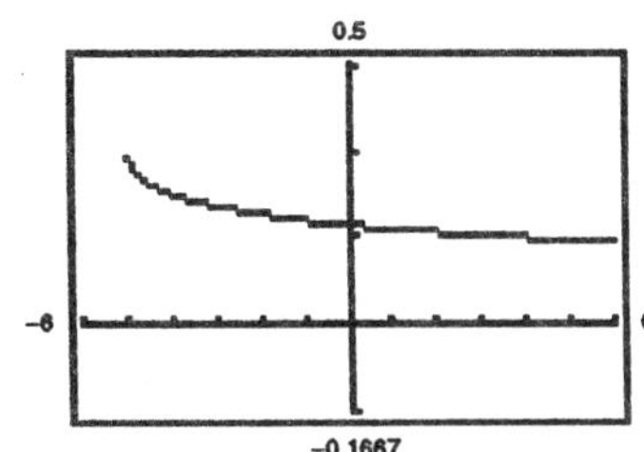

32. $\lim_{x \to 3} f(x) = \frac{1}{2}$

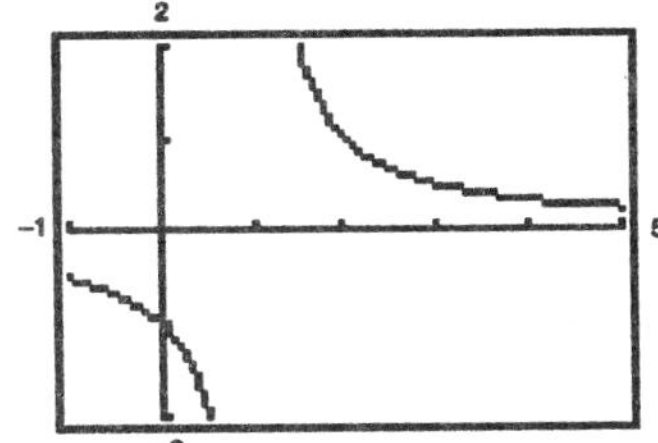

33. $\lim_{x \to 4} \frac{x^2 - x - 12}{x - 4} = 7$

n	$4 + [0.1]^n$	$f(4 + [0.1]^n)$
1	4.1	7.1
2	4.01	7.01
3	4.001	7.001
4	4.0001	7.0001

n	$4 - [0.1]^n$	$f(4 - [0.1]^n)$
1	3.9	6.9
2	3.99	6.99
3	3.999	6.999
4	3.9999	6.9999

34. If $f(2) = 4$, we cannot conclude anything about $\lim_{x \to 2} f(x)$. As $x \to 2$, the graph may have a break in it and the limit would not exist. Or the graph may approach some other value as $x \to 2$, with (2, 4) as a separate point on the graph.

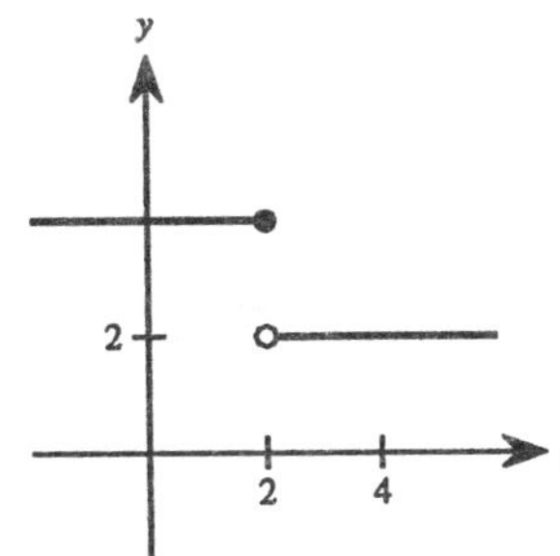

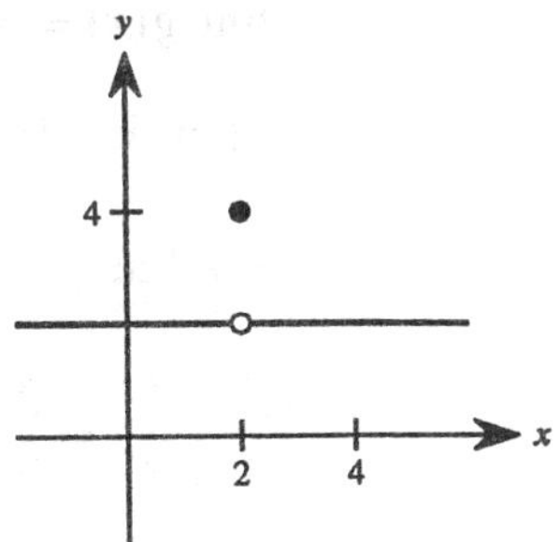

35. If $\lim_{x \to 2} f(x) = 4$, we cannot conclude anything about $f(2)$. $f(2)$ may not exist at all, or $f(2)$ may be some other value besides 4.

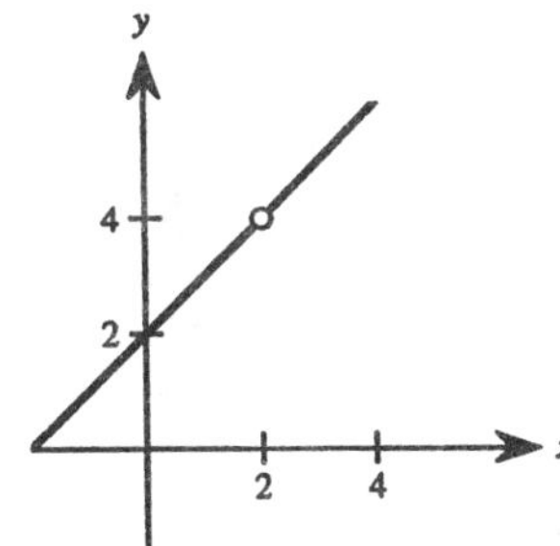

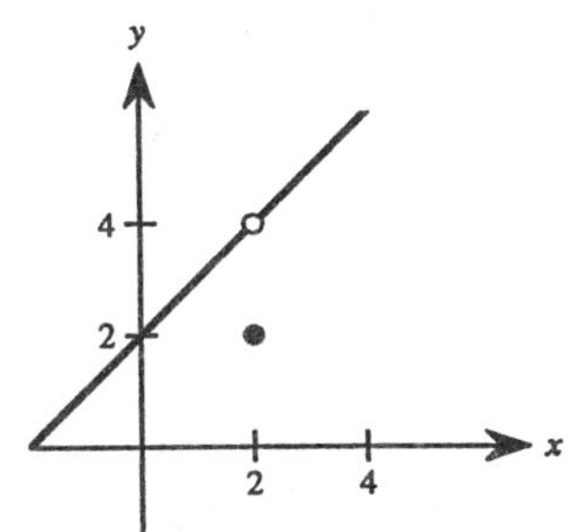

36. Let $f(x) = 1/x$ and $g(x) = -1/x$. $\lim_{x \to 0} f(x)$ and $\lim_{x \to 0} g(x)$ do not exist.

$$\lim_{x \to 0} [f(x) + g(x)] = \lim_{x \to 0} \left[\frac{1}{x} + \left(-\frac{1}{x} \right) \right] = \lim_{x \to 0} [0] = 0$$

Section 2.2 Techniques for Evaluating Limits

1. (a) $\lim_{x\to 0} g(x) = 1$

(b) $\lim_{x\to -1} g(x) = 3$

2. (a) $\lim_{x\to -2} h(x) = -5$

(b) $\lim_{x\to 0} h(x) = -3$

3. (a) $\lim_{x\to 1} g(x) = 2$

(b) $\lim_{x\to -1} g(x) = 0$

4. (a) $\lim_{x\to 1} f(x)$ does not exist.

(b) $\lim_{x\to 2} f(x) = 1$

5. $\lim_{x\to -1} \frac{x^2-1}{x+1} = \lim_{x\to -1} \frac{(x+1)(x-1)}{(x+1)} = \lim_{x\to -1} (x-1) = -2$

6. $\lim_{x\to -1} \frac{2x^2-x-3}{x+1} = \lim_{x\to -1} \frac{(2x-3)(x+1)}{x+1} = \lim_{x\to -1} (2x-3) = -5$

7. $\lim_{x\to 3} \frac{x-3}{x^2-9} = \lim_{x\to 3} \frac{(x-3)}{(x+3)(x-3)} = \lim_{x\to 3} \frac{1}{x+3} = \frac{1}{6}$

8. $\lim_{x\to -1} \frac{x^3+1}{x+1} = \lim_{x\to -1} \frac{(x+1)(x^2-x+1)}{x+1} = \lim_{x\to -1} (x^2-x+1) = 3$

9. $\lim_{x\to -2} \frac{x^3+8}{x+2} = \lim_{x\to -2} \frac{(x+2)(x^2-2x+4)}{x+2} = \lim_{x\to -2} (x^2-2x+4) = 12$

10. $\lim_{\Delta x\to 0} \frac{(x+\Delta x)^2-x^2}{\Delta x} = \lim_{\Delta x\to 0} \frac{x^2+2x\Delta x+(\Delta x)^2-x^2}{\Delta x} = \lim_{\Delta x\to 0} \frac{\Delta x(2x+\Delta x)}{\Delta x} = 2x$

11. $\lim_{\Delta x\to 0} \frac{2(x+\Delta x)-2x}{\Delta x} = \lim_{\Delta x\to 0} \frac{2x+2\Delta x-2x}{\Delta x} = \lim_{\Delta x\to 0} 2 = 2$

12. $\lim_{\Delta x\to 0} \frac{(x+\Delta x)^3-x^3}{\Delta x} = \lim_{\Delta x\to 0} \frac{x^3+3x^2\Delta x+3x(\Delta x)^2+(\Delta x)^3-x^3}{\Delta x} = \lim_{\Delta x\to 0} [3x^2+3x\Delta x+(\Delta x)^2] = 3x^2$

13.
$$\begin{aligned}\lim_{\Delta x\to 0} \frac{(x+\Delta x)^2-2(x+\Delta x)+1-(x^2-2x+1)}{\Delta x} &= \lim_{\Delta x\to 0} \frac{x^2+2x\Delta x+(\Delta x)^2-2x-2\Delta x+1-x^2+2x-1}{\Delta x}\\ &= \lim_{\Delta x\to 0} (2x+\Delta x-2)\\ &= 2x-2\end{aligned}$$

14. $\lim_{\Delta x\to 0} \frac{(1+\Delta x)^3-1}{\Delta x} = \lim_{\Delta x\to 0} \frac{1+3\Delta x+3(\Delta x)^2+(\Delta x)^3-1}{\Delta x} = \lim_{\Delta x\to 0} [3+3\Delta x+(\Delta x)^2] = 3$

15. $\lim_{x\to 5} \frac{x-5}{x^2-25} = \lim_{x\to 5} \frac{x-5}{(x+5)(x-5)} = \lim_{x\to 5} \frac{1}{x+5} = \frac{1}{10}$

16. $\lim_{x\to 2} \frac{-(x-2)}{x^2-4} = \lim_{x\to 2} \frac{-1}{x+2} = -\frac{1}{4}$

17. $\lim_{x\to 1} \frac{x^2+x-2}{x^2-1} = \lim_{x\to 1} \frac{(x+2)(x-1)}{(x+1)(x-1)} = \lim_{x\to 1} \frac{x+2}{x+1} = \frac{3}{2}$

18. $\lim_{x\to 0}\frac{\sqrt{2+x}-\sqrt{2}}{x} = \lim_{x\to 0}\frac{\sqrt{2+x}-\sqrt{2}}{x}\cdot\frac{\sqrt{2+x}+\sqrt{2}}{\sqrt{2+x}+\sqrt{2}}$

$$= \lim_{x\to 0}\frac{2+x-2}{(\sqrt{2+x}+\sqrt{2})x} = \lim_{x\to 0}\frac{1}{\sqrt{2+x}+\sqrt{2}} = \frac{1}{2\sqrt{2}} = \frac{\sqrt{2}}{4}$$

19. $\lim_{x\to 0}\frac{\sqrt{3+x}-\sqrt{3}}{x} = \lim_{x\to 0}\frac{\sqrt{3+x}-\sqrt{3}}{x}\cdot\frac{\sqrt{3+x}+\sqrt{3}}{\sqrt{3+x}+\sqrt{3}}$

$$= \lim_{x\to 0}\frac{3+x-3}{x(\sqrt{3+x}+\sqrt{3})} = \lim_{x\to 0}\frac{1}{\sqrt{3+x}+\sqrt{3}} = \frac{1}{2\sqrt{3}} = \frac{\sqrt{3}}{6}$$

20. $\lim_{x\to 0}\frac{\frac{1}{x+4}-\frac{1}{4}}{x} = \lim_{x\to 0}\frac{\frac{4-(x+4)}{4(x+4)}}{x} = \lim_{x\to 0}\frac{-1}{4(x+4)} = -\frac{1}{16}$

21. $\lim_{x\to 0}\frac{\frac{1}{2+x}-\frac{1}{2}}{x} = \lim_{x\to 0}\frac{\frac{2-(2+x)}{2(2+x)}}{x} = \lim_{x\to 0}\frac{-1}{2(2+x)} = -\frac{1}{4}$

22. $\lim_{x\to 3}\frac{\sqrt{x+1}-2}{x-3} = \lim_{x\to 3}\frac{\sqrt{x+1}-2}{x-3}\cdot\frac{\sqrt{x+1}+2}{\sqrt{x+1}+2} = \lim_{x\to 3}\frac{x-3}{(x-3)[\sqrt{x+1}+2]} = \lim_{x\to 3}\frac{1}{\sqrt{x+1}+2} = \frac{1}{4}$

23.

x	-0.1	-0.01	-0.001	0	0.001	0.01	0.1
$f(x)$	0.358	0.354	0.354	?	0.354	0.353	0.349

$$\lim_{x\to 0}\frac{\sqrt{x+2}-\sqrt{2}}{x} \approx 0.354$$

$$\lim_{x\to 0}\frac{\sqrt{x+2}-\sqrt{2}}{x} = \lim_{x\to 0}\frac{\sqrt{x+2}-\sqrt{2}}{x}\cdot\frac{\sqrt{x+2}+\sqrt{2}}{\sqrt{x+2}+\sqrt{2}}$$

$$= \lim_{x\to 0}\frac{x+2-2}{x(\sqrt{x+2}+\sqrt{2})} = \lim_{x\to 0}\frac{1}{\sqrt{x+2}+\sqrt{2}} = \frac{1}{2\sqrt{2}} = \frac{\sqrt{2}}{4}$$

24.

x	0.9	0.99	0.999	1	1.001	1.01	1.1
$f(x)$	2.130	2.013	2.001	?	1.999	1.988	1.879

$$\lim_{x\to 1}\frac{1-x}{\sqrt{5-x^2}-2} = 2$$

$$\lim_{x\to 1}\frac{1-x}{\sqrt{5-x^2}-2} = \lim_{x\to 1}\frac{1-x}{\sqrt{5-x^2}-2}\cdot\frac{\sqrt{5-x^2}+2}{\sqrt{5-x^2}+2}$$

$$= \lim_{x\to 1}\frac{(1-x)(\sqrt{5-x^2}+2)}{5-x^2-4} = \lim_{x\to 1}\frac{(1-x)(\sqrt{5-x^2}+2)}{1-x^2}$$

$$= \lim_{x\to 1}\frac{(1-x)(\sqrt{5-x^2}+2)}{(1+x)(1-x)} = \lim_{x\to 1}\frac{\sqrt{5-x^2}+2}{1+x} = \frac{4}{2} = 2$$

25.

x	-0.1	-0.01	-0.001	0	0.001	0.01	0.1
$f(x)$	-0.263	-0.251	-0.250	?	-0.250	-0.249	-0.238

$$\lim_{x\to 0} \frac{\frac{1}{2+x}-\frac{1}{2}}{x} = -\frac{1}{4}$$

$$\lim_{x\to 0} \frac{\frac{1}{2+x}-\frac{1}{2}}{x} = \lim_{x\to 0} \frac{2-(2+x)}{2(2+x)}\cdot\frac{1}{x} = \lim_{x\to 0} \frac{-x}{2(2+x)}\cdot\frac{1}{x} = \lim_{x\to 0} \frac{-1}{2(2+x)} = -\frac{1}{4}$$

26.

x	1.9	1.99	1.999	1.9999	2.0	2.0001	2.001	2.01	2.1
$f(x)$	72.39	79.20	79.92	79.99	?	80.01	80.08	80.80	88.41

$$\lim_{x\to 2} \frac{x^5-32}{x-2} = 80$$

$$\lim_{x\to 2} \frac{x^5-32}{x-2} = \lim_{x\to 2} \frac{(x-2)(x^4+2x^3+4x^2+8x+16)}{x-2} = \lim_{x\to 2} (x^4+2x^3+4x^2+8x+16) = 80$$

(**Hint:** Use long division to factor x^5-32.)

27. (a) $\lim_{x\to 3^+} f(x) = 1$

(b) $\lim_{x\to 3^-} f(x) = 1$

(c) $\lim_{x\to 3} f(x) = 1$

28. (a) $\lim_{x\to -2^+} f(x) = -2$

(b) $\lim_{x\to -2^-} f(x) = -2$

(c) $\lim_{x\to -2} f(x) = -2$

29. (a) $\lim_{x\to 3^+} f(x) = 0$

(b) $\lim_{x\to 3^-} f(x) = 0$

(c) $\lim_{x\to 3} f(x) = 0$

30. (a) $\lim_{x\to -2^+} f(x) = 2$

(b) $\lim_{x\to -2^-} f(x) = 2$

(c) $\lim_{x\to -2} f(x) = 2$

31. (a) $\lim_{x\to 3^+} f(x) = 3$

(b) $\lim_{x\to 3^-} f(x) = -3$

(c) $\lim_{x\to 3} f(x)$ does not exist.

32. (a) $\lim_{x\to -1^+} f(x) = 0$

(b) $\lim_{x\to -1^-} f(x) = 2$

(c) $\lim_{x\to -1} f(x)$ does not exist.

33. $\lim_{x\to 5^+} \frac{x-5}{x^2-25} = \lim_{x\to 5^+} \frac{1}{x+5} = \frac{1}{10}$

34. $\lim_{x\to 2^+} \frac{2-x}{x^2-4} = \lim_{x\to 2^+} -\frac{1}{x+2} = -\frac{1}{4}$

35. $\lim_{x\to 2^+} \frac{x}{\sqrt{x^2-4}}$ does not exist.

36. $$\lim_{x\to 4^-} \frac{\sqrt{x}-2}{x-4} = \lim_{x\to 4^-} \frac{\sqrt{x}-2}{x-4}\cdot\frac{\sqrt{x}+2}{\sqrt{x}+2} = \lim_{x\to 4^-} \frac{x-4}{(x-4)(\sqrt{x}+2)} = \lim_{x\to 4^-} \frac{1}{\sqrt{x}+2} = \frac{1}{4}$$

37. $$\lim_{\Delta x\to 0^+} \frac{2(x+\Delta x)-2x}{\Delta x} = \lim_{\Delta x\to 0^+} \frac{2\Delta x}{\Delta x} = \lim_{\Delta x\to 0} 2 = 2$$

38. $$\begin{aligned}\lim_{\Delta x\to 0^+} \frac{(x+\Delta x)^2+x+\Delta x-(x^2+x)}{\Delta x} &= \lim_{\Delta x\to 0} \frac{x^2+2x(\Delta x)+(\Delta x)^2+x+\Delta x-x^2-x}{\Delta x}\\ &= \lim_{\Delta x\to 0^+} \frac{2x(\Delta x)+(\Delta x)^2+\Delta x}{\Delta x}\\ &= \lim_{\Delta x\to 0^+} (2x+\Delta x+1)\\ &= 2x+0+1\\ &= 2x+1\end{aligned}$$

39. $\lim_{\Delta x \to 0^-} \dfrac{\dfrac{1}{x+\Delta x} - \dfrac{1}{x}}{\Delta x} = \lim_{\Delta x \to 0^-} \dfrac{x-(x+\Delta x)}{x(x+\Delta x)} \cdot \dfrac{1}{\Delta x}$

$= \lim_{\Delta x \to 0^-} \dfrac{-\Delta x}{x(x+\Delta x)} \cdot \dfrac{1}{\Delta x} = \lim_{\Delta x \to 0^-} \dfrac{-1}{x(x+\Delta x)} = \dfrac{-1}{x(x+0)} = -\dfrac{1}{x^2}$

40. $\lim_{x \to 1^-} \dfrac{x^2-2x+1}{x-1} = \lim_{x \to 1^-} (x-1) = 0$

41. $\lim_{x \to 0} \dfrac{|x|}{x}$ does not exist since

$\lim_{x \to 0^+} \dfrac{|x|}{x} = 1$ and $\lim_{x \to 0^-} \dfrac{|x|}{x} = -1.$

42. $\lim_{x \to 2^+} \dfrac{|x-2|}{x-2} = \lim_{x \to 2^+} \dfrac{x-2}{x-2} = 1$

$\lim_{x \to 2^-} \dfrac{|x-2|}{x-2} = \lim_{x \to 2^-} \dfrac{-(x-2)}{x-2} = -1$

$\lim_{x \to 2^+} \dfrac{|x-2|}{x-2} \neq \lim_{x \to 2^-} \dfrac{|x-2|}{x-2}$

Thus, the limit does not exist.

43. $\lim_{x \to 3^+} f(x) = \lim_{x \to 3^+} \dfrac{12-2x}{3} = 2$

$\lim_{x \to 3^-} f(x) = \lim_{x \to 3^-} \dfrac{x+2}{2} = \dfrac{5}{2}$

$\lim_{x \to 3} f(x)$ does not exist.

44. $\lim_{x \to 2^+} f(x) = \lim_{x \to 2^+} (-x^2+4x-2) = 2$

$\lim_{x \to 2^-} f(x) = \lim_{x \to 2^-} (x^2-4x+6) = 2$

$\lim_{x \to 2} f(x) = 2$

45. $\lim_{x \to 1^+} f(x) = \lim_{x \to 1^+} (x+1) = 2$

$\lim_{x \to 1^-} f(x) = \lim_{x \to 1^-} (x^3+1) = 2$

$\lim_{x \to 1} f(x) = 2$

46. $\lim_{x \to 1^+} f(x) = \lim_{x \to 1^+} (1-x) = 0$

$\lim_{x \to 1^-} f(x) = \lim_{x \to 1^-} (x) = 1$

$\lim_{x \to 1} f(x)$ does not exist.

47. $\lim_{x \to 3^-} 2[\![x-3]\!] = 2(-1) = -2$

48. $\lim_{x \to 1^+} [\![2x]\!] = 2$

49. $\lim_{x \to 1} \left(\left[\!\!\left[\dfrac{x}{4} \right]\!\!\right] + x \right) = 0 + 1 = 1$

50. $\lim_{x \to 2} \left[\!\!\left[\dfrac{x-1}{2} \right]\!\!\right] = 0$

51. $\lim_{t \to 0.5} \dfrac{s(0.5)-s(t)}{0.5-t} = \lim_{t \to 0.5} \dfrac{21-(-16t^2+25)}{0.5-t}$

$= \lim_{t \to 0.5} \dfrac{16t^2-4}{0.5-t} = \lim_{t \to 0.5} \dfrac{-16(0.25-t^2)}{0.5-t} = \lim_{t \to 0.5} \dfrac{-16(0.5+t)(0.5-t)}{0.5-t}$

$= \lim_{t \to 0.5} -16(0.5+t) = -16(0.5+0.5) = -16$ feet per second

52. $\lim_{t \to 1.1} \dfrac{s(1.1)-s(t)}{1.1-t} = \lim_{t \to 1.1} \dfrac{5.64-(-16t^2+25)}{1.1-t}$

$= \lim_{t \to 1.1} \dfrac{16t^2-19.36}{1.1-t} = \lim_{t \to 1.1} \dfrac{-16(1.21-t^2)}{1.1-t} = \lim_{t \to 1.1} \dfrac{-16(1.1+t)(1.1-t)}{1.1-t}$

$= \lim_{t \to 1.1} -16(1.1+t) = -16(1.1+1.1) = -35.2$ feet per second

53. $f(x) = \dfrac{x - 9}{\sqrt{x} - 3}$

Domain: $[0, 9) \cup (9, \infty)$

$\lim_{x \to 9} \dfrac{x - 9}{\sqrt{x} - 3} = 6$

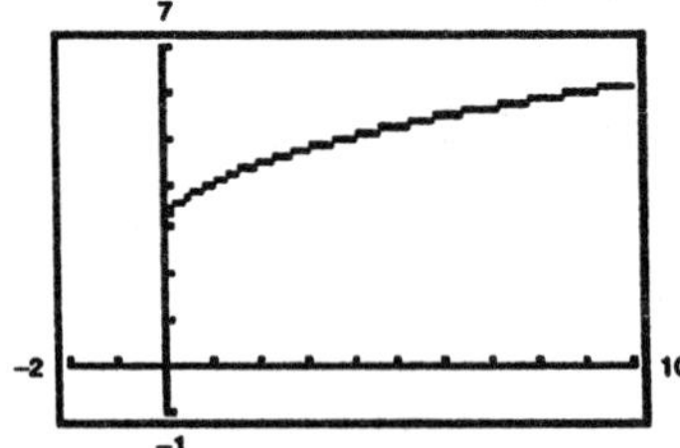

The graphing utility does not show the hole at $(9, 6)$.

54. $f(x) = \dfrac{x - 3}{x^2 - 9}$

Domain: $(-\infty, -3) \cup (-3, 3) \cup (3, \infty)$

$\lim_{x \to 3} \dfrac{x - 3}{x^2 - 9} = \dfrac{1}{6}$

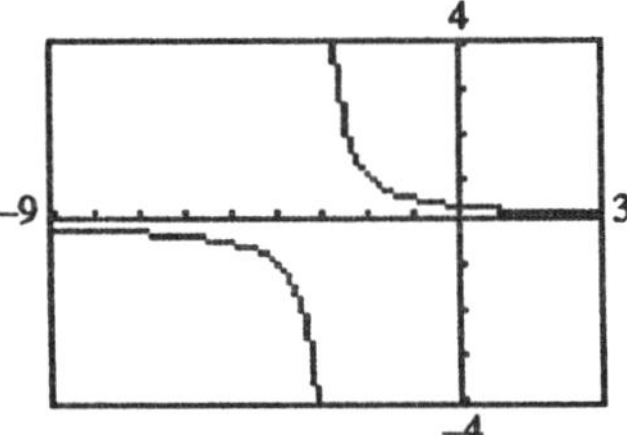

The graphing utility does not show the hole at $\left(3, \frac{1}{6}\right)$.

Section 2.3 Continuity

1. $f(x) = -\dfrac{x^3}{2}$ is continuous for all real x.

2. $f(x) = \dfrac{x^2 - 1}{x}$ has a discontinuity at $x = 0$ since $f(0)$ is not defined.

3. $f(x) = \dfrac{x^2 - 1}{x + 1}$
has a discontinuity at $x = -1$ since $f(-1)$ is not defined.

4. $f(x) = \dfrac{1}{x^2 - 4}$
has discontinuities at $x = -2$ and $x = 2$ since $f(-2)$ and $f(2)$ are not defined.

5. $f(x) = \begin{cases} x, & x < 1 \\ 2, & x = 1 \\ 2x - 1, & x > 1 \end{cases}$ has a discontinuity at $x = 1$ since $f(1) \neq \lim_{x \to 1} f(x)$.

6. $f(x) = ([\![x]\!]/2) + x$ has discontinuities at each integer k since $\lim_{x \to k^-} f(x) \neq \lim_{x \to k^+} f(x)$.

7. $f(x) = x^2 - 2x + 1$ is continuous for all real x.

8. $f(x) = 1/(x^2 + 1)$ is continuous for all real x.

9. $f(x) = 1/(x - 1)$ has a nonremovable discontinuity at $x = 1$ since $\lim_{x \to 1} f(x)$ does not exist.

10. $f(x) = x/(x^2 - 1)$ has nonremovable discontinuities at $x = 1$ and $x = -1$ since $\lim_{x \to 1} f(x)$ and $\lim_{x \to -1} f(x)$ do not exist.

11. $f(x) = \dfrac{x}{x^2 + 1}$

is continuous for all real x.

12. $f(x) = \dfrac{x - 3}{x^2 - 9}$

has a nonremovable discontinuity at $x = -3$ since $\lim_{x \to -3} f(x)$ does not exist, and has a removable discontinuity at $x = 3$ since

$$\lim_{x \to 3} f(x) = \lim_{x \to 3} \frac{1}{x + 3} = \frac{1}{6}.$$

13. $f(x) = \dfrac{x+2}{(x+2)(x-5)}$

has a nonremovable discontinuity at $x = 5$ since $\lim\limits_{x\to 5} f(x)$ does not exist, and has a removable discontinuity at $x = -2$ since

$$\lim_{x\to -2} f(x) = \lim_{x\to -2} \frac{1}{x-5} = -\frac{1}{7}.$$

14. $f(x) = \dfrac{x-1}{(x+2)(x-1)}$

has a nonremovable discontinuity at $x = -2$ since $\lim\limits_{x\to -2} f(x)$ does not exist, and has a removable discontinuity at $x = 1$ since

$$\lim_{x\to 1} f(x) = \lim_{x\to 1} \frac{1}{x+2} = \frac{1}{3}.$$

15. $f(x) = \begin{cases} x, & x \le 1 \\ x^2, & x > 1 \end{cases}$ has a **possible** discontinuity at $x = 1$.

1. $f(1) = 1$
2. $\left.\begin{array}{l} \lim\limits_{x\to 1^-} f(x) = \lim\limits_{x\to 1^-} x = 1 \\ \lim\limits_{x\to 1^+} f(x) = \lim\limits_{x\to 1^+} x^2 = 1 \end{array}\right\} \lim\limits_{x\to 1} f(x) = 1$
3. $f(1) = \lim\limits_{x\to 1} f(x)$

f is continuous at $x = 1$, therefore, f is continuous for all real x.

16. $f(x) = \begin{cases} -2x+3, & x < 1 \\ x^2, & x \ge 1 \end{cases}$ has a **possible** discontinuity at $x = 1$.

1. $f(1) = 1^2 = 1$
2. $\left.\begin{array}{l} \lim\limits_{x\to 1^-} f(x) = \lim\limits_{x\to 1^-} (-2x+3) = 1 \\ \lim\limits_{x\to 1^+} f(x) = \lim\limits_{x\to 1^+} x^2 = 1 \end{array}\right\} \lim\limits_{x\to 1} f(x) = 1$
3. $f(1) = \lim\limits_{x\to 1} f(x)$

f is continuous at $x = 1$, therefore, f is continuous for all real x.

17. $f(x) = \begin{cases} \dfrac{x}{2}+1, & x \le 2 \\ 3-x, & x > 2 \end{cases}$ has a **possible** discontinuity at $x = 2$.

1. $f(2) = \dfrac{2}{2} + 1 = 2$
2. $\left.\begin{array}{l} \lim\limits_{x\to 2^-} f(x) = \lim\limits_{x\to 2^-} \left(\dfrac{x}{2}+1\right) = 2 \\ \lim\limits_{x\to 2^+} f(x) = \lim\limits_{x\to 2^+} (3-x) = 1 \end{array}\right\} \lim\limits_{x\to 2} f(x)$ does not exist.

Therefore, f has a nonremovable discontinuity at $x = 2$.

18. $f(x) = \begin{cases} -2x, & x \le 2 \\ x^2-4x+1, & x > 2 \end{cases}$ has a **possible** discontinuity at $x = 2$.

1. $f(2) = -2(2) = -4$
2. $\left.\begin{array}{l} \lim\limits_{x\to 2^-} f(x) = \lim\limits_{x\to 2^-} (-2x) = -4 \\ \lim\limits_{x\to 2^+} f(x) = \lim\limits_{x\to 2^+} (x^2-4x+1) = -3 \end{array}\right\} \lim\limits_{x\to 2} f(x)$ does not exist.

Therefore, f has a nonremovable discontinuity at $x = 2$.

19. $f(x) = \dfrac{|x+2|}{x+2}$

has a nonremovable discontinuity at $x = -2$ since $\lim\limits_{x \to -2} f(x)$ does not exist.

20. $f(x) = \dfrac{|x-3|}{x-3}$

has a nonremovable discontinuity at $x = 3$ since $\lim\limits_{x \to 3} f(x)$ does not exist.

21. $f(x) = \begin{cases} |x-2|+3, & x < 0 \\ x+5, & x \geq 0 \end{cases}$ has a **possible** discontinuity at $x = 0$.

$\lim\limits_{x \to 0^-} f(x) = \lim\limits_{x \to 0^-} [|x-2|+3] = 2+3 = 5$

$\lim\limits_{x \to 0^+} f(x) = \lim\limits_{x \to 0^+} (x+5) = 0+5 = 5$

Continuous for all real x

22. $f(x) = \begin{cases} 3+x, & x \leq 2 \\ x^2+1, & x > 2 \end{cases}$ has a **possible** discontinuity at $x = 2$.

$\lim\limits_{x \to 2^-} f(x) = \lim\limits_{x \to 2^-} (3+x) = 5$

$\lim\limits_{x \to 2^+} f(x) = \lim\limits_{x \to 2^+} (x^2+1) = 5$

Continuous for all real x

23. $f(x) = [\![x-1]\!]$ has nonremovable discontinuities at each integer k.

24. $f(x) = x - [\![x]\!]$ has nonremovable discontinuities at each integer k.

25. $f(g(x)) = (x-1)^2$

Continuous for all real x

26. $f(g(x)) = \dfrac{1}{\sqrt{x-1}}$

Nonremovable discontinuity at $x = 1$

27. $f(g(x)) = \dfrac{1}{(x^2+5)-1} = \dfrac{1}{x^2+4}$

Continuous for all real x

28. $f(g(x)) = \sqrt{x^2} = |x|$

Continuous for all real x

29. $f(g(x)) = \dfrac{1}{1/(x-1)} = x - 1$

Removable discontinuity at $x = 1$

30. $f(g(x)) = \dfrac{1}{\sqrt{1/x}} = \sqrt{x}$

Continuous for all real x such that $x \geq 0$

31. $y = \dfrac{x^2-16}{x-4}$

Removable discontinuity at $x = 4$

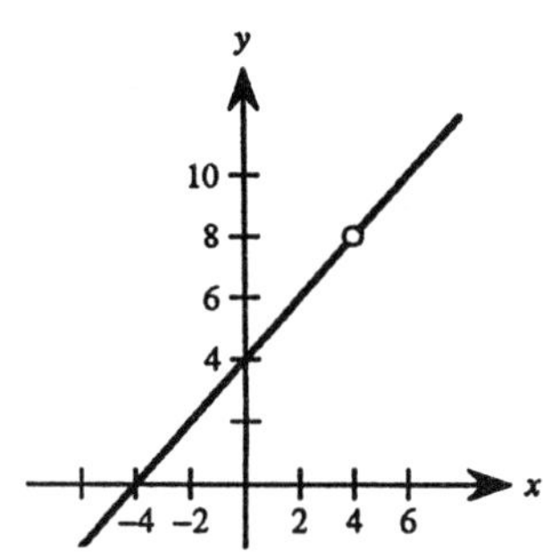

32. $y = \dfrac{x^3-8}{x-2}$

Removable discontinuity at $x = 2$

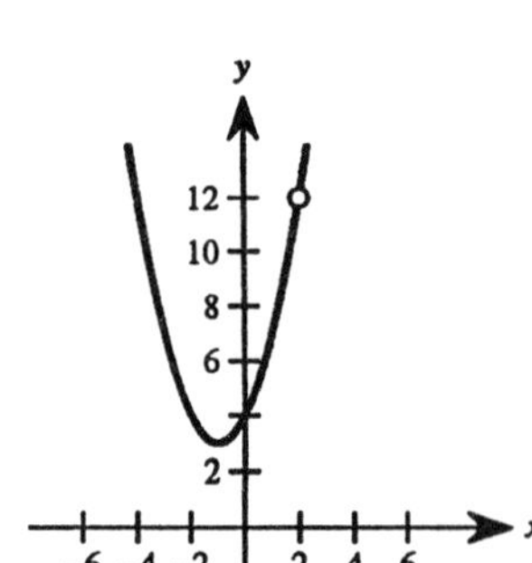

33. $y = [\![x]\!] - x$

Nonremovable discontinuity at each integer

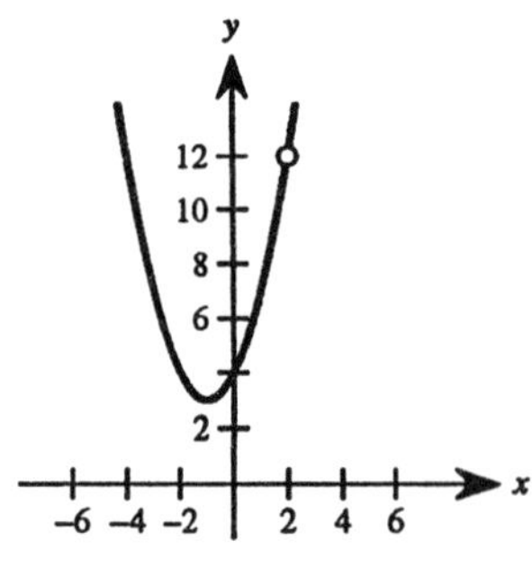

34. $f(x) = \dfrac{|x^2 - 1|}{x}$

Discontinuous at $x = 0$

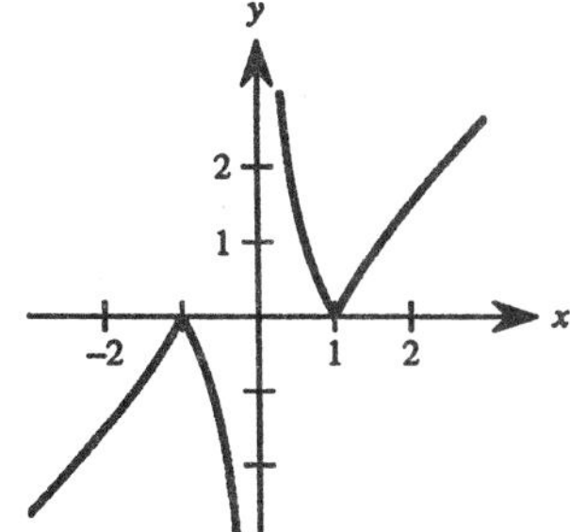

35. $f(x) = \dfrac{x^2}{x^2 - 36}$

Continuous on $(-\infty, -6)$, $(-6, 6)$, $(6, \infty)$

36. $f(x) = x\sqrt{x+3}$

Continuous on $[-3, \infty)$

37. $f(x) = \dfrac{x}{x^2+1}$

Continuous on $(-\infty, \infty)$

38. $f(x) = \dfrac{x+1}{\sqrt{x}}$

Continuous on $(0, \infty)$

39. $f(x)$ is continuous on $[2, 4]$.

$f(2) = -1$ and $f(4) = 3$

By the Intermediate Value Theorem, $f(x) = 0$ for at least one value c between 2 and 4.

40. $f(x)$ is continuous on $[0, 1]$.

$f(0) = -2$ and $f(1) = 2$

By the Intermediate Value Theorem, $f(x) = 0$ for at least one value c between 0 and 1.

41. $f(x) = x^3 + x - 1$

(a)

x	0	0.1	0.2	0.3	0.4	0.5	0.6	0.7	0.8	0.9	1
$f(x)$	−	−	−	−	−	−	−	+	+	+	+

Zero is in the subinterval (0.6, 0.7).

(b)

x	0.6	0.61	0.62	0.63	0.64	0.65	0.66	0.67	0.68	0.69	0.7
$f(x)$	−	−	−	−	−	−	−	−	−	+	+

Zero is in the subinterval (0.68, 0.69).

42. $f(x) = x^3 + 3x - 2$

(a)

x	0	0.1	0.2	0.3	0.4	0.5	0.6	0.7	0.8	0.9	1
$f(x)$	−	−	−	−	−	−	+	+	+	+	+

Zero is in the subinterval (0.5, 0.6).

(b)

x	0.5	0.51	0.52	0.53	0.54	0.55	0.56	0.57	0.58	0.59	0.6
$f(x)$	−	−	−	−	−	−	−	−	−	−	+

Zero is in the subinterval (0.59, 0.6).

43. f is a polynomial and therefore continuous; $f(3) = 11$

44. f is a polynomial and therefore continuous; $f(2) = 0$

45. f is a polynomial and therefore continuous; $f(2) = 4$

46. f has a nonremovable discontinuity at $x = 1$ outside the interval; $f(3) = 6$

47. $f(2) = 8$

Find a so that $\lim_{x\to 2^+} ax^2 = 8 \Rightarrow a = \dfrac{8}{2^2} = 2.$

48. Find a and b such that $\lim_{x\to -1^+} (ax + b) = -a + b = 2$ and $\lim_{x\to 3^-} (ax + b) = 3a + b = -2.$

$$\begin{aligned} a - b &= -2 \\ (+)\ 3a + b &= -2 \\ \hline 4a \quad &= -4 \\ a &= -1 \\ b &= 2 + (-1) = 1 \end{aligned}$$

$$f(x) = \begin{cases} 2, & x \le -1 \\ -x + 1, & -1 < x < 3 \\ -2, & x \ge 3 \end{cases}$$

49. $f(x) = \sqrt{1 - x^2}$ is continuous on $[-1, 1]$. Since $\lim_{x\to 1^-} \sqrt{1 - x^2} = 0 = f(1)$, f is continuous at $x = 1$.

50. $S(t) = 28,500(1.09)^{[\![t]\!]}$

Discontinuous at every positive integer

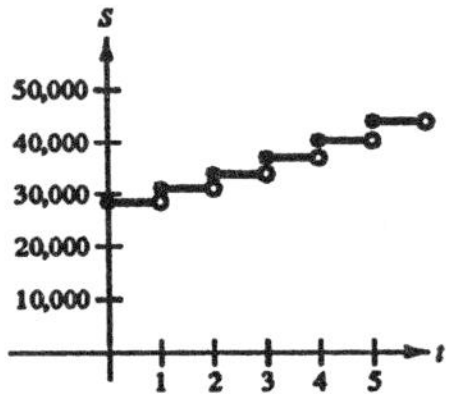

51. $C = \begin{cases} 1.04, & 0 < t \le 2 \\ 1.04 + 0.36[\![t - 1]\!], & t > 2,\ t \text{ is not an integer} \\ 1.04 + 0.36(t - 2), & t > 2,\ t \text{ is an integer} \end{cases}$

Nonremovable discontinuity at each integer greater than 2.

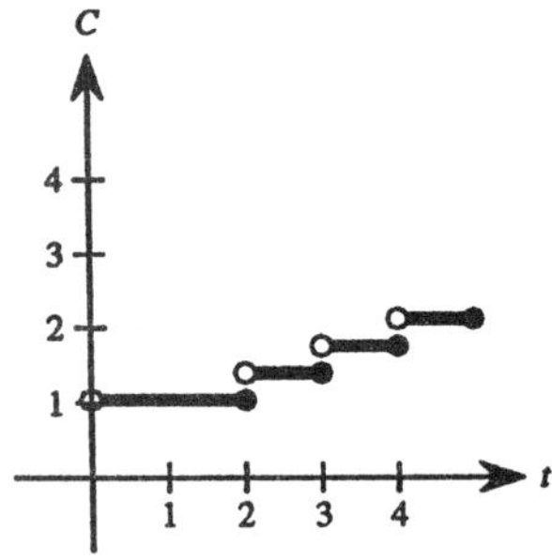

52. 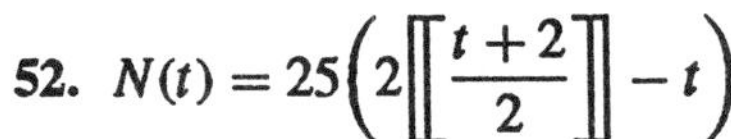 $N(t) = 25\left(2\left[\!\left[\dfrac{t+2}{2}\right]\!\right] - t\right)$

t	0	1	1.8	2	3	3.8
$N(t)$	50	25	5	50	25	5

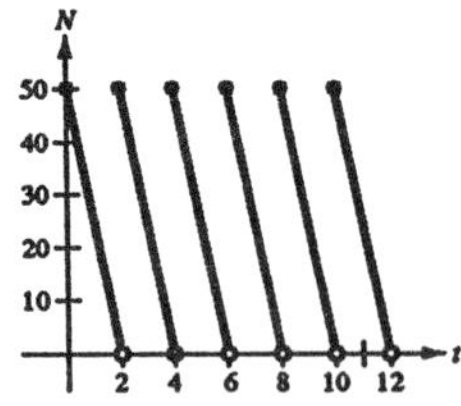

Discontinuous at every positive even integer. The company replenishes its inventory every two months.

53. $f(3) = 2(3) - 4 = 2$

$\lim_{x\to 3^-} f(x) = \lim_{x\to 3^-} (2x - 4) = 2$

$\lim_{x\to 3^+} f(x) = \lim_{x\to 3^+} (x^2 - 2x) = 3$

Therefore, $\lim_{x\to 3} f(x)$ does not exist and f is not continuous at $x = 3$.

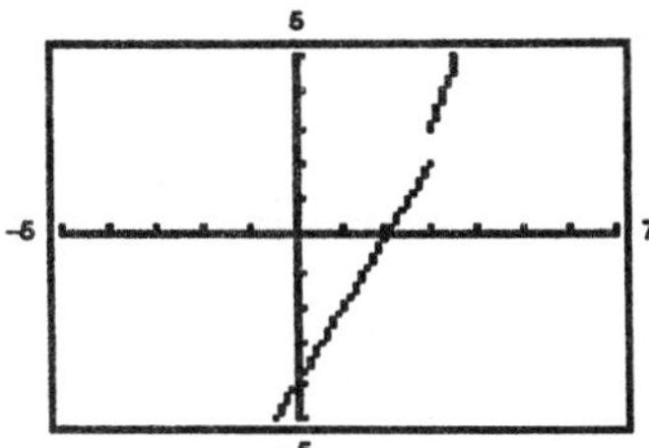

54. Let $f(t) = s(t) - r(t)$ where $s(t)$ and $r(t)$ are the position functions for the run up and down, respectively.

When $t = 0$, (8:00 A.M.):

$f(t) = s(0) - r(0)$

$= 0 -$ distance to campsite < 0.

When $t = 10$, (8:10 A.M.)

$f(t) = s(10) - r(10)$

$= \frac{1}{2}$distance to campsite $- 0 > 0$.

Since $f(0) < 0$ and $f(10) > 0$, then there must be a value t in the interval $(0, 10)$ such that $f(t) = 0$. If $f(t) = 0$, then $s(t) - r(t) = 0$, which gives us $s(t) = r(t)$. Therefore, at some time t, where $0 < t < 10$, the position functions for the run up and down are equal.

55. Apply Theorem 2.3.

56. Suppose there exists x_1 in $[a, b]$ such that $f(x_1) > 0$ and there exists x_2 in $[a, b]$ such that $f(x_2) < 0$. Then by the Intermediate Value Theorem, $f(x)$ must equal zero for some value of x in $[x_1, x_2]$ (or $[x_2, x_1]$ if $x_2 < x_1$). Thus, f would have a zero on $[a, b]$, which is a contradiction. Therefore, $f(x) > 0$ for all x in $[a, b]$ or $f(x) < 0$ for all x in $[a, b]$.

57. Let c be any real number. Then, $\lim_{x\to c} f(x)$ does not exist since there are both rational and irrational numbers arbitrarily close to c. Therefore, f is not continuous at c.

Section 2.4 Infinite Limits

1. $\lim_{x\to -2^+} \frac{1}{(x+2)^2} = \infty$

$\lim_{x\to -2^-} \frac{1}{(x+2)^2} = \infty$

2. $\lim_{x\to -2^+} \frac{1}{x+2} = \infty$

$\lim_{x\to -2^-} \frac{1}{x+2} = -\infty$

3. $\lim_{x\to -3^+} \frac{1}{x^2-9} = -\infty$

$\lim_{x\to -3^-} \frac{1}{x^2-9} = \infty$

4. $\lim_{x\to -3^+} \frac{x}{x^2-9} = \infty$

$\lim_{x\to -3^-} \frac{x}{x^2-9} = -\infty$

5. $\lim_{x\to -3^+} \frac{x^3}{x^2-9} = \infty$

$\lim_{x\to -3^-} \frac{x^3}{x^2-9} = -\infty$

6. $\lim_{x\to -3^+} \frac{x^2}{x^2-9} = -\infty$

$\lim_{x\to -3^-} \frac{x^2}{x^2-9} = \infty$

7. $\lim_{x\to 2^+} \frac{x^2-2}{(x-2)(x+1)} = \infty$

$\lim_{x\to 2^-} \frac{x^2-2}{(x-2)(x+1)} = -\infty$

Therefore, $x = 2$ is a vertical asymptote.

$\lim_{x\to -1^+} \frac{x^2-2}{(x-2)(x+1)} = \infty$

$\lim_{x\to -1^-} \frac{x^2-2}{(x-2)(x+1)} = -\infty$

Therefore, $x = -1$ is a vertical asymptote.

8. $\lim_{x\to -1^+} \frac{x^3}{x^2-1} = \infty$

$\lim_{x\to -1^-} \frac{x^3}{x^2-1} = -\infty$

Therefore, $x = -1$ is a vertical asymptote.

$\lim_{x\to 1^+} \frac{x^3}{x^2-1} = \infty$

$\lim_{x\to 1^-} \frac{x^3}{x^2-1} = -\infty$

Therefore, $x = 1$ is a vertical asymptote.

9. $\lim_{x\to 0^+} \frac{1}{x^2} = \infty = \lim_{x\to 0^-} \frac{1}{x^2}$

Therefore, $x = 0$ is a vertical asymptote.

10. $\lim_{x\to 2^+} \frac{4}{(x-2)^3} = \infty$

$\lim_{x\to 2^-} \frac{4}{(x-2)^3} = -\infty$

Therefore, $x = 2$ is a vertical asymptote.

11. $\lim_{x\to -2^+} \frac{x^2}{(x+2)(x-1)} = -\infty$

$\lim_{x\to -2^-} \frac{x^2}{(x+2)(x-1)} = \infty$

Therefore, $x = -2$ is a vertical asymptote.

$\lim_{x\to 1^+} \frac{x^2}{(x+2)(x-1)} = \infty$

$\lim_{x\to 1^-} \frac{x^2}{(x+2)(x-1)} = -\infty$

Therefore, $x = 1$ is a vertical asymptote.

12. $\lim_{x\to 1^+} \frac{2+x}{1-x} = -\infty$

$\lim_{x\to 1^-} \frac{2+x}{1-x} = \infty$

Therefore, $x = 1$ is a vertical asymptote.

13. $\lim_{x\to -2^+} \frac{x^3}{x^2-4} = \infty$

$\lim_{x\to -2^-} \frac{x^3}{x^2-4} = -\infty$

Therefore, $x = -2$ is a vertical asymptote.

$\lim_{x\to 2^+} \frac{x^3}{x^2-4} = \infty$

$\lim_{x\to 2^-} \frac{x^3}{x^2-4} = -\infty$

Therefore, $x = 2$ is a vertical asymptote.

14. No vertical asymptote since the denominator is never zero.

15. $\lim_{x\to 0^+} \left(1 - \frac{4}{x^2}\right) = -\infty = \lim_{x\to 0^-} \left(1 - \frac{4}{x^2}\right)$

Therefore, $x = 0$ is a vertical asymptote.

16. $\lim_{x\to 2^+} \frac{-2}{(x-2)^2} = -\infty = \lim_{x\to 2^-} \frac{-2}{(x-2)^2}$

Therefore, $x = 2$ is a vertical asymptote.

17. $\lim_{x \to -2^+} \frac{x}{(x+2)(x-1)} = \infty$

$\lim_{x \to -2^-} \frac{x}{(x+2)(x-1)} = -\infty$

Therefore, $x = -2$ is a vertical asymptote.

$\lim_{x \to 1^+} \frac{x}{(x+2)(x-1)} = \infty$

$\lim_{x \to 1^-} \frac{x}{(x+2)(x-1)} = -\infty$

Therefore, $x = 1$ is a vertical asymptote.

18. $\lim_{x \to -3^+} \frac{1}{(x+3)^4} = \infty = \lim_{x \to -3^-} \frac{1}{(x+3)^4}$

Therefore, $x = -3$ is a vertical asymptote.

19. $\lim_{x \to -1} \frac{x^2 - 1}{x+1} = \lim_{x \to -1} (x-1) = -2$

Removable discontinuity at $x = -1$

20. $\lim_{x \to -1} \frac{x^2 - 6x - 7}{x+1} = \lim_{x \to -1} (x-7) = -8$

Removable discontinuity at $x = -1$

21. $\lim_{x \to -1^+} \frac{x^2+1}{x+1} = \infty$

$\lim_{x \to -1^-} \frac{x^2+1}{x+1} = -\infty$

Vertical asymptote at $x = -1$

22. $\lim_{x \to -1^+} \frac{x-1}{x+1} = -\infty$

$\lim_{x \to -1^-} \frac{x-1}{x+1} = \infty$

Vertical asymptote at $x = -1$.

23. $\lim_{x \to 2^+} \frac{x-3}{x-2} = -\infty$

24. $\lim_{x \to 1^+} \frac{2+x}{1-x} = -\infty$

25. $\lim_{x \to 4^+} \frac{x^2}{x^2 - 16} = \infty$

26. $\lim_{x \to 4} \frac{x^2}{x^2+16} = \frac{1}{2}$

27. $\lim_{x \to 0^-} \left(1 + \frac{1}{x}\right) = -\infty$

28. $\lim_{x \to 0^-} \left(x^2 - \frac{1}{x}\right) = \infty$

29. $\lim_{x \to 1} \frac{x^2 - x}{(x^2+1)(x-1)} = \lim_{x \to 1} \frac{x}{x^2+1} = \frac{1}{2}$

30. $\lim_{x \to 1} \frac{x^3 - 1}{x^2 + x + 1} = \lim_{x \to 1} (x-1) = 0$

31. $\lim_{x \to 1^+} \frac{x^2+x+1}{x^3-1} = \lim_{x \to 1^+} \frac{1}{x-1} = \infty$

32. $\lim_{x \to 0^-} \frac{x^2 - 2x}{x^3} = \lim_{x \to 0^-} \frac{x-2}{x^2} = -\infty$

33. $\lim_{x \to 4} f(x) = \lim_{x \to 4} \frac{1}{(x-4)^2} = \infty$

34. $\lim_{x \to 4} g(x) = \lim_{x \to 4} (x^2 - 5x) = -4$

35. $\lim_{x \to 4} [f(x) + g(x)] = \infty + (-4) = \infty$

36. $\lim_{x \to 4} [f(x)g(x)] = (\infty)(-4) = -\infty$

37. $\lim_{x \to 4} \left[\frac{f(x)}{g(x)}\right] = \frac{\infty}{-4} = -\infty$

38. $\lim_{x \to 4} \left[\frac{g(x)}{f(x)}\right] = \frac{-4}{\infty} = 0$

39. $C = \frac{80{,}000p}{100 - p}, \quad 0 \le p < 100$

(a) $C(15) = \$14{,}117.65$

(b) $C(50) = \$80{,}000$

(c) $C(90) = \$720{,}000$

(d) $\lim_{p \to 100^-} \frac{80{,}000p}{100 - p} = \infty$

40. $C = \frac{528x}{100 - x}, \quad 0 \le x < 100$

(a) $C(25) = \$176$ million

(b) $C(50) = \$528$ milliion

(c) $C(75) = \$1584$ million

(d) $\lim_{x \to 100^-} \frac{528x}{100 - x} = \infty$

Thus, it is not possible.

41. (a) $r = \dfrac{2(7)}{\sqrt{625 - 49}} = \dfrac{7}{12}$ ft/sec

(b) $r = \dfrac{2(15)}{\sqrt{625 - 225}} = \dfrac{3}{2}$ ft/sec

(c) $\lim_{x \to 25^-} \dfrac{2x}{\sqrt{625 - x^2}} = \infty$

42.

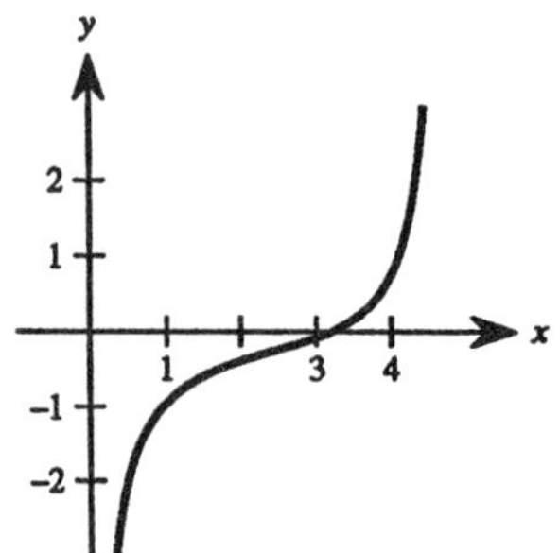

43. $\lim_{x \to 5^-} f(x) = -\infty$

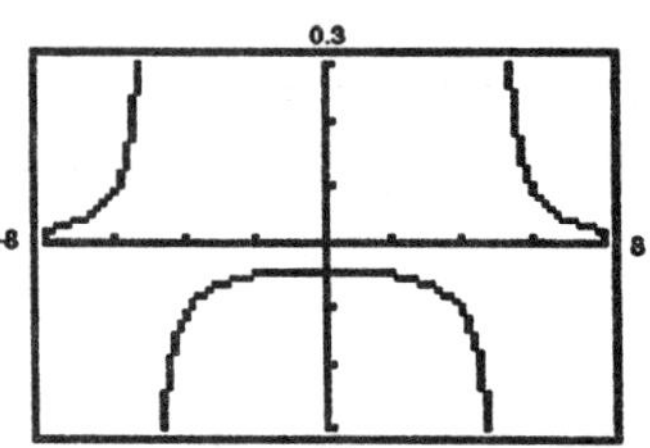

44. Let $f(x) = \dfrac{1}{x^2}$ and $g(x) = \dfrac{1}{x}$. $\lim_{x \to 0^+} \dfrac{1}{x^2} = \infty$ and $\lim_{x \to 0^+} \dfrac{1}{x} = \infty$, but $\lim_{x \to 0^+} \left(\dfrac{1}{x^2} - \dfrac{1}{x}\right) = \lim_{x \to 0^+} \left(\dfrac{1-x}{x^2}\right) = \infty \neq 0$.

Section 2.5 ε–δ Definition of Limits

1. $\lim_{x \to 2} (3x + 2) = 8$

$$|(3x + 2) - 8| < 0.01$$
$$|3x - 6| < 0.01$$
$$3|x - 2| < 0.01$$
$$0 < |x - 2| < \frac{0.01}{3}$$
$$= 0.0033 = \delta$$

2. $\lim_{x \to 4} \left[4 - \dfrac{x}{2}\right] = 2$

$$\left|\left[4 - \frac{x}{2}\right] - 2\right| < 0.01$$
$$\left|2 - \frac{x}{2}\right| < 0.01$$
$$\left|-\frac{1}{2}(x - 4)\right| < 0.01$$
$$0 < |x - 4| < 0.02 = \delta$$

3. $\lim_{x \to 2} (x^2 - 3) = 1$

$$|(x^2 - 3) - 1| < 0.01$$
$$|x^2 - 4| < 0.01$$
$$|(x + 2)(x - 2)| < 0.01$$
$$|x + 2|\,|x - 2| < 0.01$$
$$|x - 2| < \frac{0.01}{|x + 2|}$$

If we assume $1 < x < 3$, then $\delta = \dfrac{0.01}{5} = 0.002$.

4. $\lim_{x \to 5} \sqrt{x - 4} = 1$

$$|\sqrt{x - 4} - 1| < 0.01$$
$$1 < \sqrt{x - 4} < 1.01$$
$$1 < x - 4 < 1.0201$$
$$0 < x - 5 < 0.0201 = \delta$$

5. $\lim_{x \to 2} \dfrac{x^2 - 3x + 2}{x - 2} = 1$

$$\left|\frac{x^2 - 3x + 2}{x - 2} - 1\right| < 0.01$$
$$\left|\frac{(x - 2)(x - 1)}{x - 2} - 1\right| < 0.01$$
$$|x - 2| < 0.01 = \delta$$

6. $\lim_{x \to -2} \dfrac{x^2 + 5x + 6}{x + 2} = 1$

$$\left|\frac{x^2 + 5x + 6}{x + 2} - 1\right| < 0.01$$
$$\left|\frac{(x + 2)(x + 3)}{x + 2} - 1\right| < 0.01$$
$$|x + 2| < 0.01 = \delta$$

7. $\lim_{x \to 2} (x + 3) = 5$

Given $\epsilon > 0$:

$$|(x + 3) - 5| < \epsilon$$
$$|x - 2| < \epsilon = \delta$$

8. $\lim_{x \to -3} (2x + 5) = -1$

Given $\epsilon > 0$:

$$|(2x + 5) - (-1)| < \epsilon$$
$$|2x + 6| < \epsilon$$
$$2|x + 3| < \epsilon$$
$$|x + 3| < \frac{\epsilon}{2} = \delta$$

9. $\lim_{x \to 6} 3 = 3$

Given $\epsilon > 0$:

$$|3 - 3| < \epsilon$$
$$0 < \epsilon$$

Any δ will work.

10. $\lim_{x\to 2} -1 = -1$

Given $\epsilon > 0$:

$$|-1-(-1)| < \epsilon$$
$$0 < \epsilon$$

Any δ will work.

11. $\lim_{x\to 0} \sqrt[3]{x} = 0$

Given $\epsilon > 0$:

$$|\sqrt[3]{x} - 0| < \epsilon$$
$$|\sqrt[3]{x}| < \epsilon$$
$$|x| < \epsilon^3 = \delta$$

12. $\lim_{x\to 3} |x-3| = 0$

Given $\epsilon > 0$:

$$|(x-3) - 0| < \epsilon$$
$$|x-3| < \epsilon = \delta$$

13. $\lim_{x\to 0} x^2 = 0$

Given $\epsilon > 0$:

$$|x^2 - 0| < \epsilon$$
$$x^2 < \epsilon$$
$$|x| < \sqrt{\epsilon} = \delta$$

14. $\lim_{x\to 2} x^3 = 8$

Given $\epsilon > 0$:

$$|x^3 - 8| < \epsilon$$
$$|(x-2)(x^2+2x+4)| < \epsilon$$
$$|x^2+2x+4| \cdot |x-2| < \epsilon$$
$$|x-2| < \delta = \frac{\epsilon}{19} < \frac{\epsilon}{|x^2+2x+4|}$$

if we assume that x is in the interval $1 < x < 3$.

15. $\lim_{x\to 2} \frac{x^2+x-6}{x-2} = 5$

Given $\epsilon > 0$:

$$\left|\frac{x^2+x-6}{x-2} - 5\right| < \epsilon$$
$$\left|\frac{(x+3)(x-2)}{x-2} - 5\right| < \epsilon$$
$$|x-2| < \epsilon = \delta$$

16. $\lim_{x\to -10} \frac{x^2+10x}{x+10} = -10$

Given $\epsilon > 0$:

$$\left|\frac{x^2+10x}{x+10} - (-10)\right| < \epsilon$$
$$\left|\frac{x(x+10)}{x+10} + 10\right| < \epsilon$$
$$|x+10| < \epsilon = \delta$$

17. $\lim_{x\to 2} \frac{1}{x} = \frac{1}{2}$

Given $\epsilon > 0$:

$$\left|\frac{1}{x} - \frac{1}{2}\right| < \epsilon$$
$$\left|\frac{2-x}{2x}\right| < \epsilon$$
$$\frac{|x-2|}{|2x|} < \epsilon$$
$$|x-2| < \delta = 2\epsilon < |2x|\epsilon$$

if we assume that x is in the interval $1 < x < 3$.

18. $\lim_{x\to -1} \frac{3}{x+2} = 3$

Given $\epsilon > 0$:

$$\left|\frac{3}{x+2} - 3\right| < \epsilon$$
$$\frac{3}{|x+2|}|x+1| < \epsilon$$
$$|x+1| < \delta = \frac{\epsilon}{6} < \frac{|x+2|\epsilon}{3}$$

if we assume that x is in the interval $-\frac{3}{2} < x < -\frac{1}{2}$.

19. $\lim_{x\to 2} (x^2 - 2) = 2$

Given $\epsilon > 0$:

$$|(x^2 - 2) - 2| < \epsilon$$

$$|(x + 2)(x - 2)| < \epsilon$$

$$|x - 2| < \delta = \frac{\epsilon}{5} < \frac{\epsilon}{|x + 2|}$$

if we assume that x is in the interval $1 < x < 3$.

20. $\lim_{x\to -1} x^2 = 1$

Given $\epsilon > 0$:

$$|x^2 - 1| < \epsilon$$

$$|(x + 1)(x - 1)| < \epsilon$$

$$|x - 1| \cdot |x + 1| < \epsilon$$

$$|x + 1| < \delta = \frac{\epsilon}{3} < \frac{\epsilon}{|x - 1|}$$

if we assume that x is in the interval $-2 < x < 0$.

21. $\lim_{x\to 0^+} \sqrt{x} = 0$

Given $\epsilon > 0$:

$$|\sqrt{x} - 0| < \epsilon$$

$$|\sqrt{x}| < \epsilon$$

$$\sqrt{x} < \epsilon$$

$$0 < x < \epsilon^2 = \delta$$

22. $\lim_{x\to 3^-} f(x) = 2$

$$f(x) = \begin{cases} 2, & x \le 3 \\ 0, & x > 3 \end{cases}$$

Given $\epsilon > 0$:

$|2 - 2| < \epsilon$, $0 < \epsilon$

where $-\delta < x - 3 < 0$ for any $\delta > 0$.

23. $\lim_{x\to -1^+} \frac{1}{x + 1} = \infty$

Given $M > 0$, find δ such that $1/(x + 1) > M$ for $0 < x + 1 < \delta$.

$$\frac{1}{x + 1} > M$$

$$0 < x + 1 < \frac{1}{M} = \delta$$

24. $\lim_{x\to -1^-} \frac{1}{x + 1} = -\infty$

Given $N < 0$, find δ such that $1/(x + 1) < N$ for $-\delta < x + 1 < 0$.

$$\frac{1}{x + 1} < N$$

$$\frac{1}{N} < x + 1 < 0$$

Let $\delta = \frac{1}{|N|}$.

25. $\lim_{x\to 2} \frac{1}{(x - 2)^2} = \infty$

Given $M > 0$, find δ such that $1/(x - 2)^2 > M$ when $0 < |x - 2| < \delta$.

$$\frac{1}{(x - 2)^2} > M$$

$$(x - 2)^2 < \frac{1}{M}$$

$$0 < |x - 2| < \frac{1}{\sqrt{M}} = \delta$$

26. $\lim_{x\to 0} \frac{1}{x^2} = \infty$

Given $M > 0$, find $\delta > 0$ such that $1/x^2 > M$ when $0 < |x - 0| < \delta$.

$$\frac{1}{x^2} > M$$

$$x^2 < \frac{1}{M} \Rightarrow x < \sqrt{1/M}$$

Let $\delta = \sqrt{1/M}$.

27. $\lim_{x\to 3} x^2 = f(3) = 9$

28. $\lim_{x\to 1} (4 - 3x) = f(1) = 1$

29. Since $\lim_{x\to 0}(4-x^2) = 4 = \lim_{x\to 0}(4+x^2)$ and $4-x^2 \le f(x) \le 4+x^2$, then $\lim_{x\to 0} f(x) = 4$.

30. Since $\lim_{x\to a}[b-|x-a|] = b = \lim_{x\to a}[b+|x-a|]$ and $b-|x-a| \le f(x) \le b+|x-a|$, then $\lim_{x\to a} f(x) = b$.

31. If $\lim_{x\to c} f(x) = L_1$ and $\lim_{x\to c} f(x) = L_2$, then for every $\epsilon > 0$, there exists $\delta_1 > 0$ and $\delta_2 > 0$ such that $|x-c| < \delta_1 \Rightarrow |f(x)-L_1| < \epsilon$ and $|x-c| < \delta_2 \Rightarrow |f(x)-L_2| < \epsilon$. Let δ equal the smaller of δ_1 and δ_2. Then for $|x-c| < \delta$, we have

$$|L_1 - L_2| = |L_1 - f(x) + f(x) - L_2| \le |L_1 - f(x)| + |f(x) - L_2| < \epsilon + \epsilon.$$

Therefore, $|L_1 - L_2| < 2\epsilon \Rightarrow L_1 = L_2$.

32. Given $\lim_{x\to c} f(x)$ exists and $\lim_{x\to c}[f(x)+g(x)]$ does not exist, let $\lim_{x\to c} f(x) = L$ then

$$\lim_{x\to c}[f(x)+g(x)] = \lim_{x\to c} f(x) + \lim_{x\to c} g(x) = L + \lim_{x\to c} g(x).$$

The only way that $L + \lim_{x\to c} g(x)$ could not exist is for $\lim_{x\to c} g(x)$ to not exist.

33. Given $\lim_{x\to c} f(x) = 0$

For every $\epsilon > 0$, there exists $\delta > 0$ such that $|f(x) - 0| < \epsilon$ whenever $|x-c| < \delta$. Now $|f(x)-0| = |f(x)| = \big||f(x)| - 0\big| < \epsilon$ for $|x-c| < \delta$. Therefore, $\lim_{x\to c} |f(x)| = 0$.

34. If $\lim_{x\to c} |f(x)| = 0$, then $\lim_{x\to c}[-|f(x)|] = 0$.

$$-|f(x)| \le f(x) \le |f(x)|$$

$$\lim_{x\to c}[-|f(x)|] \le \lim_{x\to c} f(x) \le \lim_{x\to c} |f(x)|$$

$$0 \le \lim_{x\to c} f(x) \le 0$$

Therefore, $\lim_{x\to c} f(x) = 0$.

35. Given $\lim_{x\to c} f(x) = L$. For every $\epsilon > 0$, there exists $\delta > 0$ such that $|f(x) - L| < \epsilon$ whenever $|x-c| < \delta$. Since $\big||f(x)| - |L|\big| \le |f(x) - L| < \epsilon$ for $|x-c| < \delta$, then $\lim_{x\to c} |f(x)| = |L|$.

36. Let

$$f(x) = \begin{cases} 4, & \text{if } x \ge 0 \\ -4, & \text{if } x < 0 \end{cases}$$

$$\lim_{x\to 0} |f(x)| = \lim_{x\to 0} 4 = 4$$

$\lim_{x\to 0} f(x)$ does not exist since $\lim_{x\to 0^-} f(x) = -4$ and $\lim_{x\to 0^+} f(x) = 4$.

Chapter 2 Review Exercises

1. $\lim_{x\to 2}(5x-3) = 5(2)-3 = 7$

2. $\lim_{x\to 2}(3x+5) = 3(2)+5 = 11$

3. $\lim_{x\to 2}(5x-3)(3x+5) = [5(2)-3][3(2)+5]$
$= 7\cdot 11 = 77$

4. $\lim_{x\to 2}\left(\frac{3x+5}{5x-3}\right) = \frac{3(2)+5}{5(2)-3} = \frac{11}{7}$

5. $\lim_{x\to 3}\frac{t^2+1}{t} = \frac{3^2+1}{3} = \frac{10}{3}$

6. $\lim_{t\to 3}\frac{t^2-9}{t-3} = \lim_{t\to 3}(t+3) = 6$

7. $\lim_{t\to -2}\frac{t+2}{t^2-4} = \lim_{t\to -2}\frac{1}{t-2}$
$= -\frac{1}{4}$

8. $\lim_{x\to 0}\frac{\sqrt{4+x}-2}{x} = \lim_{x\to 0}\frac{\sqrt{4+x}-2}{x}\cdot\frac{\sqrt{4+x}+2}{\sqrt{4+x}+2} = \lim_{x\to 0}\frac{1}{\sqrt{4+x}+2} = \frac{1}{4}$

9. $\lim_{x\to 0}\frac{[1/(x+1)]-1}{x} = \lim_{x\to 0}\frac{-1}{x+1} = -1$

10. $\lim_{s\to 0}\frac{(1/\sqrt{1+s})-1}{s} = \lim_{s\to 0}\left[\frac{(1/\sqrt{1+s})-1}{s}\cdot\frac{(1/\sqrt{1+s})+1}{(1/\sqrt{1+s})+1}\right]$
$= \lim_{s\to 0}\frac{[1/(1+s)]-1}{s[1/(\sqrt{1+s})+1]} = \lim_{s\to 0}\frac{-1}{(1+s)[1/(\sqrt{1+s})+1]} = -\frac{1}{2}$

11. $\lim_{x\to -1}\frac{x^3+1}{x+1} = \lim_{x\to -1}\frac{(x+1)(x^2-x+1)}{x+1} = \lim_{x\to -1}(x^2-x+1) = 3$

12. $\lim_{x\to -2}\frac{x^2-4}{x^3+8} = \lim_{x\to -2}\frac{(x+2)(x-2)}{(x+2)(x^2-2x+4)} = \lim_{x\to -2}\frac{x-2}{x^2-2x+4} = -\frac{4}{12} = -\frac{1}{3}$

13. $\lim_{x\to 0^+}\left(x-\frac{1}{x^3}\right) = -\infty$

14. $\lim_{x\to 2^+}\frac{1}{\sqrt[3]{x^2-4}} = \infty$
$\lim_{x\to 2^-}\frac{1}{\sqrt[3]{x^2-4}} = -\infty$
Thus, $\lim_{x\to 2}\frac{1}{\sqrt[3]{x^2-4}}$ does not exist.

15. $\lim_{x\to -2^-}\frac{2x^2+x+1}{x+2} = -\infty$

16. $\lim_{x\to(1/2)}\frac{2x-1}{6x-3} = \lim_{x\to(1/2)}\frac{1}{3} = \frac{1}{3}$

17. $\lim_{x\to -1}\frac{x+1}{x^3+1} = \lim_{x\to -1}\frac{1}{x^2-x+1} = \frac{1}{3}$

18. $\lim_{x\to -1}\frac{x+1}{x^4-1} = \lim_{x\to -1}\frac{1}{(x^2+1)(x-1)} = -\frac{1}{4}$

19. $\lim_{x\to 1}\frac{x^2-2x+1}{x+1} = 0$

20. $\lim_{x\to -1^+}\frac{x^2-2x+1}{x+1} = \infty$

21. $f(x) = \dfrac{\sqrt{2x+1} - \sqrt{3}}{x-1}$

x	1.1	1.01	1.001	1.0001
$f(x)$	0.5680	0.5764	0.5773	0.5773

$$\lim_{x\to1^+} \frac{\sqrt{2x+1} - \sqrt{3}}{x-1} \approx 0.577$$

22. $f(x) = \dfrac{1 - \sqrt[3]{x}}{x-1}$

x	1.1	1.01	1.001	1.0001
$f(x)$	−0.3228	−0.3322	−0.3332	−0.3333

$$\lim_{x\to1^+} \frac{1 - \sqrt[3]{x}}{x-1} \approx -0.333$$

23.
$$\lim_{x\to1^+} \frac{\sqrt{2x+1} - \sqrt{3}}{x-1} = \lim_{x\to1^+} \frac{\sqrt{2x+1} - \sqrt{3}}{x-1} \cdot \frac{\sqrt{2x+1} + \sqrt{3}}{\sqrt{2x+1} + \sqrt{3}}$$
$$= \lim_{x\to1^+} \frac{(2x+1) - 3}{(x-1)\left[\sqrt{2x+1} + \sqrt{3}\right]} = \lim_{x\to1^+} \frac{2}{\sqrt{2x+1} + \sqrt{3}} = \frac{2}{2\sqrt{3}} = \frac{1}{\sqrt{3}}$$

24.
$$\lim_{x\to1^+} \frac{1 - \sqrt[3]{x}}{x-1} = \lim_{x\to1^+} \frac{1 - \sqrt[3]{x}}{x-1} \cdot \frac{1 + \sqrt[3]{x} + \left(\sqrt[3]{x}\right)^2}{1 + \sqrt[3]{x} + \left(\sqrt[3]{x}\right)^2}$$
$$= \lim_{x\to1^+} \frac{1-x}{(x-1)\left[1 + \sqrt[3]{x} + \left(\sqrt[3]{x}\right)^2\right]} = \lim_{x\to1^+} \frac{-1}{1 + \sqrt[3]{x} + \left(\sqrt[3]{x}\right)^2} = -\frac{1}{3}$$

25. $\lim_{x\to0} |x|/x = 1$ is false,
$\lim_{x\to0} |x|/x$ does not exist,
since $\lim_{x\to0^-} |x|/x = -1$ and
$\lim_{x\to0^+} |x|/x = 1$.

26. $\lim_{x\to0} x^3 = 0$ is true.

27. $\lim_{x\to2} f(x) = 3$ is false
since $\lim_{x\to2^-} f(x) = 3$ and
$\lim_{x\to2^+} f(x) = 0$.

28. $\lim_{x\to3} f(x) = 1$ is true.

29. $\lim_{x\to0^+} \sqrt{x} = 0$ is true.

30. $\lim_{x\to0} \sqrt[3]{x} = 0$ is true.

31. $\lim_{x\to k^+} [\![x+3]\!] = k+3$ where k is an integer.
$\lim_{x\to k^-} [\![x+3]\!] = k+2$ where k is an integer.
Nonremovable discontinuity at each integer k

32. $f(x) = \dfrac{3x^2 - x - 2}{x-1} = \dfrac{(3x+2)(x-1)}{x-1} = 3x+2$
Removable discontinuity at $x = 1$

33. $\lim_{x\to1} \dfrac{(3x+2)(x-1)}{x-1} = 5$
Removable discontinuity at $x = 1$

34. $\lim_{x\to2^-} (5-x) = 3$
$\lim_{x\to2^+} (2x-3) = 1$
Nonremovable discontinuity at $x = 2$

35. $f(x) = \dfrac{1}{(x-2)^2}$
Nonremovable discontinuity at $x = 2$

36. $f(x) = \sqrt{\dfrac{x+2}{x}}$
Nonremovable discontinuity at $x = 0$

37. $f(x) = \dfrac{3}{x+1}$
Nonremovable discontinuity at $x = -1$

38. $f(x) = \dfrac{x+1}{2x+2} = \dfrac{x+1}{2(x+1)}$

$\lim\limits_{x \to -1} \dfrac{x+1}{2(x+1)} = \dfrac{1}{2}$

Removable discontinuity at $x = -1$

39. $f(2) = 5$

Find c so that $\lim\limits_{x \to 2^+} (cx + 6) = 5$.

$$c(2) + 6 = 5$$
$$2c = -1$$
$$c = -\tfrac{1}{2}$$

40. $\lim\limits_{x \to 1^+} (x + 1) = 2$

$\lim\limits_{x \to 3^-} (x + 1) = 4$

Find b and c so that $\lim\limits_{x \to 1^-} (x^2 + bx + c) = 2$ and $\lim\limits_{x \to 3^+} (x^2 + bx + c) = 4$.

Consequently we get $1 + b + c = -2$ and $9 + 3b + c = 4$.

Solving simultaneously, $b = -3$ and $c = 4$.

41. $A = 5000(1.06)^{[\![2t]\!]}$

Nonremovable discontinuity every 6 months

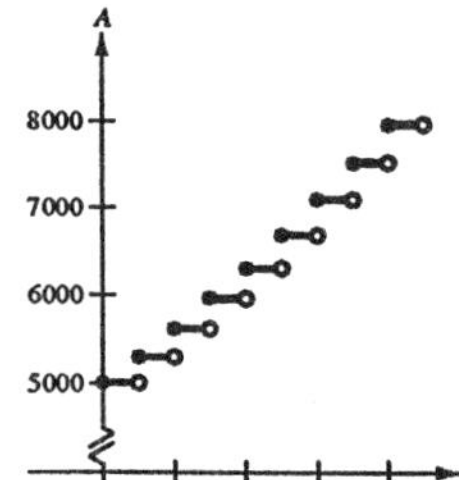

42. $A = 1000(1.14)^{[\![t]\!]}$

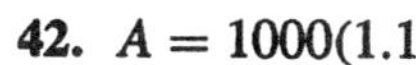

Nonremovable discontinuity every year

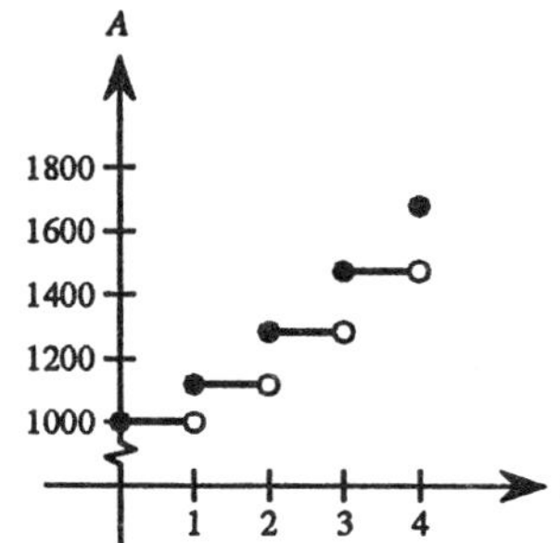

CHAPTER 3
Differentiation

Section 3.1 The Derivative and the Tangent Line Problem

1.

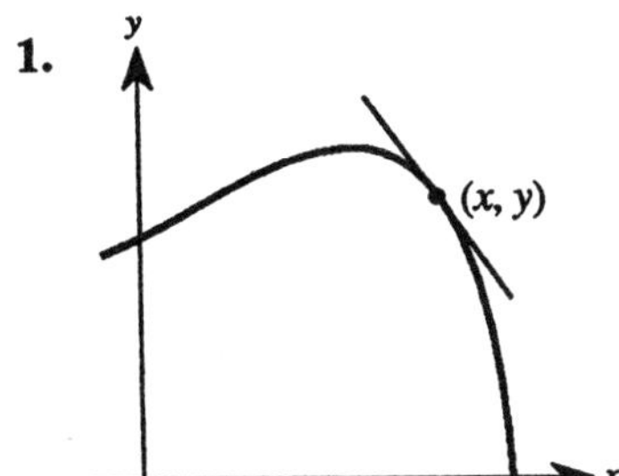

2.

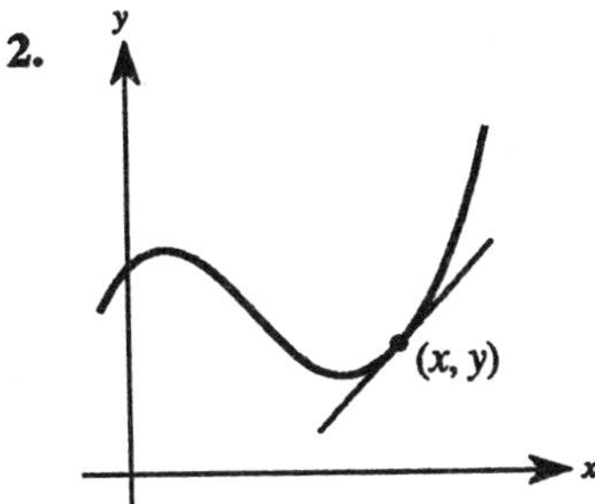

3. (a) $m = 0$

(b) $m = -3$

4. (a) $m = \frac{1}{4}$

(b) $m = 1$

5. $f(x) = 3$

$$f'(x) = \lim_{\Delta x \to 0} \frac{f(x + \Delta x) - f(x)}{\Delta x}$$
$$= \lim_{\Delta x \to 0} \frac{3 - 3}{\Delta x}$$
$$= \lim_{\Delta x \to 0} 0 = 0$$

6. $f(x) = 3x + 2$

$$f'(x) = \lim_{\Delta x \to 0} \frac{f(x + \Delta x) - f(x)}{\Delta x}$$
$$= \lim_{\Delta x \to 0} \frac{[3(x + \Delta x) + 2] - [3x + 2]}{\Delta x}$$
$$= \lim_{\Delta x \to 0} \frac{3\Delta x}{\Delta x}$$
$$= \lim_{\Delta x \to 0} 3 = 3$$

7. $f(x) = -5x$

$$f'(x) = \lim_{\Delta x \to 0} \frac{f(x + \Delta x) - f(x)}{\Delta x}$$
$$= \lim_{\Delta x \to 0} \frac{-5(x + \Delta x) - (-5x)}{\Delta x}$$
$$= \lim_{\Delta x \to 0} -5 = -5$$

8. $f(x) = 1 - x^2$

$$f'(x) = \lim_{\Delta x \to 0} \frac{f(x + \Delta x) - f(x)}{\Delta x}$$
$$= \lim_{\Delta x \to 0} \frac{[1 - (x + \Delta x)^2] - [1 - x^2]}{\Delta x}$$
$$= \lim_{\Delta x \to 0} \frac{1 - x^2 - 2x\Delta x - (\Delta x)^2 - 1 + x^2}{\Delta x}$$
$$= \lim_{\Delta x \to 0} \frac{-2x\Delta x - (\Delta x)^2}{\Delta x}$$
$$= \lim_{\Delta x \to 0} (-2x - \Delta x) = -2x$$

9. $f(x) = 2x^2 + x - 1$

$$f'(x) = \lim_{\Delta x \to 0} \frac{f(x + \Delta x) - f(x)}{\Delta x}$$

$$= \lim_{\Delta x \to 0} \frac{[2(x + \Delta x)^2 + (x + \Delta x) - 1] - [2x^2 + x - 1]}{\Delta x}$$

$$= \lim_{\Delta x \to 0} \frac{(2x^2 + 4x\Delta x + 2(\Delta x)^2 + x + \Delta x - 1) - (2x^2 + x - 1)}{\Delta x}$$

$$= \lim_{\Delta x \to 0} \frac{4x\Delta x + 2(\Delta x)^2 + \Delta x}{\Delta x} = \lim_{\Delta x \to 0} (4x + 2\Delta x + 1) = 4x + 1$$

10. $f(x) = \sqrt{x - 4}$

$$f'(x) = \lim_{\Delta x \to 0} \frac{f(x + \Delta x) - f(x)}{\Delta x} = \lim_{\Delta x \to 0} \frac{\sqrt{x + \Delta x - 4} - \sqrt{x - 4}}{\Delta x} \cdot \frac{\sqrt{x + \Delta x - 4} + \sqrt{x - 4}}{\sqrt{x + \Delta x - 4} + \sqrt{x - 4}}$$

$$= \lim_{\Delta x \to 0} \frac{(x + \Delta x - 4) - (x - 4)}{\Delta x\left[\sqrt{x + \Delta x - 4} + \sqrt{x - 4}\right]} = \lim_{\Delta x \to 0} \frac{1}{\sqrt{x + \Delta x - 4} + \sqrt{x - 4}} = \frac{1}{2\sqrt{x - 4}}$$

11. $f(x) = \dfrac{1}{x - 1}$

$$f'(x) = \lim_{\Delta x \to 0} \frac{f(x + \Delta x) - f(x)}{\Delta x}$$

$$= \lim_{\Delta x \to 0} \frac{\dfrac{1}{x + \Delta x - 1} - \dfrac{1}{x - 1}}{\Delta x}$$

$$= \lim_{\Delta x \to 0} \frac{(x - 1) - (x + \Delta x - 1)}{\Delta x(x + \Delta x - 1)(x - 1)}$$

$$= \lim_{\Delta x \to 0} \frac{-\Delta x}{\Delta x(x + \Delta x - 1)(x - 1)}$$

$$= \lim_{\Delta x \to 0} \frac{-1}{(x + \Delta x - 1)(x - 1)}$$

$$= -\frac{1}{(x - 1)^2}$$

12. $f(x) = \dfrac{1}{x^2}$

$$f'(x) = \lim_{\Delta x \to 0} \frac{f(x + \Delta x) - f(x)}{\Delta x}$$

$$= \lim_{\Delta x \to 0} \frac{\dfrac{1}{(x + \Delta x)^2} - \dfrac{1}{x^2}}{\Delta x}$$

$$= \lim_{\Delta x \to 0} \frac{x^2 - (x + \Delta x)^2}{\Delta x(x + \Delta x)^2 x^2}$$

$$= \lim_{\Delta x \to 0} \frac{-2x\Delta x - (\Delta x)^2}{\Delta x(x + \Delta x)^2 x^2}$$

$$= \lim_{\Delta x \to 0} \frac{-2x - \Delta x}{(x + \Delta x)^2 x^2}$$

$$= \frac{-2x}{x^4} = -\frac{2}{x^3}$$

13. $f(t) = t^3 - 12t$

$$f'(t) = \lim_{\Delta t \to 0} \frac{f(t + \Delta t) - f(t)}{\Delta t}$$

$$= \lim_{\Delta t \to 0} \frac{[(t + \Delta t)^3 - 12(t + \Delta t)] - [t^3 - 12t]}{\Delta t}$$

$$= \lim_{\Delta t \to 0} \frac{t^3 + 3t^2\Delta t + 3t(\Delta t)^2 + (\Delta t)^3 - 12t - 12\Delta t - t^3 + 12t}{\Delta t}$$

$$= \lim_{\Delta t \to 0} \frac{3t^2\Delta t + 3t(\Delta t)^2 + (\Delta t)^3 - 12\Delta t}{\Delta t}$$

$$= \lim_{\Delta t \to 0} (3t^2 + 3t\Delta t + (\Delta t)^2 - 12) = 3t^2 - 12$$

14. $f(t) = t^3 + t^2$

$$\begin{aligned} f'(t) &= \lim_{\Delta t \to 0} \frac{f(t+\Delta t) - f(t)}{\Delta t} \\ &= \lim_{\Delta t \to 0} \frac{[(t+\Delta t)^3 + (t+\Delta t)^2] - [t^3 + t^2]}{\Delta t} \\ &= \lim_{\Delta t \to 0} \frac{t^3 + 3t^2\Delta t + 3t(\Delta t)^2 + (\Delta t)^3 + t^2 + 2t\Delta t + (\Delta t)^2 - t^3 - t^2}{\Delta t} \\ &= \lim_{\Delta t \to 0} \frac{3t^2\Delta t + 3t(\Delta t)^2 + (\Delta t)^3 + 2t\Delta t + (\Delta t)^2}{\Delta t} \\ &= \lim_{\Delta t \to 0} (3t^2 + 3t\Delta t + (\Delta t)^2 + 2t + (\Delta t)) = 3t^2 + 2t \end{aligned}$$

15. $f(x) = x^2 + 1$

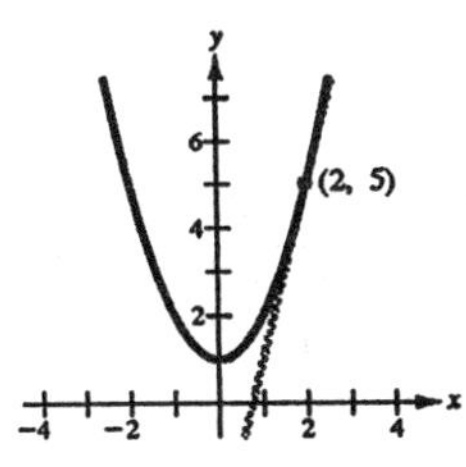

$$\begin{aligned} f'(x) &= \lim_{\Delta x \to 0} \frac{f(x+\Delta x) - f(x)}{\Delta x} \\ &= \lim_{\Delta x \to 0} \frac{[(x+\Delta x)^2 + 1] - [x^2 + 1]}{\Delta x} \\ &= \lim_{\Delta x \to 0} \frac{2x\Delta x + (\Delta x)^2}{\Delta x} \\ &= \lim_{\Delta x \to 0} (2x + \Delta x) = 2x \end{aligned}$$

At $(2, 5)$, the slope of the tangent line is $m = 2(2) = 4$. The equation of the tangent line is

$$\begin{aligned} y - 5 &= 4(x - 2) \\ y - 5 &= 4x - 8 \\ y &= 4x - 3. \end{aligned}$$

16. $f(x) = x^2 + 2x + 1$

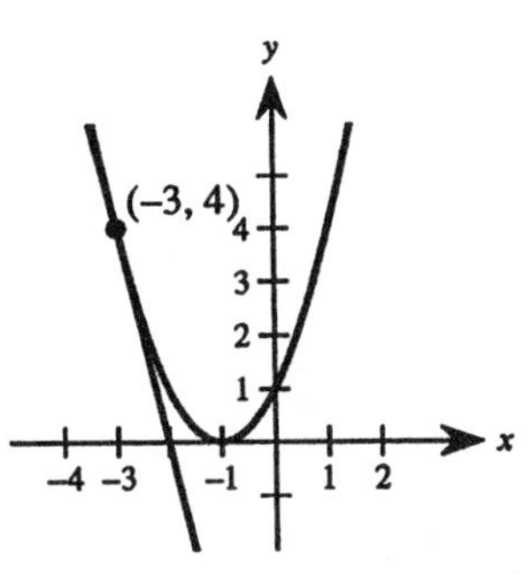

$$\begin{aligned} f'(x) &= \lim_{\Delta x \to 0} \frac{f(x+\Delta x) - f(x)}{\Delta x} \\ &= \lim_{\Delta x \to 0} \frac{[(x+\Delta x)^2 + 2(x+\Delta x) + 1] - [x^2 + 2x + 1]}{\Delta x} \\ &= \lim_{\Delta x \to 0} \frac{2x\Delta x + (\Delta x)^2 + 2\Delta x}{\Delta x} \\ &= \lim_{\Delta x \to 0} (2x + \Delta x + 2) = 2x + 2 \end{aligned}$$

At $(-3, 4)$, the slope of the tangent line is $m = 2(-3) + 2 = -4$. The equation of the tangent line is

$$\begin{aligned} y - 4 &= -4(x + 3) \\ y &= -4x - 8. \end{aligned}$$

17. $f(x) = x^3$

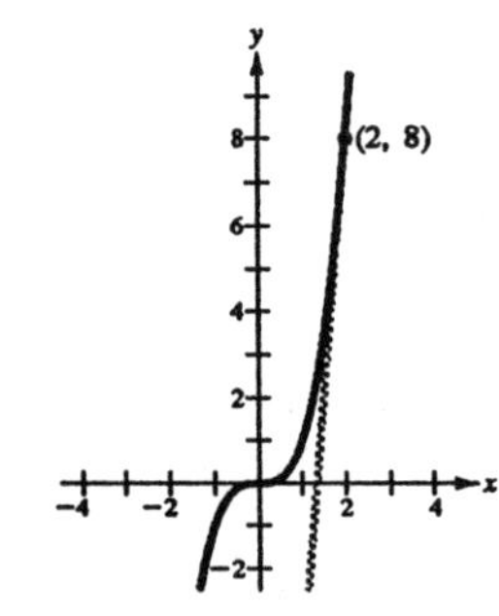

$$f'(x) = \lim_{\Delta x \to 0} \frac{f(x+\Delta x) - f(x)}{\Delta x}$$
$$= \lim_{\Delta x \to 0} \frac{(x+\Delta x)^3 - x^3}{\Delta x}$$
$$= \lim_{\Delta x \to 0} \frac{3x^2\Delta x + 3x(\Delta x)^2 + (\Delta x)^3}{\Delta x}$$
$$= \lim_{\Delta x \to 0} (3x^2 + 3x\Delta x + (\Delta x)^2) = 3x^2$$

At (2, 8), the slope of the tangent line is $m = 3(2)^2 = 12$. The equation of the tangent line is

$$y - 8 = 12(x - 2)$$
$$y = 12x - 16.$$

18. From Exercise 17 we have $f'(x) = 3x^2$. At $(-2, -8)$, the slope of the tangent line is $m = 3(-2)^2 = 12$. The equation of the tangent line is

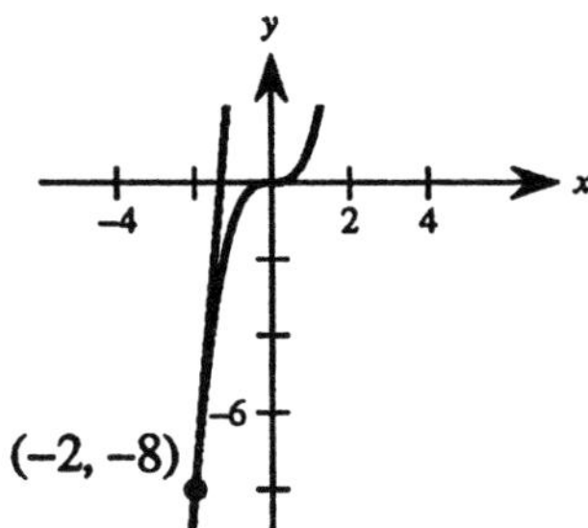

$$y - (-8) = 12[x - (-2)]$$
$$y = 12x + 16.$$

19. $f(x) = \sqrt{x+1}$

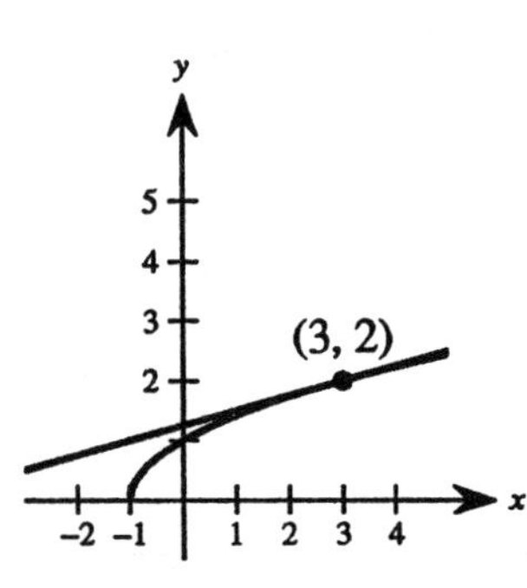

$$f'(x) = \lim_{\Delta x \to 0} \frac{f(x+\Delta x) - f(x)}{\Delta x}$$
$$= \lim_{\Delta x \to 0} \frac{\sqrt{x+\Delta x+1} - \sqrt{x+1}}{\Delta x} \cdot \frac{\sqrt{x+\Delta x+1} + \sqrt{x+1}}{\sqrt{x+\Delta x+1} + \sqrt{x+1}}$$
$$= \lim_{\Delta x \to 0} \frac{(x+\Delta x+1) - (x+1)}{\Delta x\left[\sqrt{x+\Delta x+1} + \sqrt{x+1}\right]}$$
$$= \lim_{\Delta x \to 0} \frac{1}{\sqrt{x+\Delta x+1} + \sqrt{x+1}}$$
$$= \frac{1}{2\sqrt{x+1}}$$

At (3, 2), the slope of the tangent line is

$$m = \frac{1}{2\sqrt{3+1}} = \frac{1}{4}.$$

The equation of the tangent line is

$$y - 2 = \frac{1}{4}(x - 3)$$
$$y = \frac{1}{4}x + \frac{5}{4}$$
$$4y = x + 5.$$

20. $f(x) = \dfrac{1}{x+1}$

$$f'(x) = \lim_{\Delta x\to 0}\frac{f(x+\Delta x)-f(x)}{\Delta x}$$

$$= \lim_{\Delta x\to 0}\frac{\dfrac{1}{x+\Delta x+1}-\dfrac{1}{x+1}}{\Delta x}$$

$$= \lim_{\Delta x\to 0}\frac{(x+1)-(x+\Delta x+1)}{\Delta x(x+\Delta x+1)(x+1)}$$

$$= \lim_{\Delta x\to 0} -\frac{1}{(x+\Delta x+1)(x+1)}$$

$$= -\frac{1}{(x+1)^2}$$

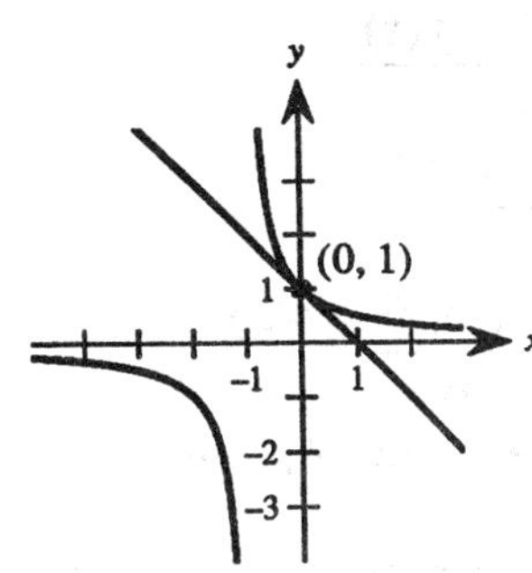

At (0, 1), the slope of the tangent line is

$$m = \frac{-1}{(0+1)^2} = -1.$$

The equation of the tangent line is $y = -x + 1$.

21. $f(x) = x^2 - 1, \quad c = 2$

$$f'(2) = \lim_{x\to 2}\frac{f(x)-f(2)}{x-2} = \lim_{x\to 2}\frac{(x^2-1)-3}{x-2} = \lim_{x\to 2}(x+2) = 4$$

22. $f(x) = x^3 + 2x, \quad c = 1$

$$f'(1) = \lim_{x\to 1}\frac{f(x)-f(1)}{x-1} = \lim_{x\to 1}\frac{x^3+2x-3}{x-1} = \lim_{x\to 1}(x^2+x+3) = 5$$

23. $f(x) = x^3 + 2x^2 + 1, \quad c = -2$

$$f'(-2) = \lim_{x\to -2}\frac{f(x)-f(-2)}{x+2} = \lim_{x\to -2}\frac{(x^3+2x^2+1)-1}{x+2} = \lim_{x\to -2} x^2 = 4$$

24. $f(x) = \dfrac{1}{x}, \quad c = 3$

$$f'(3) = \lim_{x\to 3}\frac{f(x)-f(3)}{x-3} = \lim_{x\to 3}\frac{(1/x)-(1/3)}{x-3} = \lim_{x\to 3}\frac{3-x}{3x}\cdot\frac{1}{x-3} = \lim_{x\to 3} -\frac{1}{3x} = -\frac{1}{9}$$

25. $f(x) = (x-1)^{2/3}, \quad c = 1$

$$f'(1) = \lim_{x\to 1}\frac{f(x)-f(1)}{x-1} = \lim_{x\to 1}\frac{(x-1)^{2/3}-0}{x-1} = \lim_{x\to 1}\frac{1}{(x-1)^{1/3}}$$

The limit does not exist. Thus, f is not differentiable at $x = 1$.

26. $f(x) = |x-2|, \quad c = 2$

$$f'(2) = \lim_{x\to 2}\frac{f(x)-f(2)}{x-2} = \lim_{x\to 2}\frac{|x-2|}{x-2}$$

The limit does not exist. Thus, f is not differentiable at $x = 2$.

27. $f(x)$ is differentiable everywhere except at $x = -3$. (Sharp turn in the graph)

$(-\infty, -3), (-3, \infty)$

28. $f(x)$ is differentiable everywhere except at $x = \pm 3$. (Sharp turns in the graph)
$(-\infty, -3), (-3, 3), (3, \infty)$

29. $f(x)$ is differentiable everywhere except at $x = -1$. (Discontinuity)
$(-\infty, -1), (-1, \infty)$

30. $f(x)$ is differentiable everywhere except at $x = 1$. (Discontinuity)
$(-\infty, 1), (1, \infty)$

31. $f(x)$ is differentiable everywhere except at $x = 3$. (Vertical tangent)
$(-\infty, 3), (3, \infty)$

32. $f(x)$ is differentiable everywhere except $x = 0$. (Vertical tangent)
$(-\infty, 0), (0, \infty)$

33. $f(x)$ is differentiable on the interval $(1, \infty)$. (At $x = 1$ the tangent line is vertical.)

34. $f(x)$ is differentiable everywhere except at $x = \pm 2$. (Discontinuities)
$(-\infty, -2), (-2, 2), (2, \infty)$

35. $f(x)$ is differentiable everywhere except at $x = 0$. (Discontinuity)
$(-\infty, 0), (0, \infty)$

36. $f(x)$ is differentiable everywhere except at $x = 1$. (Discontinuity)
$(-\infty, 1), (1, \infty)$

37. (a) If $f(x)$ is odd, then the graph of f is symmetric about the origin.
So, if $f'(c) = 3$, then $f'(-c) = 3$.

(b) If $f(x)$ is even, then the graph of f is symmetric about the y-axis.
So, if $f'(c) = 3$, then $f'(-c) = -3$.

38. (a) $f(x) = x^2, \quad f'(x) = 2x$

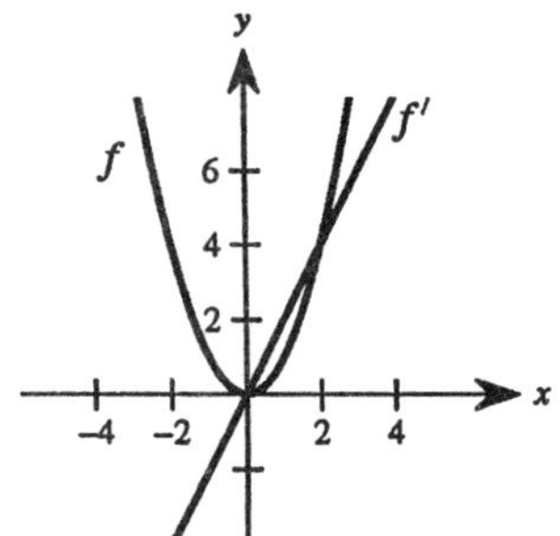

(b) $f(x) = x^3, \quad f'(x) = 3x^2$

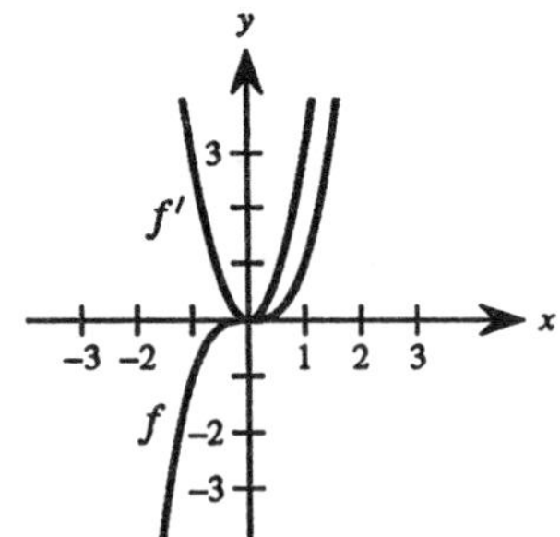

39. $f(x) = \sqrt{1 - x^2}$

The derivative from the left is

$$\lim_{x \to 1^-} \frac{f(x) - f(1)}{x - 1} = \lim_{x \to 1^-} \frac{\sqrt{1 - x^2} - 0}{x - 1} = \lim_{x \to 1^-} \frac{\sqrt{1 - x^2}}{x - 1} \cdot \frac{\sqrt{1 - x^2}}{\sqrt{1 - x^2}} = \lim_{x \to 1^-} -\frac{1 + x}{\sqrt{1 - x^2}} = -\infty.$$

The limit from the right does not exist since f is undefined for $x > 1$. Therefore, f is not differentiable at $x = 1$.

40. $f(x) = \begin{cases} x-1, & x \le 1 \\ (x-1)^2, & x > 1 \end{cases}$

The derivative from the left is

$$\lim_{x \to 1^-} \frac{f(x) - f(1)}{x - 1} = \lim_{x \to 1^-} \frac{(x-1) - 0}{x - 1} = \lim_{x \to 1^-} 1 = 1.$$

The derivative from the right is

$$\lim_{x \to 1^+} \frac{f(x) - f(1)}{x - 1} = \lim_{x \to 1^+} \frac{(x-1)^2 - 0}{x - 1} = \lim_{x \to 1^+} (x - 1) = 0.$$

The one-sided limits are not equal. Therefore, f is not differentiable at $x = 1$.

41. $f(x) = \begin{cases} (x-1)^3, & x \le 1 \\ (x-1)^2, & x > 1 \end{cases}$

The derivative from the left is

$$\lim_{x \to 1^-} \frac{f(x) - f(1)}{x - 1} = \lim_{x \to 1^-} \frac{(x-1)^3 - 0}{x - 1} = \lim_{x \to 1^-} (x - 1)^2 = 0.$$

The derivative from the right is

$$\lim_{x \to 1^+} \frac{f(x) - f(1)}{x - 1} = \lim_{x \to 1^+} \frac{(x-1)^2 - 0}{x - 1} = \lim_{x \to 1^+} (x - 1) = 0.$$

The one-sided limits are equal. Therefore, f is differentiable at $x = 1$ $(f'(1) = 0)$.

42. $f(x) = \begin{cases} x, & x \le 1 \\ x^2, & x > 1 \end{cases}$

The derivative from the left is

$$\lim_{x \to 1^-} \frac{f(x) - f(1)}{x - 1} = \lim_{x \to 1^-} \frac{x - 1}{x - 1} = \lim_{x \to 1^-} 1 = 1.$$

The derivative from the right is

$$\lim_{x \to 1^+} \frac{f(x) - f(1)}{x - 1} = \lim_{x \to 1^+} \frac{x^2 - 1}{x - 1} = \lim_{x \to 1^+} (x + 1) = 2.$$

The one-sided limits are not equal. Therefore, f is not differentiable at $x = 1$.

43. From Exercise 17 we know that $f'(x) = 3x^2$. Since the slope of the given line is 3, we have

$$3x^2 = 3$$
$$x = \pm 1.$$

Therefore, at the points (1, 1) and (−1, −1) the tangent lines are parallel to $3x - y + 1 = 0$. These lines have equations

$$y - 1 = 3(x - 1) \quad \text{and} \quad y + 1 = 3(x + 1)$$
$$y = 3x - 2 \qquad\qquad y = 3x + 2.$$

44. From the limit definition of the derivative

$$f'(x) = \frac{-1}{2x\sqrt{x}}.$$

Since the slope of the given line is $-\frac{1}{2}$, we have

$$-\frac{1}{2x\sqrt{x}} = -\frac{1}{2}$$
$$x = 1.$$

Therefore, at the point (1, 1) the tangent line is parallel to $x + 2y - 6 = 0$. The equation of this line is

$$y - 1 = -\frac{1}{2}(x - 1) = -\frac{1}{2}x + \frac{1}{2}$$
$$y = -\frac{1}{2}x + \frac{3}{2}.$$

45. Given the equation $y = 4x - x^2$ and the point $(2, 5)$, by the limit definition of the derivative, we have $dy/dx = 4 - 2x$. The equation of the tangent line is $y - 5 = (4 - 2x)(x - 2)$. Substituting the given equation for y, we have

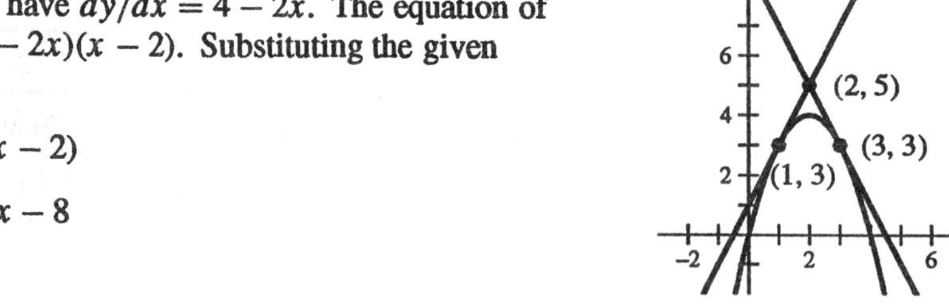

$$(4x - x^2) - 5 = (4 - 2x)(x - 2)$$
$$4x - x^2 - 5 = -2x^2 + 8x - 8$$
$$x^2 - 4x + 3 = 0$$
$$(x - 1)(x - 3) = 0.$$

Therefore, the tangent lines intersect the parabola at $(1, 3)$ and $(3, 3)$ and their equations are

$$y - 5 = 2(x - 2) \quad \text{and} \quad y - 5 = -2(x - 2)$$
$$y = 2x + 1 \qquad\qquad y = -2x + 9.$$

46. Given the equation $y = x^2$ and the point $(1, -3)$, by the limit definition of the derivative $dy/dx = 2x$. The equation of the tangent line is $y + 3 = 2x(x - 1)$. Substituting the given equation for y we have

$$x^2 + 3 = 2x(x - 1)$$
$$x^2 - 2x - 3 = 0$$
$$(x + 1)(x - 3) = 0$$

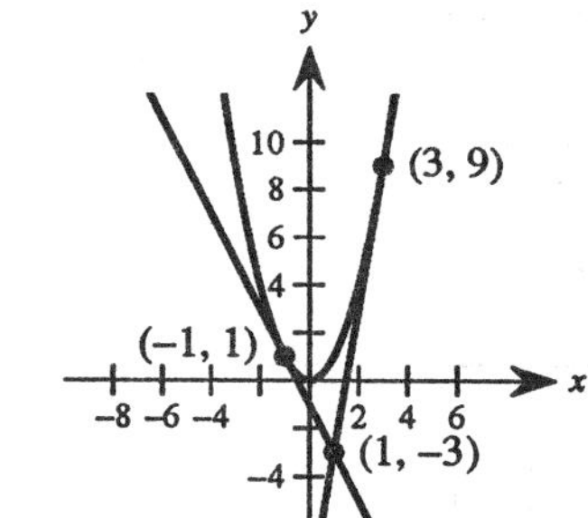

Therefore, the tangent lines intersect the curve at $(-1, 1)$ and $(3, 9)$ and their equations are

$$y - 1 = -2(x + 1) \quad \text{and} \quad y - 9 = 6(x - 3)$$
$$y - 1 = -2x - 2 \qquad\qquad y - 9 = 6x - 18$$
$$y = -2x - 1 \qquad\qquad y = 6x - 9.$$

47. $y = x^2$

Slope: $m = y' = 2x$

True, the slope is different at every point on the curve.

48. False. $y = |x - 2|$ is continuous at $x = 2$, but is not differentiable at $x = 2$. (Sharp turn in the graph)

49. True. Differentiability implies continuity.

50. False. If the derivative from the left of a point does not equal the derivative from the right of a point, then the derivative does not exist at that point. For example, if $f(x) = |x|$, then the derivative from the left of $x = 0$ is -1 and the derivative from the right of $x = 0$ is 1. At $x = 0$, the derivative does not exist.

51. $f(x) = \frac{1}{4}x^3, \quad f'(x) = \frac{3}{4}x^2$

x	-2	-1.5	-1	-0.5	0	0.5	1	1.5	2
$f(x)$	-2	-0.84375	-0.25	-0.03125	0	0.03125	0.25	0.84375	2
$f'(x)$	3	1.6875	0.75	0.1875	0	0.1875	0.75	1.6875	3

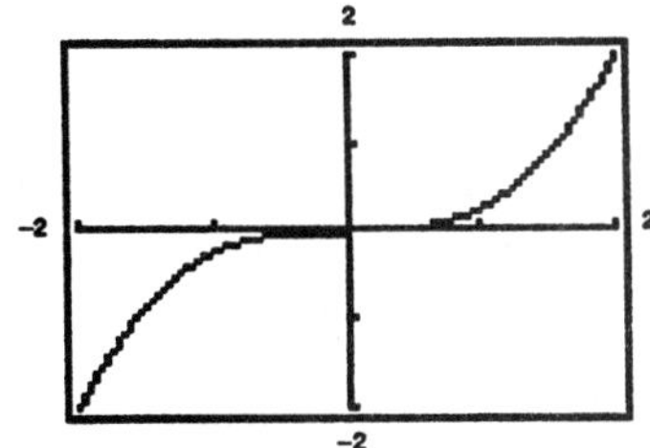

52. $f(x) = \frac{4}{x}, \quad f'(x) = -\frac{4}{x^2}$

x	-2	-1.5	-1	-0.5	0	0.5	1	1.5	2
$f(x)$	-2	-2.6667	-4	-8	undefined	8	4	2.6667	2
$f'(x)$	-1	-1.7778	-4	-16	undefined	-16	-4	-1.7778	-1

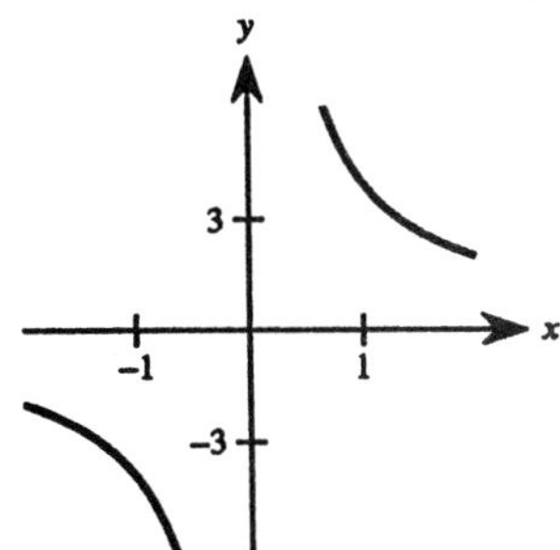

53. $f(x) = 4 - (x - 3)^2$

$$S_{\Delta x}(x) = \frac{f(2 + \Delta x) - f(2)}{\Delta x}(x - 2) + f(2)$$

$$= \frac{4 - (2 + \Delta x - 3)^2 - 3}{\Delta x}(x - 2) + 3 = \frac{1 - (\Delta x - 1)^2}{\Delta x}(x - 2) + 3 = (-\Delta x + 2)(x - 2) + 3$$

(a) $\Delta x = 1$: $S_{\Delta x} = (x - 2) + 3 = x + 1$

$\Delta x = 0.5$: $S_{\Delta x} = \left(\frac{3}{2}\right)(x - 2) + 3 = \frac{3}{2}x$

$\Delta x = 0.1$: $S_{\Delta x} = \left(\frac{19}{10}\right)(x - 2) + 3 = \frac{19}{10}x - \frac{4}{5}$

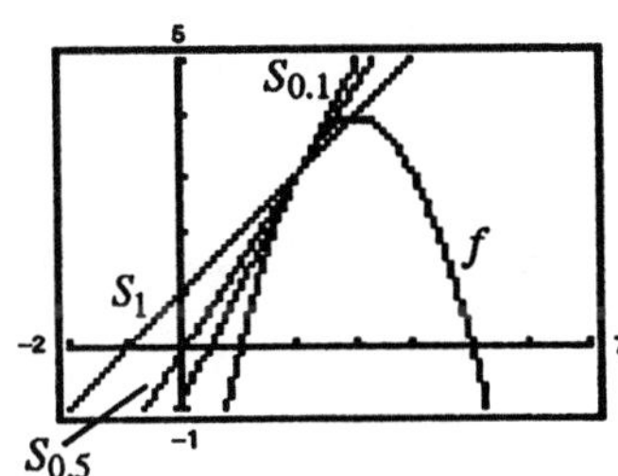

(b) As $\Delta x \to 0$, the line approaches the tangent line to f at $(2, 3)$.

54. $f(x) = x + \dfrac{1}{x}$

$$S_{\Delta x}(x) = \frac{f(2+\Delta x) - f(2)}{\Delta x}(x-2) + f(2) = \frac{(2+\Delta x) + \dfrac{1}{2+\Delta x} - \dfrac{5}{2}}{\Delta x}(x-2) + \frac{5}{2}$$

$$= \frac{2(2+\Delta x)^2 + 2 - 5(2+\Delta x)}{2(2+\Delta x)\Delta x}(x-2) + \frac{5}{2} = \frac{(2\Delta x + 3)}{2(2+\Delta x)}(x-2) + \frac{5}{2}$$

(a) $\Delta x = 1$: $\quad S_{\Delta x} = \dfrac{5}{6}(x-2) + \dfrac{5}{2} = \dfrac{5}{6}x + \dfrac{5}{6}$

$\Delta x = 0.5$: $\quad S_{\Delta x} = \dfrac{4}{5}(x-2) + \dfrac{5}{2} = \dfrac{4}{5}x + \dfrac{9}{10}$

$\Delta x = 0.1$: $\quad S_{\Delta x} = \dfrac{16}{21}(x-2) + \dfrac{5}{2} = \dfrac{16}{21}x + \dfrac{41}{42}$

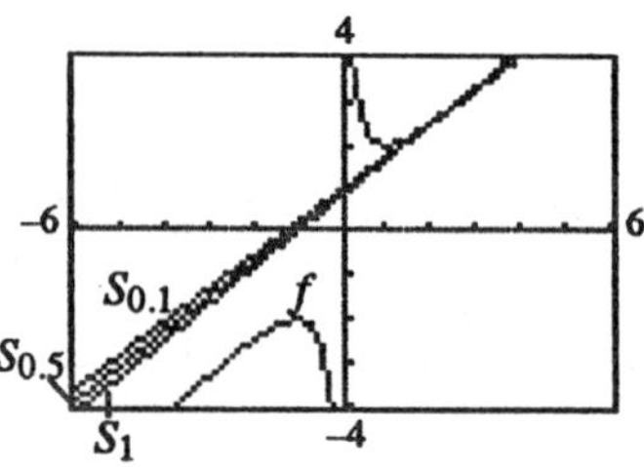

(b) As $\Delta x \to 0$, the line approaches the tangent line to f at $\left(2, \frac{5}{2}\right)$.

Section 3.2 Velocity, Acceleration, and Other Rates of Change

1. Average rate of change: $\dfrac{f(2) - f(1)}{2-1} = \dfrac{[2(2)+7] - [2(1)+7]}{1} = 2$

$$\lim_{\Delta t \to 0} \frac{f(t+\Delta t) - f(t)}{\Delta t} = \lim_{\Delta t \to 0} \frac{[2(t+\Delta t)+7] - (2t+7)}{\Delta t} = \lim_{\Delta t \to 0} 2 = 2$$

Instantaneous rate of change is the constant 2.

2. Average rate of change: $\dfrac{f(1/3) - f(0)}{(1/3) - 0} = \dfrac{[3(1/3) - 1] - [3(0) - 1]}{1/3} = 3$

$$\lim_{\Delta t \to 0} \frac{f(t+\Delta t) - f(t)}{\Delta t} = \lim_{\Delta t \to 0} \frac{[3(t+\Delta t) - 1] - (3t-1)}{\Delta t} = \lim_{\Delta t \to 0} 3 = 3$$

Instantaneous rate of change is the constant 3.

3. Average rate of change: $\dfrac{f(3) - f(0)}{3-0} = \dfrac{(1/4) - 1}{3-0} = \dfrac{-3/4}{3} = -\dfrac{1}{4}$

$$\lim_{\Delta x \to 0} \frac{f(x+\Delta x) - f(x)}{\Delta x} = \lim_{\Delta x \to 0} \frac{\dfrac{1}{x+\Delta x+1} - \dfrac{1}{x+1}}{\Delta x}$$

$$= \lim_{\Delta x \to 0} \left[-\frac{\Delta x}{(x+\Delta x+1)(x+1)} \cdot \frac{1}{\Delta x}\right] = \lim_{\Delta x \to 0} -\frac{1}{(x+\Delta x+1)(x+1)} = -\frac{1}{(x+1)^2}$$

Instantaneous rate of change: $\quad (0, 1) \Rightarrow f'(0) = \dfrac{-1}{(0+1)^2} = -1$

$$\left(3, \frac{1}{4}\right) \Rightarrow f'(3) = \frac{-1}{(3+1)^2} = -\frac{1}{16}$$

4. Average rate of change: $\dfrac{f(2) - f(1)}{2 - 1} = \dfrac{(-1/2) - (-1)}{2 - 1} = \dfrac{1}{2}$

$$\lim_{\Delta x \to 0} \frac{f(x + \Delta x) - f(x)}{\Delta x} = \lim_{\Delta x \to 0} \frac{\dfrac{-1}{x + \Delta x} - \dfrac{-1}{x}}{\Delta x} = \lim_{\Delta x \to 0} \left[\frac{\Delta x}{(x + \Delta x)x} \cdot \frac{1}{\Delta x}\right] = \lim_{\Delta x \to 0} \frac{1}{(x + \Delta x)x} = \frac{1}{x^2}$$

Instantaneous rate of change: $(1, -1) \Rightarrow f'(1) = 1$

$$\left(2, -\frac{1}{2}\right) \Rightarrow f'(2) = \frac{1}{4}$$

5. Average rate of change: $\dfrac{f(2.1) - f(2)}{2.1 - 2} = \dfrac{1.41 - 1}{0.1} = 4.1$

$$\lim_{\Delta t \to 0} \frac{f(t + \Delta t) - f(t)}{\Delta t} = \lim_{\Delta t \to 0} \frac{[(t + \Delta t)^2 - 3] - (t^2 - 3)}{\Delta t}$$

$$= \lim_{\Delta t \to 0} \frac{(t^2 + 2t(\Delta t) + (\Delta t)^2 - 3) - (t - 3)}{\Delta t}$$

$$= \lim_{\Delta t \to 0} \frac{2t(\Delta t) + (\Delta t)^2}{\Delta t} = \lim_{\Delta t \to 0} (2t + \Delta t) = 2t$$

Instantaneous rates of change: $(2, 1) \Rightarrow f'(2) = 2(2) = 4$

$$(2.1, 1.41) \Rightarrow f'(2.1) = 4.2$$

6. Average rate of change: $\dfrac{f(3) - f(-1)}{3 - (-1)} = \dfrac{-10 - 6}{3 - (-1)} = -4$

$$\lim_{\Delta x \to 0} \frac{f(x + \Delta x) - f(x)}{\Delta x} = \lim_{\Delta x \to 0} \frac{[(x + \Delta x)^2 - 6(x + \Delta x) - 1] - (x^2 - 6x - 1)}{\Delta x}$$

$$= \lim_{\Delta x \to 0} \frac{(x^2 + 2x(\Delta x) + (\Delta x)^2 - 6x - 6(\Delta x) - 1) - (x^2 - 6x - 1)}{\Delta x}$$

$$= \lim_{\Delta x \to 0} \frac{2x(\Delta x) + (\Delta x)^2 - 6(\Delta x)}{\Delta x} = \lim_{\Delta x \to 0} (2x + \Delta x - 6) = 2x - 6$$

Instantaneous rates of change: $(-1, 6) \Rightarrow f'(-1) = 2(-1) - 6 = -8$

$$(3, -10) \Rightarrow f'(3) = 2(3) - 6 = 0$$

7. $s(t) = -16t^2 + 1350$

(a) $\dfrac{s(2) - s(1)}{2 - 1} = 1286 - 1334$

$= -48$ ft/sec

(b) $s'(t) = -32t$

When $t = 1$, $s'(1) = -32$ ft/sec.

When $t = 2$, $s'(2) = -64$ ft/sec.

(c) $-16t^2 + 1350 = 0$

$$t^2 = \frac{1350}{16} \Rightarrow t = \frac{15\sqrt{6}}{4} \approx 9.2 \text{ sec}$$

(d) When $t = \dfrac{15\sqrt{6}}{4}$,

$$s'\left(\frac{15\sqrt{6}}{4}\right) = -120\sqrt{6} \approx -293.9 \text{ ft/sec.}$$

8. $v(t) = \dfrac{100t}{2t+15}$

$$a(t) = v'(t) = \lim_{\Delta t \to 0} \frac{v(t+\Delta t) - v(t)}{\Delta t} = \lim_{\Delta t \to 0} \frac{\dfrac{100(t+\Delta t)}{2(t+\Delta t)+15} - \dfrac{100t}{2t+15}}{\Delta t}$$

$$= \lim_{\Delta t \to 0} \frac{100(t+\Delta t)(2t-15) - 100t[2(t+\Delta t)+15]}{\Delta t(2t+15)[2(t+\Delta t)+15]}$$

$$= \lim_{\Delta t \to 0} \frac{1500}{(2t+15)[2(t+\Delta t)+15]} = \frac{1500}{(2t+15)^2}$$

(a) $a(5) = \dfrac{1500}{[2(5)+15]^2}$
$= 2.4 \text{ ft/sec}^2$

(b) $a(10) = \dfrac{1500}{[2(10)+15]^2}$
$= 1.2 \text{ ft/sec}^2$

(c) $a(20) = \dfrac{1500}{[2(20)+15]^2}$
$= 0.5 \text{ ft/sec}^2$

9. $s(t) = -16t^2 + 384t$

$v(t) = s'(t) = -32t + 384$

$v(5) = 224$ ft/sec

$v(10) = 64$ ft/sec

10. $s(t) = -16t^2 + 256t$

$v(t) = s'(t) = -32t + 256$

$v(5) = 96$ ft/sec

$v(10) = -64$ ft/sec

11. $s(t) = -16t^2 + 600$

$-16t^2 + 600 = 0$

$t^2 = \dfrac{600}{16}$

$t = \dfrac{10\sqrt{6}}{4} = \dfrac{5\sqrt{6}}{2}$ sec (the instant the pebble hits the ground)

$v(t) = s'(t) = -32t$

$v\left(\dfrac{5\sqrt{6}}{2}\right) = -80\sqrt{6}$ ft/sec ≈ -195.96 ft/sec

12. $s(t) = -16t^2 - 22t + 220$

$v(t) = -32t - 22$

$v(3) = -118$ ft/sec

$s(t) = -16t^2 - 22t + 220 = 112$ (height after falling 108 ft)

$s(t) = -16t^2 - 22t + 108 = 0$

$s(t) = -2(t-2)(8t+27) = 0$

$t = 2$

$v(2) = -32(2) - 22 = -86$ ft/sec

13. $s(t) = -16t^2 + s_0 = 0$ when $t = 6.8$.

$s_0 = 16(6.8)^2 \approx 740$ ft

14. Ball dropped from 100 feet:

$$s(t) = -16t^2 + v_0t + s_0 = -16t^2 + 100 = 0$$

$$t = 2.5$$

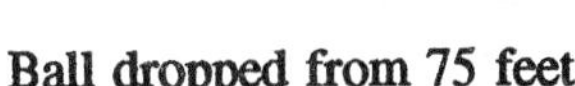

Ball dropped from 75 feet:

$$s(t) = -16t^2 + 75 = 0$$

$$t \approx 2.165$$

The ball dropped from 100 feet hits first.

15. On the interval [0, 4]:

$$v = \left(\frac{2-0}{4-0}\right)(60) = 30 \text{ mph}$$

On the interval [4, 6]:

$$v = \left(\frac{2-2}{4-6}\right)(60) = 0 \text{ mph}$$

On the interval [6, 10]:

$$v = \left(\frac{6-2}{10-6}\right)(60) = 60 \text{ mph}$$

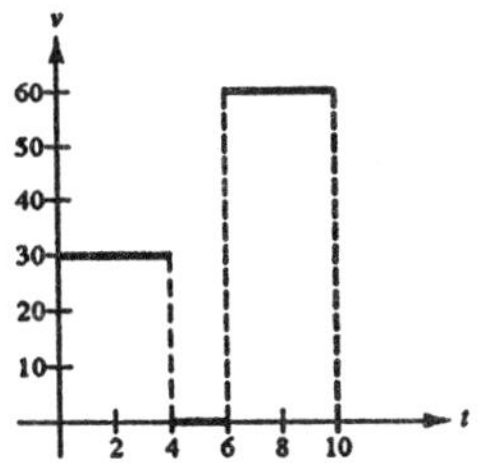

16. On the interval [0, 6]:

$$v = \left(\frac{5-0}{6-0}\right)(60) = 50 \text{ mph}$$

On the interval [6, 8]:

$$v = \left(\frac{5-5}{8-6}\right)(60) = 0 \text{ mph}$$

On the interval [8, 10]:

$$v = \left(\frac{6-5}{10-8}\right)(60) = 30 \text{ mph}$$

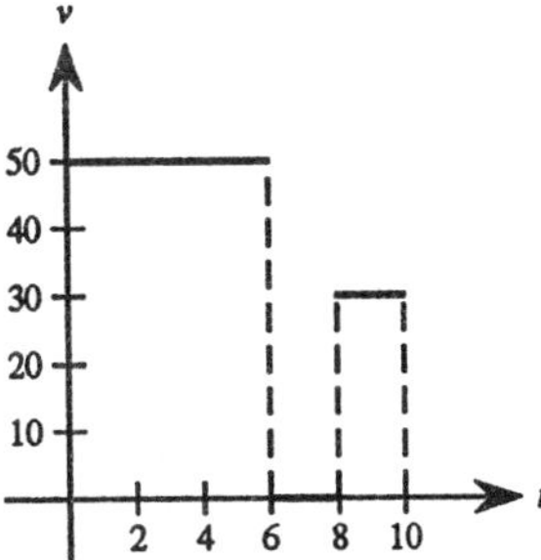

17. $v = 40$ mph $= \frac{2}{3}$ mi/min

$\left(\frac{2}{3} \text{ mi/min}\right)(6 \text{ min}) = 4$ mi

$v = 60$ mph $= 1$ mi/min

(1 mi/min)(2 min) = 2 mi

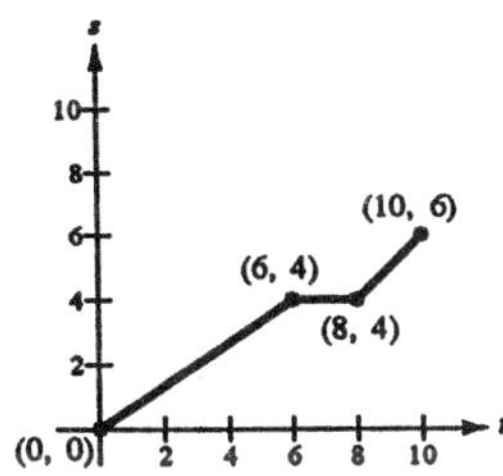

18. This graph corresponds with Exercise 15.

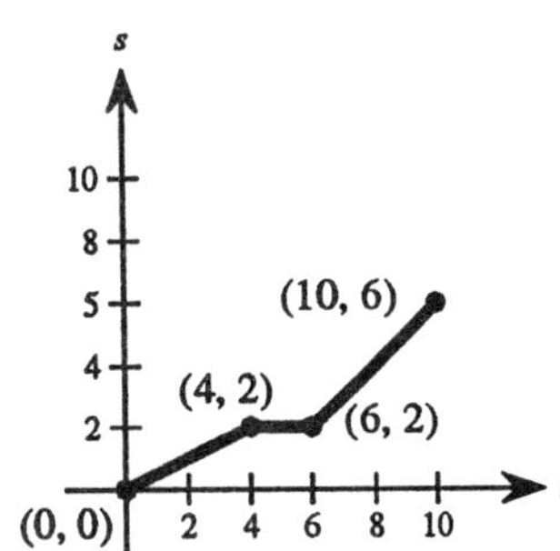

19. $f'(x) = x^2$

$$f''(x) = \lim_{\Delta x \to 0} \frac{f'(x+\Delta x) - f'(x)}{\Delta x} = \lim_{\Delta x \to 0} \left[\frac{(x+\Delta x)^2 - x^2}{\Delta x}\right] = \lim_{\Delta x \to 0} \frac{2x\Delta x + (\Delta x)^2}{\Delta x} = \lim_{\Delta x \to 0} (2x + \Delta x) = 2x$$

20. $f''(x) = x^3$

$$f'''(x) = \lim_{\Delta x \to 0} \frac{f''(x+\Delta x) - f''(x)}{\Delta x}$$

$$= \lim_{\Delta x \to 0} \frac{(x+\Delta x)^3 - x^3}{\Delta x} = \lim_{\Delta x \to 0} \frac{3x^2\Delta x + 3(\Delta x)^2 + (\Delta x)^3}{\Delta x} = \lim_{\Delta x \to 0} (3x^2 + 3\Delta x + (\Delta x)^2) = 3x^2$$

21. $f''(x) = 2 - \dfrac{2}{x} = \dfrac{2x-2}{x}$

$$f'''(x) = \lim_{\Delta x \to 0} \frac{f''(x+\Delta x) - f''(x)}{\Delta x} = \lim_{\Delta x \to 0} \frac{1}{\Delta x}\left[\frac{[2(x+\Delta x) - 2]}{x+\Delta x} - \frac{2x-2}{x}\right] = \lim_{\Delta x \to 0} \frac{2}{x(x+\Delta x)} = \frac{2}{x^2}$$

22. $f'''(x) = 2\sqrt{x-1}$

$$f^{(4)}(x) = \lim_{\Delta x \to 0} \frac{f'''(x+\Delta x) - f'''(x)}{\Delta x} = \lim_{\Delta x \to 0} \frac{2\sqrt{(x+\Delta x)-1} - 2\sqrt{x-1}}{\Delta x}$$

$$= \lim_{\Delta x \to 0} \frac{2}{\sqrt{x+\Delta x-1} + \sqrt{x-1}} = \frac{2}{\sqrt{x-1} + \sqrt{x-1}} = \frac{1}{\sqrt{x-1}}$$

23. $f^{(4)}(x) = 2x + 1$

Differentiate $f^{(4)}(x)$ twice.

$$f^{(5)}(x) = \lim_{\Delta x \to 0} \frac{f^{(4)}(x+\Delta x) - f^{(4)}(x)}{\Delta x} = \lim_{\Delta x \to 0} \frac{2(x+\Delta x) + 1 - 2x - 1}{\Delta x} = \lim_{\Delta x \to 0} 2 = 2$$

$$f^{(6)}(x) = \lim_{\Delta x \to 0} \frac{f^{(5)}(x+\Delta x) - f^{(5)}(x)}{\Delta x} = \lim_{\Delta x \to 0} \frac{2-2}{\Delta x} = \lim_{\Delta x \to 0} 0 = 0$$

24. $f(x) = 2x^2 - 2$

$$f'(x) = \lim_{\Delta x \to 0} \frac{f(x+\Delta x) - f(x)}{\Delta x}$$

$$= \lim_{\Delta x \to 0} \frac{[2(x+\Delta x)^2 - 2] - (2x^2 - 2)}{\Delta x} = \lim_{\Delta x \to 0} \frac{4x\Delta x + 2(\Delta x)^2}{\Delta x} = \lim_{\Delta x \to 0} (4x + 2\Delta x) = 4x$$

$$f''(x) = \lim_{\Delta x \to 0} \frac{f'(x+\Delta x) - f'(x)}{\Delta x} = \lim_{\Delta x \to 0} \frac{4(x+\Delta x) - 4x}{\Delta x} = \lim_{\Delta x \to 0} \frac{4\Delta x}{\Delta x} = \lim_{\Delta x \to 0} 4 = 4$$

25.

$$C = \frac{1{,}008{,}000}{Q} + 6.3Q$$

$$\frac{dC}{dQ} = -\frac{1{,}008{,}000}{Q^2} + 6.3$$

$C(351) - C(350) = 5083.095 - 5085 \approx -\$1.91$

When $Q = 350$, $dC/dQ = -\$1.93$.

26. $C =$ (gallons of fuel used)(cost per gallon) $= (15{,}000/x)(1.10)$. Thus, $C = 16{,}500/x$. By the limit definition of the derivative, we have $dC/dx = -16{,}500/x^2$.

x	10	15	20	25	30	35	40
C	\$1650	\$1100	\$825	\$660	\$550	\$471.43	\$412.50
$\dfrac{dC}{dx}$	−165	−73.33	−41.25	−26.40	−18.33	−13.47	−10.31

27. $F = 200\sqrt{T}$

By the limit definition of the derivative, we have $dF/dT = 100/\sqrt{T}$.

(a) When $T = 4$: $dF/dT = 50$

(b) When $T = 9$: $dF/dT \approx 33.33$

28. 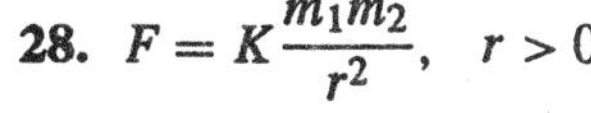

$F = K\dfrac{m_1 m_2}{r^2}, \quad r > 0$

By the limit definition of the derivative, we have

$$\frac{dF}{dr} = -2K\frac{m_1 m_2}{r^3}.$$

dF/dr is negative since the force decreases as the distance between the particles increases.

29. $\dfrac{dT}{dt} = K(T - T_a)$

30. $s = -16t^2 + 27t + 100$

When $s = 0$, we have $0 = -16t^2 + 27t + 100$. By the Quadratic Formula, we have $t \approx 3.48$ seconds until the object hits the ground.

$$v = \frac{ds}{dt}$$

By the limit definition of the derivative, we have $ds/dt = -32t + 27$. When the object hits the ground, the velocity is $v(3.48) \approx -84.4$ feet per second.

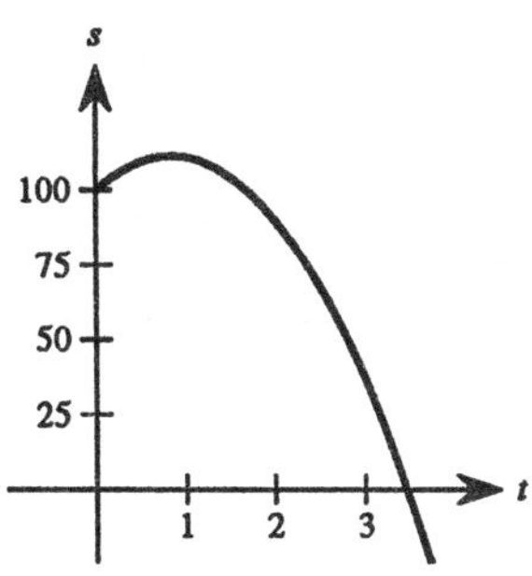

31.

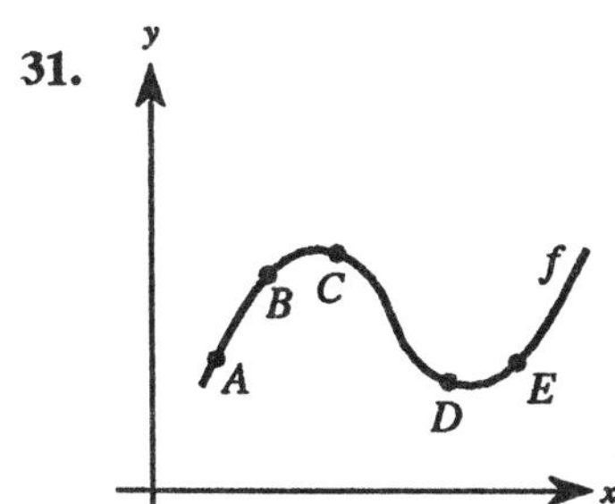

(a) The slope appears to be steepest between A and B.

(b) The average rate of change between A and B is greater than the instantaneous rate of change at B.

(c)

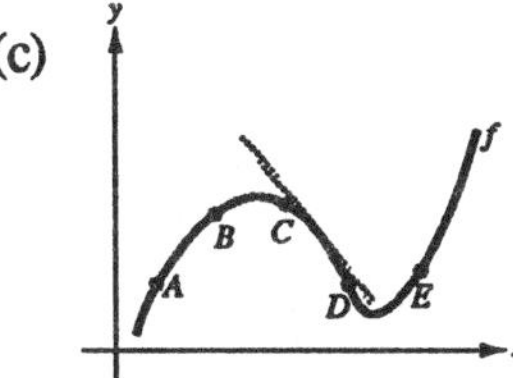

(d) The average rates of change are approximately equal between B and C and D and E.

Section 3.3 Differentiation Rules for Powers, Constant Multiples, and Sums

1. (a) $y = x^{1/2}$, $y' = \frac{1}{2}x^{-1/2}$, $y'(1) = \frac{1}{2}$

(b) $y = x^{3/2}$, $y' = \frac{3}{2}x^{1/2}$, $y'(1) = \frac{3}{2}$

(c) $y = x^2$, $y' = 2x$, $y'(1) = 2$

(d) $y = x^3$, $y' = 3x^2$, $y'(1) = 3$

2. (a) $y = x^{-1/2}$, $y' = -\frac{1}{2}x^{-3/2}$, $y(1) = -\frac{1}{2}$

(b) $y = x^{-1}$, $y' = -x^{-2}$, $y'(1) = -1$

(c) $y = x^{-3/2}$, $y' = -\frac{3}{2}x^{-5/2}$, $y'(1) = -\frac{3}{2}$

(d) $y = x^{-2}$, $y' = -2x^{-3}$, $y'(1) = -2$

3. $y = 3$, $y' = 0$

4. $f(x) = -2$, $f'(x) = 0$

5. $f(x) = x + 1$, $f'(x) = 1$

6. $g(x) = 3x - 1$, $g'(x) = 3$

7. $g(x) = x^2 + 4$, $g'(x) = 2x$

8. $y = t^2 + 2t - 3$, $y' = 2t + 2$

9. $f(t) = -2t^2 + 3t - 6$, $f'(t) = -4t + 3$

10. $y = x^3 - 9$, $y' = 3x^2$

11. $s(t) = t^3 - 2t + 4$

$s'(t) = 3t^2 - 2$

12. $f(x) = 2x^3 - x^2 + 3x$

$f'(x) = 6x^2 - 2x + 3$

13. $f(x) = \frac{1}{x}, \quad (1, 1)$

$f'(x) = -\frac{1}{x^2}$

$f'(1) = -1$

14. $f(t) = 3 - \frac{3}{5t}, \quad \left(\frac{3}{5}, 1\right)$

$f'(t) = \frac{3}{5t^2}$

$f'\left(\frac{3}{5}\right) = \frac{5}{3}$

15. $f(x) = -\frac{1}{2} + \frac{7}{5}x^3, \quad \left(0, -\frac{1}{2}\right)$

$f'(x) = \frac{21}{5}x^2$

$f'(0) = 0$

16. $y = 3x\left(x^2 - \frac{2}{x}\right), \quad (2, 18)$

$= 3x^3 - 6$

$y' = 9x^2$

$y'(2) = 36$

17. $y = (2x + 1)^2, \quad (0, 1)$

$= 4x^2 + 4x + 1$

$y' = 8x + 4$

$y'(0) = 4$

18. $f(x) = 3(5 - x)^2, \quad (5, 0)$

$= 3x^2 - 30x + 75$

$f'(x) = 6x - 30$

$f'(5) = 0$

19. $f(x) = x^2 - \frac{4}{x}$

$f'(x) = 2x + \frac{4}{x^2}$

20. $f(x) = x^2 - 3x - 3x^{-2}$

$f'(x) = 2x - 3 + 6x^{-3} = 2x - 3 + \frac{6}{x^3}$

21. $f(x) = x^3 - 3x - \frac{2}{x^4}$

$f'(x) = 3x^2 - 3 + \frac{8}{x^5}$

22. $f(x) = \frac{2x^2 - 3x + 1}{x} = 2x - 3 + \frac{1}{x}$

$f'(x) = 2 - \frac{1}{x^2} = \frac{2x^2 - 1}{x^2}$

23. $f(x) = \frac{x^3 - 3x^2 + 4}{x^2} = x - 3 + \frac{4}{x^2}$

$f'(x) = 1 - \frac{8}{x^3} = \frac{x^3 - 8}{x^3}$

24. $f(x) = (x^2 + 2x)(x + 1) = x^3 + 3x^2 + 2x$

$f'(x) = 3x^2 + 6x + 2$

25. $f(x) = x(x^2 + 1) = x^3 + x$

$f'(x) = 3x^2 + 1$

26. $f(x) = x + \frac{1}{x^2}$

$f'(x) = 1 - \frac{2}{x^3}$

27. $f(x) = x^{4/5}$

$f'(x) = \frac{4}{5}x^{-1/5} = \frac{4}{5x^{1/5}}$

28. $f(x) = x^{1/3} - 1$

$f'(x) = \frac{1}{3}x^{-2/3} = \frac{1}{3x^{2/3}}$

29. $f(x) = \sqrt[3]{x} + \sqrt[5]{x} = x^{1/3} + x^{1/5}$

$f'(x) = \frac{1}{3}x^{-2/3} + \frac{1}{5}x^{-4/5} = \frac{1}{3x^{2/3}} + \frac{1}{5x^{4/5}}$

30. $f(x) = \frac{1}{\sqrt[3]{x^2}} = x^{-2/3}$

$f'(x) = -\frac{2}{3}x^{-5/3} = -\frac{2}{3x^{5/3}}$

	Function	Rewrite	Derivative	Simplify
31.	$y = \frac{1}{3x^3}$	$y = \frac{1}{3}x^{-3}$	$y' = -x^{-4}$	$y' = -\frac{1}{x^4}$
32.	$y = \frac{2}{3x^2}$	$y = \frac{2}{3}x^{-2}$	$y' = -\frac{4}{3}x^{-3}$	$y' = -\frac{4}{3x^3}$
33.	$y = \frac{1}{(3x)^3}$	$y = \frac{1}{27}x^{-3}$	$y' = -\frac{1}{9}x^{-4}$	$y' = -\frac{1}{9x^4}$
34.	$y = \frac{\pi}{(3x)^2}$	$y = \frac{\pi}{9}x^{-2}$	$y' = -\frac{2\pi}{9}x^{-3}$	$y' = -\frac{2\pi}{9x^3}$
35.	$y = \frac{\sqrt{x}}{x}$	$y = x^{-1/2}$	$y' = -\frac{1}{2}x^{-3/2}$	$y' = -\frac{1}{2x^{3/2}}$
36.	$y = \frac{4}{x^{-3}}$	$y = 4x^3$	$y' = 12x^2$	$y' = 12x^2$

37. $y = x^4 - 3x^2 + 2$

$y' = 4x^3 - 6x$

At $(1, 0)$: $y' = 4 - 6 = -2$

Tangent line:

$$y - 0 = -2(x - 1)$$
$$2x + y - 2 = 0$$

38. $y = x^3 + x$

$y' = 3x^2 + 1$

At $(-1, -2)$: $y' = 4$

Tangent line:

$$y + 2 = 4(x + 1)$$
$$4x - y + 2 = 0$$

39. $y = x^4 - 3x^2 + 2$

$y' = 4x^3 - 6x = 2x(2x^2 - 3) = 0 \Rightarrow x = 0$ or $x = \pm\sqrt{3/2}$

At $x = 0$, $y = 2$, and at $x = \pm\sqrt{3/2}$, $y = -1/4$.

Horizontal tangents: $(0, 2)$, $\left(\pm\sqrt{3/2}, -1/4\right)$

40. $y = x^3 + x$

$y' = 3x^2 + 1 > 0$ for all x.

Therefore, there are no horizontal tangents.

41. $y = \frac{1}{x^2}$

$y' = \frac{-2}{x^3}$ cannot equal zero.

Therefore, there are no horizontal tangents.

42. $y = x^2 + 1$

$y' = 2x = 0 \Rightarrow x = 0$

At $x = 0$, $y = 1$.

Horizontal tangent: $(0, 1)$

43. Let (x_1, y_1) and (x_2, y_2) be the points of tangency. The derivatives of these functions are

$$y' = 2x \Rightarrow m = 2x_1 \quad \text{and} \quad y' = -2x + 6 \Rightarrow m = -2x_2 + 6.$$

$$m = 2x_1 = -2x_2 + 6$$
$$x_1 = -x_2 + 3$$

43. —CONTINUED—

Since $y_1 = x_2$ and $y_2 = -x_2^2 + 6x_2 - 5$,

$$m = \frac{y_2 - y_1}{x_2 - x_1} = \frac{(-x_2^2 + 6x_2 - 5) - (x_1^2)}{x_2 - x_1} = -2x_2 + 6.$$

$$\frac{(-x_2^2 + 6x_2 - 5) - (-x_2 + 3)^2}{x_2 - (-x_2 + 3)} = -2x_2 + 6$$

$$-2x_2^2 + 12x_2 - 14 = -4x_2^2 + 18x_2 - 18$$

$$2x_2^2 - 6x_2 + 4 = 0$$

$$(x_2 - 2)(x_2 - 1) = 0$$

$$x_2 = 1 \text{ or } 2$$

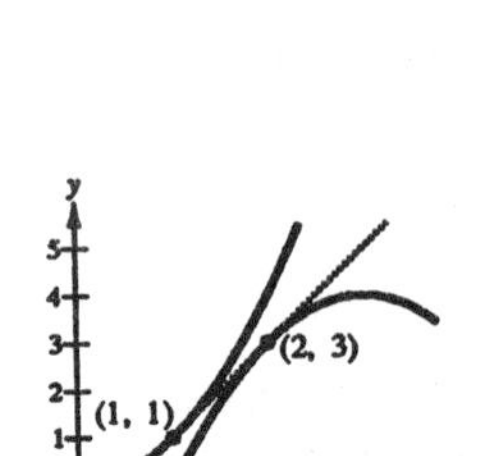

$x_2 = 1 \Rightarrow y_2 = 0, \quad x_1 = 2$ and $y_1 = 4$

Thus, the tangent line through (1, 0) and (2, 4) is

$$y - 0 = \left(\frac{4 - 0}{2 - 1}\right)(x - 1) \Rightarrow y = 4x - 4.$$

$x_2 = 2 \Rightarrow y_2 = 3, \quad x_1 = 1$ and $y_1 = 1$

Thus, the tangent line through (2, 3) and (1, 1) is

$$y - 1 = \left(\frac{3 - 2}{2 - 1}\right)(x - 1) \Rightarrow y = 2x - 1.$$

44. m_1 is the slope of the line tangent to $y = x$.

m_2 is the slope of the line tangent to $y = \dfrac{1}{x}$.

Since $y = x \Rightarrow y' = 1 \Rightarrow m_1 = 1$

and $y = \dfrac{1}{x} \Rightarrow y' = \dfrac{-1}{x^2} \Rightarrow m_2 = \dfrac{-1}{x^2}$.

The points of intersection of $y = x$ and $y = \dfrac{1}{x}$ are

$$x = \frac{1}{x} \Rightarrow x^2 = 1 \Rightarrow x = \pm 1.$$

At $x = \pm 1, \quad m_2 = -1$. Since $m_2 = -1/m_1$, these tangent lines are perpendicular at the points of intersection.

45. $A = s^2, \quad \dfrac{dA}{ds} = 2s$

When $s = 4, \ dA/ds = 8$.

46. $V = s^2, \quad \dfrac{dV}{ds} = 3s^2$

When $s = 4, \ dV/ds = 48$.

47. $R = 12{,}000p - 1000p^2$

$$\frac{dR}{dp} = 12{,}000 - 2000p$$

(a) When $p = 1, \ dR/dp = 10{,}000$.
(b) When $p = 4, \ dR/dp = 4000$.
(c) When $p = 6, \ dR/dp = 0$.
(d) When $p = 10, \ dR/dp = -8000$.

48. $P = 50\sqrt{x} - 0.5x - 500$

$$\frac{dP}{dx} = \frac{25}{\sqrt{x}} - 0.5$$

(a) When $x = 900, \ dP/dx = 1/3$.
(b) When $x = 1600, \ dP/dx = 1/8$.
(c) When $x = 2500, \ dP/dx = 0$.
(d) When $x = 3600, \ dP/dx = -1/12$.

49. $E = \frac{1}{27}(9t + 3t^2 - t^3)$

$\frac{dE}{dt} = \frac{1}{9}(3 + 2t - t^2)$

(a) When $t = 1$, $dE/dt = 4/9$.
(b) When $t = 2$, $dE/dt = 1/3$.
(c) When $t = 3$, $dE/dt = 0$.
(d) When $t = 4$, $dE/dt = -5/9$.

50. $Q(t) = 16t - 4t^2$

$Q'(t) = \frac{dQ}{dt} = 16 - 8t$

(a) $Q'\left(\frac{1}{2}\right) = 16 - 8\left(\frac{1}{2}\right) = 12$
(b) $Q'(1) = 16 - 8(1) = 8$
(c) $Q'(2) = 16 - 8(2) = 0$

51. $s(t) = -16t^2 + 5t + 30$

$0 = -16t^2 + 5t + 30$

By the Quadratic Formula, $t \approx 1.5344$ seconds until the diver hits the water.

$v = s'(t) = -32t + 5$

$v(1.53) \approx -44.1$ ft/sec

52. $s(t) = -16t^2 + 5t + h$

$v(t) = s'(t) = -32t + 5$

For $h = 40$:

$0 = -16t^2 + 5t + 40 \Rightarrow t \approx 1.7451$ sec

$v(1.7451) \approx -50.84$ ft/sec

For $h = 50$:

$0 = -16t^2 + 5t + 50 \Rightarrow t \approx 1.9309$ sec

$v(1.9309) \approx -56.79$ ft/sec

For $h = 60$:

$0 = -16t^2 + 5t + 60 \Rightarrow t \approx 2.0990$ sec

$v(2.099) \approx -62.17$ ft/sec

53. $s = -\frac{27}{10}t^2 + 27t + 6$

$v = -\frac{27}{5}t + 27$

$a = -\frac{27}{5} = -5.4$ feet per second per second

On earth, $a = -32$ feet per second per second.

54. $s = -16t^2 + 48t$

(a) $v = -32t + 48$

$a = -32$

(b) $0 = -32t + 48$

$t = \frac{48}{32} = 1.5$ sec

(c) $s(1.5) = 36$ ft

55. If $f(x) = g(x) + c$, then

$f'(x) = g'(x) + 0$.

Therefore, $f'(x) = g'(x)$.

True

56. If $y = \pi^2$, then

$y' = 0$. (π^2 is a constant.)

False

57. $f(x) = x^{3/2}$

(a)

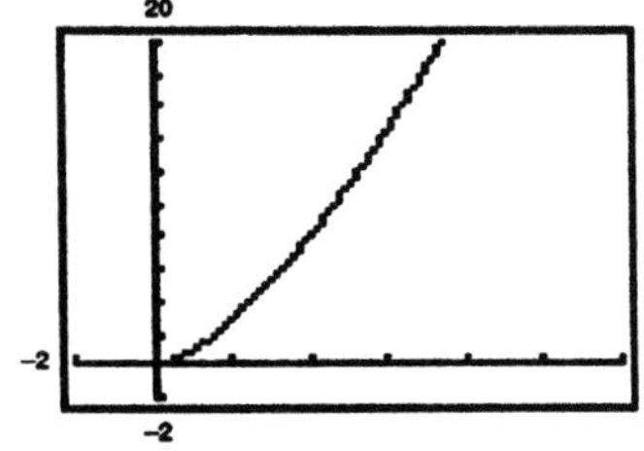

(b) $f'(x) = \frac{3}{2}x^{1/2} = \frac{3}{2}\sqrt{x}$

$T(x) = f'(4)(x - 4) + f(4)$

$= 3(x - 4) + 8$

$= 3x - 4$

(c) As you move away from (4, 8) the approximation becomes less and less accurate.

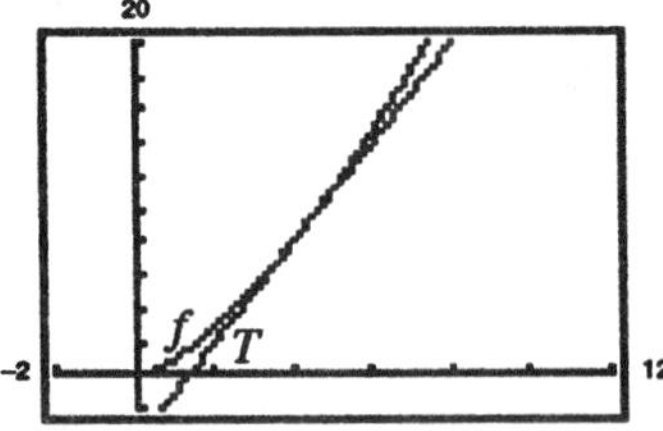

(d)

Δx	-2	-0.5	-0.1	0	0.1	0.5	2
$f(4 + \Delta x)$	2.828	6.548	7.702	8	8.302	9.546	14.697
$T(4 + \Delta x)$	2	6.5	7.7	8	8.3	9.5	14

58. $f(x) = x^3$, (1, 1)

(a)

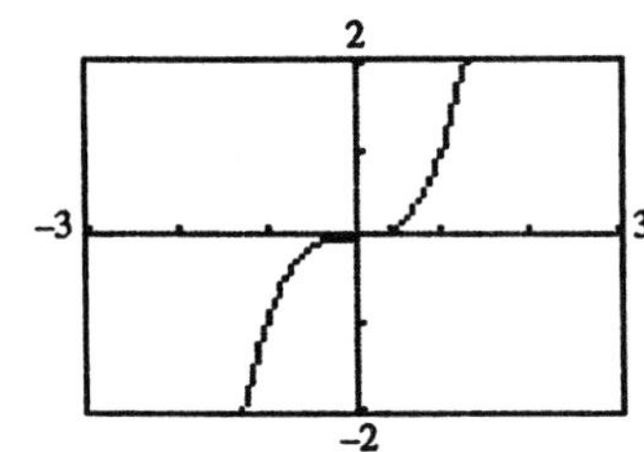

(b) $f'(x) = 3x^2$

$T(x) = f'(1)(x - 1) + f(1)$

$= 3(x - 1) + 1$

$= 3x - 2$

(c) As you move away from (1, 1), the accuracy of the approximation decreases. In Exercise 57, the graph of f was almost linear near (4, 8). Here the graph is not as "linear" near (1, 1). $T(x)$ is only a good approximation for $f(x)$ for values very close to $x = 1$.

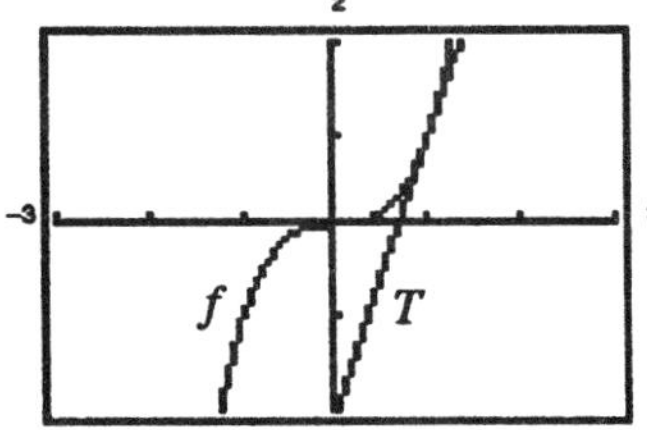

(d)

Δx	-2	-0.5	-0.1	0	0.1	0.5	2
$f(1 + \Delta x)$	-1	0.125	0.729	1	1.331	3.375	27
$T(1 + \Delta x)$	-5	-0.500	0.700	1	1.300	2.500	7

59. Let $f(x) = x$. Then,

$$f'(x) = \lim_{\Delta x \to 0} \frac{f(x + \Delta x) - f(x)}{\Delta x} = \lim_{\Delta x \to 0} \frac{(x + \Delta x) - x}{\Delta x} = \lim_{\Delta x \to 0} \frac{\Delta x}{\Delta x} = \lim_{\Delta x \to 0} 1 = 1.$$

Section 3.4 Differentiation Rules for Products and Quotients

1. $f(x) = \frac{1}{3}(2x^3 - 4)$

$f'(x) = \frac{1}{3}(6x^2) = 2x^2$

$f'(0) = 0$

2. $f(x) = \dfrac{5 - 6x^2}{7}$

$f'(x) = \dfrac{1}{7}(-12x) = \dfrac{-12x}{7}$

$f'(1) = -\dfrac{12}{7}$

3. $f(x) = 5x^{-2}(x + 3)$

$f'(x) = 5[x^{-2}(1) + (x + 3)(-2x^{-3})]$

$= 5\left[\dfrac{1}{x^2} - \dfrac{2(x + 3)}{x^3}\right]$

$= \dfrac{-5}{x^3}(x + 6)$

$f'(1) = -35$

4. $f(x) = (x^2 - 2x + 1)(x^3 - 1)$

$f'(x) = (x^2 - 2x + 1)(3x^2) + (x^3 - 1)(2x - 2)$

$= 3x^2(x - 1)^2 + 2(x - 1)^2(x^2 + x + 1)$

$= (x - 1)^2(5x^2 + 2x + 2)$

$f'(1) = 0$

5. $f(x) = (x^3 - 3x)(2x^2 + 3x + 5)$

$f'(x) = (x^3 - 3x)(4x + 3) + (2x^2 + 3x + 5)(3x^2 - 3)$

$= 10x^4 + 12x^3 - 3x^2 - 18x - 15$

$f'(0) = -15$

6. $f(x) = (x - 1)(x^2 - 3x + 2)$

$f'(x) = (x - 1)(2x - 3) + (x^2 - 3x + 2)(1)$

$= 3x^2 - 8x + 5$

$= (3x - 5)(x - 1)$

$f'(0) = 5$

7. $f(x) = (x^5 - 3x)\left(\dfrac{1}{x^2}\right)$

$f'(x) = (x^5 - 3x)\left(-\dfrac{2}{x^3}\right) + \left(\dfrac{1}{x^2}\right)(5x^4 - 3)$

$= 3x^2 + \dfrac{3}{x^2}$

$f'(-1) = 6$

8. $f(x) = \dfrac{x + 1}{x - 1}$

$f'(x) = \dfrac{(x - 1)(1) - (x + 1)(1)}{(x - 1)^2} = \dfrac{-2}{(x - 1)^2}$

$f'(2) = -2$

9. $f(x) = \dfrac{3x - 2}{2x - 3}$

$f'(x) = \dfrac{(2x - 3)(3) - (3x - 2)(2)}{(2x - 3)^2}$

$= \dfrac{-5}{(2x - 3)^2}$

10. $f(x) = \dfrac{x^3 + 3x + 2}{x^2 - 1}$

$f'(x) = \dfrac{(x^2 - 1)(3x^2 + 3) - (x^3 + 3x + 2)(2x)}{(x^2 - 1)^2}$

$= \dfrac{x^4 - 6x^2 - 4x - 3}{(x^2 - 1)^2}$

11. $f(x) = \dfrac{3 - 2x - x^2}{x^2 - 1}$

$f'(x) = \dfrac{(x^2 - 1)(-2 - 2x) - (3 - 2x - x^2)(2x)}{(x^2 - 1)^2}$

$= \dfrac{2x^2 - 4x + 2}{(x^2 - 1)^2} = \dfrac{2(x - 1)^2}{(x^2 - 1)^2}$

$= \dfrac{2}{(x + 1)^2}$

12. $f(x) = x^4\left[1 - \dfrac{2}{x + 1}\right] = x^4\left[\dfrac{x - 1}{x + 1}\right]$

$f'(x) = x^4\left[\dfrac{(x + 1) - (x - 1)}{(x + 1)^2}\right] + \left[\dfrac{x - 1}{x + 1}\right](4x^3)$

$= 2x^3\left[\dfrac{2x^2 + x - 2}{(x + 1)^2}\right]$

13. $f(x) = \dfrac{x+1}{\sqrt{x}}$

$f'(x) = \dfrac{\sqrt{x}(1) - (x+1)[1/(2\sqrt{x})]}{x} = \dfrac{x-1}{2x^{3/2}}$

14. $f(x) = \sqrt[3]{x}(\sqrt{x}+3) = x^{1/3}(x^{1/2}+3)$

$f'(x) = x^{1/3}\left(\dfrac{1}{2}x^{-1/2}\right) + (x^{1/2}+3)\left(\dfrac{1}{3}x^{-2/3}\right)$

$= \dfrac{5}{6}x^{-1/6} + x^{-2/3}$

$= \dfrac{5}{6x^{1/6}} + \dfrac{1}{x^{2/3}}$

15. $h(t) = \dfrac{t+1}{t^2+2t+2}$

$h'(t) = \dfrac{(t^2+2t+2)(1) - (t+1)(2t+2)}{(t^2+2t+2)^2}$

$= \dfrac{-t^2-2t}{(t^2+2t+2)^2}$

16. $h(x) = (x^2-1)^2 = x^4 - 2x^2 + 1$

$h'(x) = 4x^3 - 4x = 4x(x^2-1)$

17. $h(s) = (s^3-2)^2 = s^6 - 4s^3 + 4$

$h'(s) = 6s^5 - 12s^2 = 6s^2(s^3-2)$

18. $f(x) = \left(\dfrac{x^2-x-3}{x^2+1}\right)(x^2+x+1)$

$f'(x) = \left(\dfrac{x^2-x-3}{x^2+1}\right)(2x+1) + (x^2+x+1)\left[\dfrac{(x^2+1)(2x-1) - (x^2-x-3)(2x)}{(x^2+1)^2}\right]$

$= \dfrac{2x^3-x^2-7x-3}{x^2+1} + \dfrac{x^4+9x^3+8x^2+7x-1}{(x^2+1)^2}$

$= \dfrac{2x^5+4x^3+4x^2-4}{(x^2+1)^2}$

19. $g(x) = \left[\dfrac{x+1}{x+2}\right](2x-5)$

$g'(x) = \left[\dfrac{x+1}{x+2}\right](2) + (2x-5)\left[\dfrac{(x+2)(1) - (x+1)(1)}{(x+2)^2}\right] = \dfrac{2x^2+6x+4+2x-5}{(x+2)^2} = \dfrac{2x^2+8x-1}{(x+2)^2}$

20. $f(x) = (x^2-x)(x^2+1)(x^2+x+1)$

$f'(x) = (x^2-x)[(x^2+1)(2x+1) + (x^2+x+1)(2x)] + [(x^2+1)(x^2+x+1)](2x-1)$

$= (x^2-x)(4x^3+3x^2+4x+1) + (x^4+x^3+2x^2+x+1)(2x-1)$

$= 6x^5 + 4x^3 - 3x^2 - 1$

21. $f(x) = (3x^3+4x)(x-5)(x+1)$

$f'(x) = [(3x^3+4x)(x-5)](1) + (x+1)[(3x^3+4x)(1) + (x-5)(9x^2+4)]$

$= 3x^4 - 15x^3 + 4x^2 - 20x + 12x^4 - 33x^3 - 37x^2 - 12x - 20$

$= 15x^4 - 48x^3 - 33x^2 - 32x - 20$

22. $f(x) = \dfrac{x^2+c^2}{x^2-c^2}$

$f'(x) = \dfrac{(x^2-c^2)(2x) - (x^2+c^2)(2x)}{(x^2-c^2)^2} = \dfrac{-4xc^2}{(x^2-c^2)^2}$

23. $f(x) = \dfrac{c^2 - x^2}{c^2 + x^2}$

$f'(x) = \dfrac{(c^2 + x^2)(-2x) - (c^2 - x^2)(2x)}{(c^2 + x^2)^2}$

$= \dfrac{-4xc^2}{(c^2 + x^2)^2}$

24. $f(x) = \dfrac{x(x^2 - 1)}{x + 3} = \dfrac{x^3 - x}{x + 3}$

$f'(x) = \dfrac{(x + 3)(3x^2 - 1) - (x^3 - x)}{(x + 3)^2}$

$= \dfrac{2x^3 + 9x^2 - 3}{(x + 3)^2}$

	Function	*Rewrite*	*Derivative*	*Simplify*
25.	$y = \dfrac{x^2 + 2x}{x}$	$y = x + 2$	$y' = 1$	$y' = 1$
26.	$y = \dfrac{4x^{3/2}}{x}$	$y = 4\sqrt{x}$	$y' = 2x^{-1/2}$	$y' = \dfrac{2}{\sqrt{x}}$
27.	$y = \dfrac{7}{3x^3}$	$y = \dfrac{7}{3}x^{-3}$	$y' = -7x^{-4}$	$y' = -\dfrac{7}{x^4}$
28.	$y = \dfrac{4}{5x^2}$	$y = \dfrac{4}{5}x^{-2}$	$y' = -\dfrac{8}{5}x^{-3}$	$y' = -\dfrac{8}{5x^3}$
29.	$y = \dfrac{3x^2 - 5}{7}$	$y = \dfrac{1}{7}(3x^2 - 5)$	$y' = \dfrac{1}{7}(6x)$	$y' = \dfrac{6x}{7}$
30.	$y = \dfrac{x^2 - 4}{x + 2}$	$y = x - 2$	$y' = 1$	$y' = 1$

31. $f(x) = 4x^{3/2}$

$f'(x) = 6x^{1/2}$

$f''(x) = 3x^{-1/2} = \dfrac{3}{\sqrt{x}}$

32. $f(x) = \dfrac{x^2 + 2x - 1}{x} = x + 2 - \dfrac{1}{x}$

$f'(x) = 1 + \dfrac{1}{x^2}$

$f''(x) = -\dfrac{2}{x^3}$

33. $f(x) = \dfrac{x}{x - 1}$

$f'(x) = \dfrac{(x - 1)(1) - x(1)}{(x - 1)^2} = \dfrac{-1}{(x - 1)^2}$

$f''(x) = \dfrac{2}{(x - 1)^3}$

34. $f(x) = x + \dfrac{32}{x^2}$

$f'(x) = 1 - \dfrac{64}{x^3}$

$f''(x) = \dfrac{192}{x^4}$

35. $f(x) = \dfrac{x}{x - 1}$, $(2, 2)$

$f'(x) = \dfrac{(x - 1)(1) - x(1)}{(x - 1)^2} = \dfrac{-1}{(x - 1)^2}$

$f'(2) = \dfrac{-1}{(2 - 1)^2} = -1 =$ slope at $(2, 2)$

$y - 2 = -1(x - 2) \Rightarrow y = -x + 4$

36. $f(x) = (x - 1)(x^2 - 2)$, $(0, 2)$

$f'(x) = (x - 1)(2x) + (x^2 - 2)(1)$

$= 3x^2 - 2x - 2$

$f'(0) = -2 =$ slope at $(0, 2)$

$y - 2 = -2x \Rightarrow y + 2x = 2$

37. $f(x) = (x^3 - 3x + 1)(x + 2), \quad (1, -3)$

$f'(x) = (x^3 - 3x + 1)(1) + (x + 2)(3x^2 - 3)$

$= 4x^3 + 6x^2 - 6x - 5$

$f'(1) = -1 =$ slope at $(1, -3)$

$y + 3 = -1(x - 1) \Rightarrow y = -x - 2$

38. $f(x) = \dfrac{x - 1}{x + 1}, \quad \left(2, \dfrac{1}{3}\right)$

$f'(x) = \dfrac{(x + 1)(1) - (x - 1)(1)}{(x + 1)^2} = \dfrac{2}{(x + 1)^2}$

$f'(2) = \dfrac{2}{9} =$ slope at $\left(2, \dfrac{1}{3}\right)$

$y - \dfrac{1}{3} = \dfrac{2}{9}(x - 2) \Rightarrow 2x - 9y - 1 = 0$

39. $f(x) = \dfrac{x^2}{x - 1}$

$f'(x) = \dfrac{(x - 1)(2x) - x^2(1)}{(x - 1)^2}$

$= \dfrac{x^2 - 2x}{(x - 1)^2} = \dfrac{x(x - 2)}{(x - 1)^2}$

$f'(x) = 0$ when $x = 0$ or $x = 2$.

Horizontal tangents are at (0, 0) and (2, 4).

40. $f(x) = \dfrac{x^2}{x^2 + 1}$

$f'(x) = \dfrac{(x^2 + 1)(2x) - (x^2)(2x)}{(x^2 + 1)^2}$

$= \dfrac{2x}{(x^2 + 1)^2}$

$f'(x) = 0$ when $x = 0$.

Horizontal tangent is at (0, 0).

41. $f(t) = \dfrac{t^2 - t + 1}{t^2 + 1} \Rightarrow f'(t) = \dfrac{t^2 - 1}{(t^2 + 1)^2}$

(a) When $t = 0.5$, $f'(t) = -0.48$.
(b) When $t = 2$, $f'(t) = 0.12$.
(c) When $t = 8$, $f'(t) = 0.0149$.

42. $P = \dfrac{k}{V}$

$\dfrac{dP}{dV} = -\dfrac{k}{V^2}$

43. $P(t) = 500\left[1 + \dfrac{4t}{50 + t^2}\right]$

$P'(t) = 500\left[\dfrac{(50 + t^2)(4) - (4t)(2t)}{(50 + t^2)^2}\right]$

$= 500\left[\dfrac{200 - 4t^2}{(50 + t^2)^2}\right] = 2000\left[\dfrac{50 - t^2}{(50 + t^2)^2}\right]$

$P'(2) \approx 31.55$

44. $h(x) = f(x)g(x)$

$h'(x) = f(x)g'(x) + g(x)f'(x)$

$h'(c) = f(c)g'(c) + g(c)f'(c)$

$= [f(c)](0) + [g(c)](0) = 0$

45. $v(t) = \dfrac{760t}{4t^2 + 25}, \quad 0 \le t \le 4$

$a(t) = \dfrac{(4t^2 + 25)(760) - (760t)(8t)}{(4t^2 + 25)^2}$

$= \dfrac{760(4t^2 + 25 - 8t^2)}{(4t^2 + 25)^2}$

$= \dfrac{760(25 - 4t^2)}{(4t^2 + 25)^2}$

$a(2) = \dfrac{6840}{1681} \approx 4.069$

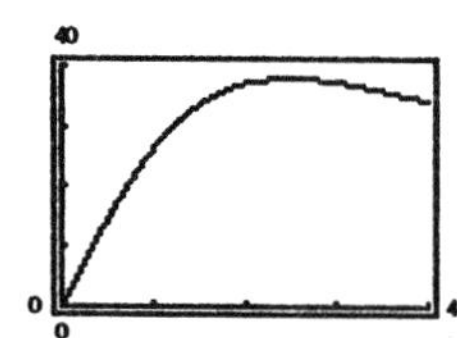

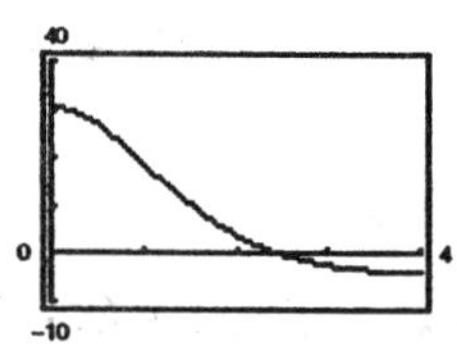

Section 3.5 The Chain Rule

	$y = f(g(x))$	$u = g(x)$	$y = f(u)$
1.	$y = (6x - 5)^4$	$u = 6x - 5$	$y = u^4$
2.	$y = \dfrac{1}{\sqrt{x+1}}$	$u = x + 1$	$y = u^{-1/2}$
3.	$y = \sqrt{x^2 - 1}$	$u = x^2 - 1$	$y = \sqrt{u}$
4.	$y = \left(\dfrac{3x}{2}\right)^2$	$u = \dfrac{3x}{2}$	$y = u^2$
5.	$y = (x^2 - 3x + 4)^6$	$u = x^2 - 3x + 4$	$y = u^6$
6.	$y = (5x - 2)^{3/2}$	$u = 5x - 2$	$y = u^{3/2}$

7. $y = (2x - 7)^3$

$y' = 3(2x - 7)^2(2) = 6(2x - 7)^2$

8. $y = (3x^2 + 1)^4$

$y' = 4(3x^2 + 1)^3(6x) = 24x(3x^2 + 1)^3$

9. $g(x) = 3(9x - 4)^4$

$g'(x) = 12(9x - 4)^3(9) = 108(9x - 4)^3$

10. $f(x) = 2(x^2 - 1)^3$

$f'(x) = 6(x^2 - 1)^2(2x) = 12x(x^2 - 1)^2$

11. $y = (x - 2)^{-1}$

$y' = -1(x - 2)^{-2}(1) = \dfrac{-1}{(x - 2)^2}$

12. $s(t) = (t^2 + 3t - 1)^{-1}$

$s'(t) = -1(t^2 + 3t - 1)^{-2}(2t + 3)$

$= \dfrac{-(2t + 3)}{(t^2 + 3t - 1)^2}$

13. $f(t) = (t - 3)^{-2}$

$f'(t) = -2(t - 3)^{-3} = \dfrac{-2}{(t - 3)^3}$

14. $y = -4(t + 2)^{-2}$

$y' = 8(t + 2)^{-3} = \dfrac{8}{(t + 2)^3}$

15. $f(x) = 3(x^3 - 4)^{-1}$

$f'(x) = -3(x^3 - 4)^{-2}(3x^2) = \dfrac{-9x^2}{(x^3 - 4)^2}$

16. $f(x) = (x^2 - 3x)^{-2}$

$f'(x) = -2(x^2 - 3x)^{-3}(2x - 3) = \dfrac{6 - 4x}{(x^2 - 3x)^3}$

17. $f(x) = x^2(x - 2)^4$

$f'(x) = x^2[4(x - 2)^3(1)] + (x - 2)^4(2x)$

$= 2x(x - 2)^3[2x + (x - 2)]$

$= 2x(x - 2)^3(3x - 2)$

18. $f(x) = x(3x - 9)^3$

$f'(x) = x[3(3x - 9)^2(3)] + (3x - 9)^3(1)$

$= (3x - 9)^2[9x + 3x - 9]$

$= 27(x - 3)^2(4x - 3)$

19. $f(t) = (1 - t)^{1/2}$

$f'(t) = \dfrac{1}{2}(1 - t)^{-1/2}(-1) = -\dfrac{1}{2\sqrt{1 - t}}$

20. $g(x) = (3 - 2x)^{1/2}$

$g'(x) = \dfrac{1}{2}(3 - 2x)^{-1/2}(-2) = -\dfrac{1}{\sqrt{3 - 2x}}$

21. $s(t) = (t^2 + 2t - 1)^{1/2}$

$$s'(t) = \frac{1}{2}(t^2 + 2t - 1)^{-1/2}(2t + 2)$$

$$= \frac{t+1}{\sqrt{t^2 + 2t - 1}}$$

22. $y = (3x^3 + 4x)^{1/3}$

$$y' = \frac{1}{3}(3x^3 + 4x)^{-2/3}(9x^2 + 4)$$

$$= \frac{9x^2 + 4}{3(3x^3 + 4x)^{2/3}}$$

23. $g(x) = (9x^2 + 4)^{1/3}$

$$g'(x) = \frac{1}{3}(9x^2 + 4)^{-2/3}(18x)$$

$$= \frac{6x}{(9x^2 + 4)^{2/3}}$$

24. $g(x) = \sqrt{x^2 - 2x + 1} = \sqrt{(x-1)^2} = |x - 1|$

$$g'(x) = \begin{cases} 1, & x > 1 \\ -1, & x < 1 \end{cases}$$

25. $y = 2(4 - x^2)^{1/2}$

$$y' = (4 - x^2)^{-1/2}(-2x) = -\frac{2x}{\sqrt{4 - x^2}}$$

26. $f(x) = -3(2 - 9x)^{1/4}$

$$f'(x) = -\frac{3}{4}(2 - 9x)^{-3/4}(-9) = \frac{27}{4(2 - 9x)^{3/4}}$$

27. $f(x) = (9 - x^2)^{2/3}$

$$f'(x) = \frac{2}{3}(9 - x^2)^{-1/3}(-2x) = -\frac{4x}{3(9 - x^2)^{1/3}}$$

28. $f(t) = (9t + 2)^{2/3}$

$$f'(t) = \frac{2}{3}(9t + 2)^{-1/3}(9) = \frac{6}{\sqrt[3]{9t + 2}}$$

29. $y = (x + 2)^{-1/2}$

$$\frac{dy}{dx} = -\frac{1}{2}(x + 2)^{-3/2} = -\frac{1}{2(x + 2)^{3/2}}$$

30. $g(t) = (t^2 - 2)^{-1/2}$

$$g'(t) = -\frac{1}{2}(t^2 - 2)^{-3/2}(2t) = -\frac{t}{(t^2 - 2)^{3/2}}$$

31. $y = x\sqrt{1 - x^2} = x(1 - x^2)^{1/2}$

$$y' = x\left[\frac{1}{2}(1 - x^2)^{-1/2}(-2x)\right] + (1 - x^2)^{1/2}(1)$$

$$= -x^2(1 - x^2)^{-1/2} + (1 - x^2)^{1/2}$$

$$= (1 - x^2)^{-1/2}[-x^2 + (1 - x^2)]$$

$$= \frac{1 - 2x^2}{\sqrt{1 - x^2}}$$

32. $y = x^2\sqrt{9 - x^2} = x^2(9 - x^2)^{1/2}$

$$y' = x^2\left[\frac{1}{2}(9 - x^2)^{-1/2}(-2x)\right] + (9 - x^2)^{1/2}(2x)$$

$$= -x^3(9 - x^2)^{-1/2} + 2x(9 - x^2)^{1/2}$$

$$= x(9 - x^2)^{-1/2}[-x^2 + 2(9 - x^2)]$$

$$= \frac{x(18 - 3x^2)}{\sqrt{9 - x^2}}$$

$$= \frac{3x(6 - x^2)}{\sqrt{9 - x^2}}$$

33. $y = \dfrac{x}{\sqrt{x^2 + 1}} = x(x^2 + 1)^{-1/2}$

$$y' = x\left[-\frac{1}{2}(x^2 + 1)^{-3/2}(2x)\right] + (x^2 + 1)^{-1/2}(1)$$

$$= -x^2(x^2 + 1)^{-3/2} + (x^2 + 1)^{-1/2}$$

$$= (x^2 + 1)^{-3/2}[-x^2 + (x^2 + 1)]$$

$$= \frac{1}{(x^2 + 1)^{3/2}}$$

34. $y = \dfrac{x^2}{\sqrt{x^2 + 9}} = x^2(x^2 + 9)^{-1/2}$

$$y' = x^2\left[-\frac{1}{2}(x^2 + 9)^{-3/2}(2x)\right] + (x^2 + 9)^{-1/2}(2x)$$

$$= -x^3(x^2 + 9)^{-3/2} + 2x(x^2 + 9)^{-1/2}$$

$$= x(x^2 + 9)^{-3/2}[-x^2 + 2(x^2 + 9)]$$

$$= \frac{x(x^2 + 18)}{(x^2 + 9)^{3/2}}$$

35. $f(t) = \dfrac{3t+2}{t-1}$

$$f'(t) = \frac{(t-1)(3)-(3t+2)}{(t-1)^2} = -\frac{5}{(t-1)^2}$$

36. $y = \left(\dfrac{2x}{x+1}\right)^{1/2}$

$$\frac{dy}{dx} = \frac{1}{2}\left(\frac{2x}{x+1}\right)^{-1/2}\left[\frac{2(x+1)-2x}{(x+1)^2}\right] = \frac{1}{\sqrt{2x}(x+1)^{3/2}}$$

37. $y = \dfrac{\sqrt{x}+1}{x^2+1}$

$$y' = \frac{1-3x^2-4x^{3/2}}{2\sqrt{x}(x^2+1)^2}$$

The zero of y' corresponds to the point on the graph of y where the tangent line is horizontal.

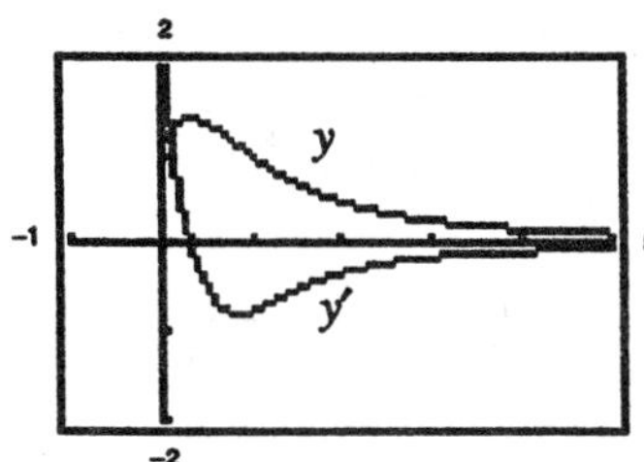

38. $y = \sqrt{\dfrac{2x}{x+1}}$

$$y' = \frac{1}{\sqrt{2x}(x+1)^{3/2}}$$

y' has no zeros.

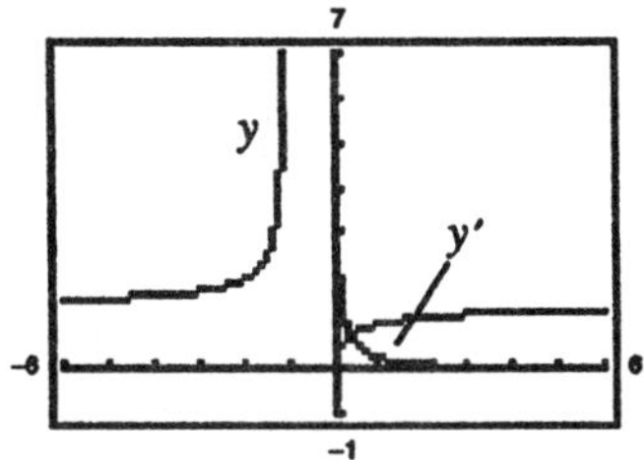

39. $g(t) = \dfrac{3t^2}{\sqrt{t^2+2t-1}}$

$$g'(t) = \frac{3t(t^2+3t-2)}{(t^2+2t-1)^{3/2}}$$

The zeros of g' correspond to the points on the graph of g where the tangent lines are horizontal.

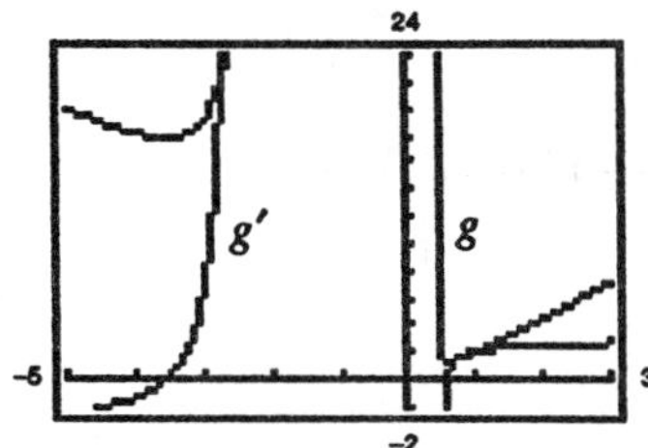

40. $f(x) = \sqrt{x}(2-x)^2$

$$f'(x) = \frac{(x-2)(5x-2)}{2\sqrt{x}}$$

The zeros of f' correspond to the points on the graph of f where the tangent lines are horizontal.

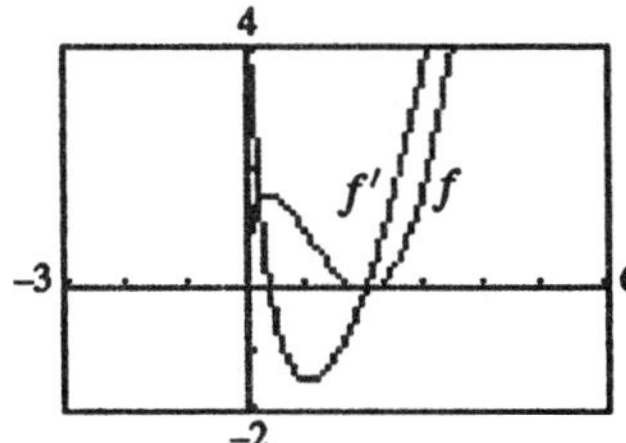

41. $y = \sqrt{\dfrac{x+1}{x}}$

$$y' = -\frac{\sqrt{(x+1)/x}}{2x(x+1)}$$

y' has no zeros.

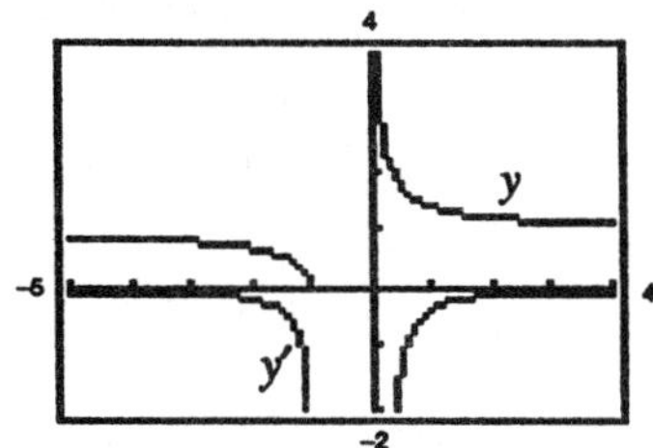

42. $y = (t^2-9)\sqrt{t+2}$

$$y' = \frac{5t^2+8t-9}{2\sqrt{t+2}}$$

The zero of y' corresponds to the point on the graph of y where the tangent line is horizontal.

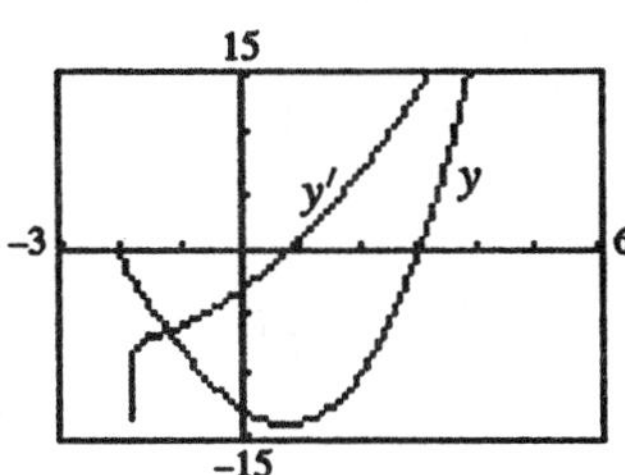

43. $s(t) = \dfrac{-2(2-t)\sqrt{1+t}}{3}$

$s'(t) = \dfrac{t}{\sqrt{1+t}}$

The zero of $s'(t)$ corresponds to the point on the graph of $s(t)$ where the tangent line is horizontal.

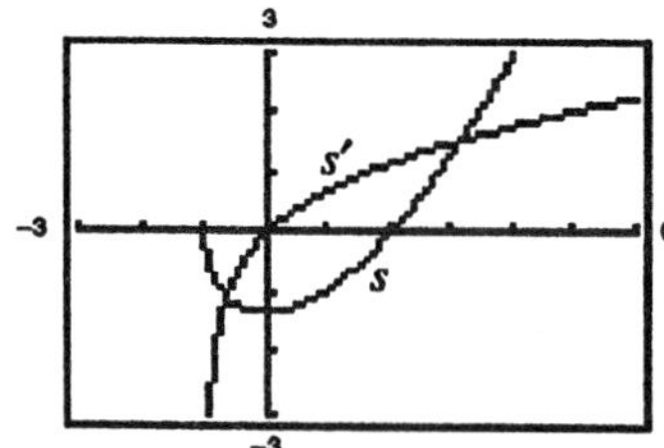

44. $g(x) = \sqrt{x-1} + \sqrt{x+1}$

$g'(x) = \dfrac{1}{2\sqrt{x-1}} + \dfrac{1}{2\sqrt{x+1}}$

g' has no zeros.

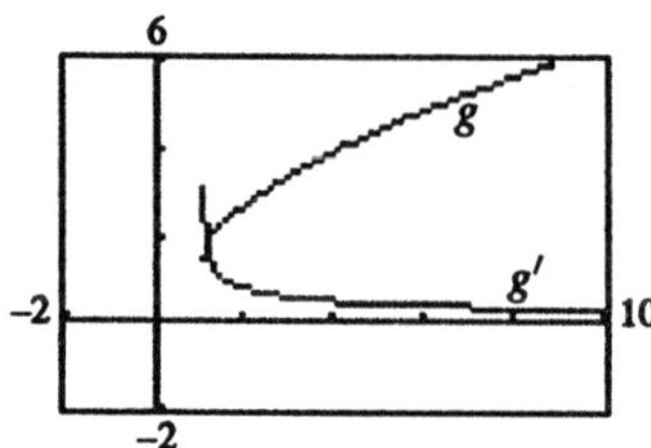

45. $f(x) = \sqrt{3x^2-2}$, $(3, 5)$

$f'(x) = \dfrac{1}{2}(3x^2-2)^{-1/2}(6x)$

$= \dfrac{3x}{\sqrt{3x^2-2}}$

$f'(3) = \dfrac{9}{5}$

Tangent line:

$y - 5 = \dfrac{9}{5}(x-3) \Rightarrow 9x - 5y - 2 = 0$

46. $f(x) = x\sqrt{x^2+5}$, $(2, 6)$

$f'(x) = x\left[\dfrac{1}{2}(x^2+5)^{-1/2}(2x)\right] + (x^2+5)^{1/2}$

$= \dfrac{x^2}{\sqrt{x^2+5}} + \sqrt{x^2+5} = \dfrac{2x^2+5}{\sqrt{x^2+5}}$

$f'(2) = \dfrac{13}{3}$

Tangent line:

$y - 6 = \dfrac{13}{3}(x-2) \Rightarrow 13x - 3y - 8 = 0$

47. $f(x) = 2(x^2-1)^3$

$f'(x) = 6(x^2-1)^2(2x)$

$= 12x(x^4 - 2x^2 + 1)$

$= 12x^5 - 24x^3 + 12x$

$f''(x) = 60x^4 - 72x^2 + 12 = 12(5x^2-1)(x^2-1)$

48. $f(x) = (x-2)^{-1}$

$f'(x) = -(x-2)^{-2} = \dfrac{-1}{(x-2)^2}$

$f''(x) = 2(x-2)^{-3} = \dfrac{2}{(x-2)^3}$

49. $f(x) = \sqrt{x^2+x+1} = (x^2+x+1)^{1/2}$

$f'(x) = \dfrac{1}{2}(x^2+x+1)^{-1/2}(2x+1)$

$f''(x) = -\dfrac{1}{4}(x^2+x+1)^{-3/2}(2x+1)^2 + \dfrac{1}{2}(x^2+x+1)^{-1/2}(2)$

$= \dfrac{1}{4}(x^2+x+1)^{-3/2}[-(4x^2+4x+1) + 4(x^2+x+1)]$

$= \dfrac{3}{4(x^2+x+1)^{3/2}}$

50. $f(t) = \dfrac{\sqrt{t^2+1}}{t}$

$f'(t) = \dfrac{[t(1/2)(t^2+1)^{-1/2}2t] - \sqrt{t^2+1}}{t^2} = \dfrac{-1}{t^2\sqrt{t^2+1}} = -(t^6+t^4)^{-1/2}$

$f''(t) = \dfrac{1}{2}(t^6+t^4)^{-3/2}(6t^5+4t^3) = \dfrac{3t^2+2}{t^3(t^2+1)^{3/2}}$

51. $|u| = \sqrt{u^2}$

$$\frac{d}{dx}[|u|] = \frac{d}{dx}\left[\sqrt{u^2}\right]$$
$$= \frac{1}{2}(u^2)^{-1/2}(2uu')$$
$$= \frac{uu'}{\sqrt{u^2}} = u'\frac{u}{|u|}, \quad u \neq 0$$

52. $f(x) = |x^2 - 4|$

$$f'(x) = 2x\left(\frac{x^2-4}{|x^2-4|}\right)$$

53. $f(x) = |x^3 + x|$

$$f'(x) = (3x^2+1)\left(\frac{x^3+x}{|x^3+x|}\right)$$

54. $f(x) = \left|\frac{4}{x}\right|$

$$f'(x) = \left(-\frac{4}{x^2}\right)\frac{4/x}{|4/x|} = \frac{-4|x|}{x^3}$$

55.

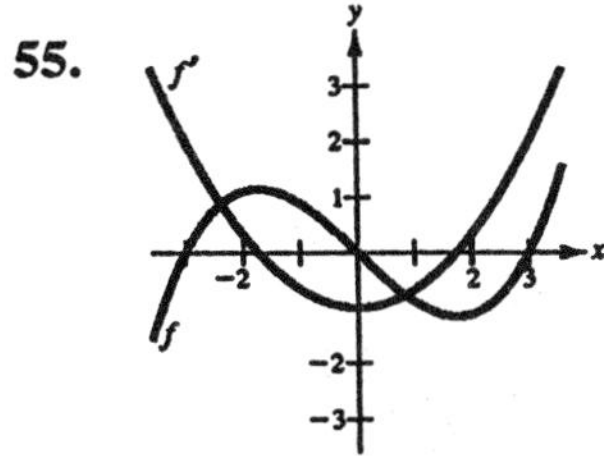

The zeros of f' correspond to the points where the graph of f has horizontal tangents.

56.

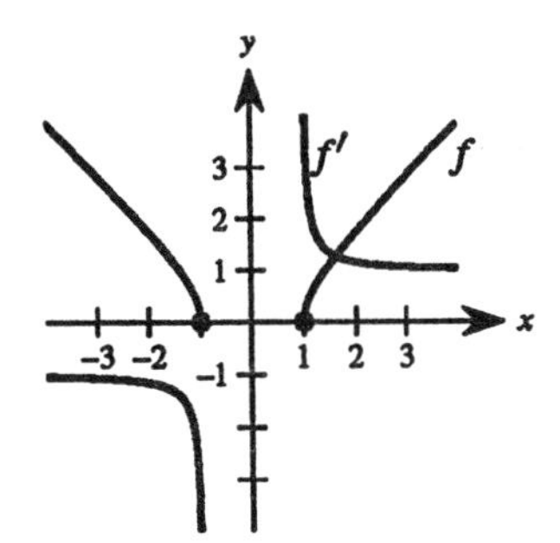

f is decreasing on $(-\infty, -1)$ so f' must be negative there. f is increasing on $(1, \infty)$ so f' must be positive there.

57. (a) $F = 132{,}400(331 - v)^{-1}$

$$F' = (-1)(132{,}400)(331-v)^{-2}(-1)$$
$$= \frac{132{,}400}{(331-v)^2}$$

When $v = 30$, $F' \approx 1.461$.

(b) $F = 132{,}400(331 + v)^{-1}$

$$F' = (-1)(132{,}400)(331+v)^{-2}(1)$$
$$= \frac{-132{,}400}{(331+v)^2}$$

When $v = 30$, $F' \approx -1.016$.

Section 3.6 Implicit Differentiation

1. $x^2 + y^2 = 16$

$$2x + 2yy' = 0$$
$$y' = \frac{-x}{y}$$

At $(3, \sqrt{7})$: $y' = \frac{-3}{\sqrt{7}} = \frac{-3\sqrt{7}}{7}$

2. $x^2 - y^2 = 16$

$$2x - 2yy' = 0$$
$$y' = \frac{x}{y}$$

At $(4, 0)$: $y' = \frac{4}{0}$

Undefined (vertical tangent)

3. $$xy = 4$$
$$xy' + y(1) = 0$$
$$xy' = -y$$
$$y' = \frac{-y}{x}$$
At $(-4, -1)$: $y' = -\frac{1}{4}$

4. $$x^2 - y^3 = 0$$
$$2x - 3y^2y' = 0$$
$$y' = \frac{2x}{3y^2}$$
At $(-1, 1)$: $y' = -\frac{2}{3}$

5. $$x^{1/2} + y^{1/2} = 9$$
$$\frac{1}{2}x^{-1/2} + \frac{1}{2}y^{-1/2}y' = 0$$
$$y' = -\frac{x^{-1/2}}{y^{-1/2}} = -\sqrt{\frac{y}{x}}$$
At $(16, 25)$: $y' = -\frac{5}{4}$

6. $$x^3 + y^3 = 8$$
$$3x^2 + 3y^2y' = 0$$
$$y' = -\frac{x^2}{y^2}$$
At $(0, 2)$: $y' = 0$

7. $$x^3 - xy + y^2 = 4$$
$$3x^2 - xy' - y + 2yy' = 0$$
$$(2y - x)y' = y - 3x^2$$
$$y' = \frac{y - 3x^2}{2y - x}$$
At $(0, -2)$: $y' = \frac{1}{2}$

8. $$x^2y + y^2x = -2$$
$$x^2y' + 2xy + y^2 + 2yxy' = 0$$
$$(x^2 + 2xy)y' = -(y^2 + 2xy)$$
$$y' = \frac{-y(y + 2x)}{x(x + 2y)}$$
At $(2, -1)$: $y' = \frac{3}{0}$

Undefined (vertical tangent)

9. $$y^2 = \frac{x^2 - 9}{x^2 + 9}$$
$$2yy' = \frac{(x^2 + 9)(2x) - (x^2 - 9)2x}{(x^2 + 9)^2}$$
$$y' = \frac{18x}{(x^2 + 9)^2 y}$$
At $(3, 0)$: $y' = \frac{54}{0}$

Undefined (vertical tangent)

10. $$(x + y)^3 = x^3 + y^3$$
$$x^3 + 3x^2y + 3xy^2 + y^3 = x^3 + y^3$$
$$3x^2y + 3xy^2 = 0$$
$$x^2y + xy^2 = 0$$
$$x^2y' + 2xy + 2xyy' + y^2 = 0$$
$$(x^2 + 2xy)y' = -(y^2 + 2xy)$$
$$y' = -\frac{y(y + 2x)}{x(x + 2y)}$$
At $(-1, 1)$: $y' = -1$

11.

$$x^3y^3 - y - x = 0$$

$$3x^3y^2y' + 3x^2y^3 - y' - 1 = 0$$

$$(3x^3y^2 - 1)y' = 1 - 3x^2y^3$$

$$y' = \frac{1 - 3x^2y^3}{3x^3y^2 - 1}$$

At $(0,\ 0)$: $y' = -1$

12.

$$(xy)^{1/2} - x + 2y = 0$$

$$\frac{1}{2}(xy)^{-1/2}(xy' + y) - 1 + 2y' = 0$$

$$\frac{x}{2\sqrt{xy}}y' + \frac{y}{2\sqrt{xy}} - 1 + 2y' = 0$$

$$xy' + y - 2\sqrt{xy} + 4\sqrt{xy}\,y' = 0$$

$$y' = \frac{2\sqrt{xy} - y}{4\sqrt{xy} + x}$$

At $(4,\ 1)$: $y' = \frac{1}{4}$

13.

$$x^{2/3} + y^{2/3} = 5$$

$$\frac{2}{3}x^{-1/3} + \frac{2}{3}y^{-1/3}y' = 0$$

$$y' = \frac{-x^{-1/3}}{y^{-1/3}} = -\sqrt[3]{\frac{y}{x}}$$

At $(8,\ 1)$: $y' = -\frac{1}{2}$

14.

$$x^3 + y^3 - 2xy = 0$$

$$3x^2 + 3y^2y' - 2xy' - 2y = 0$$

$$(3y^2 - 2x)y' = 2y - 3x^2$$

$$y' = \frac{2y - 3x^2}{3y^2 - 2x}$$

At $(1,\ 1)$: $y' = -1$

15.

$$x^3 - 2x^2y + 3xy^2 = 38$$

$$3x^2 - (2x^2y' - 4xy) + (6xyy' + 3y^2) = 0$$

$$2x(3y - x)y' = 4xy - 3x^2 - 3y^2$$

$$y' = \frac{4xy - 3x^2 - 3y^2}{2x(3y - x)}$$

At $(2,\ 3)$: $y' = \frac{-15}{28}$

16.

$$x^3 - y^3 = x - y$$

$$3x^2 - 3y^2\frac{dy}{dx} = 1 - \frac{dy}{dx}$$

$$3x^2 - 1 = \frac{dy}{dx}(3y^2 - 1)$$

$$\frac{dy}{dx} = \frac{3x^2 - 1}{3y^2 - 1}$$

At $(1,\ 1)$: $\frac{dy}{dx} = \frac{3(1)^2 - 1}{3(1)^2 - 1} = 1$

17.

$$(x^2 + 4)y = 8$$

$$(x^2 + 4)y' + y(2x) = 0$$

$$y' = \frac{-2xy}{x^2 + 4}$$

$$= \frac{-2x[8/(x^2 + 4)]}{x^2 + 4} = \frac{-16x}{(x^2 + 4)^2}$$

At $(2,\ 1)$: $y' = \frac{-32}{64} = -\frac{1}{2}$

18.

$$(4 - x)y^2 = x^3$$

$$(4 - x)(2yy') + y^2(-1) = 3x^2$$

$$y' = \frac{3x^2 + y^2}{2y(4 - x)}$$

At $(2,\ 2)$: $y' = 2$

19.

$$(x^2 + y^2)^2 = 4x^2y$$
$$2(x^2 + y^2)(2x + 2yy') = 4x^2y' + y(8x)$$
$$4x^3 + 4x^2yy' + 4xy^2 + 4y^3y' = 4x^2y' + 8xy$$
$$4x^2yy' + 4y^3y' - 4x^2y' = 8xy - 4x^3 - 4xy^2$$
$$4y'(x^2y + y^3 - x^2) = 4(2xy - x^3 - xy^2)$$
$$y' = \frac{2xy - x^3 - xy^2}{x^2y + y^3 - x^2}$$

At $(1, 1)$: $y' = 0$

20.

$$x^3 + y^3 - 6xy = 0$$
$$3x^2 + 3y^2y' - 6xy' - 6y = 0$$
$$y'(3y^2 - 6x) = 6y - 3x^2$$
$$y' = \frac{6y - 3x^2}{3y^2 - 6x} = \frac{2y - x^2}{y^2 - 2x}$$

At $\left(\frac{4}{3}, \frac{8}{3}\right)$: $y' = \frac{(16/3) - (16/9)}{(64/9) - (8/3)} = \frac{32}{40} = \frac{4}{5}$

21. $x^2 + y^2 = 16$

$$y^2 = 16 - x^2$$
$$y = \pm\sqrt{16 - x^2}$$

Explicitly: $\frac{dy}{dx} = \pm\frac{1}{2}(16 - x^2)^{-1/2}(-2x) = \frac{\mp x}{\sqrt{16 - x^2}} = -\frac{x}{y}$

Implicitly: $2x + 2yy' = 0$

$$y' = -\frac{x}{y}$$

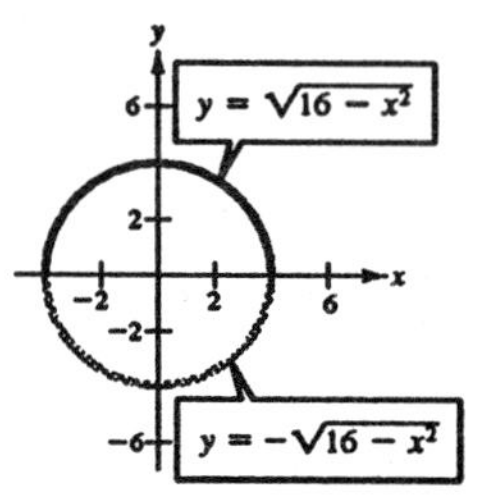

22. $(x^2 - 4x + 4) + (y^2 + 6y + 9) = -9 + 4 + 9$

$$(x - 2)^2 + (y + 3)^2 = 4$$
$$(y + 3)^2 = 4 - (x - 2)^2$$
$$y = -3 \pm \sqrt{4 - (x - 2)^2}$$

Explicitly: $\frac{dy}{dx} = \pm\frac{1}{2}[4 - (x - 2)^2]^{-1/2}(-2)(x - 2)$

$$= \frac{\mp(x - 2)}{(\sqrt{4 - (x - 2)^2} - 3) + 3} = \frac{-(x - 2)}{y + 3}$$

Implicitly: $2x + 2yy' - 4 + 6y' = 0$

$$(2y + 6)y' = -2(x - 2)$$
$$y' = \frac{-(x - 2)}{y + 3}$$

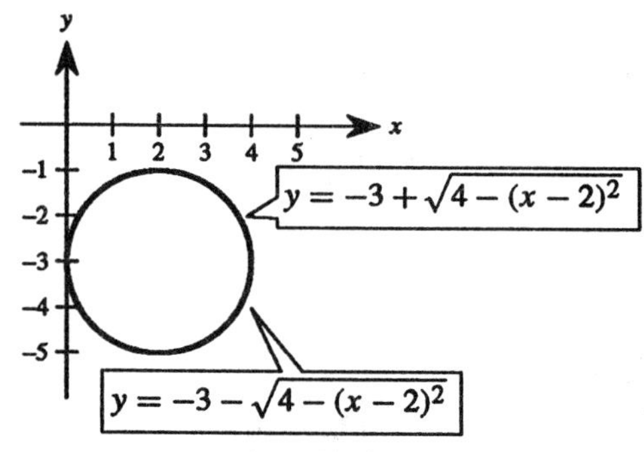

23. $16y^2 = 144 - 9x^2$

$$y^2 = \frac{1}{16}(144 - 9x^2)$$

$$y = \pm\frac{1}{4}\sqrt{144 - 9x^2}$$

Explicitly: $\dfrac{dy}{dx} = \pm\dfrac{1}{8}(144 - 9x^2)^{-1/2}(-18x)$

$$= \frac{\mp 9x}{4\sqrt{144 - 9x^2}} = \frac{\mp 9x}{16(1/4)\sqrt{144 - 9x^2}} = \frac{-9x}{16y}$$

Implicitly: $18x + 32yy' = 0$

$$y' = \frac{-9x}{16y}$$

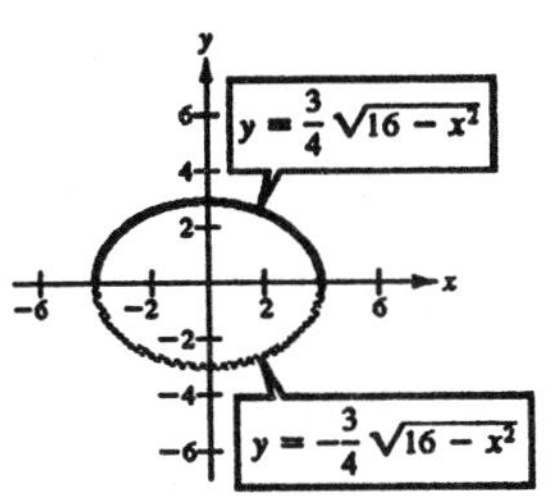

24. $y^2 = 1 + \dfrac{x^2}{4} = \dfrac{x^2 + 4}{4}$

$$y = \pm\frac{1}{2}\sqrt{x^2 + 4}$$

Explicitly: $\dfrac{dy}{dx} = \pm\dfrac{1}{4}(x^2 + 4)^{-1/2}(2x)$

$$= \frac{\pm x}{2\sqrt{x^2 + 4}} = \frac{\pm x}{4(1/2)\sqrt{x^2 + 4}} = \frac{x}{4y}$$

Implicitly: $2yy' - \dfrac{1}{2}x = 0$

$$y' = \frac{x}{4y}$$

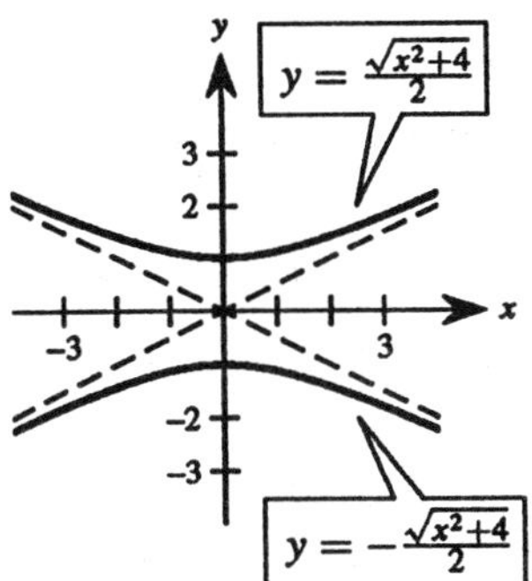

25. $x^2 + xy = 5$

$$2x + xy' + y = 0$$

$$y' = \frac{-(2x + y)}{x}$$

$$2 + xy'' + y' + y' = 0$$

$$xy'' = -2(1 + y')$$

$$y'' = \frac{-2[1 - (2x + y)/x]}{x} = \frac{2(x + y)}{x^2}$$

$$y'' = \frac{2(x + y)}{x^2} \cdot \frac{x}{x} = \frac{10}{x^3}$$

26.
$$x^2y^2 - 2x = 3$$
$$2x^2yy' + 2xy^2 - 2 = 0$$
$$x^2yy' + xy^2 - 1 = 0$$
$$y' = \frac{1 - xy^2}{x^2y}$$
$$2xyy' + x^2(y')^2 + x^2yy'' + 2xyy' + y^2 = 0$$
$$4xyy' + x^2(y')^2 + x^2yy'' + y^2 = 0$$
$$\frac{4 - 4xy^2}{x} + \frac{(1 - xy^2)^2}{x^2y^2} + x^2yy'' + y^2 = 0$$
$$4xy^2 - 4x^2y^4 + 1 - 2xy^2 + x^2y^4 + x^4y^3y'' + x^2y^4 = 0$$
$$x^4y^3y'' = 2x^2y^4 - 2xy^2 - 1$$
$$y'' = \frac{2x^2y^4 - 2xy^2 - 1}{x^4y^3}$$

27.
$$x^2 - y^2 = 16$$
$$2x - 2yy' = 0$$
$$y' = \frac{x}{y}$$
$$x - yy' = 0$$
$$1 - yy'' - (y')^2 = 0$$
$$y'' = \frac{y^2 - x^2}{y^3} = \frac{-16}{y^3}$$

28.
$$1 - xy = x - y$$
$$y - xy = x - 1$$
$$y = \frac{x - 1}{1 - x} = -1$$
$$y' = 0$$
$$y'' = 0$$

29.
$$y^2 = x^3$$
$$2yy' = 3x^2$$
$$y' = \frac{3x^2}{2y} = \frac{3x^2}{2y} \cdot \frac{xy}{xy} = \frac{3y}{2x} \cdot \frac{x^3}{y^2} = \frac{3y}{2x}$$
$$y'' = \frac{2x(3y') - 3y(2)}{4x^2} = \frac{3y}{4x^2} = \frac{3x}{4y}$$

30.
$$y^2 = 4x$$
$$2yy' = 4$$
$$y' = \frac{2}{y}$$
$$y'' = -2y^{-2}y' = \left[\frac{-2}{y^2}\right] \cdot \frac{2}{y} = \frac{-4}{y^3}$$

31. $x^2 + y^2 = 25$

$$y' = \frac{-x}{y}$$

(a) At (4, 3): Tangent line: $y - 3 = \frac{-4}{3}(x - 4) \Rightarrow 4x + 3y - 25 = 0$

Normal line: $y - 3 = \frac{3}{4}(x - 4) \Rightarrow 3x - 4y = 0$

(b) At (−3, 4): Tangent line: $y - 4 = \frac{3}{4}(x + 3) \Rightarrow 3x - 4y + 25 = 0$

Normal line: $y - 4 = \frac{-4}{3}(x + 3) \Rightarrow 4x + 3y = 0$

32. $x^2 + y^2 = 9$

$$y' = \frac{-x}{y}$$

(a) At $(0, 3)$: Tangent line: $y = 3$

Normal line: $x = 0$

(b) At $(2, \sqrt{5})$: Tangent line: $y - \sqrt{5} = \frac{-2}{\sqrt{5}}(x - 2) \Rightarrow 2x + \sqrt{5}\,y - 9 = 0$

Normal line: $y - \sqrt{5} = \frac{\sqrt{5}}{2}(x - 2) \Rightarrow \sqrt{5}\,x - 2y = 0$

33. $25x^2 + 16y^2 + 200x - 160y + 400 = 0$

$$50x + 32yy' + 200 - 160y' = 0$$

$$y' = \frac{200 + 50x}{160 - 32y}$$

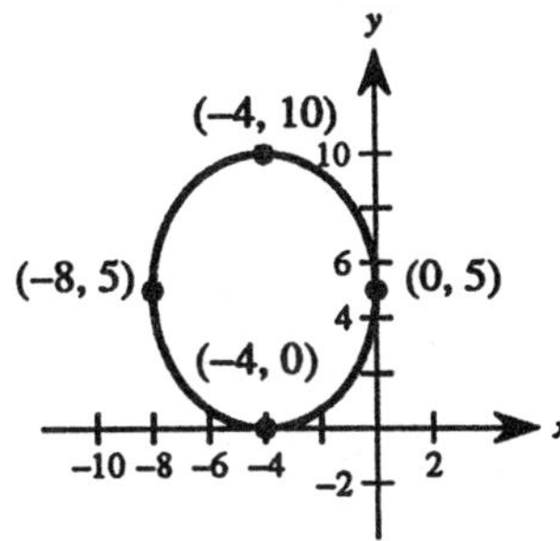

Horizontal tangents occur when $x = -4$:

$$25(16) + 16y^2 + 200(-4) - 160y + 400 = 0$$

$$y(y - 10) = 0 \Rightarrow y = 0,\ 10$$

Horizontal tangents: $(-4,\ 0),\ (-4,\ 10)$

Vertical tangents occur when $y = 5$:

$$25x^2 + 400 + 200x - 800 + 400 = 0$$

$$25x(x + 8) = 0 \Rightarrow x = 0,\ -8$$

Vertical tangents: $(0,\ 5),\ (-8,\ 5)$

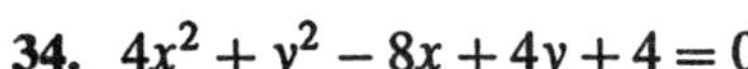

34. $4x^2 + y^2 - 8x + 4y + 4 = 0$

$$8x + 2yy' - 8 + 4y' = 0$$

$$y' = \frac{8 - 8x}{2y + 4} = \frac{4 - 4x}{y + 2}$$

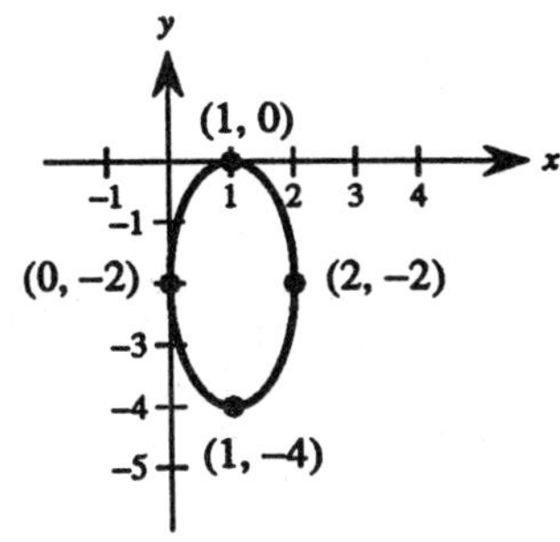

Horizontal tangents occur when $x = 1$:

$$4(1)^2 + y^2 - 8(1) + 4y + 4 = 0$$

$$y^2 + 4y = y(y + 4) = 0 \Rightarrow y = 0,\ -4$$

Horizontal tangents: $(1,\ 0),\ (1,\ -4)$

Vertical tangents occur when $y = -2$:

$$4x^2 + (-2)^2 - 8x + 4(-2) + 4 = 0$$

$$4x^2 - 8x = 4x(x - 2) = 0 \Rightarrow x = 0,\ 2$$

Vertical tangents: $(0,\ -2),\ (2,\ -2)$

35. Find the points of intersection by letting $y^2 = 4x$ in the equation

$$2x^2 + y^2 = 6: \quad 2x^2 + 4x = 6, \quad (x+3)(x-1) = 0.$$

The curves intersect at $(1, \pm 2)$.

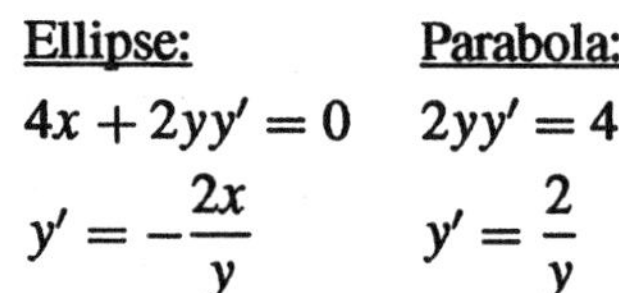

Ellipse:	Parabola:
$4x + 2yy' = 0$	$2yy' = 4$
$y' = -\dfrac{2x}{y}$	$y' = \dfrac{2}{y}$

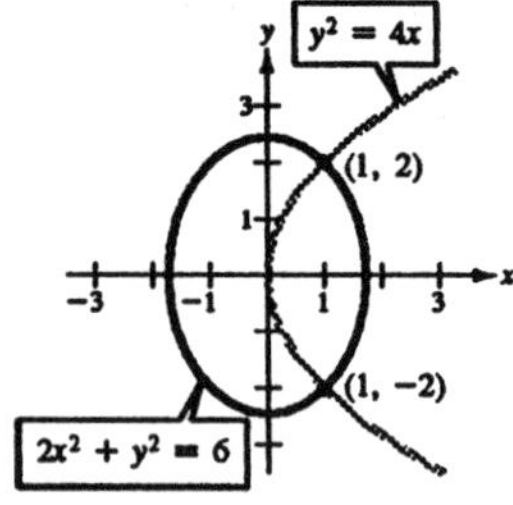

At $(1, 2)$, the slopes are $y' = -1$ and $y' = 1$.
At $(1, -2)$, the slopes are $y' = 1$ and $y' = -1$.

Tangents are perpendicular

36. Find the points of intersection by letting $y^2 = x^3$ in the equation

$$2x^2 + 3y^2 = 5: \quad 2x^2 + 3x^3 = 5, \quad 3x^3 + 2x^2 - 5 = 0.$$

Intersect when $x = 1$

Points of intersection: $(1, \pm 1)$

$y^2 = x^3$:	$2x^2 + 3y^2 = 5$:
$2yy' = 3x^2$	$4x + 6yy' = 0$
$y' = \dfrac{3x^2}{2y}$	$y' = -\dfrac{2x}{3y}$

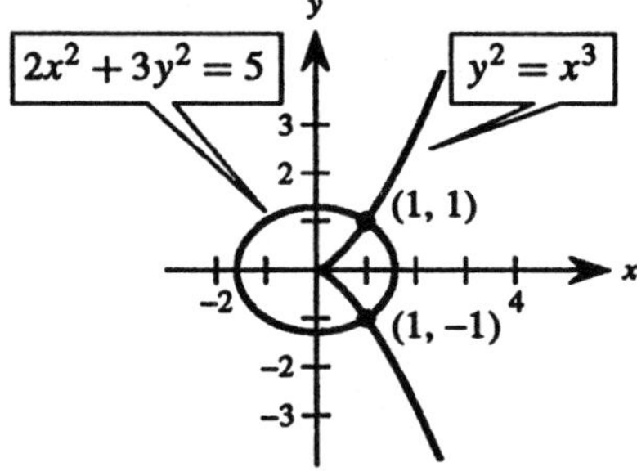

At $(1, 1)$, the slopes are $y' = \frac{3}{2}$ and $y' = -\frac{2}{3}$.
At $(1, -1)$, the slopes are $y' = -\frac{3}{2}$ and $y' = \frac{2}{3}$.

Tangents are perpendicular

37. Find the points of intersection by substituting $y = -x$ in the equation $x^2 + y^2 = 4$:

$$x^2 + (-x)^2 = 4$$
$$2x^2 = 4$$
$$x = \pm\sqrt{2}$$

Points of intersection: $(\sqrt{2}, -\sqrt{2}), (-\sqrt{2}, \sqrt{2})$

$x + y = 0$:	$x^2 + y^2 = 4$:
$y' = -1$	$2x + 2yy' = 0$
	$y' = -\dfrac{x}{y}$

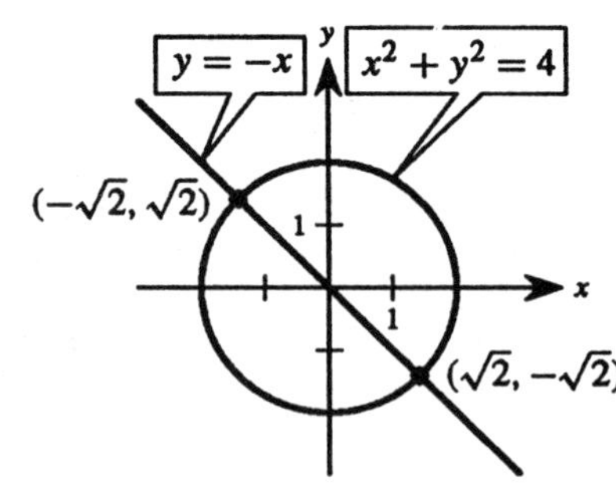

At $(\sqrt{2}, -\sqrt{2})$ the slopes are $y' = -1$ and $y' = 1$.
At $(-\sqrt{2}, \sqrt{2})$ the slopes are $y' = -1$ and $y' = 1$.

Tangents are perpendicular

38. Rewriting each equation and differentiating,

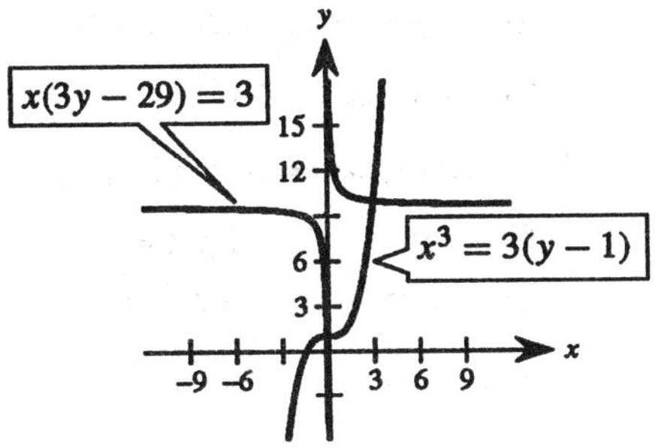

$$x^3 = 3(y-1) \qquad x(3y-29) = 3$$

$$y = \frac{x^3}{3} + 1 \qquad y = \frac{1}{3}\left(\frac{3}{x} + 29\right)$$

$$y' = x^2 \qquad y' = -\frac{1}{x^2}$$

For each value of x, the derivatives are negative reciprocals of each other. Thus, the tangent lines are orthogonal at all points of intersection.

39. $y^2 = 4x$

$$2yy' = 4$$

$$y' = \frac{2}{y} = 1 \text{ at } (1,\ 2)$$

Equation of normal at (1, 2) is $y - 2 = -1(x-1)$, $y = 3 - x$. The centers of the circles must be on the normal and at a distance of 4 units from (1, 2). Therefore,

$$(x-1)^2 + [(3-x)-2]^2 = 16$$

$$2(x-1)^2 = 16$$

$$x = 1 \pm 2\sqrt{2}.$$

Centers of the circles: $(1+2\sqrt{2},\ 2-2\sqrt{2})$ and $(1-2\sqrt{2},\ 2+2\sqrt{2})$

Equations: $(x-1-2\sqrt{2})^2 + (y-2+2\sqrt{2})^2 = 16$

$$(x-1+2\sqrt{2})^2 + (y-2-2\sqrt{2})^2 = 16$$

40. $2x + 2yy' = 0$

$$y' = \frac{-x}{y} = \text{ slope of tangent line}$$

$$\frac{y}{x} = \text{ slope of normal line}$$

Any point on the circumference of the circle will have the form $\left(x_0,\ \pm\sqrt{r^2 - x_0^2}\right)$.

Equation of normal line:

$$y \mp \sqrt{r^2 - x_0^2} = \frac{\pm\sqrt{r^2 - x_0^2}}{x_0}(x - x_0)$$

$$y = \frac{\pm\sqrt{r^2 - x_0^2}}{x_0}x \mp \frac{\sqrt{r^2 - x_0^2}}{1} \pm \sqrt{r^2 - x_0^2}$$

$$= \frac{\pm\sqrt{r^2 - x_0^2}}{x_0}x \text{ which passes through } (0,\ 0).$$

41. $\sqrt{x} + \sqrt{y} = 3$

$$\frac{1}{2}x^{-1/2} + \frac{1}{2}y^{-1/2}y' = 0$$

$$y' = -\frac{(1/2)x^{-1/2}}{(1/2)y^{-1/2}} = -\frac{\sqrt{y}}{\sqrt{x}}$$

At (4, 1): $y' = -\frac{1}{2}$

Tangent line: $y - 1 = -\frac{1}{2}(x - 4)$

$$y = -\frac{1}{2}x + 3$$

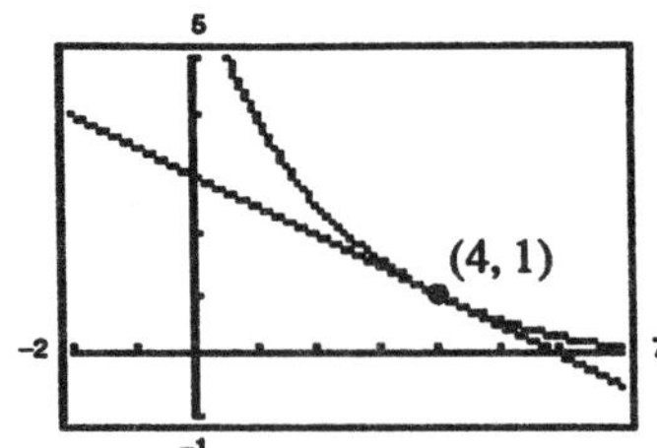

42. $x^3 + y^3 - 6xy = 0$

From Exercise 28 we have $y' = \dfrac{2y - x^2}{y^2 - 2x}$

and at $\left(\frac{4}{3}, \frac{8}{3}\right)$, $y' = \frac{4}{5}$.

Tangent line: $y - \frac{8}{3} = \frac{4}{5}\left(x - \frac{4}{3}\right)$

$$y - \frac{8}{3} = \frac{4}{5}x - \frac{16}{15}$$

$$15y - 40 = 12x - 16$$

$$0 = 12x - 15y + 24$$

43. $(x - h)^2 + (y - k)^2 = r^2$

$$2(x - h) + 2(y - k)y' = 0$$

$$y' = \frac{-2(x - h)}{2(y - k)} = -\frac{x - h}{y - k}$$

Section 3.7 Related Rates

1. $y = \sqrt{x}$

$$\frac{dy}{dt} = \left(\frac{1}{2\sqrt{x}}\right)\frac{dx}{dt}$$

$$\frac{dx}{dt} = 2\sqrt{x}\frac{dy}{dt}$$

(a) When $x = 4$ and $\frac{dx}{dt} = 3$,

$$\frac{dy}{dt} = \frac{1}{2\sqrt{4}}(3) = \frac{3}{4}.$$

(b) When $x = 25$ and $\frac{dy}{dt} = 2$,

$$\frac{dx}{dt} = 2\sqrt{25}(2) = 20.$$

2. $y = x^2 - 3x$

$$\frac{dy}{dt} = (2x - 3)\frac{dx}{dt}$$

$$\frac{dx}{dt} = \left(\frac{1}{2x - 3}\right)\frac{dy}{dt}$$

(a) When $x = 3$ and $\frac{dx}{dt} = 2$,

$$\frac{dy}{dt} = [2(3) - 3](2) = 6.$$

(b) When $x = 1$ and $\frac{dy}{dt} = 5$,

$$\frac{dx}{dt} = \frac{1}{2(1) - 3}(5) = -5.$$

3. $xy = 4$

$$x\frac{dy}{dt} + y\frac{dx}{dt} = 0$$

$$\frac{dy}{dt} = \left(-\frac{y}{x}\right)\frac{dx}{dt}$$

$$\frac{dx}{dt} = \left(-\frac{x}{y}\right)\frac{dy}{dt}$$

(a) When $x = 8$, $y = \frac{1}{2}$, and $\frac{dx}{dt} = 10$,

$$\frac{dy}{dt} = -\frac{1/2}{8}(10) = -\frac{5}{8}.$$

(b) When $x = 1$, $y = 4$, and $\frac{dy}{dt} = -6$,

$$\frac{dx}{dt} = -\frac{1}{4}(-6) = \frac{3}{2}.$$

4. $x^2 + y^2 = 25$

$$2x\frac{dx}{dt} + 2y\frac{dy}{dt} = 0$$

$$\frac{dy}{dt} = \left(-\frac{x}{y}\right)\frac{dx}{dt}$$

$$\frac{dx}{dt} = \left(-\frac{y}{x}\right)\frac{dy}{dt}$$

(a) When $x = 3$, $y = 4$, and $\frac{dx}{dt} = 8$,

$$\frac{dy}{dt} = -\frac{3}{4}(8) = -6.$$

(b) When $x = 4$, $y = 3$, and $\frac{dy}{dt} = -2$,

$$\frac{dx}{dt} = -\frac{3}{4}(-2) = \frac{3}{2}.$$

5. $A = \pi r^2$

$$\frac{dr}{dt} = 2$$

$$\frac{dA}{dt} = 2\pi r\frac{dr}{dt}$$

(a) When $r = 6$,

$$\frac{dA}{dt} = 2\pi(6)(2) = 24\pi \text{ in}^2\text{/min}.$$

(b) When $r = 24$,

$$\frac{dA}{dt} = 2\pi(24)(2) = 96\pi \text{ in}^2\text{/min}.$$

6. $V = \frac{4}{3}\pi r^3$

$$\frac{dr}{dt} = 2$$

$$\frac{dV}{dt} = 4\pi r^2\frac{dr}{dt}$$

(a) When $r = 6$,

$$\frac{dV}{dt} = 4\pi(6)^2(2) = 288\pi \text{ in}^3\text{/min}.$$

(b) When $r = 24$,

$$\frac{dV}{dt} = 4\pi(24)^2(2) = 4608\pi \text{ in}^3\text{/min}.$$

7. $A = \pi r^2$

$$\frac{dA}{dt} = 2\pi r\frac{dr}{dt}$$

If dr/dt is constant, dA/dt is proportional to r.

8. $V = \frac{4}{3}\pi r^3$

$$\frac{dV}{dt} = 4\pi r^2\frac{dr}{dt}$$

If dr/dt is constant, dV/dt is proportional to r^2.

9. $V = \frac{4}{3}\pi r^3$

$$\frac{dV}{dt} = 20$$

$$\frac{dV}{dt} = 4\pi r^2\frac{dr}{dt}$$

$$\frac{dr}{dt} = \left(\frac{1}{4\pi r^2}\right)\frac{dV}{dt}$$

(a) When $r = 1$,

$$\frac{dr}{dt} = \frac{1}{4\pi(1)^2}(20) = \frac{5}{\pi} \text{ ft/min}.$$

(b) When $r = 2$,

$$\frac{dr}{dt} = \frac{1}{4\pi(2)^2}(20) = \frac{5}{4\pi} \text{ ft/min}.$$

10. $V = \frac{1}{3}\pi r^2 h = \frac{1}{3}\pi r^2(3r) = \pi r^3$

$$\frac{dr}{dt} = 2$$

$$\frac{dV}{dt} = 3\pi r^2\frac{dr}{dt}$$

(a) When $r = 6$,

$$\frac{dV}{dt} = 3\pi(6)^2(2) = 216\pi \text{ in}^3\text{/min}.$$

(b) When $r = 24$,

$$\frac{dV}{dt} = 3\pi(24)^2(2) = 3456\pi \text{ in}^3\text{/min}.$$

11. $V = \frac{1}{3}\pi r^2 h = \frac{1}{3}\pi\left(\frac{9}{4}h^2\right)h$ [since $2r = 3h$]

$= \frac{3\pi}{4}h^3$

$\frac{dV}{dt} = 10$

$\frac{dV}{dt} = \frac{9\pi}{4}h^2\frac{dh}{dt} \Rightarrow \frac{dh}{dt} = \frac{4(dV/dt)}{9\pi h^2}$

When $h = 15$, $\frac{dh}{dt} = \frac{4(10)}{9\pi(15)^2} = \frac{8}{405\pi}$ ft/min.

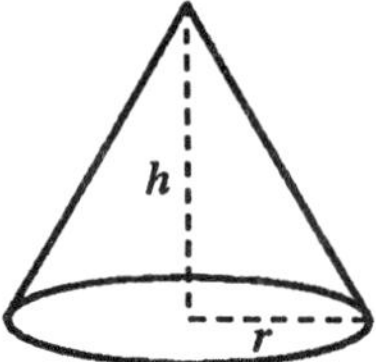

12. $V = \frac{1}{3}\pi r^2 h = \frac{1}{3}\pi\frac{25}{144}h^3 = \frac{25\pi}{3(144)}h^3$

$r = \frac{5}{12}h$

$\frac{dV}{dt} = 10$

$\frac{dV}{dt} = \frac{25\pi}{144}h^2\frac{dh}{dt} \Rightarrow \frac{dh}{dt} = \left(\frac{144}{25\pi h^2}\right)\frac{dV}{dt}$

When $h = 8$, $\frac{dh}{dt} = \frac{144}{25\pi(64)}(10) = \frac{9}{10\pi}$ ft/min.

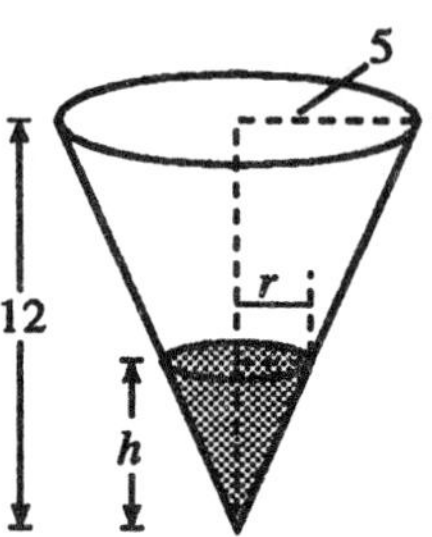

13. $V = x^3$

$\frac{dx}{dt} = 3$

$\frac{dV}{dt} = 3x^2\frac{dx}{dt}$

(a) When $x = 1$,

$\frac{dV}{dt} = 3(1)^2(3) = 9$ cm³/sec.

(b) When $x = 10$,

$\frac{dV}{dt} = 3(10)^2(3) = 900$ cm³/sec.

14. $s = 6x^2$

$\frac{dx}{dt} = 3$

$\frac{ds}{dt} = 12x\frac{dx}{dt}$

(a) When $x = 1$,

$\frac{ds}{dt} = 12(1)(3) = 36$ cm²/sec.

(b) When $x = 10$,

$\frac{ds}{dt} = 12(10)(3) = 360$ cm²/sec.

15. $y = x^2$

$\frac{dx}{dt} = 2$

$\frac{dy}{dt} = 2x\frac{dx}{dt}$

(a) When $x = 0$, $\frac{dy}{dt} = 2(0)(2) = 0$ cm/min.

(b) When $x = 3$, $\frac{dy}{dt} = 2(3)(2) = 12$ cm/min.

16. $y = x^2$

$$\frac{dx}{dt} = 2$$

$$L^2 = x^2 + y^2 = x^2 + x^4$$

$$2L\frac{dL}{dt} = (2x + 4x^3)\frac{dx}{dt}$$

$$\frac{dL}{dt} = \frac{(x + 2x^3)(dx/dt)}{L}$$

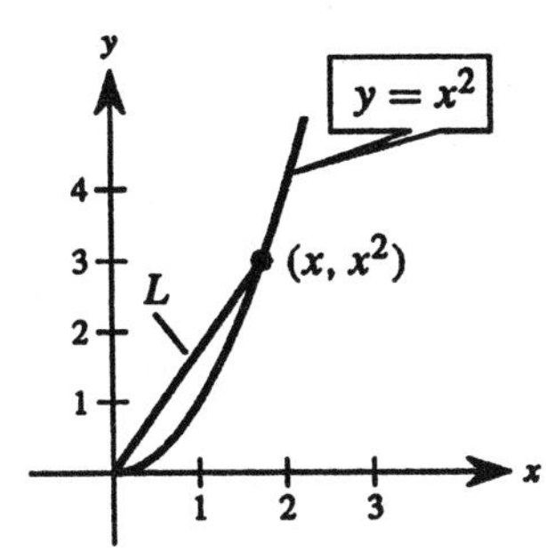

(a) When $x = 0$, $\dfrac{dL}{dt} = 0$ cm/min.

(b) When $x = 3$, $\dfrac{dL}{dt} = \dfrac{(3 + 54)(2)}{\sqrt{9 + 81}} = \dfrac{19\sqrt{10}}{5}$ cm/min.

17. $y = \dfrac{1}{1 + x^2}$

$$\frac{dx}{dt} = 2$$

$$\frac{dy}{dt} = \left[\frac{-2x}{(1 + x^2)^2}\right]\frac{dx}{dt}$$

(a) When $x = -2$,

$$\frac{dy}{dt} = \frac{-2(-2)(2)}{25} = \frac{8}{25} \text{ cm/min.}$$

(b) When $x = 0$, $\dfrac{dy}{dt} = 0$ cm/min.

(c) When $x = 2$,

$$\frac{dy}{dt} = \frac{-2(2)(2)}{25} = \frac{-8}{25} \text{ cm/min.}$$

(d) When $x = 10$,

$$\frac{dy}{dt} = \frac{-2(10)(2)}{(101)^2} = \frac{-40}{10{,}201} = -0.0039 \text{ cm/min.}$$

18. $y = x^3$

$$\frac{dx}{dt} = 2$$

$$\frac{dy}{dt} = 3x^2\frac{dx}{dt}$$

(a) When $x = -2$,

$$\frac{dy}{dt} = 3(-2)^2(2) = 24 \text{ cm/min.}$$

(b) When $x = 1$,

$$\frac{dy}{dt} = 3(1)^2(2) = 6 \text{ cm/min.}$$

(c) When $x = 0$,

$$\frac{dy}{dt} = 3(0)^2(2) = 0 \text{ cm/min.}$$

(d) When $x = 3$,

$$\frac{dy}{dt} = 3(3)^2(2) = 54 \text{ cm/min.}$$

19. (a) Total volume $= \left(\dfrac{1}{2}\right)(40)(5)(20) + (40)(4)(20) = 5200 \text{ ft}^3$

Volume of 4 ft of water $= \left(\dfrac{1}{2}\right)(32)(4)(20) = 1280 \text{ ft}^3$

% of pool filled $= \dfrac{1280}{5200}(100\%) = 24.6\%$.

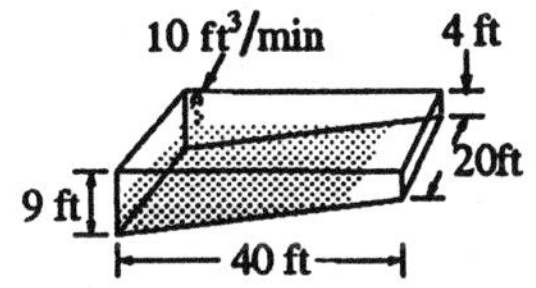

(b) $(b = 8h)$, $0 \le h \le 5$

$$V = \left(\frac{1}{2}\right)bh(20) = 10bh = 80h^2$$

$$\frac{dV}{dt} = 160h\frac{dh}{dt} \Rightarrow \frac{dh}{dt} = \left(\frac{1}{160h}\right)\frac{dV}{dt}$$

When $h = 4$ and $\dfrac{dV}{dt} = 10$, $\dfrac{dh}{dt} = \dfrac{1}{160(4)}(10) = \dfrac{1}{64}$ ft/min.

20. $V = \left(\frac{1}{2}\right)bh(12) = 6bh = 6h^2$ (since $b = h$)

$\frac{dV}{dt} = 12h\frac{dh}{dt} \Rightarrow \frac{dh}{dt} = \left(\frac{1}{12h}\right)\frac{dV}{dt}$

When $h = 1$ and $\frac{dV}{dt} = 2$, $\frac{dh}{dt} = \frac{1}{12}(2) = \frac{1}{6}$ ft/min $= 2$ in/min.

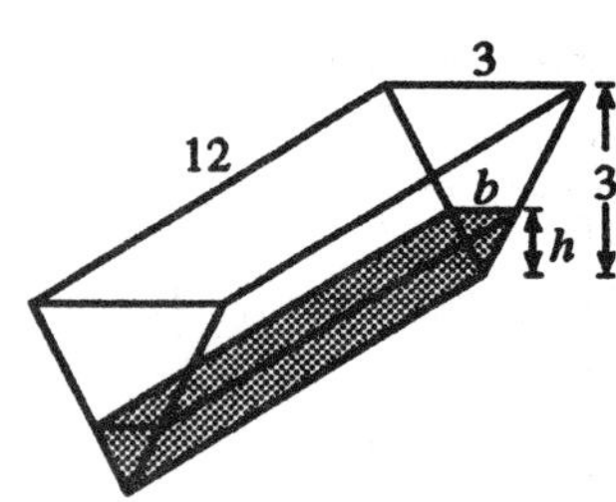

21. $x^2 + y^2 = 25^2$

$2x\frac{dx}{dt} + 2y\frac{dy}{dt} = 0$

$\frac{dy}{dt} = \frac{-x}{y} \cdot \frac{dx}{dy} = \frac{-2x}{y}$ since $\frac{dx}{dt} = 2$.

(a) When $x = 7$, $y = \sqrt{576} = 24$, $\frac{dy}{dt} = \frac{-2(7)}{24} = \frac{-7}{12}$ ft/sec.

(b) When $x = 15$, $y = \sqrt{400} = 20$, $\frac{dy}{dt} = \frac{-2(15)}{20} = \frac{-3}{2}$ ft/sec.

(c) When $x = 24$, $y = 7$, $\frac{dy}{dt} = \frac{-2(24)}{7} = \frac{-48}{7}$ ft/sec.

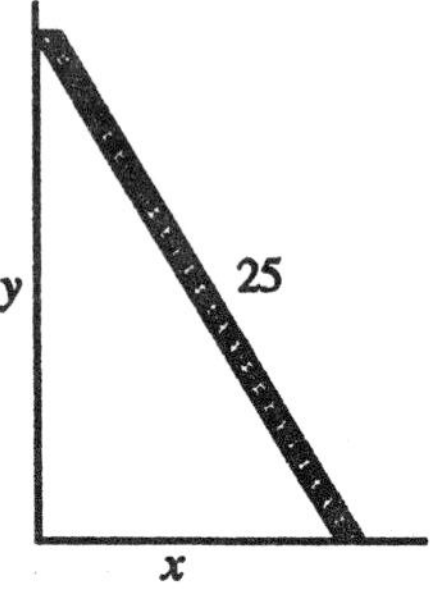

22. $x^2 + y^2 = 16^2$

$2x\frac{dx}{dt} + 2y\frac{dy}{dt} = 0$

$\frac{dx}{dt} = -\frac{y}{x} \cdot \frac{dy}{dt} = -\frac{0.5y}{x}$ since $\frac{dy}{dt} = 0.5$

When $x = 8$, $y = \sqrt{192} = 8\sqrt{3}$, $\frac{dx}{dt} = -\frac{0.5(8\sqrt{3})}{8} = -\frac{\sqrt{3}}{2}$ ft/sec.

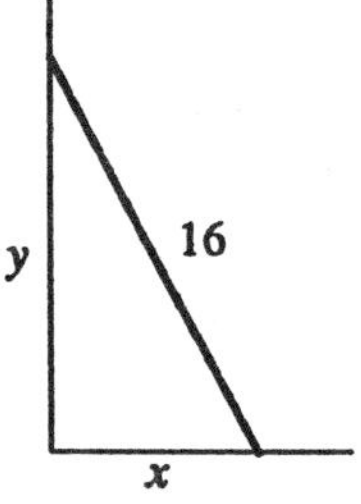

23. $A = \frac{1}{2}xy$

$\frac{dA}{dt} = \frac{1}{2}\left(x\frac{dy}{dt} + y\frac{dx}{dt}\right)$

From Exercise 21(a) we have $x = 7$, $y = 24$, $\frac{dx}{dt} = 2$, and $\frac{dy}{dt} = -\frac{7}{12}$.

Thus, $\frac{dA}{dt} = \frac{1}{2}\left[7\left(-\frac{7}{12}\right) + 24(2)\right] = \frac{527}{24} \approx 21.96$ ft^2/sec.

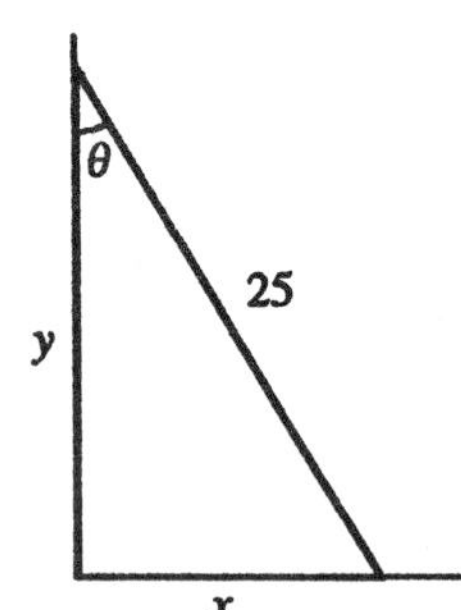

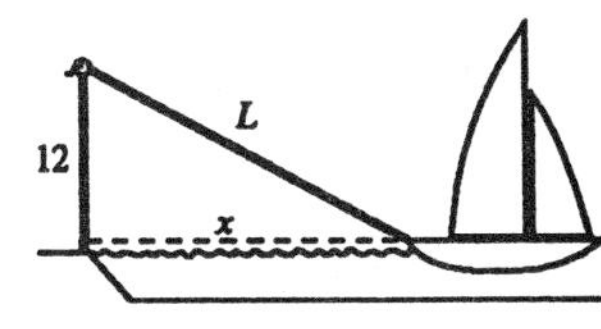

24. $L^2 = 144 + x^2$

$$2L\frac{dL}{dt} = 2x\frac{dx}{dt}$$

$$\frac{dx}{dt} = \frac{L}{x} \cdot \frac{dL}{dt} = -\frac{4L}{x} \text{ since } \frac{dL}{dt} = -4 \text{ ft/sec}$$

When $L = 13$, $x = \sqrt{L^2 - 144} = \sqrt{169 - 144} = 5$

$$\frac{dx}{dt} = -\frac{4(13)}{5} = -\frac{52}{5} = -10.4 \text{ ft/sec.}$$

Speed of the boat increases as it approaches the dock.

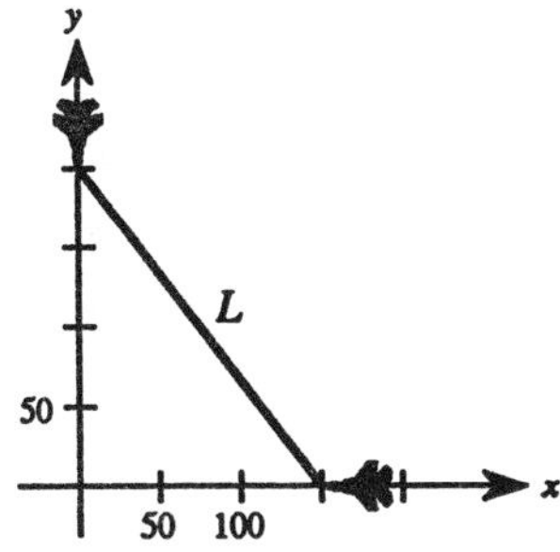

25. (a) $L^2 = x^2 + y^2$

$$\frac{dx}{dt} = -450$$

$$\frac{dy}{dt} = -600$$

$$\frac{dL}{dt} = \frac{x(dx/dt) + y(dy/dt)}{L}$$

When $x = 150$ and $y = 200$,

$$\frac{dL}{dt} = \frac{150(-450) + 200(-600)}{250} = -750 \text{ mph.}$$

(b) $t = \dfrac{250}{750} = \dfrac{1}{3}$ hr $= 20$ min

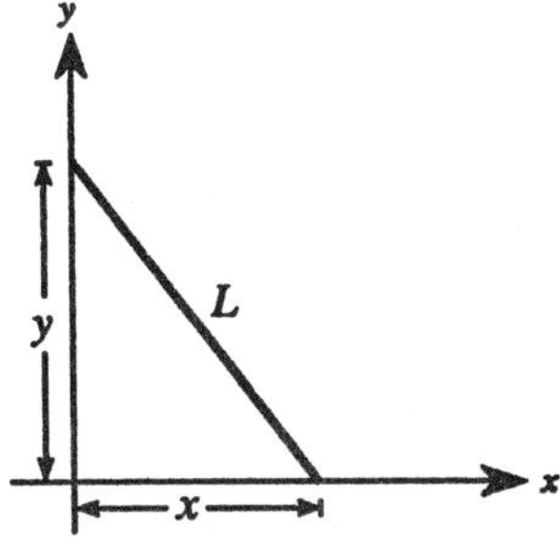

26.

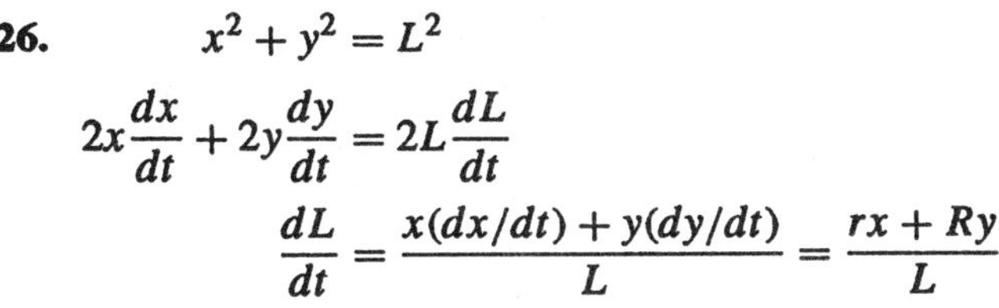

$$x^2 + y^2 = L^2$$

$$2x\frac{dx}{dt} + 2y\frac{dy}{dt} = 2L\frac{dL}{dt}$$

$$\frac{dL}{dt} = \frac{x(dx/dt) + y(dy/dt)}{L} = \frac{rx + Ry}{L}$$

27.

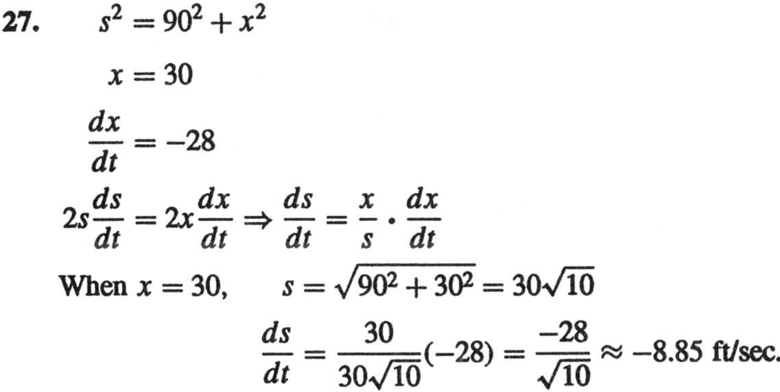

$$s^2 = 90^2 + x^2$$

$$x = 30$$

$$\frac{dx}{dt} = -28$$

$$2s\frac{ds}{dt} = 2x\frac{dx}{dt} \Rightarrow \frac{ds}{dt} = \frac{x}{s} \cdot \frac{dx}{dt}$$

When $x = 30$, $s = \sqrt{90^2 + 30^2} = 30\sqrt{10}$

$$\frac{ds}{dt} = \frac{30}{30\sqrt{10}}(-28) = \frac{-28}{\sqrt{10}} \approx -8.85 \text{ ft/sec.}$$

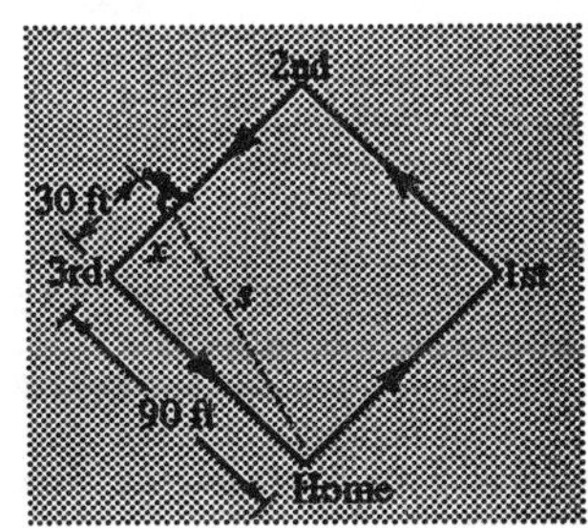

28. $s^2 = 90^2 + x^2$

$x = 60$

$\frac{dx}{dt} = 28$

$\frac{ds}{dt} = \frac{x}{s} \cdot \frac{dx}{dt}$

When $x = 60$, $s = \sqrt{90^2 + 60^2} = 30\sqrt{13}$

$$\frac{ds}{dt} = \frac{60}{30\sqrt{13}}(28) = \frac{56}{\sqrt{13}} \approx 15.5 \text{ ft/sec.}$$

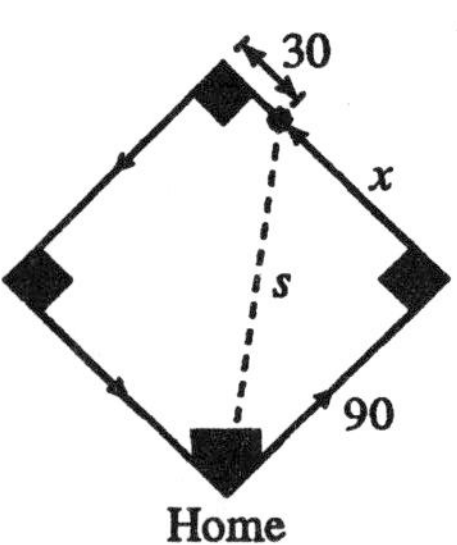

29. (a) $\frac{15}{6} = \frac{y}{y - x}$

$\frac{dx}{dt} = 5$

$y = \frac{5}{3}x$

$$\frac{dy}{dt} = \frac{5}{3} \cdot \frac{dx}{dt} = \frac{5}{3}(5) = \frac{25}{3} \text{ ft/sec}$$

(b) $\frac{d(y - x)}{dt} = \frac{dy}{dt} - \frac{dx}{dt} = \frac{25}{3} - 5 = \frac{10}{3}$ ft/sec

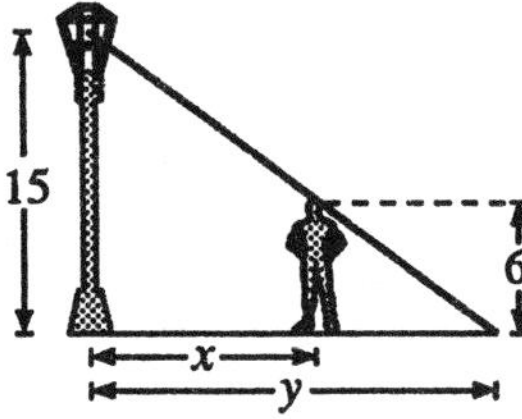

30. $x^2 + y^2 = s^2$

$x = \sqrt{s^2 - y^2}$

$\frac{ds}{dt} = -240$ mph

$y = 6$ mi

$x = \sqrt{s^2 - 36}$

$$\frac{dx}{dt} = \left(\frac{1}{2}\right)(s^2 - 36)^{-1/2}\left(2s\frac{ds}{dt}\right) = \frac{s}{\sqrt{s^2 - 36}} \cdot \frac{ds}{dt}$$

When $s = 10$, $\frac{dx}{dt} = \frac{10}{\sqrt{10^2 - 36}}(-240)$

$$\frac{dx}{dt} = \frac{10}{8}(-240) = -300 \text{ mph.}$$

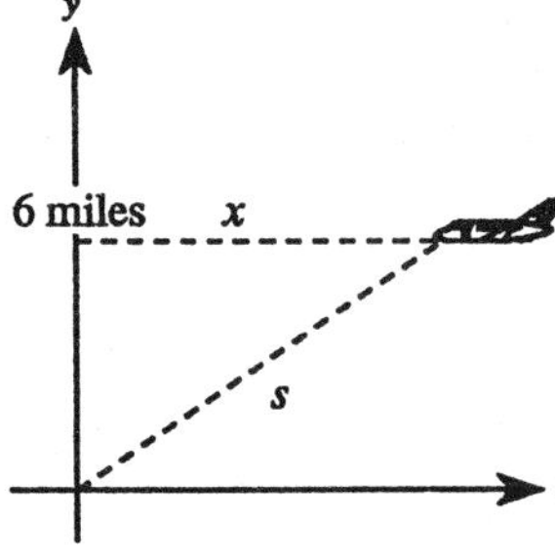

31. Since the evaporation rate is proportional to the surface area, $dV/dt = k(4\pi r^2)$. However, since $V = (4/3)\pi r^3$, we have $(dV/dt) = 4\pi r^2(dr/dt)$. Therefore,

$$k(4\pi r^2) = 4\pi r^2\frac{dr}{dt} \Rightarrow k = \frac{dr}{dt}.$$

32. $$\frac{1}{R} = \frac{1}{R_1} + \frac{1}{R_2}$$

$$\frac{dR_1}{dt} = 1$$

$$\frac{dR_2}{dt} = 1.5$$

$$\frac{1}{R^2} \cdot \frac{dR}{dt} = \frac{1}{R_1{}^2} \cdot \frac{dR_1}{dt} + \frac{1}{R_2{}^2} \cdot \frac{dR_2}{dt}$$

When $R_1 = 50$ and $R_2 = 75$,

$$R = 30$$

$$\frac{dR}{dt} = (30)^2\left[\frac{1}{(50)^2}(1) + \frac{1}{(75)^2}(1.5)\right] = 0.6 \text{ ohms/sec.}$$

33. $$pv^{1.3} = k$$

$$1.3pv^{0.3}\frac{dv}{dt} + v^{1.3}\frac{dp}{dt} = 0$$

$$v^{0.3}\left(1.3p\,\frac{dv}{dt} + v\,\frac{dp}{dt}\right) = 0$$

34. When $y = 20$, $x = \sqrt{40^2 - 20^2} = 20\sqrt{3}$ and $s = 40$.

$$x^2 + (40 - y)^2 = s^2$$

$$2x\frac{dx}{dt} + 2(40 - y)(-1)\frac{dy}{dt} = 2s\frac{ds}{dt}$$

$$x\frac{dx}{dt} + (y - 40)\frac{dy}{dt} = s\frac{ds}{dt}$$

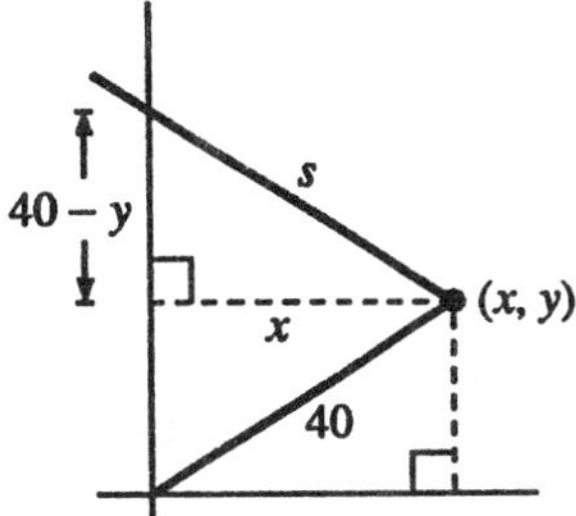

Also,

$$x^2 + y^2 = 40^2$$

$$2x\frac{dx}{dt} + 2y\frac{dy}{dt} = 0 \Rightarrow \frac{dy}{dt} = -\frac{x}{y}\frac{dx}{dt}.$$

Thus,

$$x\frac{dx}{dt} + (y - 40)\left(-\frac{x}{y}\frac{dx}{dt}\right) = s\frac{ds}{dt}$$

$$\frac{dx}{dt}\left[x - x + \frac{40x}{y}\right] = s\frac{ds}{dt}$$

$$\frac{dx}{dt} = \frac{sy}{40x}\frac{ds}{dt} = \frac{(40)(20)}{(40)(20\sqrt{3})}(-0.5) = -\frac{\sqrt{3}}{6} \text{ ft/sec}$$

$$\frac{dy}{dt} = -\frac{x}{y}\frac{dx}{dt} = -\frac{20\sqrt{3}}{20}\left(-\frac{\sqrt{3}}{6}\right) = \frac{1}{2} \text{ ft/sec.}$$

Chapter 3 Review Exercises

1. $f(x) = x^3 - 3x^2$

$f'(x) = 3x^2 - 6x = 3x(x-2)$

2. $f(x) = 2x - x^{-2}$

$f'(x) = 2 + 2x^{-3} = 2\left(1 + \frac{1}{x^3}\right) = \frac{2(x^3+1)}{x^3}$

3. $f(x) = x^{1/2} - x^{-1/2}$

$f'(x) = \frac{1}{2}x^{-1/2} + \frac{1}{2}x^{-3/2} = \frac{x+1}{2x^{3/2}}$

4. $f(x) = \frac{x+1}{x-1}$

$f'(x) = \frac{(x-1)(1) - (x+1)(1)}{(x-1)^2} = \frac{-2}{(x-1)^2}$

5. $g(t) = \frac{2}{3}t^{-2}$

$g'(t) = \frac{-4}{3}t^{-3} = \frac{-4}{3t^3}$

6. $h(x) = \frac{2}{9}x^{-2}$

$h'(x) = \frac{-4}{9}x^{-3} = \frac{-4}{9x^3}$

7. $f(x) = (x^3+1)^{1/2}$

$f'(x) = \frac{1}{2}(x^3+1)^{-1/2}(3x^2) = \frac{3x^2}{2\sqrt{x^3+1}}$

8. $f(x) = (x^2-1)^{1/3}$

$f'(x) = \frac{1}{3}(x^2-1)^{-2/3}(2x) = \frac{2x}{3(x^2-1)^{2/3}}$

9. $f(x) = (3x^2+7)(x^2-2x+3)$

$f'(x) = (3x^2+7)(2x-2) + (x^2-2x+3)(6x)$

$= 2(6x^3 - 9x^2 + 16x - 7)$

10. $f(x) = \left(x^2 + \frac{1}{x}\right)^5$

$f'(x) = 5\left(x^2 + \frac{1}{x}\right)^4\left(2x - \frac{1}{x^2}\right)$

11. $f(s) = (s^2-1)^{5/2}(s^3+5)$

$f'(s) = (s^2-1)^{5/2}(3s^2) + (s^3+5)^{5/2}(s^2-1)^{3/2}(2s)$

$= s(s^2-1)^{3/2}[3s(s^2-1) + 5(s^3+5)] = s(s^2-1)^{3/2}(8s^3 - 3s + 25)$

12. $h(\theta) = \frac{\theta}{(1-\theta)^3}$

$h'(\theta) = \frac{(1-\theta)^3 - \theta[3(1-\theta)^2(-1)]}{(1-\theta)^6} = \frac{(1-\theta)^2(1-\theta+3\theta)}{(1-\theta)^6} = \frac{2\theta+1}{(1-\theta)^4}$

13. $f(x) = \frac{-2x^2}{x-1}$

$f'(x) = \frac{(x-1)(-4x) - (-2x^2)(1)}{(x-1)^2}$

$= \frac{4x - 2x^2}{(x-1)^2} = \frac{2x(2-x)}{(x-1)^2}$

14. $f(x) = \frac{6x-5}{x^2+1}$

$f'(x) = \frac{(x^2+1)(6) - (6x-5)(2x)}{(x^2+1)^2}$

$= \frac{2(3+5x-3x^2)}{(x^2+1)^2}$

15. $f(x) = \frac{x^2+x-1}{x^2-1}$

$f'(x) = \frac{(x^2-1)(2x+1) - (x^2+x-1)(2x)}{(x^2-1)^2}$

$= \frac{-(x^2+1)}{(x^2-1)^2}$

16. $f(t) = t^2(t-1)^5$

$f'(t) = t^2[5(t-1)^4(1)] + (t-1)^5(2t)$

$= t(t-1)^4[5t + 2(t-1)]$

$= t(t-1)^4(7t-2)$

17. $f(x) = -2(1-4x^2)^2$

$$f'(x) = -4(1-4x^2)(-8x) = 32x(1-4x^2) = 32x - 128x^3$$

18. $f(x) = (x^2+2x-8)^2$

$$f'(x) = 2(x^2+2x-8)(2x+2) = 4(x^3+3x^2-6x-8)$$

19. $f(x) = (4-3x^2)^{-1}$

$$f'(x) = -(4-3x^2)^{-2}(-6x) = \frac{6x}{(4-3x^2)^2}$$

20. $f(x) = 9(3x^2-2x)^{-1}$

$$f'(x) = -9(3x^2-2x)^{-2}(6x-2) = \frac{18(1-3x)}{(3x^2-2x)^2}$$

21. $g(x) = (2x)(x+1)^{-1/2}$

$$g'(x) = (2x)\left[-\frac{1}{2}(x+1)^{-3/2}(1)\right] + (x+1)^{-1/2}(2) = \frac{-x}{(x+1)^{3/2}} + \frac{2}{(x+1)^{1/2}} \cdot \frac{(x+1)}{(x+1)} = \frac{x+2}{(x+1)^{3/2}}$$

22. $g(x) = x\sqrt{x^2+1} = x(x^2+1)^{1/2}$

$$g'(x) = x\left[\frac{1}{2}(x^2+1)^{-1/2}(2x)\right] + (x^2+1)^{1/2}(1)$$

$$= x^2(x^2+1)^{-1/2} + (x^2+1)^{1/2} = (x^2+1)^{-1/2}[x^2+(x^2+1)] = \frac{2x^2+1}{\sqrt{x^2+1}}$$

23. $f(t) = (t+1)^{1/2}(t+1)^{1/3} = (t+1)^{5/6}$

$$f'(t) = \frac{5}{6}(t+1)^{-1/6}(1) = \frac{5}{6(t+1)^{1/6}}$$

24. $y = \sqrt{3x}\,(x+2)^3$

$$y' = \sqrt{3x}\,[3(x+2)^2(1)] + (x+2)^3\left[\frac{1}{2}(3x)^{-1/2}(3)\right]$$

$$= \frac{3}{2}(3x)^{-1/2}(x+2)^2[6x+(x+2)]$$

$$= \frac{3(x+2)^2(7x+2)}{2\sqrt{3x}}$$

25. $f(x) = \sqrt{x^2+9}$

$$f'(x) = \frac{1}{2}(x^2+9)^{-1/2}(2x) = \frac{x}{\sqrt{x^2+9}}$$

$$f''(x) = \frac{\sqrt{x^2+9} - \left(x^2/\sqrt{x^2+9}\right)}{x^2+9} = \frac{x^2+9-x^2}{(x^2+9)^{3/2}} = \frac{9}{(x^2+9)^{3/2}}$$

26. $h(x) = x\sqrt{x^2-1}$

$$h'(x) = x\left(\frac{1}{2}\right)(x^2-1)^{-1/2}(2x) + (x^2-1)^{1/2}$$

$$= \frac{x^2}{\sqrt{x^2-1}} + \sqrt{x^2-1} = \frac{2x^2-1}{\sqrt{x^2-1}}$$

$$h''(x) = \frac{4x\sqrt{x^2-1} - (2x^2-1)\left(x/\sqrt{x^2-1}\right)}{x^2-1}$$

$$= \frac{4x^3-4x-2x^3+x}{(x^2-1)\sqrt{x^2-1}} = \frac{x(2x^2-3)}{(x^2-1)^{3/2}}$$

27. $f(t) = \dfrac{t}{(1-t)^2}$

$f'(t) = \dfrac{(1-t)^2 - 2t(1-t)(-1)}{(1-t)^4}$

$= \dfrac{t+1}{(1-t)^3}$

$f''(t) = \dfrac{(1-t)^3 - (t+1)(3)(1-t)^2(-1)}{(1-t)^6}$

$= \dfrac{2(t+2)}{(1-t)^4}$

28. $h(x) = x^2 + \dfrac{3}{x}$

$h'(x) = 2x - \dfrac{3}{x^2}$

$h''(x) = 2 + \dfrac{6}{x^3}$

29. $g(x) = \dfrac{6x-5}{x^2+1}$

$g'(x) = \dfrac{6(x^2+1) - (6x-5)(2x)}{(x^2+1)^2} = \dfrac{2(-3x^2+5x+3)}{(x^2+1)^2}$

$g''(x) = 2\left[\dfrac{(x^2+1)^2(-6x+5) - (-3x^2+5x+3)(2)(x^2+1)(2x)}{(x^2+1)^4}\right] = \dfrac{2(6x^3-15x^2-18x+5)}{(x^2+1)^3}$

30. $f(x) = 3x^4 - 6x^3 + 16x^2 - 14x + 21$

$f'(x) = 12x^3 - 18x^2 + 32x - 14$

$f''(x) = 36x^2 - 36x + 32 = 4(9x^2 - 9x + 8)$

31. $x^2 + 3xy + y^3 = 10$

$2x + 3xy' + 3y + 3y^2y' = 0$

$3(x + y^2)y' = -(2x + 3y)$

$y' = \dfrac{-(2x+3y)}{3(x+y^2)}$

32. $x^2 + 9y^2 - 4x + 3y - 7 = 0$

$2x + 18yy' - 4 + 3y' = 0$

$3(6y+1)y' = 4 - 2x$

$y' = \dfrac{4-2x}{3(6y+1)}$

33. $y\sqrt{x} - x\sqrt{y} = 16$

$y\left(\dfrac{1}{2}x^{-1/2}\right) + x^{1/2}y' - x\left(\dfrac{1}{2}y^{-1/2}y'\right) - y^{1/2} = 0$

$\left(\sqrt{x} - \dfrac{x}{2\sqrt{y}}\right)y' = \sqrt{y} - \dfrac{y}{2\sqrt{x}}$

$\dfrac{2\sqrt{xy} - x}{2\sqrt{y}}y' = \dfrac{2\sqrt{xy} - y}{2\sqrt{x}}$

$y' = \dfrac{2\sqrt{xy} - y}{2\sqrt{x}} \cdot \dfrac{2\sqrt{y}}{2\sqrt{xy} - x} = \dfrac{2y\sqrt{x} - y\sqrt{y}}{2x\sqrt{y} - x\sqrt{x}}$

34. $y^2 + x^2 - 6y - 2x - 5 = 0$

$2yy' + 2x - 6y' - 2 = 0$

$y' = \dfrac{1-x}{y-3}$

35. $y^2 - x^2 = 25$

$2yy' - 2x = 0$

$y' = \dfrac{x}{y}$

36.

$$y^2 = x^3 - x^2y + xy - y^2$$
$$0 = x^3 - x^2y + xy - 2y^2$$
$$0 = 3x^2 - x^2y' - 2xy + xy' + y - 4yy'$$
$$(x^2 - x + 4y)y' = 3x^2 - 2xy + y$$
$$y' = \frac{3x^2 - 2xy + y}{x^2 - x + 4y}$$

37. $y = (x + 3)^3$

$y' = 3(x + 3)^2$

At $(-2, 1)$: $y' = 3$

Tangent line: $y - 1 = 3(x + 2)$

$3x - y + 7 = 0$

Normal line: $y - 1 = -\frac{1}{3}(x + 2)$

$x + 3y - 1 = 0$

38. $y = (x - 2)^2$

$y' = 2(x - 2)$

At $(2, 0)$: $y' = 0$

Tangent line: $y = 0$

Normal line: $x = 2$

39. $x^2 + y^2 = 20$

$2x + 2yy' = 0$

$y' = -\frac{x}{y}$

At $(2, 4)$: $y' = -\frac{1}{2}$

Tangent line: $y - 4 = -\frac{1}{2}(x - 2)$

$x + 2y - 10 = 0$

Normal line: $y - 4 = 2(x - 2)$

$2x - y = 0$

40. $x^2 - y^2 = 16$

$2x - 2yy' = 0$

$y' = \frac{x}{y}$

At $(5, 3)$: $y' = \frac{5}{3}$

Tangent line: $y - 3 = \frac{5}{3}(x - 5)$

$5x - 3y - 16 = 0$

Normal line: $y - 3 = -\frac{3}{5}(x - 5)$

$3x + 5y - 30 = 0$

41. $y = (x - 2)^{2/3}$

$y' = \frac{2}{3}(x - 2)^{-1/3} = \frac{2}{3\sqrt[3]{x - 2}}$

At $(3, 1)$: $y' = \frac{2}{3}$

Tangent line: $y - 1 = \frac{2}{3}(x - 3)$

$2x - 3y - 3 = 0$

Normal line: $y - 1 = -\frac{3}{2}(x - 3)$

$3x + 2y - 11 = 0$

42. $y = \frac{2x}{1 - x^2}$

$y' = \frac{2(1 - x^2) - 2x(-2x)}{(1 - x^2)^2} = \frac{2(x^2 + 1)}{(1 - x^2)^2}$

At $(0, 0)$: $y' = 2$

Tangent line: $y = 2x$

Normal line: $y = -\frac{1}{2}x$

43. $f(x) = \frac{1}{3}x^3 + x^2 - x - 1$

$f'(x) = x^2 + 2x - 1$

(a) $x^2 + 2x - 1 = -1$

$x(x + 2) = 0$

$(0, -1), \left(-2, \frac{7}{3}\right)$

(b) $x^2 + 2x - 1 = 2$

$(x + 3)(x - 1) = 0$

$(-3, 2), \left(1, -\frac{2}{3}\right)$

(c) $x^2 + 2x - 1 = 0$

$(x + 1)^2 = 2$

$x = -1 \pm \sqrt{2}$

$\left(-1 + \sqrt{2}, \frac{2(1 - 2\sqrt{2})}{3}\right)$,

$\left(-1 - \sqrt{2}, \frac{2(1 + 2\sqrt{2})}{3}\right)$

44. $f(x) = x^2 + 1$

$f'(x) = 2x$

(a) $2x = -1$

$x = -\frac{1}{2}$

$\left(-\frac{1}{2}, \frac{5}{4}\right)$

(b) $2x = 0$

$x = 0$

$(0, 1)$

(c) $2x = 1$

$x = \frac{1}{2}$

$\left(\frac{1}{2}, \frac{5}{4}\right)$

45. $f(x) = \frac{1}{x^2}$

$$f'(x) = \lim_{\Delta x \to 0} \frac{f(x + \Delta x) - f(x)}{\Delta x} = \lim_{\Delta x \to 0} \frac{\frac{1}{(x + \Delta x)^2} - \frac{1}{x^2}}{\Delta x} = \lim_{\Delta x \to 0} \frac{x^2 - (x - \Delta x)^2}{\Delta x(x + \Delta x)^2 x^2}$$

$$= \lim_{\Delta x \to 0} \frac{-2x(\Delta x) - (\Delta x)^2}{\Delta x(x + \Delta x)^2 x^2} = \lim_{\Delta x \to 0} \frac{-2x - \Delta x}{(x + \Delta x)^2 x^2} = -\frac{2x}{x^4} = -\frac{2}{x^3}$$

46. $f(x) = \frac{x + 1}{x - 1}$

$$f'(x) = \lim_{\Delta x \to 0} \frac{f(x + \Delta x) - f(x)}{\Delta x} = \lim_{\Delta x \to 0} \frac{\frac{x + \Delta x + 1}{x + \Delta x - 1} - \frac{x + 1}{x - 1}}{\Delta x}$$

$$= \lim_{\Delta x \to 0} \frac{(x + \Delta x + 1)(x - 1) - (x + \Delta x - 1)(x + 1)}{\Delta x(x + \Delta x - 1)(x - 1)}$$

$$= \lim_{\Delta x \to 0} \frac{(x^2 + x\Delta x + x - x - \Delta x - 1) - (x^2 + x\Delta x - x + x + \Delta x - 1)}{\Delta x(x + \Delta x - 1)(x - 1)}$$

$$= \lim_{\Delta x \to 0} \frac{-2\Delta x}{\Delta x(x + \Delta x - 1)(x - 1)} = \lim_{\Delta x \to 0} \frac{-2}{(x + \Delta x - 1)(x - 1)} = \frac{-2}{(x - 1)^2}$$

47. $f(x) = \sqrt{x + 2}$

$$f'(x) = \lim_{\Delta x \to 0} \frac{f(x + \Delta x) - f(x)}{\Delta x} = \lim_{\Delta x \to 0} \frac{\sqrt{x + \Delta x + 2} - \sqrt{x + 2}}{\Delta x} \cdot \frac{\sqrt{x + \Delta x + 2} + \sqrt{x + 2}}{\sqrt{x + \Delta x + 2} + \sqrt{x + 2}}$$

$$= \lim_{\Delta x \to 0} \frac{(x + \Delta x + 2) - (x + 2)}{\Delta x\left(\sqrt{x + \Delta x + 2} + \sqrt{x + 2}\right)} = \lim_{\Delta x \to 0} \frac{1}{\sqrt{x + \Delta x + 2} + \sqrt{x + 2}} = \frac{1}{2\sqrt{x + 2}}$$

48. $f(x) = \dfrac{1}{\sqrt{x}}$

$$f'(x) = \lim_{\Delta x \to 0} \frac{f(x+\Delta x) - f(x)}{\Delta x} = \lim_{\Delta x \to 0} \frac{\dfrac{1}{\sqrt{x+\Delta x}} - \dfrac{1}{\sqrt{x}}}{\Delta x}$$

$$= \lim_{\Delta x \to 0} \frac{\sqrt{x} - \sqrt{x+\Delta x}}{\Delta x\sqrt{x}\sqrt{x+\Delta x}} \cdot \frac{\sqrt{x} + \sqrt{x+\Delta x}}{\sqrt{x} + \sqrt{x+\Delta x}} = \lim_{\Delta x \to 0} \frac{x - (x+\Delta x)}{\Delta x\sqrt{x}\sqrt{x+\Delta x}\left(\sqrt{x} + \sqrt{x+\Delta x}\right)}$$

$$= \lim_{\Delta x \to 0} \frac{-1}{\sqrt{x}\sqrt{x+\Delta x}\left(\sqrt{x} + \sqrt{x+\Delta x}\right)} = -\frac{1}{\sqrt{x}\sqrt{x}\left(\sqrt{x} + \sqrt{x}\right)} = -\frac{1}{2x\sqrt{x}} = -\frac{1}{2x^{3/2}}$$

49. $s(t) = t + \dfrac{1}{t+1}$

$v(t) = s'(t) = 1 - \dfrac{1}{(t+1)^2}$

$a(t) = s''(t) = \dfrac{2}{(t+1)^3}$

50. $s(t) = \dfrac{1}{t^2 + 2t + 1} \Rightarrow s(t) = (t+1)^{-2}$

$v(t) = s'(t) = -\dfrac{2}{(t+1)^3}$

$a(t) = s''(t) = \dfrac{6}{(t+1)^4}$

51. $T = 700(t^2 + 4t + 10)^{-1}$

$T' = \dfrac{-1400(t+2)}{(t^2 + 4t + 10)^2}$

(a) When $t = 1$, $T' = \dfrac{-1400(1+2)}{(1+4+10)^2} \approx -18.667$

(b) When $t = 3$, $T' = \dfrac{-1400(3+2)}{(9+12+10)^2} \approx -7.284$

(c) When $t = 5$, $T' = \dfrac{-1400(5+2)}{(25+20+10)^2} \approx -3.240$

(d) When $t = 10$, $T' = \dfrac{-1400(10+2)}{(100+40+10)^2} \approx -0.747$

52. $v = \sqrt{2gh} = \sqrt{2(32)h} = 8\sqrt{h}$

$\dfrac{dv}{dh} = \dfrac{4}{\sqrt{h}}$

(a) When $h = 9$, $\dfrac{dv}{dh} = \dfrac{4}{3}$ ft/sec.

(b) When $h = 4$, $\dfrac{dv}{dh} = 2$ ft/sec.

53. Use the equation for a free-falling object from Section 3.2.

$s(t) = -16t^2 + v_0 t + s_0$

Consider $s_0 = 0$, then $s(t) = -16t^2 + v_0 t$. This is a parabola opening downward. Therefore, the vertex will be the highest point on the graph. Completing the square, we have

$$s(t) = -16\left(t^2 - \frac{v_0}{16}t + \frac{v_0^2}{32^2}\right) + \frac{v_0^2}{64}$$

$$= -16\left(t - \frac{v_0}{32}\right)^2 + \frac{v_0^2}{64}.$$

Thus, the vertex is $(v_0/32,\ v_0^2/64)$ and we let $v_0^2/64 = 49 \Rightarrow v_0 = 56$ ft/sec.

54. $s(t) = -16t^2 + 14{,}400 = 0$

$16t^2 = 14{,}400$

$t = 30$ sec

Since 600 mph $= \frac{1}{6}$ mi/sec, in 30 seconds the bomb will move horizontally $\left(\frac{1}{6}\right)(30) = 5$ miles.

55. (a)

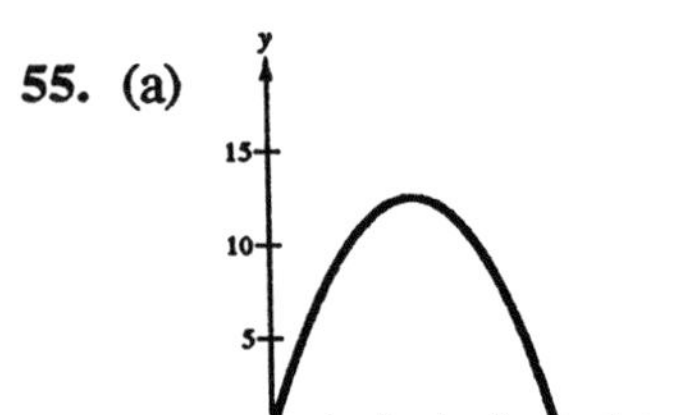

(b) Total horizontal distance: 50 ft

(c) Ball reaches its maximum height when x is 25 ft.

(d)
$$y = x - 0.02x^2$$
$$y' = 1 - 0.04x$$
$$y'(0) = 1$$
$$y'(10) = 0.6$$
$$y'(25) = 0$$
$$y'(30) = -0.2$$
$$y'(50) = -1$$

(e) $y'(25) = 0$

56. (a)

(b) $y = x - \dfrac{32}{v_0^2}x^2$

$= x\left(1 - \dfrac{32}{v_0^2}x\right) = 0$ if $x = 0$ or $x = \dfrac{v_0^2}{32}$.

Projectile strikes the ground when $x = \dfrac{v_0^2}{32}$.

Projectile reaches its maximum height at

$x = \dfrac{v_0^2}{64}$.

(c)
$$y' = 1 - \frac{64}{v_0^2}x$$
$$f'\left(\frac{v_0^2}{64}\right) = 0$$

57. $y = x - \dfrac{32}{v_0^2}x^2 = x\left(1 - \dfrac{32}{v_0^2}x\right) = 0$ when $x = 0$ and $x = \dfrac{x_0^2}{32}$. Therefore, the range is $x = \dfrac{v_0^2}{32}$. When the initial velocity is doubled the range is $x = \dfrac{(2v_0)^2}{32} = \dfrac{4v_0^2}{32}$ or four times the initial range. The maximum height occurs when $x = \dfrac{v_0^2}{64}$. The maximum height is

$$y\left(\frac{v_0^2}{64}\right) = \frac{v_0^2}{64} - \frac{32}{v_0^2}\left(\frac{v_0^2}{64}\right)^2 = \frac{v_0^2}{64} - \frac{v_0^2}{128} = \frac{v_0^2}{128}.$$

If the initial velocity is doubled, the maximum height is $y\left[\dfrac{(2v_0)^2}{64}\right] = \dfrac{(2v_0)^2}{128} = 4\left(\dfrac{v_0^2}{32}\right)$ or four times the original maximum height.

58. $v_0 = 70$ ft/sec

Range: $x = \dfrac{v_0^2}{32} = \dfrac{(70)^2}{32} = 153.125$ ft

Maximum height: $y = \dfrac{v_0^2}{128} = \dfrac{(70)^2}{128} \approx 38.28$ ft

59. $y = \sqrt{x}$

$$\frac{dy}{dt} = 2 \text{ units/sec}$$

$$\frac{dy}{dt} = \frac{1}{2\sqrt{x}}\frac{dx}{dt} \Rightarrow \frac{dx}{dt} = 2\sqrt{x}\frac{dy}{dt} = 4\sqrt{x}$$

(a) When $x = \frac{1}{2}$, $\frac{dx}{dt} = 2\sqrt{2}$ units/sec.

(b) When $x = 1$, $\frac{dx}{dt} = 4$ units/sec.

(c) When $x = 4$, $\frac{dx}{dt} = 8$ units/sec.

60. $y = \sqrt{x}$

$$L^2 = x^2 + y^2$$

$$\frac{dy}{dt} = 2 \text{ units/sec}$$

$$L^2 = y^4 + y^2$$

$$2L\frac{dL}{dt} = (4y^3 + 2y)\frac{dy}{dt}$$

$$\frac{dL}{dt} = \frac{4y^3 + 2y}{2L}\frac{dy}{dt} = \frac{4y^3 + 2y}{L} = \frac{(4x + 2)\sqrt{x}}{L}$$

(a) When $x = \frac{1}{2}$, $L = \sqrt{\left(\frac{1}{2}\right)^2 + \left(\frac{1}{\sqrt{2}}\right)^2} = \frac{\sqrt{3}}{2}$ and $\frac{dL}{dt} = \frac{(2+2)(1/\sqrt{2})}{\sqrt{3}/2} = \frac{8}{\sqrt{6}}$ units/sec.

(b) When $x = 1$, $L = \sqrt{(1)^2 + (1)^2} = \sqrt{2}$ and $\frac{dL}{dt} = \frac{(4+2)(1)}{\sqrt{2}} = 3\sqrt{2}$ units/sec.

(c) When $x = 4$, $L = \sqrt{(4)^2 + (2)^2} = 2\sqrt{5}$ and $\frac{dL}{dt} = \frac{(16+2)(2)}{2\sqrt{5}} = \frac{18}{\sqrt{5}}$ units/sec.

61. $\frac{s}{h} = \frac{1/2}{2}$

$$s = \frac{1}{4}h$$

$$\frac{dV}{dt} = 1$$

Width of water when at depth h is $w = 2 + 2\left(\frac{1}{4}h\right) = \frac{4+h}{2}$.

$$V = \frac{5}{2}\left(2 + \frac{4+h}{2}\right)h = \frac{5}{4}(8+h)h$$

$$\frac{dV}{dt} = \frac{5}{2}(4+h)\frac{dh}{dt}$$

$$\frac{dh}{dt} = \frac{2(dV/dt)}{5(4+h)}$$

When $h = 1$, $\frac{dh}{dt} = \frac{2}{25}$ ft/min.

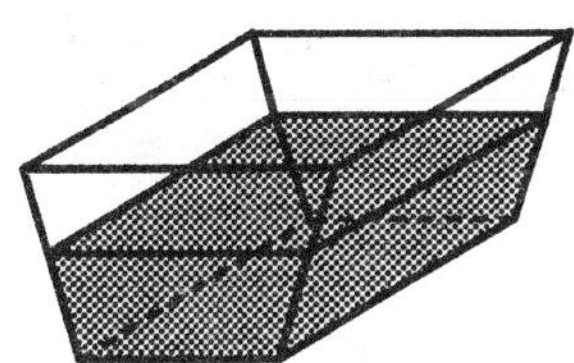

62. $g = \sqrt{x(x+n)} = (x^2 + nx)^{1/2}$

$$\frac{dg}{dx} = \frac{1}{2}(x^2 + nx)^{-1/2}(2x + n) = \frac{2x + n}{2\sqrt{x(x+n)}}$$

$$\frac{a}{g} = \frac{[x + (x+n)]/2}{\sqrt{x(x+n)}} = \frac{2x+n}{2\sqrt{x(x+n)}}$$

Thus, $\dfrac{dg}{dx} = \dfrac{a}{g}$.

63. $y = \dfrac{1}{x}$

$$y' = -x^{-2} = \frac{-1}{x^2}$$

$$y'' = 2 \cdot 1x^{-3} = \frac{2 \cdot 1}{x^3}$$

$$y''' = -3 \cdot 2 \cdot 1x^{-4} = \frac{-(3 \cdot 2 \cdot 1)}{x^4}$$

$$y^{(4)} = 4 \cdot 3 \cdot 2 \cdot 1x^{-5} = \frac{4 \cdot 3 \cdot 2 \cdot 1}{x^5}$$

$$\vdots$$

$$y^{(n)} = (-1)^n n(n-1) \ldots 3 \cdot 2 \cdot 1x^{-(n+1)} = \frac{(-1)^n n!}{x^{n+1}}$$

64. $f(x) = 4 - |x - 2|$

(a) Continuous at $x = 2$

(b) Not differentiable at $x = 2$ because of sharp turn in graph

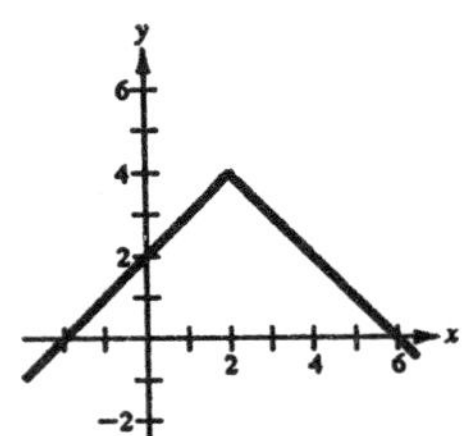

65. $f(x) = \begin{cases} x^2 + 4x + 2, & \text{if } x < -2 \\ 1 - 4x - x^2, & \text{if } x \geq -2 \end{cases}$

(a) Nonremovable discontinuity at $x = -2$

(b) Not differentiable at $x = -2$ because function is discontinuous there

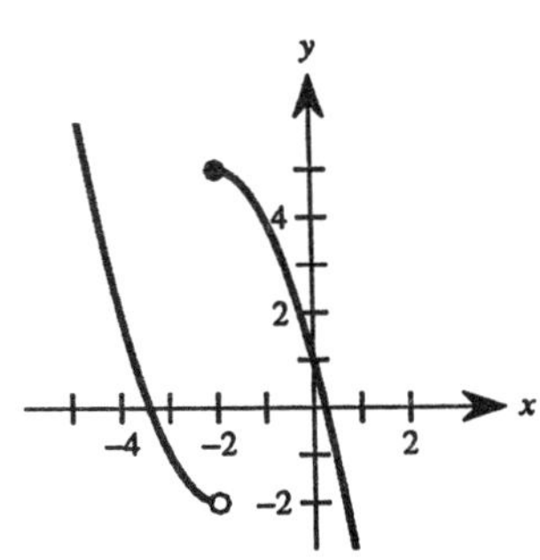

CHAPTER 4
Applications of Differentiation

Section 4.1 Extrema on an Interval

1. $f(x) = \dfrac{x^2}{x^2+4}$

$f'(x) = \dfrac{(x^2+4)(2x) - (x^2)(2x)}{(x^2+4)^2}$

$= \dfrac{8x}{(x^2+4)^2}$

$f'(0) = 0$

2. $f(x) = -x^2 + 4x$

$f'(x) = -2x + 4$

$f'(2) = 0$

3. $f(x) = x + \dfrac{32}{x^2}$

$f'(x) = 1 - \dfrac{64}{x^3}$

$f'(4) = 0$

4. $f(x) = -3x\sqrt{x+1}$

$f'(x) = -3x\left[\frac{1}{2}(x+1)^{-1/2}\right] + \sqrt{x+1}(-3) = -\frac{3}{2}(x+1)^{-1/2}[x + 2(x+1)] = -\frac{3}{2}(x+1)^{-1/2}(3x+2)$

$f'\left(-\frac{2}{3}\right) = 0$

5. $f(x) = (x+2)^{2/3}$

$f'(x) = \frac{2}{3}(x+2)^{-1/3}$

$f'(-2)$ is undefined.

6. Using the limit definition of the derivative,

$$\lim_{x\to 0^-} \frac{f(x) - f(0)}{x - 0} = \lim_{x\to 0^-} \frac{(4 - |x|) - 4}{x} = 1$$

$$\lim_{x\to 0^+} \frac{f(x) - f(0)}{x - 0} = \lim_{x\to 0^+} \frac{(4 - |x|) - 4}{x - 0} = -1$$

$f'(0)$ does not exist, since the one-sided derivatives are not equal.

7. $f(x) = 2(3 - x)$

$f'(x) = -2$

Absolute minimum: $(2, 2)$

Absolute maximum: $(-1, 8)$

8. $f(x) = \dfrac{2x+5}{3}$

$f'(x) = \dfrac{2}{3}$

Absolute minimum: $\left(0, \frac{5}{3}\right)$

Absolute maximum: $(5, 5)$

9. $f(x) = -x^2 + 3x$

$f'(x) = -2x + 3$

Absolute minima: $(0, 0)$, $(3, 0)$

Absolute maximum: $\left(\frac{3}{2}, \frac{9}{4}\right)$

10. $f(x) = x^2 + 2x - 4$

$f'(x) = 2x + 2$

Absolute minimum: $(-1, -5)$

Absolute maximum: $(1, -1)$

11. $f(x) = x^3 - 3x^2$

$f'(x) = 3x^2 - 6x$

Absolute minima:

$(-1, -4)$, $(2, -4)$

Absolute maxima:

$(0, 0)$, $(3, 0)$

12. $f(x) = x^3 - 12x$

$f'(x) = 3x^2 - 12$

Absolute minimum: $(2, -16)$

Absolute maximum: $(4, 16)$

13. $f(x) = 3x^{2/3} - 2x$

$f'(x) = 2x^{-1/3} - 2$

Absolute minimum: (0, 0)

Absolute maximum: (−1, 5)

14. $g(x) = \sqrt[3]{x}$

$$g'(x) = \frac{1}{3\sqrt[3]{x^2}}$$

Absolute minimum: (−1, −1)

Absolute maximum: (1, 1)

15. $h(t) = 4 - |t - 4|$

$$h'(t) = \frac{t-4}{|t-4|}$$

Absolute minimum: (1, 1)

Absolute maximum: (4, 4)

16. $g(t) = \dfrac{t^2}{t^2 + 3}$

$$g'(t) = \frac{6t}{(t^2+3)^2}$$

Absolute minimum: (0, 0)

Absolute maxima: $\left(-1, \frac{1}{4}\right), \left(1, \frac{1}{4}\right)$

17. $h(s) = \dfrac{1}{s-2}$

$$h'(s) = \frac{-1}{(s-2)^2}$$

Absolute minimum: (1, −1)

Absolute maximum: $\left(0, -\frac{1}{2}\right)$

18. $h(t) = \dfrac{t}{t-2}$

$$h'(t) = \frac{-2}{(t-2)^2}$$

Absolute minimum: $\left(5, \frac{5}{3}\right)$

Absolute maximum: (3, 3)

19. $f(x) = \dfrac{1}{x^2}$

f is continuous on [1, 2] but not on [0, 2].

20. $y = f(x) = \dfrac{1}{x+1}$

f is continuous on [0, 2] but not on [−2, 0].

21. (a) Yes
(b) No

22. (a) Yes
(b) No

23. (a) No
(b) Yes

24. (a) No
(b) Yes

25. (a) Minimum: (0, −3)
Maximum: (2, 1)
(b) Minimum: (0, −3)
(c) Maximum: (2, 1)
(d) No extrema

26. (a) Minimum: (4, 1)
Maximum: (1, 4)
(b) Maximum: (1, 4)
(c) Minimum: (4, 1)
(d) No extrema

27. $f(x) = \dfrac{1}{x^2+1}, \quad [0, 3]$

$$f'(x) = \frac{-2x}{(x^2+1)^2}$$

$$f''(x) = \frac{-2(1-3x^2)}{(x^2+1)^3}$$

$$f'''(x) = \frac{24x - 24x^3}{(x^2+1)^4}$$

Setting $f''' = 0$, we have $x = 0, \pm 1$.

$|f''(0)| = 2$ is the maximum value.

28. Using the results of Exercise 27 on the interval $\left[\frac{1}{2}, 3\right]$, $|f''(1)| = \frac{1}{2}$ is the maximum value.

29. $f(x) = (1+x^3)^{1/2}$, $[0, 2]$

$f'(x) = \frac{3}{2}x^2(1+x^3)^{-1/2}$

$f''(x) = \frac{3}{4}(x^4+4x)(1+x^3)^{-3/2}$

$f'''(x) = -\frac{3}{8}(x^6+20x^3-8)(1+x^3)^{-5/2}$

Setting $f''' = 0$, we have $x^6+20x^3-8=0$.

$$x^3 = \frac{-20 \pm \sqrt{400-4(1)(-8)}}{2} \Rightarrow x = \sqrt[3]{-10 \pm \sqrt{108}}$$

$\left|f''\left(\sqrt[3]{-10+\sqrt{108}}\right)\right| \approx 1.47$ is the maximum value.

30. $f(x) = 3x^5 - 10x^3$, $[0, 1]$

$f'(x) = 15x^4 - 30x^2$

$f''(x) = 60x^3 - 60x$

$f'''(x) = 180x^2 - 60 = 60(3x^2-1)$

$f'''(x) = 0$ when $x = \frac{\sqrt{3}}{3}$.

$\left|f''\left(\frac{\sqrt{3}}{3}\right)\right| = \frac{40\sqrt{3}}{3}$ is the maximum value.

31. $f(x) = 15x^4 - \left(\frac{2x-1}{2}\right)^6$, $[0, 1]$

$f'(x) = 60x^3 - 6\left(\frac{2x-1}{2}\right)^5$

$f''(x) = 180x^2 - 30\left(\frac{2x-1}{2}\right)^4$

$f'''(x) = 360x - 120\left(\frac{2x-1}{2}\right)^3$

$f^{(4)}(x) = 360 - 360\left(\frac{2x-1}{2}\right)^2$

$f^{(5)}(x) = -720\left(\frac{2x-1}{2}\right)$

$f^{(5)}(x) = 0$ when $x = \frac{1}{2}$.

$\left|f^{(4)}\left(\frac{1}{2}\right)\right| = 360$ is the maximum value.

32. $f(x) = x^5 - 5x^4 + 20x^3 + 600$, $[0, \frac{3}{2}]$

$f'(x) = 5x^4 - 20x^3 + 60x^2$

$f''(x) = 20x^3 - 60x^2 + 120x$

$f'''(x) = 60x^2 - 120x + 120$

$f^{(4)}(x) = 120x - 120$

$f^{(5)}(x) = 120$

$|f^{(4)}(0)| = 120$ is the maximum value.

33. $f(x) = (x+1)^{2/3}$, $[0, 2]$

$f'(x) = \frac{2}{3}(x+1)^{-1/3}$

$f''(x) = -\frac{2}{9}(x+1)^{-4/3}$

$f'''(x) = \frac{8}{27}(x+1)^{-7/3}$

$f^{(4)}(x) = -\frac{56}{81}(x+1)^{-10/3}$

$f^{(5)}(x) = \frac{560}{243}(x+1)^{-13/3}$

$|f^{(4)}(0)| = \frac{56}{81}$ is the maximum value.

34. $f(x) = \dfrac{1}{x^2+1}$, $[-1, 1]$

$f'''(x) = \dfrac{24x - 24x^3}{(x^2+1)^4}$ (See Exercise 29.)

$f^{(4)}(x) = \dfrac{24(5x^4 - 10x^2 + 1)}{(x^2+1)^5}$

$f^{(5)}(x) = \dfrac{-240(3x^4 - 10x^3 + 3)}{(x^2+1)^6}$

$|f^{(4)}(0)| = 24$ is the maximum value.

35. $P = VI - RI^2 = 12I - 0.5I^2$, $0 \le I \le 15$

$P = 0$ when $I = 0$.

$P = 67.5$ when $I = 15$.

$P' = 12 - I = 0$

Critical number: $I = 12$ amps

When $I = 12$ amps, $P = 72$, the maximum output.

36. $C = 2x + \dfrac{300{,}000}{x}$, $0 \le x \le 300$

$C(0)$ is undefined.

$C(300) = 1600$

$C' = 2 - \dfrac{300{,}000}{x^2} = 0$

$2x^2 = 300{,}000$

$x^2 = 150{,}000$

$x = 100\sqrt{15} \approx 387 > 300$

C is minimized when $x = 300$ units.

37. $f(x) = 3.2x^5 + 5x^3 - 3.5x$, $[0, 1]$

$f'(x) = 16x^4 + 15x^2 - 3.5$

$16x^4 + 15x^2 - 3.5 = 0$

$$x^2 = \frac{-15 \pm \sqrt{(15)^2 - 4(16)(-3.5)}}{2(16)} = \frac{-15 \pm \sqrt{449}}{32}$$

$$x = \sqrt{\frac{-15 + \sqrt{449}}{32}} \approx 0.4398$$

$f(0) = 0$

$f(1) = 4.7$

$$f\left(\sqrt{\frac{-15 + \sqrt{449}}{32}}\right) \approx -1.0613$$

Relative minimum: $(0.4398, -1.0613)$

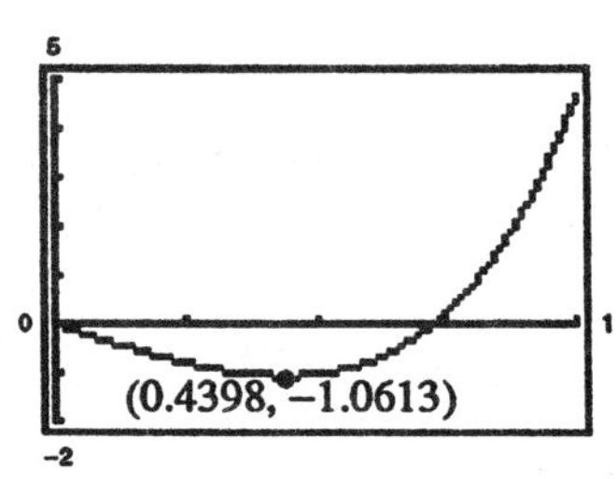

38. $f(x) = \frac{4}{3}x\sqrt{3-x}, \quad [0, 3]$

$$f'(x) = \frac{4}{3}\left[x\left(\frac{1}{2}\right)(3-x)^{-1/2}(-1) + (3-x)^{1/2}(1)\right]$$

$$= \frac{4}{3}(3-x)^{-1/2}\left(\frac{1}{2}\right)[-x + 2(3-x)] = \frac{2(6-3x)}{3\sqrt{3-x}} = \frac{6(2-x)}{3\sqrt{3-x}} = \frac{2(2-x)}{\sqrt{3-x}}$$

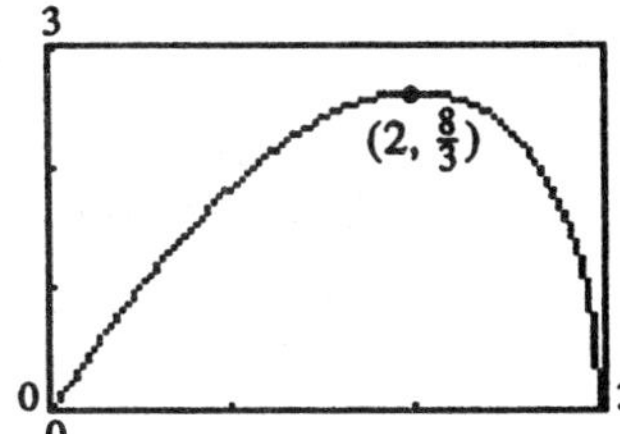

Critical number: $x = 2$

$f(0) = 0$

$f(3) = 0$

$f(2) = \frac{8}{3}$

Relative maximum: $\left(2, \frac{8}{3}\right)$

Section 4.2 Rolle's Theorem and the Mean Value Theorem

1. Rolle's Theorem does not apply to $f(x) = 1 - |x - 1|$ over $[0, 2]$ since f is not differentiable at $x = 1$.

2. $f(x) = \frac{x^2 - 4}{x^2}$

$f(-2) = f(2) = 0$

However, Rolle's Theorem does not apply since the function is not differentiable at $x = 0$.

3. $f(x) = x^2 - 2x, \quad [0, 2]$

$f(0) = f(2) = 0$

f is continuous on $[0, 2]$. f is differentiable on $(0, 2)$. Rolle's Theorem applies.

$f'(x) = 2x - 2$

$2x - 2 = 0 \Rightarrow x = 1$

c-value: 1

4. $f(x) = x^2 - 3x + 2, \quad [1, 2]$

$f(1) = f(2) = 0$

f is continuous on $[1, 2]$. f is differentiable on $(1, 2)$. Rolle's Theorem applies.

$f'(x) = 2x - 3$

$2x - 3 = 0 \Rightarrow x = \frac{3}{2} = 1.5$

c-value: $\frac{3}{2} = 1.5$

5. $f(x) = (x-1)(x-2)(x-3), \quad [1, 3]$

$f(1) = f(2) = f(3) = 0$

f is continuous on $[1, 2]$ and $[2, 3]$. f is differentiable on $(1, 2)$ and $(2, 3)$. Rolle's Theorem applies.

$$f(x) = x^3 - 6x^2 + 11x - 6$$

$$f'(x) = 3x^2 - 12x + 11$$

$3x^2 - 12x + 11 = 0 \Rightarrow x = \frac{6 \pm \sqrt{3}}{3}$

In $(1, 2)$, $c = \left(6 - \sqrt{3}\right)/3$.

In $(2, 3)$, $c = \left(6 + \sqrt{3}\right)/3$.

6. $f(x) = (x-3)(x+1)^2, \quad [-1, 3]$

$f(-1) = f(3) = 0$

f is continuous on $[-1, 3]$. f is differentiable on $(-1, 3)$. Rolle's Theorem applies.

$$f'(x) = (x-3)(2)(x+1) + (x+1)^2$$

$$= (x+1)[2x - 6 + x + 1]$$

$$= (x+1)(3x-5)$$

c-value: $\frac{5}{3}$

7. $f(x) = |x| - 1, \quad [-1, 1]$

$f(-1) = f(1) = 0$

f is continuous on $[-1, 1]$.
f is not differentiable on$(-1, 1)$ since $f'(0)$ does not exist. Rolle's Theorem does not apply.

8. $f(x) = 3 - |x - 3|, \quad [0, 6]$

$f(0) = f(6) = 0$

f is continuous on $[0, 6]$. f is not differentiable on $(0, 6)$ since $f'(3)$ does not exist. Rolle's Theorem does not apply.

9. $f(x) = x^{2/3} - 1, \quad [-8, 8]$

$f(-8) = f(8) = 3$

f is continuous on $[-8, 8]$. f is not differentiable on $(-8, 8)$ since $f'(0)$ does not exist. Rolle's Theorem does not apply.

10. $f(x) = x - x^{1/3}, \quad [0, 1]$

$f(0) = f(1) = 0$

f is continuous on $[0, 1]$. f is differentiable on $(0, 1)$. (*Note:* f is not differentiable at $x = 0$.) Rolle's Theorem applies.

$$f'(x) = 1 - \frac{1}{3\sqrt[3]{x^2}}$$

$$1 = \frac{1}{3\sqrt[3]{x^2}}$$

$$\sqrt[3]{x^2} = \frac{1}{3}$$

$$x^2 = \frac{1}{27}$$

$$x = \sqrt{\frac{1}{27}} = \frac{\sqrt{3}}{9}$$

c-value: $\dfrac{\sqrt{3}}{9}$

11. $f(x) = \dfrac{x^2 - 2x - 3}{x + 2}, \quad [-1, 3]$

$f(-1) = f(3) = 0$

f is continuous on $[-1, 3]$. (*Note:* The discontinuity, $x = -2$, is not in the interval.) f is differentiable on $(-1, 3)$. Rolle's Theorem applies.

$$f'(x) = \frac{(x+2)(2x-2) - (x^2 - 2x - 3)(1)}{(x+2)^2}$$

$$= \frac{x^2 + 4x - 1}{(x+2)^2}$$

$$x = \frac{-4 \pm 2\sqrt{5}}{2} = -2 \pm \sqrt{5}$$

c-value: $-2 + \sqrt{5}$

12. $f(x) = \dfrac{x^2 - 1}{x}, \quad [-1, 1]$

$f(-1) = f(1) = 0$

f is not continous on $[-1, 1]$ since $f(0)$ does not exist. Rolle's Theorem does not apply.

13. $f(x) = x^2$ is continuous and differentiable on $[-2, 1]$.

$$\frac{f(1) - f(-2)}{1 - (-2)} = \frac{1 - 4}{3}$$

$$= -1$$

$f'(x) = 2x = -1$ when $x = -\frac{1}{2}$.
Therefore, $c = -\frac{1}{2}$.

14. $f(x) = x(x^2 - x - 2)$ is continuous and differentiable on $[-1, 1]$.

$$\frac{f(1) - f(-1)}{1 - (-1)} = -1$$

$$f'(x) = 3x^2 - 2x - 2$$

$$= -1$$

$$(3x + 1)(x - 1) = 0$$

$$c = -\frac{1}{3}$$

15. $f(x) = x^{2/3}$ is continuous on $[0, 1]$ and differentiable on $(0, 1)$.

$$\frac{f(1) - f(0)}{1 - 0} = 1$$

$$f'(x) = \frac{2}{3}x^{-1/3} = 1$$

$$x = \left(\frac{2}{3}\right)^3 = \frac{8}{27}$$

$$c = \frac{8}{27}$$

16. $f(x) = (x + 1)/x$ is continuous and differentiable on $[1/2, 2]$.

$$\frac{f(2) - f(1/2)}{2 - (1/2)} = \frac{(3/2) - 3}{3/2} = -1$$

$$f'(x) = \frac{-1}{x^2} = -1$$

$$x^2 = 1$$

$$c = 1$$

17. $f(x) = x/(x + 1)$ is continuous and differentiable on $[-1/2, 2]$.

$$\frac{f(2) - f(-1/2)}{2 - (-1/2)} = \frac{(2/3) - (-1)}{5/2} = \frac{2}{3}$$

$$f'(x) = \frac{1}{(x + 1)^2} = \frac{2}{3}$$

$$(x + 1)^2 = \frac{3}{2}$$

$$x = -1 \pm \frac{\sqrt{6}}{2}$$

In the interval $\left(-\frac{1}{2}, 2\right)$: $c = -1 + \frac{\sqrt{6}}{2}$

18. $f(x) = \sqrt{x - 2}$ is continuous on $[2, 6]$ and differentiable on $(2, 6)$.

$$\frac{f(6) - f(2)}{6 - 2} = \frac{2 - 0}{4} = \frac{1}{2}$$

$$f'(x) = \frac{1}{2\sqrt{x - 2}} = \frac{1}{2}$$

$$\sqrt{x - 2} = 1$$

$$c = 3$$

19. $f(x) = x^3$ is continuous and differentiable on $[0, 1]$.

$$\frac{f(1) - f(0)}{1 - 0} = \frac{1 - 0}{1} = 1$$

$$f'(x) = 3x^2 = 1$$

$$x = \pm\frac{\sqrt{3}}{2}$$

In the interval $(0, 1)$: $c = \frac{\sqrt{3}}{3}$

20. $f(x) = x^3 - 2x$ is continuous and differentiable on $[0, 2]$.

$$\frac{f(2) - f(0)}{2 - 0} = \frac{4 - 0}{2} = 2$$

$$f'(x) = 3x^2 - 2 = 2$$

$$3x^2 = 4$$

$$x = \pm\frac{2\sqrt{3}}{3}$$

In the interval $(0, 2)$: $c = \frac{2\sqrt{3}}{3}$

21. $f(t) = -16t^2 + 48t + 32$

(a) $f(1) = f(2) = 64$

(b) $v = f'(t)$ must be 0 at some time in [1, 2].

$$f'(t) = -32t + 48 = 0$$

$$t = \frac{3}{2} \text{ seconds}$$

22. $C(x) = 10\left(\frac{1}{x} + \frac{x}{x+3}\right)$

(a) $C(3) = C(6) = \frac{25}{3}$

(b) $$C'(x) = 10\left(-\frac{1}{x^2} + \frac{3}{(x+3)^2}\right) = 0$$

$$\frac{3}{x^2 + 6x + 9} = \frac{1}{x^2}$$

$$2x^2 - 6x - 9 = 0$$

$$x = \frac{6 \pm \sqrt{108}}{4} = \frac{6 \pm 6\sqrt{3}}{4} = \frac{3 \pm 3\sqrt{3}}{2}$$

In the interval [3, 6]: $c = \frac{3 + 3\sqrt{3}}{2} \approx 4.098$

23. $s(t) = -16t^2 + 500$

(a) $V_{\text{avg}} = \frac{s(3) - s(0)}{3 - 0} = -\frac{144}{3} = -48$ ft/sec

(b) $s(t)$ is continuous on [0, 3] and differentiable on (0, 3). Therefore, the Mean Value Theorem applies.

$$v(t) = s'(t) = -32t = -48$$

$$t = \frac{3}{2} \text{ seconds}$$

In the interval [0, 3]: $c = \frac{3}{2}$

24. $s(t) = 200\left(5 - \frac{9}{2+t}\right)$

(a) $\frac{s(12) - s(0)}{12 - 0} = \frac{200[5 - (9/14)] - 200[5 - (9/2)]}{12} = \frac{450}{7}$

(b) $$s'(t) = 200\left(\frac{9}{(2+t)^2}\right) = \frac{450}{7}$$

$$\frac{1}{(2+t)^2} = \frac{1}{28}$$

$$2 + t = 2\sqrt{7}$$

$$t = 2\sqrt{7} - 2 \approx 3.2915 \text{ months}$$

$s'(t)$ is equal to the average value on April 10th (April 9th in a leap year).

25. $$f(x) = \frac{1}{x-4}$$

$$\frac{f(6) - f(2)}{6 - 2} = \frac{(1/2) - (-1/2)}{4} = \frac{1}{4}$$

$f'(c) = -\frac{1}{(x-4)^2} < 0$ for all real numbers except 4.

Therefore, $f'(c) \neq \frac{1}{4}$ for any real number c in the interval (2, 6). This does not contradict the Mean Value Theorem since it does not apply to $f(x)$ on the interval [2, 6]. f has a discontinuity at $x = 4$.

26. Let f be continuous on $[a, b]$ with $f(a) = f(b)$.

Case 1: If f is differentiable on (a, b), then Rolle's Theorem says that there exists c in (a, b) such that $f'(c) = 0$. Thus, c is a critical number.

Case 2: If f is not differentiable on (a, b), then there exists c in (a, b) such that $f'(c)$ is undefined. Thus, c is a critical number.

27. Suppose that $p(x) = x^{2n+1} + ax + b$ has two real roots x_1 and x_2. Then by Rolle's Theorem, since $p(x_1) = p(x_2) = 0$, there exists c in (x_1, x_2) such that $p'(c) = 0$, but $p'(x) = (2n+1)x^{2n} + a = 0$, $n > 0$, $a > 0$. Thus, $p'(x) \neq 0$ for any value of x. Therefore, $p(x)$ cannot have two real roots.

28. (a) Consider two consecutive zeros of $p'(x)$ and suppose that between them there are two zeros of $p(x)$. Then by Rolle's Theorem, there would be another zero of $p'(x)$ between the zeros of $p(x)$. This contradicts the fact that the zeros of $p'(x)$ are consecutive.

(b) Let x_1, x_2, and x_3 be the zeros of $p(x)$ where $x_1 < x_2 < x_3$. Then by Rolle's Theorem, there exists c_1 and c_2, where $x_1 < c_1 < x_2$ and $x_2 < c_2 < x_3$, such that $p'(c_1) = p'(c_2) = 0$. Then, by Rolle's Theorem again, there exists d, where $c_1 < d < c_2$ such that $p''(d) = 0$.

29. Suppose $f(x)$ is not constant on $[a, b]$. Then there exists x_1 and x_2 in $[a, b]$ such that $f(x_1) \neq f(x_2)$. Then by the Mean Value Theorem, there exists c in (a, b) such that

$$f'(c) = \frac{f(x_2) - f(x_1)}{x_2 - x_1} \neq 0.$$

This contradicts the fact that $f'(x) = 0$ for all x in $[a, b]$.

30. If $p(x) = Ax^2 + Bx + C$, then

$$\begin{aligned}
p'(x) = 2Ax + B = \frac{f(b) - f(a)}{b - a} &= \frac{(Ab^2 + Bb + C) - (Aa^2 + Ba + C)}{b - a} \\
&= \frac{A(b^2 - a^2) + B(b - a)}{b - a} \\
&= \frac{(b - a)[A(b + a) + B]}{b - a} \\
&= A(b + a) + B.
\end{aligned}$$

Thus, $2Ax = A(b + a)$ and $x = (b + a)/2$ which is the midpoint of $[a, b]$.

31. Let $h(x) = f(x) - g(x)$. Then $h'(x) = f'(x) - g'(x) = 0$; $h(x)$ is constant. Therefore, $h(x) = f(x) - g(x) = C$ and $f(x) = g(x) + C$.

32. Suppose $f(x)$ has two fixed points c_1 and c_2. Then by the Mean Value Theorem, there exists c such that

$$f'(c) = \frac{f(c_2) - f(c_1)}{c_2 - c_1} = \frac{c_2 - c_1}{c_2 - c_1} = 1.$$

This contradicts the fact that $f'(x) < 1$ for all x.

33. $f(x) = \sqrt{x}$, $[1, 9]$

(a) $(1, 1)$, $(9, 3)$

$$m = \frac{3-1}{9-1} = \frac{1}{4}$$

$$y - 1 = \frac{1}{4}(x - 1)$$

$$y = \frac{1}{4}x + \frac{3}{4}$$

$$0 = x - 4y + 3$$

3 tangent f secant 1 1 9

(b) $f'(x) = \dfrac{1}{2\sqrt{x}}$

$$\frac{f(9) - f(1)}{9 - 1} = \frac{1}{4}$$

$$\frac{1}{2\sqrt{c}} = \frac{1}{4}$$

$$\sqrt{c} = 2$$

$$c = 4$$

$$(c, f(c)) = (4, 2)$$

$$m = f'(4) = \frac{1}{4}$$

$$y - 2 = \frac{1}{4}(x - 4)$$

$$y = \frac{1}{4}x + 1$$

$$0 = x - 4y + 4$$

Section 4.3 Increasing and Decreasing Functions and the First Derivative Test

1. $f(x) = x^2 - 6x + 8$
Increasing on: $(3, \infty)$
Decreasing on: $(-\infty, 3)$

2. $y = -(x+1)^2$
Increasing on: $(-\infty, -1)$
Decreasing on: $(-1, \infty)$

3. $y = \dfrac{x^3}{4} - 3x$
Increasing on:
$(-\infty, -2)$, $(2, \infty)$
Decreasing on: $(-2, 2)$

4. $f(x) = x^4 - 2x^2$
Increasing on:
$(-1, 0)$, $(1, \infty)$
Decreasing on:
$(-\infty, -1)$, $(0, 1)$

5. $f(x) = \dfrac{1}{x^2}$
Increasing on: $(-\infty, 0)$
Decreasing on: $(0, \infty)$

6. $y = \dfrac{x^2}{x+1}$
Increasing on:
$(-\infty, -2)$, $(0, \infty)$
Decreasing on:
$(-2, -1)$, $(-1, 0)$

7. $f(x) = -2x^2 + 4x + 3$

$f'(x) = -4x + 4 = 0$

Critical number: $x = 1$

Test intervals:	$-\infty < x < 1$	$1 < x < \infty$
Sign of $f'(x)$:	$f' > 0$	$f' < 0$
Conclusion:	Increasing	Decreasing

Increasing on: $(-\infty, 1)$
Decreasing on: $(1, \infty)$
Relative maximum: $(1, 5)$

8. $f(x) = x^2 + 8x + 10$

$f'(x) = 2x + 8 = 0$

Critical number: $x = -4$

Test intervals:	$-\infty < x < -4$	$-4 < x < \infty$
Sign of $f'(x)$:	$f' < 0$	$f' > 0$
Conclusion:	Decreasing	Increasing

Increasing on: $(-4, \infty)$
Decreasing on: $(-\infty, -4)$
Relative minimum: $(-4, -6)$

9. $f(x) = x^2 - 6x$

$f'(x) = 2x - 6 = 0$

Critical number: $x = 3$

Test intervals:	$-\infty < x < 3$	$3 < x < \infty$
Sign of $f'(x)$:	$f' < 0$	$f' > 0$
Conclusion:	Decreasing	Increasing

Increasing on: $(3, \infty)$
Decreasing on: $(-\infty, 3)$
Relative minimum: $(3, -9)$

10. $f(x) = (x-1)^2(x+2)$

$f'(x) = (x-1)^2(1) + (x+2)(2)(x-1) = (x-1)[(x-1) + 2(x+2)] = 3(x-1)(x+1) = 0$

Critical numbers: $x = -1, 1$

Test intervals:	$-\infty < x < -1$	$-1 < x < 1$	$1 < x < \infty$
Sign of $f'(x)$:	$f' > 0$	$f' < 0$	$f' > 0$
Conclusion:	Increasing	Decreasing	Increasing

Increasing on: $(-\infty, -1), (1, \infty)$
Decreasing on: $(-1, 1)$
Relative maximum: $(-1, 4)$
Relative minimum: $(1, 0)$

11. $f(x) = 2x^3 + 3x^2 - 12x$

$f'(x) = 6x^2 + 6x - 12 = 6(x+2)(x-1) = 0$

Critical numbers: $x = -2, 1$

Test intervals:	$-\infty < x < -2$	$-2 < x < 1$	$1 < x < \infty$
Sign of $f'(x)$:	$f' > 0$	$f' < 0$	$f' > 0$
Conclusion:	Increasing	Decreasing	Increasing

Increasing on: $(-\infty, -2), (1, \infty)$
Decreasing on: $(-2, 1)$
Relative maximum: $(-2, 20)$
Relative minimum: $(1, -7)$

12. $f(x) = (x-3)^3$

$f'(x) = 3(x-3)^2 = 0$

Critical number: $x = 3$

Test intervals:	$-\infty < x < 3$	$3 < x < \infty$
Sign of $f'(x)$:	$f' > 0$	$f' > 0$
Conclusion:	Increasing	Increasing

Increasing on: $(-\infty, \infty)$
No relative extrema

13. $f(x) = x^{1/3} + 1$

$f'(x) = \frac{1}{3}x^{-2/3} = \frac{1}{3x^{2/3}}$

Critical number: $x = 0$

Test intervals:	$-\infty < x < 0$	$0 < x < \infty$
Sign of $f'(x)$:	$f' > 0$	$f' > 0$
Conclusion:	Increasing	Increasing

Increasing on: $(-\infty, \infty)$
No relative extrema

14. $f(x) = x^{2/3}(x-5) = x^{5/3} - 5x^{2/3}$

$f'(x) = \frac{5}{3}x^{2/3} - \frac{10}{3}x^{-1/3} = \frac{5}{3}x^{-1/3}(x-2)$

Critical numbers: $x = 0, 2$

Test intervals:	$-\infty < x < 0$	$0 < x < 2$	$2 < x < \infty$
Sign of $f'(x)$:	$f' > 0$	$f' < 0$	$f' > 0$
Conclusion:	Increasing	Decreasing	Increasing

Increasing on: $(-\infty, 0)$, $(2, \infty)$
Decreasing on: $(0, 2)$
Relative maximum: $(0, 0)$
Relative minimum: $(2, -3\sqrt[3]{4})$

15. $f(x) = \dfrac{x^2}{x^2 - 9}$

$f'(x) = \dfrac{(x^2 - 9)(2x) - (x^2)(2x)}{(x^2 - 9)^2} = \dfrac{-18x}{(x^2 - 9)^2}$

Critical number: $x = 0$

Discontinuities: $x = -3,\ 3$

Test intervals:	$-\infty < x < -3$	$-3 < x < 0$	$0 < x < 3$	$3 < x < \infty$
Sign of $f'(x)$:	$f' > 0$	$f' > 0$	$f' < 0$	$f' < 0$
Conclusion:	Increasing	Increasing	Decreasing	Decreasing

Increasing on: $(-\infty, -3),\ (-3, 0)$
Decreasing on: $(0, 3),\ (3, \infty)$
Relative maximum: $(0, 0)$

16. $f(x) = \dfrac{x+3}{x^2} = \dfrac{1}{x} + \dfrac{3}{x^2}$

$f'(x) = -\dfrac{1}{x^2} - \dfrac{6}{x^3} = \dfrac{-(x+6)}{x^3}$

Critical number: $x = -6$
Discontinuity: $x = 0$

Test intervals:	$-\infty < x < -6$	$-6 < x < 0$	$0 < x < \infty$
Sign of $f'(x)$:	$f' < 0$	$f' > 0$	$f' < 0$
Conclusion:	Decreasing	Increasing	Decreasing

Increasing on : $(-6, 0)$
Decreasing on: $(-\infty, -6),\ (0, \infty)$
Relative minimum: $\left(-6, -\frac{1}{12}\right)$

17. $f(x) = \dfrac{x^5 - 5x}{5}$

$f'(x) = x^4 - 1$

Critical numbers: $x = -1,\ 1$

Test intervals:	$-\infty < x < -1$	$-1 < x < 1$	$1 < x < \infty$
Sign of $f'(x)$:	$f' > 0$	$f' < 0$	$f' > 0$
Conclusion:	Increasing	Decreasing	Increasing

Increasing on: $(-\infty, -1),\ (1, \infty)$
Decreasing on: $(-1, 1)$
Relative maximum: $\left(-1, \frac{4}{5}\right)$
Relative minimum: $\left(1, -\frac{4}{5}\right)$

18. $f(x) = x^4 - 32x + 4$

$f'(x) = 4x^3 - 32 = 4(x^3 - 8)$

Critical number: $x = 2$

Test intervals:	$-\infty < x < 2$	$2 < x < \infty$
Sign of $f'(x)$:	$f' < 0$	$f' > 0$
Conclusion:	Decreasing	Increasing

Increasing on: $(2, \infty)$
Decreasing on: $(-\infty, 2)$
Relative minimum: $(2, -44)$

19. $f(x) = x^3 - 6x^2 + 15$

$f'(x) = 3x^2 - 12x = 3x(x - 4) = 0$

Critical numbers: $x = 0, 4$

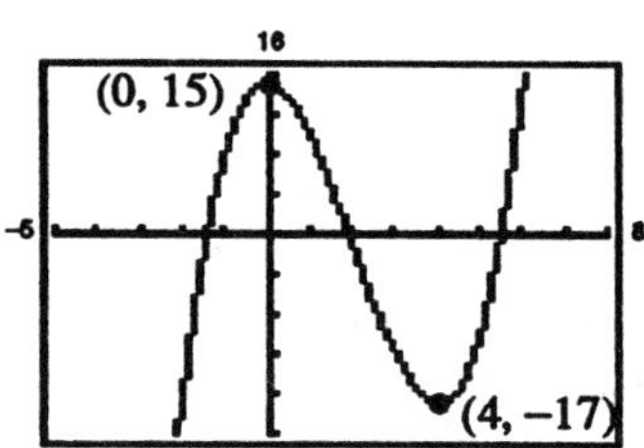

Test intervals:	$-\infty < x < 0$	$0 < x < 4$	$4 < x < \infty$
Sign of $f'(x)$:	$f' > 0$	$f' < 0$	$f' > 0$
Conclusion:	Increasing	Decreasing	Increasing

Increasing on: $(-\infty, 0), (4, \infty)$
Decreasing on: $(0, 4)$
Relative maximum: $(0, 15)$
Relative minimum: $(4, -17)$

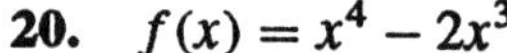

20. $f(x) = x^4 - 2x^3$

$f'(x) = 4x^3 - 6x^2 = 2x^2(2x - 3) = 0$

Critical numbers: $x = 0, \frac{3}{2}$

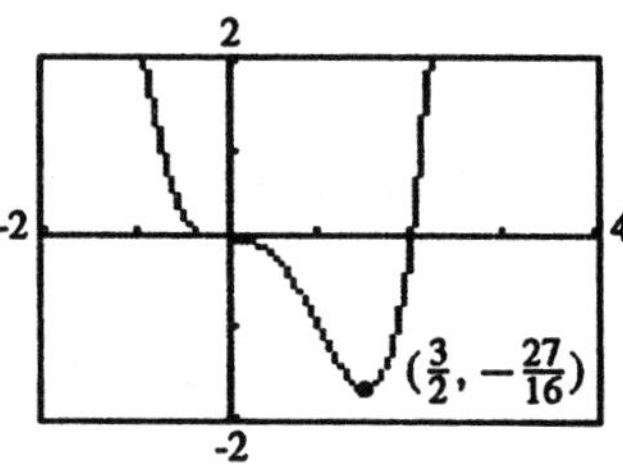

Test intervals:	$-\infty < x < 0$	$0 < x < \frac{3}{2}$	$\frac{3}{2} < x < \infty$
Sign of $f'(x)$:	$f' < 0$	$f' < 0$	$f' > 0$
Conclusion:	Decreasing	Decreasing	Increasing

Decreasing on: $(-\infty, \frac{3}{2})$
Increasing on: $(\frac{3}{2}, \infty)$
Relative minimum: $(\frac{3}{2}, -\frac{27}{16})$

21. $f(x) = (x-1)^{2/3}$

$f'(x) = \dfrac{2}{3(x-1)^{1/3}}$

Critical number: $x = 1$

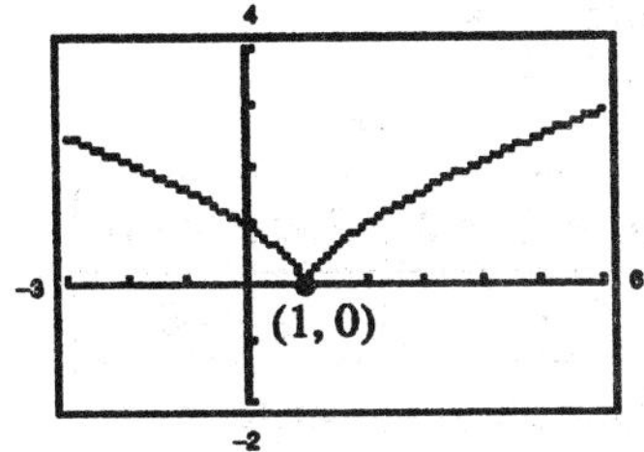

Test intervals:	$-\infty < x < 1$	$1 < x < \infty$
Sign of $f'(x)$:	$f' < 0$	$f' > 0$
Conclusion:	Decreasing	Increasing

Increasing on: $(1, \infty)$
Decreasing on: $(-\infty, 1)$
Relative minimum: $(1, 0)$

22. $f(x) = (x-1)^{1/3}$

$f'(x) = \dfrac{1}{3(x-1)^{2/3}}$

Critical number: $x = 1$

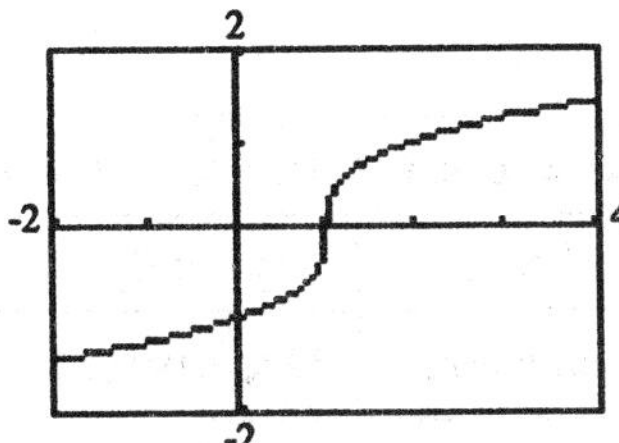

Test intervals:	$-\infty < x < 1$	$1 < x < \infty$
Sign of $f'(x)$:	$f' > 0$	$f' > 0$
Conclusion:	Increasing	Increasing

Increasing on: $(-\infty, \infty)$
No relative extrema

23. $f(x) = x + \dfrac{1}{x}$

$f'(x) = 1 - \dfrac{1}{x^2} = \dfrac{x^2-1}{x^2}$

Critical numbers: $x = -1, 1$
Discontinuity: $x = 0$

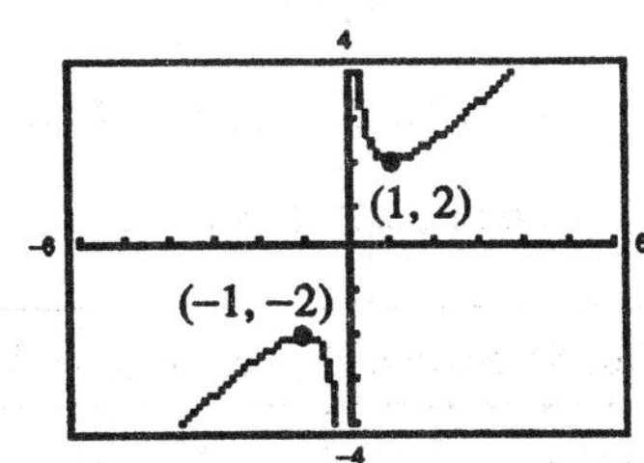

Test intervals:	$-\infty < x < -1$	$-1 < x < 0$	$0 < x < 1$	$1 < x < \infty$
Sign of $f'(x)$:	$f' > 0$	$f' < 0$	$f' < 0$	$f' > 0$
Conclusion:	Increasing	Decreasing	Decreasing	Increasing

Increasing on: $(-\infty, -1)$, $(1, \infty)$
Decreasing on: $(-1, 0)$, $(0, 1)$
Relative maximum: $(-1, -2)$
Relative minimum: $(1, 2)$

24. $f(x) = \dfrac{x}{x+1}$

$$f'(x) = \frac{(x+1)(1) - (x)(1)}{(x+1)^2} = \frac{1}{(x+1)^2}$$

Discontinuity: $x = -1$

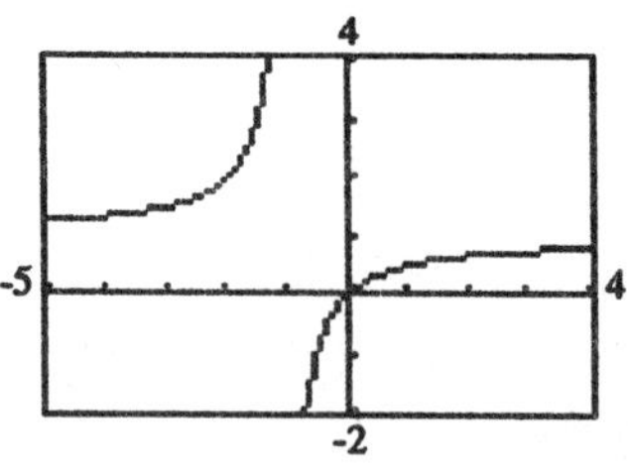

Test intervals:	$-\infty < x < -1$	$-1 < x < \infty$
Sign of $f'(x)$:	$f' > 0$	$f' > 0$
Conclusion:	Increasing	Increasing

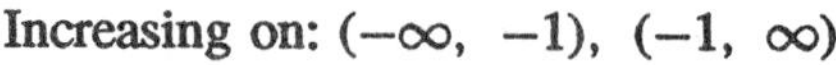
Increasing on: $(-\infty, -1)$, $(-1, \infty)$
No relative extrema

25. $f(x) = \dfrac{x^2 - 2x + 1}{x+1}$

$$f'(x) = \frac{(x+1)(2x-2) - (x^2 - 2x + 1)(1)}{(x+1)^2} = \frac{x^2 + 2x - 3}{(x+1)^2} = \frac{(x+3)(x-1)}{(x+1)^2}$$

Critical numbers: $x = -3, 1$
Discontinuity: $x = -1$

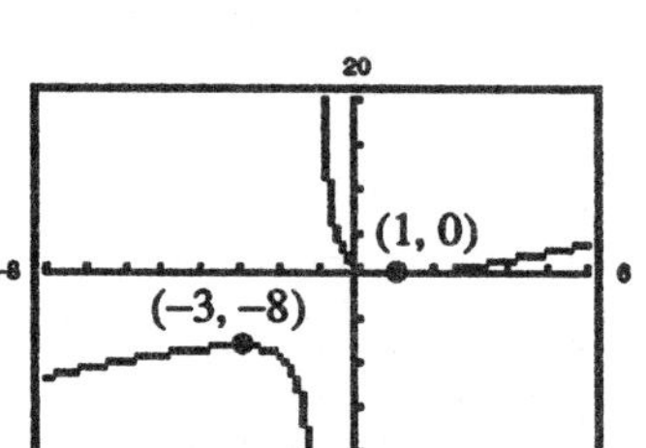

Test intervals:	$-\infty < x < -3$	$-3 < x < -1$	$-1 < x < 1$	$1 < x < \infty$
Sign of $f'(x)$:	$f' > 0$	$f' < 0$	$f' < 0$	$f' > 0$
Conclusion:	Increasing	Decreasing	Decreasing	Increasing

Increasing on: $(-\infty, -3)$, $(1, \infty)$
Decreasing on: $(-3, -1)$, $(-1, 1)$
Relative maximum: $(-3, -8)$
Relative minimum: $(1, 0)$

26. $f(x) = \dfrac{x^2 - 3x - 4}{x-2}$

$$f'(x) = \frac{(x-2)(2x-3) - (x^2 - 3x - 4)(1)}{(x-2)^2} = \frac{x^2 - 4x + 10}{(x-2)^2}$$

Discontinuity: $x = 2$

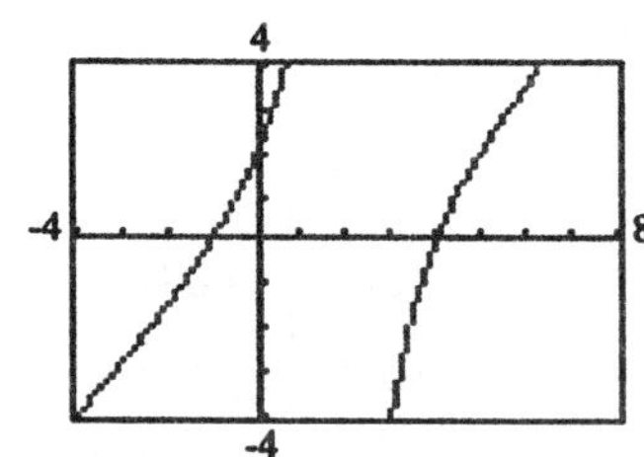

Test intervals:	$-\infty < x < 2$	$2 < x < \infty$
Sign of $f'(x)$:	$f' > 0$	$f' > 0$
Conclusion:	Increasing	Increasing

Increasing on: $(-\infty, 2)$, $(2, \infty)$
No relative extrema

27. $f(x) = x^2$

$f'(x) = 2x = 0 \Rightarrow x = 0$ is a critical number

$f'(x) < 0$ on $(-\infty, 0) \Rightarrow f$ is decreasing

$f'(x) > 0$ on $(0, \infty) \Rightarrow f$ is increasing

(a) f is not strictly monotonic on $(-\infty, \infty)$.
(b) f is strictly monotonic on $(-\infty, 0)$.
(c) f is strictly monotonic on $(0, \infty)$.

28. $f(x) = x^3 - x$

$f'(x) = 3x^2 - 1 = 0 \Rightarrow x = \pm\dfrac{1}{\sqrt{3}} \approx \pm 0.57735$

$\left(-\infty, -\frac{1}{\sqrt{3}}\right)$	$\left(-\frac{1}{\sqrt{3}}, \frac{1}{\sqrt{3}}\right)$	$\left(\frac{1}{\sqrt{3}}, \infty\right)$
$f'(x) > 0$	$f'(x) < 0$	$f'(x) > 0$
Increasing	Decreasing	Increasing

(a) f is not strictly monotonic on $(-1, 0)$.
(b) f is strictly monotonic on $\left(-1, -\frac{1}{2}\right)$.
(c) f is not strictly monotonic on $(-1, 1)$.

29. $s(t) = 96t - 16t^2 = 16t(6 - t) = 0, \quad 0 \le t \le 6$

$s'(t) = 96 - 32t = 0$

$t = 3$

Test intervals:	$0 < t < 3$	$3 < t < 6$
Sign of $s'(t)$:	$s' > 0$	$s' < 0$
Conclusion:	Increasing	Decreasing

Moving upward when $0 < t < 3$
Moving downward when $3 < t < 6$
Maximum height: $s(3) = 144$ ft

30. $s(t) = -16t^2 + 64t = 16t(4 - t) = 0, \quad 0 \le t \le 4$

$s'(t) = -32t + 64 = 0$

Test intervals:	$0 < t < 2$	$2 < t < 4$
Sign of $s'(t)$:	$s' > 0$	$s' < 0$
Conclusion:	Increasing	Decreasing

Moving upward when $0 < t < 2$
Moving downward when $2 < t < 4$
Maximum height: $s(2) = 64$ ft

31. $v = k(R - r)r^2 = k(Rr^2 - r^3)$

$v' = k(2Rr - 3r^2) = kr(2R - 3r) = 0$

$r = 0$ or $\frac{2}{3}R$

Maximum when $r = \frac{2}{3}R$

32. $C = \dfrac{3t}{27 + t^3}$

$$C' = \frac{(27 + t^3)(3) - (3t)(3t^2)}{(27 + t^3)^2}$$

$$= \frac{3(27 - 2t^3)}{(27 + t^3)^2} = 0$$

$$t = \frac{3}{\sqrt[3]{2}} \approx 2.38 \text{ hours}$$

33. $C = 0.29483t + 0.04253t^2 - 0.00035t^3, \quad 0 \le t \le 120$

$C' = 0.29483 + 0.08506t - 0.00105t^2 = 0$

$$t = \frac{-0.08506 \pm \sqrt{(0.08506)^2 - 4(-0.00105)(0.29483)}}{2(-0.00105)} \approx 84.3388$$

Test intervals:	$0 < t < 84.3388$	$84.3388 < t < 120$
Sign of C':	$C' > 0$	$C' < 0$
Conclusion:	Increasing	Decreasing

Increasing when $0 < t < 84.3388$ min
Decreasing when $84.3388 < t < 120$ min

34. $P = 2.44x - \dfrac{x^2}{20{,}000} - 5000, \quad 0 \le x \le 35{,}000$

$P' = 2.44 - \dfrac{x}{10{,}000} = 0$

$x = 24{,}400$

Test intervals:	$0 < x < 24{,}400$	$24{,}400 < x < 35{,}000$
Sign of P':	$P' > 0$	$P' < 0$
Conclusion:	Increasing	Decreasing

Increasing when $0 < x < 24{,}400$ hamburgers
Decreasing when $24{,}400 < x < 35{,}000$ hamburgers

35. $W = 0.033t^2 - 0.3974t + 7.3032, \quad 0 \le t \le 14$

$W' = 0.066t - 0.3974 = 0$

$t \approx 6.02$

Test intervals:	$0 < t < 6.02$	$6.02 < t < 14$
Sign of W':	$W' < 0$	$W' > 0$
Conclusion:	Decreasing	Increasing

Increasing when $6.02 < t < 14$ days
Decreasing when $0 < t < 6.02$ days

36. $P = \dfrac{vR_1R_2}{(R_1+R_2)^2}$, v and R_1 are constant

$$\frac{dP}{dR_2} = \frac{(R_1+R_2)^2(vR_1) - vR_1R_2[2(R_1+R_2)(1)]}{(R_1+R_2)^4}$$

$$= \frac{vR_1(R_1-R_2)}{(R_1+R_2)^3} = 0 \Rightarrow R_2 = R_1$$

Maximum when $R_1 = R_2$

37. $R = \sqrt{0.001T^4 - 4T + 100}$

$$R' = \frac{0.004T^3 - 4}{2\sqrt{0.001T^4 - 4T + 100}}$$

$= 0$

$T = 10°$

38. $f(x) = x,\ g(x) = x^3,$
$h(x) = x - x^3$

(a) $h(x) = x - x^3 = x(1 - x^2)$
On the interval $(0, 1)$, both x and $(1 - x^2)$ are positive and therefore, $h(x)$ is positive. Hence, $f(x) > g(x)$ on the interval $(0, 1)$.

(b)

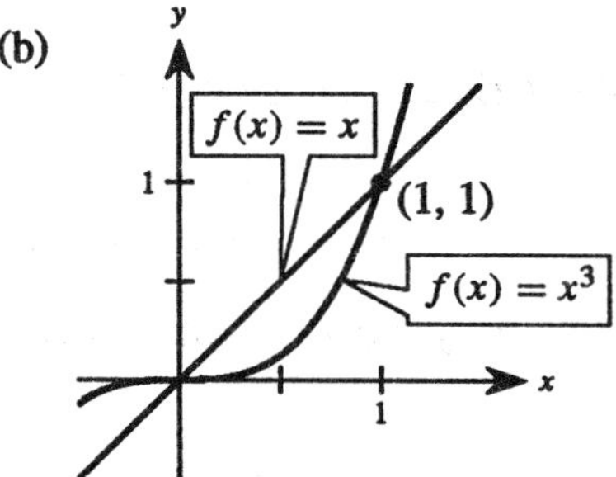

39. $f(x) = ax^3 + bx^2 + cx + d$
Relative minimum: $(0, 0)$
Relative maximum: $(2, 2)$
Since $f(0) = 0 \Rightarrow d = 0,\ f'(x) = 3ax^2 + 2bx + c.$

Since $(0, 0)$ is a relative minimum
$\Rightarrow f'(0) = 0 \Rightarrow c = 0.$

Since $(2, 2)$ is a relative maximum,
$f'(2) = 0 \Rightarrow 12a + 4b = 0.$
$f(2) = 2 \Rightarrow 8a + 4b = 2$

Solving this system of equations simultaneously, we have $a = -\frac{1}{2}$ and $b = \frac{3}{2}$. Therefore, $f(x) = -\frac{1}{2}x^3 + \frac{3}{2}x^2$.

40. $f(x) = ax^2 + bx + c$ passes through $(2, 10)$
Relative maximum: $(5, 20)$
$f'(x) = 2ax + b = 0$
Critical number: $x = 5$

$$x = \frac{-b}{2a} = 5 \Rightarrow b = -10a$$

$f(x) = ax^2 - 10ax + c$

$f(5) = 25a - 50a + c = 20 \Rightarrow -25a + c = 20$

$f(2) = 4a - 20a + c = 10 \Rightarrow \underline{-16a + c = 10}$

$-9a = 10$

$a = -\frac{10}{9},\quad b = \frac{100}{9},\quad c = -\frac{70}{9}$

$f(x) = -\frac{10}{9}(x^2 - 10x + 7)$

In Exercises 41–46, $f'(x) > 0$ on $(-\infty, -4)$, $f'(x) < 0$ on $(-4, 6)$ and $f'(x) > 0$ on $(6, \infty)$.

41. $g(x) = f(x) + 5$

$g'(x) = f'(x)$

$g'(0) = f'(0) < 0$

42. $g(x) = 3f(x) - 3$

$g'(x) = 3f'(x)$

$g'(-5) = 3f'(-5) > 0$

43. $g(x) = -f(x)$

$g'(x) = -f'(x)$

$g'(-6) = -f'(-6) < 0$

44. $g(x) = -f(x)$

$g'(x) = -f'(x)$

$g'(0) = -f'(0) > 0$

45. $g(x) = f(x - 10)$

$g'(x) = f'(x - 10)$

$g'(0) = f'(-10) > 0$

46. $g(x) = f(x - 10)$

$g'(x) = f'(x - 10)$

$g'(8) = f'(-2) < 0$

47. $f'(x) = \begin{cases} > 0, & x < 4 \Rightarrow f \text{ is increasing on } (-\infty, 4). \\ \text{undefined}, & x = 4 \\ < 0, & x > 4 \Rightarrow f \text{ is decreasing on } (4, \infty). \end{cases}$

Two possibilities for $f(x)$ are given below.

(a)

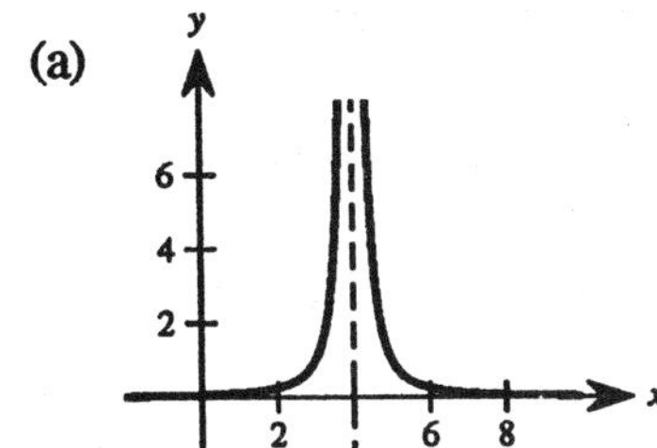

(b)

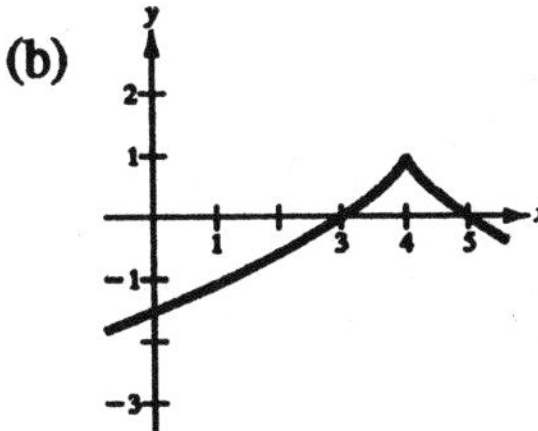

48. Critical number: $x = 5$

$f'(4) = -2.5 \Rightarrow f$ is decreasing at $x = 4$.

$f'(6) = 3 \Rightarrow f$ is increasing at $x = 6$.

$(5, f(5))$ is a relative minimum.

49. $f(x) = 2x\sqrt{9 - x^2}$, $[-3, 3]$

$$f'(x) = 2x\left[\frac{1}{2}(9 - x^2)^{-1/2}(-2x)\right] + 2\sqrt{9 - x^2} = 2(9 - x^2)^{-1/2}[-x^2 + (9 - x^2)] = \frac{2(9 - 2x^2)}{\sqrt{9 - x^2}}$$

(a)

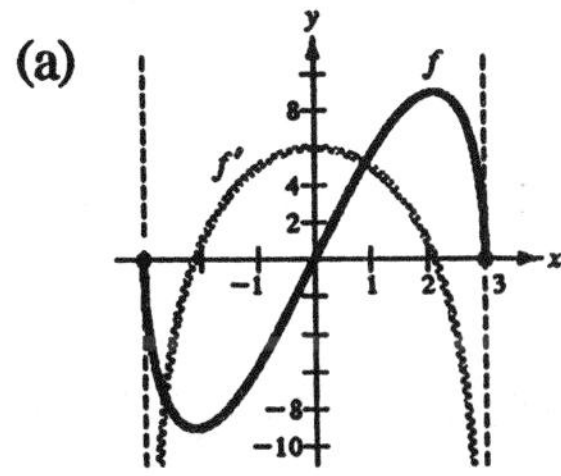

(b) 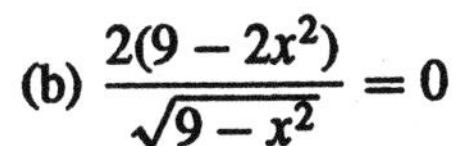

$$\frac{2(9 - 2x^2)}{\sqrt{9 - x^2}} = 0$$

Critical numbers:

$$x = \pm\frac{3}{\sqrt{2}} = \pm\frac{3\sqrt{2}}{2}$$

(c)

Intervals:	$\left(-3, -\frac{3\sqrt{2}}{2}\right)$	$\left(-\frac{3\sqrt{2}}{2}, \frac{3\sqrt{2}}{2}\right)$	$\left(\frac{3\sqrt{2}}{2}, 3\right)$
	$f'(x) < 0$	$f'(x) > 0$	$f'(x) < 0$
	Decreasing	Increasing	Decreasing

50. $f(x) = 10(5 - \sqrt{x^2 - 3x + 16}),\ [0, 5]$

$$f'(x) = 10\left(-\frac{1}{2}\right)(x^2 - 3x + 16)^{-1/2}(2x - 3) = -\frac{5(2x - 3)}{\sqrt{x^2 - 3x + 16}}$$

(a)

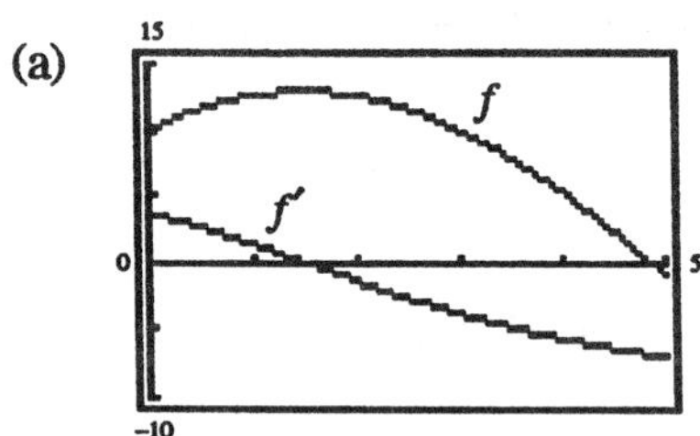

(b) $-\dfrac{5(2x - 3)}{\sqrt{x^2 - 3x + 16}} = 0$

Critical number: $x = \dfrac{3}{2}$

(c)

Intervals:	$(0, \frac{3}{2})$	$(\frac{3}{2}, 5)$
	$f'(x) > 0$	$f'(x) < 0$
	Increasing	Decreasing

51. Assume that $f'(x) < 0$ for all x in the interval (a, b) and let $x_1 < x_2$ be any two points in the interval. By the Mean Value Theorem, we know there exists a number c such that $x_1 < c < x_2$, and

$$f'(c) = \frac{f(x_2) - f(x_1)}{x_2 - x_1}.$$

Since $f'(c) < 0$ and $x_2 - x_1 > 0$, then $f(x_2) - f(x_1) < 0$, which implies that $f(x_2) < f(x_1)$. Thus, f is decreasing on the interval.

52. *Case 2:* Suppose f' changes from positive to negative at c. Then there exists a and b in I such that $f'(x) > 0$ for all x in (a, c) and $f'(x) < 0$ for all x in (c, b). By Theorem 4.5, f is increasing on (a, c) and decreasing on (c, b). Therefore, $f(c)$ is a maximum of f on (a, b) and thus, a relative maximum of f.

Case 3: Suppose f' does not change signs at c. Then there exists a and b in I such that

(i) $f'(x) > 0$ for all x in (a, c) and (c, b) which implies that f is increasing on (a, c) and (c, b) and $f(c)$ is not a relative minimum or a relative maximum.

(ii) $f'(x) < 0$ for all x in (a, c) and (c, b) which implies that f is decreasing on (a, c) and (c, b) and $f(c)$ is not a relative minimum or a relative maximum.

Section 4.4 Concavity and the Second Derivative Test

1. $y = x^2 - x - 2$

Concave upward: $(-\infty, \infty)$

2. $f(x) = \dfrac{24}{x^2 + 12}$

Concave upward: $(-\infty, -2), (2, \infty)$

Concave downward: $(-2, 2)$

3. $y = -x^3 + 3x^2 - 2$

Concave upward: $(-\infty, 1)$

Concave downward: $(1, \infty)$

4. $f(x) = \dfrac{x^2 - 1}{2x + 1}$

Concave upward: $\left(-\infty, -\frac{1}{2}\right)$

Concave downward: $\left(-\frac{1}{2}, \infty\right)$

5. $f(x) = \dfrac{x^2 + 1}{x^2 - 1}$

Concave upward: $(-\infty, -1), (1, \infty)$

Concave downward: $(-1, 1)$

6. $y = \dfrac{1}{270}(-3x^5 + 40x^3 + 135x)$

Concave upward: $(-\infty, -2), (0, 2)$

Concave downward: $(-2, 0), (2, \infty)$

7. $f(x) = 6x - x^2$

$f'(x) = 6 - 2x$

$f''(x) = -2$

Critical number: $x = 3$
$f''(3) < 0$
Therefore, $(3, 9)$ is a relative maximum.

8. $f(x) = x^2 + 3x - 8$

$f'(x) = 2x + 3$

$f''(x) = 2$

Critical number: $x = -\frac{3}{2}$
$f''\left(-\frac{3}{2}\right) > 0$
Therefore, $\left(-\frac{3}{2}, -\frac{41}{4}\right)$ is a relative minimum.

9. $f(x) = (x - 5)^2$

$f'(x) = 2(x - 5)$

$f''(x) = 2$

Critical number: $x = 5$
$f''(5) > 0$
Therefore, $(5, 0)$ is a relative minimum.

10. $f(x) = -(x - 5)^2$

$f'(x) = -2(x - 5)$

$f''(x) = -2$

Critical number: $x = 5$
$f''(5) < 0$
Therefore, $(5, 0)$ is a relative maximum.

11. $f(x) = x^3 - 3x^2 + 3$

$f'(x) = 3x^2 - 6x = 3x(x - 2)$

$f''(x) = 6x - 6 = 6(x - 1)$

Critical numbers: $x = 0, \quad x = 2$
$f''(0) = -6 < 0$
Therefore, $(0, 3)$ is a relative maximum.
$f''(2) = 6 > 0$
Therefore, $(2, -1)$ is a relative minimum.

12. $f(x) = 5 + 3x^2 - x^3$

$f'(x) = 6x - 3x^2 = 3x(2 - x)$

$f''(x) = 6 - 6x = 6(1 - x)$

Critical numbers: $x = 0, \quad x = 2$
$f''(0) = 6 > 0$
Therefore, $(0, 5)$ is a relative minimum.
$f''(2) = -6 < 0$
Therefore, $(2, 9)$ is a relative maximum.

13. $f(x) = x^4 - 4x^3 + 2$

$f'(x) = 4x^3 - 12x^2 = 4x^2(x - 3)$

$f''(x) = 12x^2 - 24x = 12x(x - 2)$

Critical numbers: $x = 0, \ x = 3$
$f''(0) = 0$, so we must use the First Derivative Test.
$f'(x) < 0$ on the intervals $(-\infty, 0)$ and $(0, 3)$.
Hence, $(0, 2)$ is not an extremum.
$f''(3) > 0$ so $(3, -25)$ is a relative minimum.

14. $f(x) = x^3 - 9x^2 + 27x - 26$

$f'(x) = 3(x - 3)^2$

$f''(x) = 6(x - 3)$

Critical number: $x = 3$
$f''(3) = 0$, so we must use the First Derivative Test.
$f'(x) \geq 0$ for all x and, therefore, there are no relative extrema.

15. $f(x) = x^{2/3} - 3$

$$f'(x) = \frac{2}{3x^{1/3}}$$

$$f''(x) = \frac{-2}{9x^{4/3}}$$

Critical number: $x = 0$
However, $f''(0)$ is undefined, so we must use the First Derivative Test. Since $f'(x) < 0$ on $(-\infty, 0)$ and $f'(x) > 0$ on $(0, \infty)$, $(0, -3)$ is a relative minimum.

16. $f(x) = \sqrt{x^2 + 1}$

$$f'(x) = \frac{x}{\sqrt{x^2 + 1}}$$

Critical number: $x = 0$

$$f''(x) = -\frac{x^2}{(x^2 + 1)^{3/2}} + \frac{1}{\sqrt{x^2 + 1}}$$

$f''(0) = 1 > 0$
Therefore, $(0, 1)$ is a relative minimum.

17. $f(x) = x + \dfrac{4}{x}$

$f'(x) = 1 - \dfrac{4}{x^2}$

$f''(x) = \dfrac{8}{x^3}$

Critical numbers: $x = \pm 2$

$f''(-2) < 0$

Therefore, $(-2, -4)$ is a relative maximum.

$f''(2) > 0$

Therefore, $(2, 4)$ is a relative minimum.

18. $f(x) = \dfrac{x}{x-1}$

$f'(x) = \dfrac{-1}{(x-1)^2}$

There are no critical numbers and $x = 1$ is not in the domain. There are no relative extrema.

19. $f(x) = x^3 - 12x$

$f'(x) = 3x^2 - 12 = 3(x+2)(x-2) = 0$ when $x = \pm 2$.

$f''(x) = 6x$

$f''(-2) = -12 < 0 \Rightarrow (-2, 16)$ is a relative maximum.

$f''(2) = 12 > 0 \Rightarrow (2, -16)$ is a relative minimum.

$f''(x) = 6x = 0$ when $x = 0$.

Test interval	$-\infty < x < 0$	$0 < x < \infty$
Sign of $f''(x)$	$f''(x) < 0$	$f''(x) > 0$
Conclusion	Concave downward	Concave upward

Point of inflection: $(0, 0)$

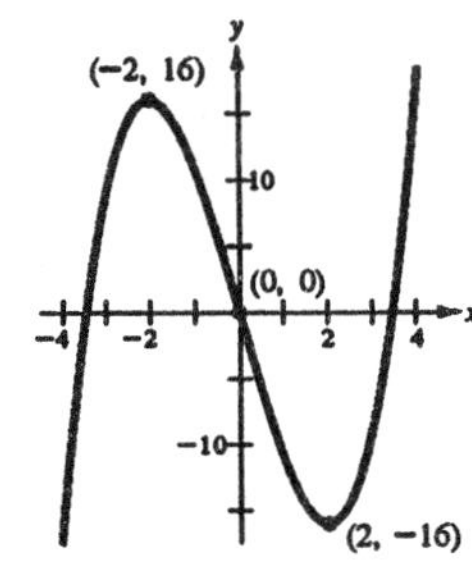

20. $f(x) = x^3 + 1$

$f'(x) = 3x^2 = 0$ when $x = 0$.

$f''(x) = 6x = 0$ when $x = 0$.

Since $f'(x) \geq 0$ for all x and the concavity changes at $x = 0$. $(0, 1)$ is a point of inflection.

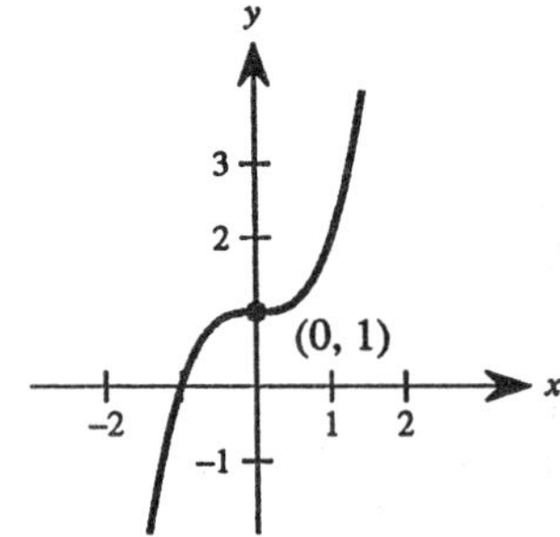

21. $f(x) = x^3 - 6x^2 + 12x - 8$

$f'(x) = 3x^2 - 12x + 12$

$= 3(x-2)^2 = 0$ when $x = 2$.

$f''(x) = 6(x-2) = 0$ when $x = 2$.

Since $f'(x) > 0$ when $x \neq 2$ and the concavity changes at $x = 2$. $(2, 0)$ is a point of inflection.

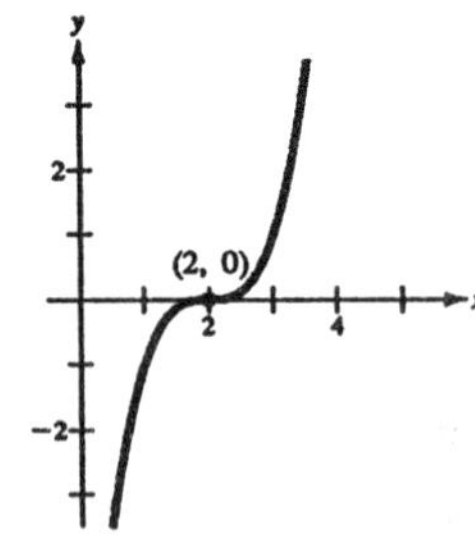

22. $f(x) = 2x^3 - 3x^2 - 12x + 8$

$f'(x) = 6x^2 - 6x - 12 = 6(x - 2)(x + 1) = 0$ when $x = -1, \ 2.$

$f''(x) = 12x - 6$

$f''(-1) = -18 < 0 \Rightarrow (-1, \ 15)$ is a relative maximum.

$f''(2) = 18 > 0 \Rightarrow (2, \ -12)$ is a relative minimum.

$f''(x) = 12x - 6 = 0$ when $x = \frac{1}{2}$.

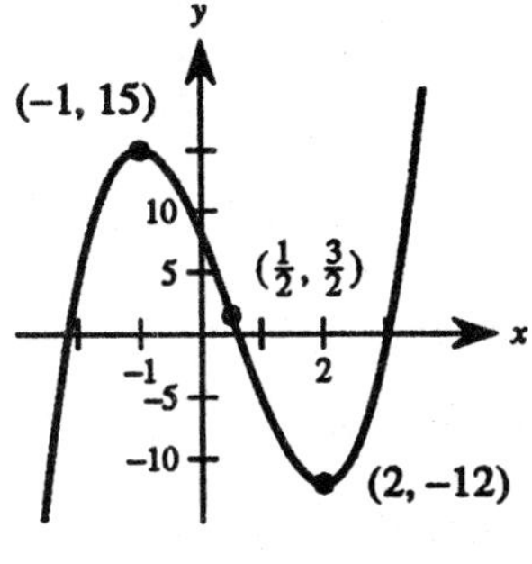

Test interval	$-\infty < x < \frac{1}{2}$	$\frac{1}{2} < x < \infty$
Sign of $f''(x)$	$f''(x) < 0$	$f''(x) > 0$
Conclusion	Concave downward	Concave upward

Point of inflection: $(\frac{1}{2}, \ \frac{3}{2})$

23. $f(x) = \frac{1}{4}x^4 - 2x^2$

$f'(x) = x^3 - 4x = x(x + 2)(x - 2) = 0$ when $x = 0, \ \pm 2.$

$f''(x) = 3x^2 - 4$

$f''(-2) = 8 > 0 \Rightarrow (-2, \ -4)$ is a relative minimum.

$f''(0) = -4 < 0 \Rightarrow (0, \ 0)$ is a relative maximum.

$f''(2) = 8 > 0 \Rightarrow (2, \ -4)$ is a relative minimum.

$f''(x) = 3x^2 - 4 = 0$ when $x = \pm\frac{2}{\sqrt{3}}$.

Test interval	$-\infty < x < -\frac{2}{\sqrt{3}}$	$-\frac{2}{\sqrt{3}} < x < \frac{2}{\sqrt{3}}$	$\frac{2}{\sqrt{3}} < x < \infty$
Sign of $f''(x)$	$f''(x) > 0$	$f''(x) < 0$	$f''(x) > 0$
Conclusion	Concave upward	Concave downward	Concave upward

Point of inflection: $\left(\pm\frac{2}{\sqrt{3}}, \ -\frac{20}{9}\right)$

24. $f(x) = 2x^4 - 8x + 3$

$f'(x) = 8x^3 - 8 = 8(x - 1)(x^2 + x + 1) = 0$ when $x = 1.$

$f''(x) = 24x^2 = 0$ when $x = 0.$

Since $f''(1) > 0, \ (1, \ -3)$ is a relative minimum. However, $(0, 3)$ is not a point of inflection since $f''(x) \geq 0$ for all x.

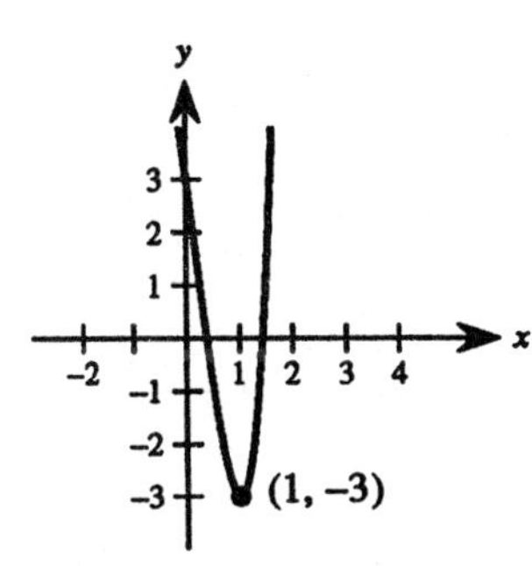

25. $f(x) = x(x-4)^3$

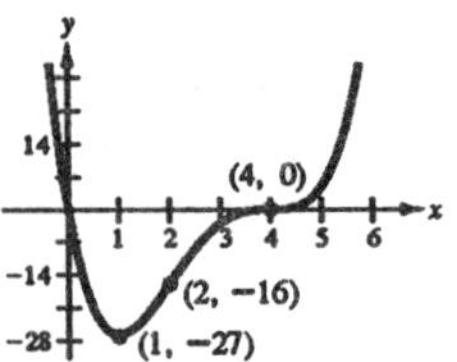

$f'(x) = x[3(x-4)^2] + (x-4)^3$

$= (x-4)^2(4x-4) = 4(x-1)(x-4)^2 = 0$ when $x = 1,\ 4$.

$f''(x) = 4(x-1)[2(x-4)] + 4(x-4)^2$

$= 4(x-4)[2(x-1)+(x-4)]$

$= 4(x-4)(3x-6) = 12(x-4)(x-2)$

$f''(1) = 36 > 0 \Rightarrow (1,\ -27)$ is a relative minimum.

$f''(4) = 0 \Rightarrow$ test fails

By the First Derivative Test we see that $x = 4$ does not yield a relative extrema.

$f''(x) = 12(x-4)(x-2) = 0$ when $x = 2,\ 4$.

Test interval	$-\infty < x < 2$	$2 < x < 4$	$4 < x < \infty$
Sign of $f''(x)$	$f''(x) > 0$	$f''(x) < 0$	$f''(x) > 0$
Conclusion	Concave upward	Concave downward	Concave upward

Points of inflection: $(2,\ -16)$, $(4,\ 0)$

26. $f(x) = x^3(x-4)$

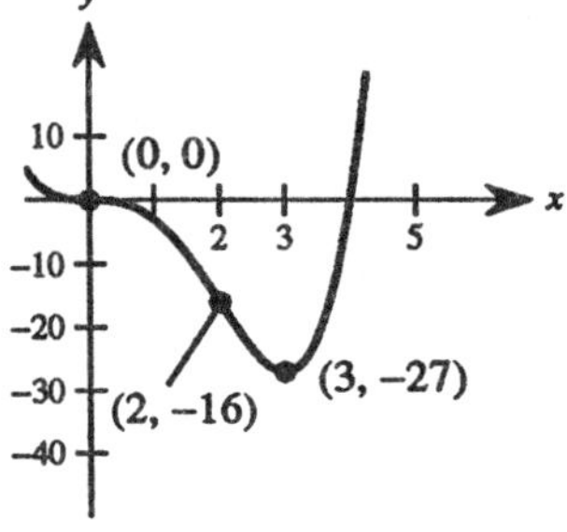

$f'(x) = x^3 + 3x^2(x-4)$

$= x^2[x + 3(x-4)] = 4x^2(x-3) = 0$ when $x = 0,\ 3$.

$f''(x) = 4x^2 + 8x(x-3) = 4x[x + 2(x-3)] = 12x(x-2) = 0$

$f''(0) = 0 \Rightarrow$ test fails

$f''(3) = 36 > 0 \Rightarrow (3,\ -27)$ is a relative minimum.

By the First Derivative Test we see that $x = 0$ does not yield a relative extrema.

$f''(x) = 12x(x-2) = 0$ when $x = 0,\ 2$.

Test interval	$-\infty < x < 0$	$0 < x < 2$	$2 < x < \infty$
Sign of $f''(x)$	$f''(x) > 0$	$f''(x) < 0$	$f''(x) > 0$
Conclusion	Concave upward	Concave downward	Concave upward

Points of inflection: $(0,\ 0)$, $(2,\ -16)$

27. $f(x) = x^2 + \dfrac{1}{x^2}$, Domain: $(-\infty,\ 0) \cup (0,\ \infty)$

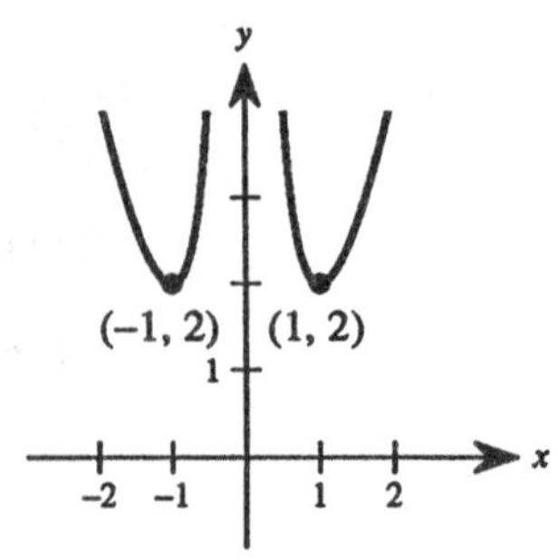

$f'(x) = 2x - \dfrac{2}{x^3} = \dfrac{2(x^4-1)}{x^3} = 0$ when $x = \pm 1$.

$f''(x) = 2 + \dfrac{6}{x^4} = \dfrac{2(x^4+3)}{x^4} > 0$ for all x in the domain.

$f''(\pm 1) = 8 > 0 \Rightarrow (\pm 1,\ 2)$ are relative minima.

28. $f(x) = \dfrac{x^2}{x^2 - 1}$, Domain: $(-\infty, -1) \cup (-1, 1) \cup (1, \infty)$

$$f'(x) = \frac{(x^2-1)(2x) - x^2(2x)}{(x^2-1)^2} = \frac{-2x}{(x^2-1)^2} = 0 \text{ when } x = 0.$$

$$f''(x) = \frac{(x^2-1)^2(-2) + 2x(2)(x^2-1)(2x)}{(x^2-1)^4}$$

$$= \frac{2(3x^2+1)}{(x^2-1)^3}$$

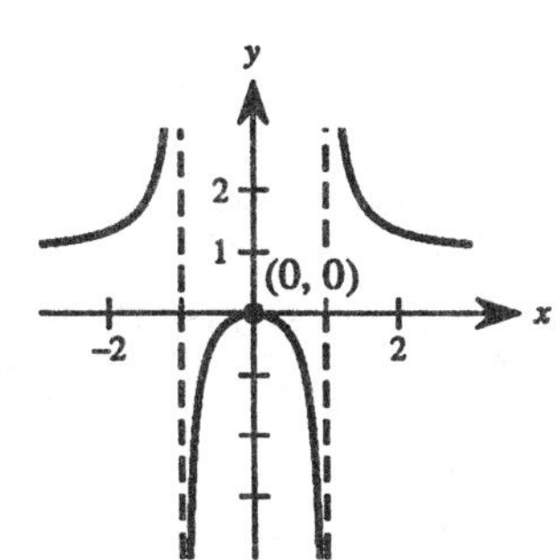

Since $f''(0) < 0$, then $(0, 0)$ is a relative maximum. Since $f''(x) \neq 0$, nor is it undefined in the domain of f, there are no points of inflection.

29. $f(x) = x\sqrt{x+3}$, Domain: $[-3, \infty)$

$$f'(x) = x\left(\frac{1}{2}\right)(x+3)^{-1/2} + \sqrt{x+3} = \frac{3(x+2)}{2\sqrt{x+3}} = 0 \text{ when } x = -2.$$

$$f''(x) = \frac{6\sqrt{x+3} - 3(x+2)(x+3)^{-1/2}}{4(x+3)} = \frac{3(x+4)}{4(x+3)^{3/2}}$$

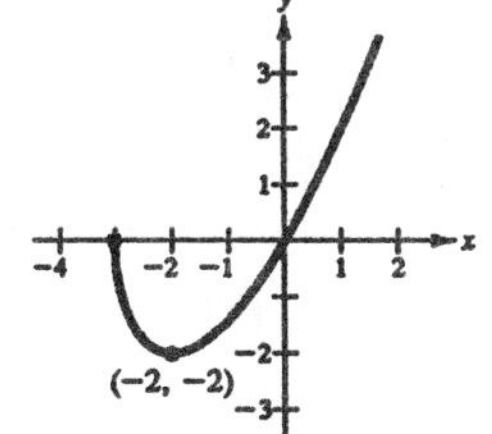

Since $f''(-2) > 0$, then $(-2, -2)$ is a relative minimum. $f''(x) > 0$ on the entire domain of f (except for $x = -3$, for which $f''(x)$ is undefined). There are no points of inflection.

30. $f(x) = x\sqrt{x+1}$, Domain: $[-1, \infty)$

$$f'(x) = (x)\frac{1}{2}(x+1)^{-1/2} + \sqrt{x+1} = \frac{3x+2}{2\sqrt{x+1}} = 0 \text{ when } x = -\frac{2}{3}.$$

$$f''(x) = \frac{6\sqrt{x+1} - (3x+2)(x+1)^{-1/2}}{4(x+1)} = \frac{3x+4}{4(x+1)^{3/2}}$$

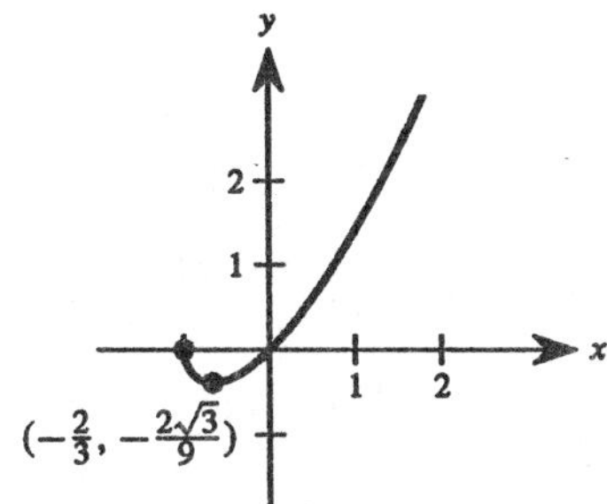

Since $f'(2/3) > 0$, then $\left(-2/3, -(2\sqrt{3})/9\right)$ is a relative minimum. $f''(x) > 0$ on the entire domain of f (except for $x = -1$, for which $f''(x)$ is undefined). There are no points of inflection.

31. $f(x) = \dfrac{x}{x^2 - 4}$, Domain: $(-\infty, -2) \cup (-2, 2) \cup (2, \infty)$

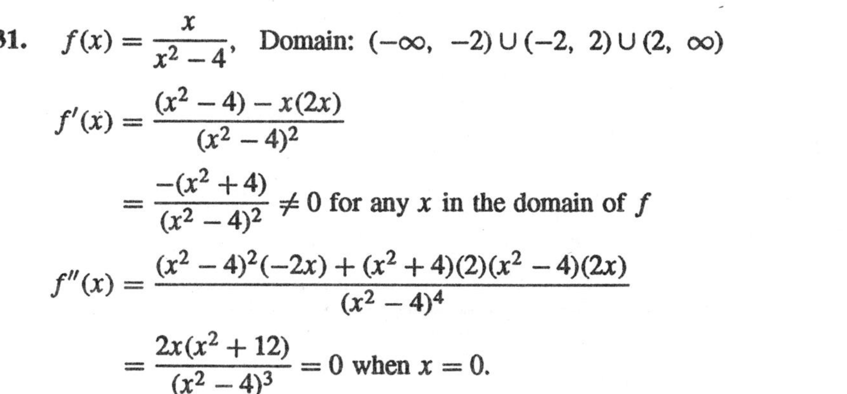

$$f'(x) = \frac{(x^2-4) - x(2x)}{(x^2-4)^2}$$

$$= \frac{-(x^2+4)}{(x^2-4)^2} \neq 0 \text{ for any } x \text{ in the domain of } f$$

$$f''(x) = \frac{(x^2-4)^2(-2x) + (x^2+4)(2)(x^2-4)(2x)}{(x^2-4)^4}$$

$$= \frac{2x(x^2+12)}{(x^2-4)^3} = 0 \text{ when } x = 0.$$

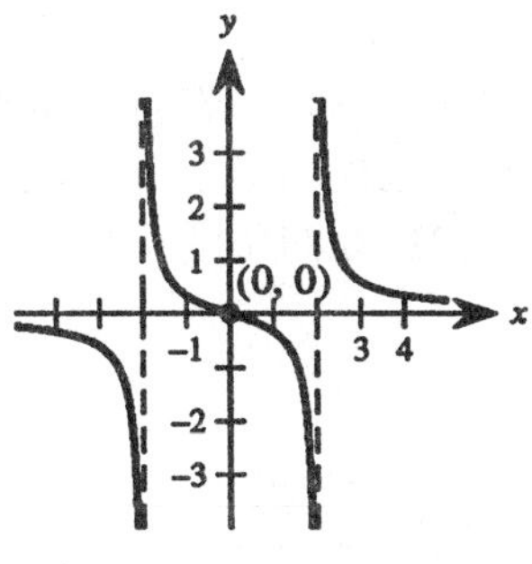

Since $f''(x) > 0$ on $(-2, 0)$ and $f''(x) < 0$ on $(0, 2)$, then $(0, 0)$ is an inflection point.

32. $f(x) = \dfrac{1}{x^2 - x - 2}$, Domain: $(-\infty, -1) \cup (-1, 2) \cup (2, \infty)$

$$f'(x) = \frac{-(2x-1)}{(x^2-x-2)^2} = 0 \text{ when } x = \frac{1}{2}.$$

$$f''(x) = \frac{(x^2-x-2)^2(-2) + (2x-1)(2)(x^2-x-2)(2x-1)}{(x^2-x-2)^4}$$

$$= \frac{6(x^2-x+1)}{(x^2-x-2)^3}$$

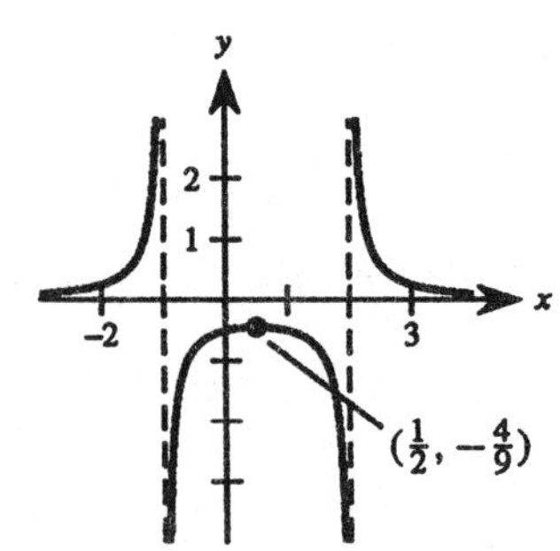

Since $f''\left(\frac{1}{2}\right) < 0$, then $\left(\frac{1}{2}, -\frac{4}{9}\right)$ is a relative maximum. Since $f''(x) \neq 0$, nor is it undefined in the domain of f, there are no points of inflection.

33. $f(x) = \dfrac{x-2}{x^2 - 4x + 3}$, Domain: $(-\infty, 1) \cup (1, 3) \cup (3, \infty)$

$$f'(x) = \frac{(x^2-4x+3) - (x-2)(2x-4)}{(x^2-4x+3)^2} = \frac{-x^2+4x-5}{(x^2-4x+3)^2} \neq 0$$

$$f''(x) = \frac{(x^2-4x+3)^2(-2x+4) - (-x^2+4x-5)(2)(x^2-4x+3)(2x-4)}{(x^2-4x+3)^4}$$

$$= \frac{2(x^3-6x^2+15x-14)}{(x^2-4x+3)^3} = 0 \text{ when } x = 2.$$

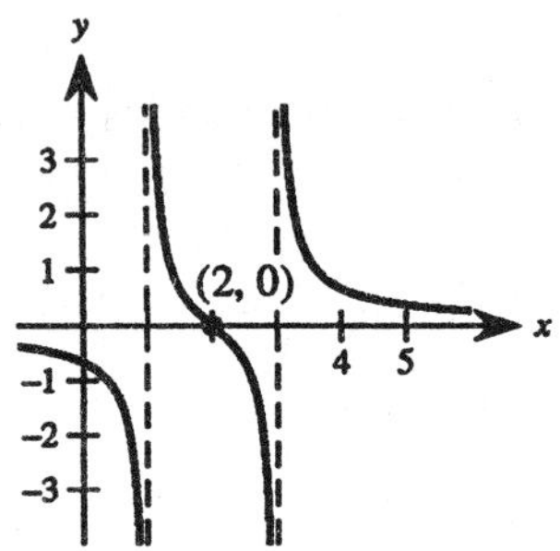

Since $f''(x) > 0$ on $(1, 2)$ and $f''(x) < 0$ on $(2, 3)$, then $(2, 0)$ is a point of inflection.

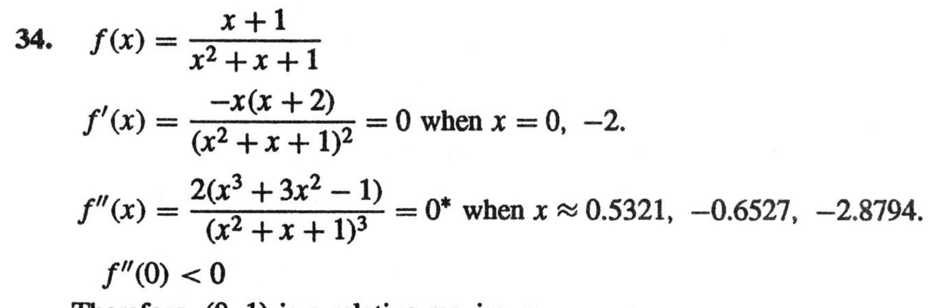

34. $f(x) = \dfrac{x+1}{x^2+x+1}$

$$f'(x) = \frac{-x(x+2)}{(x^2+x+1)^2} = 0 \text{ when } x = 0, -2.$$

$$f''(x) = \frac{2(x^3+3x^2-1)}{(x^2+x+1)^3} = 0^* \text{ when } x \approx 0.5321, -0.6527, -2.8794.$$

$f''(0) < 0$

Therefore, $(0, 1)$ is a relative maximum.

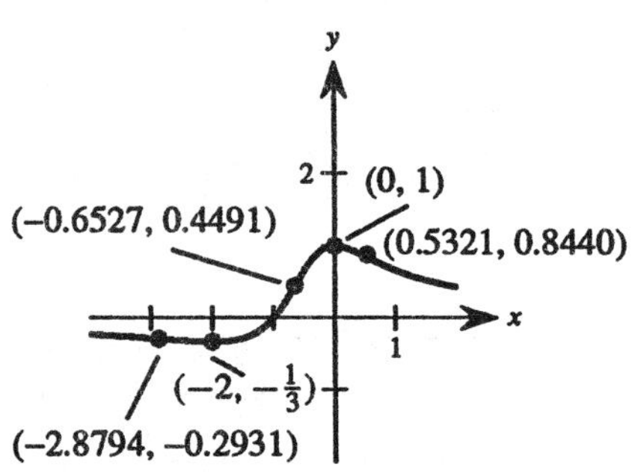

$f''(-2) > 0$

Therefore, $\left(-2, -\frac{1}{3}\right)$ is a relative minimum.

Points of inflection to four decimal places:

$(0.5321, 0.8440)$, $(-0.6527, 0.4491)$ and $(-2.8794, -0.2931)$

34. —CONTINUED—

*We can solve $x^3 + 3x^2 - 1 = 0$ by using an interpolation method on a programmable calculator or by the Bisection Method written below in BASIC.

```
10   REM DEFINE THE FUNCTION
20      DEF FNY(X) = X^3 + 3 * X^2 - 1
30   REM A = LEFT ENDPOINT, B = RIGHT ENDPOINT
40      INPUT " ENTER LEFT ENDPOINT:", A
50      INPUT "ENTER RIGHT ENDPOINT:", B
60   REM DETERMINE MIDPOINT = C
70      C = (A + B)/2
80   REM M = MAXIMUM ERROR
90      M = (B - A)/2
100  REM TEST THE SIZE OF THE ERROR
110     IF M < .00001 GOTO 160
120  REM TEST THE SIGN OF F(C)
130     IF FNY(A) * FNY(C) < 0 THEN B = C : GOTO 70
140     IF FNY(B) * FNY(C) < 0 THEN A = C : GOTO 70
150  REM PRINT THE APPROXIMATION
160     PRINT "A";A, "C =";C, "B =";B, "MAXIMUM ERROR =";M
170  END
```

The output from the program is:

ENTER LEFT ENDPOINT: −3
ENTER RIGHT ENDPOINT: −2
$A = -2.87939, \quad C = -2.87939, \quad B = -2.87938$, MAXIMUM ERROR $= 7.62939E - 06$

ENTER LEFT ENDPOINT: −1
ENTER RIGHT ENDPOINT: 0
$A = -.65271, \quad C = -.652702, \quad B = -.652695$, MAXIMUM ERROR $= 7.62939E - 06$

ENTER LEFT ENDPOINT: 0
ENTER RIGHT ENDPOINT: 1
$A = .532074, \quad C = .532082, \quad B = .532089$, MAXIMUM ERROR $= 7.62939E - 06$

35. $f(x) = 0.2x^2(x-3)^3, \quad [-1, 4]$

(a) $f'(x) = 0.2x(5x-6)(x-3)^2$

$f''(x) = (x-3)(4x^2 - 9.6x + 3.6)$

$= 0.4(x-3)(10x^2 - 24x + 9)$

(b) $f''(0) < 0 \Rightarrow (0,\ 0)$ is a relative maximum.

$f''\left(\frac{6}{5}\right) > 0 \Rightarrow (1.2,\ -1.6796)$ is a relative minimum.

Points of inflection:

$(3,\ 0), (0.4652,\ -0.7049), (1.9348,\ -0.9049)$

(c)

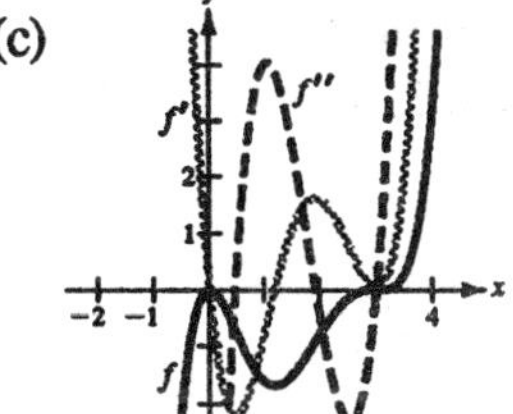

f is increasing when $f' > 0$ and decreasing when $f' < 0$. f is concave upward when $f'' > 0$ and concave downward when $f'' < 0$.

36. $f(x) = x^2\sqrt{6 - x^2}, \quad [-\sqrt{6}, \sqrt{6}]$

(a) $f'(x) = \dfrac{3x(4 - x^2)}{\sqrt{6 - x^2}}$

$f'(x) = 0$ when $x = 0, \ x = \pm 2$.

$f''(x) = \dfrac{6(x^4 - 9x^2 + 12)}{(6 - x^2)^{3/2}}$

$f''(x) = 0$ when $x = \pm\sqrt{\dfrac{9 - \sqrt{33}}{2}}$.

(b) $f''(0) > 0 \Rightarrow (0, 0)$ is a relative minimum.

$f''(\pm 2) < 0 \Rightarrow (\pm 2, 4\sqrt{2})$ are relative maxima.

Points of inflection: $(\pm 1.2758, 3.4035)$

(c)

The graph of f is increasing when $f' > 0$ and decreasing when $f' < 0$. f is concave upward when $f'' > 0$ and concave downward when $f'' < 0$.

37. (a)

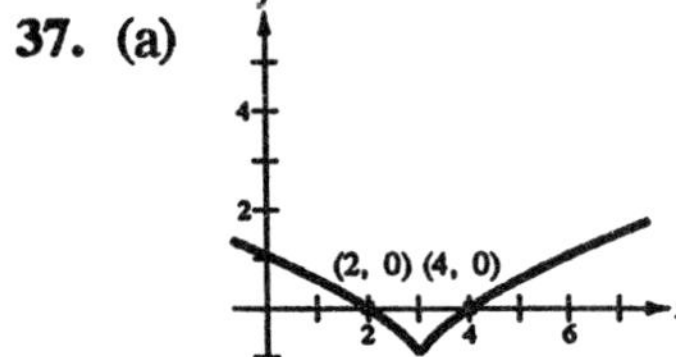

38. (a)

37. (b)

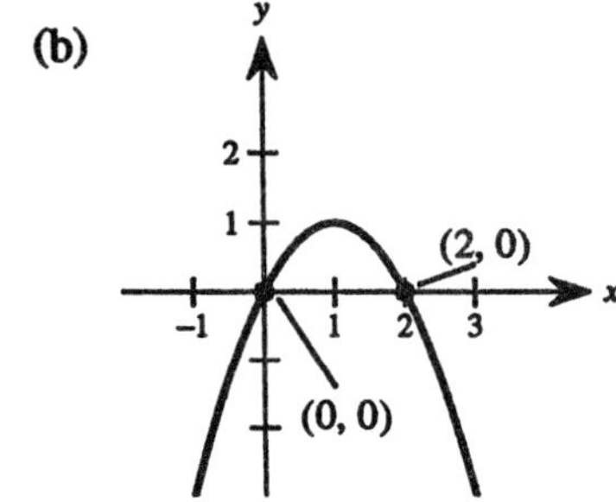

38. (b)

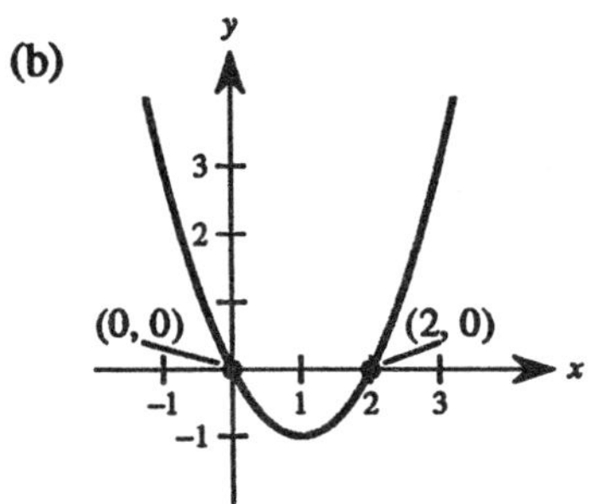

39.

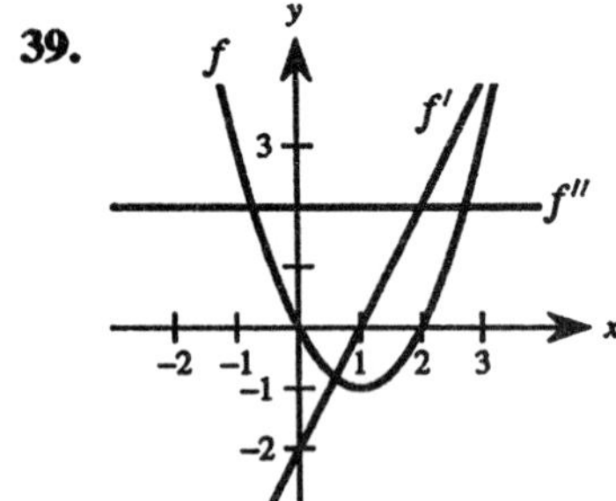

40.

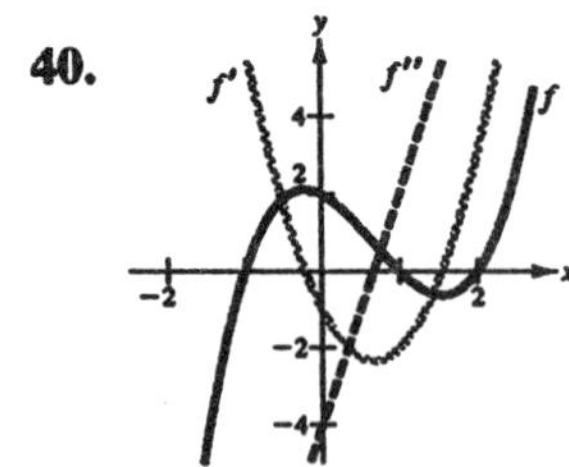

41. $n = 1$:

$f(x) = x - c$

$f'(x) = 1$

$f''(x) = 0$

No inflection points

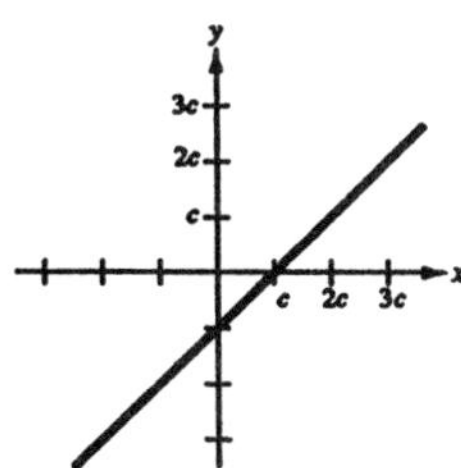

$n = 2$:

$f(x) = (x - c)^2$

$f'(x) = 2(x - c)$

$f''(x) = 2$

No inflection points

Relative minimum: $(c, 0)$

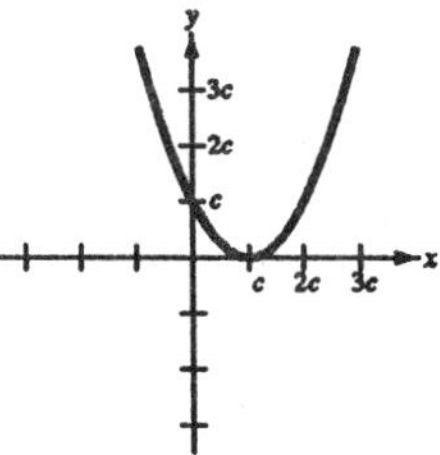

$n = 3$:

$f(x) = (x - c)^3$

$f'(x) = 3(x - c)^2$

$f''(x) = 6(x - c)$

Inflection point at $(c, 0)$

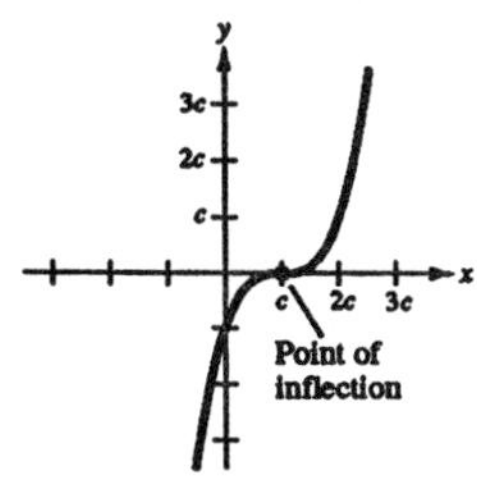

$n = 4$:

$f(x) = (x - c)^4$

$f'(x) = 4(x - c)^3$

$f''(x) = 12(x - c)^2$

No inflection points

Relative minimum: $(c, 0)$

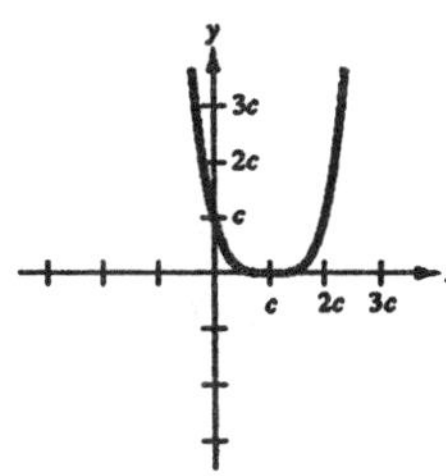

Conclusion: If $n \geq 3$ and n is odd, then $(c, 0)$ is an inflection point.
If $n \geq 2$ and n is even, then $(c, 0)$ is a relative minimum.

42. (a) $f(x) = \sqrt[3]{x}$

$f'(x) = \frac{1}{3}x^{-2/3}$

$f''(x) = -\frac{2}{9}x^{-5/3}$

Inflection point: $(0, 0)$

(b) $f''(x)$ does not exist at $x = 0$.

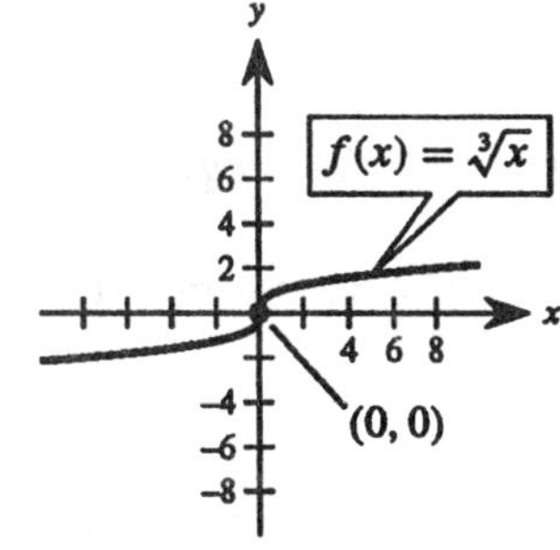

43. $f(x) = x(x-6)^2 = x^3 - 12x^2 + 36x$

$$f'(x) = 3x^2 - 24x + 36 = 3(x-2)(x-6) = 0$$

Relative extrema: (2, 32) and (6, 0)
Point of inflection (4, 16) is midway between the relative extrema of f.

44. Assume the zeros of f are all real. Then express the function as $f(x) = a(x-r_1)(x-r_2)(x-r_3)$ where r_1, r_2, and r_3 are the zeros of f. From the Product Rule for a function involving three factors we have

$$f'(x) = a[(x-r_1)(x-r_2) + (x-r_1)(x-r_3) + (x-r_2)(x-r_3)]$$

$$f''(x) = a[(x-r_1) + (x-r_2) + (x-r_1) + (x-r_3) + (x-r_2) + (x-r_3)]$$

$$= a[6x - 2(r_1 + r_2 + r_3)].$$

Consequently, $f''(x) = 0$ if

$$x = \frac{2(r_1 + r_2 + r_3)}{6} = \frac{r_1 + r_2 + r_3}{3} = \text{(average of } r_1, r_2, \text{ and } r_3).$$

45. $f(x) = ax^3 + bx^2 + cx + d$
Maximum: $(-4, 1)$
Minimum: $(0, 0)$
Point of inflection: $\left(-2, \frac{1}{2}\right)$

(a) $f'(x) = 3ax^2 + 2bx + c, \quad f''(x) = 6ax + 2b$

$$f(0) = 0 \Rightarrow d = 0$$

$$f(-4) = 1 \Rightarrow -64a + 16b - 4c = 1$$

$$f'(-4) = 0 \Rightarrow 48a - 8b + c = 0$$

$$f'(0) = 0 \Rightarrow c = 0$$

$$f''(-2) = 0 \Rightarrow -12a + 2b = 0$$

Solving this system yields $a = \frac{1}{32}$ and $b = 6a = \frac{3}{16}$.

$$f(x) = \frac{1}{32}x^3 + \frac{3}{16}x^2$$

(b) The plane would be descending at the greatest rate at the point of inflection, or two miles from touchdown.

46. $E = kqx(x^2 + a^2)^{-3/2}$

$$E' = kqx\left[-\frac{3}{2}(x^2 + a^2)^{-5/2}(2x)\right] + kq(x^2 + a^2)^{-3/2} = kq(x^2 + a^2)^{-5/2}[-3x^2 + (x^2 + a^2)] = \frac{kq(a^2 - 2x^2)}{(x^2 + a^2)^{5/2}}$$

By the First Derivative Test, E is maximized when $x = (a\sqrt{2})/2$.

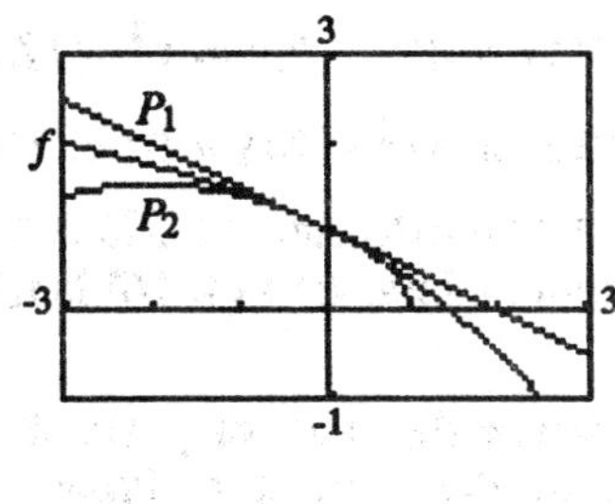

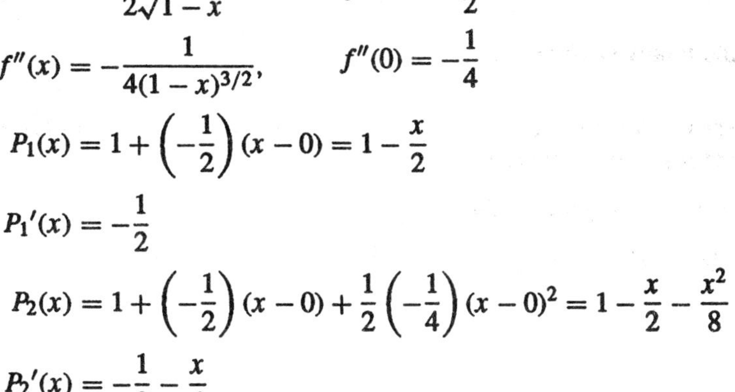

47. $f(x) = \sqrt{1-x}, \qquad f(0) = 1$

$$f'(x) = -\frac{1}{2\sqrt{1-x}}, \qquad f'(0) = -\frac{1}{2}$$

$$f''(x) = -\frac{1}{4(1-x)^{3/2}}, \qquad f''(0) = -\frac{1}{4}$$

$$P_1(x) = 1 + \left(-\frac{1}{2}\right)(x-0) = 1 - \frac{x}{2}$$

$$P_1'(x) = -\frac{1}{2}$$

$$P_2(x) = 1 + \left(-\frac{1}{2}\right)(x-0) + \frac{1}{2}\left(-\frac{1}{4}\right)(x-0)^2 = 1 - \frac{x}{2} - \frac{x^2}{8}$$

$$P_2'(x) = -\frac{1}{2} - \frac{x}{4}$$

$$P_2''(x) = -\frac{1}{4}$$

The values of f, P_1, P_2, and their first derivatives are equal at $x = 0$. The values of the second derivatives of f and P_2 are equal at $x = 0$.

48. $f(x) = \dfrac{\sqrt{x}}{x-1}, \qquad f(2) = \sqrt{2}$

$$f'(x) = \frac{-(x+1)}{2\sqrt{x}(x-1)^2}, \qquad f'(2) = -\frac{3}{2\sqrt{2}} = -\frac{3\sqrt{2}}{4}$$

$$f''(x) = \frac{3x^2+6x-1}{4x^{3/2}(x-1)^3}, \qquad f''(2) = \frac{23}{8\sqrt{2}} = \frac{23\sqrt{2}}{16}$$

$$P_1(x) = \sqrt{2} + \left(-\frac{3\sqrt{2}}{4}\right)(x-2) = -\frac{3\sqrt{2}}{4}x + \frac{5\sqrt{2}}{2}$$

$$P_1'(x) = -\frac{3\sqrt{2}}{4}$$

$$P_2(x) = \sqrt{2} + \left(-\frac{3\sqrt{2}}{4}\right)(x-2) + \frac{1}{2}\left(\frac{23\sqrt{2}}{16}\right)(x-2)^2 = \sqrt{2} - \frac{3\sqrt{2}}{4}(x-2) + \frac{23\sqrt{2}}{32}(x-2)^2$$

$$P_2'(x) = -\frac{3\sqrt{2}}{4} + \frac{23\sqrt{2}}{16}(x-2)$$

$$P_2''(x) = \frac{23\sqrt{2}}{16}$$

The values of f, P_1, P_2, and their first derivatives are equal at $x = 2$. The values of the second derivatives of f and P_2 are equal at $x = 2$.

49. (a) The rate of change of sales is increasing.
$S'' > 0$

(b) The rate of change of sales is decreasing.
$S'' < 0$

(c) The rate of change of sales is constant.
$S' = C, \quad S'' = 0$

(d) Sales are steady.
$S = C, \quad S' = 0, \quad S'' = 0$

(e) Sales are declining, but at a lower rate.
$S' < 0, \quad S'' > 0$

(f) Sales have bottomed out and have started to rise.
$S' > 0$

Section 4.5 Limits at Infinity

1. $f(x) = \dfrac{3x^2}{x^2+2}$

No vertical asymptotes
Horizontal asymptote: $y = 3$
Matches (h)

2. $f(x) = \dfrac{2x}{\sqrt{x^2+2}}$

No vertical asymptotes
Horizontal asymptotes: $y = \pm 2$
Matches (c)

3. $f(x) = \dfrac{x}{x^2+2}$

No vertical asymptotes
Horizontal asymptote: $y = 0$
Matches (e)

4. $f(x) = 2 + \dfrac{x^2}{x^4+1}$

No vertical asymptotes
Horizontal asymptote: $y = 2$
Matches (a)

5. $f(x) = \dfrac{-6x}{\sqrt{4x^2+5}}$

No vertical asymptotes
Horizontal asymptotes: $y = \pm 3$
Matches (d)

6. $f(x) = 5 - \dfrac{1}{x^2+1}$

No vertical asymptotes
Horizontal asymptote: $y = 5$
Matches (g)

7. $f(x) = \dfrac{4}{x^2+1}$

No vertical asymptotes
Horizontal asymptote: $y = 0$
Matches (b)

8. $f(x) = \dfrac{2x^2-3x+5}{x^2+1}$

No vertical asymptotes
Horizontal asymptote: $y = 2$
Matches (f)

9. $$\lim_{x\to\infty} \frac{2x-1}{3x+2} = \lim_{x\to\infty} \frac{2-(1/x)}{3+(2/x)} = \frac{2-0}{3+0} = \frac{2}{3}$$

10. $$\lim_{x\to\infty} \frac{5x^3+1}{10x^3-3x^2+7} = \lim_{x\to\infty} \frac{5+(1/x^3)}{10-(3/x)+(7/x^3)} = \frac{1}{2}$$

11. $$\lim_{x\to\infty} \frac{x}{x^2-1} = \lim_{x\to\infty} \frac{1/x}{1-(1/x^2)} = \frac{0}{1} = 0$$

12. $$\lim_{x\to\infty} \frac{2x^{10}-1}{10x^{11}-3} = \lim_{x\to\infty} \frac{(2/x)-(1/x^{11})}{10-(3/x^{11})} = \frac{0}{10} = 0$$

13. $$\lim_{x\to-\infty} \frac{5x^2}{x+3} = \lim_{x\to-\infty} \frac{5x}{1+(3/x)} = -\infty$$

Limit does not exist.

14. $$\lim_{x\to\infty} \frac{x^3-2x^2+3x+1}{x^2-3x+2} = \lim_{x\to\infty} \frac{x-2+(3/x)+(1/x^2)}{1-(3/x)+(2/x^2)} = \infty$$

Limit does not exist.

15. $$\lim_{x\to\infty} \left(2x - \frac{1}{x^2}\right) = \infty - 0 = \infty$$

Limit does not exist.

16. $$\lim_{x\to\infty} \frac{1}{(x+3)^2} = 0$$

17. $$\lim_{x\to-\infty} \left(\frac{2x}{x-1} + \frac{3x}{x+1}\right) = \lim_{x\to-\infty} \left(\frac{2}{1-(1/x)} + \frac{3}{1+(1/x)}\right) = 2+3 = 5$$

18. $\lim_{x\to\infty}\left(\frac{2x^2}{x-1}+\frac{3x}{x+1}\right)=\lim_{x\to\infty}\left(\frac{2x}{1-(1/x)}+\frac{3}{1+(1/x)}\right)=\infty+3=\infty$

Limit does not exist.

19. $\lim_{x\to-\infty}\frac{x}{\sqrt{x^2-x}}=\lim_{x\to-\infty}\frac{1}{\frac{\sqrt{x^2-x}}{-\sqrt{x^2}}},\quad \left(\text{for } x<0 \text{ we have } x=-\sqrt{x^2}\right)$

$=\lim_{x\to-\infty}\frac{-1}{\sqrt{1-(1/x)}}=-1$

20. $\lim_{x\to\infty}\frac{x}{\sqrt{x^2+1}}=\lim_{x\to\infty}\frac{1}{\sqrt{1+(1/x^2)}}=1$

21. $\lim_{x\to\infty}\frac{2x+1}{\sqrt{x^2-x}}=\lim_{x\to\infty}\frac{2+(1/x)}{\sqrt{1-(1/x)}}=2$

22. $\lim_{x\to-\infty}\frac{-3x+1}{\sqrt{x^2+x}}=\lim_{x\to-\infty}\frac{-3+(1/x)}{\frac{\sqrt{x^2+x}}{-\sqrt{x^2}}},\quad \left(\text{for } x<0 \text{ we have } -\sqrt{x^2}=x\right)$

$=\lim_{x\to-\infty}\frac{3-(1/x)}{\sqrt{1+(1/x)}}=3$

23. $\lim_{x\to\infty}\frac{x^2-x}{\sqrt{x^4+x}}=\lim_{x\to\infty}\frac{1-(1/x)}{\sqrt{1+(1/x^3)}}=1$

24. $\lim_{x\to\infty}\frac{2x}{\sqrt{4x^2+1}}=\lim_{x\to\infty}\frac{2}{\sqrt{4+(1/x^2)}}=1$

25. $\lim_{x\to-\infty}\left(x+\sqrt{x^2+3}\right)=\lim_{x\to-\infty}\left[\left(x+\sqrt{x^2+3}\right)\cdot\frac{x-\sqrt{x^2+3}}{x-\sqrt{x^2+3}}\right]=\lim_{x\to-\infty}\frac{-3}{x-\sqrt{x^2+3}}=0$

26. $\lim_{x\to\infty}\left(2x-\sqrt{4x^2+1}\right)=\lim_{x\to\infty}\left[\left(2x-\sqrt{4x^2+1}\right)\cdot\frac{2x+\sqrt{4x^2+1}}{2x+\sqrt{4x^2+1}}\right]=\lim_{x\to\infty}\frac{-1}{2x+\sqrt{4x^2+1}}=0$

27. $\lim_{x\to\infty}\left(x-\sqrt{x^2+x}\right)=\lim_{x\to\infty}\left[\left(x-\sqrt{x^2+x}\right)\cdot\frac{x+\sqrt{x^2+x}}{x+\sqrt{x^2+x}}\right]$

$=\lim_{x\to\infty}\frac{-x}{x+\sqrt{x^2+x}}=\lim_{x\to\infty}\frac{-1}{1+\sqrt{1+(1/x)}}=-\frac{1}{2}$

28. $\lim_{x\to-\infty}\left(3x+\sqrt{9x^2-x}\right)=\lim_{x\to-\infty}\left[\left(3x+\sqrt{9x^2-x}\right)\cdot\frac{3x-\sqrt{9x^2-x}}{3x-\sqrt{9x^2-x}}\right]$

$=\lim_{x\to-\infty}\frac{x}{3x-\sqrt{9x^2-x}}$

$=\lim_{x\to-\infty}\frac{1}{3-\frac{\sqrt{9x^2-x}}{-\sqrt{x^2}}}\quad \left(\text{for } x<0 \text{ we have } x=-\sqrt{x^2}\right)$

$=\lim_{x\to-\infty}\frac{1}{3+\sqrt{9-(1/x)}}=\frac{1}{6}$

29. $y = \dfrac{2+x}{1-x}$

Intercepts: $(-2, 0)$, $(0, 2)$

Symmetry: none

Horizontal asymptote: $y = -1$ since

$$\lim_{x\to-\infty} \frac{2+x}{1-x} = -1 = \lim_{x\to\infty} \frac{2+x}{1-x}.$$

Discontinuity: $x = 1$ (Vertical asymptote)

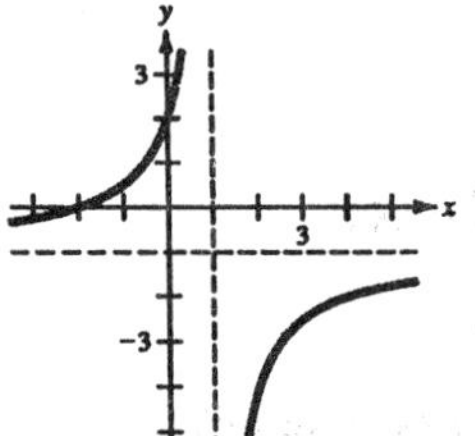

30. $y = \dfrac{x-3}{x-2}$

Intercepts: $(3, 0)$, $\left(0, \frac{3}{2}\right)$

Symmetry: none

Horizontal asymptote: $y = 1$ since

$$\lim_{x\to-\infty} \frac{x-3}{x-2} = 1 = \lim_{x\to\infty} \frac{x-3}{x-2}.$$

Discontinuity: $x = 2$ (Vertical asymptote)

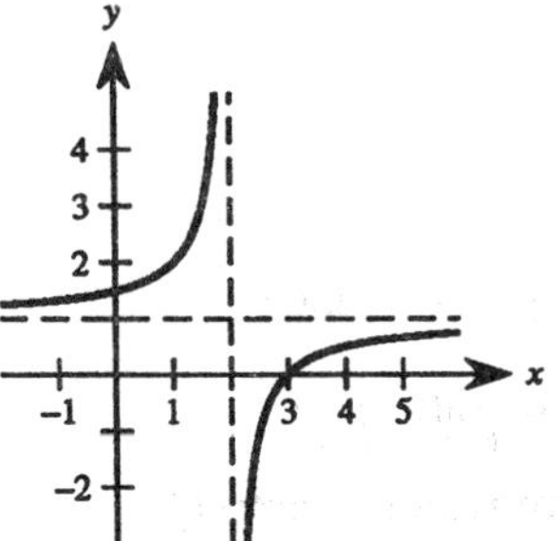

31. $y = \dfrac{x^2}{x^2+9}$

Intercept: $(0, 0)$

Symmetry: y-axis

Horizontal asymptote: $y = 1$ since

$$\lim_{x\to-\infty} \frac{x^2}{x^2+9} = 1 = \lim_{x\to\infty} \frac{x^2}{x^2+9}.$$

Relative minimum: $(0, 0)$

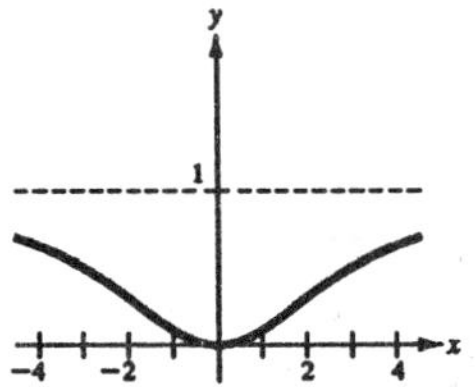

32. $y = \dfrac{x^2}{x^2-9}$

Intercept: $(0, 0)$

Symmetry: y-axis

Horizontal asymptote: $y = 1$ since

$$\lim_{x\to-\infty} \frac{x^2}{x^2-9} = 1 = \lim_{x\to\infty} \frac{x^2}{x^2-9}.$$

Discontinuities: $x = \pm 3$ (Vertical asymptotes)

Relative maximum: $(0, 0)$

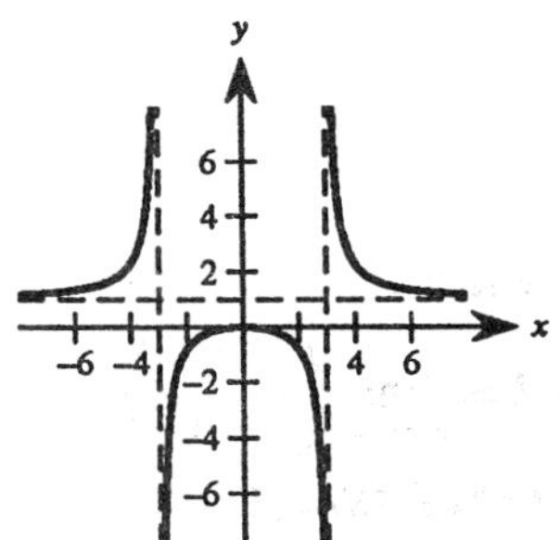

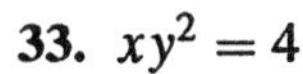

33. $xy^2 = 4$

Domain: $x > 0$

Intercepts: none

Symmetry: x-axis

Horizontal asymptote: $y = 0$ since

$$\lim_{x\to\infty} \frac{2}{\sqrt{x}} = 0 = \lim_{x\to\infty} -\frac{2}{\sqrt{x}}.$$

Discontinuity: $x = 0$ (Vertical asymptote)

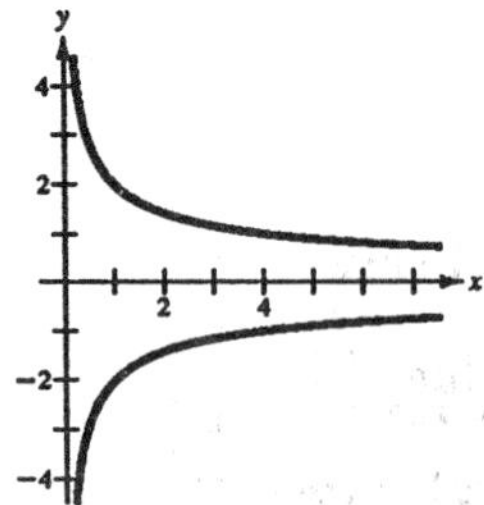

34. $x^2y = 4$

Intercepts: none

Symmetry: y-axis

Horizontal asymptote: $y = 0$ since

$$\lim_{x\to-\infty} \frac{4}{x^2} = 0 = \lim_{x\to\infty} \frac{4}{x^2}.$$

Discontinuity: $x = 0$ (Vertical asymptote)

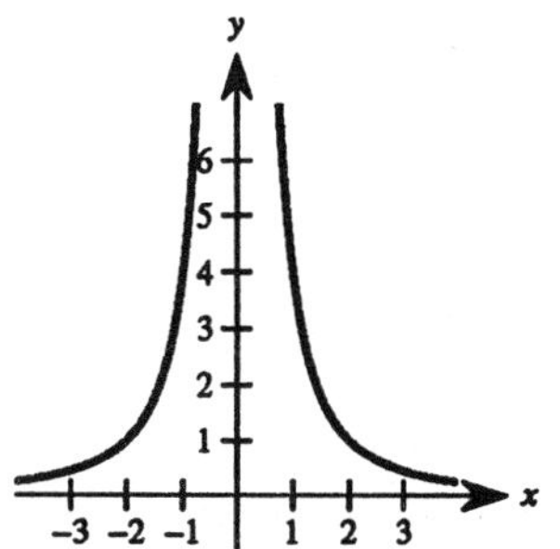

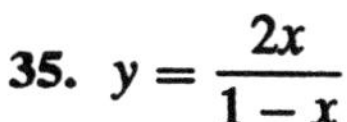

35. $y = \dfrac{2x}{1 - x}$

Intercept: (0, 0)

Symmetry: none

Horizontal asymptote: $y = -2$ since

$$\lim_{x\to-\infty} \frac{2x}{1-x} = -2 = \lim_{x\to\infty} \frac{2x}{1-x}.$$

Discontinuity: $x = 1$ (Vertical asymptote)

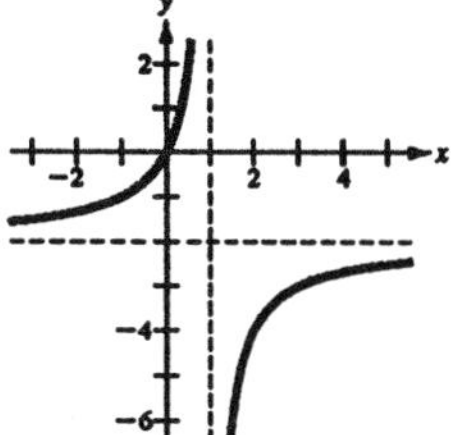

36. $y = \dfrac{2x}{1 - x^2}$

Intercept: (0, 0)

Symmetry: origin

Horizontal asymptote: $y = 0$ since

$$\lim_{x\to-\infty} \frac{2x}{1-x^2} = 0 = \lim_{x\to\infty} \frac{2x}{1-x^2}$$

Discontinuities: $x = \pm 1$ (Vertical asymptotes)

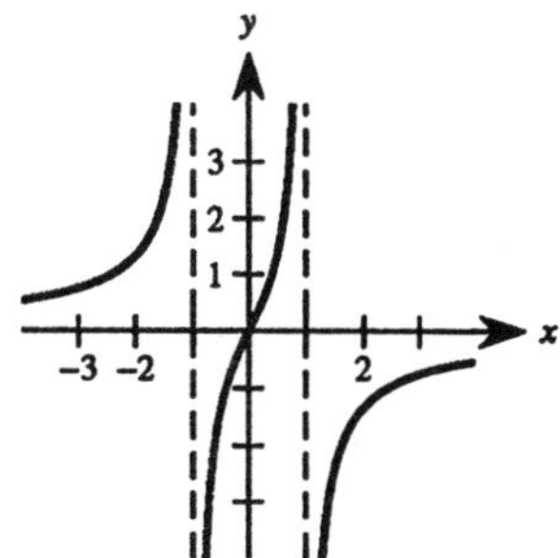

37. $y = 2 - \dfrac{3}{x^2}$

Intercepts: $\left(\pm\sqrt{3/2},\ 0\right)$

Symmetry: y-axis

Horizontal asymptote: $y = 2$ since

$$\lim_{x\to-\infty} \left(2 - \frac{3}{x^2}\right) = 2 = \lim_{x\to\infty} \left(2 - \frac{3}{x^2}\right).$$

Discontinuity: $x = 0$ (Vertical asymptote)

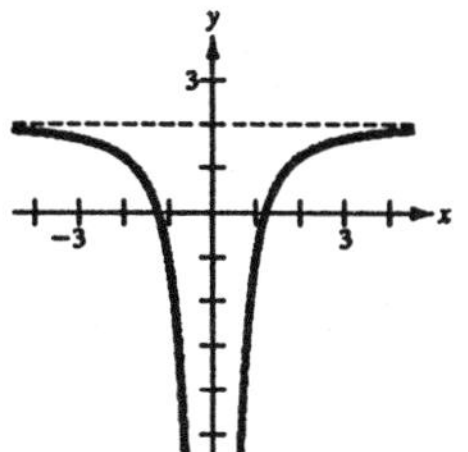

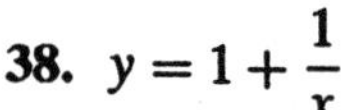

38. $y = 1 + \dfrac{1}{x}$

Intercept: $(-1,\ 0)$

Symmetry: none

Horizontal asymptote: $y = 1$ since

$$\lim_{x\to-\infty} \left(1 + \frac{1}{x}\right) = 1 = \lim_{x\to\infty} \left(1 + \frac{1}{x}\right).$$

Discontinuity: $x = 0$ (Vertical asymptote)

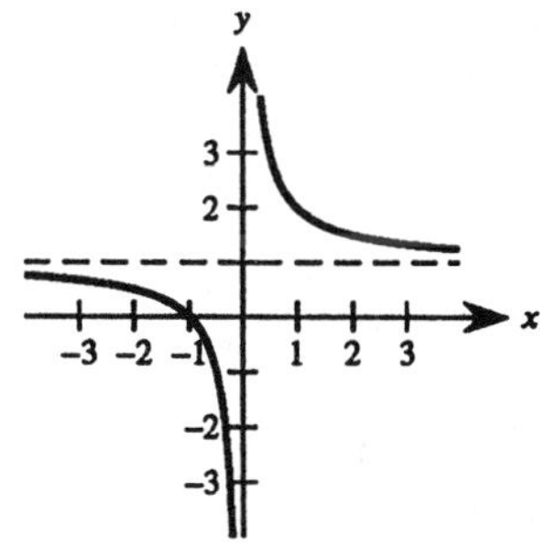

39. $y = \dfrac{x^3}{\sqrt{x^2-4}}$

Domain: $(-\infty, -2)$, $(2, \infty)$

Intercepts: none

Symmetry: origin

Horizontal asymptote: none

Vertical asymptotes:

$x = \pm 2$ (discontinuities)

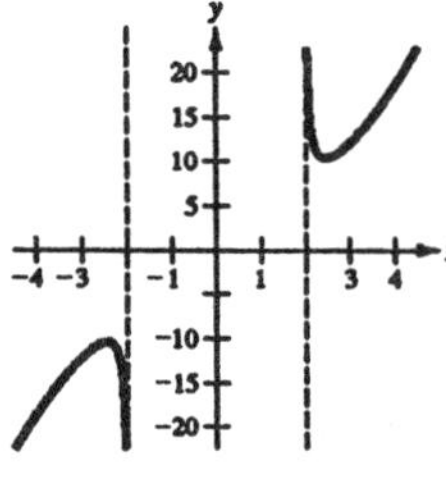

40. $y = \dfrac{x}{\sqrt{x^2-4}}$

Domain: $(-\infty, -2)$, $(2, \infty)$

Intercepts: none

Symmetry: origin

Horizontal asymptotes: $y = \pm 1$ since

$$\lim_{x\to\infty} \frac{x}{\sqrt{x^2-4}} = 1, \quad \lim_{x\to-\infty} \frac{x}{\sqrt{x^2-4}} = -1.$$

Vertical asymptotes:

$x = \pm 2$ (discontinuities)

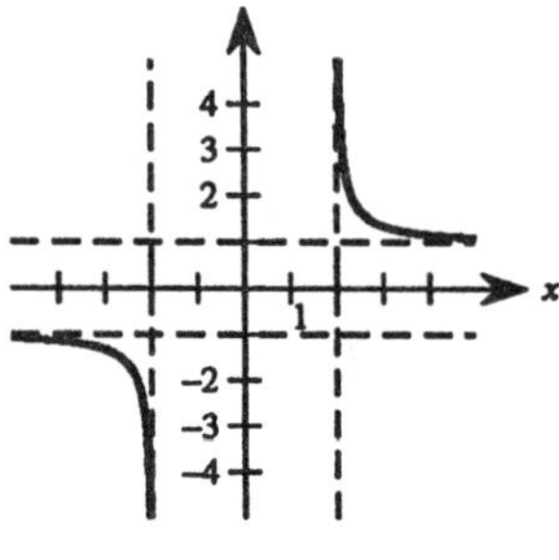

41. $f(x) = 5 - \dfrac{1}{x^2} = \dfrac{5x^2-1}{x^2}$

Domain: $(-\infty, 0)$, $(0, \infty)$

$f'(x) = \dfrac{2}{x^3} \Rightarrow$ No relative extrema

$f''(x) = -\dfrac{6}{x^4} \Rightarrow$ No points of inflection

Vertical asymptote: $x = 0$

Horizontal asymptote: $y = 5$

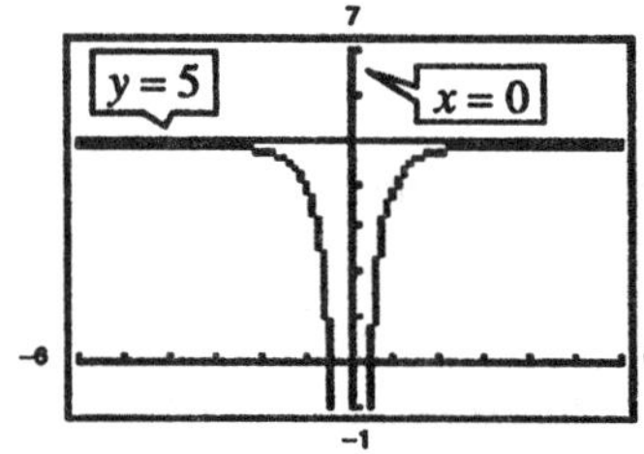

42. $f(x) = \dfrac{x^2}{x^2-1}$

$$f'(x) = \frac{(x^2-1)(2x) - x^2(2x)}{(x^2-1)^2} = \frac{-2x}{(x^2-1)^2} = 0 \text{ when } x = 0.$$

$$f''(x) = \frac{(x^2-1)^2(-2) + 2x(2)(x^2-1)(2x)}{(x^2-1)^4} = \frac{2(3x^2+1)}{(x^2-1)^3}$$

Since $f''(0) < 0$, then $(0, 0)$ is a relative maximum. Since $f''(x) \neq 0$, nor is it undefined in the domain of f, there are no points of inflection.

Vertical asymptotes: $x = \pm 1$

Horizontal asymptote: $y = 1$

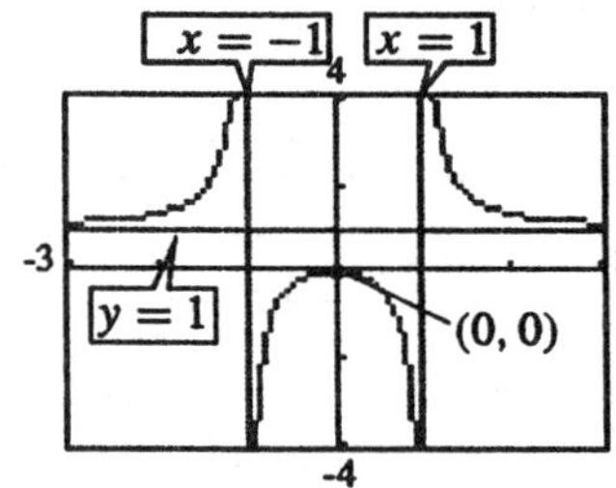

43. $f(x) = \dfrac{x}{x^2 - 4}$

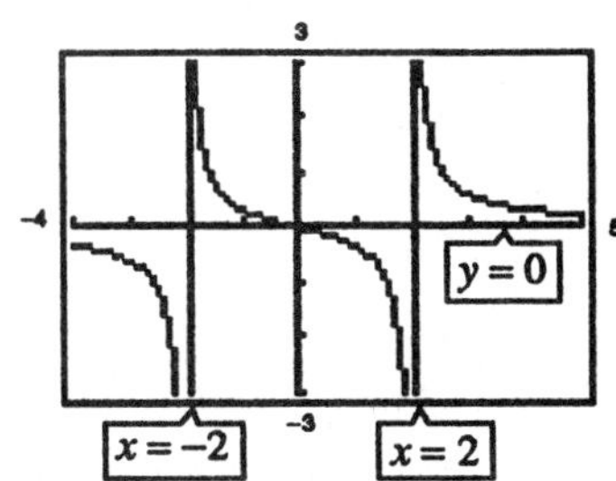

$$f'(x) = \frac{(x^2 - 4) - x(2x)}{(x^2 - 4)^2}$$

$$= \frac{-(x^2 + 4)}{(x^2 - 4)^2} \neq 0 \text{ for any } x \text{ in the domain of } f$$

$$f''(x) = \frac{(x^2 - 4)^2(-2x) + (x^2 + 4)(2)(x^2 - 4)(2x)}{(x^2 - 4)^4}$$

$$= \frac{2x(x^2 + 12)}{(x^2 - 4)^3} = 0 \text{ when } x = 0.$$

Since $f''(x) > 0$ on $(-2,\ 0)$ and $f''(x) < 0$ on $(0, 2)$, then $(0, 0)$ is a point of inflection.
Vertical asymptotes: $x = \pm 2$
Horizontal asymptote: $y = 0$

44. $f(x) = \dfrac{1}{x^2 - x - 2} = \dfrac{1}{(x + 1)(x - 2)}$

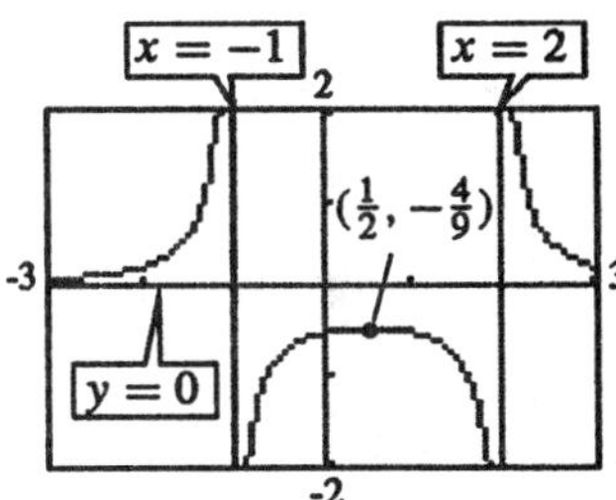

$$f'(x) = \frac{-(2x - 1)}{(x^2 - x - 2)^2} = 0 \text{ when } x = \frac{1}{2}$$

$$f''(x) = \frac{(x^2 - x - 2)^2(-2) + (2x - 1)(2)(x^2 - x - 2)(2x - 1)}{(x^2 - x - 2)^4}$$

$$= \frac{6(x^2 - x + 1)}{(x^2 - x - 2)^3}$$

Since $f''\left(\frac{1}{2}\right) < 0$, then $\left(\frac{1}{2},\ -\frac{4}{9}\right)$ is a relative maximum. Since $f''(x) \neq 0$, nor is it undefined in the domain of f, there are no points of inflection.
Vertical asymptotes: $x = -1,\ x = 2$
Horizontal asymptote: $y = 0$

45. $f(x) = \dfrac{x - 2}{x^2 - 4x + 3} = \dfrac{x - 2}{(x - 1)(x - 3)}$

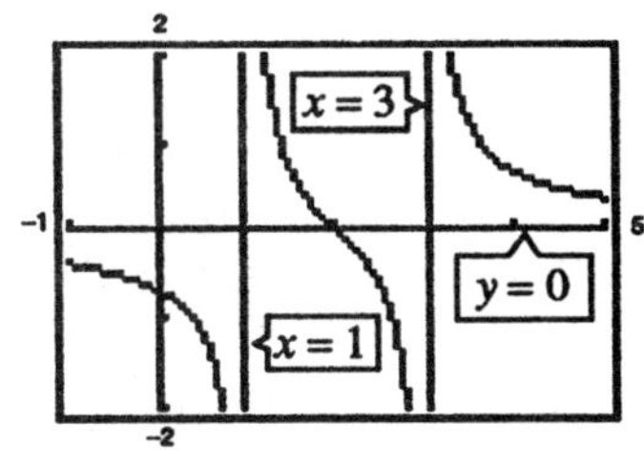

$$f'(x) = \frac{(x^2 - 4x + 3) - (x - 2)(2x - 4)}{(x^2 - 4x + 3)^2} = \frac{-x^2 + 4x - 5}{(x^2 - 4x + 3)^2} \neq 0$$

$$f''(x) = \frac{(x^2 - 4x + 3)^2(-2x + 4) - (-x^2 + 4x - 5)(2)(x^2 - 4x + 3)(2x - 4)}{(x^2 - 4x + 3)^4}$$

$$= \frac{2(x^3 - 6x^2 + 15x - 14)}{(x^2 - 4x + 3)^3} = 0 \text{ when } x = 2.$$

Since $f''(x) > 0$ on $(1, 2)$ and $f''(x) < 0$ on $(2, 3)$, then $(2, 0)$ is a point of inflection.
Vertical asymptote: $x = 1,\ x = 3$
Horizontal asymptote: $y = 0$

46. $f(x) = \dfrac{x+1}{x^2+x+1}$

$f'(x) = \dfrac{-x(x+2)}{(x^2+x+1)^2} = 0$ when $x = 0, \ -2.$

$f''(x) = \dfrac{2(x^3+3x^2-1)}{(x^2+x+1)^3} = 0$ when $x \approx 0.5321, \ -0.6527, \ -2.8794$

$f''(0) < 0$

Therefore, $(0, 1)$ is a relative maximum.

$f''(-2) > 0$

Therefore, $\left(-2, \ -\frac{1}{3}\right)$ is a relative minimum.

Points of inflection: $(0.5321, \ 0.8440)$, $(-0.6527, \ 0.4491)$ and $(-2.8794, \ -0.2931)$

Horizontal asymptote: $y = 0$

(−0.6527, 0.4491)
(0.5321, 0.8440)
(0, 1)
$\left(-2, -\frac{1}{3}\right)$
(−2.8794, −0.2931)

47. $f(x) = \dfrac{x^3-3x^2+2}{x(x-3)}, \quad g(x) = x + \dfrac{2}{x(x-3)}$

(a)

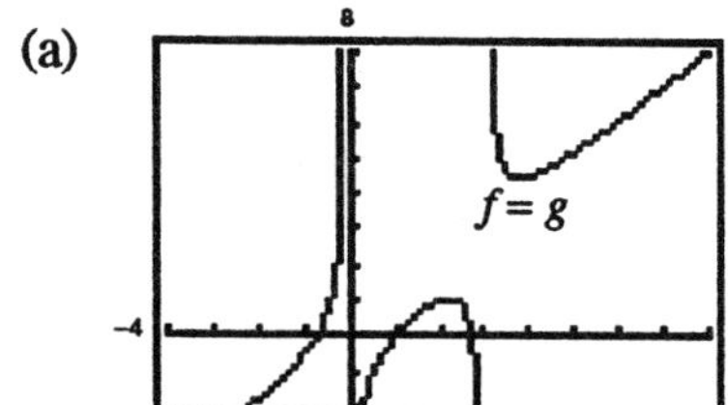

(b) $f(x) = \dfrac{x^3-3x^2+2}{x(x-3)}$

$= \dfrac{x^2(x-3)}{x(x-3)} + \dfrac{2}{x(x-3)}$

$= x + \dfrac{2}{x(x-3)} = g(x)$

(c)

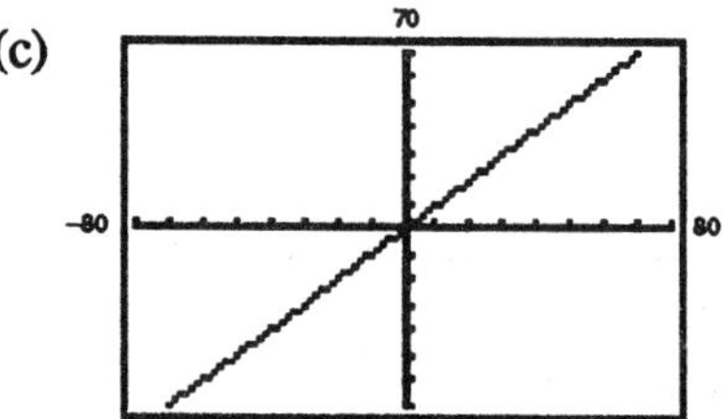

The graph appears as the slant asymptote $y = x$.

48. $f(x) = -\dfrac{x^3-2x^2+2}{2x^2}, \quad g(x) = -\dfrac{1}{2}x + 1 - \dfrac{1}{x^2}$

(a)

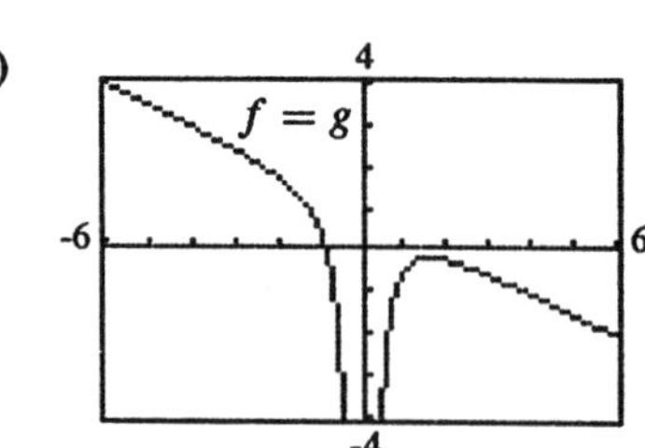

(b) $f(x) = -\dfrac{x^3-2x^2+2}{2x^2}$

$= -\left[\dfrac{x^3}{2x^2} - \dfrac{2x^2}{2x^2} + \dfrac{2}{2x^2}\right]$

$= -\dfrac{1}{2}x + 1 - \dfrac{1}{x^2} = g(x)$

(c)

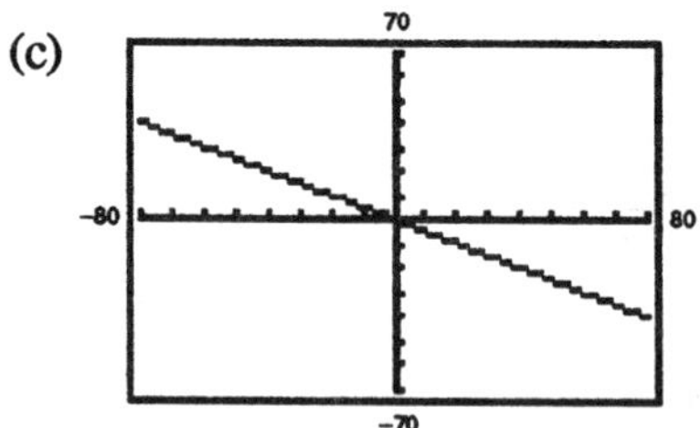

The graph appears as the slant asymptote $y = -\frac{1}{2}x + 1$.

49.

x	1	10	10^2	10^3	10^4	10^5	10^6
$f(x)$	2.000	0.348	0.101	0.032	0.010	0.003	0.001

$$\lim_{x\to\infty} \frac{x+1}{x\sqrt{x}} = 0$$

50.

x	1	10	10^2	10^3	10^4	10^5	10^6
$f(x)$	1	0.513	0.501	0.500	0.500	0.500	0.500

$$\lim_{x\to\infty} \left[x - \sqrt{x(x-1)}\right] = \tfrac{1}{2}$$

51.

x	1	10	10^2	10^3	10^4	10^5	10^6
$f(x)$	−0.2361	−0.0250	−0.0025	−0.0002	−0.000025	−0.000002	0.0

$$\lim_{x\to\infty} \left(2x - \sqrt{4x^2+1}\right) = 0$$

52.

x	1	10	10^2	10^3	10^4	10^5	10^6
$f(x)$	1.0	5.1	50.1	500.1	5000.1	50000.2	500000.2

$$\lim_{x\to\infty} \left[x^2 - x\sqrt{x(x-1)}\right] = \infty$$

Limit does not exist

53. $C = 0.5x + 500$

$$\overline{C} = \frac{C}{x}$$

$$\overline{C} = 0.5 + \frac{500}{x}$$

$$\lim_{x\to\infty} \left(0.5 + \frac{500}{x}\right) = 0.5$$

54. $\displaystyle\lim_{v\to c} \frac{m_0}{\sqrt{1-(v^2/c^2)}} = \frac{m_0}{0} \to \infty$

Limit does not exist

55. $\displaystyle\lim_{v_1/v_2\to\infty} 100\left[1 - \frac{1}{(v_1/v_2)^c}\right] = 100[1-0] = 100\%$

56. (a) $f(x) = \dfrac{|x|}{x+1}$

$$\lim_{x\to\infty} \frac{|x|}{x+1} = 1$$

$$\lim_{x\to-\infty} \frac{|x|}{x+1} = -1$$

Therefore, $y = 1$ and $y = -1$ are both horizontal asymptotes.

(b) $f(x) = \dfrac{2x}{\sqrt{x^2+1}}$

$$\lim_{x\to\infty} \frac{2x}{\sqrt{x^2+1}} = 2$$

$$\lim_{x\to-\infty} \frac{2x}{\sqrt{x^2+1}} = -2$$

Therefore, $y - 2$ and $y = -2$ are both horizontal asymptotes.

57. $f(x) = \dfrac{3x}{\sqrt{4x^2+1}}$

Horizontal asymptotes: $y = -\frac{3}{2}$, $y = \frac{3}{2}$

No vertical asymptotes

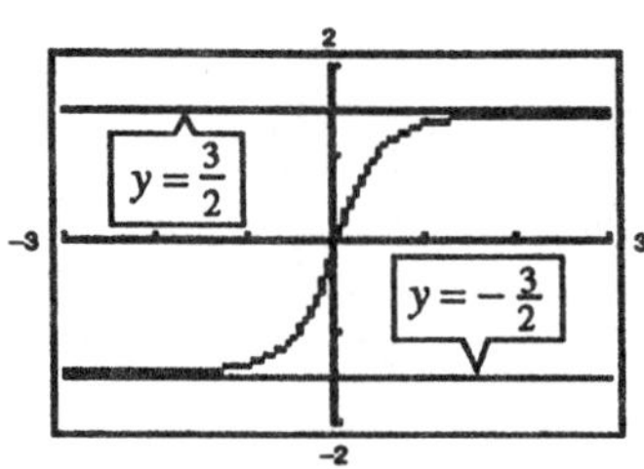

58. $\displaystyle\lim_{x\to\infty} \frac{p(x)}{q(x)} = \lim_{x\to\infty} \frac{a_n x^n + \cdots + a_1 x + a_0}{b_m x^m + \cdots + b_1 x + b_0}$

Divide $p(x)$ and $q(x)$ by x^m.

Case 1: If $n < m$,

$$\lim_{x\to\infty} \frac{p(x)}{q(x)} = \lim_{x\to\infty} \frac{\dfrac{a_n}{x^{m-n}} + \cdots + \dfrac{a_1}{x^{m-1}} + \dfrac{a_0}{x^m}}{b_m + \cdots + \dfrac{b_1}{x^{m-1}} + \dfrac{b_0}{x^m}} = \frac{0 + \cdots + 0 + 0}{b_m + \cdots + 0 + 0} = \frac{0}{b_m} = 0.$$

Case 2: If $m = n$,

$$\lim_{x\to\infty} \frac{p(x)}{q(x)} = \lim_{x\to\infty} \frac{a_n + \cdots + \dfrac{a_1}{x^{m-1}} + \dfrac{a_0}{x^m}}{b_m + \cdots + \dfrac{b_1}{x^{m-1}} + \dfrac{b_0}{x^m}} = \frac{a_n + \cdots + 0 + 0}{b_m + \cdots + 0 + 0} = \frac{a_n}{b_m}.$$

Case 3: If $n > m$,

$$\lim_{x\to\infty} \frac{p(x)}{q(x)} = \lim_{x\to\infty} \frac{a_n x^{n-m} + \cdots + \dfrac{a_1}{x^{m-1}} + \dfrac{a_0}{x^m}}{b_m + \cdots + \dfrac{b_1}{x^{m-1}} + \dfrac{b_0}{x^m}} = \frac{\pm\infty + \cdots + 0}{b_m + \cdots + 0} = \pm\infty.$$

Section 4.6 A Summary of Curve Sketching

1. $y = x^3 - 3x^2 + 3$

$y' = 3x^2 - 6x = 3x(x-2) = 0$ when $x = 0$, $x = 2$.

$y'' = 6x - 6 = 6(x-1) = 0$ when $x = 1$.

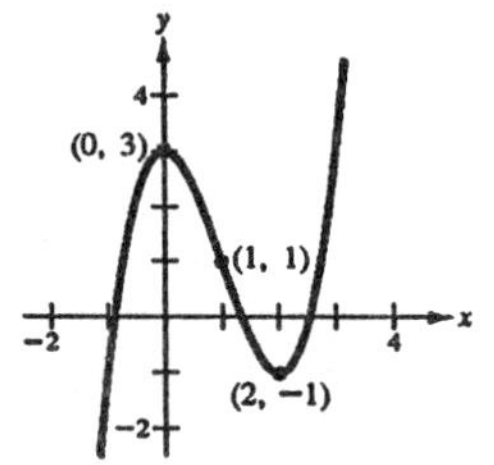

	y	y'	y''	Conclusion
$-\infty < x < 0$		+	−	Increasing, concave down
$x = 0$	3	0	−	Relative maximum
$0 < x < 1$		−	−	Decreasing, concave down
$x = 1$	1	−	0	Point of inflection
$1 < x < 2$		−	+	Decreasing, concave up
$x = 2$	−1	0	+	Relative minimum
$2 < x < \infty$		+	+	Increasing, concave up

2. $y = -\frac{1}{3}(x^3 - 3x + 2)$

$y' = -x^2 + 1 = 0$ when $x = \pm 1$.

$y'' = -2x = 0$ when $x = 0$.

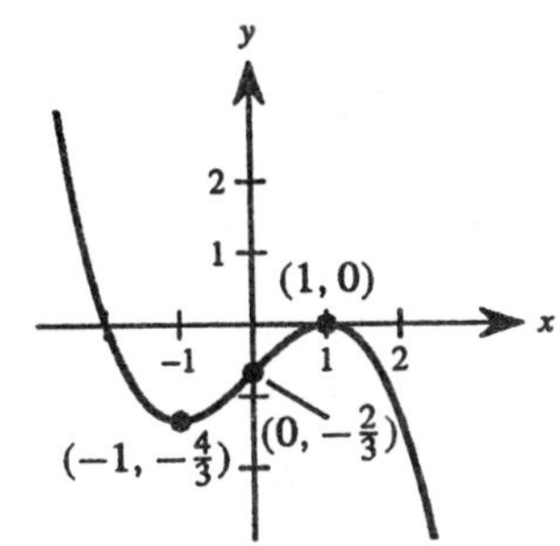

	y	y'	y''	Conclusion
$-\infty < x < -1$		$-$	$+$	Decreasing, concave up
$x = -1$	$-\frac{4}{3}$	0	$+$	Relative minimum
$-1 < x < 0$		$+$	$+$	Increasing, concave up
$x = 0$	$-\frac{2}{3}$	$+$	0	Point of inflection
$0 < x < 1$		$+$	$-$	Increasing, concave down
$x = 1$	0	0	$-$	Relative maximum
$1 < x < \infty$		$-$	$-$	Decreasing, concave down

3. $y = 2 - x - x^3$

$y' = -1 - 3x^2$ No critical numbers

$y'' = -6x = 0$ when $x = 0$.

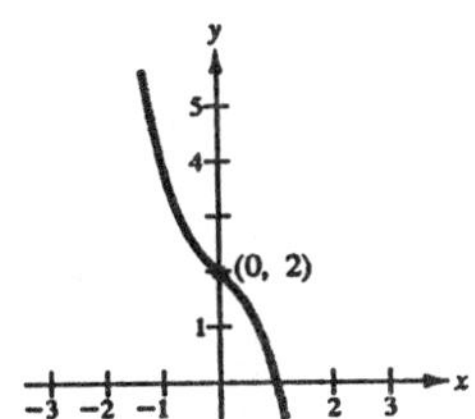

	y	y'	y''	Conclusion
$-\infty < x < 0$		$-$	$+$	Decreasing, concave up
$x = 0$	2	$-$	0	Point of inflection
$0 < x < \infty$		$-$	$-$	Decreasing, concave down

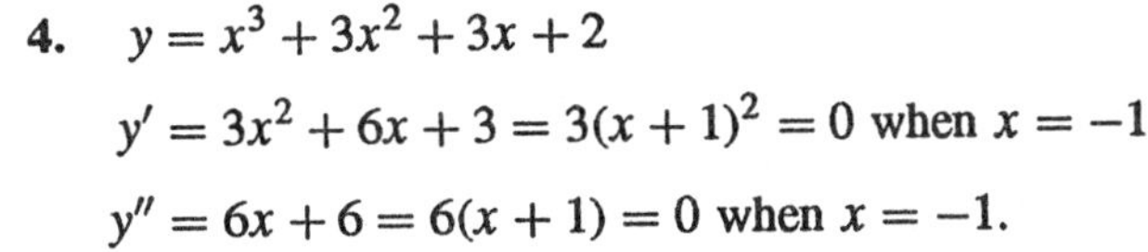

4. $y = x^3 + 3x^2 + 3x + 2$

$y' = 3x^2 + 6x + 3 = 3(x + 1)^2 = 0$ when $x = -1$.

$y'' = 6x + 6 = 6(x + 1) = 0$ when $x = -1$.

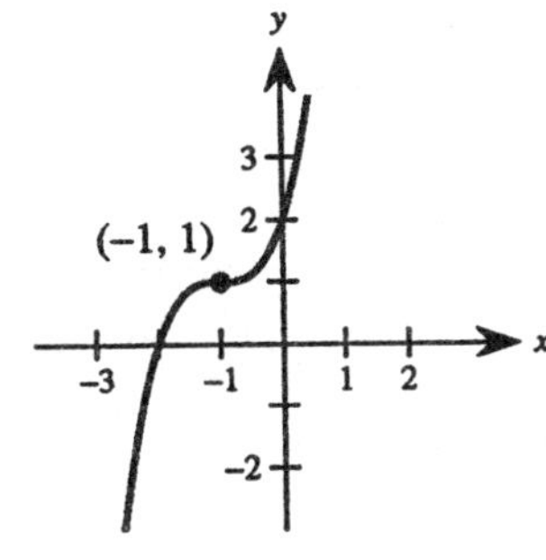

	y	y'	y''	Conclusion
$-\infty < x < -1$		$+$	$-$	Increasing, concave down
$x = -1$	1	0	0	Point of inflection
$-1 < x < \infty$		$+$	$+$	Increasing, concave up

5. $f(x) = 3x^3 - 9x + 1$

$f'(x) = 9x^2 - 9 = 9(x^2 - 1) = 0$ when $x = \pm 1$.

$f''(x) = 18x = 0$ when $x = 0$.

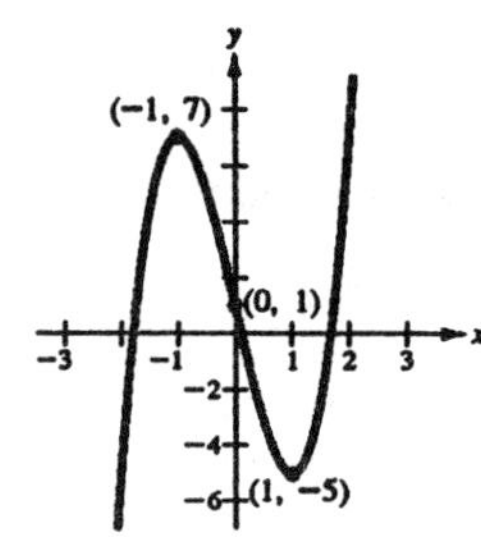

	$f(x)$	$f'(x)$	$f''(x)$	Conclusion
$-\infty < x < -1$		$+$	$-$	Increasing, concave down
$x = -1$	7	0	$-$	Relative maximum
$-1 < x < 0$		$-$	$-$	Decreasing, concave down
$x = 0$	1	$-$	0	Point of inflection
$0 < x < 1$		$-$	$+$	Decreasing, concave up
$x = 1$	-5	0	$+$	Relative minimum
$1 < x < \infty$		$+$	$+$	Increasing, concave up

6. $f(x) = (x + 1)(x - 2)(x - 5)$

$f'(x) = (x + 1)(x - 2) + (x + 1)(x - 5) + (x - 2)(x - 5)$

$= 3(x^2 - 4x + 1) = 0$ when $x = 2 \pm \sqrt{3}$.

$f''(x) = 6(x - 2) = 0$ when $x = 2$.

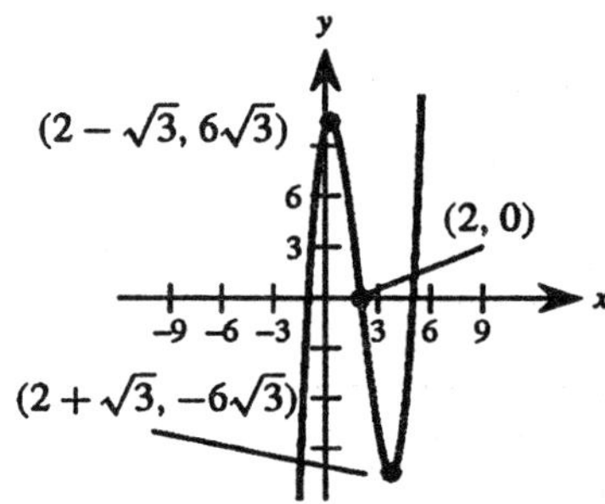

	$f(x)$	$f'(x)$	$f''(x)$	Conclusion
$-\infty < x < 2 - \sqrt{3}$		$+$	$-$	Increasing, concave down
$x = 2 - \sqrt{3}$	$6\sqrt{3}$	0	$-$	Relative maximum
$2 - \sqrt{3} < x < 2$		$-$	$-$	Decreasing, concave down
$x = 2$	0	$-$	0	Point of inflection
$2 < x < 2 + \sqrt{3}$		$-$	$+$	Decreasing, concave up
$x = 2 + \sqrt{3}$	$-6\sqrt{3}$	0	$+$	Relative minimum
$2 + \sqrt{3} < x < \infty$		$+$	$+$	Increasing, concave up

7. $f(x) = -x^3 + 3x^2 + 9x - 2$

$f'(x) = -3x^2 + 6x + 9 = -3(x+1)(x-3) = 0$ when $x = -1,\ 3$.

$f''(x) = -6x + 6 = -6(x-1) = 0$ when $x = 1$.

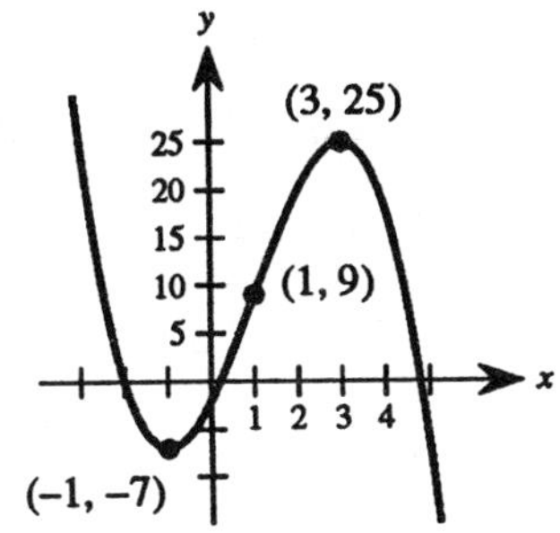

	$f(x)$	$f'(x)$	$f''(x)$	Conclusion
$-\infty < x < -1$		$-$	$+$	Decreasing, concave up
$x = -1$	-7	0	$+$	Relative minimum
$-1 < x < 1$		$+$	$+$	Increasing, concave up
$x = 1$	9	$+$	0	Point of inflection
$1 < x < 3$		$+$	$-$	Increasing, concave down
$x = 3$	25	0	$-$	Relative maximum
$3 < x < \infty$		$-$	$-$	Decreasing, concave down

8. $f(x) = \frac{1}{3}(x-1)^3 + 2$

$f'(x) = (x-1)^2 = 0$ when $x = 1$.

$f''(x) = 2(x-1) = 0$ when $x = 1$.

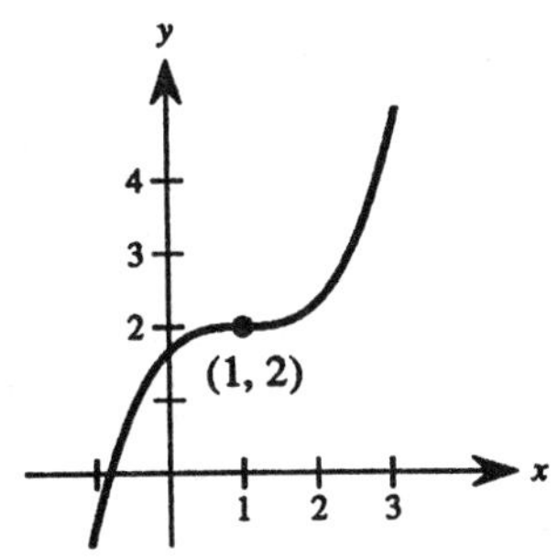

	$f(x)$	$f'(x)$	$f''(x)$	Conclusion
$-\infty < x < 1$		$+$	$-$	Increasing, concave down
$x = 1$	2	0	0	Point of inflection
$1 < x < \infty$		$+$	$+$	Increasing, concave up

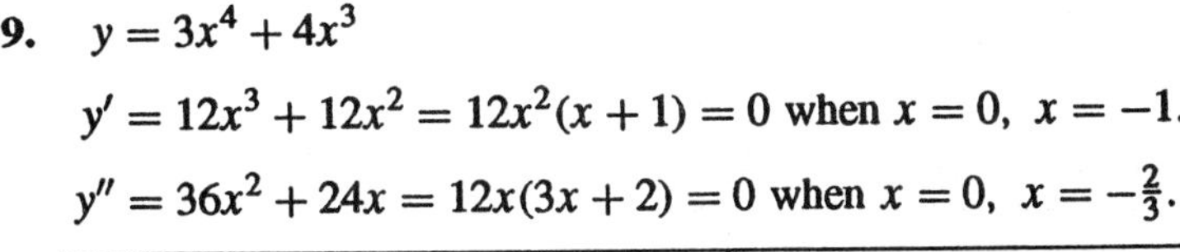

9. $y = 3x^4 + 4x^3$

$y' = 12x^3 + 12x^2 = 12x^2(x+1) = 0$ when $x = 0,\ x = -1$.

$y'' = 36x^2 + 24x = 12x(3x+2) = 0$ when $x = 0,\ x = -\frac{2}{3}$.

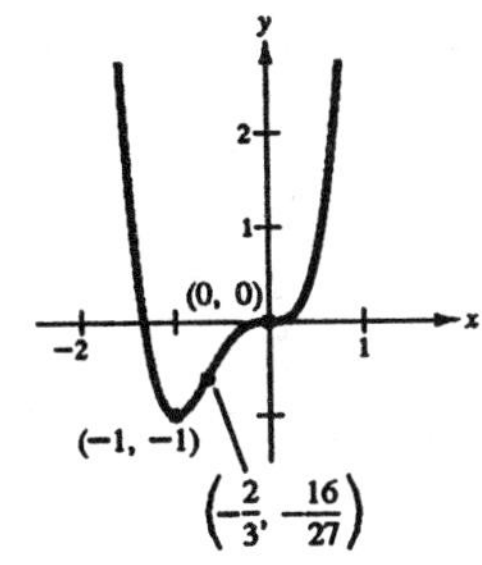

	y	y'	y''	Conclusion
$-\infty < x < -1$		$-$	$+$	Decreasing, concave up
$x = -1$	-1	0	$+$	Relative minimum
$-1 < x < -\frac{2}{3}$		$+$	$+$	Increasing, concave up
$x = -\frac{2}{3}$	$-\frac{16}{27}$	$+$	0	Point of inflection
$-\frac{2}{3} < x < 0$		$+$	$-$	Increasing, concave down
$x = 0$	0	0	0	Point of inflection
$0 < x < \infty$		$+$	$+$	Increasing, concave up

10. $y = 3x^4 - 6x^2$

$y' = 12x^3 - 12x = 12x(x^2 - 1) = 0$ when $x = 0,\ x = \pm 1.$

$y'' = 36x^2 - 12 = 12(3x^2 - 1) = 0$ when $x = \pm\dfrac{\sqrt{3}}{3}.$

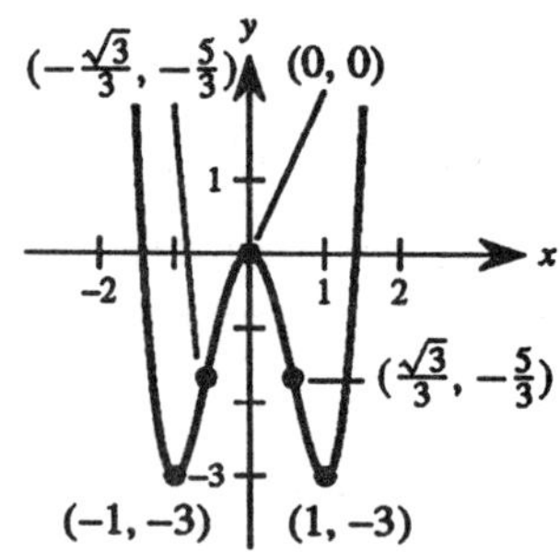

	y	y'	y''	Conclusion
$-\infty < x < -1$		$-$	$+$	Decreasing, concave up
$x = -1$	-3	0	$+$	Relative minimum
$-1 < x < -\frac{\sqrt{3}}{3}$		$+$	$+$	Increasing, concave up
$x = -\frac{\sqrt{3}}{3}$	$-\frac{5}{3}$	$+$	0	Point of inflection
$-\frac{\sqrt{3}}{3} < x < 0$		$+$	$-$	Increasing, concave down
$x = 0$	0	0	$-$	Relative maximum
$0 < x < \frac{\sqrt{3}}{3}$		$-$	$-$	Decreasing, concave down
$x = \frac{\sqrt{3}}{3}$	$-\frac{5}{3}$	$-$	0	Point of inflection
$\frac{\sqrt{3}}{3} < x < 1$		$-$	$+$	Decreasing, concave up
$x = 1$	-3	0	$+$	Relative minimum
$1 < x < \infty$		$+$	$+$	Increasing, concave up

11. $f(x) = x^4 - 4x^3 + 16x$

$f'(x) = 4x^3 - 12x^2 + 16 = 4(x+1)(x-2)^2 = 0$ when $x = -1,\ x = 2.$

$f''(x) = 12x^2 - 24x = 12x(x-2) = 0$ when $x = 0,\ x = 2.$

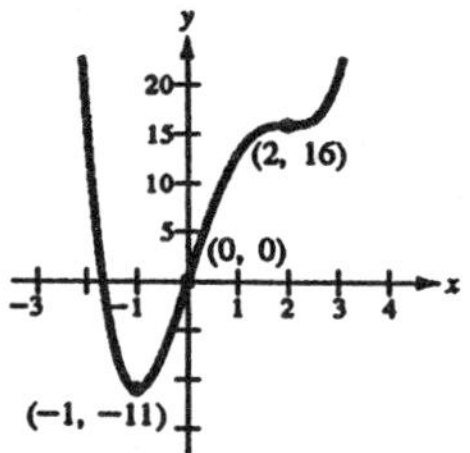

	$f(x)$	$f'(x)$	$f''(x)$	Conclusion
$-\infty < x < -1$		$-$	$+$	Decreasing, concave up
$x = -1$	-11	0	$+$	Relative minimum
$-1 < x < 0$		$+$	$+$	Increasing, concave up
$x = 0$	0	$+$	0	Point of inflection
$0 < x < 2$		$+$	$-$	Increasing, concave down
$x = 2$	16	0	0	Point of inflection
$2 < x < \infty$		$+$	$+$	Increasing, concave up

12. $f(x) = x^4 - 8x^3 + 18x^2 - 16x + 5$

$f'(x) = 4x^3 - 24x^2 + 36x - 16 = 4(x-4)(x-1)^2 = 0$ when $x = 1,\ x = 4$.

$f''(x) = 12x^2 - 48x + 36 = 12(x-3)(x-1) = 0$ when $x = 3,\ x = 1$.

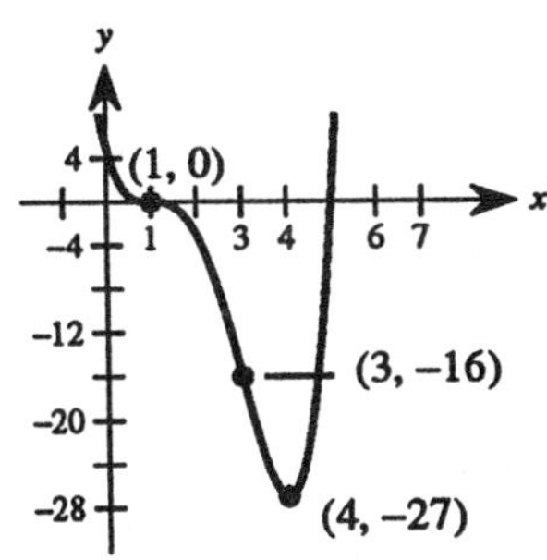

	$f(x)$	$f'(x)$	$f''(x)$	Conclusion
$-\infty < x < 1$		$-$	$+$	Decreasing, concave up
$x = 1$	0	0	0	Point of inflection
$1 < x < 3$		$-$	$-$	Decreasing, concave down
$x = 3$	-16	$-$	0	Point of inflection
$3 < x < 4$		$-$	$+$	Decreasing, concave up
$x = 4$	-27	0	$+$	Relative minimum
$4 < x < \infty$		$+$	$+$	Increasing, concave up

13. $f(x) = x^4 - 4x^3 + 16x - 16$

$f'(x) = 4x^3 - 12x^2 + 16 = 4(x+1)(x-2)^2 = 0$ when $x = -1,\ 2$.

$f''(x) = 12x^2 - 24x = 12x(x-2) = 0$ when $x = 0,\ 2$.

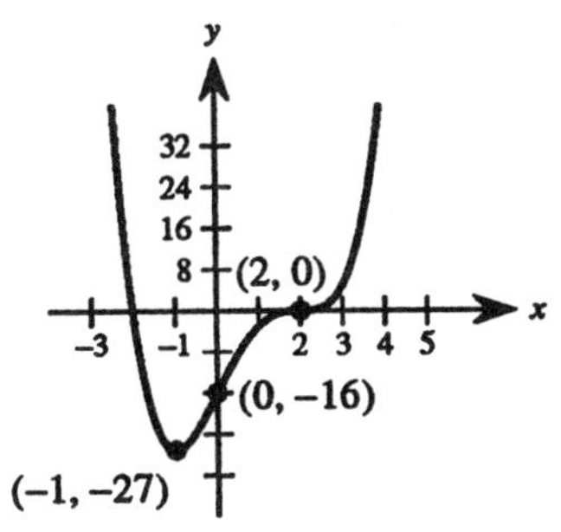

	$f(x)$	$f'(x)$	$f''(x)$	Conclusion
$-\infty < x < -1$		$-$	$+$	Decreasing, concave up
$x = -1$	-27	0	$+$	Relative minimum
$-1 < x < 0$		$+$	$+$	Increasing, concave up
$x = 0$	-16	$+$	0	Point of inflection
$0 < x < 2$		$+$	$-$	Increasing, concave down
$x = 2$	0	0	0	Point of inflection
$2 < x < \infty$		$+$	$+$	Increasing, concave up

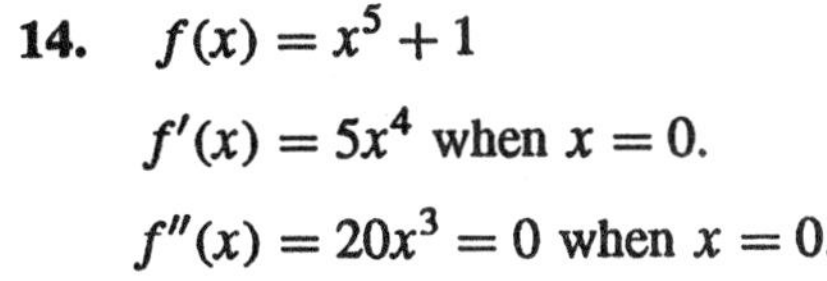

14. $f(x) = x^5 + 1$

$f'(x) = 5x^4$ when $x = 0$.

$f''(x) = 20x^3 = 0$ when $x = 0$.

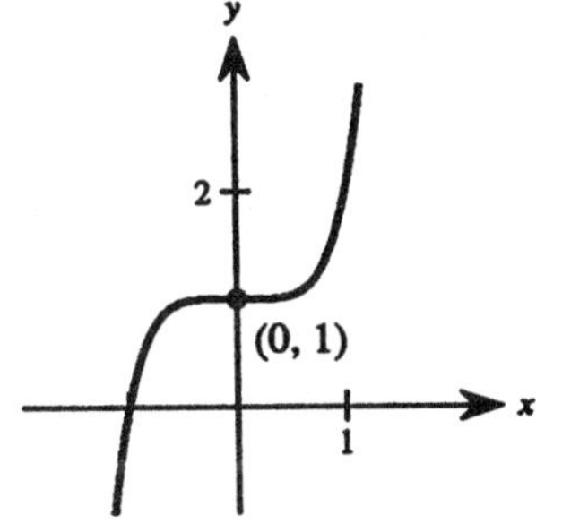

	$f(x)$	$f'(x)$	$f''(x)$	Conclusion
$-\infty < x < 0$		$+$	$-$	Increasing, concave down
$x = 0$	1	0	0	Point of inflection
$0 < x < \infty$		$+$	$+$	Increasing, concave up

15. $y = x^5 - 5x$

$y' = 5x^4 - 5 = 5(x^4 - 1) = 0$ when $x = \pm 1$.

$y'' = 20x^3 = 0$ when $x = 0$.

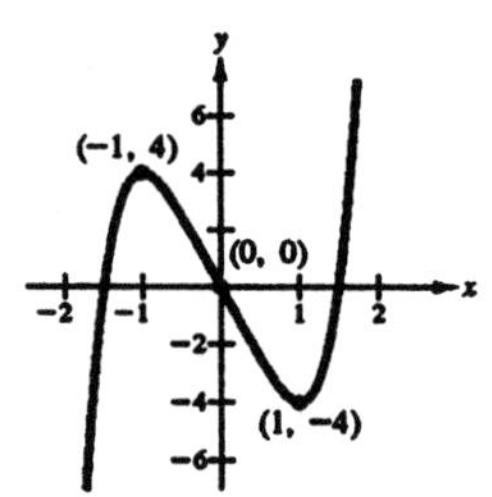

	y	y'	y''	Conclusion
$-\infty < x < -1$		$+$	$-$	Increasing, concave down
$x = -1$	4	0	$-$	Relative maximum
$-1 < x < 0$		$-$	$-$	Decreasing, concave down
$x = 0$	0	$-$	0	Point of inflection
$0 < x < 1$		$-$	$+$	Decreasing, concave up
$x = 1$	-4	0	$+$	Relative minimum
$1 < x < \infty$		$+$	$+$	Increasing, concave up

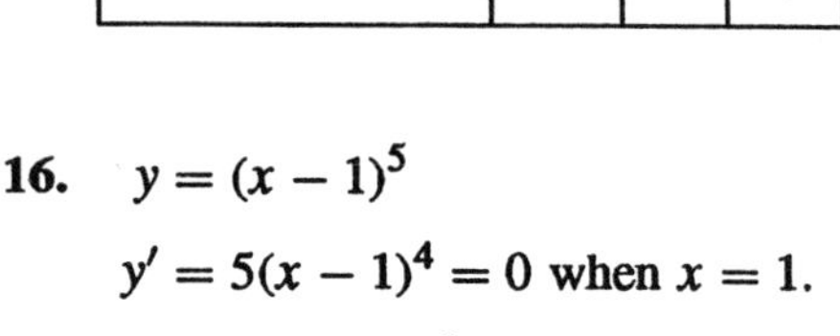

16. $y = (x - 1)^5$

$y' = 5(x - 1)^4 = 0$ when $x = 1$.

$y'' = 20(x - 1)^3 = 0$ when $x = 1$.

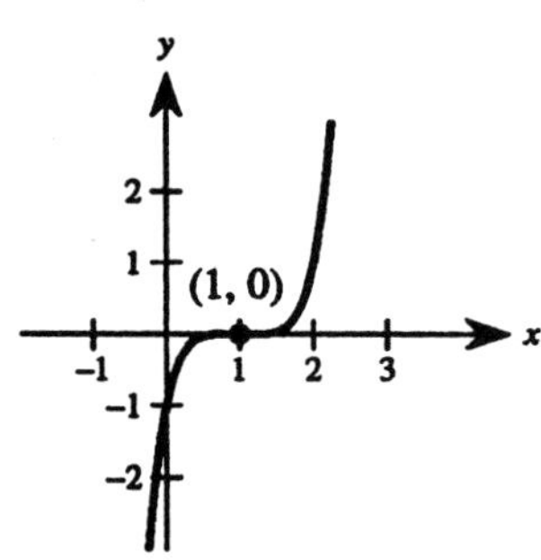

	y	y'	y''	Conclusion
$-\infty < x < 1$		$+$	$-$	Increasing, concave down
$x = 1$	0	0	0	Point of inflection
$1 < x < \infty$		$+$	$+$	Increasing, concave up

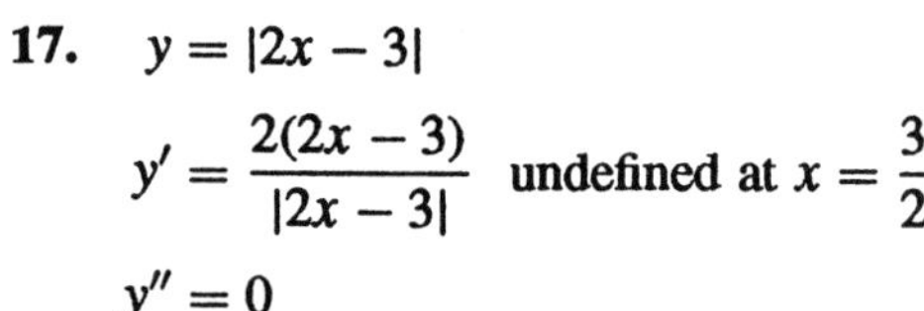

17. $y = |2x - 3|$

$y' = \dfrac{2(2x - 3)}{|2x - 3|}$ undefined at $x = \dfrac{3}{2}$

$y'' = 0$

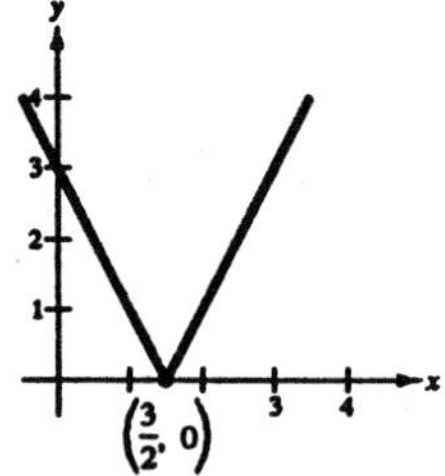

	y	y'	Conclusion
$-\infty < x < \frac{3}{2}$		$-$	Decreasing
$x = \frac{3}{2}$	0	Undefined	Relative minimum
$\frac{3}{2} < x < \infty$		$+$	Increasing

18. $y = |x^2 - 6x + 5|$

$$y' = \frac{2(x-3)(x^2-6x+5)}{|x^2-6x+5|} = \frac{2(x-3)(x-5)(x-1)}{|(x-5)(x-1)|}$$

$= 0$ when $x = 3$ and undefined when $x = 1,\ x = 5.$

$$y'' = \frac{2(x^2-6x+5)}{|x^2-6x+5|} = \frac{2(x-5)(x-1)}{|(x-5)(x-1)|} \text{ undefined when } x = 1,\ x = 5.$$

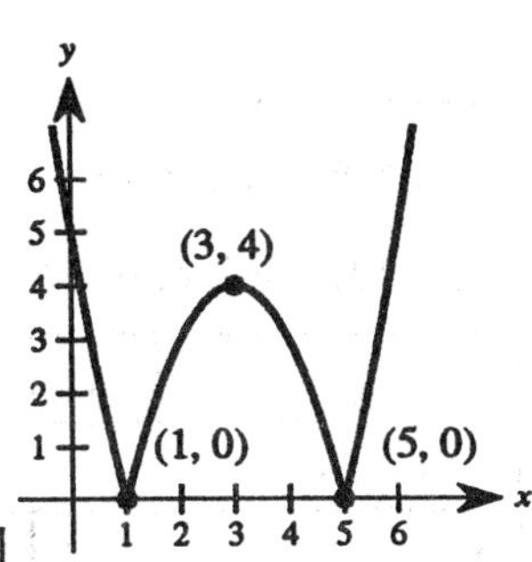

	y	y'	y''	Conclusion
$-\infty < x < 1$		$-$	$+$	Decreasing, concave up
$x = 1$	0	Undefined	Undefined	Relative minimum, point of inflection
$1 < x < 3$		$+$	$-$	Increasing, concave down
$x = 3$	4	0	$-$	Relative maximum
$3 < x < 5$		$-$	$-$	Decreasing, concave down
$x = 5$	0	Undefined	Undefined	Relative minimum, point of inflection
$5 < x < \infty$		$+$	$+$	Increasing, concave up

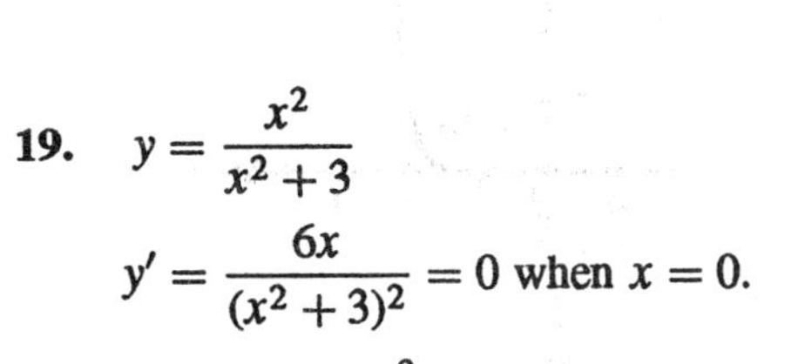

19. $y = \dfrac{x^2}{x^2 + 3}$

$$y' = \frac{6x}{(x^2+3)^2} = 0 \text{ when } x = 0.$$

$$y'' = \frac{18(1-x^2)}{(x^2+3)^3} = 0 \text{ when } x = \pm 1.$$

Horizontal asymptote: $y = 1$

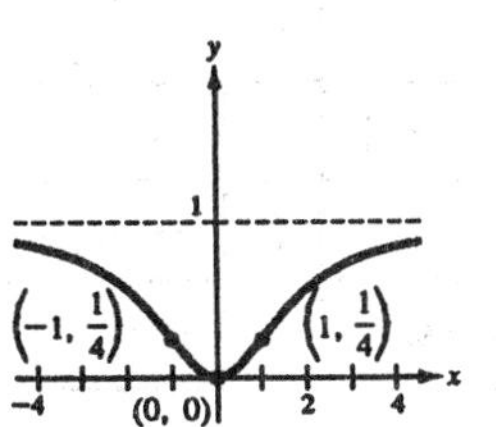

	y	y'	y''	Conclusion
$-\infty < x < -1$		$-$	$-$	Decreasing, concave down
$x = -1$	$\frac{1}{4}$	$-$	0	Point of inflection
$-1 < x < 0$		$-$	$+$	Decreasing, concave up
$x = 0$	0	0	$+$	Relative minimum
$0 < x < 1$		$+$	$+$	Increasing, concave up
$x = 1$	$\frac{1}{4}$	$+$	0	Point of inflection
$1 < x < \infty$		$+$	$-$	Increasing, concave down

20. $y = \dfrac{x}{x^2+1}$

$y' = \dfrac{1-x^2}{(x^2+1)^2} = \dfrac{(1-x)(1+x)}{(x^2+1)^2} = 0$ when $x = \pm 1$.

$y'' = -\dfrac{2x(3-x^2)}{(x^2+1)^3} = 0$ when $x = 0, \ \pm\sqrt{3}$.

Horizontal asymptote: $y = 0$

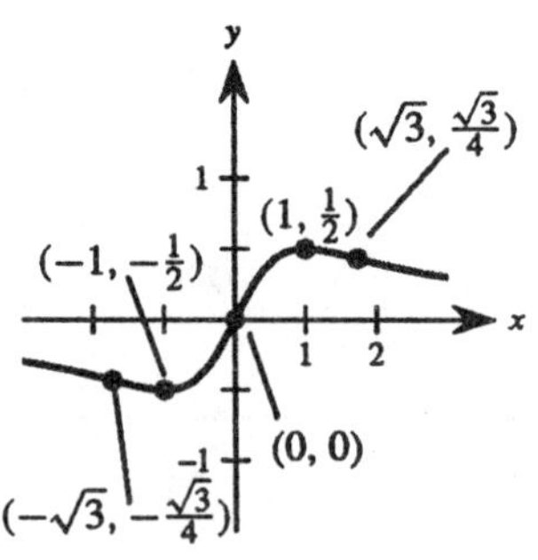

	y	y'	y''	Conclusion
$-\infty < x < -\sqrt{3}$		$-$	$-$	Decreasing, concave down
$x = -\sqrt{3}$	$-\frac{\sqrt{3}}{4}$	$-$	0	Point of inflection
$-\sqrt{3} < x < -1$		$-$	$+$	Decreasing, concave up
$x = -1$	$-\frac{1}{2}$	0	$+$	Relative minimum
$-1 < x < 0$		$+$	$+$	Increasing, concave up
$x = 0$	0	$+$	0	Point of inflection
$0 < x < 1$		$+$	$-$	Increasing, concave down
$x = 1$	$\frac{1}{2}$	0	$-$	Relative maximum
$1 < x < \sqrt{3}$		$-$	$-$	Decreasing, concave down
$x = \sqrt{3}$	$\frac{\sqrt{3}}{4}$	$-$	0	Point of inflection
$\sqrt{3} < x < \infty$		$-$	$+$	Decreasing, concave up

21. $y = x\sqrt{4-x}$, Domain: $(-\infty, 4]$

$y' = \dfrac{8-3x}{2\sqrt{4-x}} = 0$ when $x = \dfrac{8}{3}$ and undefined when $x = 4$.

$y'' = \dfrac{3x-16}{4(4-x)^{3/2}} = 0$ when $x = \dfrac{16}{3}$ and undefined when $x = 4$.

Note: $x = \frac{16}{3}$ is not in the domain.

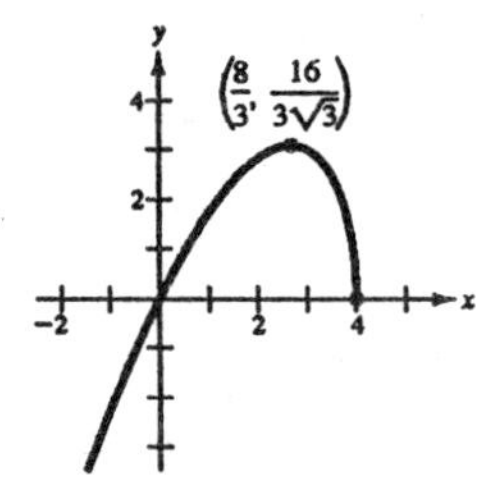

	y	y'	y''	Conclusion
$-\infty < x < \frac{8}{3}$		$+$	$-$	Increasing, concave down
$x = \frac{8}{3}$	$\frac{16}{3\sqrt{3}}$	0	$-$	Relative maximum
$\frac{8}{3} < x < 4$		$-$	$-$	Decreasing, concave down
$x = 4$	0	Undefined	Undefined	Endpoint

22. $y = x\sqrt{4 - x^2}$, Domain: $[-2, 2]$

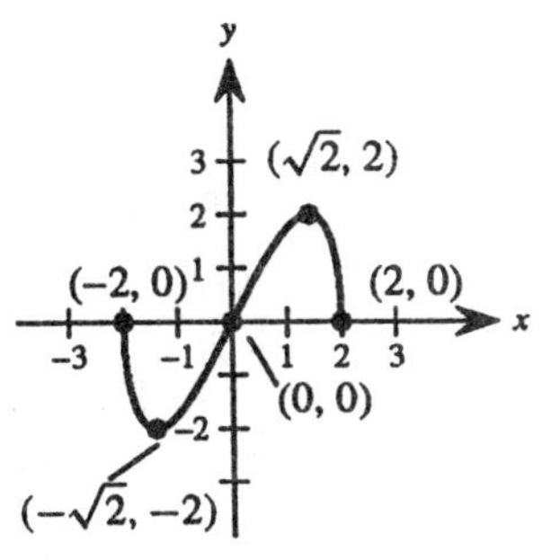

$y' = \dfrac{4 - 2x^2}{\sqrt{4 - x^2}} = 0$ when $x = \pm\sqrt{2}$ and undefined when $x = \pm 2$.

$y'' = \dfrac{2x(x^2 - 6)}{(4 - x^2)^{3/2}} = 0$ when $x = 0$, ± 6 and undefined when $x = \pm 2$.

Note: $x = \pm\sqrt{6}$ are not in the domain.

	y	y'	y''	Conclusion
$x = -2$	0	Undefined	Undefined	Endpoint
$-2 < x < -\sqrt{2}$		$-$	$+$	Decreasing, concave up
$x = -\sqrt{2}$	-2	0	$+$	Relative minimum
$-\sqrt{2} < x < 0$		$+$	$+$	Increasing, concave up
$x = 0$	0	$+$	0	Point of inflection
$0 < x < \sqrt{2}$		$+$	$-$	Increasing, concave down
$x = \sqrt{2}$	2	0	$-$	Relative maximum
$\sqrt{2} < x < 2$		$-$	$-$	Decreasing, concave down
$x = 2$	0	Undefined	Undefined	Endpoint

23. $y = 3x^{2/3} - 2x$

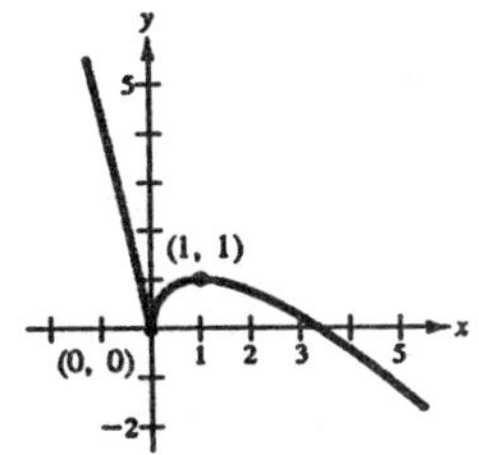

$y' = 2x^{-1/3} - 2 = \dfrac{2(1 - x^{1/3})}{x^{1/3}}$

$= 0$ when $x = 1$ and undefined when $x = 0$.

$y'' = \dfrac{-2}{3x^{4/3}} < 0$ when $x \neq 0$.

	y	y'	y''	Conclusion
$-\infty < x < 0$		$-$	$-$	Decreasing, concave down
$x = 0$	0	Undefined	Undefined	Relative minimum
$0 < x < 1$		$+$	$-$	Increasing, concave down
$x = 1$	1	0	$-$	Relative maximum
$1 < x < \infty$		$-$	$-$	Decreasing, concave down

24. $y = 3x^{2/3} - x^2$

$y' = \dfrac{2}{x^{1/3}} - 2x = \dfrac{2(1 - x^{4/3})}{x^{1/3}} = 0$ when $x = \pm 1$, undefined when $x = 0$.

$y'' = -2\left(\dfrac{1}{3x^{4/3}} + 1\right) < 0$ when $x \neq 0$.

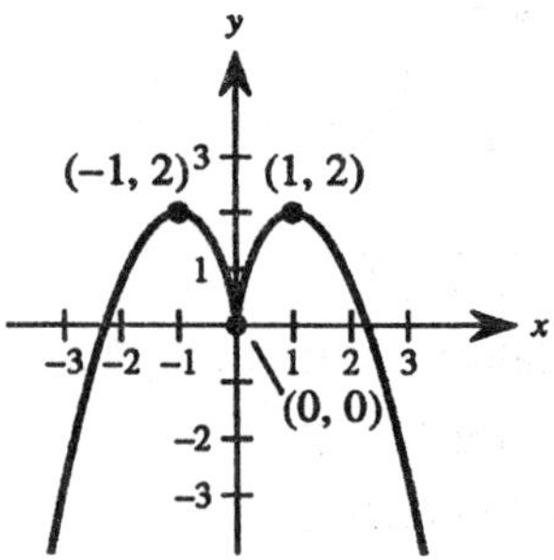

	y	y'	y''	Conclusion
$-\infty < x < -1$		$+$	$-$	Increasing, concave down
$x = -1$	2	0	$-$	Relative maximum
$-1 < x < 0$		$-$	$-$	Decreasing, concave down
$x = 0$	0	Undefined	Undefined	Relative minimum
$0 < x < 1$		$+$	$-$	Increasing, concave down
$x = 1$	2	0	$-$	Relative maximum
$1 < x < \infty$		$-$	$-$	Decreasing, concave down

25. $y = \dfrac{x}{\sqrt{x^2 + 7}}$

$y' = \dfrac{7}{(x^2 + 7)^{3/2}} > 0$ for all x.

$y'' = \dfrac{-21x}{(x^2 + 7)^{5/2}} = 0$ when $x = 0$.

Horizontal asymptotes: $y = \pm 1$

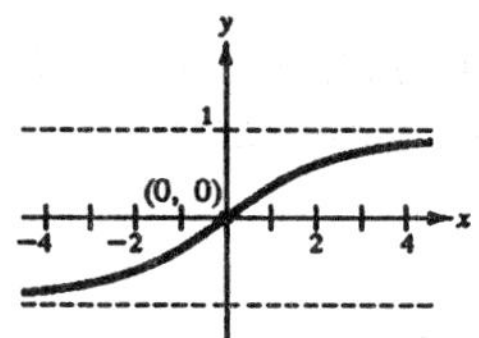

	$f(x)$	$f'(x)$	$f''(x)$	Conclusion
$-\infty < x < 0$		$+$	$+$	Increasing, concave up
$x = 0$	0	$+$	0	Point of inflection
$0 < x < \infty$		$+$	$-$	Increasing, concave down

26. $f(x) = \dfrac{4x}{\sqrt{x^2 + 15}}$

$f'(x) = \dfrac{60}{(x^2 + 15)^{3/2}} > 0$ for all x.

$f''(x) = \dfrac{-180x}{(x^2 + 15)^{5/2}} = 0$ when $x = 0$.

Horizontal asymptotes: $y = \pm 4$

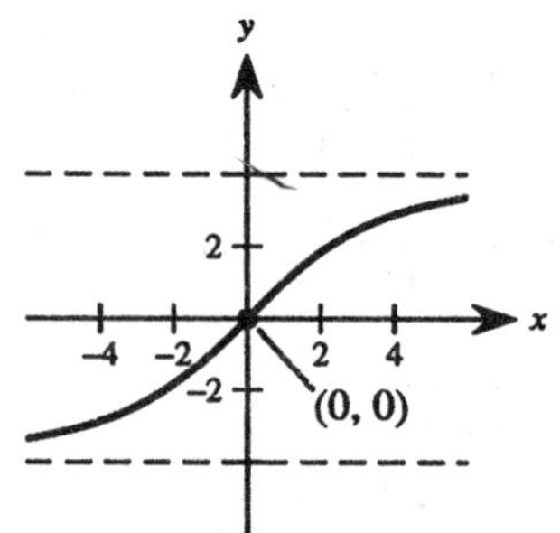

	$f(x)$	$f'(x)$	$f''(x)$	Conclusion
$-\infty < x < 0$		$+$	$+$	Increasing, concave up
$x = 0$	0	$+$	0	Point of inflection
$0 < x < \infty$		$+$	$-$	Increasing, concave down

27. $y = \dfrac{1}{x-2} - 3$

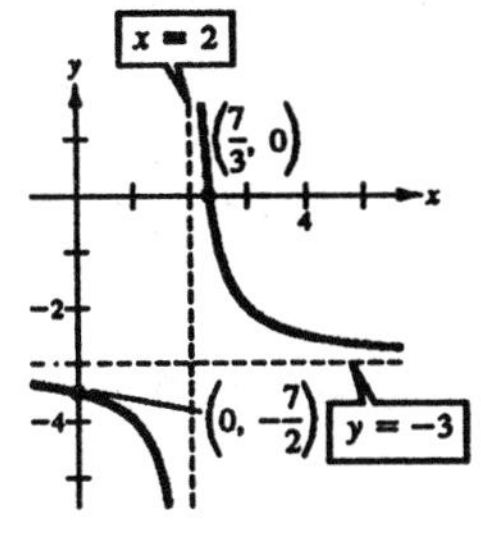

$y' = -\dfrac{1}{(x-2)^2} < 0$ when $x \neq 2$.

$y'' = \dfrac{2}{(x-2)^3}$

No relative extrema, no points of inflection

Intercepts: $\left(\frac{7}{3}, 0\right)$, $\left(0, -\frac{7}{2}\right)$

Vertical asymptote: $x = 2$

Horizontal asymptote: $y = -3$

28. $y = \dfrac{x^2+1}{x^2-2}$

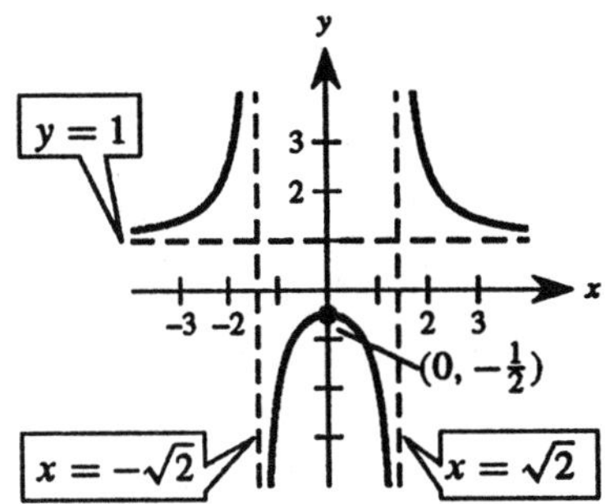

$y' = \dfrac{-6x}{(x^2-2)^2} = 0$ when $x = 0$.

$y'' = \dfrac{6(3x^2+2)}{(x^2-2)^3} < 0$ when $x = 0$.

Therefore, $\left(0, -\frac{1}{2}\right)$ is a relative maximum.

Intercept: $\left(0, -\frac{1}{2}\right)$

Vertical asymptotes: $x = \pm\sqrt{2}$

Horizontal asymptote: $y = 1$

29. $y = \dfrac{2x}{x^2-1}$

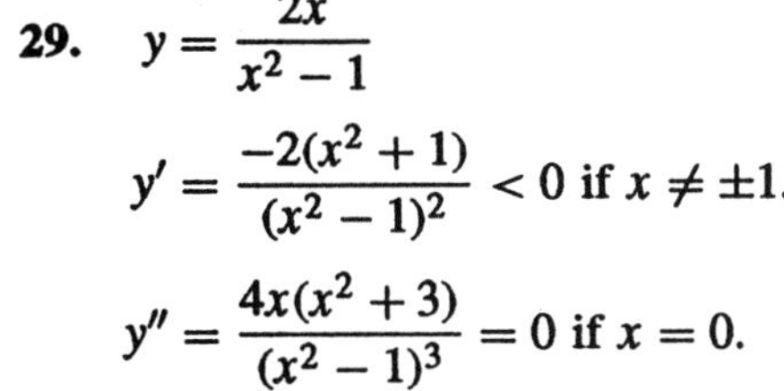

$y' = \dfrac{-2(x^2+1)}{(x^2-1)^2} < 0$ if $x \neq \pm 1$.

$y'' = \dfrac{4x(x^2+3)}{(x^2-1)^3} = 0$ if $x = 0$.

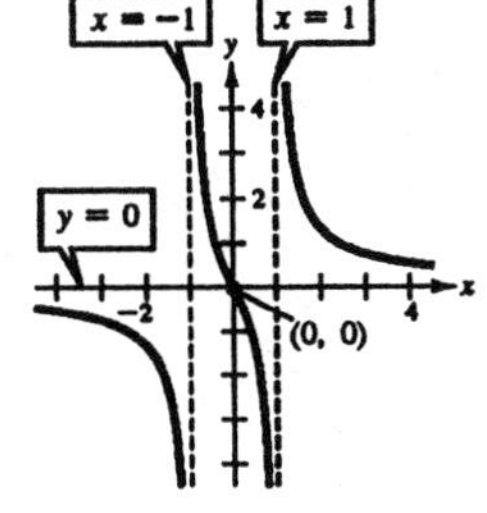

Inflection point: (0, 0)

Intercept: (0, 0)

Vertical asymptotes: $x = \pm 1$

Horizontal asymptote: $y = 0$

Symmetry with respect to the origin

30. $y = \dfrac{x^2-6x+12}{x-4} = x - 2 + \dfrac{4}{x-4}$

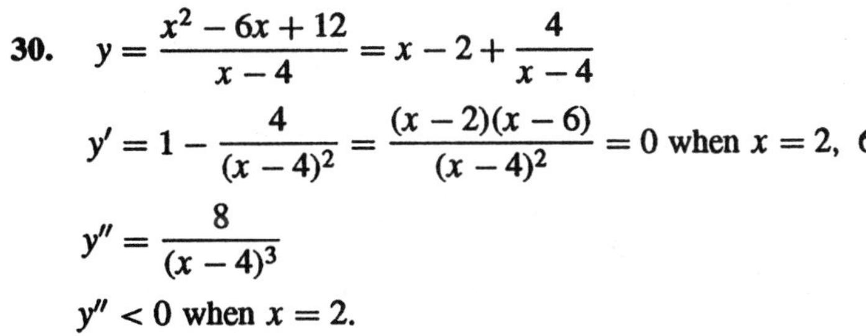

$y' = 1 - \dfrac{4}{(x-4)^2} = \dfrac{(x-2)(x-6)}{(x-4)^2} = 0$ when $x = 2,\ 6$.

$y'' = \dfrac{8}{(x-4)^3}$

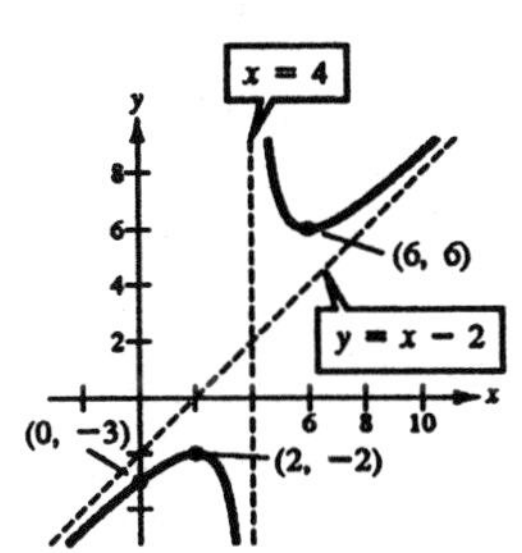

$y'' < 0$ when $x = 2$.

Therefore, $(2, -2)$ is a relative maximum.

$y'' > 0$ when $x = 6$.

Therefore, $(6, 6)$ is a relative minimum.

Vertical asymptote: $x = 4$

Slant asymptote: $y = x - 2$

31. $f(x) = \dfrac{x+2}{x} = 1 + \dfrac{2}{x}$

$f'(x) = \dfrac{-2}{x^2} < 0$ when $x \neq 0$.

$f''(x) = \dfrac{4}{x^3} \neq 0$

Intercept: $(-2,\ 0)$
Vertical asymptote: $x = 0$
Horizontal asymptote: $y = 1$

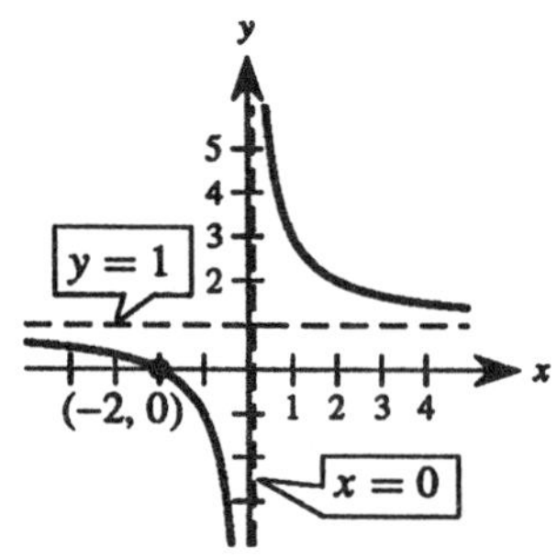

32. $f(x) = x + \dfrac{32}{x^2}$

$f'(x) = 1 - \dfrac{64}{x^3} = \dfrac{(x-4)(x^2+4x+16)}{x^3} = 0$ when $x = 4$.

$f''(x) = \dfrac{192}{x^4} > 0$ if $x \neq 0$.

Therefore, (4, 6) is a relative minimum.
Intercept: $\left(-2\sqrt[3]{4},\ 0\right)$
Vertical asymptote: $x = 0$
Slant asymptote: $y = x$

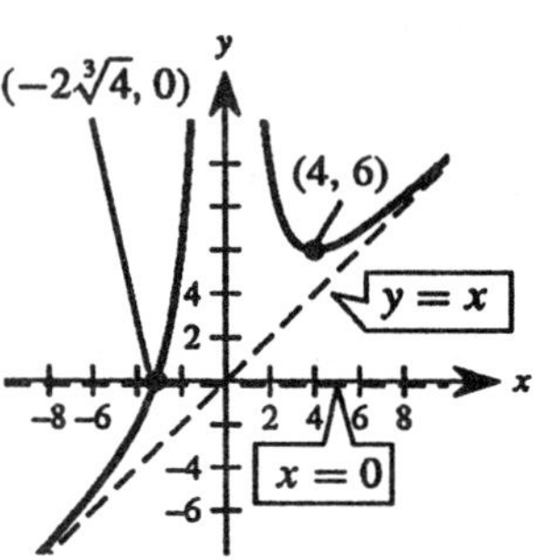

33. $f(x) = \dfrac{x^2+1}{x} = x + \dfrac{1}{x}$

$f'(x) = 1 - \dfrac{1}{x^2} = 0$ when $x = \pm 1$.

$f''(x) = \dfrac{2}{x^3} \neq 0$

Relative maximum: $(-1,\ -2)$
Relative minimum: $(1, 2)$
Vertical asymptote: $x = 0$
Slant asymptote: $y = x$

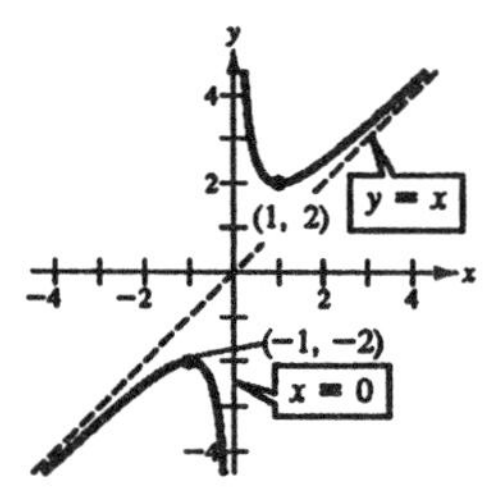

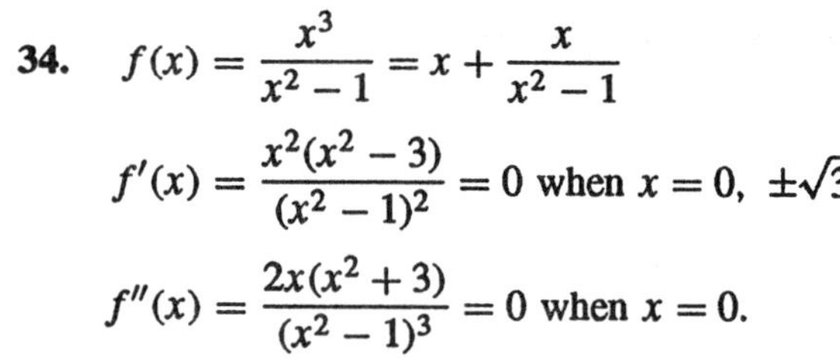

34. $f(x) = \dfrac{x^3}{x^2-1} = x + \dfrac{x}{x^2-1}$

$f'(x) = \dfrac{x^2(x^2-3)}{(x^2-1)^2} = 0$ when $x = 0,\ \pm\sqrt{3}$.

$f''(x) = \dfrac{2x(x^2+3)}{(x^2-1)^3} = 0$ when $x = 0$.

Relative maximum: $\left(-\sqrt{3},\ -\dfrac{3\sqrt{3}}{2}\right)$

Relative minimum: $\left(\sqrt{3},\ \dfrac{3\sqrt{3}}{2}\right)$

Inflection point: $(0, 0)$
Vertical asymptotes: $x = \pm 1$
Slant asymptote: $y = x$
Origin symmetry

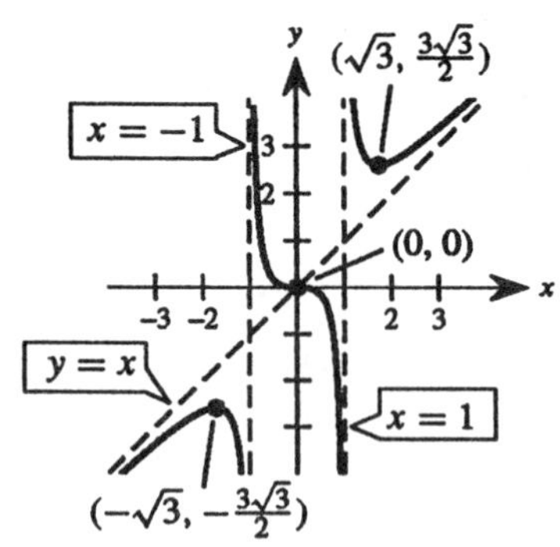

35. $y = \dfrac{x^3}{2x^2 - 8} = \dfrac{x}{2} + \dfrac{2x}{x^2 - 4}$

$y' = \dfrac{x^2(x^2 - 12)}{2(x^2 - 4)^2} = 0$ when $x = 0, \ \pm 2\sqrt{3}$.

$y'' = \dfrac{4x(x^2 + 12)}{(x^2 - 4)^3} = 0$ when $x = 0$.

Relative maximum: $\left(-2\sqrt{3}, \ -\dfrac{3\sqrt{3}}{2}\right)$

Relative minimum: $\left(2\sqrt{3}, \ \dfrac{3\sqrt{3}}{2}\right)$

Inflection point: (0, 0)

Vertical asymptotes: $x = \pm 2$

Slant asymptote: $y = \dfrac{x}{2}$

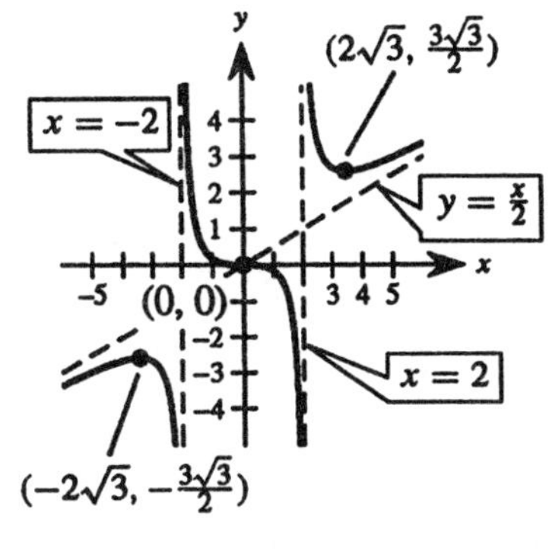

36. $y = \dfrac{2x^2 - 5x + 5}{x - 2} = 2x - 1 + \dfrac{3}{x - 2}$

$y' = 2 - \dfrac{3}{(x - 2)^2} = \dfrac{2x^2 - 8x + 5}{(x - 2)^2} = 0$ when $x = \dfrac{4 \pm \sqrt{6}}{2}$.

$y'' = \dfrac{6}{(x - 2)^3} \neq 0.$

Relative maximum: $\left(\dfrac{4 - \sqrt{6}}{2}, \ -1.8990\right)$

Relative minimum: $\left(\dfrac{4 + \sqrt{6}}{2}, \ 7.8990\right)$

Vertical asymptote: $x = 2$

Slant asymptote: $y = 2x - 1$

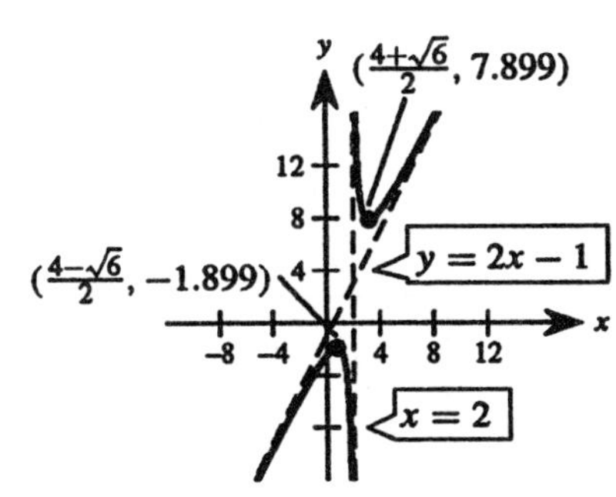

37. $f(x) = \dfrac{20x}{x^2 + 1} - \dfrac{1}{x} = \dfrac{19x^2 - 1}{x(x^2 + 1)}$

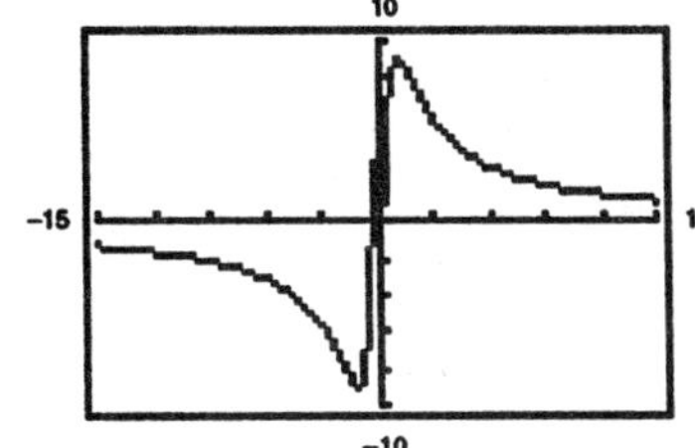

38. $f(x) = 5\left(\dfrac{1}{x - 4} - \dfrac{1}{x + 2}\right)$

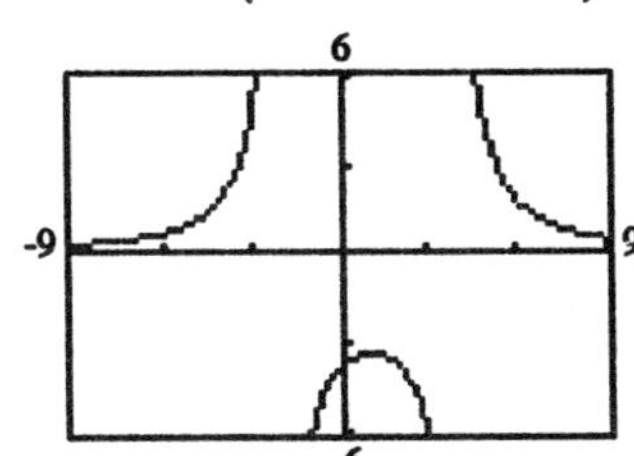

39. $f(x) = \dfrac{4(x-1)^2}{x^2-4x+5}$

Vertical asymptote: none
Horizontal asymptote: $y = 4$

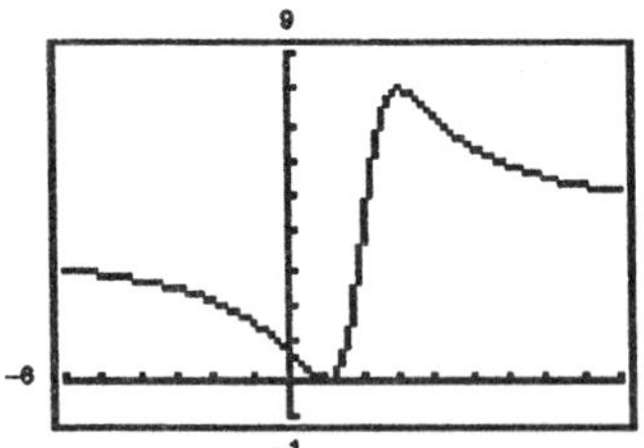

The graph crosses the horizontal asymptote $y = 4$. If a function has a vertical asymptote at $x = c$, the graph would not cross it since $f(c)$ is undefined.

40. $g(x) = \dfrac{3x^4-5x+3}{x^4+1}$

Vertical asymptote: none
Horizontal asymptote: $y = 3$

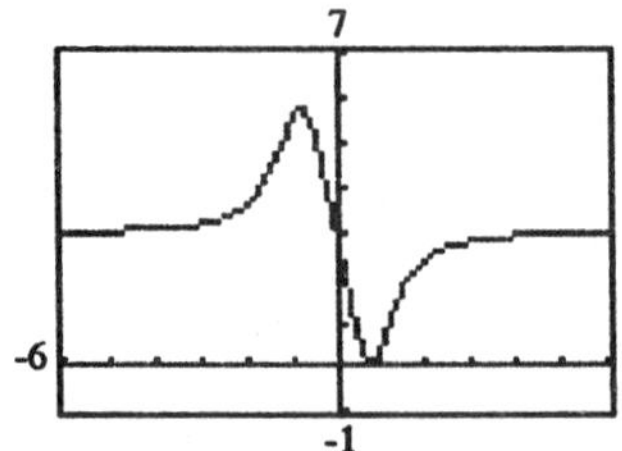

The graph crosses the horizontal asymptote $y = 3$. If a function has a vertical asymptote at $x = c$, the graph would not cross it since $f(c)$ is undefined.

41. $h(x) = \dfrac{6-2x}{3-x}$

$$= \frac{2(3-x)}{3-x} = \begin{cases} 2, & \text{if } x \neq 3 \\ \text{Undefined}, & \text{if } x = 3 \end{cases}$$

The rational function is not reduced to lowest terms.

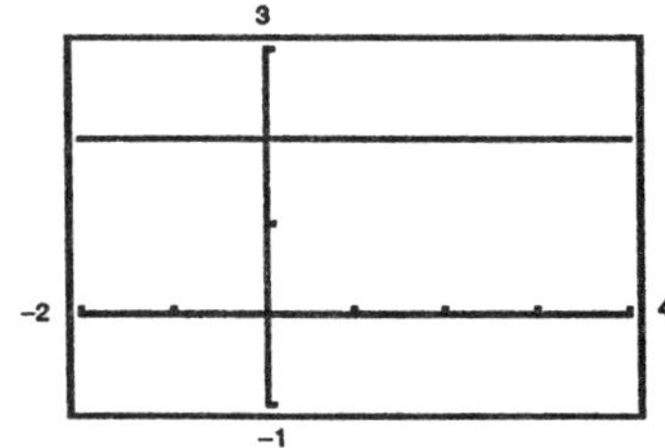

42. $g(x) = \dfrac{x^2+x-2}{x-1}$

$$= \frac{(x+2)(x-1)}{x-1} = \begin{cases} x+2, & \text{if } x \neq 1 \\ \text{Undefined}, & \text{if } x = 1 \end{cases}$$

The rational function is not reduced to lowest terms.

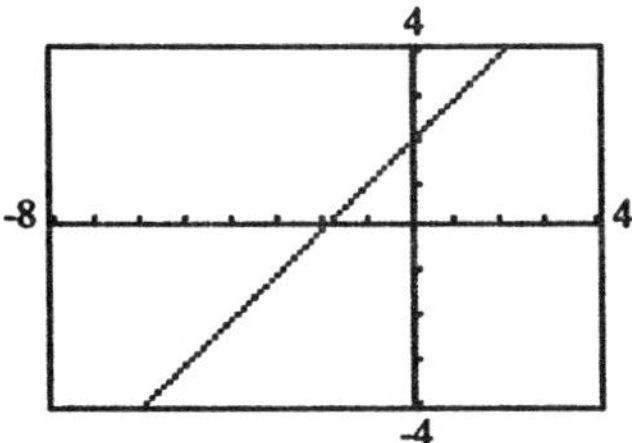

43. $f(x) = -\dfrac{x^2-3x-1}{x-2} = -x+1+\dfrac{3}{x-2}$

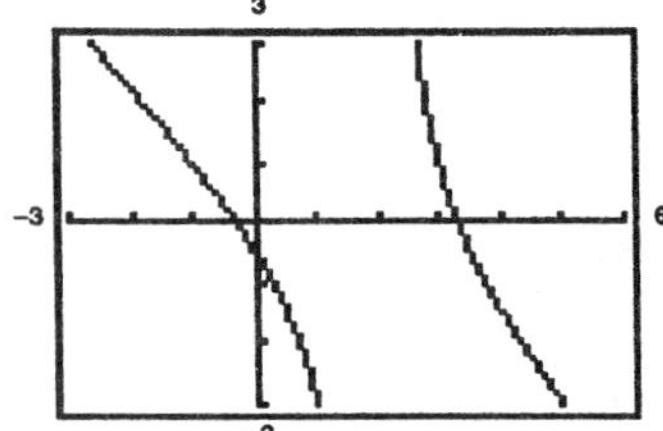

The graph appears to approach the slant asymptote $y = -x + 1$.

44. $g(x) = \dfrac{2x^2-8x-15}{x-5} = 2x+2-\dfrac{5}{x-5}$

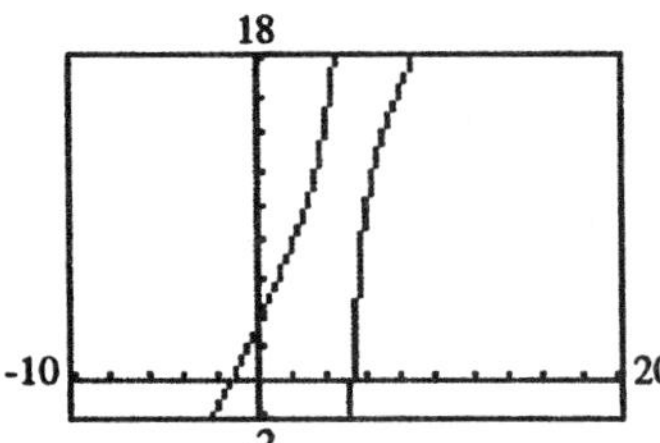

The graph appears to approach the slant asymptote $y = 2x + 2$.

45. Vertical asymptote: $x = 5$
Horizontal asymptote: $y = 0$

$$y = \frac{1}{x-5}$$

46. Vertical asymptote: $x = -3$
Horizontal asymptote: none

$$y = \frac{x^2}{x+3}$$

47. Vertical asymptote: $x = 5$
Slant asymptote: $y = 3x + 2$

$$y = 3x + 2 + \frac{1}{x-5} = \frac{3x^2-13x-9}{x-5}$$

48. Vertical asymptote: $x = 0$
Slant asymptote: $y = -x$

$$y = -x + \frac{1}{x} = \frac{1-x^2}{x}$$

49. $f'(x) = -x(x-1)(x-3) = -x^2 + 4x - 3$
Relative minimum when $x = 1$
Relative maximum when $x = 3$

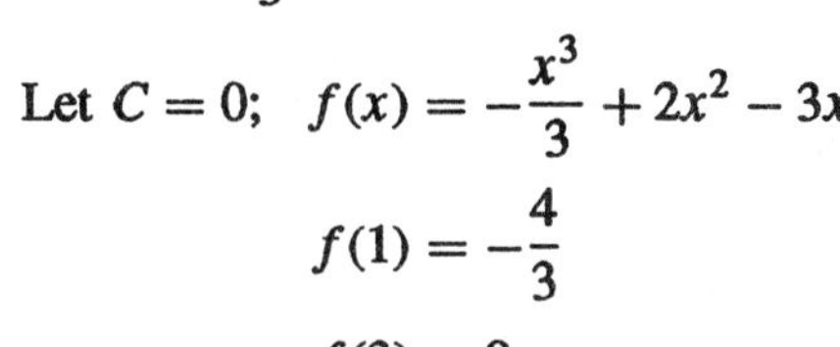

$$f(x) = -\frac{x^3}{3} + 2x^2 - 3x + C, \quad C \text{ is any constant}$$

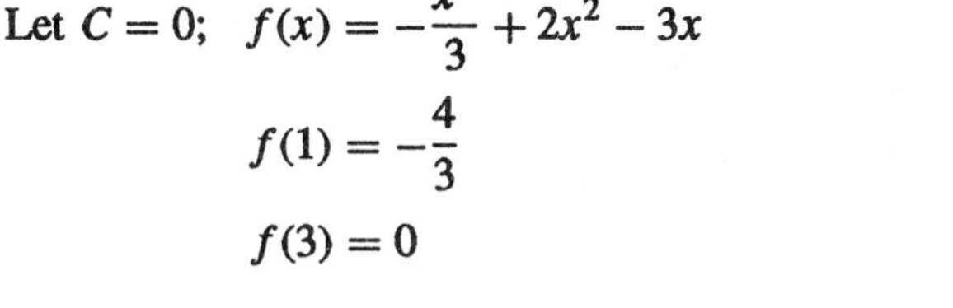

$$\text{Let } C = 0; \quad f(x) = -\frac{x^3}{3} + 2x^2 - 3x$$

$$f(1) = -\frac{4}{3}$$

$$f(3) = 0$$

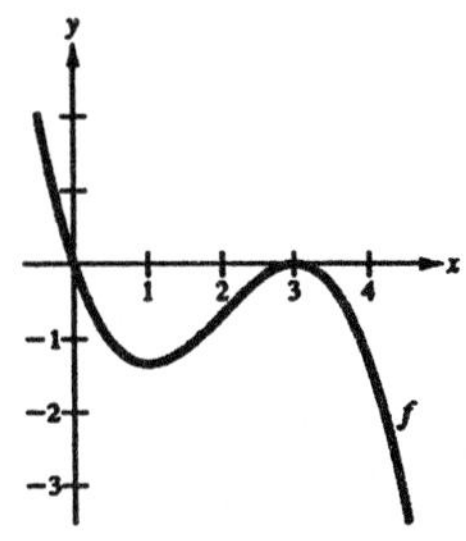

50. $f' > 0$ for $x > 0 \Rightarrow f(x)$ is increasing.
$f' < 0$ for $x < 0 \Rightarrow f(x)$ is decreasing.
f' is undefined when $x = 0$.
One such function is

$$f(x) = -\frac{1}{x^2}.$$

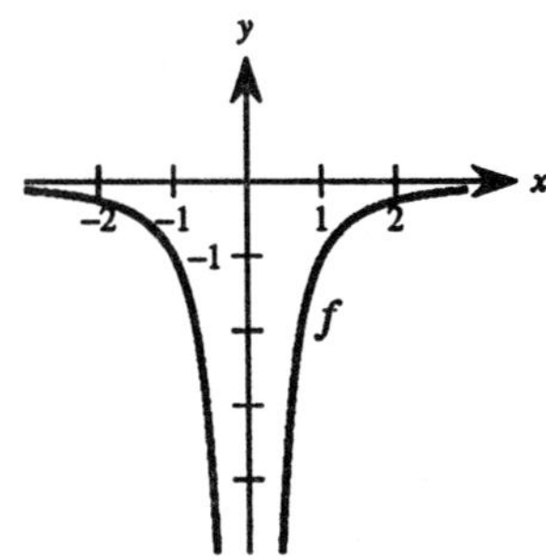

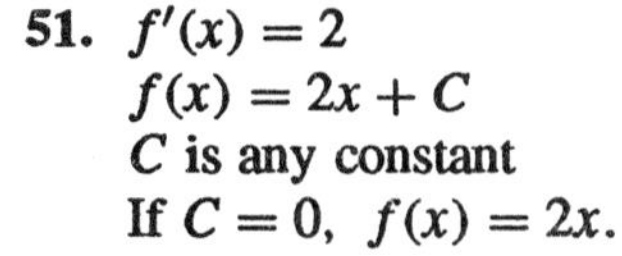

51. $f'(x) = 2$
$f(x) = 2x + C$
C is any constant
If $C = 0$, $f(x) = 2x$.

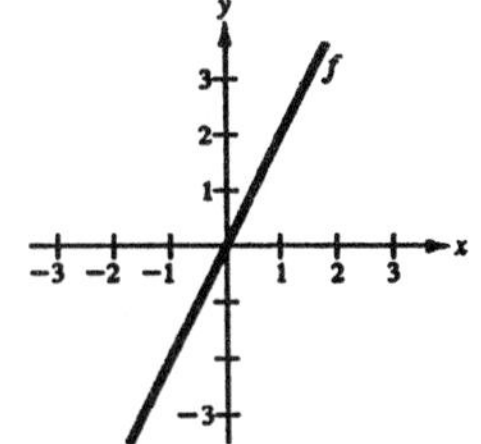

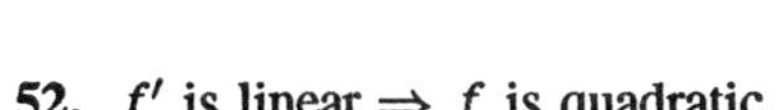

52. f' is linear $\Rightarrow f$ is quadratic.

$f' < 0$ when $x < 0 \Rightarrow f$ is decreasing.
$f' > 0$ when $x > 0 \Rightarrow f$ is increasing.
Therefore, $(0, f(0))$ is a relative minimum.
$f(x) = ax^2 + b, \ a > 0$
If $a = \frac{5}{2}$ and $b = 0$, then $f(x) = \frac{5}{2}x^2$.

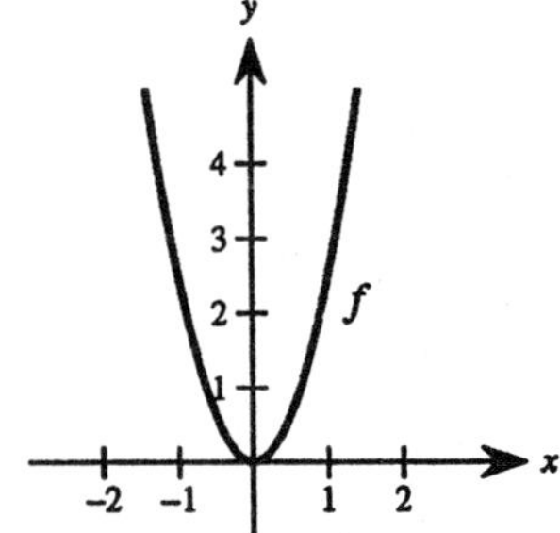

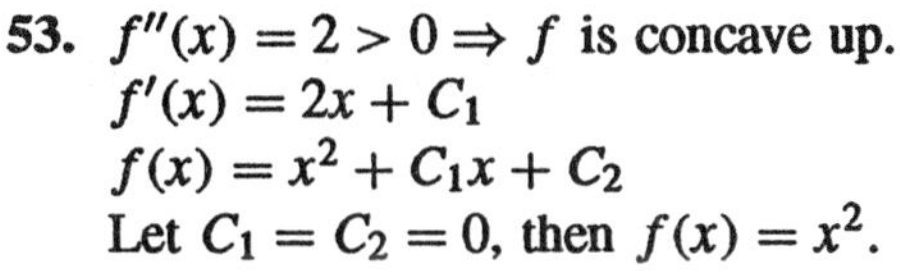

53. $f''(x) = 2 > 0 \Rightarrow f$ is concave up.
$f'(x) = 2x + C_1$
$f(x) = x^2 + C_1x + C_2$
Let $C_1 = C_2 = 0$, then $f(x) = x^2$.

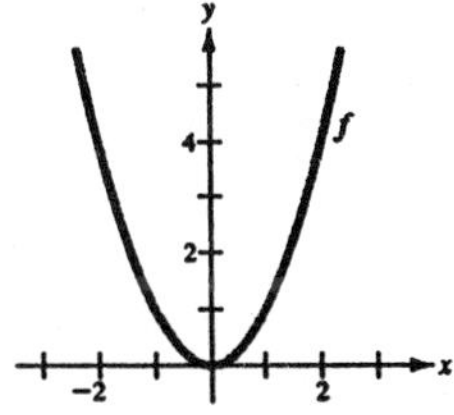

54. f'' is linear $\Rightarrow f$ is cubic.
$f(x) = ax^3 + bx^2 + cx + d$
$f'' < 0$ when $x < 0 \Rightarrow$ concave down.
$f'' > 0$ when $x > 0 \Rightarrow$ concave up.
Therefore, $(0, f(0))$ is a point of inflection.
Since f is concave up to the right and concave down to the left $\Rightarrow a > 0$.
Let $f(x) = x^3$.

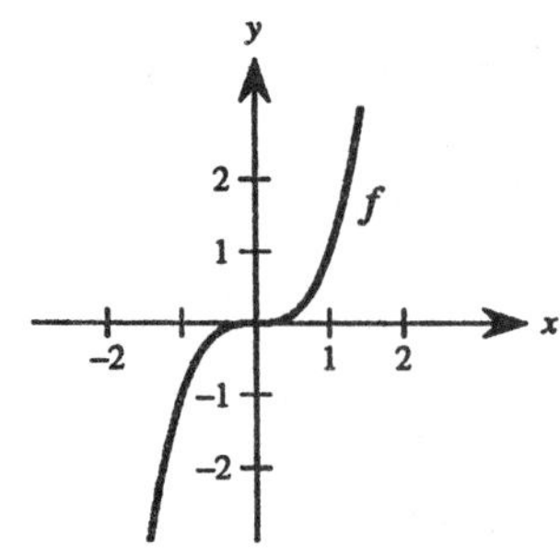

Section 4.7 Optimization Problems

1. (a)

First number, x	Second number	Product, P
10	$110 - 10$	$10(110 - 10) = 1000$
20	$110 - 20$	$20(110 - 20) = 1800$
30	$110 - 30$	$30(110 - 30) = 2400$
40	$110 - 40$	$40(110 - 40) = 2800$
50	$110 - 50$	$50(110 - 50) = 3000$
60	$110 - 60$	$60(110 - 60) = 3000$

(b) $P = x(110 - x) = 110x - x^2$

(c) $\dfrac{dP}{dx} = 110 - 2x = 0$ when $x = 55$.

$\dfrac{d^2P}{dx^2} = -2 < 0$

P is a maximum when $x = 110 - x = 55$.

(d)

First number, x	Second number	Product, P
10	$110 - 10$	$10(110 - 10) = 1000$
20	$110 - 20$	$20(110 - 20) = 1800$
30	$110 - 30$	$30(110 - 30) = 2400$
40	$110 - 40$	$40(110 - 40) = 2800$
50	$110 - 50$	$50(110 - 50) = 3000$
60	$110 - 60$	$60(110 - 60) = 3000$
70	$110 - 70$	$70(110 - 70) = 2800$
80	$110 - 80$	$80(110 - 80) = 2400$
90	$110 - 90$	$90(110 - 90) = 1800$
100	$110 - 100$	$100(110 - 100) = 1000$

(e)

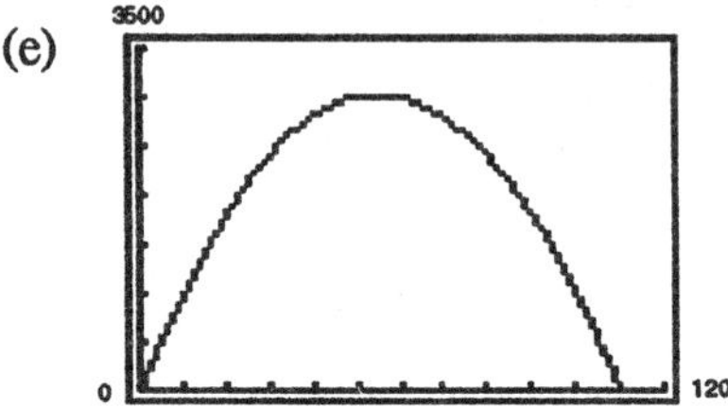

2. Let x and y be two positive numbers such that $x + y = S$.

$P = xy = x(S - x) = Sx - x^2$

$\dfrac{dP}{dx} = S - 2x = 0$ when $x = \dfrac{S}{2}$.

$\dfrac{d^2P}{dx^2} = -2$

P is a maximum when $x = y = S/2$.

3. Let x and y be two positive numbers such that $xy = 192$.

$$S = x + y = x + \frac{192}{x}$$

$$\frac{dS}{dx} = 1 - \frac{192}{x^2} = 0 \text{ when } x = \sqrt{192}.$$

$$\frac{d^2S}{dx^2} = \frac{384}{x^3} > 0 \text{ when } x > 0.$$

S is a minimum when $x = y = \sqrt{192}$.

4. Let x and y be two positive numbers such that $xy = 192$.

$$S = x + 3y = 3y + \frac{192}{y}$$

$$\frac{dS}{dy} = 3 - \frac{192}{y^2} = 0 \text{ when } y = 8.$$

$$\frac{d^2S}{dy^2} = \frac{384}{y^3} > 0 \text{ when } y > 0.$$

S is a minimum when $y = 8$ and $x = 24$.

5. Let x be the positive number.

$$S = x + \frac{1}{x}$$

$$\frac{dS}{dx} = 1 - \frac{1}{x^2} = 0 \text{ when } x = 1.$$

$$\frac{d^2S}{dx^2} = \frac{2}{x^3} > 0 \text{ when } x > 0.$$

The sum is a minimum when $x = 1$ and $1/x = 1$.

6. Let x and y be two positive numbers such that $x + 2y = 100$.

$$P = xy = y(100 - 2y) = 100y - 2y^2$$

$$\frac{dP}{dy} = 100 - 4y = 0 \text{ when } y = 25.$$

$$\frac{d^2P}{dy^2} = -4$$

P is a maximum when $x = 50$ and $y = 25$.

7. Let x be the length and y the width of the rectangle.

$$2x + 2y = 100$$

$$y = 50 - x$$

$$A = xy = x(50 - x)$$

$$\frac{dA}{dx} = 50 - 2x = 0 \text{ when } x = 25.$$

$$\frac{d^2A}{dx^2} = -2$$

A is maximum when $x = y = 25$ feet.

8. Let x be the length and y the width of the rectangle.

$$2x + 2y = P$$

$$y = \frac{P - 2x}{2} = \frac{P}{2} - x$$

$$A = xy = x\left(\frac{P}{2} - x\right) = \frac{P}{2}x - x^2$$

$$\frac{dA}{dx} = \frac{P}{2} - 2x = 0 \text{ when } x = \frac{P}{4}.$$

$$\frac{d^2A}{dx^2} = -2$$

A is maximum when $x = y = P/4$ units.

9. Let x be the length and y the width of the rectangle.

$$xy = 64$$

$$y = \frac{64}{x}$$

$$P = 2x + 2y = 2x + 2\left(\frac{64}{x}\right) = 2x + \frac{128}{x}$$

$$\frac{dP}{dx} = 2 - \frac{128}{x^2} = 0 \text{ when } x = 8.$$

$$\frac{d^2P}{dx^2} = \frac{256}{x^3} > 0 \text{ when } x = 8.$$

P is minimum when $x = y = 8$ feet.

10. Let x be the length and y the width of the rectangle.

$$xy = A$$

$$y = \frac{A}{x}$$

$$P = 2x + 2y = 2x + 2\left(\frac{A}{x}\right) = 2x + \frac{2A}{x}$$

$$\frac{dP}{dx} = 2 - \frac{2A}{x^2} = 0 \text{ when } x = \sqrt{A}.$$

$$\frac{d^2P}{dx^2} = \frac{4A}{x^3} > 0 \text{ when } x = \sqrt{A}.$$

P is minimum when $x = y = \sqrt{A}$ feet.

11. $d = \sqrt{(x-4)^2 + (\sqrt{x} - 0)^2}$

$$d' = \frac{2(x-4)+1}{\sqrt{(x-4)^2 + x}} = 0 \text{ when } x = \frac{7}{2}.$$

By the First Derivative Test, the point nearest to (4, 0) is $\left(7/2, \sqrt{7/2}\right)$.

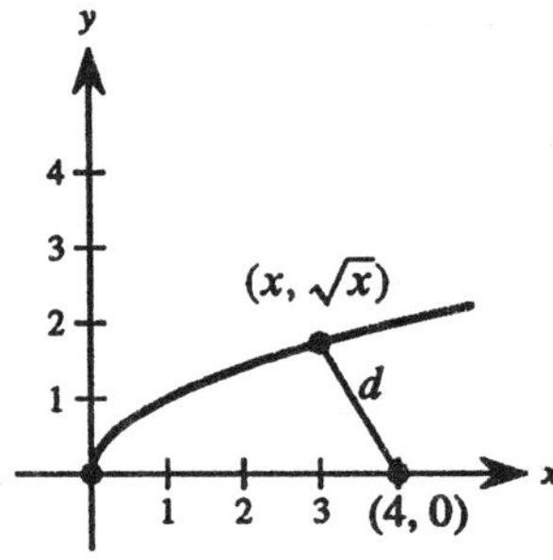

12. $d = \sqrt{(x-2)^2 + [y - (1/2)]^2} = \sqrt{(x-2)^2 + [x^2 - (1/2)]^2}$

$$d' = \frac{2(x^3 - 1)}{\sqrt{(x-2)^2 + [x^2 - (1/2)]^2}} = 0$$

$2(x^3 - 1) = 0$ when $x = 1$ and $y = 1$.

By the First Derivative Test, the point nearest to $\left(2, \frac{1}{2}\right)$ is (1, 1).

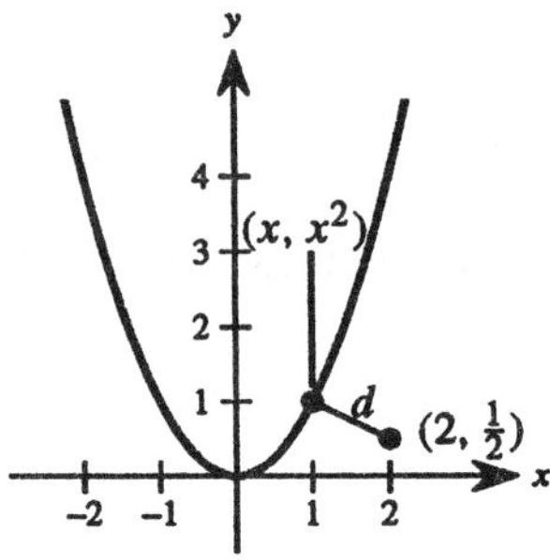

13. $xy = 180{,}000$ (See figure.)

$S = x + 2y = \left(x + \dfrac{360{,}000}{x}\right)$ where

S is the length of fence needed.

$$\frac{dS}{dx} = 1 - \frac{360{,}000}{x^2} = 0 \text{ when } x = 600.$$

$$\frac{d^2S}{dx^2} = \frac{720{,}000}{x^3} > 0 \text{ when } x > 0.$$

S is a minimum when $x = 600$ and $y = 300$.

14. $4x + 3y = 200$ is the perimeter. (See figure.)

$$A = 2xy = 2x\left(\frac{200 - 4x}{3}\right) = \frac{8}{3}(50x - x^2)$$

$$\frac{dA}{dx} = \frac{8}{3}(50 - 2x) = 0 \text{ when } x = 25.$$

$$\frac{d^2A}{dx^2} = -\frac{16}{3}$$

A is a maximum when $x = 25$ and $y = \frac{100}{3}$.

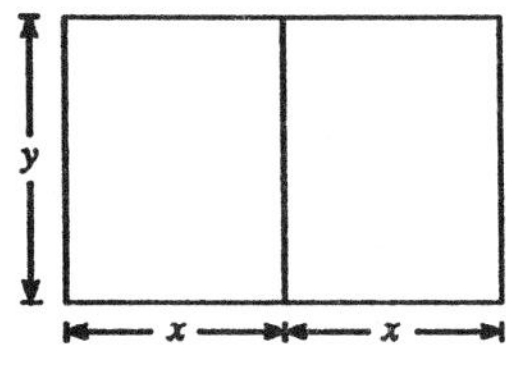

15. $V = x(12 - 2x)^2$ (See figure.)

$$\frac{dV}{dx} = 2x(12 - 2x)(-2) + (12 - 2x)^2 = 12(6 - x)(2 - x) = 0 \text{ when } x = 2, 6.$$

$$\frac{d^2V}{dx^2} = 12(2x - 8)$$

$$\frac{d^2V}{dx^2} = 48 > 0 \text{ when } x = 6.$$

$$\frac{d^2V}{dx^2} = -48 < 0 \text{ when } x = 2.$$

$V = 128$ is maximum when $x = 2$.

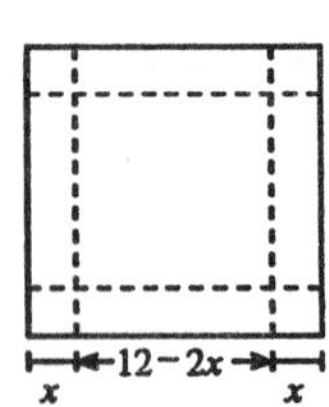

16. (a) $V = x(s - 2x)^2$

$$\frac{dV}{dx} = 2x(s - 2x)(-2) + (s - 2x)^2$$

$$= (s - 2x)(s - 6x) = 0 \text{ when } x = \frac{s}{2}, \frac{s}{6}.$$

$$\frac{d^2V}{dx^2} = 24x - 8s$$

$$\frac{d^2V}{dx^2} > 0 \text{ when } x = \frac{s}{2}.$$

$$\frac{d^2V}{dx^2} < 0 \text{ when } x = \frac{s}{6}.$$

$$V = \frac{2s^3}{27} \text{ is maximum when } x = \frac{s}{6}.$$

(b) If the length is doubled,

$$V = \frac{2}{27}(2s)^3 = 8\left(\frac{2}{27}s^3\right).$$

Volume is increased by a factor of 8.

17. $V = x(3 - 2x)(2 - 2x)$ (See figure.)

$$\frac{dV}{dx} = (3 - 2x)(2 - 2x) + x(2 - 2x)(-2) + x(3 - 2x)(-2)$$

$$\frac{dV}{dx} = 12x^2 - 20x + 6 = 0 \text{ when } x = \frac{5 \pm \sqrt{7}}{6}.$$

$$\frac{d^2V}{dx^2} = 24x - 20 < 0 \text{ when } x = \frac{5 - \sqrt{7}}{6}.$$

The dimensions are $\left(\frac{5 - \sqrt{7}}{6}\text{ ft}\right)$ by $\left(\frac{1 + \sqrt{7}}{3}\text{ ft}\right)$ by $\left(\frac{4 + \sqrt{7}}{3}\text{ ft}\right)$ or (0.392 ft) by (1.215 ft) by (2.215 ft) yield a maximum volume.

18. $S = x^2 + 3xy$ (See figure.)

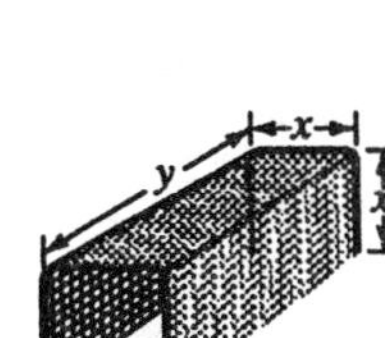

$$x^2y = \frac{250}{3}$$

$$y = \frac{250}{3x^2}$$

$$S = x^2 + \frac{250}{x}$$

$$\frac{dS}{dx} = 2x - \frac{250}{x^2} = \frac{2x^3 - 250}{x^2} = 0 \text{ when } x = 5.$$

$$\frac{d^2S}{dx^2} = 2 + \frac{500}{x^3} > 0 \text{ when } x > 0.$$

S is a minimum when $x = 5$ and $y = \frac{10}{3}$.

19. $16 = 2y + x + \pi\left(\dfrac{x}{2}\right)$

$$32 = 4y + 2x + \pi x$$

$$x = \frac{32 - 2x - \pi x}{4}$$

$$A = xy + \frac{\pi}{2}\left(\frac{x}{2}\right)^2 = \left(\frac{32 - 2x + \pi x}{4}\right)x + \frac{\pi x^2}{8} = 8x - \frac{1}{2}x^2 - \frac{\pi}{4}x^2 + \frac{\pi}{8}x^2$$

$$\frac{dA}{dx} = 8 - x - \frac{\pi}{2}x + \frac{\pi}{4}x = 8 - x\left(1 + \frac{\pi}{4}\right) = 0 \text{ when } x = \frac{8}{1 + (\pi/4)} = \frac{32}{4 + \pi}.$$

$$\frac{d^2A}{dx^2} = -\left(1 + \frac{\pi}{4}\right) < 0$$

$$x = \frac{32 - 2[32/(4+\pi)] - \pi[32/(4+\pi)]}{4} = \frac{16}{4+\pi}$$

The area is maximum when $y = 16/(4+\pi)$ feet and $x = 32/(4+\pi)$ feet.

20. $P = 2x + 2\pi r = 2x + 2\pi\left(\dfrac{y}{2}\right)$

$$P = 2x + \pi y = 200$$

$$y = \frac{200 - 2x}{\pi}$$

$$A = xy = \frac{2x}{\pi}(100 - x) = \frac{2}{\pi}(100x - x^2)$$

$$\frac{dA}{dx} = \frac{2}{\pi}(100 - 2x) = 0 \text{ when } x = 50 \text{ meters and } y = \frac{100}{\pi} \text{ meters.}$$

21. $A = \dfrac{xy}{2} = \dfrac{1}{2}x\left(3 + \dfrac{6}{x-2}\right) = \dfrac{3x^2}{2(x-2)}$

$$\frac{dA}{dx} = \frac{3x(x-4)}{2(x-2)^2} = 0 \text{ when } x = 0,\ 4.$$

By the First Derivative Test, A is a minimum when $x = 4$ and $y = 6$. The vertices of the triangle are $(0,\ 0)$, $(4,\ 0)$, and $(0,\ 6)$.

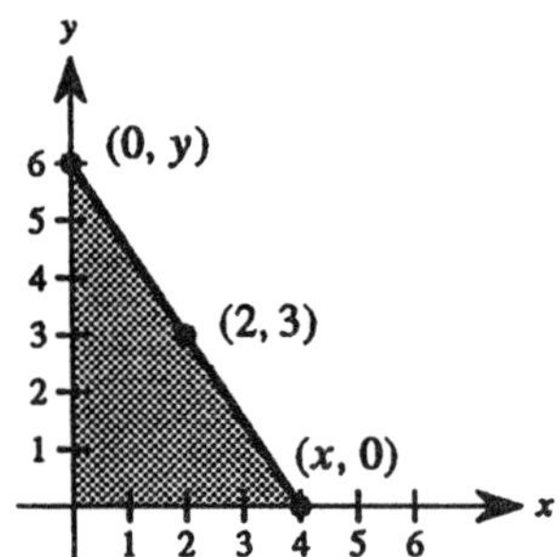

22. $A = \dfrac{1}{2}\left[2\sqrt{16 - x^2}\right](x + 4) = (x + 4)\sqrt{16 - x^2}$

$$\frac{dA}{dx} = (x+4)\left[\frac{1}{2}(16 - x^2)^{-1/2}(-2x)\right] + \sqrt{16 - x^2}$$

$$= \frac{16 - 4x - 2x^2}{\sqrt{16 - x^2}} = \frac{2(2-x)(4+x)}{\sqrt{16 - x^2}} = 0 \text{ when } x = 2.$$

By the First Derivative Test, A is a maximum at the critical point $x = 2$. The vertices are at $(-4,\ 0)$, $(2,\ 2\sqrt{3})$, and $(2,\ -2\sqrt{3})$. The triangle is equilateral with sides of length $4\sqrt{3}$.

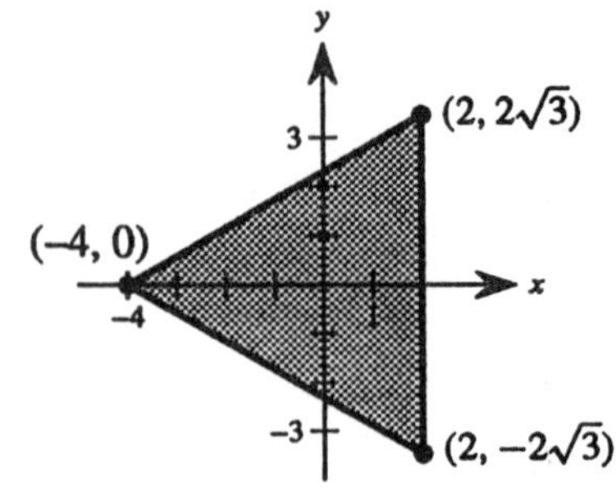

23. $A = 2x\sqrt{25-x^2}$ (See figure.)

$$\frac{dA}{dx} = 2x\left(\frac{1}{2}\right)\left(\frac{-2x}{\sqrt{25-x^2}}\right) + 2\sqrt{25-x^2}$$

$$= 2\left(\frac{25-2x^2}{\sqrt{25-x^2}}\right) = 0 \text{ when } x = y = \frac{5\sqrt{2}}{2} \approx 3.54.$$

By the First Derivative Test, the inscribed rectangle of maximum area has vertices

$$\left(\pm\frac{5\sqrt{2}}{2}, 0\right), \left(\pm\frac{5\sqrt{2}}{2}, \frac{5\sqrt{2}}{2}\right).$$

Length: $5\sqrt{2}$; Width: $\frac{5\sqrt{2}}{2}$

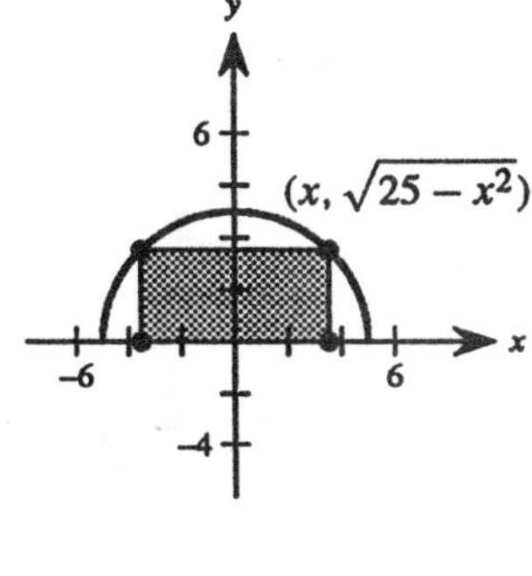

24. $A = 2x\sqrt{r^2-x^2}$ (See figure.)

$$\frac{dA}{dx} = \frac{2(r^2-2x^2)}{\sqrt{r^2-x^2}} = 0 \text{ when } x = \frac{\sqrt{2}r}{2}.$$

By the First Derivative Test, A is maximum when the rectangle has dimensions $\sqrt{2}r$ by $(\sqrt{2}r)/2$.

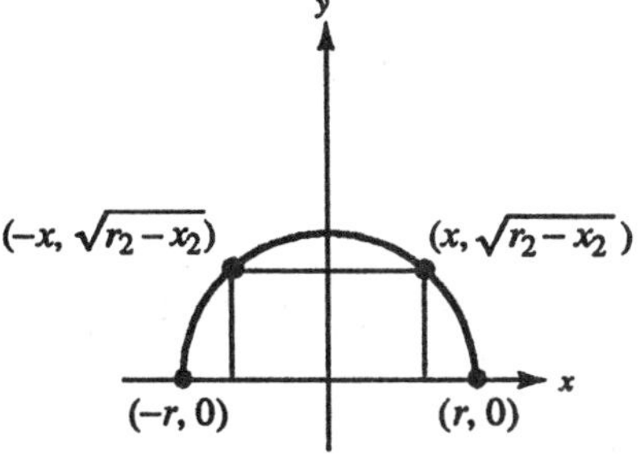

25. $A = \frac{1}{2}(2r+2x)\sqrt{r^2-x^2} = (r+x)\sqrt{r^2-x^2}$ (See figure.)

$$\frac{dA}{dx} = (r+x)\left(\frac{1}{2}\right)(r^2-x^2)^{-1/2}(-2x) + \sqrt{r^2-x^2}$$

$$= \frac{-x(r+x)}{\sqrt{r^2-x^2}} + \sqrt{r^2-x^2} = \frac{r^2-rx-2x^2}{\sqrt{r^2-x^2}}$$

$$= \frac{(r-2x)(r+x)}{\sqrt{r^2-x^2}} = 0 \text{ when } x = \frac{r}{2}.$$

By the First Derivative Test, A will be a maximum when the trapezoid bases are r and $2r$, and the altitude is $(\sqrt{3}r)/2$.

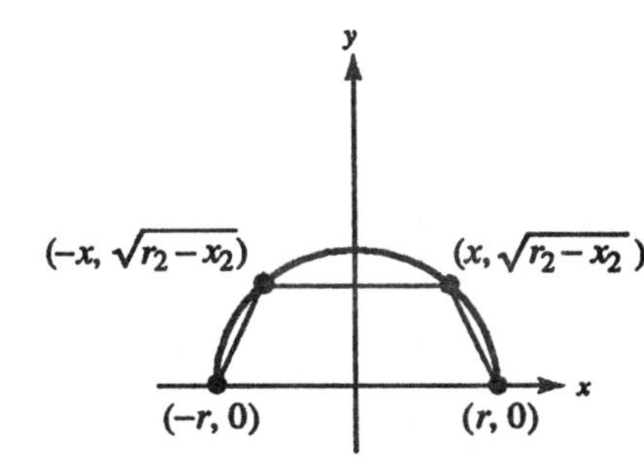

26. $xy = 30 \Rightarrow y = \frac{30}{x}$

$$A = (x+2)\left(\frac{30}{x}+4\right) \text{ (See figure.)}$$

$$\frac{dA}{dx} = (x+2)\left(\frac{-30}{x^2}\right) + \left(\frac{30}{x}+4\right) = \frac{4(x^2-15)}{x^2} = 0 \text{ when } x = \sqrt{15}.$$

$$y = \frac{30}{\sqrt{15}} = 2\sqrt{15}$$

By the First Derivative Test, the dimensions $(x+2)$ by $(y+4)$ or $(2+\sqrt{15})$ by $2(2+\sqrt{15})$ or approximately (5.873) by (11.746) yield a maximum area.

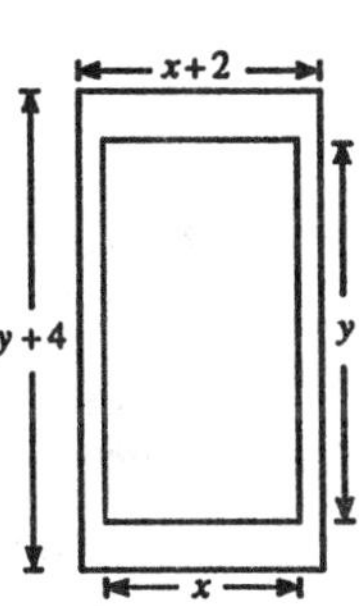

27. $V = \pi r^2 h = 22$ cubic inches or $h = \dfrac{22}{\pi r^2}$

(a)

Radius, r	Height	Surface area
0.2	$\dfrac{22}{\pi(0.2)^2}$	$2\pi(0.2)\left[0.2 + \dfrac{22}{\pi(0.2)^2}\right] \approx 220.3$
0.4	$\dfrac{22}{\pi(0.4)^2}$	$2\pi(0.4)\left[0.4 + \dfrac{22}{\pi(0.4)^2}\right] \approx 111.0$
0.6	$\dfrac{22}{\pi(0.6)^2}$	$2\pi(0.6)\left[0.6 + \dfrac{22}{\pi(0.6)^2}\right] \approx 75.6$
0.8	$\dfrac{22}{\pi(0.8)^2}$	$2\pi(0.8)\left[0.8 + \dfrac{22}{\pi(0.8)^2}\right] \approx 59.0$

(b) $S = 2\pi r^2 + 2\pi rh = 2\pi r(r+h) = 2\pi r\left[r + \dfrac{22}{\pi r^2}\right] = 2\pi r^2 + \dfrac{44}{r}$

(c) $\dfrac{dS}{dr} = 4\pi r - \dfrac{44}{r^2} = 0$ when $r = \sqrt[3]{11/\pi} \approx 1.52$ in.

$h = \dfrac{22}{\pi r^2} \approx 3.04$ in.

Note: Notice that $h = \dfrac{22}{\pi r^2} = \dfrac{22}{\pi(11/\pi)^{2/3}} = 2 \cdot \dfrac{11^{1/3}}{\pi^{1/3}} = 2r.$

(d)

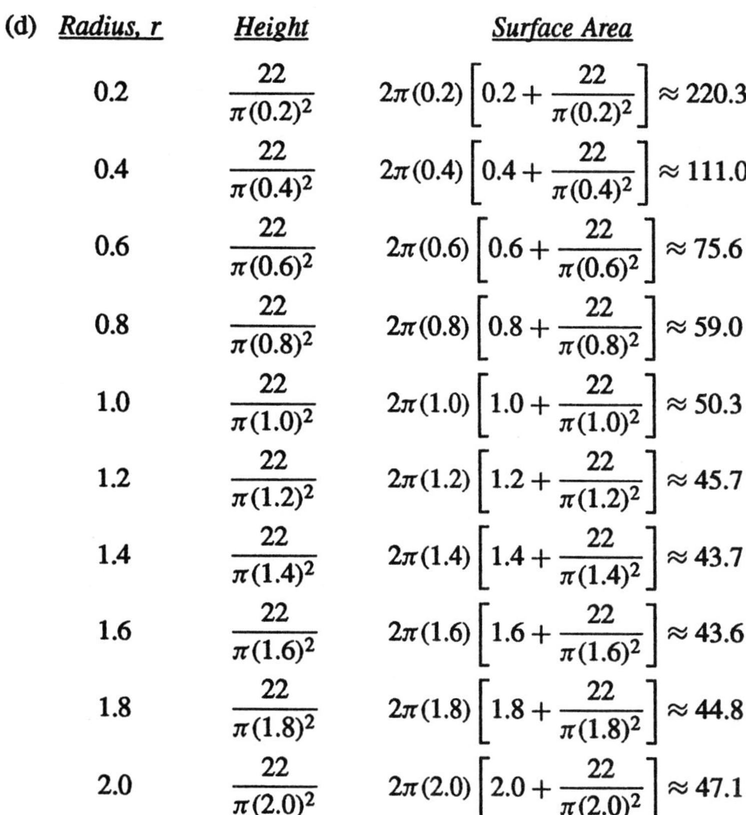

Radius, r	Height	Surface Area
0.2	$\dfrac{22}{\pi(0.2)^2}$	$2\pi(0.2)\left[0.2 + \dfrac{22}{\pi(0.2)^2}\right] \approx 220.3$
0.4	$\dfrac{22}{\pi(0.4)^2}$	$2\pi(0.4)\left[0.4 + \dfrac{22}{\pi(0.4)^2}\right] \approx 111.0$
0.6	$\dfrac{22}{\pi(0.6)^2}$	$2\pi(0.6)\left[0.6 + \dfrac{22}{\pi(0.6)^2}\right] \approx 75.6$
0.8	$\dfrac{22}{\pi(0.8)^2}$	$2\pi(0.8)\left[0.8 + \dfrac{22}{\pi(0.8)^2}\right] \approx 59.0$
1.0	$\dfrac{22}{\pi(1.0)^2}$	$2\pi(1.0)\left[1.0 + \dfrac{22}{\pi(1.0)^2}\right] \approx 50.3$
1.2	$\dfrac{22}{\pi(1.2)^2}$	$2\pi(1.2)\left[1.2 + \dfrac{22}{\pi(1.2)^2}\right] \approx 45.7$
1.4	$\dfrac{22}{\pi(1.4)^2}$	$2\pi(1.4)\left[1.4 + \dfrac{22}{\pi(1.4)^2}\right] \approx 43.7$
1.6	$\dfrac{22}{\pi(1.6)^2}$	$2\pi(1.6)\left[1.6 + \dfrac{22}{\pi(1.6)^2}\right] \approx 43.6$
1.8	$\dfrac{22}{\pi(1.8)^2}$	$2\pi(1.8)\left[1.8 + \dfrac{22}{\pi(1.8)^2}\right] \approx 44.8$
2.0	$\dfrac{22}{\pi(2.0)^2}$	$2\pi(2.0)\left[2.0 + \dfrac{22}{\pi(2.0)^2}\right] \approx 47.1$

(e)

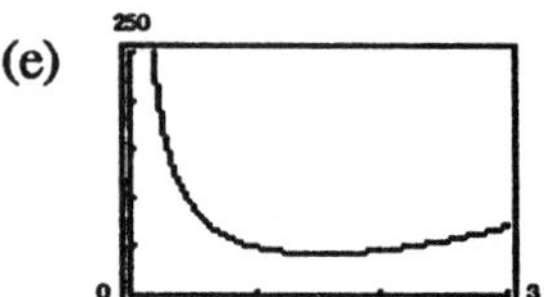

28. $V = \pi r^2 h = V_0$ cubic units or $h = \dfrac{V_0}{\pi r^2}$

$$S = 2\pi r^2 + 2\pi r h = 2\left(\pi r^2 + \frac{V_0}{r}\right)$$

$$\frac{dS}{dr} = 2\left(2\pi r - \frac{V_0}{r^2}\right) = 0 \text{ when } r = \sqrt[3]{\frac{V_0}{2\pi}} \text{ units.}$$

$$h = \frac{V_0}{\pi\left(\sqrt[3]{V_0/2\pi}\right)^2} = \frac{V_0\sqrt[3]{4\pi^2}}{\pi\sqrt[3]{V_0^2}} \cdot \frac{\sqrt[3]{V_0}}{\sqrt[3]{V_0}} = \frac{\sqrt[3]{4\pi^2 V_0}}{\pi} = 2r$$

By the First Derivative Test, this will yield the minimum surface area.

29. Let x be the sides of the square ends and y the length of the package.

$$P = 4x + y = 108$$

$$V = x^2 y = x^2(108 - 4x) = 108x^2 - 4x^3$$

$$\frac{dV}{dx} = 216x - 12x^2 = 12x(18 - x) = 0 \text{ when } x = 18.$$

$$\frac{d^2V}{dx^2} = 216 - 24x = -216 \text{ when } x = 18.$$

The volume is maximum when $x = 18$ inches and $y = 108 - 4(18) = 36$ inches.

30. $V = \pi r^2 x, \quad x + 2\pi r = 108 \Rightarrow x = 108 - 2\pi r$ (See figure.)

$$V = \pi r^2(108 - 2\pi r) = \pi(108r^2 - 2\pi r^3)$$

$$\frac{dV}{dr} = \pi(216r - 6\pi r^2) = 6\pi r(36 - \pi r) = 0 \text{ when } r = \frac{36}{\pi} \text{ and } x = 36.$$

Volume is maximum when $x = 36$ inches and $r = 36/\pi \approx 11.459$ inches.

31. $V = \dfrac{1}{3}\pi x^2 h = \dfrac{1}{3}\pi x^2\left(r + \sqrt{r^2 - x^2}\right)$ (See figure.)

$$\frac{dV}{dx} = \frac{1}{3}\pi\left[\frac{-x^3}{\sqrt{r^2 - x^2}} + 2x\left(r + \sqrt{r^2 - x^2}\right)\right]$$

$$= \frac{\pi x}{3\sqrt{r^2 - x^2}}\left(2r^2 + 2r\sqrt{r^2 - x^2} - 3x^2\right) = 0 \text{ when } x = 0, \ \frac{2\sqrt{2}r}{3}.$$

By the First Derivative Test, the volume is a maximum when

$$x = \frac{2\sqrt{2}r}{3} \text{ and } h = r + \sqrt{r^2 - x^2} = \frac{4r}{3}.$$

Thus, the maximum volume is $V = \dfrac{1}{3}\pi\left(\dfrac{8r^2}{9}\right)\left(\dfrac{4r}{3}\right) = \dfrac{32\pi r^3}{81}$ cubic units.

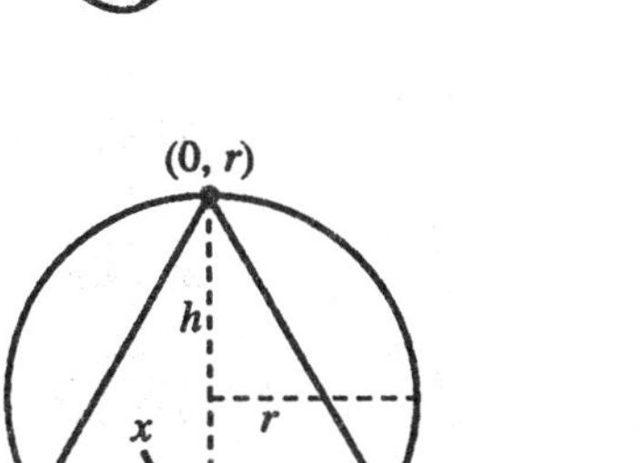

32. $V = \pi r^2 h = \pi x^2\left(2\sqrt{r^2 - x^2}\right) = 2\pi x^2\sqrt{r^2 - x^2}$ (See figure.)

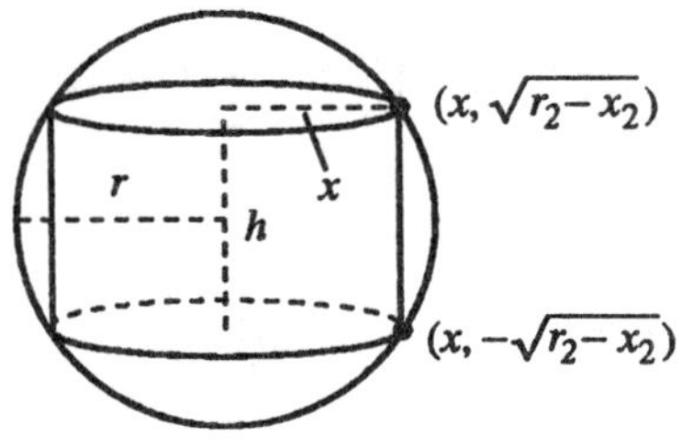

$$\frac{dV}{dx} = 2\pi\left[x^2\left(\frac{1}{2}\right)(r^2 - x^2)^{-1/2}(-2x) + 2x\sqrt{r^2 - x^2}\right]$$

$$= \frac{2\pi x}{\sqrt{r^2 - x^2}}(2r^2 - 3x^2) = 0 \text{ when } x = 0 \text{ and}$$

$$x^2 = \frac{2r^2}{3} \Rightarrow x = \frac{\sqrt{6}r}{3}.$$

By the First Derivative Test, the volume is a maximum when

$$x = \frac{\sqrt{6}r}{3} \text{ and } h = \frac{2r}{\sqrt{3}}.$$

Thus, the maximum volume is $V = \pi\left(\frac{2}{3}r^2\right)\left(\frac{2r}{\sqrt{3}}\right) = \frac{4\pi r^3}{3\sqrt{3}}$.

33. $12 = \frac{4}{3}\pi r^3 + \pi r^2 h$

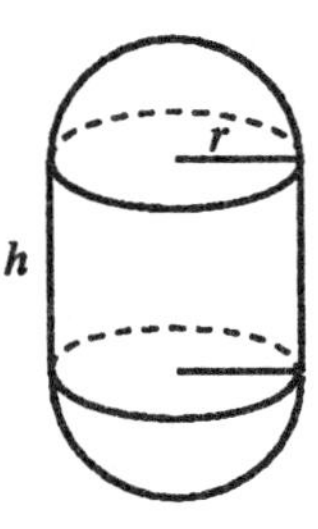

$$h = \frac{12 - (4/3)\pi r^3}{\pi r^2} = \frac{12}{\pi r^2} - \frac{4}{3}r$$

$$S = 4\pi r^2 + 2\pi rh = 4\pi r^2 + 2\pi r\left(\frac{12}{\pi r^2} - \frac{4}{3}r\right)$$

$$= 4\pi r^2 + \frac{24}{r} - \frac{8}{3}\pi r^2 = \frac{4}{3}\pi r^2 + \frac{24}{r}$$

$$\frac{dS}{dr} = \frac{8}{3}\pi r - \frac{24}{r^2} = 0 \text{ when } r = \sqrt[3]{9/\pi} \approx 1.42 \text{ inches.}$$

$$\frac{d^2S}{dr^2} = \frac{8}{3}\pi + \frac{48}{r^3} > 0 \text{ when } r = \sqrt[3]{9/\pi} \text{ inches.}$$

The surface area is minimum when $r = \sqrt[3]{9/\pi}$ inches and $h = 0$. The resulting solid is a sphere of radius $r = 1.42$ inches.

34. $3000 = \frac{4}{3}\pi r^3 + \pi r^2 h$

$$h = \frac{3000}{\pi r^2} - \frac{4}{3}r$$

Let k = cost per square foot of the surface area of the sides, then $2k$ = cost per square foot of the hemispherical ends.

$$C = 2k(4\pi r^2) + k(2\pi rh) = k\left[8\pi r^2 + 2\pi r\left(\frac{3000}{\pi r^2} - \frac{4}{3}r\right)\right] = k\left[\frac{16}{3}\pi r^2 + \frac{6000}{r}\right]$$

$$\frac{dC}{dr} = k\left[\frac{32}{3}\pi r - \frac{6000}{r^2}\right] = 0 \text{ when } r = \sqrt[3]{\frac{1125}{2\pi}} \approx 5.636 \text{ feet and } h \approx 22.545 \text{ feet.}$$

By the Second Derivative Test, we have

$$\frac{d^2C}{dr^2} = k\left[\frac{32}{3}\pi + \frac{12000}{r^3}\right] > 0.$$

Therefore, these dimensions will produce a minimum cost.

35. Let x the the length of a side of the square and y the length of a side of the triangle.

$$4x + 3y = 10$$

$$A = x^2 + \frac{1}{2}y\left(\frac{\sqrt{3}}{2}y\right)$$

$$= \frac{(10-3y)^2}{16} + \frac{\sqrt{3}}{4}y^2$$

$$\frac{dA}{dy} = \frac{1}{8}(10-3y)(-3) + \frac{\sqrt{3}}{2}y = 0$$

$$-30 + 9y + 4\sqrt{3}\,y = 0$$

$$y = \frac{30}{9+4\sqrt{3}}$$

$$\frac{d^2A}{dy^2} = \frac{9+4\sqrt{3}}{8} > 0$$

A is minimum when

$$y = \frac{30}{9+4\sqrt{3}} \text{ and } x = \frac{10\sqrt{3}}{9+4\sqrt{3}}.$$

36. Let x be the length of a side of the square and r the radius of the circle.

$$4x + 2\pi r = 16$$

$$x = 4 - \frac{1}{2}\pi r$$

$$A = x^2 + \pi r^2 = \left(4 - \frac{1}{2}\pi r\right)^2 + \pi r^2$$

$$\frac{dA}{dr} = 2\left(4 - \frac{1}{2}\pi r\right)\left(-\frac{1}{2}\pi\right) + 2\pi r$$

$$-8\pi + (\pi^2 + 4\pi)r = 0 \text{ when } r = \frac{8}{4+\pi}.$$

$$\frac{d^2A}{dr^2} = \frac{\pi^2}{2} > 0$$

A is minimum when $r = \dfrac{8}{4+\pi}$ and $x = \dfrac{16}{4+\pi}$.

37. Let x be the length of the sides of the isosceles triangle (the hypotenuse will be $\sqrt{2}x$) and r the radius of the circle.

$$(2x + \sqrt{2}x) + 2\pi r = 10$$

$$x = \frac{10 - 2\pi r}{2+\sqrt{2}}$$

$$A = \frac{1}{2}x^2 + \pi r^2 = \frac{1}{2}\left(\frac{10-2\pi r}{2+\sqrt{2}}\right)^2 + \pi r^2 = \frac{1}{3+2\sqrt{2}}(25 - 10\pi r + \pi^2 r^2) + \pi r^2$$

$$\frac{dA}{dr} = \frac{1}{3+2\sqrt{2}}(-10\pi + 2\pi^2 r) + 2\pi r = 0 \Rightarrow r = \frac{5}{\pi + 3 + 2\sqrt{2}}$$

(a) Minimum: $2\pi r = 2\pi\left(\dfrac{5}{\pi+3+2\sqrt{2}}\right) = \dfrac{10\pi}{\pi+3+2\sqrt{2}} \approx 3.5$ feet

(b) Maximum: $2\pi r = 10$ feet (endpoint extrema)

38. (a) Let x be the side of the triangle and y the side of the square.

$$A = \frac{\sqrt{3}}{4}x^2 + \left(5 - \frac{3}{4}x\right)^2, \quad 0 \le x \le \frac{20}{3}$$

$$\frac{dA}{dx} = \frac{\sqrt{3}}{2}x + 2\left(5 - \frac{3}{4}x\right)\left(-\frac{3}{4}\right) = 0$$

$$x = \frac{60}{4\sqrt{3}+9}$$

When $x = 0$, $A = 25$, when $x = 60/(4\sqrt{3}+9)$, $A \approx 10.847$, and when $x = 20/3$, $A \approx 19.245$. Area is maximum when all 20 feet are used on the square.

38. —CONTINUED—

(b) Let x be the side of the square and y the side of the pentagon.

$$A = x^2 + 1.7204774\left(4 - \frac{4}{5}x\right)^2, \quad 0 \le x \le 5$$

$$A' = 2x - 2.75276384\left(4 - \frac{4}{5}x\right) = 0$$

$$x \approx 2.62$$

When $x = 0$, $A \approx 27.528$, when $x \approx 2.62$, $A \approx 13.102$, and when $x = 5$, $A \approx 25$. Area is maximum when all 20 feet are used on the pentagon.

(c) Let x be the side of the pentagon and y the side of the hexagon.

$$A = 1.7204774x^2 + \frac{3}{2}(\sqrt{3})\left(\frac{20 - 5x}{6}\right)^2, \quad 0 \le x \le 4$$

$$A' = 3.4409548x + 3\sqrt{3}\left(-\frac{5}{6}\right)\left(\frac{20 - 5x}{6}\right) = 0$$

$$x \approx 2.0475$$

When $x = 0$, $A \approx 28.868$, when $x \approx 2.0475$, $A \approx 14.091$, and when $x = 4$, $A \approx 27.528$. Area is maximum when all 20 feet are used on the hexagon.

39. Let S be the strength and k the constant of proportionality. Given $h^2 + w^2 = 24^2$, $h^2 = 24^2 - w^2$,

$$S = kwh^2$$

$$S = kw(576 - w^2) = k(576w - w^3)$$

$$\frac{dS}{dw} = k(576 - 3w^2) = 0 \text{ when } w = 8\sqrt{3}, \quad h = 8\sqrt{6}.$$

$$\frac{d^2S}{dw^2} = -6kw < 0 \text{ for } w > 0$$

These values yield a maximum.

40. Let F be the illumination at point P which is x units from source 1.

$$F = \frac{kI_1}{x^2} + \frac{kI_2}{(d - x)^2}$$

$$\frac{dF}{dx} = \frac{-2kI_1}{x^3} + \frac{2kI_2}{(d - x)^3} = 0 \text{ when } \frac{2kI_1}{x^3} = \frac{2kI_2}{(d - x)^3}.$$

$$\frac{\sqrt[3]{I_1}}{\sqrt[3]{I_2}} = \frac{x}{d - x}$$

$$x = \frac{d\sqrt[3]{I_1}}{\sqrt[3]{I_1} + \sqrt[3]{I_2}}$$

$$\frac{d^2F}{dx^2} = \frac{6kI_1}{x^4} + \frac{6kI_2}{(d - x)^4} > 0 \text{ when } x > 0.$$

This is the minimum point.

41. $$S = \sqrt{x^2+4}, \quad L = \sqrt{1+(3-x)^2}$$

$$T = \frac{\sqrt{x^2+4}}{2} + \frac{\sqrt{x^2-6x+10}}{4}$$

$$\frac{dT}{dx} = \frac{x}{2\sqrt{x^2+4}} + \frac{x-3}{4\sqrt{x^2-6x+10}} = 0$$

$$\frac{x^2}{x^2+4} = \frac{9-6x+x^2}{4(x^2-6x+10)}$$

$$x^4 - 6x^3 + 9x^2 + 8x - 12 = 0$$

We can find the roots of this equation in the interval [0, 3] with the bisection method (see Section 4.4, Exercise 34). This equation has only one root in this interval ($x = 1$). Testing at this value and at the endpoints, we see that $x = 1$ yields the minimum time. Thus, the man should row to a point 1 mile from the nearest point on the coast.

42. $$T = \frac{\sqrt{x^2+4}}{4} + \frac{\sqrt{x^2-6x+10}}{4}$$

$$\frac{dT}{dx} = \frac{1}{4}\left[\frac{x}{\sqrt{x^2+4}} + \frac{x-3}{\sqrt{x^2-6x+10}}\right] = 0 \text{ when } \frac{x^2}{x^2+4} = \frac{(x-3)^2}{x^2-6x+10}.$$

Thus, $x^2 - 8x + 12 = 0 \Rightarrow x = 2, 6$. In the interval [0, 3], we have $x = 2$ miles. Testing at $x = 2$ and at the endpoints, we see that $x = 2$ yields a minimum time.

43. $$A = 2x\sqrt{100-x^2}$$

$$\frac{dA}{dx} = \frac{2(100-2x^2)}{\sqrt{100-x^2}} = 0 \text{ (From Exercise 24)}$$

$$x = 5\sqrt{2}$$

Dimensions: $10\sqrt{2} \times 5\sqrt{2} \approx 14.1421 \times 7.071$

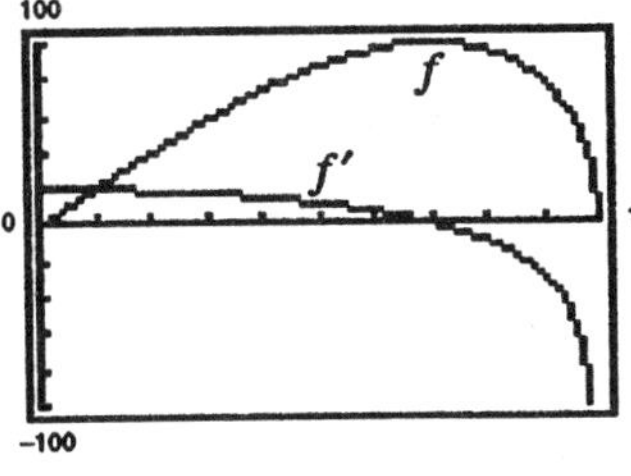

44. $$\frac{y-8}{0-5} = \frac{8-0}{5-x} \Rightarrow y = \frac{8x}{x-5}$$

$$L^2 = x^2 + y^2 = x^2 + \left(\frac{8x}{x-5}\right)^2$$

$$2LL' = 2x + 2\left(\frac{8x}{x-5}\right)\left[\frac{8x}{(x-5)^2}\right]$$

$$LL' = x - \frac{320x}{(x-5)^3}$$

$$x[(x-5)^3 - 320] = 0$$

$$x = \sqrt[3]{320} + 5$$

$$L = \sqrt{(\sqrt[3]{320}+5)^2 + \left[\frac{8(\sqrt[3]{320}+5)^2}{\sqrt[3]{320}}\right]^2} \approx 18.22 \text{ feet}$$

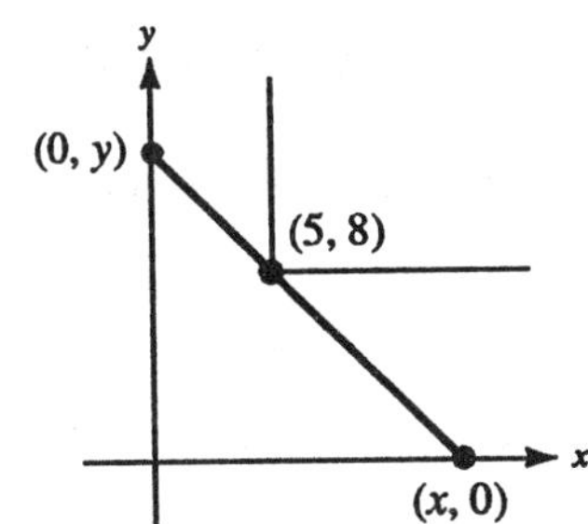

Section 4.8 Newton's Method

1. $f(x) = x^2 - 3$

$f'(x) = 2x$

$x_1 = 1.7$

n	x_n	$f(x_n)$	$f'(x_n)$	$\frac{f(x_n)}{f'(x_n)}$	$x_n - \frac{f(x_n)}{f'(x_n)}$
1	1.700	−0.110	3.400	−0.032	1.732

2. $f(x) = 3x^2 - 2$

$f'(x) = 6x$

$x_1 = 1$

n	x_n	$f(x_n)$	$f'(x_n)$	$\frac{f(x_n)}{f'(x_n)}$	$x_n - \frac{f(x_n)}{f'(x_n)}$
1	1.000	1.000	6.000	0.167	0.833

3. $f(x) = 3x^3 - 2$

$f'(x) = 9x^2$

$x_1 = 1$

n	x_n	$f(x_n)$	$f'(x_n)$	$\frac{f(x_n)}{f'(x_n)}$	$x_n - \frac{f(x_n)}{f'(x_n)}$
1	1.000	1.000	9.000	0.111	0.889

4. $f(x) = x^3 - x^2 - 2x - 2$

$f'(x) = 3x^2 - 2x - 2$

$x_1 = 2.5$

n	x_n	$f(x_n)$	$f'(x_n)$	$\frac{f(x_n)}{f'(x_n)}$	$x_n - \frac{f(x_n)}{f'(x_n)}$
1	2.500	2.375	11.750	0.202	2.298

5. $f(x) = x^3 + x - 1$

$f'(x) = 3x^2 + 1$

n	x_n	$f(x_n)$	$f'(x_n)$	$\frac{f(x_n)}{f'(x_n)}$	$x_n - \frac{f(x_n)}{f'(x_n)}$
1	0.5000	−0.3750	1.7500	−0.2143	0.7143
2	0.7143	0.0787	2.5306	0.0311	0.6832
3	0.6832	0.0021	2.4002	0.0009	0.6823

Approximation to the zero: $x = 0.682$

6. $f(x) = x^5 + x - 1$

$f'(x) = 5x^4 + 1$

n	x_n	$f(x_n)$	$f'(x_n)$	$\frac{f(x_n)}{f'(x_n)}$	$x_n - \frac{f(x_n)}{f'(x_n)}$
1	0.5000	−0.4687	1.3125	−0.3571	0.8571
2	0.8571	0.3198	3.6983	0.0865	0.7707
3	0.7707	0.0426	2.7640	0.0154	0.7553
4	0.7553	0.0011	2.6272	0.0004	0.7549

Approximation to the zero: $x = 0.755$

7. $f(x) = 3\sqrt{x-1} - x$

$f'(x) = \dfrac{3}{2\sqrt{x-1}} - 1$

n	x_n	$f(x_n)$	$f'(x_n)$	$\dfrac{f(x_n)}{f'(x_n)}$	$x_n - \dfrac{f(x_n)}{f'(x_n)}$
1	1.2000	0.1416	2.3541	0.0602	1.1398
2	1.1398	−0.0181	3.0118	−0.0060	1.1458
3	1.1458	−0.0003	2.9283	−0.0001	1.1459

Approximation to the zero: $x = 1.146$

8. $f(x) = x^3 - 3.9x^2 + 4.79x - 1.881$

$f'(x) = 3x^2 - 7.8x + 4.79$

n	x_n	$f(x_n)$	$f'(x_n)$	$\dfrac{f(x_n)}{f'(x_n)}$	$x_n - \dfrac{f(x_n)}{f'(x_n)}$
1	0.5000	−0.3360	1.6400	−0.2049	0.7049
2	0.7049	−0.0921	0.7824	−0.1177	0.8226
3	0.8226	−0.0231	0.4037	−0.0572	0.8799
4	0.8799	−0.0045	0.2494	−0.0181	0.8980
5	0.8980	−0.0004	0.2048	−0.0019	0.8999
6	0.8999	0.0000	0.2002	0.0000	0.8999

Approximation to the zero: $x = 0.900$

n	x_n	$f(x_n)$	$f'(x_n)$	$\dfrac{f(x_n)}{f'(x_n)}$	$x_n - \dfrac{f(x_n)}{f'(x_n)}$
1	1.1	0.001	−0.16	−0.0062	1.1062
2	1.1062	0.000	−0.1673	0.0000	1.1062

Approximation to the zero: $x = 1.106$

n	x_n	$f(x_n)$	$f'(x_n)$	$\dfrac{f(x_n)}{f'(x_n)}$	$x_n - \dfrac{f(x_n)}{f'(x_n)}$
1	1.9	0	0.8000	0	1.9

Approximation to the zero: $x = 1.9$

9. $f(x) = x^4 - 10x^2 - 11$

$f'(x) = 4x^3 - 20x$

n	x_n	$f(x_n)$	$f'(x_n)$	$\dfrac{f(x_n)}{f'(x_n)}$	$x_n - \dfrac{f(x_n)}{f'(x_n)}$
1	3.5000	16.5625	101.5000	0.1632	3.3368
2	3.3368	1.6307	81.8748	0.0199	3.3169
3	3.3169	0.0224	79.6298	0.0003	3.3166

Approximation to the zero: $x = 3.317$

10. $f(x) = x^3 + 3$

$f'(x) = 3x^2$

n	x_n	$f(x_n)$	$f'(x_n)$	$\frac{f(x_n)}{f'(x_n)}$	$x_n - \frac{f(x_n)}{f'(x_n)}$
1	−1.5000	−0.3750	6.7500	−0.0556	−1.4444
2	−1.4444	−0.0137	6.2589	−0.0022	−1.4422
3	−1.4422	−0.0000	6.2398	0.0000	−1.4422

Approximation to the zero: $x = -1.442$

11. $f(x) = 2x + 1 - \sqrt{x+4}$

$f'(x) = 2 - \frac{1}{2\sqrt{x+4}}$

n	x_n	$f(x_n)$	$f'(x_n)$	$\frac{f(x_n)}{f'(x_n)}$	$x_n - \frac{f(x_n)}{f'(x_n)}$
1	0.6	0.0552	1.7669	0.0312	0.5688
2	0.5688	0.0001	1.7661	0.0001	0.5687
3	0.5687	−0.0001	1.7661	−0.0001	0.5688

Approximation to the zero: $x = 0.569$

12. $f(x) = 3 - x - \frac{1}{x^2+1}$

$f'(x) = -1 + \frac{2x}{(x^2+1)^2}$

n	x_n	$f(x_n)$	$f'(x_n)$	$\frac{f(x_n)}{f'(x_n)}$	$x_n - \frac{f(x_n)}{f'(x_n)}$
1	2.9	−0.0063	−0.9345	0.0067	2.8933
2	2.8933	0.0000	−0.9341	0.0000	2.8933

Approximation to the zero: $x = 2.893$

13. $y = 2x^3 - 6x^2 + 6x - 1$

$y' = 6x^2 - 12x + 6$

$x_1 = 1$

n	x_n	$f(x_n)$	$f'(x_n)$
1	1	1	0

$f'(x_1) = 0$; therefore, the method fails.

14. $y = 4x^3 - 12x^2 + 12x - 3$

$y' = 12x^2 - 24x + 12$

$x_1 = \frac{3}{2}$

n	x_n	$f(x_n)$	$f'(x_n)$	$\frac{f(x_n)}{f'(x_n)}$	$x_n - \frac{f(x_n)}{f'(x_n)}$
1	$\frac{3}{2}$	$\frac{3}{2}$	3	$\frac{1}{2}$	1
2	1	1	0	–	–

$f'(x_2) = 0$; therefore, the method fails.

15. $y = -x^3 + 3x^2 - x + 1$

$y' = -3x^2 + 6x - 1$

$x_1 = 1$

n	x_n	$f(x_n)$	$f'(x_n)$	$\frac{f(x_n)}{f'(x_n)}$	$x_n - \frac{f(x_n)}{f'(x_n)}$
1	1	2	2	1	0
2	0	1	−1	−1	1
3	1	–	–	–	–

Fails to converge

16. $f(x) = \sqrt[3]{x-1}$

$f'(x) = \dfrac{1}{3(x-1)^{2/3}}$

n	x_n	$f(x_n)$	$f'(x_n)$	$\dfrac{f(x_n)}{f'(x_n)}$	$x_n - \dfrac{f(x_n)}{f'(x_n)}$
1	2.0000	1.0000	0.3333	3.0000	−1.0000
2	−1.0000	−1.2599	0.2100	−6.0000	5.0000
3	5.0000	1.5874	0.1323	12.0000	−7.0000
4	−7.0000	−2.0000	0.0833	−24.0000	17.0000

Fails to converge

17. $f(x) = x^2 - a = 0$

$f'(x) = 2x$

$$x_{i+1} = x_i - \frac{x_i^2 - a}{2x_i}$$

$$= \frac{2x_i^2 - x_i^2 + a}{2x_i} = \frac{x_i^2 + a}{2x_i}$$

18. $f(x) = x^n - a = 0$

$f'(x) = nx^{n-1}$

$$x_{i+1} = x_i - \frac{x_i^n - a}{nx_i^{n-1}}$$

$$= \frac{nx_i^n - x_i^n + a}{nx_i^{n-1}} = \frac{(n-1)x_i^n + a}{nx_i^{n-1}}$$

19. $x_{i+1} = \dfrac{x_i^2 + 7}{2x_i}$

i	1	2	3	4	5
x_i	2	2.75	2.6477	2.6458	2.6458

$\sqrt{7} \approx 2.646$

20. $x_{i+1} = \dfrac{x_i^2 + 5}{2x_i}$

i	1	2	3	4	5	6	7
x_i	2	2.25	2.23611	2.23610	2.2361	2.2361	2.2361

$\sqrt{5} \approx 2.236$

21. $x_{i+1} = \dfrac{3x_i^4 + 6}{4x_i^3}$

i	1	2	3	4
x_i	1.5000	1.5694	1.5651	1.5651

$\sqrt[4]{6} \approx 1.565$

22. $x_{i+1} = \dfrac{2x_i^3 + 15}{3x_i^2}$

i	1	2	3	4
x_i	2.5	2.4667	2.4662	2.4662

$\sqrt[3]{15} \approx 2.466$

23. $f(x) = \dfrac{1}{x} - a = 0$

$f'(x) = -\dfrac{1}{x^2}$

$$x_{n+1} = x_n - \frac{(1/x_n) - a}{-1/x_n^2} = x_n + x_n^2\left(\frac{1}{x_n} - a\right) = x_n + x_n - x_n^2 a = 2x_n - x_n^2 a = x_n(2 - ax_n)$$

24. (a) $x_{n+1} = x_n(2 - 3x_n)$

i	1	2	3	4
x_i	0.3	0.33	0.3333	0.3333

$\frac{1}{3} \approx 0.333$

(b) $x_{n+1} = x_n(2 - 11x_n)$

i	1	2	3	4
x_i	0.1	0.09	0.0909	0.0909

$\frac{1}{11} \approx 0.091$

25. $y = f(x) = 4 - x^2, \quad (1, 0)$

$d = \sqrt{(x-1)^2 + (y-0)^2} = \sqrt{(x-1)^2 + (4-x^2)^2} = \sqrt{x^4 - 7x^2 - 2x + 17}$

d is minimized when $d_i = x^4 - 7x^2 - 2x + 17$ is a minimum.

$g(x) = d_i' = 4x^3 - 14x - 2$

$g'(x) = 12x^2 - 14$

n	x_n	$g(x_n)$	$g'(x_n)$	$\frac{g(x_n)}{g'(x_n)}$	$x_n - \frac{g(x_n)}{g'(x_n)}$
1	2	2	34	0.0588	1.9412
2	1.9412	0.0830	31.2191	0.0026	1.9386
3	1.9386	0.0020	31.0980	0.0001	1.9385

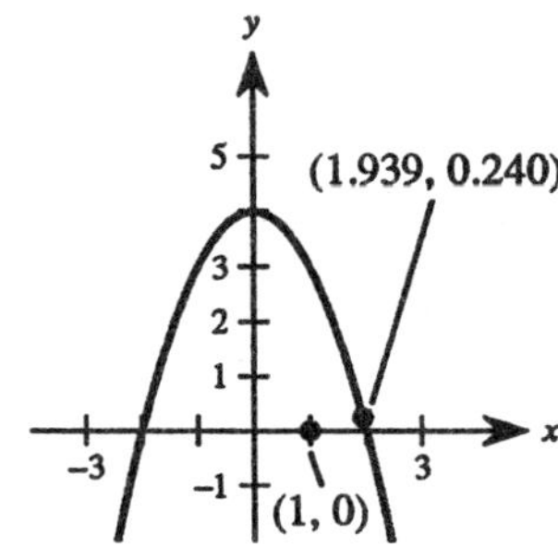

$x \approx 1.939$

Point closest to (1, 0) is approximately (1.939, 0.240).

26. $y = f(x) = x^2, \quad (4, -3)$

$d = \sqrt{(x-4)^2 + (y+3)^2} = \sqrt{(x-4)^2 + (x^2+3)^2} = \sqrt{x^4 + 7x^2 - 8x + 25}$

d is minimum when $d_i = x^4 + 7x^2 - 8x + 25$ is minimum.

$g(x) = d_i = 4x^3 + 14x - 8 = 2(2x^3 + 7x - 4) = 0$

$g'(x) = 12x^2 + 14$

n	x_n	$g(x_n)$	$g'(x_n)$	$\frac{g(x_n)}{g'(x_n)}$	$x_n - \frac{g(x_n)}{g'(x_n)}$
1	0.5	−0.5000	17	−0.0294	0.5294
2	0.5294	0.0051	17.3632	0.0003	0.5291
3	0.5291	−0.0001	17.3594	0.0000	0.5291

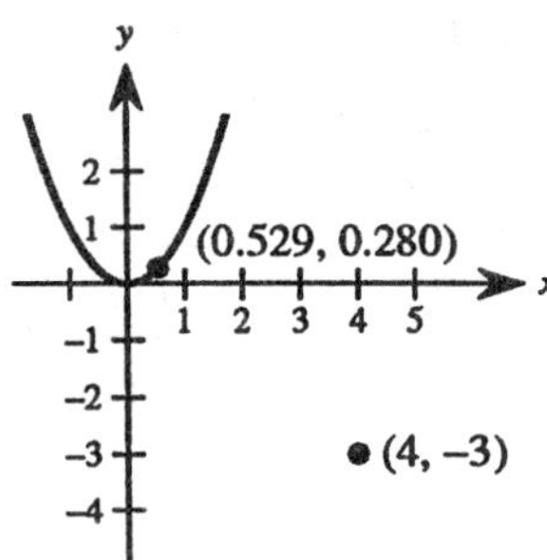

$x \approx 0.529$

Point closest to (4, −3) is approximately (0.529, 0.280).

27. Minimize:

$$T = \frac{\text{Distance rowed}}{\text{Rate rowed}} + \frac{\text{Distance walked}}{\text{Rate walked}}$$

$$T = \frac{\sqrt{x^2+4}}{3} + \frac{\sqrt{x^2-6x+10}}{4}$$

$$T' = \frac{x}{3\sqrt{x^2+4}} + \frac{x-3}{4\sqrt{x^2-6x+10}} = 0$$

$$4x\sqrt{x^2-6x+10} = -3(x-3)\sqrt{x^2+4}$$

$$16x^2(x^2-6x+10) = 9(x-3)^2(x^2+4)$$

$$7x^4 - 42x^3 + 43x^2 + 216x - 324 = 0$$

Let $f(x) = 7x^4 - 42x^3 + 43x^2 + 216x - 324$ and $f'(x) = 28x^3 - 126x^2 + 86x + 216$. Since $f(1) = -100$ and $f(2) = 56$, the solution is in the interval $(1, 2)$.

n	x_n	$f(x_n)$	$f'(x_n)$	$\frac{f(x_n)}{f'(x_n)}$	$x_n - \frac{f(x_n)}{f'(x_n)}$
1	1.7000	19.5887	135.6240	0.1444	1.5556
2	1.5556	−0.0481	150.2780	0.0019	1.5575
3	1.5575	−0.7627	150.0828	−0.0051	1.5626

Approximation: $x \approx 1.563$ miles

28. Maximize:

$$C = \frac{3t^2+t}{50+t^3}$$

$$C' = \frac{-3t^4 - 2t^3 + 300t + 50}{(50+t^3)^2} = 0$$

Let $f(x) = 3t^4 + 2t^3 - 300t - 50$ and $f'(x) = 12t^3 + 6t^2 - 300$. Since $f(4) = -354$ and $f(5) = 575$, the solution is in the interval $(4, 5)$.

n	x_n	$f(x_n)$	$f'(x_n)$	$\frac{f(x_n)}{f'(x_n)}$	$x_n - \frac{f(x_n)}{f'(x_n)}$
1	4.5000	12.4374	915.0000	0.0136	4.4864
2	4.4864	0.0658	904.3822	0.0001	4.4863

Approximation: $t \approx 4.486$ hours

29. $f(x) = x^3 - 3x^2 + 3, \quad f'(x) = 3x^2 - 6x$

(a)

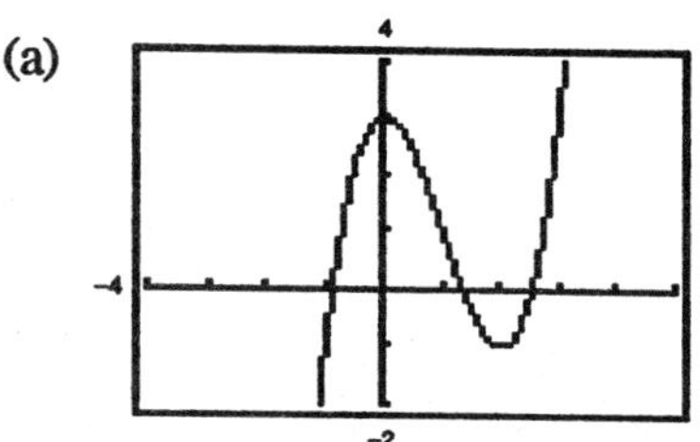

(b) $x_1 = 1$

$$x_2 = x_1 - \frac{f(x_1)}{f'(x_1)} \approx 1.333$$

(c) $x_1 = \frac{1}{4}$

$$x_2 = x_1 - \frac{f(x_1)}{f'(x_1)} \approx 2.405$$

(d)

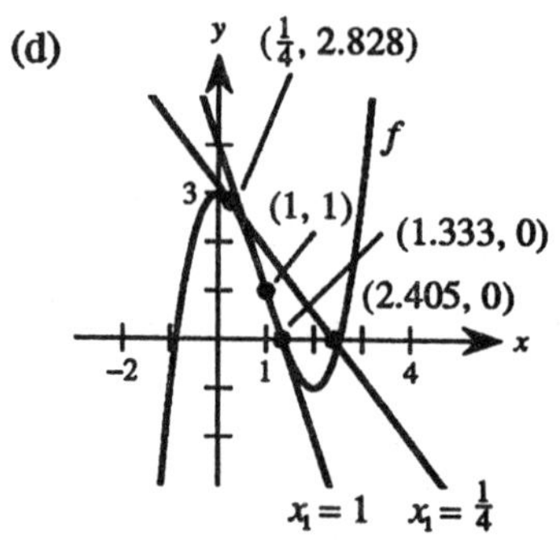

The x-intercepts correspond to the values resulting from the first iteration of Newton's Method.

(e) If the initial guess x_1 is not "close to" the desired zero of the function, the x-intercept of the tangent line may approximate another zero of the function.

30. $f(x) = \frac{1}{4}x^3 - 3x^2 + \frac{3}{4}x - 2$

$f'(x) = \frac{3}{4}x^2 - 6x + \frac{3}{4}$

Let $x_1 = 12$.

n	x_n	$f(x_n)$	$f'(x_n)$	$\frac{f(x_n)}{f'(x_n)}$	$x_n - \frac{f(x_n)}{f'(x_n)}$
1	12.0000	7.0000	36.7500	0.1905	11.8095
2	11.8095	0.2151	34.4912	0.0062	11.8033
3	11.8033	0.0015	34.4186	0.0000	11.8033

Approximation: $x \approx 11.803$

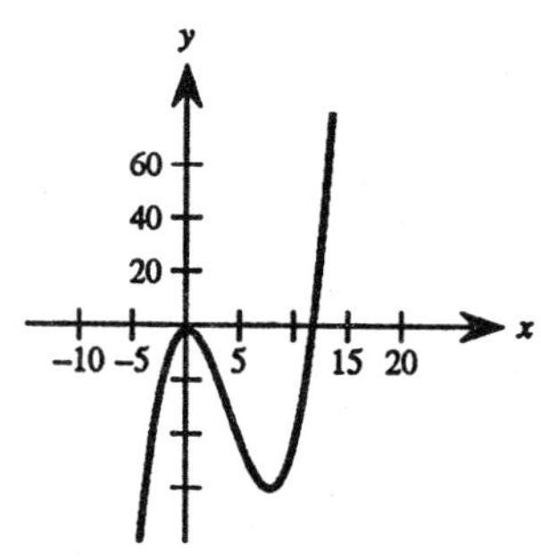

Section 4.9 Differentials

1. $y = 3x^2 - 4$

$dy = 6x\,dx$

2. $y = 2x^{3/2}$

$dy = 3\sqrt{x}\,dx$

3. $y = 4x^3$

$dy = 12x^2\,dx$

4. $y = 3$

$dy = 0$

5. $y = \frac{x+1}{2x-1}$

$$dy = \frac{-3}{(2x-1)^2}\,dx$$

6. $y = \frac{x}{x+5}$

$$dy = \frac{5}{(x+5)^2}\,dx$$

7. $y = \sqrt{x}$

$$dy = \frac{1}{2\sqrt{x}}\,dx$$

8. $y = \sqrt{x^2 - 4}$

$$dy = \frac{x}{\sqrt{x^2-4}}\,dx$$

9. $y = x\sqrt{1 - x^2}$

$$dy = \left(x\frac{-x}{\sqrt{1-x^2}} + \sqrt{1-x^2}\right)dx$$

$$= \frac{1-2x^2}{\sqrt{1-x^2}}\,dx$$

10. $y = \sqrt{x} + \frac{1}{\sqrt{x}}$

$$dy = \left(\frac{1}{2\sqrt{x}} - \frac{1}{2x\sqrt{x}}\right)dx$$

$$= \frac{x-1}{2x\sqrt{x}}\,dx$$

11. $y = x - 1$

$dy = dx$

$x = 2$

$dx = \Delta x$	dy	Δy	$\Delta y - dy$	$\frac{dy}{\Delta y}$
1.0000	1.0000	1.0000	0.0000	1.0000
0.5000	0.5000	0.5000	0.0000	1.0000
0.1000	0.1000	0.1000	0.0000	1.0000
0.0100	0.0100	0.0100	0.0000	1.0000
0.0010	0.0010	0.0010	0.0000	1.0000

12. $y = 2x$

$dy = 2\,dx$

$x = 2$

$dx = \Delta x$	dy	Δy	$\Delta y - dy$	$\frac{dy}{\Delta y}$
1.0000	2.0000	2.0000	0.0000	1.0000
0.5000	1.0000	1.0000	0.0000	1.0000
0.1000	0.2000	0.2000	0.0000	1.0000
0.0100	0.0200	0.0200	0.0000	1.0000
0.0010	0.0020	0.0020	0.0000	1.0000

13. $y = x^2$

$dy = 2x\,dx$

$x = 2$

$dx = \Delta x$	dy	Δy	$\Delta y - dy$	$\frac{dy}{\Delta y}$
1.0000	4.0000	5.0000	1.0000	0.8000
0.5000	2.0000	2.2500	0.2500	0.8889
0.1000	0.4000	0.4100	0.0100	0.9756
0.0100	0.0400	0.0401	0.0001	0.9975
0.0010	0.0040	0.0040	0.0000	1.0000

14. $y = \frac{1}{x^2}$

$dy = \frac{-2}{x^3}\,dx$

$x = 2$

$dx = \Delta x$	dy	Δy	$\Delta y - dy$	$\frac{dy}{\Delta y}$
1.0000	−0.2500	−0.1389	0.1111	1.8000
0.5000	−0.1250	−0.0900	0.0350	1.3889
0.1000	−0.0250	−0.0232	0.0018	1.0756
0.0100	−0.0025	−0.0025	0.0000	1.0075
0.0010	−0.0003	−0.0003	0.0000	1.0009

15. $y = x^5$

$dy = 5x^4\,dx$

$x = 2$

$\Delta x = dx$	dy	Δy	$\Delta y - dy$	$\frac{dy}{\Delta y}$
1.0000	80.0000	211.0000	131.0000	0.3791
0.5000	40.0000	65.6562	25.6562	0.6092
0.1000	8.0000	8.8410	0.8410	0.9049
0.0100	0.8000	0.8080	0.0080	0.9901
0.0010	0.0800	0.0801	0.0001	0.9990

16. $y = \sqrt{x}$

$dy = \frac{dx}{2\sqrt{x}}$

$x = 2$

$\Delta x = dx$	dy	Δy	$\Delta y - dy$	$\frac{dy}{\Delta y}$
1.0000	0.3536	0.3178	−0.0357	1.1124
0.5000	0.1768	0.1669	−0.0099	1.0590
0.1000	0.0354	0.0349	−0.0004	1.0124
0.0100	0.0035	0.0035	0.0000	1.0012
0.0010	0.0004	0.0004	0.0000	1.0001

17. $A(x) = x^2$

(a) $dA = 2x\,dx = 2x\Delta x$

$\Delta A = (x + \Delta x)^2 - x^2$

$= 2x\Delta x + (\Delta x)^2$

(b)

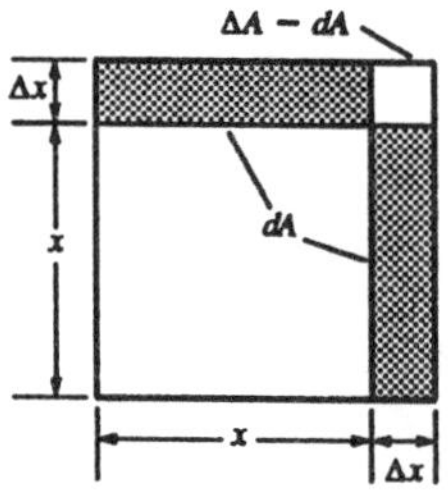

(c) $\Delta A - dA = (\Delta x)^2$

18. $A = x^2$

$x = 12$

$\Delta x = dx = \pm\frac{1}{64}$

$dA = 2x\,dx$

$\Delta A \approx dA = 2(12)\left(\pm\frac{1}{64}\right)$

$= \pm\frac{3}{8}$ square inches

19. $A = \pi r^2$

$r = 14$

$\Delta r = dr = \pm\frac{1}{4}$

$\Delta A \approx dA = 2\pi r\,dr$

$= \pi(28)\left(\pm\frac{1}{4}\right)$

$= \pm 7\pi$ square inches

20. $x = 12$ inches

$\Delta x = dx = \pm 0.03$ inch

(a) $V = x^3$

$dV = 3x^2\,dx$

$= 3(12)^2(\pm 0.03)$

$= \pm 12.96$ cubic inches

(b) $S = 6x^2$

$dS = 12x\,dx$

$= 12(12)(\pm 0.03)$

$= \pm 4.32$ square inches

21. (a) $x = 15$ centimeters

$\Delta x = dx = \pm 0.05$ centimeters

$A = x^2$

$dA = 2x\,dx = 2(15)(\pm 0.05)$

$= \pm 1.5$ square centimeters

Percentage error:

$$\frac{dA}{A} = \frac{\pm 1.5}{(15)^2} = 0.00666\ldots = \frac{2}{3}\%$$

(b) $\dfrac{dA}{A} = \dfrac{2x\,dx}{x^2} = \dfrac{2\,dx}{x} \le 0.025$

$\dfrac{dx}{x} \le \dfrac{0.025}{2} = 0.0125 = 1.25\%$

22. (a) $C = 56$ centimeters

$\Delta C = dC = \pm 1.2$ centimeters

$C = 2\pi r \Rightarrow r = \dfrac{C}{2\pi}$

$A = \pi r^2 = \pi\left(\dfrac{C}{2\pi}\right)^2 = \dfrac{1}{4\pi}C^2$

$dA = \dfrac{1}{2\pi}C\,dC = \dfrac{1}{2\pi}(56)(\pm 1.2) = \dfrac{33.6}{\pi}$

$\dfrac{dA}{A} = \dfrac{33.6/\pi}{(1/4\pi)(56)^2} \approx 0.042857 = 4.2857\%$

(b) $\dfrac{dA}{A} = \dfrac{(1/2\pi)C\,dC}{(1/4\pi)C^2} = \dfrac{2\,dC}{C} \le 0.03$

$\dfrac{dC}{C} \le \dfrac{0.03}{2} = 0.015 = 1.5\%$

23. $r = 6$ inches

$\Delta r = dr = \pm 0.02$ inches

(a) $V = \frac{4}{3}\pi r^3$

$dV = 4\pi r^2\, dr = 4\pi(6)^2(\pm 0.02) = \pm 2.88\pi$ cubic inches

(b) $S = 4\pi r^2$

$dS = 8\pi r\, dr = 8\pi(6)(\pm 0.02) = \ 0.96\pi$ square inches

(c) Relative error: $\frac{dV}{V} = \frac{4\pi r^2\, dr}{(4/3)\pi r^3} = \frac{3\,dr}{r} = \frac{3}{6}(0.02) = 0.01 = 1\%$

Relative error: $\frac{dS}{S} = \frac{8\pi r\, dr}{4\pi r^2} = \frac{2\,dr}{r} = \frac{2(0.02)}{6} = 0.0067 = \frac{2}{3}\%$

24. $P = (500x - x^2) - \left(\frac{1}{2}x^2 - 77x + 3000\right)$, x changes from 115 to 120

$dP = (500 - 2x - x + 77)\, dx = (577 - 3x)\, dx = [577 - 3(115)](120 - 115) = 1160$

Approximate percentage change: $\frac{dP}{P}(100) = \frac{1160}{43{,}517.50}(100) = 2.7\%$

25. (a) $T = 2\pi\sqrt{L/g}$

$dT = \frac{\pi}{g\sqrt{L/g}}dL$

Relative error:

$\frac{dT}{T} = \frac{(\pi\, dL)/(g\sqrt{L/g})}{2\pi\sqrt{L/g}}$

$= \frac{dL}{2L} = \frac{1}{2}(\text{relative error in } L)$

$= \frac{1}{2}(0.005) = 0.0025$

Percentage error: $\frac{dT}{T}(100) = 0.25\% = \frac{1}{4}\%$

(b) $(0.0025)(3600)(24) = 216$ seconds

$= 3.6$ minutes

26. $T = 2\pi\sqrt{L/g}$

$T^2 = 4\pi^2\left(\frac{L}{g}\right)$

$g = \frac{4\pi^2 L}{T^2}$

$dg = 4\pi^2 L\frac{(-2)}{T^3}dT$

Relative error in g:

$\frac{dg}{g} = \frac{(-8\pi^2 L\, dT)/T^3}{(4\pi^2 L)/T^2}$

$= -2\frac{dT}{T} = -2$ (relative error in period)

$= -2(0.001) = -0.002$

Percentage error in g: $\left|\frac{dg}{g}\right|(100) = \frac{1}{5}\%$

27. $E = IR$

$R = \frac{E}{I}$

$dR = -\frac{E}{I^2}dI$

$\frac{dR}{R} = \frac{-(E/I^2)\, dI}{E/I} = -\frac{dI}{I}$

$\left|\frac{dR}{R}\right| = \left|-\frac{dI}{I}\right| = \left|\frac{dI}{I}\right|$

28. $C = \frac{80{,}000p}{100 - p}$, $0 \le p < 100$

(a) $p = 40$

$\Delta p = dp = 2$

$dC = \frac{8{,}000{,}000}{(100 - p)^2}dp$

$= \frac{8{,}000{,}000}{(100 - 40)^2}(2) = \$4444.44$

(b) $p = 75$

$\Delta p = dp = 2$

$dC = \frac{8{,}000{,}000}{(100 - 75)^2}(2) = \$25{,}600.00$

29. Let $\epsilon = \dfrac{\Delta y}{\Delta x} - f'(x)$ then $\Delta x e = \Delta y - f'(x)\Delta x$. Since $\Delta x = dx$, we have $\Delta y - f'(x)\,dx = \Delta x e$ or $\Delta y - dy = \epsilon\Delta x$ where $\epsilon \to 0$ as $\Delta x \to 0$.

Section 4.10 Business and Economics Applications

1. $R = 900x - 0.1x^2$

$\dfrac{dR}{dx} = 900 - 0.2x = 0$ when $x = 4500$.

By the First Derivative Test, $x = 4500$ is a maximum.

2. $R = 600x^2 - 0.02x^3$

$\dfrac{dR}{dx} = 1200x - 0.06x^2 = 6x(200 - 0.01x) = 0$ when $x = 0,\ 20{,}000$.

By the First Derivative Test, $x = 20{,}000$ is a maximum.

3.
$$R = \frac{1{,}000{,}000x}{0.02x^2 + 1800}$$
$$\frac{dR}{dx} = 1{,}000{,}000\frac{0.02x^2 + 1800 - x(0.04x)}{(0.02x^2 + 1800)^2} = 0$$

$1800 - 0.02x^2 = 0$ when $x = 300$.

By the First Derivative Test, $x = 300$ is a maximum.

4. $R = 30x^{2/3} - 2x$

$\dfrac{dR}{dx} = 20x^{-1/3} - 2 = \dfrac{20}{x^{1/3}} - 2 = 0$ when $x = 1000$.

By the First Derivative Test, $x = 1000$ is a maximum.

5. $\overline{C} = 0.125x + 20 + \dfrac{5000}{x}$

$\dfrac{d\overline{C}}{dx} = 0.125 - \dfrac{5000}{x^2} = 0$ when $x = 200$.

By the First Derivative Test, $x = 200$ yields the minimum average cost.

6. $\overline{C} = 0.001x^2 - 5 + \dfrac{250}{x}$

$\dfrac{d\overline{C}}{dx} = 0.002x - \dfrac{250}{x^2} = 0$ when $x = 50$.

By the First Derivative Test, $x = 50$ yields the minimum average cost.

7. $\overline{C} = 3000 - x(300 - x)^{1/2}$

$\dfrac{d\overline{C}}{dx} = -x\left(\dfrac{1}{2}\right)(300 - x)^{-1/2}(-1) - (300 - x)^{1/2} = 0$ when $x = 200$.

By the First Derivative Test, $x = 200$ yields the minimum average cost.

8. $\overline{C} = \dfrac{2x^2 - x + 5000}{x^2 + 2500}$

$$\frac{d\overline{C}}{dx} = \frac{(x^2+2500)(4x-1) - (2x^2 - x + 5000)(2x)}{(x^2+2500)^2} = \frac{x^2 - 2500}{(x^2+2500)^2} = 0 \text{ when } x = 50.$$

By the First Derivative Test, $x = 50$ yields the minimum average cost.

9. $C = 100 + 30x$

$p = 90 - x$

$P = xp - C = 90x - x^2 - 30x - 100 = -x^2 + 60x - 100$

$\dfrac{dP}{dx} = -2x + 60 = 0$ when $x = 30$, so $p = 60$.

By the First Derivative Test, $x = 30$ is a maximum.

10. $C = 2400x + 5200$

$p = \dfrac{2}{5}(15{,}000 - x^2)$

$P = \dfrac{2}{5}(15{,}000x - x^3) - 2400x - 5200 = -\dfrac{2x^3}{5} + 3600x - 5200$

$\dfrac{dP}{dx} = -\dfrac{6x^2}{5} + 3600 = 0$ when $x \approx 55$, so $p \approx 4800$.

By the First Derivative Test, $x \approx 55$ is a maximum.

11. $C = 4000 - 40x + 0.02x^2$

$p = 50 - 0.01x$

$P = 50x - 0.01x^2 - 4000 + 40x - 0.02x^2 = -0.03x^2 + 90x - 4000$

$\dfrac{dP}{dx} = -0.06x + 90 = 0$ when $x = 1500$, so $p = 35$.

By the First Derivative Test, $x = 1500$ is a maximum.

12. $C = 35x + 2\sqrt{x-1}$

$p = 40 - \sqrt{x-1}$

$P = 40x - x\sqrt{x-1} - 35x - 2\sqrt{x-1} = 5x - (x+2)\sqrt{x-1}$

$$\frac{dP}{dx} = 5 - \left[(x+2)\left(\frac{1}{2}\right)(x-1)^{-1/2} + (x-1)^{1/2}\right]$$

$$= 5 - \left[\frac{x + 2 + 2(x-1)}{2\sqrt{x-1}}\right] = 5 - \frac{3x}{2\sqrt{x-1}} = \frac{10\sqrt{x-1} - 3x}{2\sqrt{x-1}} = 0 \text{ when } x = 10, \text{ so } p = 37.$$

By the First Derivative Test, $x = 10$ is a maximum.

13. $C = 800 - 10x + \dfrac{1}{4}x^2$

$\dfrac{dC}{dx} = -10 + \dfrac{1}{2}x = 0$ when $x = 20$.

C is minimum when $x = 20$ since $\dfrac{d^2C}{dx^2} = \dfrac{1}{2} > 0$.

14. $P = 230 + 20s - \frac{1}{2}s^2$

$\frac{dP}{ds} = 20 - s = 0$ when $s = 20$.

P is maximum when advertising is \$2000 since $\frac{d^2P}{ds^2} = -1 < 0$.

15. Let x be the number of units purchased, p be the price per unit, and P be the total profit.

$$p = 90 - (0.10)(x - 100) = 100 - 0.1x$$

Since each radio costs the manufacturer \$60, the profit per radio is $p - 60$ and the total profit is $P = x(p - 60) = 40x - 0.1x^2$.

$$\frac{dP}{dx} = 40 - 0.2x = 0 \text{ when } x = \frac{40}{0.2} = 200 \left(\text{maximum since } \frac{d^2P}{dx^2} = -0.02 < 0\right)$$

Manufacturer's largest ordering size for maximum profit is 200 units

16. Let

$$\begin{aligned}
x &= \text{the number of \$30 increases in rent.} \\
\text{Profit} &= \text{Revenue} - \text{Cost} \\
P &= (\text{number of apartments rented})(\text{rent/unit}) - (\text{number of apartments rented})(\$36) \\
&= (50 - x)(540 + 30x) - (50 - x)(36) = (50 - x)(504 + 30x) \\
\frac{dP}{dx} &= (50 - x)(30) + (504 + 30x)(-1) \\
&= 996 - 60x = 0 \text{ when } x = 16.6 \approx 17 \left(\text{maximum since } \frac{d^2P}{dx^2} = -60 < 0\right)
\end{aligned}$$

Maximum profit occurs when there are 17 increases of \$30 in rent and the rent is $540 + 30(17) = \$1050$.

17. Let T be the total cost.

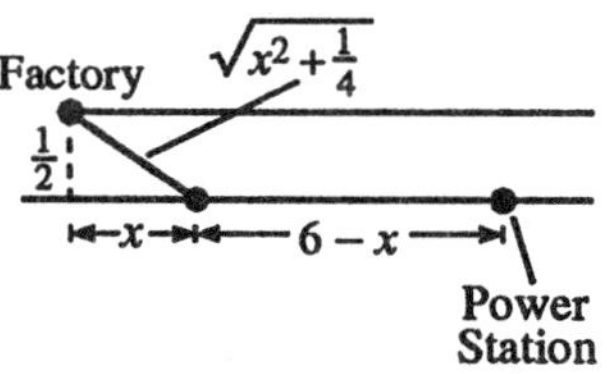

$$\begin{aligned}
T &= 8(5280)\sqrt{x^2 + (1/4)} + 6(5280)(6 - x) \\
&= 2(5280)[4\sqrt{x^2 + (1/4)} + 18 - 3x] \\
\frac{dT}{dx} &= 2(5280)\left[\frac{4(2x)}{2\sqrt{x^2 + (1/4)}} - 3\right] \\
&= 2(5280)\left[\frac{4x - 3\sqrt{x^2 - (1/4)}}{\sqrt{x^2 + (1/4)}}\right] = 0
\end{aligned}$$

When $4x - 3\sqrt{x^2 + (1/4)} = 0$, $x = 3/(2\sqrt{7}) \approx 0.57$ mile. By the First Derivative Test, $x \approx 0.57$ is a minimum.

18. Let K be the cost per mile for laying pipe on land and T be the total cost.

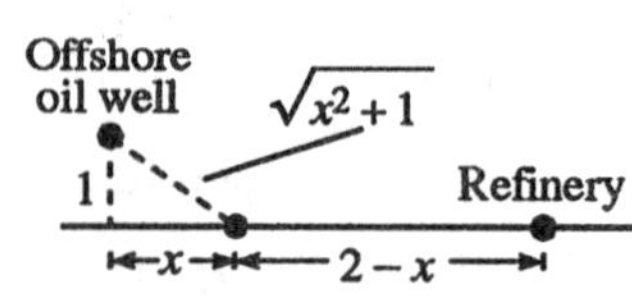

$$\begin{aligned}
T &= 2K\sqrt{x^2 + 1} + K(2 - x) = K\left(2\sqrt{x^2 + 1} + 2 - x\right) \\
\frac{dT}{dx} &= K\left[\frac{2x}{\sqrt{x^2 + 1}} - 1\right] \\
&= 0
\end{aligned}$$

When $2x - \sqrt{x^2 + 1} = 0$, $x = 1/\sqrt{3} \approx 0.58$ mile. By the First Derivative Test, $x = 1/\sqrt{3}$ is a minimum.

19. Let d be the amount deposited in the bank, i be the interest rate paid by the bank, and P be the profit.

$$P = (0.12)d - id$$

$$d = ki^2 \text{ (since } d \text{ is proportional to } i^2)$$

$$P = (0.12)(ki^2) - i(ki^2) = k(0.12i^2 - i^3)$$

$$\frac{dP}{di} = k(0.24i - 3i^2) = 0 \text{ when } i = \frac{0.24}{3} = 0.08.$$

$$\frac{d^2P}{di^2} = 0.24 - 6i < 0 \text{ when } i = 0.08.$$

The profit is a maximum when $i = 8\%$.

20. Average cost $= \overline{C} = \frac{C}{s} = \frac{f(x)}{x}$

$$\frac{d\overline{C}}{dx} = \frac{xf'(x) - f(x)}{x^2} = 0 \Rightarrow xf'(x) - f(x) = 0 \text{ when } f'(x) = \frac{f(x)}{x}$$

Marginal cost = average cost

This condition will yield a minimum (if it exists).

21.

$$C = 2x^2 + 5x + 18$$

Average cost $= \frac{C}{x} = \overline{C} = 2x + 5 + \frac{18}{x}$

$$\frac{d\overline{C}}{dx} = 2 - \frac{18}{x^2} = 0 \text{ when } x = 3.$$

$$\overline{C}(3) = 6 + 5 + 6 = 17$$

By the First Derivative Test, $x = 3$ is a minimum.

Marginal cost: $\frac{dC}{dx} = 4x + 5$

At $x = 3$: $\frac{dC}{dx} = 17 = \overline{C}(3)$

22.

$$C = x^3 - 6x^2 + 13x$$

Average cost $= \frac{C}{x} = \overline{C} = x^2 - 6x + 13$

$$\frac{d\overline{C}}{dx} = 2x - 6 = 0 \text{ when } x = 3.$$

$$\frac{dC}{dx} = 3x^2 - 12x + 13 \text{ when } x = 3.$$

Marginal cost: $\frac{dC}{dx} = 27 - 36 + 13 = 4$

Average cost: $\overline{C} = 9 - 18 + 13 = 4$

23. (a) $P = R - C = xp - C = x\left(100 - \frac{1}{2}x^2\right) - (4x + 375)$

$$\frac{dP}{dx} = x(-x) + \left(100 - \frac{1}{2}x^2\right) - 4 = -\frac{3}{2}x^2 + 96 = 0 \text{ when } x = 8.$$

$$p = 100 - \frac{1}{2}(64) = \$68 \text{ (maximum by the First Derivative Test)}$$

(b) $\overline{C}(x) = \frac{C}{x} = 4 + \frac{375}{x}$

$$\overline{C}(8) = 4 + \frac{375}{8} \approx \$50.88$$

24. For the linear demand function, the rate of change (slope) is $m = -\frac{25}{5}$. Therefore, the demand x is

$$x = 800 - \frac{25}{5}(p - 25) = 925 - 5p$$

$$R = xp = -5p^2 + 925p$$

$$\frac{dR}{dp} = -10p + 925 = 0 \text{ when } p = \$92.50.$$

25. $C = 100\left(\dfrac{200}{x^2} + \dfrac{x}{x+30}\right), \quad 1 \le x$

$C' = 100\left(-\dfrac{400}{x^3} + \dfrac{30}{(x+30)^2}\right) = 0$

$f(x) = 10(3x^3 - 40x^2 - 2400x - 36{,}000)$

$f'(x) = 10(9x^2 - 80x - 2400)$

n	x_n	$f(x_n)$	$f'(x_n)$	$\dfrac{f(x_n)}{f'(x_n)}$	$x_n - \dfrac{f(x_n)}{f'(x_n)}$
1	40	−40,000	88,000	−0.4545	40.4545
2	40.4545	659.841	90,927.40	0.0073	40.4472
3	40.4472	−3.756	90,880.10	0.0000	40.4472

Approximation: $x \approx 40$ units

26. $\overline{C} = \dfrac{C}{x} = \dfrac{800}{x} + 0.4 + 0.02x + 0.0001x^2$

$\overline{C}' = -\dfrac{800}{x^2} + 0.02 + 0.0002x = 0$

$f(x) = 0.0002x^3 + 0.02x^2 - 800 = 0$

$f'(x) = 0.0006x^2 + 0.04x$

n	x_n	$f(x_n)$	$f'(x_n)$	$\dfrac{f(x_n)}{f'(x_n)}$	$x_n - \dfrac{f(x_n)}{f'(x_n)}$
1	130	−22.6	15.3400	−1.4733	131.4733
2	131.4733	0.2138	15.6301	0.0137	131.4597
3	131.4597	0.0012	15.6274	0.0000	131.4597

Approximation: $x \approx 131$ units

27. $R = 900x - 0.1x^2$

$x = 3000$

$dx = 100$

$dR = (900 - 0.2x)\,dx$

$= [900 - 0.2(3000)](100)$

$= \$30{,}000$

28. $P = (500x - x^2) - \left(\dfrac{1}{2}x^2 - 77x + 3000\right)$

$x = 175$

$dx = 5$

$dP = [(500 - 2x) - (x - 77)]\,dx$

$= (-3x + 577)\,dx$

$= [-3(175) + 577](5) = \$260$

$\dfrac{dP}{P} = \dfrac{260}{52037.5} \approx 0.004996 \approx 0.5\%$ or $\dfrac{1}{2}\%$

29. $S_1 = (4m - 1)^2 + (5m - 6)^2 + (10m - 3)^2$

$\dfrac{dS_1}{dm} = 2(4m-1)(4) + 2(5m-6)(5) + 2(10m-3)(10) = 282m - 128 = 0$ when $m = \dfrac{64}{141}$.

Line: $y = \dfrac{64}{141}x$

$S = \left|4\left(\dfrac{64}{141}\right) - 1\right| + \left|5\left(\dfrac{64}{141}\right) - 6\right| + \left|10\left(\dfrac{64}{141}\right) - 3\right|$

$= \left|\dfrac{256}{141} - 1\right| + \left|\dfrac{320}{141} - 6\right| + \left|\dfrac{640}{141} - 3\right| = \dfrac{858}{141} \approx 6.1$ mi

30. $S_2 = |4m - 1| + |5m - 6| + |10m - 3|$

Using a graphing utility, you can see that the minimum occurs when $m = 0.3$.

Line $y = 0.3x$

$$S_2 = |4(0.3) - 1| + |5(0.3) - 6| + |10(0.3) - 3| = 4.7 \text{ mi.}$$

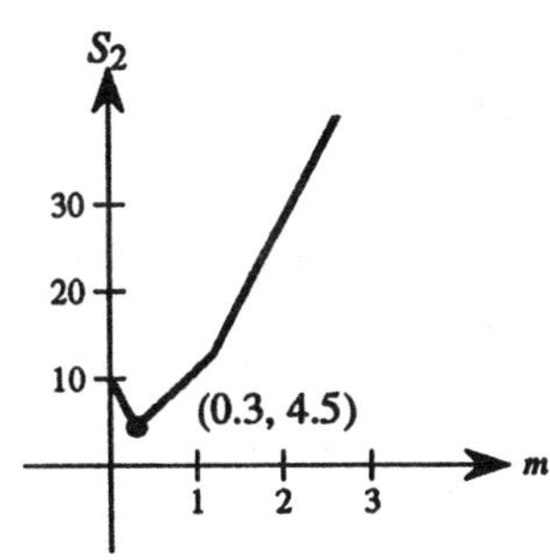

Chapter 4 Review Exercises

1. $f(x) = 4x - x^2 = x(4 - x)$; Domain: $(-\infty, \infty)$, Range: $(-\infty, 4]$

$f'(x) = 4 - 2x = 0$ when $x = 2$.

$f''(x) = -2$

Therefore, $(2, 4)$ is a relative maximum.

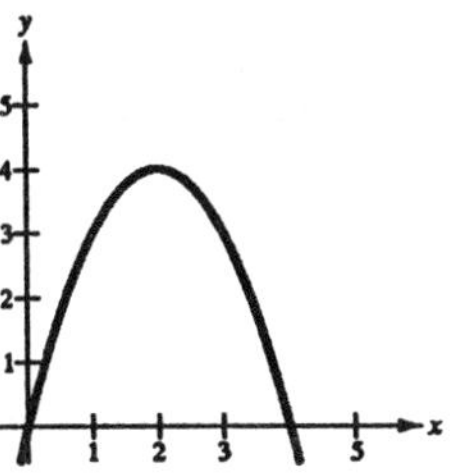

2. $f(x) = 4x^3 - x^4 = x^3(4 - x)$; Domain: $(-\infty, \infty)$, Range: $(-\infty, 27]$

$f'(x) = 12x^2 - 4x^3 = 4x^2(3 - x) = 0$ when $x = 0, 3$.

$f''(x) = 24x - 12x^2 = 12x(2 - x) = 0$ when $x = 0, 2$.

$f''(3) < 0$

Therefore, $(3, 27)$ is a relative maximum.
Points of inflection: $(0, 0)$, $(2, 16)$

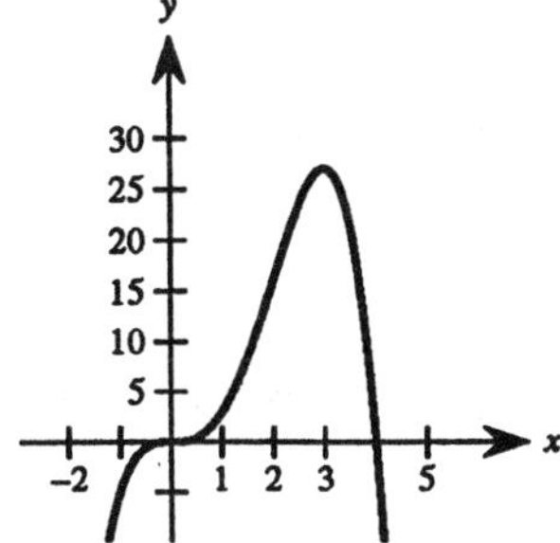

3. $f(x) = x\sqrt{16 - x^2}$; Domain: $[-4, 4]$, Range: $[-8, 8]$

$f'(x) = \dfrac{16 - 2x^2}{\sqrt{16 - x^2}} = 0$ when $x = \pm 2\sqrt{2}$ and undefined when $x = \pm 4$.

$f''(x) = \dfrac{2x(x^2 - 24)}{(16 - x^2)^{3/2}}$

$f''\left(-2\sqrt{2}\right) > 0$

Therefore, $\left(-2\sqrt{2}, -8\right)$ is a relative minimum.

$f''\left(2\sqrt{2}\right) < 0$

Therefore, $\left(2\sqrt{2}, 8\right)$ is a relative maximum. $\check{4}$ Point of inflection: $(0, 0)$
Intercepts: $(-4, 0)$, $(0, 0)$, $(4, 0)$
Symmetry with respect to origin

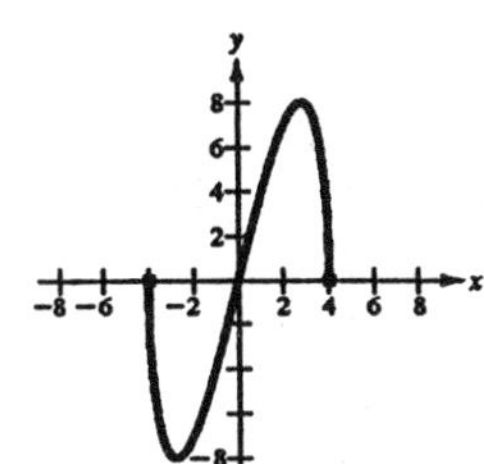

4. $f(x) = x + \dfrac{4}{x^2}$; Domain: $(-\infty, 0) \cup (0, \infty)$, Range: $(-\infty, \infty)$

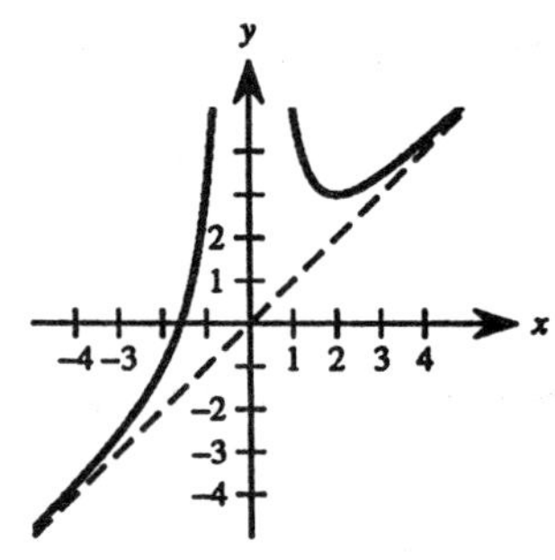

$f'(x) = 1 - \dfrac{8}{x^3} = \dfrac{x^3 - 8}{x^3} = 0$ when $x = 2$.

$f''(x) = \dfrac{24}{x^4} > 0$

$f''(2) > 0$

Therefore, $(2, 3)$ is a relative minimum.

Intercept: $(-\sqrt[3]{4}, 0)$

Vertical asymptote: $x = 0$

Slant asymptote: $y = x$

5. $f(x) = \dfrac{x+1}{x-1}$; Domain: $(-\infty, 1) \cup (1, \infty)$, Range: $(-\infty, 1) \cup (1, \infty)$

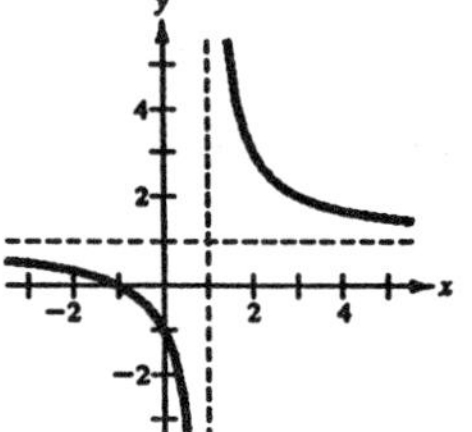

$f'(x) = \dfrac{-2}{(x-1)^2} < 0$ if $x \neq 1$.

$f''(x) = \dfrac{4}{(x-1)^3}$

Horizontal asymptote: $y = 1$

Vertical asymptote: $x = 1$

Intercepts: $(-1, 0)$, $(0, -1)$

6. $f(x) = x^2 + \dfrac{1}{x} = \dfrac{x^3 + 1}{x}$; Domain: $(-\infty, 0) \cup (0, \infty)$, Range: $(-\infty, \infty)$

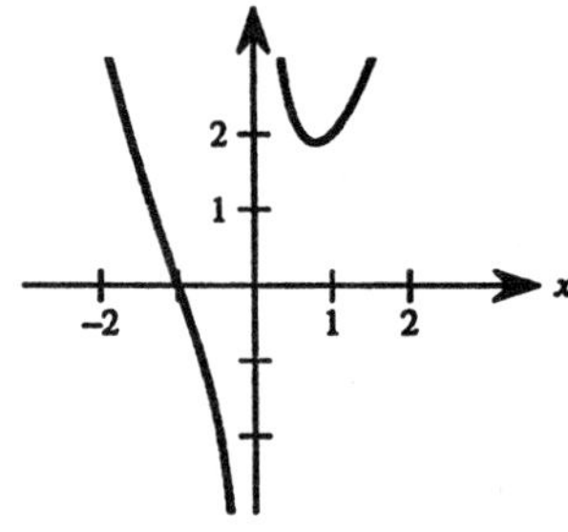

$f'(x) = 2x - \dfrac{1}{x^2} = \dfrac{2x^3 - 1}{x^2} = 0$ when $x = \dfrac{1}{\sqrt[3]{2}}$.

$f''(x) = 2 + \dfrac{2}{x^3} = \dfrac{2(x^3 + 1)}{x^3} = 0$ when $x = -1$.

$f''(1/\sqrt[3]{2}) > 0$

Therefore, $(1/\sqrt[3]{2}, 3/\sqrt[3]{4})$ is a relative minimum.

Point of inflection: $(-1, 0)$

Intercept: $(-1, 0)$

Vertical asymptote: $x = 0$

7. $f(x) = x^3 + x + \dfrac{4}{x}$; Domain: $(-\infty, \infty)$, Range: $(-\infty, -6] \cup [6, \infty)$

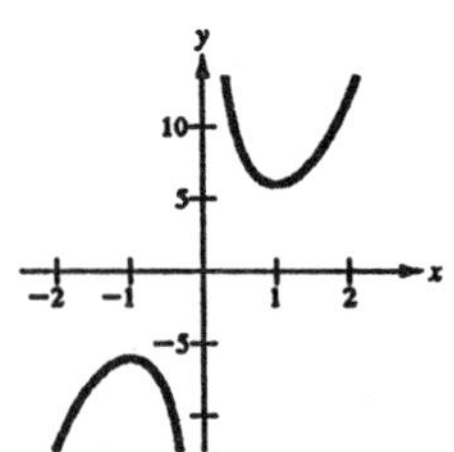

$f'(x) = 3x^2 + 1 - \dfrac{4}{x^2} = \dfrac{3x^4 + x^2 - 4}{x^2} = 0$ when $x = \pm 1$.

$f''(x) = 6x + \dfrac{8}{x^3} = \dfrac{6x^4 + 8}{x^3} \neq 0$

$f''(-1) < 0$

Therefore, $(-1, -6)$ is a relative maximum.

$f''(1) > 0$

Therefore, $(1, 6)$ is a relative minimum. ŏ Vertical asymptote: $x = 0$

Symmetric with respect to origin

8. $f(x) = x^3(x+1)$; Domain: $(-\infty, \infty)$, Range: $\left(-\frac{27}{256}, \infty\right)$

$f'(x) = x^3 + 3x^2(x+1) = x^2(4x+3) = 0$ when $x = 0, -\frac{3}{4}$.

$f''(x) = 12x^2 + 6x = 6x(2x+1) = 0$ when $x = 0, -\frac{1}{2}$.

$f''\left(-\frac{3}{4}\right) > 0$

Therefore, $\left(-\frac{3}{4}, -\frac{27}{256}\right)$ is a relative minimum.

Points of inflection: $\left(-\frac{1}{2}, -\frac{1}{16}\right)$, $(0, 0)$

Intercepts: $(-1, 0)$, $(0, 0)$

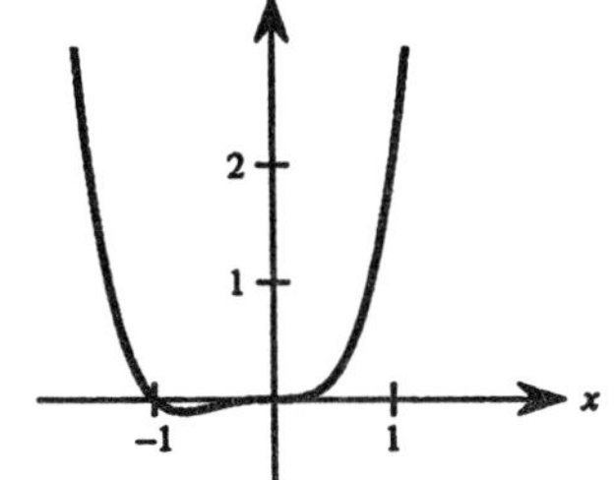

9. $f(x) = (x-1)^3(x-3)^2$; Domain: $(-\infty, \infty)$, Range: $(-\infty, \infty)$

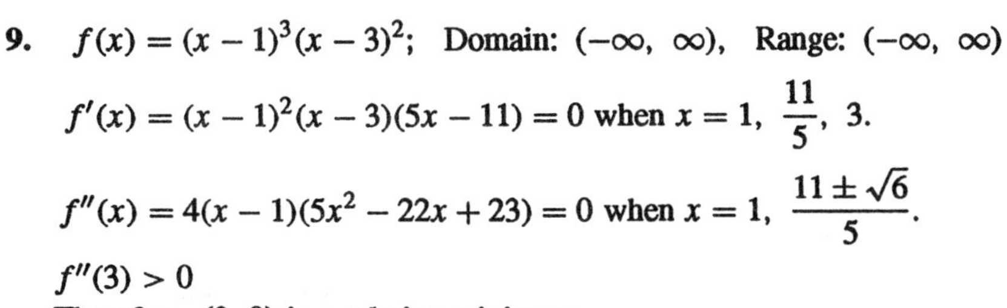

$f'(x) = (x-1)^2(x-3)(5x-11) = 0$ when $x = 1, \dfrac{11}{5}, 3$.

$f''(x) = 4(x-1)(5x^2 - 22x + 23) = 0$ when $x = 1, \dfrac{11 \pm \sqrt{6}}{5}$.

$f''(3) > 0$

Therefore, $(3, 0)$ is a relative minimum.

$f''\left(\dfrac{11}{5}\right) < 0$

Therefore, $\left(\frac{11}{5}, \frac{3456}{3125}\right)$ is a relative maximum.

Points of inflection: $(1, 0)$, $\left(\dfrac{11-\sqrt{6}}{5}, 0.60\right)$, $\left(\dfrac{11+\sqrt{6}}{5}, 0.46\right)$

Intercepts: $(0, -9)$, $(1, 0)$, $(3, 0)$

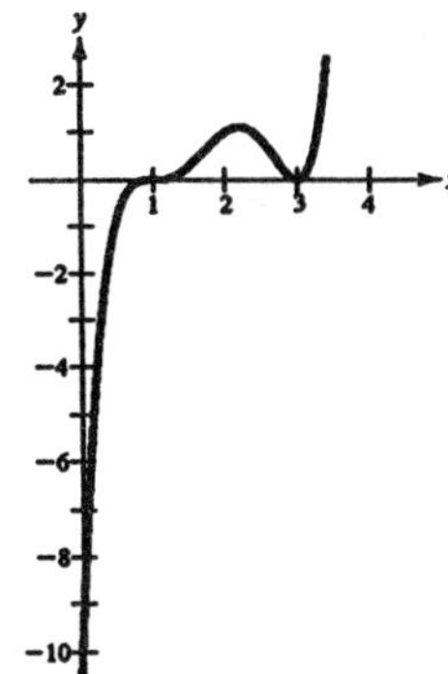

10. $f(x) = (x-3)(x+2)^3$; Domain: $(-\infty, \infty)$, Range: $\left[-\frac{16,875}{256}, \infty\right)$

$f'(x) = (x-3)(3)(x+2)^2 + (x+2)^3$

$= (4x-7)(x+2)^2 = 0$ when $x = -2, \frac{7}{4}$.

$f''(x) = (4x-7)(2)(x+2) + (x+2)^2(4)$

$= 6(2x-1)(x+2) = 0$ when $x = -2, \frac{1}{2}$.

$f''\left(\frac{7}{4}\right) > 0$

Therefore, $\left(\frac{7}{4}, -\frac{16,875}{256}\right)$ is a relative minimum.

Points of inflection: $(-2, 0)$, $\left(\frac{1}{2}, -\frac{625}{16}\right)$

Intercepts: $(-2, 0)$, $(0, -24)$, $(3, 0)$

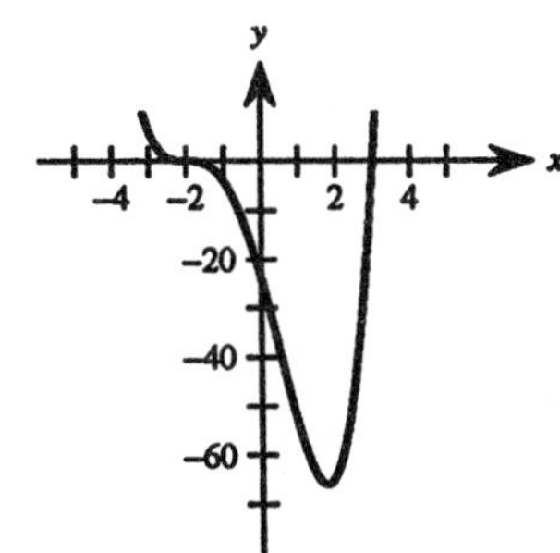

11. $f(x) = (5-x)^3$; Domain: $(-\infty, \infty)$, Range: $(-\infty, \infty)$

$f'(x) = -3(5-x)^2 < 0$ if $x \neq 5$.

$f''(x) = 6(5-x) = 0$ if $x = 5$.

Point of inflection: $(5, 0)$

Intercepts: $(5, 0)$, $(0, 125)$

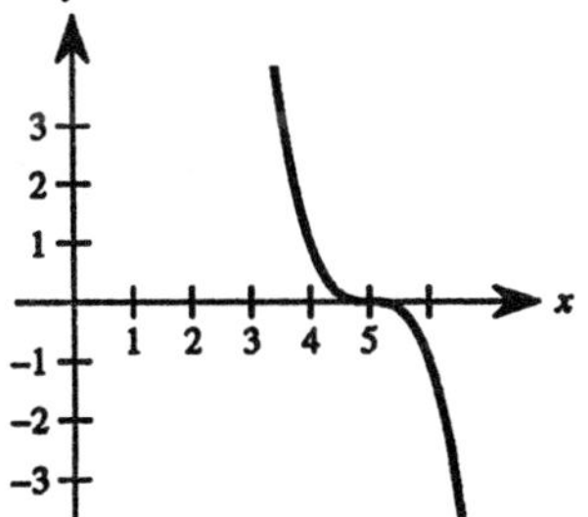

12. $f(x) = (x^2 - 4)^2$; Domain: $(-\infty, \infty)$, Range: $[0, \infty)$

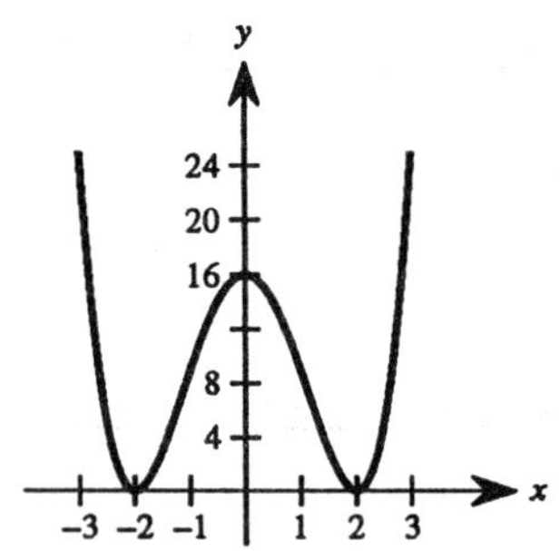

$f'(x) = 4x(x^2 - 4) = 0$ when $x = 0, \pm 2$.

$f''(x) = 4(3x^2 - 4) = 0$ when $x = \pm\dfrac{2\sqrt{3}}{3}$.

$f''(0) < 0$

Therefore, $(0, 16)$ is a relative maximum.

$f''(\pm 2) > 0$

Therefore, $(\pm 2, 0)$ are relative minima.

Points of inflection: $\left(\pm\dfrac{2\sqrt{3}}{3}, \dfrac{64}{9}\right)$

Intercepts: $(-2, 0)$, $(0, 16)$, $(2, 0)$

Symmetry with respect to y-axis

13. $f(x) = x^{1/3}(x + 3)^{2/3}$; Domain: $(-\infty, \infty)$, Range: $(-\infty, \infty)$

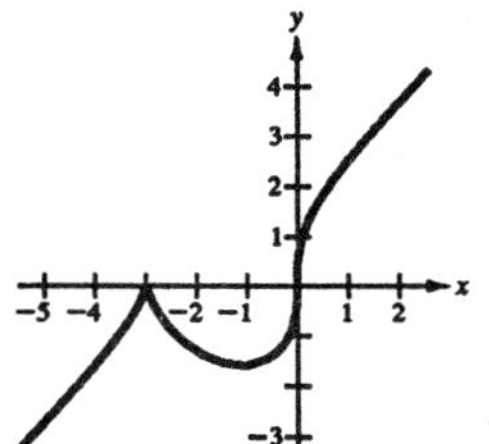

$f'(x) = \dfrac{x + 1}{(x + 3)^{1/3}x^{2/3}} = 0$ when $x = -1$ and undefined when $x = -3, 0$.

$f''(x) = \dfrac{-2}{x^{5/3}(x + 3)^{4/3}}$ is undefined when $x = 0, -3$.

$f'(-4) > 0$, $f'(-2) < 0$

Therefore, $(-3, 0)$ is a relative maximum.

$f''(-1) > 0$

Therefore, $\left(-1, -\sqrt[3]{4}\right)$ is a relative minimum.

$f''(-1) > 0$, $f''(1) < 0$

Therefore, $(0, 0)$ is a point of inflection.

Intercepts: $(-3, 0)$, $(0, 0)$

14. $f(x) = (x - 2)^{1/3}(x + 1)^{2/3}$

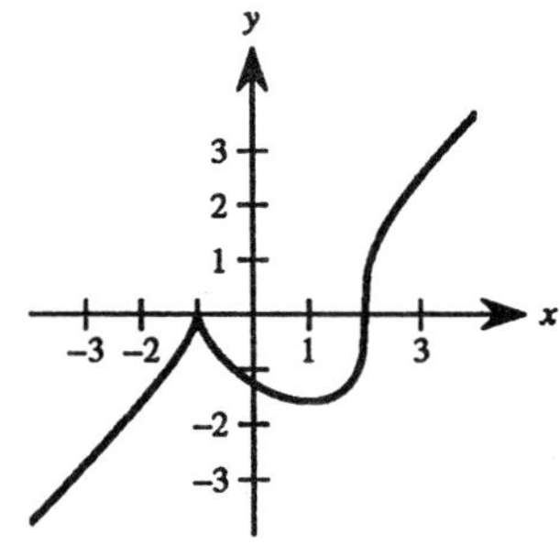

Domain: $(-\infty, \infty)$

Range: $(-\infty, \infty)$

Graph of Exercise 13 translated two units to the right

(x replaced by $x - 2$)

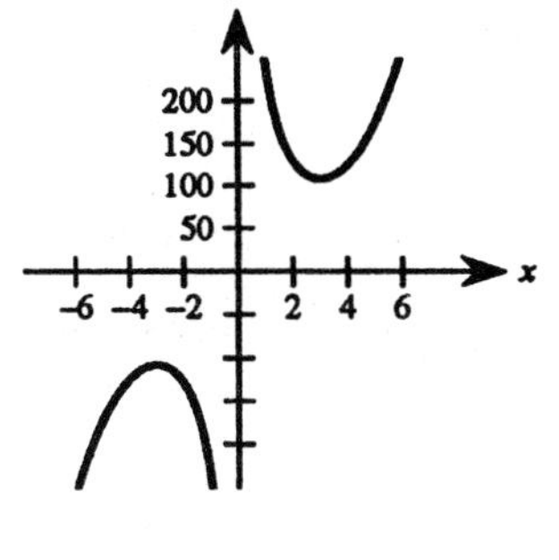

15. $f(x) = x^3 + \dfrac{243}{x} = \dfrac{x^4 + 243}{x}$;

Domain: $(-\infty,\ 0) \cup (0,\ \infty)$, Range: $(-\infty,\ -108] \cup [108,\ \infty)$

$f'(x) = 3x^2 - \dfrac{243}{x^2} = \dfrac{3(x^4 - 81)}{x^2} = 0$ when $x = \pm 3$.

$f''(x) = 6x + \dfrac{2(243)}{x^3} = \dfrac{6(x^4 + 81)}{x^3}$

$f''(-3) < 0$

Therefore, $(-3,\ -108)$ is a relative maximum.

$f''(3) > 0$

Therefore, $(3,\ 108)$ is a relative minimum.

Symmetric with respect to the origin

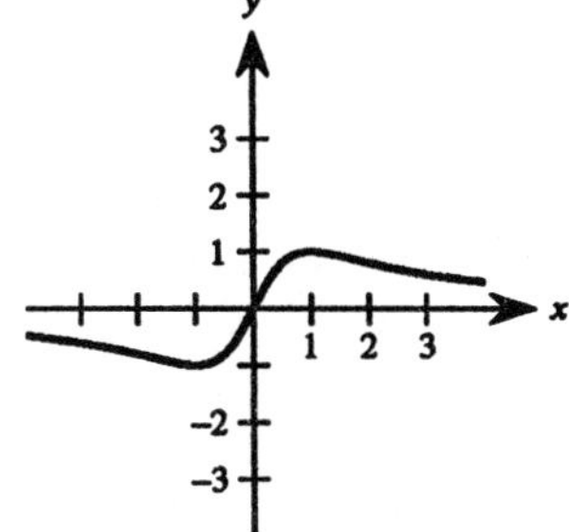

16. $f(x) = \dfrac{2x}{1 + x^2}$; Domain: $(-\infty,\ \infty)$, Range: $[-1,\ 1]$

$f'(x) = \dfrac{2(1 - x)(1 + x)}{(1 + x^2)^2} = 0$ when $x = \pm 1$.

$f''(x) = \dfrac{-2x(3 - x^2)}{(1 + x^2)^3} = 0$ when $x = 0,\ \pm\sqrt{3}$.

$f''(1) < 0$

Therefore, $(1,\ 1)$ is a relative maximum.

$f''(-1) > 0$

Therefore, $(-1,\ -1)$ is a relative minimum.

Points of inflection: $\left(-\sqrt{3},\ -\dfrac{\sqrt{3}}{2}\right)$, $(0,\ 0)$, $\left(\sqrt{3},\ \dfrac{\sqrt{3}}{2}\right)$

Intercept: $(0,\ 0)$

Symmetric with respect to the origin

Horizontal asymptote: $y = 0$

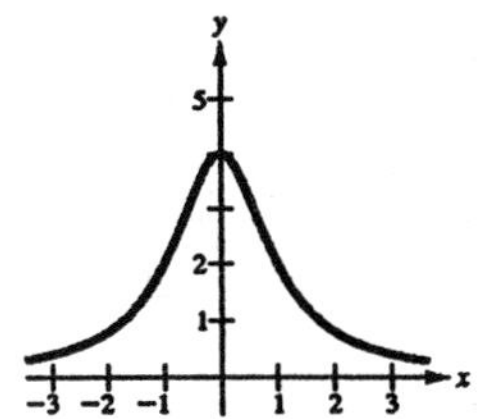

17. $f(x) = \dfrac{4}{1 + x^2}$; Domain: $(-\infty,\ \infty)$, Range: $(0,\ 4]$

$f'(x) = \dfrac{-8x}{(1 + x^2)^2} = 0$ when $x = 0$.

$f''(x) = \dfrac{-8(1 - 3x^2)}{(1 + x^2)^3} = 0$ when $x = \pm\dfrac{\sqrt{3}}{3}$.

$f''(0) < 0$

Therefore, $(0,\ 4)$ is a relative maximum.

Points of inflection: $(\pm\sqrt{3}/3,\ 3)$

Intercept: $(0,\ 4)$

Symmetric to the y-axis

Horizontal asymptote: $y = 0$

18. $f(x) = \dfrac{x^2}{1+x^4}$; Domain: $(-\infty, \infty)$, Range: $\left[0, \dfrac{1}{2}\right]$

$$f'(x) = \frac{(1+x^4)(2x) - x^2(4x^3)}{(1+x^4)^2} = \frac{2x(1-x)(1+x)(1+x^2)}{(1+x^4)^2} = 0 \text{ when } x = 0, \ \pm 1.$$

$$f''(x) = \frac{(1+x^4)^2(2-10x^4) - (2x-2x^5)(2)(1+x^4)(4x^3)}{(1+x^4)^4} = \frac{2(1-12x^4+3x^8)}{(1+x^4)^3} = 0 \text{ when } x = \pm\sqrt[4]{\frac{6 \pm \sqrt{33}}{3}}.$$

$f''(\pm 1) < 0$

Therefore, $(\pm 1, \ 1/2)$ are relative maxima.

$f''(0) > 0$

Therefore, $(0, 0)$ is a relative minimum.

Points of inflection: $\left(\pm\sqrt[4]{\dfrac{6-\sqrt{33}}{3}}, \ 0.29\right)$, $\left(\pm\sqrt[4]{\dfrac{6+\sqrt{33}}{3}}, \ 0.40\right)$

Intercept: $(0, 0)$

Symmetric to the y-axis

Horizontal asymptote: $y = 0$

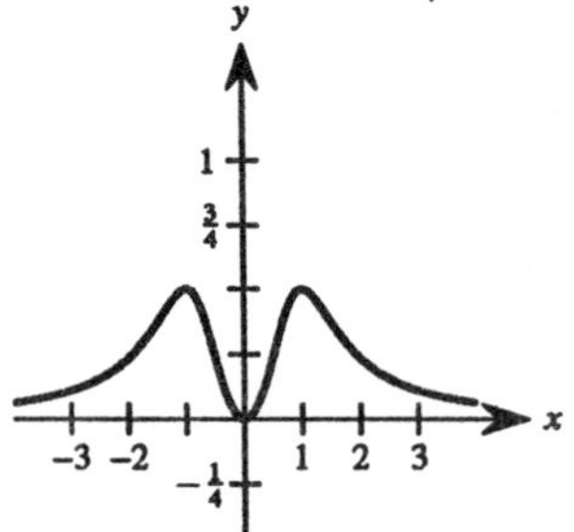

19. $f(x) = |x^2 - 9|$; Domain: $(-\infty, \infty)$, Range: $[0, \infty)$

$$f'(x) = \frac{2x(x^2-9)}{|x^2-9|} = 0 \text{ when } x = 0 \text{ and is undefined when } x = \pm 3.$$

$$f''(x) = \frac{2(x^2-9)}{|x^2-9|} \text{ is undefined at } x = \pm 3.$$

$f''(0) < 0$

Therefore, $(0, 9)$ is a relative maximum.

Relative minima: $(\pm 3, \ 0)$

Points of inflection: $(\pm 3, \ 0)$

Intercepts: $(\pm 3, \ 0)$, $(0, \ 0)$

Symmetric to the y-axis

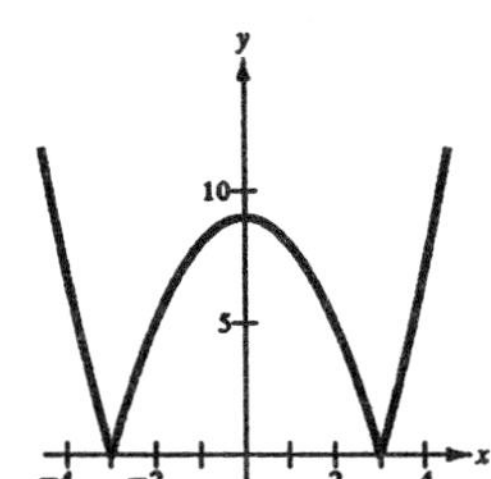

20. $f(x) = |9 - x^2| = |x^2 - 9|$

Same as Exercise 19.

21. $f(x) = |x^3 - 3x^2 + 2x| = |x(x-2)(x-1)|$; Domain: $(-\infty, \infty)$, Range: $[0, \infty)$

$$f'(x) = \frac{(3x^2-6x+2)(x^3-3x^2+2x)}{|x^3-3x^2+2x|} = 0 \text{ when } x = \frac{3 \pm \sqrt{3}}{3} \text{ and undefined when } x = 0, \ 2, \ 1.$$

$$f''(x) = \frac{(6x-6)(x^3-3x^2+2x)}{|x^3-3x^2+2x|}$$

$f''\left(\dfrac{3 \pm \sqrt{3}}{3}\right) < 0$

Therefore, $\left[(3 \pm \sqrt{3})/3, \ 0.38\right]$ are relative maxima.

By the First Derivative Test, $(1, 0)$, $(0, 0)$, and $(2, 0)$ are relative minima.

Intercepts: $(0, 0)$, $(1, 0)$, $(2, 0)$

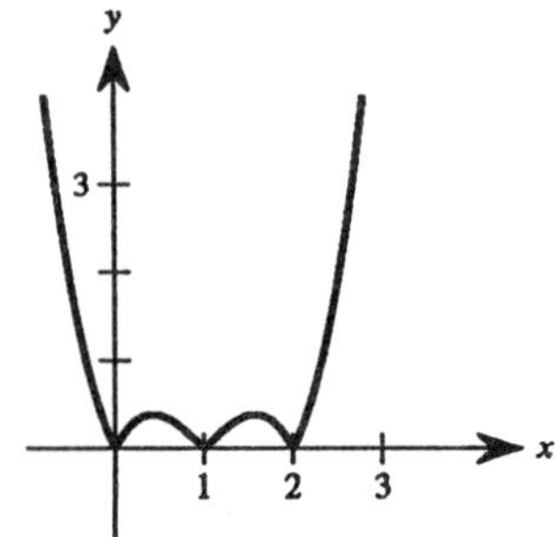

22. $f(x) = |x-1| + |x-3| = \begin{cases} -2x+4, & x \le 1 \\ 2, & 1 < x \le 3 \\ 2x-4, & x > 3 \end{cases}$

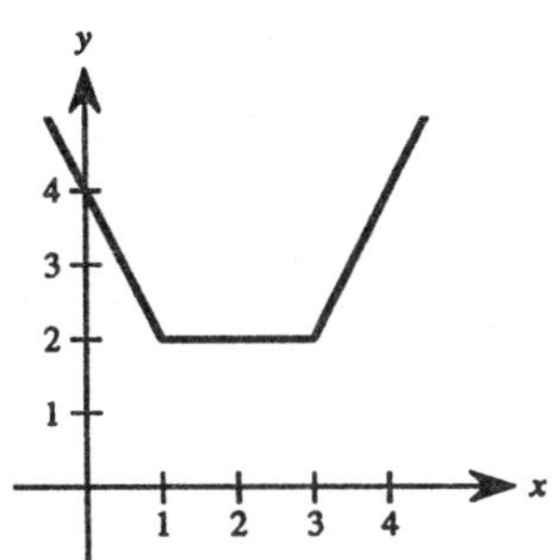

Domain: $(-\infty, \infty)$
Range: $[2, \infty)$
Intercept: $(0, 4)$

23. $f(x) = \dfrac{1}{|x-1|} = \begin{cases} \dfrac{1}{x-1}, & x > 1 \\ \dfrac{1}{1-x}, & x < 1 \end{cases}$

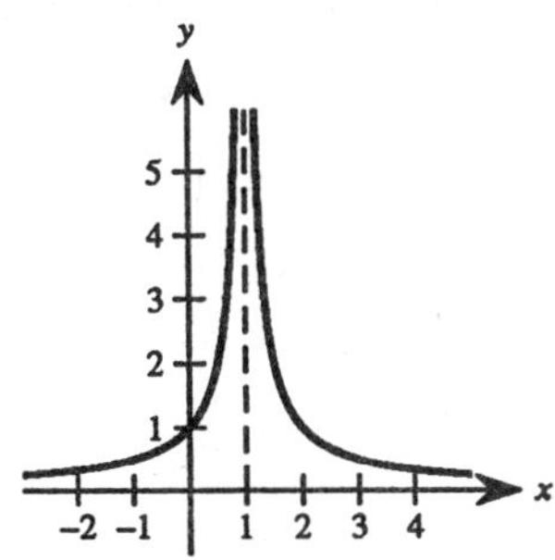

Domain: $(-\infty, 1) \cup (1, \infty)$
Range: $(0, \infty)$
Intercept: $(0, 1)$
Vertical asymptote: $x = 1$
Horizontal asymptote: $y = 0$

24. $f(x) = \dfrac{x-1}{1+3x^2}$; Domain: $(-\infty, \infty)$, Range: $\left(\dfrac{-3-2\sqrt{3}}{6}, \dfrac{2\sqrt{3}-3}{6}\right)$

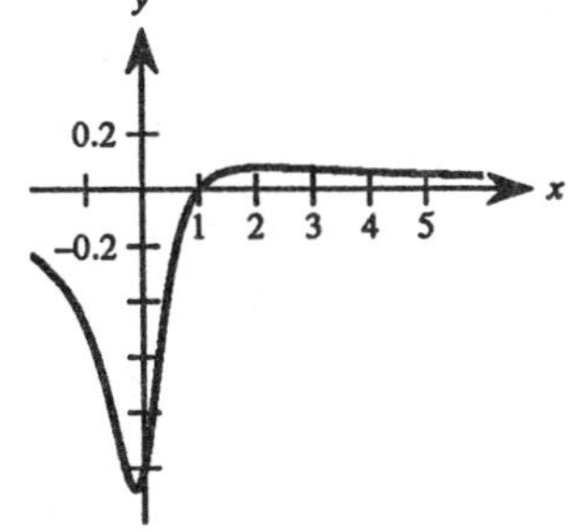

$$f'(x) = \frac{1+6x-3x^2}{(1+3x^2)^2} = 0 \text{ when } x = \frac{3 \pm 2\sqrt{3}}{3}.$$

$$f''(x) = \frac{6(1-3x-9x^2+3x^3)}{(1+3x^2)^3}$$

By the First Derivative Test, $\left(\dfrac{3-2\sqrt{3}}{3}, \dfrac{-3-2\sqrt{3}}{6}\right)$ is a relative minimum

and $\left(\dfrac{3+2\sqrt{3}}{3}, \dfrac{2\sqrt{3}-3}{6}\right)$ is a relative maximum.

Using Newton's Method, we find that $f''(x) = 0$ when $x = -0.484, 0.210, 3.274$.

Points of inflection: $(-0.484, -0.872)$, $(0.210, -0.698)$, $(3.274, 0.069)$
Intercepts: $(1, 0)$, $(0, -1)$
Horizontal asymptote: $y = 0$

25. (a) $x^2 + 4y^2 - 2x - 16y + 13 = 0$

$$\frac{(x-1)^2}{4} + \frac{(y-2)^2}{1} = 1$$

Maximum and minimum are at endpoints of minor axis: (1, 3), (1, 1)

(b) $2x + 8yy' - 2 - 16y' = 0$

$$8(y-2)y' = 2(1-x)$$

$$y' = \frac{1-x}{4(y-2)}$$

Ellipse has horizontal tangents when $x = 1$.

$$1 + 4y^2 - 2 - 16y + 13 = 0$$

$$4y^2 - 16y + 12 = 0$$

$$(y-3)(y-1) = 0$$

Maximum: (1, 3)
Minimum: (1, 1)

26. (a) n is even

(b) n is odd. If n is odd, then $n = 2k + 1$ and

$$f(x) = x^{2k+1}$$

$$f'(x) = (2k+1)x^{2k}$$

$$f''(x) = 2k(2k+1)x^{2k-1}$$

Hence, $2k - 1$ is odd and f'' changes sign as x increases through $x = 0$.

27.
$$f(x) = \frac{2x+3}{3x+2}, \quad 1 \le x \le 5$$

$$\frac{f(b) - f(a)}{b - a} = \frac{(13/17) - 1}{5 - 1} = -\frac{1}{17}$$

$$f'(x) = \frac{-5}{(3x+2)^2} = -\frac{1}{17}$$

$$(3x+2)^2 = 85$$

$$x = \frac{-2 \pm \sqrt{85}}{3}$$

$$c = \frac{-2 + \sqrt{85}}{3}$$

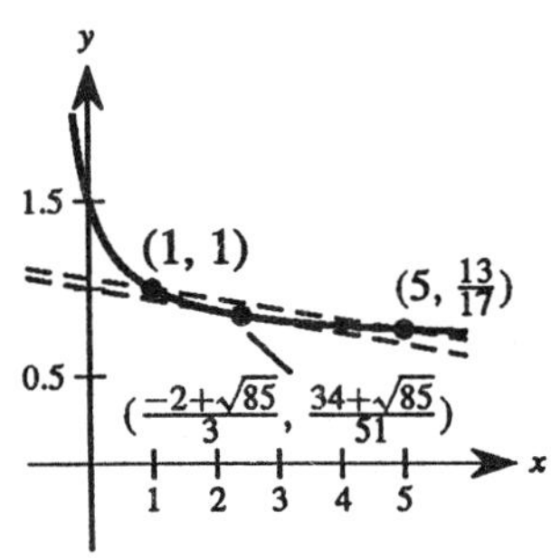

28.
$$f(x) = \frac{1}{x}, \quad 1 \le x \le 4$$

$$\frac{f(b) - f(a)}{b - a} = \frac{(1/4) - 1}{4 - 1} = \frac{-3/4}{3} = -\frac{1}{4}$$

$$f'(x) = \frac{-1}{x^2} = -\frac{1}{4}$$

$$x = \pm 2$$

$$c = 2$$

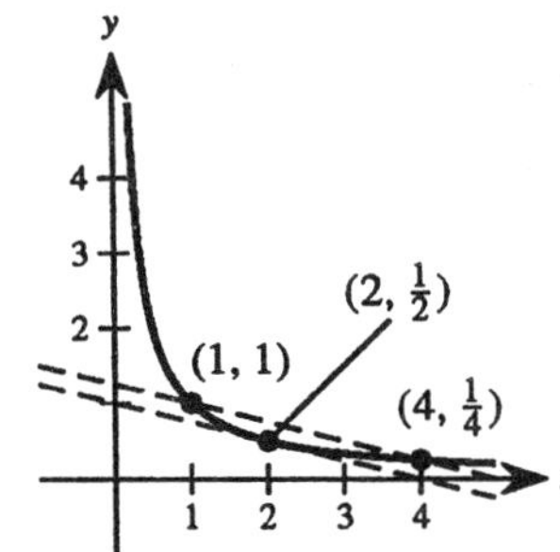

29. $f(x) = x^{2/3}, \quad 1 \le x \le 8$

$$\frac{f(b) - f(a)}{b - a} = \frac{4 - 1}{8 - 1} = \frac{3}{7}$$

$$f'(x) = \frac{2}{3}x^{-1/3} = \frac{3}{7}$$

$$x = \left(\frac{14}{9}\right)^3$$

$$c = \left(\frac{14}{9}\right)^3$$

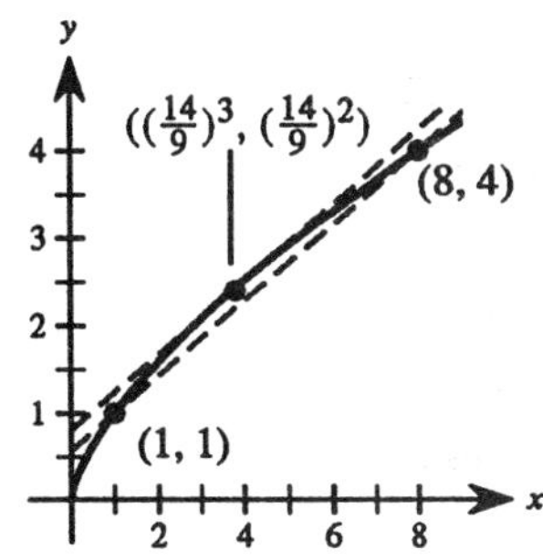

30. $f(x) = |x^2 - 9|, \quad 0 \le x \le 2$

$$\frac{f(b) - f(a)}{b - a} = \frac{5 - 9}{2 - 0} = -2$$

On the interval [0, 2]:

$f(x) = 9 - x^2$

$f'(x) = -2x \Rightarrow -2x = -2$

$x = 1$

$c = 1$

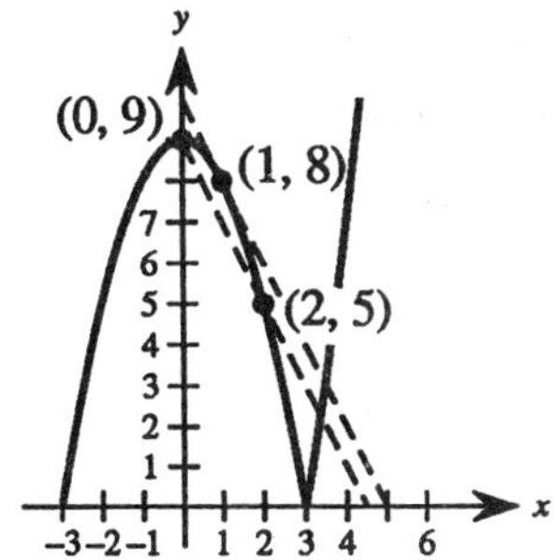

31. $f(x) = x - \frac{1}{x}, \quad 1 \le x \le 4$

$$\frac{f(b) - f(a)}{b - a} = \frac{(15/4) - 0}{4 - 1} = \frac{5}{4}$$

$$f'(x) = 1 + \frac{1}{x^2} = \frac{5}{4}$$

$$x = \pm 2$$

$$c = 2$$

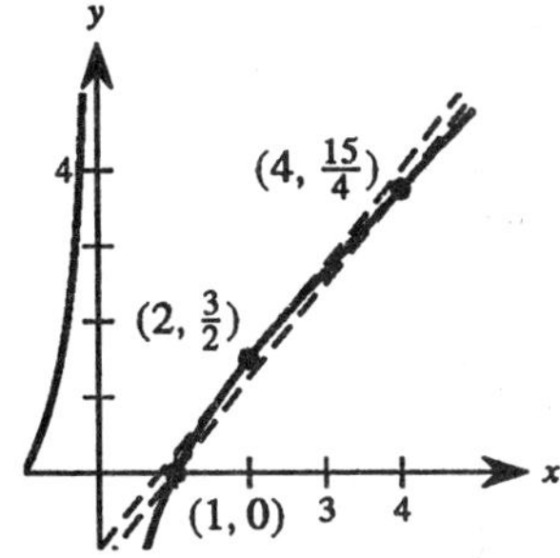

32. $f(x) = \sqrt{x} - 2x, \quad 0 \le x \le 4$

$$\frac{f(b) - f(a)}{b - a} = \frac{-6 - 0}{4 - 0} = -\frac{3}{2}$$

$$f'(x) = \frac{1}{2\sqrt{x}} - 2 = -\frac{3}{2}$$

$$x = 1$$

$$c = 1$$

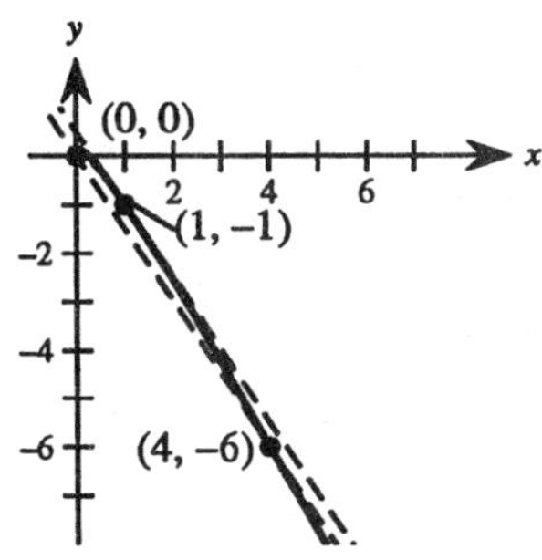

33. No; the function is discontinuous on [−2, 1].

34. $f(x) = 3 - |x - 4|$

$f(1) = 3 - |1 - 4| = 0$

$f(7) = 3 - |7 - 4| = 0$

$f'(x) = -\dfrac{x-4}{|x-4|} \neq 0$ on $[1,\ 7]$

Rolle's Theorem requires that the function be differentiable in $(1,\ 7)$, but this function is not differentiable at $x = 4$.

35.

$$f(x) = Ax^2 + Bx + C$$

$$\frac{f(x_2) - f(x_1)}{x_2 - x_1} = \frac{A(x_2{}^2 - x_1{}^2) + B(x_2 - x_1)}{x_2 - x_1}$$

$$= A(x_1 + x_2) + B$$

$$f'(x) = 2Ax + B = A(x_1 + x_2) + B$$

$$2Ax = A(x_1 + x_2)$$

$$x = \frac{x_1 + x_2}{2} = \text{midpoint of } [x_1,\ x_2]$$

36.

$$f(x) = 2x^2 - 3x + 1$$

$$\frac{f(b) - f(a)}{b - a} = \frac{21 - 1}{4 - 0} = 5$$

$$f'(x) = 4x - 3 = 5$$

$$x = 2 = \text{midpoint of } [0,\ 4]$$

37. Let $t = 0$ at noon.

$L = d^2 = (100 - 12t)^2 + (-10t)^2 = 10{,}000 - 2400t + 244t^2$

$\dfrac{dL}{dt} = -2400 + 488t = 0$ when $t = \dfrac{300}{61} \approx 4.92$ hr.

Ship A at $(40.98,\ 0)$; Ship B at $(0,\ -49.18)$

$d^2 = 10,000 - 2400t + 244t^2$

$= 4098.36$ when $t = 4.92 \approx$ 4:55 P.M.

$d \approx 64$ miles

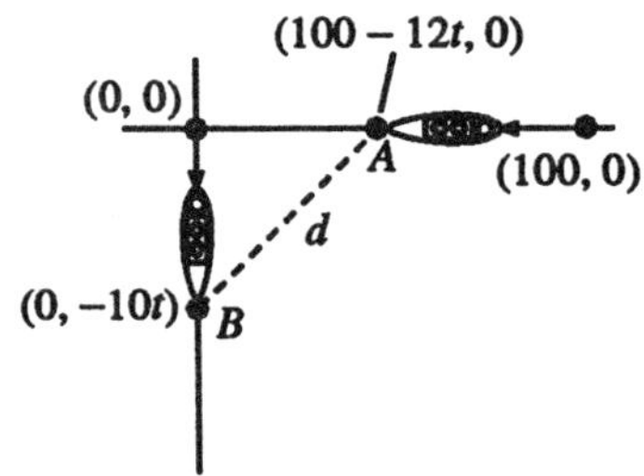

38. $C = \dfrac{1}{4}x^2 + 62x + 125$

$p = 75 - \dfrac{1}{3}x$

$R = xp$

(a) $P = R - C = x\left(75 - \dfrac{1}{3}x\right) - \left(\dfrac{1}{4}x^2 + 62x + 125\right)$

$\dfrac{dP}{dx} = -\dfrac{1}{3}x + 75 - \dfrac{1}{3}x - \dfrac{1}{2}x - 62 = -\dfrac{7}{6}x + 13 = 0$ when $x = \dfrac{78}{7}$ (11 units).

(b) $\overline{C}(x) = \dfrac{C}{x} = \dfrac{1}{4}x + 62 + \dfrac{125}{x}$ (average cost)

$\dfrac{d\overline{C}}{dx} = \dfrac{1}{4} - \dfrac{125}{x^2} = 0$

$x^2 - 500 = 0$ when $x = 10\sqrt{5}$ (22 units).

(c) $\eta = \dfrac{p/x}{dp/dx} = \dfrac{(75/x) - (1/3)}{-1/3} = 1 - \dfrac{225}{x} = \dfrac{x - 255}{x}$

39. $p = 36 - 4x$

$C = 2x^2 + 6$

$R = xp$

$P = R - C = x(36 - 4x) - (2x^2 + 6) = -6(x^2 - 6x + 1)$

$\dfrac{dP}{dx} = -6(2x - 6)$ when $x = 3$, $P = \$48$

40. Ellipse: $\dfrac{x^2}{144} + \dfrac{y^2}{16} = 1$

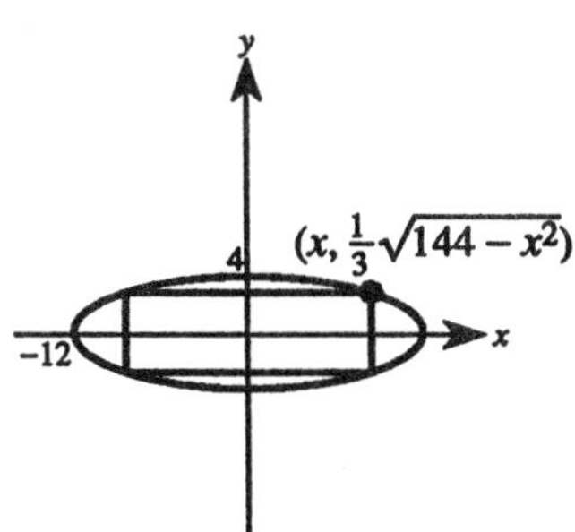

$$A = (2x)\left(\frac{2}{3}\sqrt{144 - x^2}\right) = \frac{4}{3}x\sqrt{144 - x^2}$$

$$\frac{dA}{dx} = \frac{4}{3}\left[\frac{-x^2}{\sqrt{144 - x^2}} + \sqrt{144 - x^2}\right]$$

$$= \frac{4}{3}\left[\frac{144 - 2x^2}{\sqrt{144 - x^2}}\right] = 0 \text{ when } x = \sqrt{72} \text{ or } x = 6\sqrt{2}$$

The dimensions of the rectangle are $12\sqrt{2}$ by $4\sqrt{2}$.

41. We have points $(0,\ y)$, $(x,\ 0)$, and $(1, 8)$.

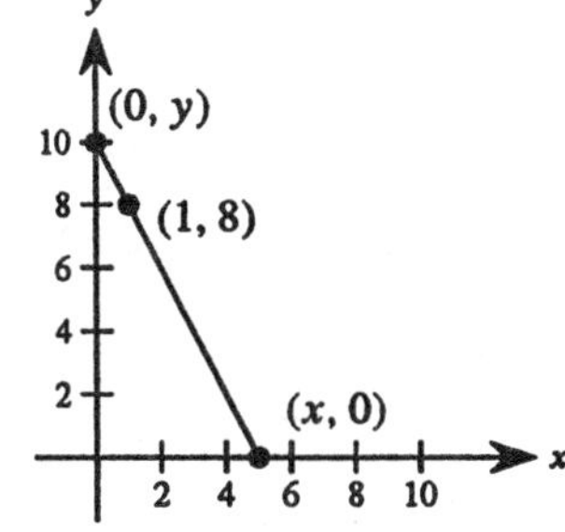

$$m = \frac{y - 8}{1} = \frac{8}{x - 1} \text{ or } y = \frac{8x}{x - 1}$$

$$L^2 = x^2 + \left(\frac{8x}{x - 1}\right)^2$$

$$2LL' = 2x + 128\left(\frac{x}{x - 1}\right)\left[\frac{(x - 1) - x}{(x - 1)^2}\right] = 0$$

$$x - \frac{64x}{(x - 1)^3} = 0$$

$x[(x - 1)^3 - 64] = 0$ when $x = 0,\ 5$ (minimum).

Vertices of triangle: $(0, 0)$, $(5, 0)$, $(0, 10)$

42. We have points $(0,\ y)$, $(x,\ 0)$, and $(4, 5)$.

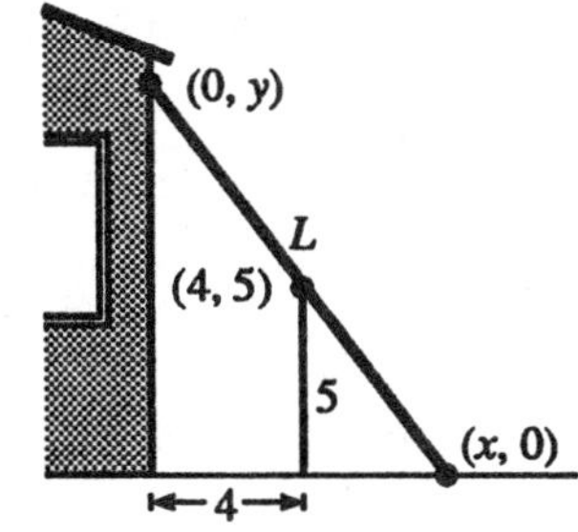

$$m = \frac{y - 5}{0 - 4} = \frac{5 - 0}{4 - x} \text{ or } y = \frac{5x}{x - 4}$$

$$L^2 = x^2 + \left(\frac{5x}{x - 4}\right)^2$$

$$2LL' = 2x + 50\left(\frac{x}{x - 4}\right)\left[\frac{x - 4 - x}{(x - 4)^2}\right] = 0$$

$$x - \frac{100x}{(x - 4)^3} = 0$$

$x[(x - 4)^3 - 100] = 0$ when $x = 0$ or $x = 4 + \sqrt[3]{100}$.

$$L = \sqrt{x^2 + \frac{25x^2}{(x - 4)^2}} = \frac{x}{x - 4}\sqrt{(x - 4)^2 + 25} = \frac{\sqrt[3]{100} + 4}{\sqrt[3]{100}}\sqrt{100^{2/3} + 25} = 12.7 \text{ feet}$$

43. Let x be the number of people in excess of eighty.

$$R = (\text{number of people})(\text{price per person}) = (80 + x)(8 - 0.05x)$$

$$\frac{dR}{dx} = (80 + x)(-0.05) + (8 - 0.05x) = 4 - 0.10x = 0 \text{ when } x = 40$$

Revenue will be maximum when 120 people go on the bus.

44. Width of rectangle: $a - \dfrac{a-b}{b}x - x$

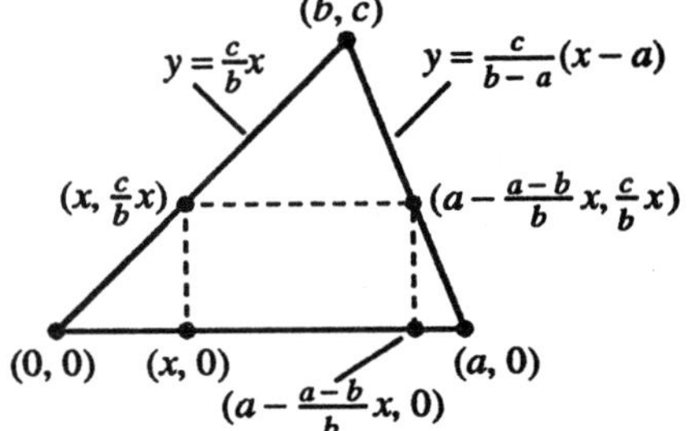

Height of rectangle: $\dfrac{c}{b}x$ (See figure.)

$$A = (\text{width})(\text{height}) = \left(a - \frac{a-b}{b}x - x\right)\left(\frac{c}{b}x\right) = \left(a - \frac{a}{b}x\right)\frac{c}{b}x$$

$$\frac{dA}{dx} = \left(a - \frac{a}{b}x\right)\frac{c}{b} + \left(\frac{c}{b}x\right)\left(-\frac{a}{b}\right) = \frac{ac}{b} - \frac{2ac}{b^2}x = 0 \text{ when } x = \frac{b}{2}.$$

$$A\left(\frac{b}{2}\right) = \left(a - \frac{a}{b}\frac{b}{2}\right)\left(\frac{c}{b}\frac{b}{2}\right)$$

$$= \left(\frac{a}{2}\right)\left(\frac{c}{2}\right) = \frac{1}{4}ac = \frac{1}{2}\left(\frac{1}{2}ac\right) = \frac{1}{2}(\text{area of triangle})$$

45. $A = (\text{average of bases})(\text{height})$

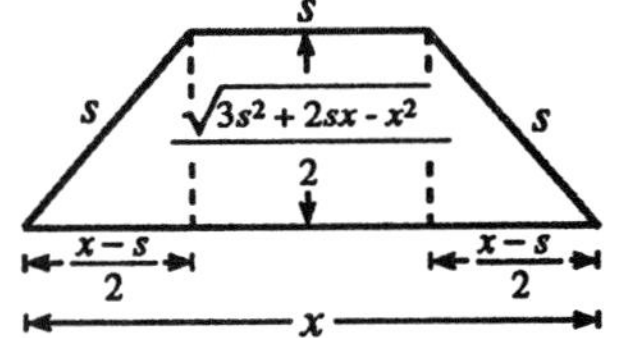

$$= \left(\frac{x+s}{2}\right)\frac{\sqrt{3s^2 + 2sx - x^2}}{2} \quad \text{(See figure.)}$$

$$\frac{dA}{dx} = \frac{1}{4}\left[\frac{(s-x)(s+x)}{\sqrt{3s^2 + 2sx - x^2}} + \sqrt{3s^2 + 2sx - x^2}\right]$$

$$= \frac{2(2s - x)(s + x)}{4\sqrt{3s^2 + 2sx - x^2}} = 0 \text{ when } x = 2s.$$

A is a maximum when $x = 2s$.

46. The cost C is proportional to $s^{3/2}$, $C = 50$ when $s = 25$, and fixed costs are \$100.

$$C = ks^{3/2} \text{ where } 50 = k(25)^{3/2} \Rightarrow k = \frac{2}{5}.$$

$$\overline{C} = \frac{(2/5)s^{3/2} + 100}{s} = \frac{2}{5}s^{1/2} + \frac{100}{s}$$

$$\overline{C}' = \frac{1}{5s^{1/2}} - \frac{100}{s^2} = 0 \text{ when } s^{3/2} = 500.$$

$$s = 500^{2/3} \approx 63 \text{ mi/hr}$$

47. $m = \dfrac{y - 6}{0 - 4} = \dfrac{6 - 0}{4 - x}$

$$LL' = x[(x - 4)^3 - 144] = 0$$

$$L = 14.05 \text{ feet}$$

(See solution to Exercise 41.)

48. We can form a right triangle with vertices $(0, y)$, $(0, 0)$, and $(x, 0)$. Choosing a point (a, b) on the hypotenuse (assuming the triangle is in the first quadrant), the slope is

$$m = \frac{b-y}{0-a} = \frac{b-0}{a-x} \Rightarrow y = \frac{-bx}{a-x}$$

$$L^2 = x^2 + y^2 = x^2 + \left(\frac{-bx}{a-x}\right)^2$$

$$2LL' = 2x + 2\left[\frac{-bx}{a-x}\;\frac{-ab}{(a-x)^2}\right]$$

$$\frac{2x[(a-x)^3 + ab^2]}{(a-x)^3} = 0 \text{ when } x = 0,\ a + \sqrt[3]{ab^2}.$$

Choosing the nonzero value, we have $y = b + \sqrt[3]{a^2b}$.

$$L = \sqrt{\left(a + \sqrt[3]{ab^2}\right)^2 + \left(b + \sqrt[3]{a^2b}\right)^2} = (a^2 + 3a^{4/3}b^{2/3} + 3a^{2/3}b^{4/3} + b^2)^{1/2} = (a^{2/3} + b^{2/3})^{3/2} \text{ feet}$$

49. $C = \left(\frac{Q}{x}\right)s + \left(\frac{x}{2}\right)r$

$$\frac{dC}{dx} = -\frac{Qs}{x^2} + \frac{r}{2} = 0$$

$$\frac{Qs}{x^2} = \frac{r}{2}$$

$$x^2 = \frac{2Qs}{r}$$

$$x = \sqrt{\frac{2Qs}{r}}$$

50. $p = 600 - 3x$

$$C = 0.3x^2 + 6x + 600$$

$$P = xp - C - xt$$

$$P = -3.3x^2 + (594 - t)x - 600$$

$$\frac{dP}{dx} = -6.6x + 594 - t = 0$$

$$x = \frac{594 - t}{6.6}$$

(a) $t = 5$, $x = 89$, $P = 25681.70$
(b) $t = 10$, $x = 88$, $P = 25236.80$
(c) $t = 20$, $x = 87$, $P = 24360.30$

51. $f(x) = x^3 - 3x - 1$, $[-1, 0]$

$$f'(x) = 3x^2 - 3$$

$$x_{n+1} = x_n - \frac{{x_n}^3 - 3x_n - 1}{3{x_n}^2 - 3}$$

$$x \approx -0.347$$

n	x_n	$f(x_n)$
1	–0.5000	0.3750
2	–0.3333	–0.0370
3	–0.3472	–0.0002
4	–0.3473	0.0000

52. Find the zeros to $f(x) = x^4 - x - 3$.

$f'(x) = 4x^3 - 1$

$$x_{n+1} = x_n - \frac{{x_n}^4 - x_n - 3}{4{x_n}^3 - 1}$$

f changes sign in $[-2, -1]$.

n	x_n	$f(x_n)$
1	−1.2000	0.2736
2	−1.1654	0.0101
3	−1.1640	0.0000

On the interval $[-2, -1]$: $x \approx -1.164$

f changes sign in $[1, 2]$.

n	x_n	$f(x_n)$
1	1.2000	−2.1264
2	1.5597	1.3578
3	1.4639	0.1285
4	1.4528	0.0016
5	1.4526	0.0000

On the interval $[1, 2]$: $x \approx 1.453$

53. $f(x) = x^3 + 2x + 1$

$f'(x) = 3x^2 + 2$

$$x_{n+1} = x_n - \frac{{x_n}^3 + 2x_n + 1}{3{x_n}^2 + 2}$$

f changes sign in $[-1, 0]$.

n	x_n	$f(x_n)$
1	−0.5000	−0.1250
2	−0.4545	−0.0030
3	−0.4534	0.0000

On the interval $[-1, 0]$: $x \approx -0.453$

54. $s = 4\pi r^2$

$$ds = 8\pi r\,dr$$
$$= 8\pi(9)(0.025) \approx 1.8\pi \text{ square inches}$$
$$V = \frac{4}{3}\pi r^3$$
$$dV = 4\pi r^2\,dr$$
$$= 4\pi(9)^2(0.025) \approx 8.1\pi \text{ cubic inches}$$

55. $s = 6x^2$

$$ds = 12x\,dx$$
$$\%\text{ error} = \frac{ds}{s}(100) = \frac{12x\,dx}{6x^2}(100) = 2\left[\frac{dx}{x}(100)\right] = 2(1\%) = 2\%$$
$$V = x^3$$
$$dV = 3x^2\,dx$$
$$\%\text{ error} = \frac{dV}{V}(100) = \frac{3x^2\,dx}{x^3}(100) = 3\left[\frac{dx}{x}(100)\right] = 3(1\%) = 3\%$$

56. $p = 75 - \frac{1}{4}x$

$$\Delta p = p(8) - p(7) = \left(75 - \tfrac{8}{4}\right) - \left(75 - \tfrac{7}{4}\right) = -\tfrac{1}{4}$$
$$dp = -\tfrac{1}{4}\,dx = -\tfrac{1}{4}(1) = -\tfrac{1}{4}$$

CHAPTER 5
Integration

Section 5.1 Antiderivatives and Indefinite Integration

	Given	*Rewrite*	*Integrate*	*Simplify*
1.	$\int \sqrt[3]{x}\,dx$	$\int x^{1/3}\,dx$	$\dfrac{x^{4/3}}{4/3}+C$	$\dfrac{3}{4}x^{4/3}+C$
2.	$\int \dfrac{1}{x^2}\,dx$	$\int x^{-2}\,dx$	$\dfrac{x^{-1}}{-1}+C$	$-\dfrac{1}{x}+C$
3.	$\int \dfrac{1}{x\sqrt{x}}\,dx$	$\int x^{-3/2}\,dx$	$\dfrac{x^{-1/2}}{-1/2}+C$	$-\dfrac{2}{\sqrt{x}}+C$
4.	$\int x(x^2+3)\,dx$	$\int (x^3+3x)\,dx$	$\dfrac{x^4}{4}+3\left(\dfrac{x^2}{2}\right)+C$	$\dfrac{1}{4}x^4+\dfrac{3}{2}x^2+C$
5.	$\int \dfrac{1}{2x^3}\,dx$	$\dfrac{1}{2}\int x^{-3}\,dx$	$\dfrac{1}{2}\left(\dfrac{x^{-2}}{-2}\right)+C$	$-\dfrac{1}{4x^2}+C$
6.	$\int \dfrac{1}{(2x)^3}\,dx$	$\dfrac{1}{8}\int x^{-3}\,dx$	$\dfrac{1}{8}\left(\dfrac{x^{-2}}{-2}\right)+C$	$-\dfrac{1}{16x^2}+C$

7. $\int (x^3+2)\,dx = \dfrac{1}{4}x^4+2x+C$

Check: $\dfrac{d}{dx}\left(\dfrac{1}{4}x^4+2x+C\right)=x^3+2$

8. $\int (x^2-2x+3)\,dx = \dfrac{1}{3}x^3-x^2+3x+C$

Check: $\dfrac{d}{dx}\left(\dfrac{1}{3}x^3-x^2+3x+C\right)=x^2-2x+3$

9. $\int (x^{3/2}+2x+1)\,dx = \dfrac{2}{5}x^{5/2}+x^2+x+C$

Check: $\dfrac{d}{dx}\left(\dfrac{2}{5}x^{5/2}+x^2+x+C\right)=x^{3/2}+2x+1$

10. $\int \left(\sqrt{x}+\dfrac{1}{2\sqrt{x}}\right)dx = \int \left(x^{1/2}+\dfrac{1}{2}x^{-1/2}\right)dx = \dfrac{x^{3/2}}{3/2}+\dfrac{1}{2}\left(\dfrac{x^{1/2}}{1/2}\right)+C=\dfrac{2}{3}x^{3/2}+x^{1/2}+C$

Check: $\dfrac{d}{dx}\left(\dfrac{2}{3}x^{3/2}+x^{1/2}+C\right)=x^{1/2}+\dfrac{1}{2}x^{-1/2}=\sqrt{x}+\dfrac{1}{2\sqrt{x}}$

11. $\int \sqrt[3]{x^2}\,dx = \int x^{2/3}\,dx = \dfrac{x^{5/3}}{5/3}+C=\dfrac{3}{5}x^{5/3}+C$

Check: $\dfrac{d}{dx}\left(\dfrac{3}{5}x^{5/3}+C\right)=x^{2/3}=\sqrt[3]{x^2}$

12. $\int \left(\sqrt[4]{x^3}+1\right)dx = \int (x^{3/4}+1)\,dx = \dfrac{4}{7}x^{7/4}+x+C$

Check: $\dfrac{d}{dx}\left(\dfrac{4}{7}x^{7/4}+x+C\right)=x^{3/4}+1=\sqrt[4]{x^3}+1$

13. $\int \dfrac{1}{x^3}\,dx = \int x^{-3}\,dx = \dfrac{x^{-2}}{-2}+C=-\dfrac{1}{2x^2}+C$

Check: $\dfrac{d}{dx}\left(-\dfrac{1}{2x^2}+C\right)=\dfrac{1}{x^3}$

14. $\int \dfrac{1}{x^4}\,dx = \int x^{-4}\,dx = \dfrac{x^{-3}}{-3}+C=-\dfrac{1}{3x^3}+C$

Check: $\dfrac{d}{dx}\left(-\dfrac{1}{3x^3}+C\right)=\dfrac{1}{x^4}$

15. $\displaystyle\int \frac{1}{4x^2}\,dx = \frac{1}{4}\int x^{-2}\,dx$

$\displaystyle = \frac{1}{4}\left(\frac{x^{-1}}{-1}\right) + C = -\frac{1}{4x} + C$

Check: $\displaystyle\frac{d}{dx}\left(-\frac{1}{4x} + C\right) = \frac{1}{4x^2}$

16. $\displaystyle\int (2x + x^{-1/2})\,dx = x^2 + 2x^{1/2} + C$

Check: $\displaystyle\frac{d}{dx}(x^2 + 2x^{1/2} + C) = 2x + x^{-1/2}$

17. $\displaystyle\int \frac{x^2+x+1}{\sqrt{x}}\,dx = \int (x^{3/2} + x^{1/2} + x^{-1/2})\,dx = \frac{2}{5}x^{5/2} + \frac{2}{3}x^{3/2} + 2x^{1/2} + C = \frac{2}{15}x^{1/2}(3x^2 + 5x + 15) + C$

Check: $\displaystyle\frac{d}{dx}\left(\frac{2}{5}x^{5/2} + \frac{2}{3}x^{3/2} + 2x^{1/2} + C\right) = x^{3/2} + x^{1/2} + x^{-1/2} = \frac{x^2+x+1}{\sqrt{x}}$

18. $\displaystyle\int \frac{x^2+1}{x^2}\,dx = \int (1 + x^{-2})\,dx = x + \frac{x^{-1}}{-1} + C = x - \frac{1}{x} + C$

Check: $\displaystyle\frac{d}{dx}\left(x - \frac{1}{x} + C\right) = 1 + \frac{1}{x^2} = \frac{x^2+1}{x^2}$

19. $\displaystyle\int (x+1)(3x-2)\,dx = \int (3x^2 + x - 2)\,dx$

$\displaystyle = x^3 + \frac{1}{2}x^2 - 2x + C$

Check:

$\displaystyle\frac{d}{dx}\left(x^3 + \frac{1}{2}x^2 - 2x + C\right) = 3x^2 + x - 2$

$= (x+1)(3x-2)$

20. $\displaystyle\int (2t^2 - 1)^2\,dt = \int (4t^4 - 4t^2 + 1)\,dt$

$\displaystyle = \frac{4}{5}t^5 - \frac{4}{3}t^3 + t + C$

Check:

$\displaystyle\frac{d}{dt}\left(\frac{4}{5}t^5 - \frac{4}{3}t^3 + t + C\right) = 4t^4 - 4t^2 + 1$

$= (2t^2 - 1)^2$

21. $\displaystyle\int \frac{t^2+2}{t^2}\,dt = \int (1 + 2t^{-2})\,dt$

$\displaystyle = t + 2\left(\frac{t^{-1}}{-1}\right) + C = t - \frac{2}{t} + C$

Check: $\displaystyle\frac{d}{dt}\left(t - \frac{2}{t} + C\right) = 1 + \frac{2}{t^2} = \frac{t^2+2}{t^2}$

22. $\displaystyle\int (1 - 2y + 3y^2)\,dy = y - y^2 + y^3 + C$

Check: $\displaystyle\frac{d}{dy}\left(y - y^2 + y^3 + C\right) = 1 - 2y + 3y^2$

23. $\displaystyle\int y^2\sqrt{y}\,dy = \int y^{5/2}\,dy = \frac{2}{7}y^{7/2} + C$

Check: $\displaystyle\frac{d}{dy}\left(\frac{2}{7}y^{7/2} + C\right) = y^{5/2} = y^2\sqrt{y}$

24. $\displaystyle\int (1+3t)t^2\,dt = \int (t^2 + 3t^3)\,dt = \frac{1}{3}t^3 + \frac{3}{4}t^4 + C$

Check: $\displaystyle\frac{d}{dt}\left(\frac{1}{3}t^3 + \frac{3}{4}t^4 + C\right) = t^2 + 3t^3 = (1+3t)t^2$

25. $\displaystyle\int dx = \int 1\,dx = x + C$

Check: $\displaystyle\frac{d}{dx}(x + C) = 1$

26. $\displaystyle\int 3\,dt = 3t + C$

Check: $\displaystyle\frac{d}{dt}(3t + C) = 3$

27. $f'(x) = 2$

$f(x) = 2x + C$

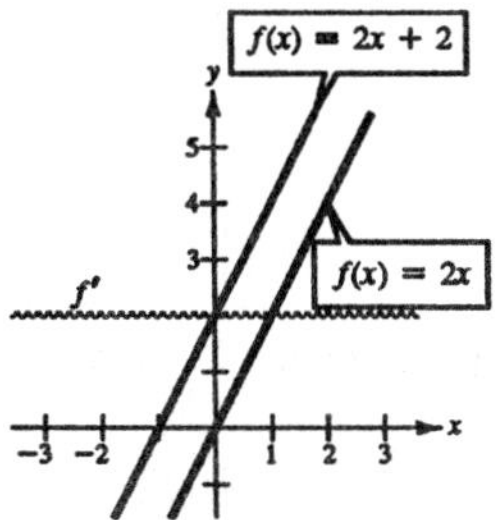

28. $f'(x) = x$

$f(x) = \frac{x^2}{2} + C$

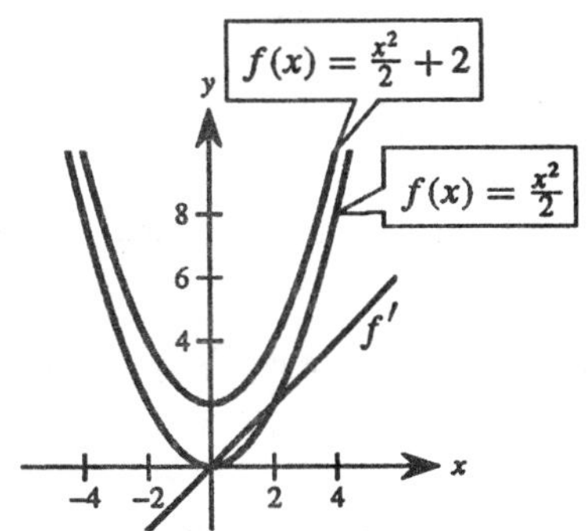

29. $f'(x) = 1 - x^2$

$f(x) = x - \frac{x^3}{3} + C$

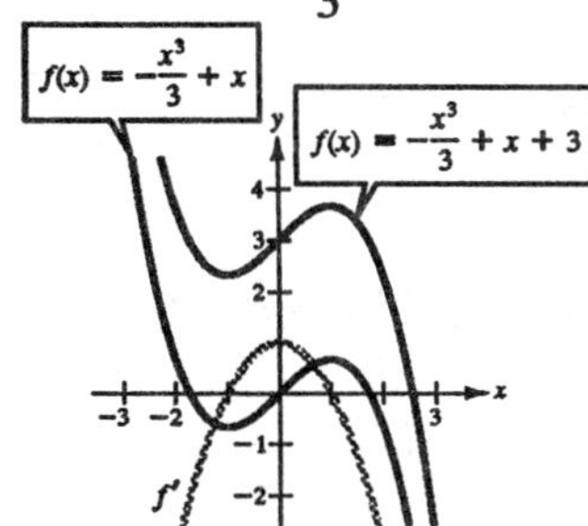

30. $f'(x) = \frac{1}{x^2}$

$f(x) = -\frac{1}{x} + C$

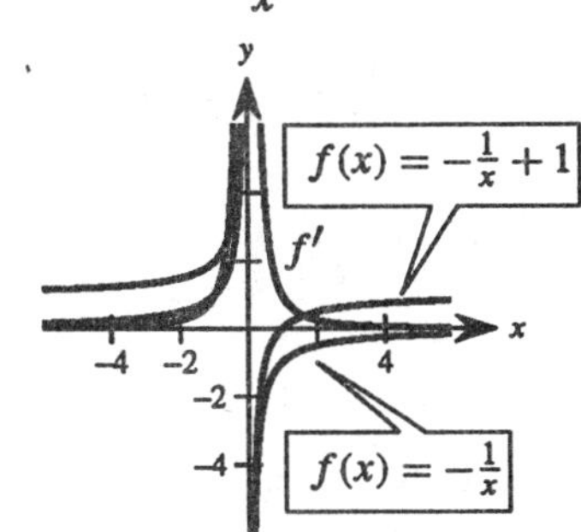

31. $\frac{dy}{dx} = 2x - 1, \quad (1, 1)$

$y = \int (2x - 1)\,dx = x^2 - x + C$

$1 = (1)^2 - (1) + C \Rightarrow C = 1$

$y = x^2 - x + 1$

32. $\frac{dy}{dx} = 2(x - 1) = 2x - 2, \quad (3, 2)$

$y = \int 2(x - 1)\,dx = x^2 - 2x + C$

$2 = (3)^2 - 2(3) + C \Rightarrow C = -1$

$y = x^2 - 2x - 1$

33. $\frac{dy}{dx} = 3x^2 - 1, \quad (0, 2)$

$y = \int (3x^2 - 1)\,dx = x^3 - x + C$

$2 = 0^3 - 0 + C \Rightarrow C = 2$

$y = x^3 - x + 2$

34. $\frac{dy}{dx} = -\frac{1}{x^2} = -x^{-2}, \quad (1, 3)$

$y = \int -x^{-2}\,dx = \frac{1}{x} + C$

$3 = \frac{1}{1} + C \Rightarrow C = 2$

$y = \frac{1}{x} + 2$

35. $f''(x) = 2$

$f'(2) = 5$

$f(2) = 10$

$$f'(x) = \int 2\,dx = 2x + C_1$$

$$f'(2) = 4 + C_1 = 5 \Rightarrow C_1 = 1$$

$$f'(x) = 2x + 1$$

$$f(x) = \int (2x + 1)\,dx = x^2 + x + C_2$$

$$f(2) = 6 + C_2 = 10 \Rightarrow C_2 = 4$$

$$f(x) = x^2 + x + 4$$

36. $f''(x) = x^2$

$f'(0) = 6$

$f(0) = 3$

$$f'(x) = \int x^2\,dx = \tfrac{1}{3}x^3 + C_1$$

$$f'(0) = 0 + C_1 = 6 \Rightarrow C_1 = 6$$

$$f'(x) = \tfrac{1}{3}x^3 + 6$$

$$f(x) = \int \left(\tfrac{1}{3}x^3 + 6\right) dx = \tfrac{1}{12}x^4 + 6x + C_2$$

$$f(0) = 0 + 0 + C_2 = 3 \Rightarrow C_2 = 3$$

$$f(x) = \tfrac{1}{12}x^4 + 6x + 3$$

37. $f''(x) = x^{-3/2}$

$f'(4) = 2$

$f(0) = 0$

$$f'(x) = \int x^{-3/2}\,dx = -2x^{-1/2} + C_1 = -\frac{2}{\sqrt{x}} + C_1$$

$$f'(4) = -\frac{2}{2} + C_1 = 2 \Rightarrow C_1 = 3$$

$$f'(x) = -\frac{2}{\sqrt{x}} + 3$$

$$f(x) = \int (-2x^{-1/2} + 3)\,dx = -4x^{1/2} + 3x + C_2$$

$$f(0) = 0 + 0 + C_2 = 0 \Rightarrow C_2 = 0$$

$$f(x) = -4x^{1/2} + 3x = -4\sqrt{x} + 3x$$

38. $f''(x) = x^{-3/2}$

$f'(1) = 2$

$f(9) = -4$

$$f'(x) = \int x^{-3/2}\,dx = -2x^{-1/2} + C_1 = -\frac{2}{\sqrt{x}} + C_1$$

$$f'(1) = -2 + C_1 = 2 \Rightarrow C_1 = 4$$

$$f'(x) = -\frac{2}{\sqrt{x}} + 4$$

$$f(x) = \int (-2x^{-1/2} + 4)\,dx = -4x^{1/2} + 4x + C_2$$

$$= -4\sqrt{x} + 4x + C_2$$

$$f(9) = -12 + 36 + C_2 = -4 \Rightarrow C_2 = -28$$

$$f(x) = -4\sqrt{x} + 4x - 28 = -4(\sqrt{x} - x + 7)$$

39. $s_0 = 1600$

$v_0 = 0$

$$v(t) = \int -32\,dt = -32t + C_1$$

$$v(0) = C_1 = 0 \Rightarrow v(t) = -32t$$

$$s(t) = \int -32t\,dt = -16t^2 + C_2$$

$$s(0) = C_2 = 1600 \Rightarrow s(t) = -16t^2 + 1600$$

$$-16t^2 + 1600 = 0$$

$t = 10$ seconds

40. $s_0 = 0$

$v_0 = 60$

$$v(t) = \int -32\,dt = -32t + C_1$$

$$v(0) = C_1 = 60$$

$$v(t) = -32t + 60$$

$$s(t) = \int (-32t + 60)\,dt = -16t^2 + 60t + C_2$$

$$s(0) = C_2 = 0$$

$$s(t) = -16t^2 + 60t$$

$s'(t) = v(t) = -32t + 60 = 0$ when $t = 1.875$.

$s''(t) = -32 < 0$

Maximum height at $s(1.875)$:

$-56.25 + 112.50 = 56.25$ feet

41. $v(t) = \int -32\,dt = -32t + C_1$

$v_0 = C_1$

$s(t) = \int (-32t + v_0)\,dt = -16t^2 + v_0t + C_2$

$s(0) = C_2 = 0 \Rightarrow s(t) = -16t^2 + v_0t$

$s'(t) = -32t + v_0 = 0$ when

$t = \frac{v_0}{32}$ = time to reach maximum height.

$$s\left(\frac{v_0}{32}\right) = -16\left(\frac{v_0}{32}\right)^2 + v_0\left(\frac{v_0}{32}\right) = 550$$

$$-\frac{v_0^2}{64} + \frac{v_0^2}{32} = 550$$

$v_0^2 = 35{,}200$

$v_0 \approx 187.617$ ft/sec

42. $s''(t) = a(t) = -32$ ft/sec^2

$s'(0) = v_0$

$s(0) = s_0$

$s'(t) = v(t) = \int -32\,dt = -32t + C_1$

$s'(0) = 0 + C_1 = v_0 \Rightarrow C_1 = v_0$

$s'(t) = -32t + v_0$

$s(t) = \int (-32t + v_0)\,dt = -16t^2 + v_0t + C_2$

$s(0) = 0 + 0 + C_2 = s_0 \Rightarrow C_2 = s_0$

$s(t) = -16t^2 + v_0t + s_0$

43. $v_0 = 16$ ft/sec

$s_0 = 64$ ft

(a) $s(t) = -16t^2 + 16t + 64 = 0$

$-16(t^2 - t - 4) = 0$

$$t = \frac{1 \pm \sqrt{17}}{2}$$

Choosing the positive value,

$$t = \frac{1 + \sqrt{17}}{2} \approx 2.562 \text{ seconds.}$$

(b) $v(t) = s'(t) = -32t + 16$

$$v\left(\frac{1+\sqrt{17}}{2}\right) = -32\left(\frac{1+\sqrt{17}}{2}\right) + 16$$

$$= -16\sqrt{17} \approx -65.970 \text{ ft/sec}$$

44. $a(t) = k$

$v(t) = kt$

$s(t) = \frac{k}{2}t^2$ since $v(0) = s(0) = 0$.

At the time of lift-off, $kt = 160$ and $(k/2)t^2 = 0.7$. Since $(k/2)t^2 = 0.7$,

$$t = \sqrt{\frac{1.4}{k}}$$

$$k\sqrt{\frac{1.4}{k}} = 160$$

$$1.4k = 160^2 \Rightarrow k = \frac{160^2}{1.4} \approx 18285.714 \text{ mi/hr}^2$$

$$\approx 7.45 \text{ ft/sec}^2.$$

45. $v(0) = 15$ mph $= 22$ ft/sec

$v(13) = 50$ mph $= \frac{220}{3}$ ft/sec

$s(0) = 0$

(a) $a(t) = a$(constant acceleration)

$v(t) = at + C$

$v(0) = C = 22$ ft/sec

$v(t) = at + 22$

$v(13) = 13a + 22$

$= \frac{220}{3}$ when $a = \frac{154}{39} \approx 3.95$ ft/sec^2.

(b) $v(t) = \frac{154}{39}t + 22$

$s(t) = \left(\frac{77}{39}\right)t^2 + 22t$ since $s(0) = 0$.

$s(13) = \frac{77}{39}(13)^2 + 22(13)$

$= \frac{1859}{3}$ ft ≈ 619.67 ft.

46. $v(0) = 45 \text{ mph} = 66 \text{ ft/sec}$

$30 \text{ mph} = 44 \text{ ft/sec}$

$15 \text{ mph} = 22 \text{ ft/sec}$

$a(t) = -a$

$v(t) = -at + 66$

$s(t) = -\frac{a}{2}t^2 + 66t$ (Let $s(0) = 0$.)

$v(t) = 0$ after car moves 132 ft.

$-at + 66 = 0$ when $t = \frac{66}{a}$.

$$s\left(\frac{66}{a}\right) = -\frac{a}{2}\left(\frac{66}{a}\right)^2 + 66\left(\frac{66}{a}\right)$$

$$= 132 \text{ when } a = \frac{33}{2} = 16.5.$$

$a(t) = -16.5$

$v(t) = -16.5t + 66$

$s(t) = -8.25t^2 + 66t$

(a) $-16.5t + 66 = 44$

$$t = \frac{22}{16.5}$$

$$s\left(\frac{22}{16.5}\right) \approx 73.33 \text{ ft}$$

(b) $-16.5t + 66 = 22$

$$t = \frac{44}{16.5}$$

$$s\left(\frac{44}{16.5}\right) \approx 117.33 \text{ ft}$$

(c)

20 60 100 140 x

47. Truck: $v(t) = 30$

$s(t) = 30t$ (Let $s(0) = 0$.)

Automobile: $a(t) = 6$

$v(t) = 6t$ (Let $v(0) = 0$.)

$s(t) = 3t^2$ (Let $s(0) = 0$.)

At the point where the automobile overtakes the truck,

$30t = 3t^2$

$0 = 3t^2 - 30t$

$0 = 3t(t - 10)$ when $t = 10$ sec.

(a) $s(10) = 3(10)^2 = 300$ ft

(b) $v(10) = 6(10) = 60 \text{ ft/sec} \approx 41 \text{ mph}$

48. $a(t) = a$ (since acceleration is constant)

$v(t) = at$ (Let $v(0) = 0$.)

$s(t) = \frac{at^2}{2}$ (Let $s(0) = 0$.)

$s(4) = \frac{a(4)^2}{2} = 8a = 100$

$a = 12.5 \text{ cm/sec}^2$

49. Let d be the distance traversed and a be the uniform acceleration. Furthermore, note that $v(0) = 0$ and let $s(0) = 0$.

$$a(t) = a$$

$$v(t) = at$$

$$s(t) = \frac{at^2}{2} = d \text{ when } t = \sqrt{\frac{2d}{a}}.$$

The highest speed is

$$v\left(\sqrt{\frac{2d}{a}}\right) = a\sqrt{\frac{2d}{a}} = \sqrt{2ad}$$

and the mean speed is $\sqrt{2ad}/2 = \sqrt{ad/2}$. The time necessary to traverse the distance at the mean speed must satisfy the equation

$$\sqrt{\frac{ad}{2}}t = d \text{ or } t = \sqrt{\frac{2d}{a}}.$$

This is the same time as under uniform acceleration.

50. If $C'(x) = k$, then $C(x) = kx + C_1$ where k and C_1 are constants. Thus, the cost function is linear.

Section 5.2 Area

1. $\displaystyle\sum_{i=1}^{5}(2i+1) = 2\sum_{i=1}^{5} i + \sum_{i=1}^{5} 1 = 2(1+2+3+4+5)+5 = 35$

2. $\displaystyle\sum_{i=1}^{6} 2i = 2+4+6+8+10+12 = 42$

3. $\displaystyle\sum_{k=0}^{4}\frac{1}{k^2+1} = 1+\frac{1}{2}+\frac{1}{5}+\frac{1}{10}+\frac{1}{17} = \frac{158}{85}$

4. $\displaystyle\sum_{j=3}^{5}\frac{1}{j} = \frac{1}{3}+\frac{1}{4}+\frac{1}{5} = \frac{47}{60}$

5. $\displaystyle\sum_{k=1}^{4} c = c+c+c+c = 4c$

6. $\displaystyle\sum_{n=1}^{10}\frac{3}{n+1} = 3\left(\frac{1}{2}+\frac{1}{3}+\frac{1}{4}+\frac{1}{5}+\frac{1}{6}+\frac{1}{7}+\frac{1}{8}+\frac{1}{9}+\frac{1}{10}+\frac{1}{11}\right) = \frac{55{,}991}{9240}$

7. $\displaystyle\sum_{i=1}^{4}[(i-1)^2+(i+1)^3] = (0+8)+(1+27)+(4+64)+(9+125) = 238$

8. $\displaystyle\sum_{k=2}^{5}(k+1)(k-3) = (3)(-1)+(4)(0)+(5)(1)+(6)(2) = 14$

9. $\displaystyle\sum_{i=1}^{9}\frac{1}{3i}$

10. $\displaystyle\sum_{i=1}^{15}\frac{5}{1+i}$

11. $\displaystyle\sum_{j=1}^{8}\left[2\left(\frac{j}{8}\right)+3\right]$

12. $\displaystyle\sum_{j=1}^{4}\left[1-\left(\frac{j}{4}\right)^2\right]$

13. $\displaystyle\sum_{k=1}^{6}\left[\left(\frac{k}{6}\right)^2+2\right]\left(\frac{1}{6}\right) = \frac{1}{6}\sum_{k=1}^{6}\left[\left(\frac{k}{6}\right)^2+2\right]$

14. $\displaystyle\sum_{k=1}^{n}\left[\left(\frac{k}{n}\right)^2+2\right]\left(\frac{1}{n}\right) = \frac{1}{n}\sum_{k=1}^{n}\left[\left(\frac{k}{n}\right)^2+2\right]$

15. $\displaystyle\frac{2}{n}\sum_{i=1}^{n}\left[\left(\frac{2i}{n}\right)^3-\left(\frac{2i}{n}\right)\right]$

16. $\displaystyle\frac{2}{n}\sum_{i=1}^{n}\left[1-\left(\frac{2i}{n}-1\right)^2\right]$

17. $\frac{3}{n}\sum_{i=1}^{n}\left[2\left(1+\frac{3i}{n}\right)^2\right]$

18. $\frac{1}{n}\sum_{i=0}^{n-1}\sqrt{1-\left(\frac{i}{n}\right)^2}$

19. $\sum_{i=1}^{20} 2i = 2\sum_{i=1}^{20} i = 2\left[\frac{20(21)}{2}\right] = 420$

20. $\sum_{i=1}^{10} i(i^2+1) = \sum_{i=1}^{10} i^3 + \sum_{i=1}^{10} i$

$= \left[\frac{10(11)}{2}\right]^2 + \left[\frac{10(11)}{2}\right] = 3080$

21. $\sum_{i=1}^{20} (i-1)^2 = \sum_{i=1}^{19} i^2 = \left[\frac{19(20)(39)}{6}\right] = 2470$

22. $\sum_{i=1}^{15} (2i-3) = 2\sum_{i=1}^{15} i - 3(15)$

$= 2\left[\frac{15(16)}{2}\right] - 45 = 195$

23. $\sum_{i=1}^{15} \frac{1}{n^3}(i-1)^2 = \frac{1}{n^3}\sum_{i=1}^{14} i^2 = \frac{1}{n^3}\left[\frac{14(15)(29)}{6}\right] = \frac{1015}{n^3}$

24. $\sum_{i=1}^{10} (i^2-1) = \sum_{i=1}^{10} i^2 - \sum_{i=1}^{10} 1 = \left[\frac{10(11)(21)}{6}\right] - 10 = 375$

25. $\lim_{n\to\infty}\left[\left(\frac{4}{3n^3}\right)(2n^3+3n^2+n)\right] = \lim_{n\to\infty}\left[\frac{8}{3}+\frac{4}{n}+\frac{4}{3n^2}\right] = \frac{8}{3}$

26. $\lim_{n\to\infty}\left(\frac{8}{3}+\frac{4}{n}+\frac{4}{3n^2}\right) = \frac{8}{3}$

27. $\lim_{n\to\infty}\left[\left(\frac{81}{n^4}\right)\frac{n^2(n+1)^2}{4}\right] = \frac{81}{4}\lim_{n\to\infty}\left[\frac{n^4+2n^3+n^2}{n^4}\right] = \frac{81}{4}(1) = \frac{81}{4}$

28. $\lim_{n\to\infty}\left[\left(\frac{64}{n^3}\right)\frac{n(n+1)(2n+1)}{6}\right] = \frac{64}{6}\lim_{n\to\infty}\left[\frac{2n^3+3n^2+n}{n^3}\right] = \frac{64}{6}(2) = \frac{64}{3}$

29. $\lim_{n\to\infty}\left[\left(\frac{18}{n^2}\right)\frac{n(n+1)}{2}\right] = \frac{18}{2}\lim_{n\to\infty}\left[\frac{n^2+n}{n^2}\right] = \frac{18}{2}(1) = 9$

30. $\lim_{n\to\infty}\left[\left(\frac{1}{n^2}\right)\frac{n(n+1)}{2}\right] = \frac{1}{2}\lim_{n\to\infty}\left[\frac{n^2+n}{n^2}\right] = \frac{1}{2}(1) = \frac{1}{2}$

31. $\lim_{n\to\infty}\sum_{i=1}^{n}\frac{1}{n^3}(i-1)^2 = \lim_{n\to\infty}\frac{1}{n^3}\sum_{i=1}^{n-1} i^2 = \lim_{n\to\infty}\frac{1}{n^3}\left[\frac{(n-1)(n)(2n-1)}{6}\right]$

$= \lim_{n\to\infty}\frac{1}{6}\left[\frac{2n^3-3n^2+n}{n^3}\right] = \lim_{n\to\infty}\left[\frac{1}{6}\left(\frac{2-(3/n)+(1/n^2)}{1}\right)\right] = \frac{1}{3}$

32. $\lim_{n\to\infty}\sum_{i=1}^{n}\left(1+\frac{2i}{n}\right)^2\left(\frac{2}{n}\right) = \lim_{n\to\infty}\frac{2}{n^3}\sum_{i=1}^{n}(n+2i)^2 = \lim_{n\to\infty}\frac{2}{n^3}\left[\sum_{i=1}^{n} n^2 + 4n\sum_{i=1}^{n} i + 4\sum_{i=1}^{n} i^2\right]$

$= \lim_{n\to\infty}\frac{2}{n^3}\left[n^3 + (4n)\left(\frac{n(n+1)}{2}\right) + \frac{4(n)(n+1)(2n+1)}{6}\right]$

$= 2\lim_{n\to\infty}\left[1+2+\frac{2}{n}+\frac{4}{3}+\frac{2}{n}+\frac{2}{3n^2}\right]$

$= 2\left(1+2+\frac{4}{3}\right) = \frac{26}{3}$

33. $\lim_{n\to\infty} \sum_{i=1}^{n} \left(\frac{16i}{n^2}\right) = \lim_{n\to\infty} \frac{16}{n^2} \sum_{i=1}^{n} i = \lim_{n\to\infty} \frac{16}{n^2}\left(\frac{n(n+1)}{2}\right) = \lim_{n\to\infty} \left[8\left(\frac{n^2+n}{n^2}\right)\right] = 8 \lim_{n\to\infty} \left(1+\frac{1}{n}\right) = 8$

34. $\lim_{n\to\infty} \sum_{i=1}^{n} \left(\frac{2i}{n}\right)\left(\frac{2}{n}\right) = \lim_{n\to\infty} \frac{4}{n^2} \sum_{i=1}^{n} i = \lim_{n\to\infty} \frac{4}{n^2}\left(\frac{n(n+1)}{2}\right) = \lim_{n\to\infty} \frac{4}{2}\left(1+\frac{1}{n}\right) = 2$

35.
$$\lim_{n\to\infty} \sum_{i=1}^{n} \left(1+\frac{2i}{n}\right)^3\left(\frac{2}{n}\right) = 2 \lim_{n\to\infty} \frac{1}{n^4} \sum_{i=1}^{n} (n+2i)^3$$
$$= 2 \lim_{n\to\infty} \frac{1}{n^4} \sum_{i=1}^{n} (n^3 + 6n^2 i + 12ni^2 + 8i^3)$$
$$= 2 \lim_{n\to\infty} \frac{1}{n^4} \left[n^4 + 6n^2\left(\frac{n(n+1)}{2}\right) + 12n\left(\frac{n(n+1)(2n+1)}{6}\right) + 8\left(\frac{n(n+1)}{2}\right)^2\right]$$
$$= 2 \lim_{n\to\infty} \left(1 + 3 + \frac{3}{n} + 4 + \frac{6}{n} + \frac{2}{n^2} + 2 + \frac{4}{n} + \frac{2}{n^2}\right)$$
$$= 2 \lim_{n\to\infty} \left(10 + \frac{13}{n} + \frac{4}{n^2}\right) = 20$$

36.
$$\lim_{n\to\infty} \sum_{i=1}^{n} \left(1+\frac{i}{n}\right)\left(\frac{2}{n}\right) = 2 \lim_{n\to\infty} \frac{1}{n}\left[\sum_{i=1}^{n} 1 + \frac{1}{n}\sum_{i=1}^{n} i\right] = 2 \lim_{n\to\infty} \frac{1}{n}\left[n + \frac{1}{n}\left(\frac{n(n+1)}{2}\right)\right]$$
$$= 2 \lim_{n\to\infty} \left[1 + \frac{n^2+n}{2n^2}\right] = 2\left(1+\frac{1}{2}\right) = 3$$

37. $S(\Delta) = \sqrt{\frac{1}{4}}\left(\frac{1}{4}\right) + \sqrt{\frac{1}{2}}\left(\frac{1}{4}\right) + \sqrt{\frac{3}{4}}\left(\frac{1}{4}\right) + \sqrt{1}\left(\frac{1}{4}\right) = \frac{3+\sqrt{2}+\sqrt{3}}{8} \approx 0.768$

$s(\Delta) = 0\left(\frac{1}{4}\right) + \sqrt{\frac{1}{4}}\left(\frac{1}{4}\right) + \sqrt{\frac{1}{2}}\left(\frac{1}{4}\right) + \sqrt{\frac{3}{4}}\left(\frac{1}{4}\right) = \frac{1+\sqrt{2}+\sqrt{3}}{8} \approx 0.518$

38.
$$S(\Delta) = \left(\sqrt{\frac{1}{4}}+1\right)\frac{1}{4} + \left(\sqrt{\frac{1}{2}}+1\right)\frac{1}{4} + \left(\sqrt{\frac{3}{4}}+1\right)\frac{1}{4} + (\sqrt{1}+1)\frac{1}{4}$$
$$+ \left(\sqrt{\frac{5}{4}}+1\right)\frac{1}{4} + \left(\sqrt{\frac{3}{2}}+1\right)\frac{1}{4} + \left(\sqrt{\frac{7}{4}}+1\right)\frac{1}{4} + (\sqrt{2}+1)\frac{1}{4}$$
$$= \frac{1}{4}\left(8 + \frac{1}{2} + \frac{\sqrt{2}}{2} + \frac{\sqrt{3}}{2} + 1 + \frac{\sqrt{5}}{2} + \frac{\sqrt{6}}{2} + \frac{\sqrt{7}}{2} + \sqrt{2}\right) \approx 4.038$$
$$s(\Delta) = (0+1)\frac{1}{4} + \left(\sqrt{\frac{1}{4}}+1\right)\frac{1}{4} + \left(\sqrt{\frac{1}{2}}+1\right)\frac{1}{4} + \cdots + \left(\sqrt{\frac{7}{4}}+1\right)\frac{1}{4} \approx 3.685$$

39. $S(\Delta) = 1\left(\frac{1}{5}\right) + \frac{1}{6/5}\left(\frac{1}{5}\right) + \frac{1}{7/5}\left(\frac{1}{5}\right) + \frac{1}{8/5}\left(\frac{1}{5}\right) + \frac{1}{9/5}\left(\frac{1}{5}\right) = \frac{1}{5} + \frac{1}{6} + \frac{1}{7} + \frac{1}{8} + \frac{1}{9} \approx 0.746$

$s(\Delta) = \frac{1}{6/5}\left(\frac{1}{5}\right) + \frac{1}{7/5}\left(\frac{1}{5}\right) + \frac{1}{8/5}\left(\frac{1}{5}\right) + \frac{1}{9/5}\left(\frac{1}{5}\right) + \frac{1}{2}\left(\frac{1}{5}\right) = \frac{1}{6} + \frac{1}{7} + \frac{1}{8} + \frac{1}{9} + \frac{1}{10} \approx 0.646$

40. $S(\Delta) = \frac{1}{4-2}\left(\frac{1}{2}\right) + \frac{1}{4.5-2}\left(\frac{1}{2}\right) + \frac{1}{5-2}\left(\frac{1}{2}\right) + \frac{1}{5.5-2}\left(\frac{1}{2}\right) = \frac{1}{2}\left(\frac{1}{2} + \frac{2}{5} + \frac{1}{3} + \frac{2}{7}\right) \approx 0.760$

$s(\Delta) = \frac{1}{4.5-2}\left(\frac{1}{2}\right) + \frac{1}{5-2}\left(\frac{1}{2}\right) + \frac{1}{5.5-2}\left(\frac{1}{2}\right) + \frac{1}{6-2}\left(\frac{1}{2}\right) = \frac{1}{2}\left(\frac{2}{5} + \frac{1}{3} + \frac{2}{7} + \frac{1}{4}\right) \approx 0.635$

41. $S(\Delta) = 1\left(\frac{1}{5}\right) + \sqrt{1-\left(\frac{1}{5}\right)^2}\left(\frac{1}{5}\right) + \sqrt{1-\left(\frac{2}{5}\right)^2}\left(\frac{1}{5}\right) + \sqrt{1-\left(\frac{3}{5}\right)^2}\left(\frac{1}{5}\right) + \sqrt{1-\left(\frac{4}{5}\right)^2}\left(\frac{1}{5}\right)$

$$= \frac{1}{5}\left[1 + \frac{\sqrt{24}}{5} + \frac{\sqrt{21}}{5} + \frac{\sqrt{16}}{5} + \frac{\sqrt{9}}{5}\right] \approx 0.859$$

$$s(\Delta) = \sqrt{1-\left(\frac{1}{5}\right)^2}\left(\frac{1}{5}\right) + \sqrt{1-\left(\frac{2}{5}\right)^2}\left(\frac{1}{5}\right) + \sqrt{1-\left(\frac{3}{5}\right)^2}\left(\frac{1}{5}\right) + \sqrt{1-\left(\frac{4}{5}\right)^2}\left(\frac{1}{5}\right) + 0 \approx 0.659$$

42. $S(\Delta) = \sqrt{\frac{1}{4}+1}\left(\frac{1}{4}\right) + \sqrt{\frac{1}{2}+1}\left(\frac{1}{4}\right) + \sqrt{\frac{3}{4}+1}\left(\frac{1}{4}\right) + \sqrt{2}\left(\frac{1}{4}\right) = \frac{\sqrt{5}+\sqrt{6}+\sqrt{7}+\sqrt{8}}{8} \approx 1.270$

$$s(\Delta) = 1\left(\frac{1}{4}\right) + \sqrt{\frac{1}{4}+1}\left(\frac{1}{4}\right) + \sqrt{\frac{1}{2}+1}\left(\frac{1}{4}\right) + \sqrt{\frac{3}{4}+1}\left(\frac{1}{4}\right) = \frac{2+\sqrt{5}+\sqrt{6}+\sqrt{7}}{8} \approx 1.166$$

43. (a)

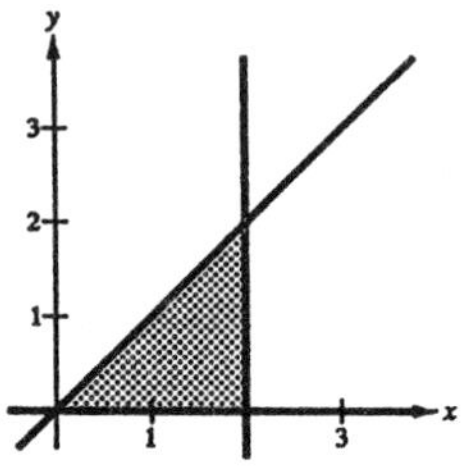

(b) $\Delta x = \frac{2-0}{n} = \frac{2}{n}$

Endpoints: $0 < 1\left(\frac{2}{n}\right) < 2\left(\frac{2}{n}\right) < \cdots < (n-1)\left(\frac{2}{n}\right) < n\left(\frac{2}{n}\right) = 2$

(c) Since $y = x$ is increasing, $f(m_i) = f(x_{i-1})$ on $[x_{i-1}, x_i]$.

$$s(n) = \sum_{i=1}^{n} f(x_i)\Delta x = \sum_{i=1}^{n} f\left(\frac{2i-2}{n}\right)\left(\frac{2}{n}\right) = \sum_{i=1}^{n}\left[(i-1)\left(\frac{2}{n}\right)\right]\left(\frac{2}{n}\right)$$

(d) $f(M_i) = f(x_i)$ on $[x_{i-1}, x_i]$

$$S(n) = \sum_{i=1}^{n} f(x_i)\Delta x = \sum_{i=1}^{n} f\left(\frac{2i}{n}\right)\frac{2}{n} = \sum_{i=1}^{n}\left[i\left(\frac{2}{n}\right)\right]\left(\frac{2}{n}\right)$$

(e)

n	5	10	50	100
$s(n)$	1.6	1.8	1.96	1.98
$S(n)$	2.4	2.2	2.04	2.02

(f) $\lim_{n\to\infty} \sum_{i=1}^{n}\left[(i-1)\left(\frac{2}{n}\right)\right]\left(\frac{2}{n}\right) = \lim_{n\to\infty}\frac{4}{n^2}\sum_{i=1}^{n}(i-1) = \lim_{n\to\infty}\frac{4}{n^2}\left[\frac{n(n+1)}{2} - n\right]$

$$= \lim_{n\to\infty}\left[\frac{2(n+1)}{n} - \frac{4}{n}\right] = 2$$

$$\lim_{n\to\infty} \sum_{i=1}^{n}\left[i\left(\frac{2}{n}\right)\right]\left(\frac{2}{n}\right) = \lim_{n\to\infty}\frac{4}{n^2}\sum_{i=1}^{n} i = \lim_{n\to\infty}\left(\frac{4}{n^2}\right)\frac{n(n+1)}{2} = \lim_{n\to\infty}\frac{2(n+1)}{n} = 2$$

44. (a)

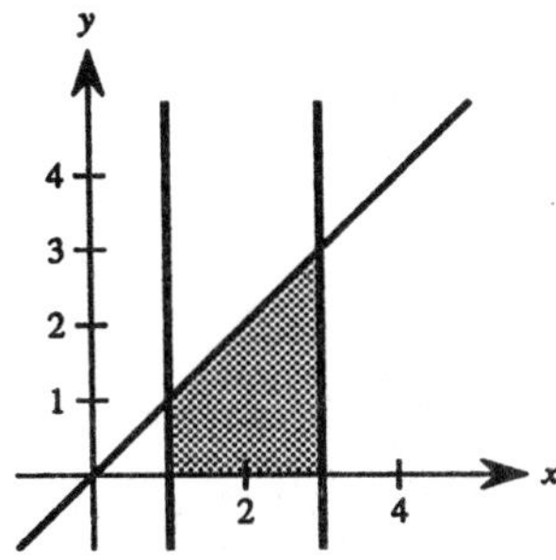

(b) $\Delta x = \dfrac{3-1}{n} = \dfrac{2}{n}$

Endpoints: $1 < 1 + \dfrac{2}{n} < 1 + \dfrac{4}{n} < \cdots < 1 + \dfrac{2n}{n} = 3$

$$1 < 1 + 1\left(\frac{2}{n}\right) < 1 + 2\left(\frac{2}{n}\right) < \cdots < 1 + (n-1)\left(\frac{2}{n}\right) < 1 + n\left(\frac{2}{n}\right)$$

(c) Since $y = x$ is increasing, $f(m_i) = f(x_{i-1})$ on $[x_{i-1},\ x_i]$.

$$s(n) = \sum_{i=1}^{n} f(x_{i-1})\Delta x = \sum_{i=1}^{n} f\left[1 + (i-1)\left(\frac{2}{n}\right)\right]\left(\frac{2}{n}\right) = \sum_{i=1}^{n}\left[1 + (i-1)\left(\frac{2}{n}\right)\right]\left(\frac{2}{n}\right)$$

(d) $f(M_i) = f(x_i)$ on $[x_{i-1},\ x_i]$

$$S(n) = \sum_{i=1}^{n} f(x_i)\Delta x = \sum_{i=1}^{n} f\left[1 + i\left(\frac{2}{n}\right)\right]\left(\frac{2}{n}\right) = \sum_{i=1}^{n}\left[1 + i\left(\frac{2}{n}\right)\right]\left(\frac{2}{n}\right)$$

(e)

n	5	10	50	100
$s(n)$	3.6	3.8	3.96	3.98
$S(n)$	4.4	4.2	4.04	4.02

(f) $$\lim_{n\to\infty} \sum_{i=1}^{n}\left[1 + (i-1)\left(\frac{2}{n}\right)\right]\left(\frac{2}{n}\right) = \lim_{n\to\infty}\left(\frac{2}{n}\right)\left[n + \frac{2}{n}\left(\frac{n(n-1)}{2} - n\right)\right]$$

$$= \lim_{n\to\infty}\left[2 + \frac{2n+2}{n} - \frac{4}{n}\right] = \lim_{n\to\infty}\left[4 - \frac{2}{n}\right] = 4$$

$$\lim_{n\to\infty} \sum_{i=1}^{n}\left[1 + i\left(\frac{2}{n}\right)\right]\left(\frac{2}{n}\right) = \lim_{n\to\infty}\frac{2}{n}\left[n + \left(\frac{2}{n}\right)\frac{n(n+1)}{2}\right]$$

$$= \lim_{n\to\infty}\left[2 + \frac{2(n+1)}{n}\right] = \lim_{n\to\infty}\left[4 + \frac{2}{n}\right] = 4$$

45. $y = -2x + 3$ on $[0,\ 1]$ $\left(\textit{Note: } \Delta x = \dfrac{1-0}{n} = \dfrac{1}{n}\right)$

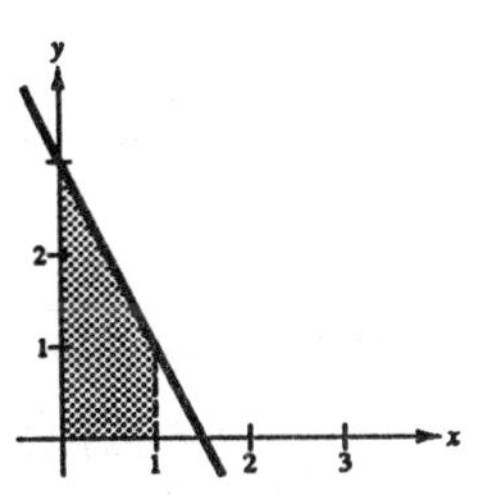

$$S(n) = \sum_{i=1}^{n} f\left(\frac{i}{n}\right)\left(\frac{1}{n}\right) = \sum_{i=1}^{n}\left[-2\left(\frac{i}{n}\right) + 3\right]\left(\frac{1}{n}\right)$$

$$= 3 - \frac{2}{n^2}\sum_{i=1}^{n} i = 3 - \frac{2(n+1)n}{2n^2} = 2 - \frac{1}{n}$$

Area $= \displaystyle\lim_{n\to\infty} S(n) = 2$

46. $y = 3x - 4$ on $[2, 5]$ $\left(\textit{Note: } \Delta x = \frac{5-2}{n} = \frac{3}{n}\right)$

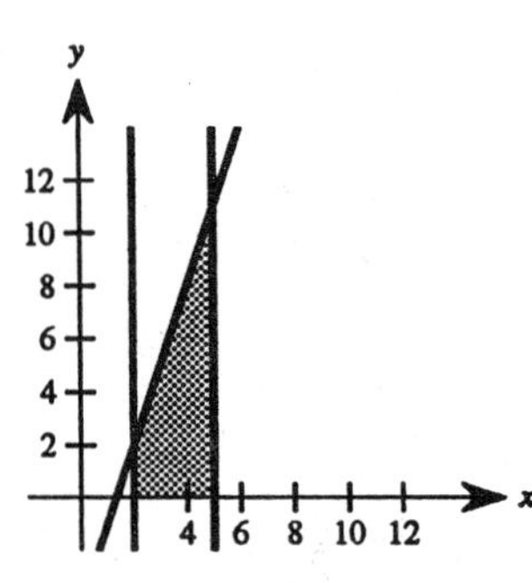

$$S(n) = \sum_{i=1}^{n} f\left(2 + \frac{3i}{n}\right)\left(\frac{3}{n}\right) = \sum_{i=1}^{n}\left[3\left(2 + \frac{3i}{n}\right) - 4\right]\left(\frac{3}{n}\right)$$

$$= 18 + 3\left(\frac{3}{n}\right)^2 \sum_{i=1}^{n} i - 12 = 6 + \frac{27}{n^2}\left(\frac{(n+1)n}{2}\right) = 6 + \frac{27}{2}\left(1 + \frac{1}{n}\right)$$

$$\text{Area} = \lim_{n\to\infty} S(n) = 6 + \frac{27}{2} = \frac{39}{2}$$

47. $y = x^2 + 2$ on $[0, 1]$ $\left(\textit{Note: } \Delta x = \frac{1}{n}\right)$

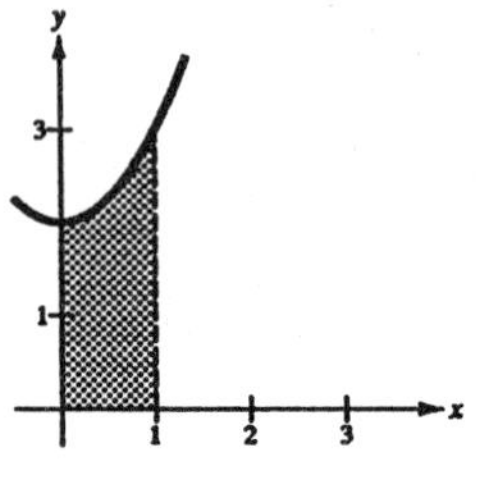

$$S(n) = \sum_{i=1}^{n} f\left(\frac{i}{n}\right)\left(\frac{1}{n}\right) = \sum_{i=1}^{n}\left[\left(\frac{i}{n}\right)^2 + 2\right]\left(\frac{1}{n}\right)$$

$$= \left[\frac{1}{n^3}\sum_{i=1}^{n} i^2\right] + 2 = \frac{n(n+1)(2n+1)}{6n^3} + 2 = \frac{1}{6}\left(2 + \frac{3}{n} + \frac{1}{n^2}\right) + 2$$

$$\text{Area} = \lim_{n\to\infty} S(n) = \frac{7}{3}$$

48. $y = 1 - x^2$ on $[-1, 1]$

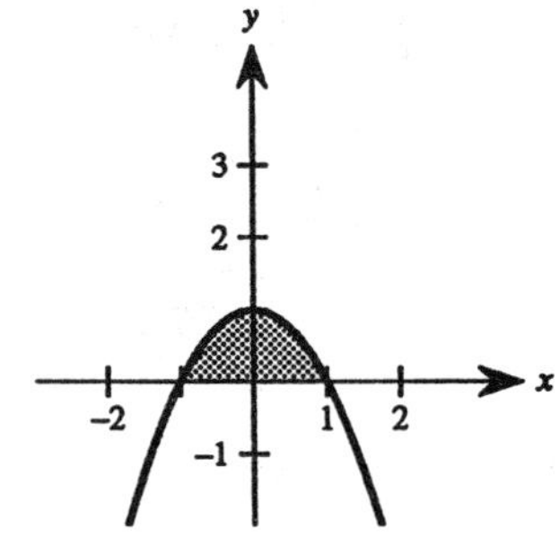

Find area of region over the interval [0, 1]. $\left(\textit{Note: } \Delta x = \frac{1}{n}\right)$

$$S(n) = \sum_{i=1}^{n} f\left(\frac{i}{n}\right)\left(\frac{1}{n}\right) = \sum_{i=1}^{n}\left[1 - \left(\frac{i}{n}\right)^2\right]\frac{1}{n}$$

$$= 1 - \frac{1}{n^3}\sum_{i=1}^{n} i^2 = 1 - \frac{n(n+1)(2n+1)}{6n^3} = 1 - \frac{1}{6}\left(2 + \frac{3}{n} + \frac{1}{n^2}\right)$$

$$\frac{1}{2}\text{Area} = \lim_{n\to\infty} S(n) = 1 - \frac{1}{3} = \frac{2}{3}$$

$$\text{Area} = \frac{4}{3}$$

49. $y = 2x^2$ on $[1, 3]$ $\left(\textit{Note: } \Delta x = \frac{3-1}{n} = \frac{2}{n}\right)$

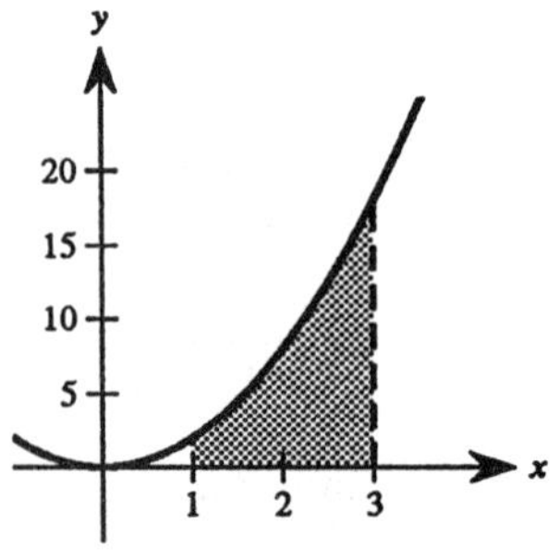

$$S(n) = \sum_{i=1}^{n} f\left(1 + \frac{2i}{n}\right)\left(\frac{2}{n}\right)$$

$$= \sum_{i=1}^{n}\left[2\left(1 + \frac{2i}{n}\right)^2\right]\left(\frac{2}{n}\right) = \frac{4}{n}\sum_{i=1}^{n}\left(1 + \frac{4i}{n} + \frac{4i^2}{n^2}\right)$$

$$= \frac{4}{n}\left[n + \frac{4}{n}\left(\frac{n(n+1)}{2}\right) + \left(\frac{4}{n^2}\right)\frac{n(n+1)(2n+1)}{6}\right]$$

$$= \frac{52}{3} + \frac{16}{n} + \frac{8}{3n^2}$$

$$\text{Area} = \lim_{n\to\infty} S(n) = \frac{52}{3}$$

50. $y = 2x^2 - x + 1$ on $[0, 2]$ $\left(\textit{Note: } \Delta x = \dfrac{2-0}{n} = \dfrac{2}{n}\right)$

$$\begin{aligned}
S(n) &= \sum_{i=1}^{n} f\left(\frac{2i}{n}\right)\left(\frac{2}{n}\right) \\
&= \sum_{i=1}^{n} \left[2\left(\frac{2i}{n}\right)^2 - \left(\frac{2i}{n}\right) + 1\right]\left(\frac{2}{n}\right) \\
&= \frac{16}{n^3}\sum_{i=1}^{n} i^2 - \frac{4}{n^2}\sum_{i=1}^{n} i + \frac{2}{n}\sum_{i=1}^{n} 1 \\
&= \frac{16}{n^3}\left[\frac{n(n+1)(2n+1)}{6}\right] - \frac{4}{n^2}\left[\frac{n(n+1)}{2}\right] + 2 \\
&= \frac{8}{3}\left(2 + \frac{3}{n} + \frac{1}{n^2}\right) - 2\left(1 + \frac{1}{n}\right) + 2
\end{aligned}$$

$$\text{Area} = \lim_{n\to\infty} S(n) = \frac{16}{3} - 2 + 2 = \frac{16}{3}$$

51. $y = 1 - x^3$ on $[0, 1]$ $\left(\textit{Note: } \Delta x = \dfrac{1-0}{n} = \dfrac{1}{n}\right)$

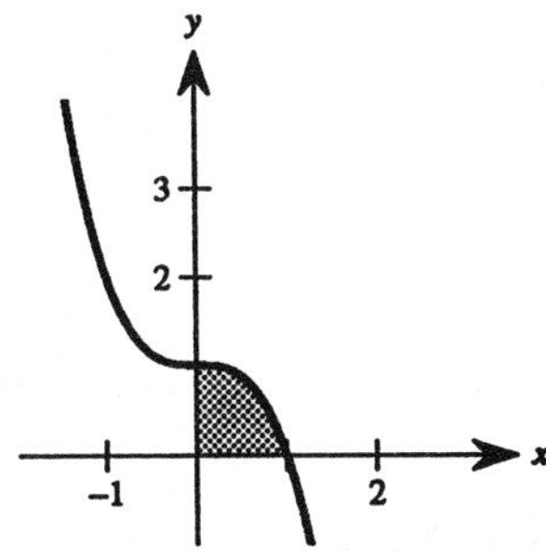

$$\begin{aligned}
S(n) &= \sum_{i=1}^{n} f\left(\frac{i}{n}\right)\left(\frac{1}{n}\right) \\
&= \sum_{i=1}^{n} \left[1 - \left(\frac{i}{n}\right)^3\right]\left(\frac{1}{n}\right) = 1 - \frac{1}{n^4}\sum_{i=1}^{n} i^3 \\
&= 1 - \frac{1}{n^4}\left[\frac{n(n+1)}{2}\right]^2 = 1 - \frac{1}{4} - \frac{1}{2n} - \frac{1}{4n^2}
\end{aligned}$$

$$\text{Area} = \lim_{n\to\infty} S(n) = 1 - \frac{1}{4} = \frac{3}{4}$$

52. $y = 2x - x^3$ on $[0, 1]$ $\left(\textit{Note: } \Delta x = \dfrac{1-0}{n} = \dfrac{1}{n}\right)$

Since y both increases and decreases on $[0, 1]$, $T(n)$ is neither an upper nor a lower sum.

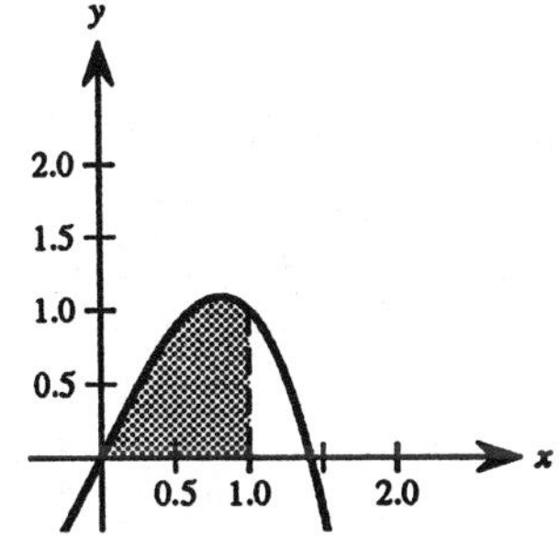

$$\begin{aligned}
T(n) &= \sum_{i=1}^{n} f\left(\frac{i}{n}\right)\left(\frac{1}{n}\right) \\
&= \sum_{i=1}^{n} \left[2\left(\frac{i}{n}\right) - \left(\frac{i}{n}\right)^3\right]\left(\frac{1}{n}\right) = \frac{2}{n^2}\sum_{i=1}^{n} i - \frac{1}{n^4}\sum_{i=1}^{n} i^3 \\
&= \frac{n(n+1)}{n^2} - \frac{1}{n^4}\left[\frac{n(n+1)}{2}\right]^2 = 1 - \frac{1}{n} - \frac{1}{4} - \frac{2}{4n} - \frac{1}{4n^2}
\end{aligned}$$

$$\text{Area} = \lim_{n\to\infty} T(n) = 1 - \frac{1}{4} = \frac{3}{4}$$

53. $y = x^2 - x^3$ on $[-1, 1]$ $\left(\textit{Note: } \Delta x = \frac{1-(-1)}{n} = \frac{2}{n}\right)$

Again, $T(n)$ is neither an upper nor a lower sum.

$$T(n) = \sum_{i=1}^{n} f\left(-1 + \frac{2i}{n}\right)\left(\frac{2}{n}\right)$$

$$= \sum_{i=1}^{n} \left[\left(-1 + \frac{2i}{n}\right)^2 - \left(-1 + \frac{2i}{n}\right)^3\right]\left(\frac{2}{n}\right)$$

$$= \sum_{i=1}^{n} \left[\left(1 - \frac{4i}{n} + \frac{4i^2}{n^2}\right) - \left(-1 + \frac{6i}{n} - \frac{12i^2}{n^2} + \frac{8i^3}{n^3}\right)\right]\left(\frac{2}{n}\right)$$

$$= \sum_{i=1}^{n} \left[2 - \frac{10i}{n} + \frac{16i^2}{n^2} - \frac{8i^3}{n^3}\right]\left(\frac{2}{n}\right)$$

$$= \frac{4}{n}\sum_{i=1}^{n} 1 - \frac{20}{n^2}\sum_{i=1}^{n} i + \frac{32}{n^3}\sum_{i=1}^{n} i^2 - \frac{16}{n^4}\sum_{i=1}^{n} i^3$$

$$= \frac{4}{n}(n) - \frac{20}{n^2} \cdot \frac{n(n+1)}{2} + \frac{32}{n^3} \cdot \frac{n(n+1)(2n+1)}{6} - \frac{16}{n^4} \cdot \frac{n^2(n+1)^2}{4}$$

$$= 4 - 10\left(1 + \frac{1}{n}\right) + \frac{16}{3}\left(2 + \frac{3}{n} + \frac{1}{n^2}\right) - 4\left(1 + \frac{2}{n} + \frac{1}{n^2}\right)$$

$$\text{Area} = \lim_{n\to\infty} T(n) = 4 - 10 + \frac{32}{3} - 4 = \frac{2}{3}$$

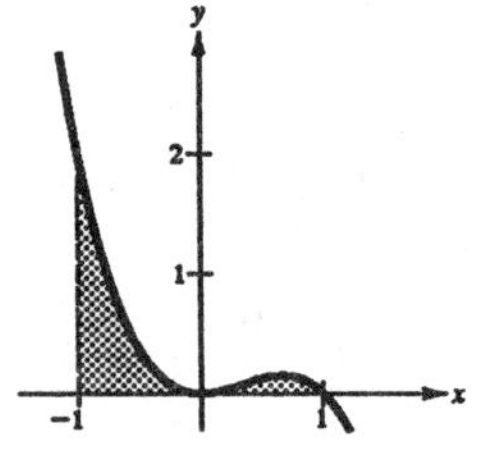

54. $y = x^2 - x^3$ on $[-1, 0]$ $\left(\textit{Note: } \Delta x = \frac{0-(-1)}{n} = \frac{1}{n}\right)$

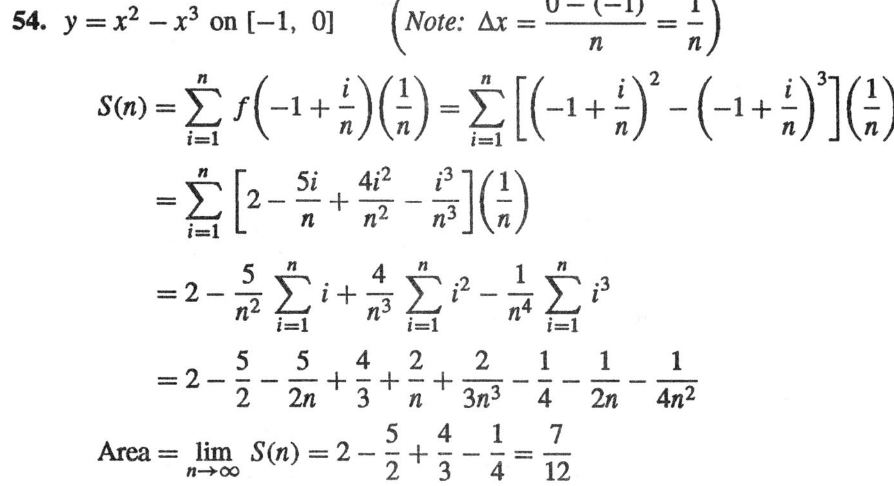

$$S(n) = \sum_{i=1}^{n} f\left(-1 + \frac{i}{n}\right)\left(\frac{1}{n}\right) = \sum_{i=1}^{n} \left[\left(-1 + \frac{i}{n}\right)^2 - \left(-1 + \frac{i}{n}\right)^3\right]\left(\frac{1}{n}\right)$$

$$= \sum_{i=1}^{n} \left[2 - \frac{5i}{n} + \frac{4i^2}{n^2} - \frac{i^3}{n^3}\right]\left(\frac{1}{n}\right)$$

$$= 2 - \frac{5}{n^2}\sum_{i=1}^{n} i + \frac{4}{n^3}\sum_{i=1}^{n} i^2 - \frac{1}{n^4}\sum_{i=1}^{n} i^3$$

$$= 2 - \frac{5}{2} - \frac{5}{2n} + \frac{4}{3} + \frac{2}{n} + \frac{2}{3n^3} - \frac{1}{4} - \frac{1}{2n} - \frac{1}{4n^2}$$

$$\text{Area} = \lim_{n\to\infty} S(n) = 2 - \frac{5}{2} + \frac{4}{3} - \frac{1}{4} = \frac{7}{12}$$

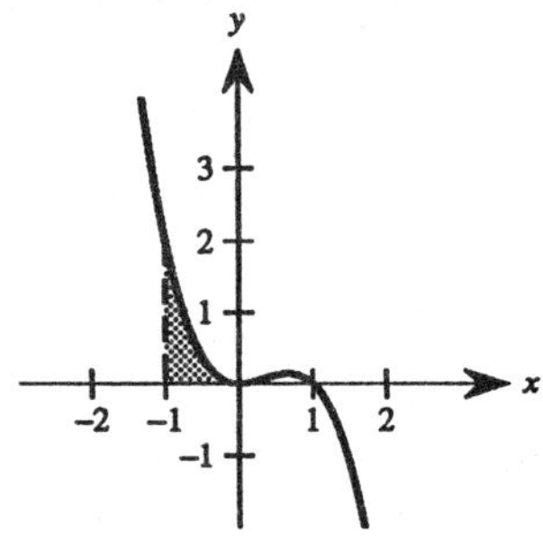

55. $f(y) = 3y$ on $[0, 2]$ $\left(\textit{Note: } \Delta y = \frac{2-0}{n} = \frac{2}{n}\right)$

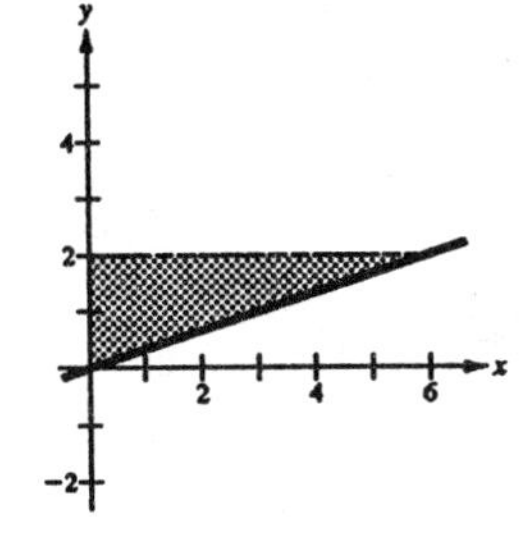

$$s(n) = \sum_{i=1}^{n} f(m_i)\Delta y = \sum_{i=1}^{n} f\left(\frac{2i}{n}\right)\left(\frac{2}{n}\right)$$

$$= \sum_{i=1}^{n} 3\left(\frac{2i}{n}\right)\left(\frac{2}{n}\right) = \frac{12}{n^2}\sum_{i=1}^{n} i$$

$$= \left(\frac{12}{n^2}\right)\cdot\frac{n(n+1)}{2} = \frac{6(n+1)}{n} = 6 + \frac{6}{n}$$

$$\text{Area} = \lim_{n\to\infty} s(n) = \lim_{n\to\infty}\left(6 + \frac{6}{n}\right) = 6$$

56. $f(y) = y^2$ on $[0, 3]$ $\left(\textit{Note: } \Delta y = \frac{3-0}{n} = \frac{3}{n}\right)$

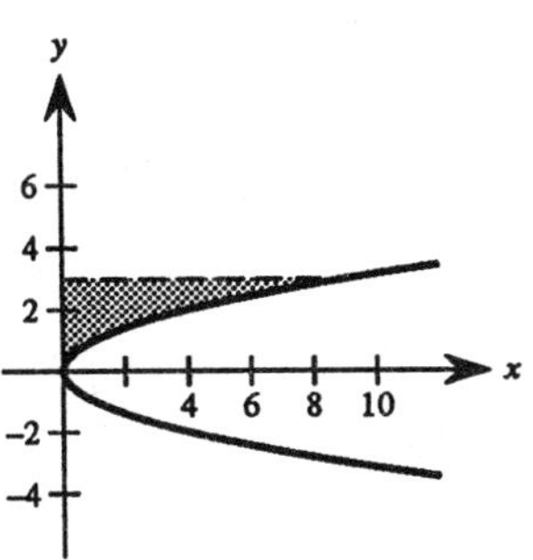

$$S(n) = \sum_{i=1}^{n} f\left(\frac{3i}{n}\right)\left(\frac{3}{n}\right) = \sum_{i=1}^{n}\left(\frac{3i}{n}\right)^2\left(\frac{3}{n}\right)$$

$$= \frac{27}{n^3}\sum_{i=1}^{n} i^2 = \frac{27}{n^3}\cdot\frac{n(n+1)(2n+1)}{6}$$

$$= \frac{9}{n^2}\left(\frac{2n^2+3n+1}{2}\right) = 9 + \frac{27}{2n} + \frac{9}{2n^2}$$

$$\text{Area} = \lim_{n\to\infty} S(n) = \lim_{n\to\infty}\left(9 + \frac{27}{2n} + \frac{9}{2n^2}\right) = 9$$

57. $f(x) = x^2 + 3, \quad 0 \le x \le 2, \quad n = 4$

$$\Delta x = \frac{1}{2}, \quad c_1 = \frac{1}{4}, \quad c_2 = \frac{3}{4}, \quad c_3 = \frac{5}{4}, \quad c_4 = \frac{7}{4}$$

$$\text{Area} \approx \sum_{i=1}^{n} f(c_i)\Delta x = \sum_{i=1}^{4}[c_i^2 + 3]\left(\frac{1}{2}\right) = \frac{1}{2}\left[\left(\frac{1}{16}+3\right) + \left(\frac{9}{16}+3\right) + \left(\frac{25}{16}+3\right) + \left(\frac{49}{16}+3\right)\right] = \frac{69}{8}$$

58. $f(x) = x^2 + 4x, \quad 0 \le x \le 4, \quad n = 4$

$$\Delta x = 1, \quad c_1 = \frac{1}{2}, \quad c_2 = \frac{3}{2}, \quad c_3 = \frac{5}{2}, \quad c_4 = \frac{7}{2}$$

$$\text{Area} \approx \sum_{i=1}^{n} f(c_i)\Delta x = \sum_{i=1}^{4}[c_i^2 + 4c_i](1) = \left[\left(\frac{1}{4}+2\right) + \left(\frac{9}{4}+6\right) + \left(\frac{25}{4}+10\right) + \left(\frac{49}{4}+14\right)\right] = 53$$

59. $f(x) = \sqrt{x-1}, \quad 1 \le x \le 2, \quad n = 4$

$$\Delta x = \frac{1}{4}, \quad c_1 = \frac{9}{8}, \quad c_2 = \frac{11}{8}, \quad c_3 = \frac{13}{8}, \quad c_4 = \frac{15}{8}$$

$$\text{Area} \approx \sum_{i=1}^{n} f(c_i)\Delta x = \sum_{i=1}^{4}\sqrt{c_i - 1}\left(\frac{1}{4}\right) = \frac{1}{4}\left[\sqrt{\frac{9}{8}-1} + \sqrt{\frac{11}{8}-1} + \sqrt{\frac{13}{8}-1} + \sqrt{\frac{15}{8}-1}\right] \approx 0.6730$$

60. $f(x) = \dfrac{1}{x^2+1}, \quad 0 \le x \le 2, \quad n = 4$

$\Delta x = \dfrac{1}{2}, \quad c_1 = \dfrac{1}{4}, \quad c_2 = \dfrac{3}{4}, \quad c_3 = \dfrac{5}{4}, \quad c_4 = \dfrac{7}{4}$

$$\text{Area} \approx \sum_{i=1}^{n} f(c_i)\Delta x = \sum_{i=1}^{4} \frac{1}{c_i{}^2+1}\left(\frac{1}{2}\right)$$

$$= \frac{1}{2}\left[\frac{1}{(1/16)+1} + \frac{1}{(9/16)+1} + \frac{1}{(25/16)+1} + \frac{1}{(49/16)+1}\right] = 8\left[\frac{1}{17} + \frac{1}{25} + \frac{1}{41} + \frac{1}{65}\right] \approx 1.1088$$

61. $f(x) = \sqrt{x}$ on $[0, 4]$.

n	4	8	12	16	20
Approximate area	5.3838	5.3523	5.3439	5.3403	5.3384

62. $f(x) = \dfrac{8}{x^2+1}$ on $[2, 6]$.

n	4	8	12	16	20
Approximate area	2.3397	2.3755	2.3824	2.3848	2.3860

63. $N = \dfrac{10(5+3t)}{1+0.04t}$

$N(5) \approx 167, \quad N(10) \approx 250, \quad N(25) \approx 400$

$$\lim_{t\to\infty} \frac{10(5+3t)}{1+0.04t} = \lim_{t\to\infty} \frac{30t+50}{0.04t+1} = \frac{30}{0.04} = 750$$

64. $\displaystyle\lim_{n\to\infty} \frac{b+\theta a(n-1)}{1+\theta(n-1)} = \frac{\theta a}{\theta} = a$

Section 5.3 Riemann Sums and the Definite Integral

1. $\displaystyle\int_0^5 3\,dx$

2. $\displaystyle\int_0^2 (4-2x)\,dx$

3. $\displaystyle\int_{-4}^4 (4-|x|)\,dx$

4. $\displaystyle\int_0^2 x^2\,dx$

5. $\displaystyle\int_{-2}^2 (4-x^2)\,dx$

6. $\displaystyle\int_0^2 (y-2)^2\,dy$

7. $\displaystyle\int_0^2 y^3\,dy$

8. $\displaystyle\int_{-1}^1 (x^2+1)^3\,dx$

9. $\displaystyle\int_0^2 \sqrt{x+1}\,dx$

10. $\displaystyle\int_{-1}^1 \frac{1}{x^2+1}\,dx$

11. Rectangle

$A = bh = 3(4)$

$A = \displaystyle\int_0^3 4\,dx = 12$

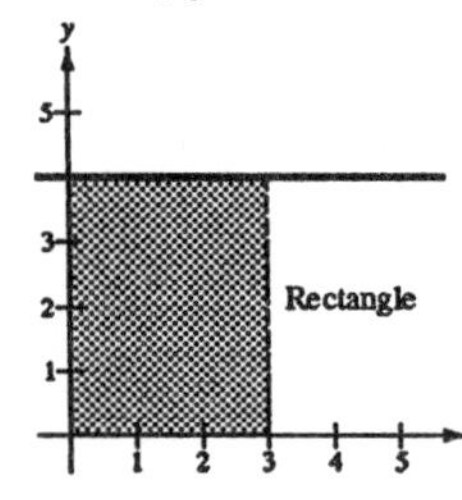

12. Rectangle

$A = bh = 2(4)(a)$

$A = \displaystyle\int_{-a}^a 4\,dx = 8a$

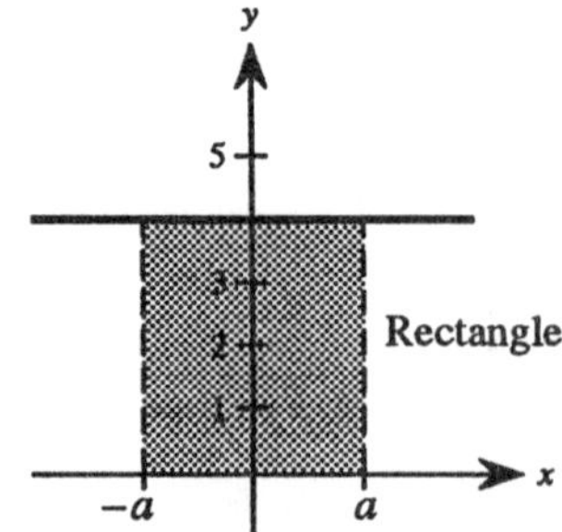

13. Triangle

$A = \frac{1}{2}bh = \frac{1}{2}(4)(4)$

$A = \displaystyle\int_0^4 x\,dx = 8$

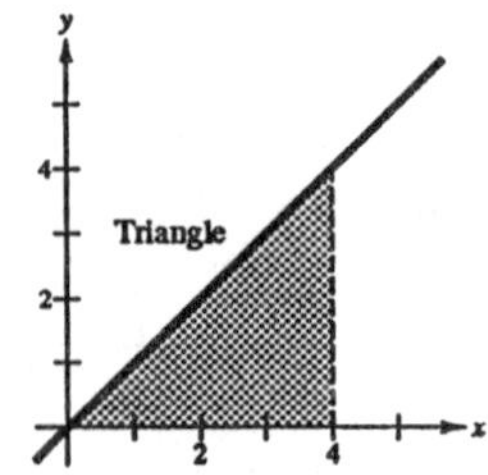

14. Triangle

$$A = \frac{1}{2}bh = \frac{1}{2}(4)(2)$$

$$A = \int_0^4 \frac{x}{2}\,dx = 4$$

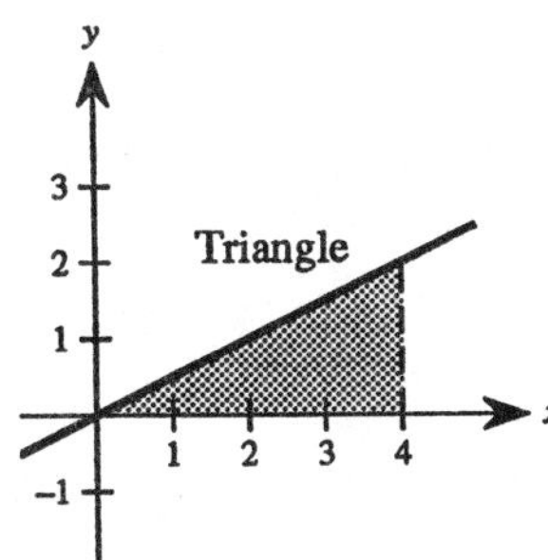

15. Trapezoid

$$A = \frac{b_1 + b_2}{2}h = 2\left(\frac{5+9}{2}\right)$$

$$A = \int_0^2 (2x+5)\,dx = 14$$

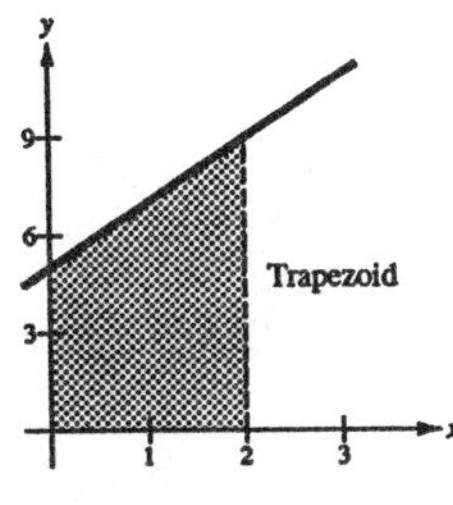

16. Triangle

$$A = \frac{1}{2}bh = \frac{1}{2}(5)(5)$$

$$A = \int_0^5 (5-x)\,dx = \frac{25}{2}$$

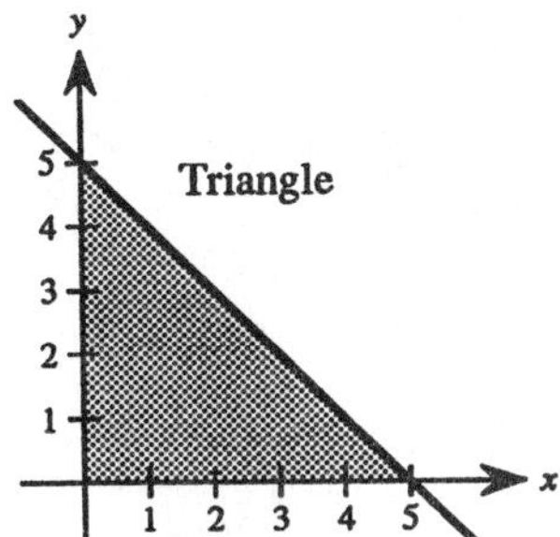

17. Triangle

$$a = \frac{1}{2}bh = \frac{1}{2}(2)(1)$$

$$A = \int_{-1}^{1} (1-|x|)\,dx = 1$$

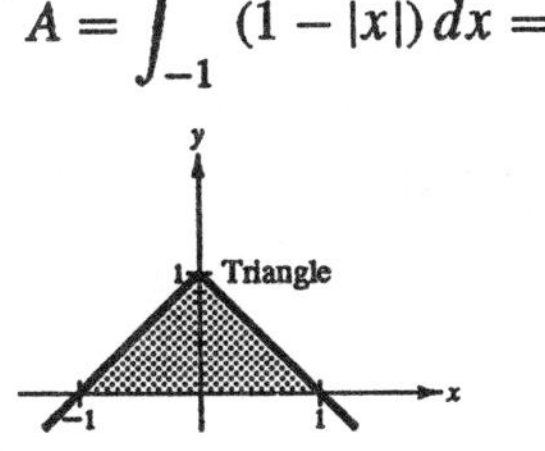

18. Triangle

$$A = \frac{1}{2}bh = \frac{1}{2}(2a)a$$

$$A = \int_{-a}^{a} (a-|x|)\,dx = a^2$$

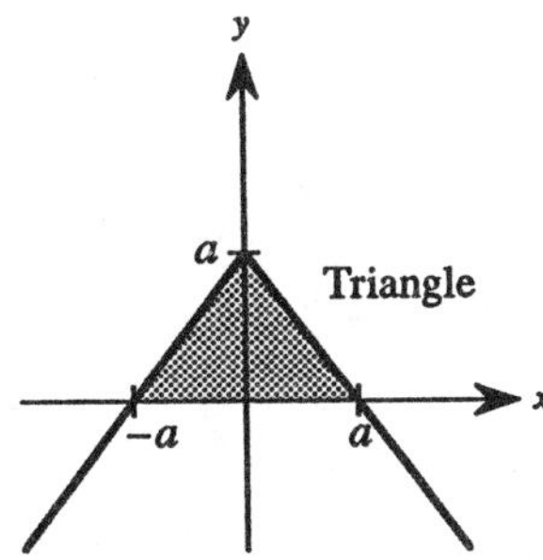

19. Semicircle

$$A = \frac{1}{2}\pi r^2 = \frac{1}{2}\pi(3)^2$$

$$A = \int_{-3}^{3} \sqrt{9-x^2}\,dx = \frac{9\pi}{2}$$

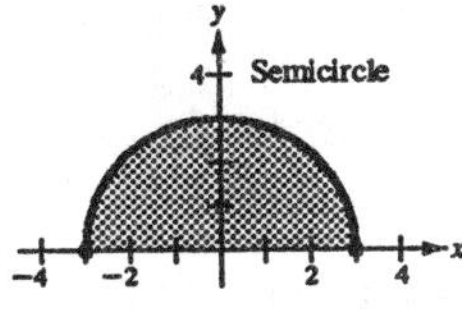

20. Semicircle

$$A = \tfrac{1}{2}\pi r^2$$

$$A = \int_{-r}^{r} \sqrt{r^2-x^2}\,dx = \tfrac{1}{2}\pi r^2$$

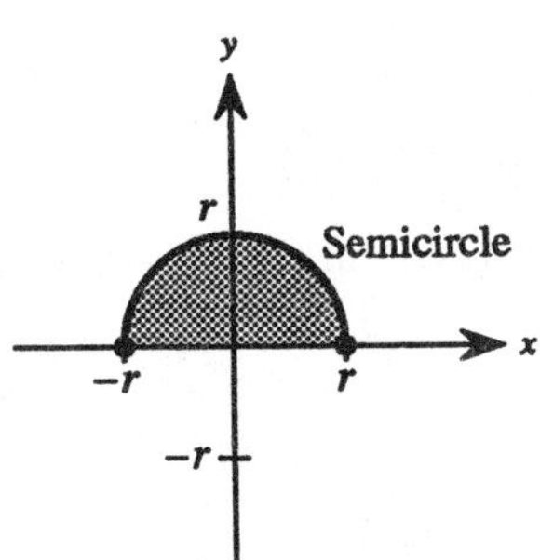

21. (a) $\displaystyle\int_0^7 f(x)\,dx = \int_0^5 f(x)\,dx + \int_5^7 f(x)\,dx = 10 + 3 = 13$

(b) $\displaystyle\int_5^0 f(x)\,dx = -\int_0^5 f(x)\,dx = -10$

(c) $\displaystyle\int_5^5 f(x)\,dx = 0$

(d) $\displaystyle\int_0^5 3f(x)\,dx = 3\int_0^5 f(x)\,dx = 3(10) = 30$

22. (a) $\int_0^6 f(x)\,dx = \int_0^3 f(x)\,dx + \int_3^6 f(x)\,dx = 4 + (-1) = 3$

(b) $\int_6^3 f(x)\,dx = -\int_3^6 f(x)\,dx = -(-1) = 1$

(c) $\int_4^4 f(x)\,dx = 0$

(d) $\int_3^6 -5f(x)\,dx = -5\int_3^6 f(x)\,dx = -5(-1) = 5$

23. (a) $\int_2^6 [f(x) + g(x)]\,dx = \int_2^6 f(x)\,dx + \int_2^6 g(x)\,dx = 10 + (-2) = 8$

(b) $\int_2^6 [g(x) - f(x)]\,dx = \int_2^6 g(x)\,dx - \int_2^6 f(x)\,dx = -2 - 10 = -12$

(c) $\int_2^6 2g(x)\,dx = 2\int_2^6 g(x)\,dx = 2(-2) = -4$

(d) $\int_2^6 3f(x)\,dx = 3\int_2^6 f(x)\,dx = 3(10) = 30$

24. (a) $\int_{-1}^0 f(x)\,dx = \int_{-1}^1 f(x)\,dx - \int_0^1 f(x)\,dx = 0 - 5 = -5$

(b) $\int_0^1 f(x)\,dx - \int_{-1}^0 f(x)\,dx = 5 - (-5) = 10$

(c) $\int_{-1}^1 3f(x)\,dx = 3\int_{-1}^1 f(x)\,dx = 3(0) = 0$

(d) $\int_0^1 3f(x)\,dx = 3\int_0^1 f(x)\,dx = 3(5) = 15$

25. $y = 6$ on $[4, 10]$ $\left(\textit{Note: } \Delta x = \dfrac{10-4}{n} = \dfrac{6}{n},\ \Delta x \to 0 \text{ as } n \to \infty\right)$

$$S(n) = \sum_{i=1}^{n} f\left(4 + \frac{6i}{n}\right)\left(\frac{6}{n}\right) = \sum_{i=1}^{n} 6\left(\frac{6}{n}\right) = \sum_{i=1}^{n} \frac{36}{n} = 36$$

$$\int_4^{10} 6\,dx = \lim_{\Delta x \to \infty} S(n) = \lim_{n\to\infty} 36 = 36$$

26. $y = x$ on $[-2, 3]$ $\left(\textit{Note: } \Delta x = \dfrac{3-(-2)}{n} = \dfrac{5}{n},\ \Delta x \to 0 \text{ as } n \to \infty\right)$

$$S(n) = \sum_{i=1}^{n} f\left(-2 + \frac{5i}{n}\right)\left(\frac{5}{n}\right) = \sum_{i=1}^{n} \left(-2 + \frac{5i}{n}\right)\left(\frac{5}{n}\right) = -10 + \frac{25}{n^2}\sum_{i=1}^{n} i$$

$$= -10 + \left(\frac{25}{n^2}\right)\frac{n(n+1)}{2} = -10 + \frac{25}{2}\left(1 + \frac{1}{n}\right) = \frac{5}{2} + \frac{25}{2n}$$

$$\int_{-2}^3 x\,dx = \lim_{\Delta x \to 0} S(n) = \lim_{n\to\infty}\left(\frac{5}{2} + \frac{25}{2n}\right) = \frac{5}{2}$$

27. $y = x^3$ on $[-1, 1]$ $\left(\textit{Note: } \Delta x = \dfrac{1-(-1)}{n} = \dfrac{2}{n},\ \Delta x \to 0 \text{ as } n \to \infty\right)$

$$S(n) = \sum_{i=1}^{n} f\left(-1 + \frac{2i}{n}\right)\left(\frac{2}{n}\right) = \sum_{i=1}^{n}\left(-1 + \frac{2i}{n}\right)^3\left(\frac{2}{n}\right) = \sum_{i=1}^{n}\left[-1 + \frac{6i}{n} - \frac{12i^2}{n^2} + \frac{8i^3}{n^3}\right]\left(\frac{2}{n}\right)$$

$$= -2 + \frac{12}{n^2}\sum_{i=1}^{n} i - \frac{24}{n^3}\sum_{i=1}^{n} i^2 + \frac{16}{n^4}\sum_{i=1}^{n} i^3$$

$$= -2 + 6\left(1 + \frac{1}{n}\right) - 4\left(2 + \frac{3}{n} + \frac{1}{n^2}\right) + 4\left(1 + \frac{2}{n} + \frac{1}{n^2}\right) = \frac{2}{n}$$

$$\int_{-1}^{1} x^3\,dx = \lim_{\Delta x \to 0} S(n) = \lim_{n \to \infty} \frac{2}{n} = 0$$

28. $y = x^3$ on $[0, 1]$ $\left(\textit{Note: } \Delta x = \dfrac{1-0}{n} = \dfrac{1}{n},\ \Delta x \to 0 \text{ as } n \to \infty\right)$

$$S(n) = \sum_{i=1}^{n} f\left(\frac{i}{n}\right)\left(\frac{1}{n}\right) = \sum_{i=1}^{n}\left(\frac{i}{n}\right)^3\left(\frac{1}{n}\right) = \frac{1}{n^4}\sum_{i=1}^{n} i^3 = \frac{1}{n^4}\left[\frac{n(n+1)}{2}\right]^2 = \frac{1}{4}\left(1 + \frac{2}{n} + \frac{1}{n^2}\right)$$

$$\int_{0}^{1} x^3\,dx = \lim_{\Delta x \to 0} S(n) = \frac{1}{4}\lim_{n \to \infty}\left(1 + \frac{2}{n} + \frac{1}{n^2}\right) = \frac{1}{4}$$

29. $y = x^2 + 1$ on $[1, 2]$ $\left(\textit{Note: } \Delta x = \dfrac{2-1}{n} = \dfrac{1}{n},\ \Delta x \to 0 \text{ as } n \to \infty\right)$

$$S(n) = \sum_{i=1}^{n} f\left(1 + \frac{i}{n}\right)\left(\frac{1}{n}\right) = \sum_{i=1}^{n}\left[\left(1 + \frac{i}{n}\right)^2 + 1\right]\left(\frac{1}{n}\right) = \sum_{i=1}^{n}\left[1 + \frac{2i}{n} + \frac{i^2}{n^2} + 1\right]\left(\frac{1}{n}\right)$$

$$= 2 + \frac{2}{n^2}\sum_{i=1}^{n} i + \frac{1}{n^3}\sum_{i=1}^{n} i^2 = 2 + \left(1 + \frac{1}{n}\right) + \frac{1}{6}\left(2 + \frac{3}{n} + \frac{1}{n^2}\right) = \frac{10}{3} + \frac{3}{2n} + \frac{1}{6n^2}$$

$$\int_{1}^{2} (x^2 + 1)\,dx = \lim_{\Delta x \to 0} S(n) = \lim_{n \to \infty}\left(\frac{10}{3} + \frac{3}{2n} + \frac{1}{6n^2}\right) = \frac{10}{3}$$

30. $y = 4x^2$ on $[1, 2]$ $\left(\textit{Note: } \Delta x = \dfrac{2-1}{n} = \dfrac{1}{n},\ \Delta x \to 0 \text{ as } n \to \infty\right)$

$$S(n) = \sum_{i=1}^{n} f\left(1 + \frac{i}{n}\right)\left(\frac{1}{n}\right) = \sum_{i=1}^{n}\left[4\left(1 + \frac{i}{n}\right)^2\right]\left(\frac{1}{n}\right) = 4\sum_{i=1}^{n}\left[1 + \frac{2i}{n} + \frac{i^2}{n^2}\right]\left(\frac{1}{n}\right)$$

$$= 4\left[1 + \frac{1}{n^2}\sum_{i=1}^{n} i + \frac{1}{n^3}\sum_{i=1}^{n} i^2\right] = 4\left[1 + \left(\frac{2}{n^2}\right)\frac{n(n+1)}{2} + \left(\frac{1}{n^3}\right)\frac{n(n+1)(2n+1)}{6}\right]$$

$$= 4\left[1 + \left(1 + \frac{1}{n}\right) + \frac{1}{6}\left(2 + \frac{3}{n} + \frac{1}{n^2}\right)\right] = 4\left(\frac{7}{3} + \frac{3}{2n} + \frac{1}{6n^2}\right)$$

$$\int_{1}^{2} 4x^2\,dx = \lim_{\Delta x \to 0} S(n) = 4\lim_{n \to \infty}\left(\frac{7}{3} + \frac{3}{2n} + \frac{1}{6n^2}\right) = \frac{28}{3}$$

31. $f(x) = \sqrt{x}, \quad y = 0, \quad x = 0, \quad x = 2, \quad c_i = \dfrac{2i^2}{n^2}$

$$\Delta x_i = \frac{2i^2}{n^2} - \frac{2(i-1)^2}{n^2} = \frac{2(2i-1)}{n^2}$$

$$\sum_{i=1}^{n} f(c_i)\Delta x_i = \sum_{i=1}^{n} \sqrt{\frac{2i^2}{n^2}}\left[\frac{2(2i-1)}{n^2}\right] = \frac{2\sqrt{2}}{n^3}\sum_{i=1}^{n}(2i^2 - i)$$

$$= \frac{2\sqrt{2}}{n^3}\left[2\left(\frac{n(n+1)(2n+1)}{6}\right) - \frac{n(n+1)}{2}\right]$$

$$= \frac{2\sqrt{2}}{n^3}\left[\frac{4n^3 + 3n^2 - n}{6}\right] = \sqrt{2}\left[\frac{4}{3} + \frac{1}{n} - \frac{2}{n^2}\right]$$

$$\lim_{n\to\infty}\sum_{i=1}^{n} f(c_i)\Delta x_i = \lim_{n\to\infty}\sqrt{2}\left[\frac{4}{3} + \frac{1}{n} - \frac{2}{n^2}\right] = \frac{4\sqrt{2}}{3}$$

32. $f(x) = \sqrt[3]{x}, \quad y = 0, \quad x = 0, \quad x = 1, \quad c_i = \dfrac{i^3}{n^3}$

$$\Delta x_i = \frac{i^3}{n^3} - \frac{(i-1)^3}{n^3} = \frac{3i^2 - 3i + 1}{n^3}$$

$$\sum_{i=1}^{n} f(c_i)\Delta x_i = \sum_{i=1}^{n} \sqrt[3]{\frac{i^3}{n^3}}\left[\frac{3i^2 - 3i + 1}{n^3}\right] = \frac{1}{n^4}\sum_{i=1}^{n}(3i^3 - 3i^2 + i)$$

$$= \frac{1}{n^4}\left[3\left(\frac{n^2(n+1)^2}{4}\right) - 3\left(\frac{n(n+1)(2n+1)}{6}\right) + \frac{n(n+1)}{2}\right]$$

$$= \frac{1}{n^4}\left[\frac{3n^4 + 6n^3 + 3n^2}{4} - \frac{2n^3 + 3n^2 + n}{2} + \frac{n^2 + n}{2}\right]$$

$$= \frac{1}{n^4}\left[\frac{3n^4}{4} + \frac{n^3}{2} - \frac{n^2}{4}\right] = \left[\frac{3}{4} + \frac{1}{2n} - \frac{1}{4n^2}\right]$$

$$\lim_{n\to\infty}\sum_{i=1}^{n} f(c_i)\Delta x_i = \lim_{n\to\infty}\left[\frac{3}{4} + \frac{1}{2n} - \frac{1}{4n^2}\right] = \frac{3}{4}$$

33. $\displaystyle\lim_{\|\Delta\|\to 0}\sum_{i=1}^{n}(3c_i + 10)\Delta x_i = \int_{-1}^{5}(3x + 10)\,dx$ on the interval $[-1, 5]$.

34. $\displaystyle\lim_{\|\Delta\|\to 0}\sum_{i=1}^{n} 6c_i(4 - c_i)^2\Delta x_i = \int_{0}^{4} 6x(4 - x)^2\,dx$ on the interval $[0, 4]$.

35. $\displaystyle\int_0^3 x\sqrt{3 - x}\,dx$

n	4	8	12	16	20
$L(n)$	3.6830	3.9956	4.0707	4.1016	4.1177
$M(n)$	4.3082	4.2076	4.1838	4.1740	4.1690
$R(n)$	3.6830	3.9956	4.0707	4.1016	4.1177

36. $\int_0^3 \frac{5}{x^2+1}\,dx$

n	4	8	12	16	20
$L(n)$	7.9224	7.0855	6.8062	6.6662	6.5822
$M(n)$	6.2485	6.2470	6.2460	6.2457	6.2455
$R(n)$	4.5474	5.3980	5.6812	5.8225	5.9072

37. $f(x) = \begin{cases} 1, & x \text{ is rational.} \\ 0, & x \text{ is irrational.} \end{cases}$

is not integrable on the interval [0, 1]. As $\|\Delta\| \to 0$, $f(c_i) = 1$ or $f(c_i) = 0$ in each subinterval since there are an infinite number of both rational and irrational numbers in any interval, no matter how small.

38. $f(x) = 1/\sqrt[3]{x}$ is integrable on $[-1, 1]$, but is not continuous on $[-1, 1]$. There is a discontinuity at $x = 0$.

$$\int_{-1}^{1} \frac{1}{\sqrt{x}}\,dx = \int_{-1}^{0} \frac{1}{\sqrt[3]{x}}\,dx + \int_0^1 \frac{1}{\sqrt[3]{x}}\,dx = \left[\frac{3}{2}x^{2/3}\right]_{-1}^{0} + \left[\frac{3}{2}x^{2/3}\right]_0^1 = \left(0 - \frac{3}{2}\right) + \left(\frac{3}{2} - 0\right) = 0$$

Section 5.4 The Fundamental Theorem of Calculus

1. $\int_0^1 2x\,dx = x^2\Big]_0^1 = 1 - 0 = 1$

2. $\int_2^7 3\,dv = 3v\Big]_2^7 = 3(7) - 3(2) = 15$

3. $\int_{-1}^{0} (x-2)\,dx = \left[\frac{x^2}{2} - 2x\right]_{-1}^{0} = 0 - \left(\frac{1}{2} + 2\right) = -\frac{5}{2}$

4. $\int_2^5 (-3v+4)\,dv = \left[-\frac{3}{2}v^2 + 4v\right]_2^5 = \left(-\frac{75}{2} + 20\right) - (-6+8) = -\frac{39}{2}$

5. $\int_{-1}^{1} (t^2 - 2)\,dt = \left[\frac{t^3}{3} - 2t\right]_{-1}^{1} = \left(\frac{1}{3} - 2\right) - \left(-\frac{1}{3} + 2\right) = -\frac{10}{3}$

6. $\int_0^3 (3x^2 + x - 2)\,dx = \left[x^3 + \frac{x^2}{2} - 2x\right]_0^3 = \left(27 + \frac{9}{2} - 6\right) - 0 = \frac{51}{2}$

7. Let $u = 2t - 1$, $u' = 2$.

$$\int_0^1 (2t-1)^2\,dt = \frac{1}{2}\int_0^1 (2t-1)^2(2)\,dt = \frac{1}{6}(2t-1)^3\Big]_0^1 = \frac{1}{6}[1 - (-1)] = \frac{1}{3}$$

8. $\int_{-1}^{1} (t^3 - 9t)\,dt = \left[\frac{1}{4}t^4 - \frac{9}{2}t^2\right]_{-1}^{1} = \left(\frac{1}{4} - \frac{9}{2}\right) - \left(\frac{1}{4} - \frac{9}{2}\right) = 0$

9. $\int_1^2 \left(\frac{3}{x^2} - 1\right)dx = \left[-\frac{3}{x} - x\right]_1^2 = \left(-\frac{3}{2} - 2\right) - (-3-1) = \frac{1}{2}$

10. $\int_0^1 (3x^3 - 9x + 7)\,dx = \left[\frac{3x^4}{4} - \frac{9}{2}x^2 + 7x\right]_0^1 = \left(\frac{3}{4} - \frac{9}{2} + 7\right) - 0 = \frac{13}{4}$

11. $\int_1^2 (5x^4+5)\,dx = \left[x^5+5x\right]_1^2 = (32+10)-(1+5) = 36$

12. $\int_{-3}^3 v^{1/3}\,dv = \frac{3}{4}v^{4/3}\Big]_{-3}^3 = \frac{3}{4}\left[(\sqrt[3]{3})^4 - (\sqrt[3]{-3})^4\right] = 0$

13. $\int_{-1}^1 (\sqrt[3]{t}-2)\,dt = \left[\frac{3}{4}t^{4/3}-2t\right]_{-1}^1 = \left(\frac{3}{4}-2\right)-\left(\frac{3}{4}+2\right) = -4$

14. $\int_1^2 \sqrt{\frac{2}{x}}\,dx = \int_1^2 \sqrt{2}x^{-1/2}\,dx = \left[\sqrt{2}(2x^{1/2})\right]_1^2 = 4-2\sqrt{2}$

15. $\int_1^4 \frac{u-2}{\sqrt{u}}\,du = \int_1^4 (u^{1/2}-2u^{-1/2})\,du = \left[\frac{2}{3}u^{3/2}-4u^{1/2}\right]_1^4 = \left[\frac{2}{3}(\sqrt{4})^3-4\sqrt{4}\right]-\left[\frac{2}{3}-4\right] = \frac{2}{3}$

16. $\int_{-2}^{-1} \left(u-\frac{1}{u^2}\right)du = \left[\frac{u^2}{2}+\frac{1}{u}\right]_{-2}^{-1} = \left(\frac{1}{2}-1\right)-\left(2-\frac{1}{2}\right) = -2$

17. $\int_0^1 \frac{x-\sqrt{x}}{3}\,dx = \frac{1}{3}\int_0^1 (x-x^{1/2})\,dx = \frac{1}{3}\left(\frac{x^2}{2}-\frac{2}{3}x^{3/2}\right)\Big]_0^1 = \frac{1}{3}\left(\frac{1}{2}-\frac{2}{3}\right) = -\frac{1}{18}$

18. $\int_0^2 (2-t)\sqrt{t}\,dt = \int_0^2 (2t^{1/2}-t^{3/2})\,dt = \left[\frac{4}{3}t^{3/2}-\frac{2}{5}t^{5/2}\right]_0^2 = \left[\frac{t\sqrt{t}}{15}(20-6t)\right]_0^2 = \frac{2\sqrt{2}}{15}(20-12) = \frac{16\sqrt{2}}{15}$

19. $\int_{-1}^0 (t^{1/3}-t^{2/3})\,dt = \left[\frac{3}{4}t^{4/3}-\frac{3}{5}t^{5/3}\right]_{-1}^0 = 0-\left(\frac{3}{4}+\frac{3}{5}\right) = -\frac{27}{20}$

20. $\int_{-8}^{-1} \frac{x-x^2}{2\sqrt[3]{x}}\,dx = \frac{1}{2}\int_{-8}^{-1} (x^{2/3}-x^{5/3})\,dx$

$= \frac{1}{2}\left(\frac{3}{5}x^{5/3}-\frac{3}{8}x^{8/3}\right)\Big]_{-8}^{-1} = \left[\frac{x^{5/3}}{80}(24-15x)\right]_{-8}^{-1} = -\frac{1}{80}(39)+\frac{32}{80}(144) = \frac{4569}{80}$

21. $\int_{-1}^1 |x|\,dx = \int_{-1}^0 (-x)\,dx + \int_0^1 x\,dx = \left[-\frac{x^2}{2}\right]_{-1}^0 + \left[\frac{x^2}{2}\right]_0^1 = \frac{1}{2}+\frac{1}{2} = 1$

22. $\int_0^3 |2x-3|\,dx = \int_0^{3/2} (3-2x)\,dx + \int_{3/2}^3 (2x-3)\,dx$

$= \left[3x-x^2\right]_0^{3/2} + \left[x^2-3x\right]_{3/2}^3 = \left(\frac{9}{2}-\frac{9}{4}\right)-0+(9-9)-\left(\frac{9}{4}-\frac{9}{2}\right) = 2\left(\frac{9}{2}-\frac{9}{4}\right) = \frac{9}{2}$

23. $\int_0^4 |x^2-4x+3|\,dx = \int_0^1 (x^2-4x+3)\,dx - \int_1^3 (x^2-4x+3)\,dx + \int_3^4 (x^2-4x+3)\,dx$

$= \left[\frac{x^3}{3}-2x^2+3x\right]_0^1 - \left[\frac{x^3}{3}-2x^2+3x\right]_1^3 + \left[\frac{x^3}{3}-2x^2+3x\right]_3^4$

$= \left(\frac{1}{3}-2+3\right)+\left(\frac{1}{3}-2+3\right)+\left(\frac{64}{3}-32+12\right)-(9-18+9)$

$= \frac{8}{3}+\frac{64}{3}-20 = 4$

24. $\int_{-1}^1 |x^3|\,dx = -\int_{-1}^0 x^3\,dx + \int_0^1 x^3\,dx = \left[-\frac{x^4}{4}\right]_{-1}^0 + \left[\frac{x^4}{4}\right]_0^1 = \frac{1}{4}+\frac{1}{4} = \frac{1}{2}$

25. $A = \int_0^1 (x - x^2)\,dx = \left[\frac{x^2}{2} - \frac{x^3}{3}\right]_0^1 = \frac{1}{6}$

26. $A = -\int_{-1}^3 (x^2 - 2x - 3)\,dx = -\left[\frac{x^3}{3} - x^2 - 3x\right]_{-1}^3 = -(9 - 9 - 9) + \left(-\frac{1}{3} - 1 + 3\right) = \frac{32}{3}$

27. $A = \int_{-1}^1 (1 - x^4)\,dx = \left[x - \frac{1}{5}x^5\right]_{-1}^1 = \frac{8}{5}$

28. $A = \int_1^2 \frac{1}{x^2}\,dx = -\frac{1}{x}\Big]_1^2 = -\frac{1}{2} + 1 = \frac{1}{2}$

29. Let $u = 2x$, $u' = 2$.

$$A = \int_0^4 \sqrt[3]{2x}\,dx = \frac{1}{2}\int_0^4 (2x)^{1/3} 2\,dx = \frac{3}{8}(2x)^{4/3}\Big]_0^4 = 6$$

30. $A = \int_0^3 (3 - x)\sqrt{x}\,dx = \int_0^3 (3x^{1/2} - x^{3/2})\,dx = \left[2x^{3/2} - \frac{2}{5}x^{5/2}\right]_0^3 = \frac{x\sqrt{x}}{5}(10 - 2x)\Big]_0^3 = \frac{12\sqrt{3}}{5}$

31. Since $y \geq 0$ on $[0, 2]$, $A = \int_0^2 (3x^2 + 1)\,dx = \left[x^3 + x\right]_0^2 = 8 + 2 = 10.$

32. Since $y \geq 0$ on $[0, 4]$, $A = \int_0^4 (1 + \sqrt{x})\,dx = \left[x + \frac{2}{3}x^{3/2}\right]_0^4 = 4 + \frac{2}{3}(8) = \frac{28}{3}.$

33. Since $y \geq 0$ on $[0, 2]$, $A = \int_0^2 (x^3 + x)\,dx = \left[\frac{x^4}{4} + \frac{x^2}{2}\right]_0^2 = 4 + 2 = 6.$

34. Since $y \geq 0$ on $[0, 3]$, $A = \int_0^3 (3x - x^2)\,dx = \left[\frac{3}{2}x^2 - \frac{x^3}{3}\right]_0^3 = \frac{9}{2}.$

35. $\int_0^2 x^3\,dx = \frac{x^4}{4}\Big]_0^2 = 4$

$f(c)(2 - 0) = 4$

$c^3 = 2$

$c = \sqrt[3]{2} \approx 1.2599$

36. $\int_1^3 \frac{9}{x^3}\,dx = -\frac{9}{2x^2}\Big]_1^3 = -\frac{1}{2} + \frac{9}{2} = 4$

$f(c)(3 - 1) = 4$

$\frac{9}{c^3} = 2$

$c^3 = \frac{9}{2}$

$c = \sqrt[3]{\frac{9}{2}} \approx 1.6510$

37. $\frac{1}{3 - 0}\int_0^3 (-x^2 + 4x)\,dx = \frac{1}{3}\left[-\frac{x^3}{3} + 2x^2\right]_0^3 = 3$

Average $= 3$

$-x^2 + 4x = 3$, $0 = x^2 - 4x + 3$, $0 = (x - 1)(x - 3)$

In the interval $[0, 3]$, $x = 1$ or $x = 3$.

38. $\frac{1}{9 - 1}\int_1^9 \sqrt{x}\,dx = \frac{1}{8}\left[\frac{2}{3}x^{3/2}\right]_1^9 = \frac{13}{6}$

Average $= \frac{13}{6}$

$\sqrt{x} = \frac{13}{6}$

$x = \frac{169}{36}$

39. $\dfrac{1}{2-(-2)}\displaystyle\int_{-2}^{2}(4-x^2)\,dx = \frac{1}{4}\left[4x-\frac{1}{3}x^3\right]_{-2}^{2}$

$= \dfrac{1}{4}\left[\left(8-\dfrac{8}{3}\right)-\left(-8+\dfrac{8}{3}\right)\right] = \dfrac{8}{3}$

Average $= \dfrac{8}{3}$

$4-x^2 = \dfrac{8}{3}$ when $x = \pm\dfrac{2\sqrt{3}}{3} \approx \pm 1.155.$

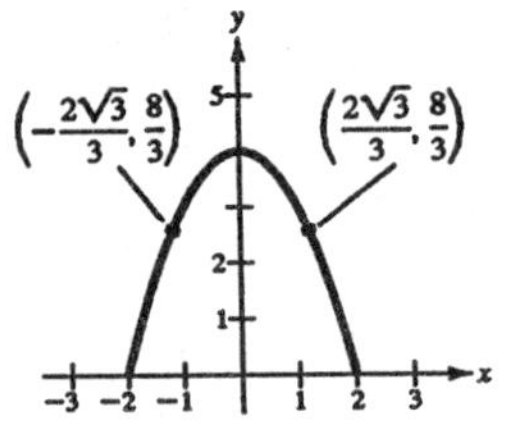

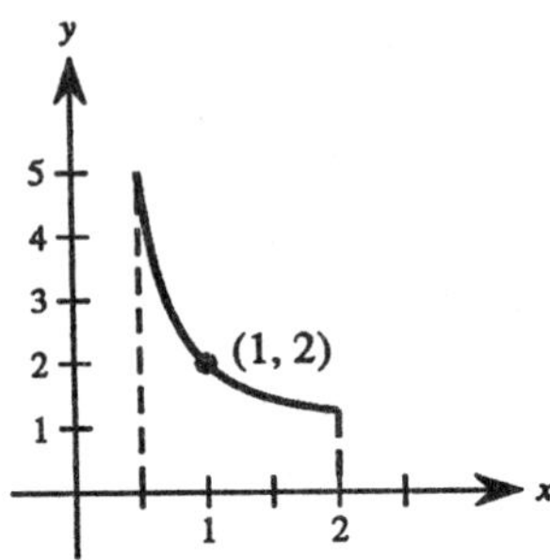

40. $\dfrac{1}{2-(1/2)}\displaystyle\int_{1/2}^{2}\frac{x^2+1}{x^2}\,dx = \frac{2}{3}\int_{1/2}^{2}(1+x^{-2})\,dx$

$= \dfrac{2}{3}\left[x-\dfrac{1}{x}\right]_{1/2}^{2} = \dfrac{2}{3}\left(\dfrac{3}{2}+\dfrac{3}{2}\right) = 2$

Average $= 2$

In the interval $\left[\dfrac{1}{2}, 2\right]$, $\dfrac{x^2+1}{x^2} = 2$ when $x = 1$.

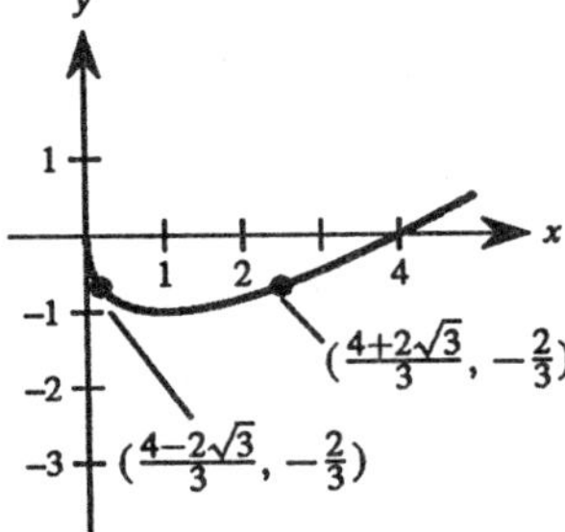

41. $\dfrac{1}{4-0}\displaystyle\int_{0}^{4}(x-2\sqrt{x})\,dx = \frac{1}{4}\left[\frac{x^2}{2}-\frac{4}{3}x^{3/2}\right]_{0}^{4} = \frac{1}{4}\left(8-\frac{32}{3}\right) = -\frac{2}{3}$

Average $= -\dfrac{2}{3}$

$x-2\sqrt{x} = -\dfrac{2}{3}$

$3x-6\sqrt{x}+2 = 0$

$\sqrt{x} = \dfrac{6\pm\sqrt{36-24}}{6} = \dfrac{3\pm\sqrt{3}}{3}$

$x = \left(\dfrac{3\pm\sqrt{3}}{3}\right)^2 = \dfrac{4\pm 2\sqrt{3}}{3} \approx 2.488,\ 0.179$

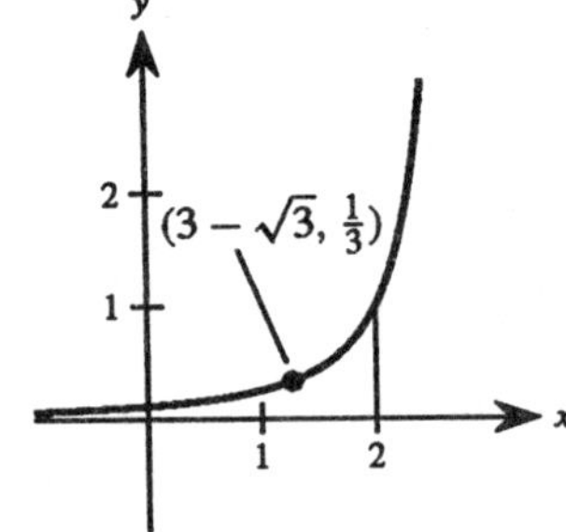

42. $\dfrac{1}{2-0}\displaystyle\int_{0}^{2}\frac{1}{(x-3)^2}\,dx = -\frac{1}{2(x-3)}\bigg]_{0}^{2} = \frac{1}{3}$

Average $= \dfrac{1}{3}$

$\dfrac{1}{(x-3)^2} = \dfrac{1}{3}$

$(x-3)^2 = 3$

$x-3 = \pm\sqrt{3},\quad x = 3\pm\sqrt{3}$

In the interval $[0, 2]$, $x = 3-\sqrt{3} \approx 1.27$.

$(3-\sqrt{3}, \tfrac{1}{3})$

43. (a) $\displaystyle\int_{0}^{x}(t+2)\,dt = \left[\frac{t^2}{2}+2t\right]_{0}^{x} = \frac{1}{2}x^2+2x$

(b) $\dfrac{d}{dx}\left[\dfrac{1}{2}x^2+2x\right] = x+2$

44. (a) $\displaystyle\int_0^x t(t^2+1)\,dt = \frac{1}{2}\int_0^x (t^2+1)(2t)\,dt = \frac{1}{4}(t^2+1)^2\Big]_0^x = \frac{1}{4}(x^2+1)^2 - \frac{1}{4}$

(b) $\displaystyle\frac{d}{dx}\left[\frac{1}{4}(x^2+1)^2 - \frac{1}{4}\right] = \frac{1}{4}(2)(x^2+1)(2x) - 0 = x(x^2+1)$

45. (a) $\displaystyle\int_8^x \sqrt[3]{t}\,dt = \frac{3}{4}t^{4/3}\Big]_8^x$

$\displaystyle= \frac{3}{4}(x^{4/3} - 16) = \frac{3}{4}x^{4/3} - 12$

(b) $\displaystyle\frac{d}{dx}\left[\frac{3}{4}x^{4/3} - 12\right] = x^{1/3} = \sqrt[3]{x}$

46. (a) $\displaystyle\int_4^x \sqrt{t}\,dt = \frac{2}{3}t^{3/2}\Big]_4^x$

$\displaystyle= \frac{2}{3}x^{3/2} - \frac{16}{3} = \frac{2}{3}(x^{3/2} - 8)$

(b) $\displaystyle\frac{d}{dx}\left[\frac{2}{3}x^{3/2} - \frac{16}{3}\right] = x^{1/2} = \sqrt{x}$

47. (a) $\displaystyle\int_1^x \frac{1}{t^2}\,dt = -\frac{1}{t}\Big]_1^x = 1 - \frac{1}{x}$

(b) $\displaystyle\frac{d}{dx}\left[1 - \frac{1}{x}\right] = \frac{1}{x^2}$

48. (a) $\displaystyle\int_0^x t^{3/2}\,dt = \frac{2}{5}t^{5/2}\Big]_0^x = \frac{2}{5}x^{5/2}$

(b) $\displaystyle\frac{d}{dx}\left[\frac{2}{5}x^{5/2}\right] = x^{3/2}$

49. $\displaystyle F(x) = \int_{-2}^x (t^2 - 2t + 5)\,dt$

$F'(x) = x^2 - 2x + 5$

50. $\displaystyle F(x) = \int_1^x \sqrt[4]{t}\,dt$

$F'(x) = \sqrt[4]{x}$

51. $\displaystyle F(x) = \int_{-1}^x \sqrt{t^4+1}\,dt$

$F'(x) = \sqrt{x^4+1}$

52. $\displaystyle F(x) = \int_1^x \frac{t^2}{t^2+1}\,dt$

$\displaystyle F'(x) = \frac{x^2}{x^2+1}$

53. $\displaystyle\frac{1}{5-0}\int_0^5 (0.1729t + 0.1522t^2 - 0.0374t^3)\,dt \approx \frac{1}{5}(0.08645t^2 + 0.05073t^3 - 0.00935t^4)\Big]_0^5 \approx 0.5318$ liter

54. $\displaystyle\frac{1}{R-0}\int_0^R k(R^2 - r^2)\,dr = \frac{k}{R}\left(R^2r - \frac{r^3}{3}\right)\Big]_0^R = \frac{2kR^2}{3}$

55. (a) $\displaystyle\int_1^7 f(x)\,dx$

= Sum of the areas

$= A_1 + A_2 + A_3 + A_4$

$\displaystyle= \frac{1}{2}(3+1) + \frac{1}{2}(1+2) + \frac{1}{2}(2+1) + (3)(1)$

$= 8$

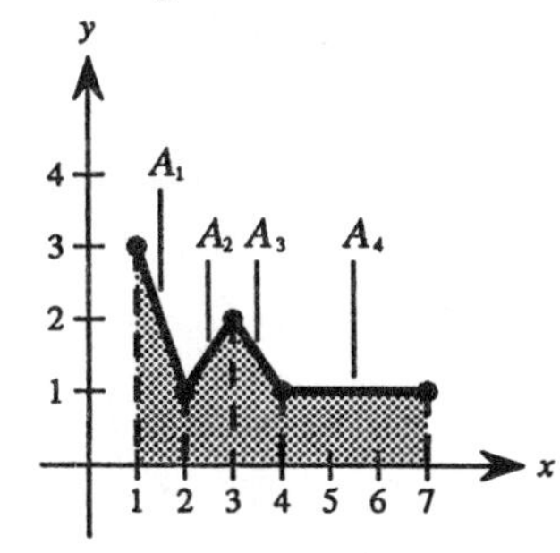

(b) Average value $\displaystyle= \frac{\int_1^7 f(x)\,dx}{7-1} = \frac{8}{6} = \frac{4}{3}$

(c) $A = 8 + (6)(2) = 20$

Average value $\displaystyle= \frac{20}{6} = \frac{10}{3}$

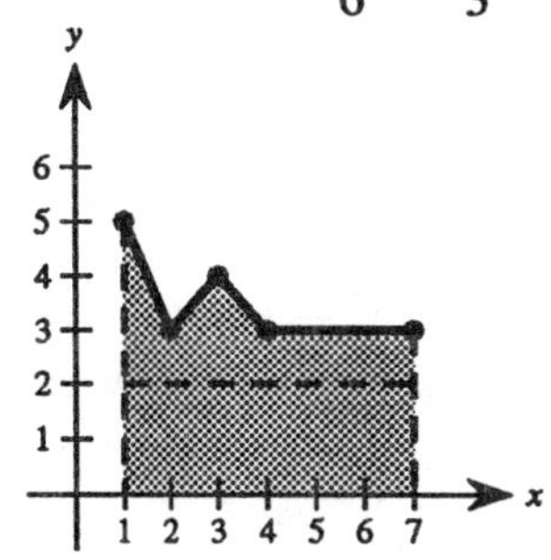

56. $P = 5(\sqrt{t} + 30)$

(a)

t	1	2	3	4	5	6
P	155	157.071	158.660	160	161.180	162.247

Average profit $\approx \frac{1}{6}(155 + 157.071 + 158.660 + 160 + 161.180 + 162.247) = \frac{954.158}{6} \approx 159.026$

(b) $\frac{1}{6}\int_{0.5}^{6.5} 5(\sqrt{t} + 30)\,dt = \frac{1}{6}\left[5\left(\frac{2}{3}t^{3/2} + 30t\right)\right]_{0.5}^{6.5} \approx \frac{954.061}{6} \approx 159.010$

(c) The definite integral yields a better approximation.

Section 5.5 Integration by Substitution

$\int f(g(x))g'(x)\,dx$	$u = g(x)$	$du = g'(x)\,dx$
1. $\int (5x^2 + 1)^2(10x)\,dx$	$5x^2 + 1$	$10x\,dx$
2. $\int x^2\sqrt{x^3 + 1}\,dx$	$x^3 + 1$	$3x^2\,dx$
3. $\int \frac{x}{\sqrt{x^2 + 1}}\,dx$	$x^2 + 1$	$2x\,dx$
4. $\int (x^3 + 3)3x^2\,dx$	$x^3 + 3$	$3x^2\,dx$

5. $\int (1 + 2x)^4 2\,dx = \frac{(1 + 2x)^5}{5} + C$

Check: $\frac{d}{dx}\left[\frac{(1 + 2x)^5}{5} + C\right] = 2(1 + 2x)^4$

6. $\int (x^2 - 1)^3 2x\,dx = \frac{(x^2 - 1)^4}{4} + C$

Check: $\frac{d}{dx}\left[\frac{(x^2 - 1)^4}{4} + C\right] = 2x(x^2 - 1)^3$

7. $\int (9 - x^2)^{1/2}(-2x)\,dx = \frac{(9 - x^2)^{3/2}}{3/2} + C = \frac{2}{3}(9 - x^2)^{3/2} + C$

Check: $\frac{d}{dx}\left[\frac{2}{3}(9 - x^2)^{3/2} + C\right] = \frac{2}{3} \cdot \frac{3}{2}(9 - x^2)^{1/2}(-2x) = \sqrt{9 - x^2}(-2x)$

8. $\int (1 - 2x^2)^3(-4x)\,dx = \frac{(1 - 2x^2)^4}{4} + C$

Check: $\frac{d}{dx}\left[\frac{(1 - 2x^2)^4}{4} + C\right] = \frac{4(1 - 2x^2)^3(-4x)}{4} = (1 - 2x^2)^3(-4x)$

9. $\int x^2(x^3 - 1)^4\,dx = \frac{1}{3}\int (x^3 - 1)^4(3x^2)\,dx = \frac{1}{3}\left[\frac{(x^3 - 1)^5}{5}\right] + C = \frac{(x^3 - 1)^5}{15} + C$

Check: $\frac{d}{dx}\left[\frac{(x^3 - 1)^5}{15} + C\right] = x^2(x^3 - 1)^4$

10. $\int x(4x^2+3)^3\,dx = \frac{1}{8}\int (4x^2+3)^3(8x)\,dx = \frac{1}{8}\left[\frac{(4x^2+3)^4}{4}\right] + C = \frac{(4x^2+3)^4}{32} + C$

Check: $\frac{d}{dx}\left[\frac{(4x^2+3)^4}{32} + C\right] = \frac{4(4x^2+3)^3(8x)}{32} = x(4x^2+3)^3$

11. $\int 5x(1-x^2)^{1/3}\,dx = -\frac{5}{2}\int (1-x^2)^{1/3}(-2x\,dx) = -\frac{5}{2}\cdot\frac{(1-x^2)^{4/3}}{4/3} + C = -\frac{15}{8}(1-x^2)^{4/3} + C$

Check: $\frac{d}{dx}\left[-\frac{15}{8}(1-x^2)^{4/3} + C\right] = -\frac{15}{8}\cdot\frac{4}{3}(1-x^2)^{1/3}(-2x) = 5x(1-x^2)^{1/3} = 5x\sqrt[3]{1-x^2}$

12. $\int u^3\sqrt{u^4+2}\,du = \frac{1}{4}\int (u^4+2)^{1/2}(4u^3)\,du = \frac{1}{4}\left[\frac{(u^4+2)^{3/2}}{3/2}\right] + C = \frac{1}{6}(u^4+2)^{3/2} + C$

Check: $\frac{d}{dx}\left[\frac{1}{6}(u^4+2)^{3/2} + C\right] = u^3\sqrt{u^4+2}$

13. $\int \frac{x^2}{(1+x^3)^2}\,dx = \frac{1}{3}\int (1+x^3)^{-2}(3x^2)\,dx = \frac{1}{3}\left[\frac{(1+x^3)^{-1}}{-1}\right] + C = -\frac{1}{3(1+x^3)} + C$

Check: $\frac{d}{dx}\left[-\frac{1}{3(1+x^3)} + C\right] = \frac{x^2}{(1+x^3)^2}$

14. $\int \frac{x^2}{(16-x^3)^2}\,dx = -\frac{1}{3}\int (16-x^3)^{-2}(-3x^2)\,dx = -\frac{1}{3}\left[\frac{(16-x^3)^{-1}}{-1}\right] + C = \frac{1}{3(16-x^3)} + C$

Check: $\frac{d}{dx}\left[\frac{1}{3(16-x^3)} + C\right] = \frac{1}{3}(-1)(16-x^3)^{-2}(-3x^2) = \frac{x^2}{(16-x^3)^2}$

15. $\int \frac{4x}{\sqrt{16-x^2}}\,dx = -2\int (16-x^2)^{-1/2}(-2x)\,dx = -2\left[\frac{(16-x^2)^{1/2}}{1/2}\right] + C = -4\sqrt{16-x^2} + C$

Check: $\frac{d}{dx}\left[-4\sqrt{16-x^2} + C\right] = -4\left(\frac{1}{2}\right)(16-x^2)^{-1/2}(-2x) = \frac{4x}{\sqrt{16-x^2}}$

16. $\int \frac{10x^2}{\sqrt{1+x^3}}\,dx = \frac{10}{3}\int (1+x^3)^{-1/2}(3x^2)\,dx = \frac{10}{3}\left[\frac{(1+x^3)^{1/2}}{1/2}\right] + C = \frac{20}{3}\sqrt{1+x^3} + C$

Check: $\frac{d}{dx}\left[\frac{20}{3}\sqrt{1+x^3} + C\right] = \frac{20}{3}\cdot\frac{1}{2}(1+x^3)^{-1/2}(3x^2) = \frac{10x^2}{\sqrt{1+x^3}}$

17. $\int \frac{x+1}{(x^2+2x-3)^2}\,dx = \frac{1}{2}\int (x^2+2x-3)^{-2}(2x+2)\,dx = \frac{1}{2}\left[\frac{(x^2+2x-3)^{-1}}{-1}\right] + C = -\frac{1}{2(x^2+2x-3)} + C$

Check: $\frac{d}{dx}\left[-\frac{1}{2(x^2+2x-3)} + C\right] = \frac{x+1}{(x^2+2x-3)^2}$

18. $\int \frac{x-4}{\sqrt{x^2-8x+1}}\,dx = \frac{1}{2}\int (x^2-8x+1)^{-1/2}(2x-8)\,dx = \frac{1}{2}\left[\frac{(x^2-8x+1)^{1/2}}{1/2}\right] + C = \sqrt{x^2-8x+1} + C$

Check: $\frac{d}{dx}\left[\sqrt{x^2-8x+1} + C\right] = \frac{1}{2}(x^2-8x+1)^{-1/2}(2x-8) = \frac{x-4}{\sqrt{x^2-8x+1}}$

19. $\displaystyle\int\left(1+\frac{1}{t}\right)^3\left(\frac{1}{t^2}\right)dt = -\int\left(1+\frac{1}{t}\right)^3\left(-\frac{1}{t^2}\right)dt$

$$= -\frac{[1+(1/t)]^4}{4} + C$$

Check: $\displaystyle\frac{d}{dt}\left[-\frac{[1+(1/t)]^4}{4}+C\right] = \frac{1}{t^2}\left(1+\frac{1}{t}\right)^3$

20. $\displaystyle\int\frac{1}{(3x)^2}\,dx = \frac{1}{9}\int x^{-2}\,dx$

$$= \frac{1}{9}\left(\frac{x^{-1}}{-1}\right) + C = -\frac{1}{9x} + C$$

Check: $\displaystyle\frac{d}{dx}\left[-\frac{1}{9x}+C\right] = \frac{1}{9x^2} = \frac{1}{(3x)^2}$

21. $\displaystyle\int\frac{1}{\sqrt{2x}}\,dx = \frac{1}{2}\int(2x)^{-1/2}2\,dx$

$$= \frac{1}{2}\left[\frac{(2x)^{1/2}}{1/2}\right] + C = \sqrt{2x} + C$$

Check: $\displaystyle\frac{d}{dx}[\sqrt{2x}+C] = \frac{1}{\sqrt{2x}}$

22. $\displaystyle\int\frac{1}{2\sqrt{x}}\,dx = \frac{1}{2}\int x^{-1/2}\,dx$

$$= \frac{1}{2}\left(\frac{x^{1/2}}{1/2}\right) + C = \sqrt{x} + C$$

Check: $\displaystyle\frac{d}{dx}[\sqrt{x}+C] = \frac{1}{2\sqrt{x}}$

23. $\displaystyle\int\frac{x^2+3x+7}{\sqrt{x}}\,dx = \int(x^{3/2}+3x^{1/2}+7x^{-1/2})\,dx$

$$= \frac{2}{5}x^{5/2} + 2x^{3/2} + 14x^{1/2} + C = \frac{2}{5}\sqrt{x}(x^2+5x+35) + C$$

Check: $\displaystyle\frac{d}{dx}\left[\frac{2}{5}\sqrt{x}(x^2+5x+35)+C\right] = \frac{2}{5}\sqrt{x}(2x+5) + \frac{x^2+5x+35}{5\sqrt{x}} = \frac{5x^2+15x+35}{5\sqrt{x}} = \frac{x^2+3x+7}{\sqrt{x}}$

24. $\displaystyle\int\frac{t+2t^2}{\sqrt{t}}\,dt = \int(t^{1/2}+2t^{3/2})\,dt = \frac{2}{3}t^{3/2} + \frac{4}{5}t^{5/2} + C = \frac{2}{15}t^{3/2}(5+6t) + C$

Check: $\displaystyle\frac{d}{dt}\left[\frac{2}{15}t^{3/2}(5+6t)+C\right] = t^{1/2} + 2t^{3/2} = \frac{t+2t^2}{\sqrt{t}}$

25. $\displaystyle\int t^2\left(t-\frac{2}{t}\right)dt = \int(t^3-2t)\,dt = \frac{1}{4}t^4 - t^2 + C$

Check: $\displaystyle\frac{d}{dt}\left[\frac{1}{4}t^4 - t^2 + C\right] = t^3 - 2t = t^2\left(t-\frac{2}{t}\right)$

26. $\displaystyle\int\left(\frac{t^3}{3}+\frac{1}{4t^2}\right)dt = \int\left(\frac{1}{3}t^3+\frac{1}{4}t^{-2}\right)dt = \frac{1}{3}\left(\frac{t^4}{4}\right) + \frac{1}{4}\left(\frac{t^{-1}}{-1}\right) + C = \frac{1}{12}t^4 - \frac{1}{4t} + C$

Check: $\displaystyle\frac{d}{dt}\left[\frac{1}{12}t^4 - \frac{1}{4t} + C\right] = \frac{1}{3}t^3 + \frac{1}{4t^2}$

27. $\displaystyle\int(9-y)\sqrt{y}\,dy = \int(9y^{1/2}-y^{3/2})\,dy = 9\left(\frac{2}{3}y^{3/2}\right) - \frac{2}{5}y^{5/2} + C = \frac{2}{5}y^{3/2}(15-y) + C$

Check: $\displaystyle\frac{d}{dy}\left[\frac{2}{5}y^{3/2}(15-y)+C\right] = 9y^{1/2} - y^{3/2} = (9-y)\sqrt{y}$

28. $\displaystyle\int 2\pi y(8-y^{3/2})\,dy = 2\pi\int(8y-y^{5/2})\,dy = 2\pi\left(4y^2-\frac{2}{7}y^{7/2}\right) + C = \frac{4\pi y^2}{7}(14-y^{3/2}) + C$

Check: $\displaystyle\frac{d}{dy}\left[\frac{4\pi y^2}{7}(14-y^{3/2})+C\right] = 16\pi y - 2\pi y^{5/2} = (8-y^{3/2})$

29. $u = \sqrt{x+2},\quad x = u^2 - 2,\quad dx = 2u\,du$

$$\int x\sqrt{x+2}\,dx = \int (u^2-2)u(2u\,du)$$

$$= 2\int (u^4 - 2u^2)\,du = 2\left[\frac{u^5}{5} - \frac{2u^3}{3}\right] + C$$

$$= \frac{2u^3}{15}[3u^2 - 10] + C = \frac{2}{15}(x+2)^{3/2}[3(x+2) - 10] + C = \frac{2}{15}(x+2)^{3/2}(3x-4) + C$$

30. $u = \sqrt{2x+1},\quad x = \dfrac{u^2-1}{2},\quad dx = u\,du$

$$\int x\sqrt{2x+1}\,dx = \frac{1}{2}\int (u^2-1)u(u\,du)$$

$$= \frac{1}{2}\int (u^4 - u^2)\,du = \frac{1}{2}\left(\frac{u^5}{5} - \frac{u^3}{3}\right) + C$$

$$= \frac{u^3}{30}(3u^2 - 5) + C = \frac{(2x+1)^{3/2}}{30}(6x-2) + C = \frac{(2x+1)^{3/2}}{15}(3x-1) + C$$

31. $u = \sqrt{1-x},\quad x = 1 - u^2,\quad dx = -2u\,du$

$$\int x^2\sqrt{1-x}\,dx = \int (1-u^2)^2 u(-2u\,du)$$

$$= -2\int (u^2 - 2u^4 + u^6)\,du = -2\left(\frac{u^3}{3} - \frac{2u^5}{5} + \frac{u^7}{7}\right) + C$$

$$= \frac{-2u^3}{105}(35 - 42u^2 + 15u^4) + C = \frac{-2}{105}(1-x)^{3/2}(15x^2 + 12x + 8) + C$$

32. $u = \sqrt{x+2},\quad x = u^2 - 2,\quad dx = 2u\,du$

$$\int x^3\sqrt{x+2}\,dx = \int (u^2-2)^3 u(2u\,du) = 2\int (u^8 - 6u^6 + 12u^4 - 8u^2)\,du$$

$$= 2\left(\frac{u^9}{9} - \frac{6u^7}{7} + \frac{12u^5}{5} - \frac{8u^3}{3}\right) + C$$

$$= \frac{2u^3}{315}(35u^6 - 270u^4 + 756u^2 - 840) + C$$

$$= \frac{2}{315}(x+2)^{3/2}[35(x+2)^3 - 270(x+2)^2 + 756(x+2) - 840] + C$$

$$= \frac{2}{315}(x+2)^{3/2}(35x^3 - 60x^2 + 96x - 128) + C$$

33. $u = \sqrt{2x-1},\quad x = \dfrac{u^2+1}{2},\quad dx = u\,du$

$$\int \frac{x^2-1}{\sqrt{2x-1}}\,dx = \frac{1}{4}\int \frac{u^4 + 2u^2 - 3}{u}u\,du = \frac{1}{4}\left(\frac{1}{5}u^5 + \frac{2}{3}u^3 - 3u\right) + C$$

$$= \frac{u}{60}(3u^4 + 10u^2 - 45) + C = \frac{\sqrt{2x-1}}{15}(3x^2 + 2x - 13) + C$$

34. $u = \sqrt{x+3},\quad x = u^2 - 3,\quad dx = 2u\,du$

$$\int \frac{2x-1}{\sqrt{x+3}}\,dx = \int \frac{2u^2-7}{u}(2u)\,du$$

$$= 2\left(\frac{2}{3}u^3 - 7u\right) + C = \frac{2u}{3}(2u^2 - 21) + C = \frac{2}{3}\sqrt{x+3}(2x-15) + C$$

35. $u = \sqrt{x+1}, \quad x = u^2 - 1, \quad dx = 2u\,du$

$$\int \frac{-x}{(x+1) - \sqrt{x+1}}\,dx = \int \frac{-(u^2-1)}{u^2 - u}(2u)\,du$$
$$= -2\int \frac{u^2-1}{u-1}\,du = -2\int (u+1)\,du$$
$$= -2\left(\frac{u^2}{2} + u\right) + C = -u^2 - 2u + C$$
$$= -[(x+1) + 2\sqrt{x+1}] + C = -(x + 2\sqrt{x+1}) + C_1$$

36. $u = \sqrt[3]{t-4}, \quad t = u^3 + 4, \quad dt = 3u^2\,du$

$$\int t\sqrt[3]{t-4}\,dt = \int (u^3+4)u(3u^2\,du)$$
$$= 3\int (u^6 + 4u^3)\,du = 3\left(\frac{u^7}{7} + u^4\right) + C = \frac{3u^4}{7}(u^3+7) + C = \frac{3}{7}(t-4)^{4/3}(t+3) + C$$

37. $u = \sqrt{2x+1}, \quad x = \dfrac{u^2-1}{2}, \quad dx = u\,du$

$$\int \frac{x}{\sqrt{2x+1}}\,dx = \int \left(\frac{u^2-1}{2}\right)u^{-1}u\,du = \frac{1}{2}\int (u^2-1)\,du$$
$$= \frac{1}{2}\left(\frac{u^3}{3} - u\right) + C = \frac{u}{6}(u^2-3) + C = \frac{1}{6}\sqrt{2x+1}(2x+1-3) + C = \frac{1}{3}\sqrt{2x+1}(x-1) + C$$

38. $u = \sqrt{2-x}, \quad x = 2 - u^2, \quad dx = -2u\,du$

$$\int (x+1)\sqrt{2-x}\,dx = \int (3-u^2)u(-2u)\,du = -2\int (3u^2 - u^4)\,du$$
$$= -2\left(u^3 - \frac{u^5}{5}\right) + C = -\frac{2u^3}{5}(5-u^2) + C$$
$$= -\frac{2}{5}(2-x)^{3/2}[5 - (2-x)] + C = -\frac{2}{5}(2-x)^{3/2}(x+3) + C$$

39. Let $u = x^2 + 1, \quad du = 2x\,dx$

$$\int_{-1}^{1} x(x^2+1)^3\,dx = \frac{1}{2}\int_{-1}^{1} (x^2+1)^3(2x)\,dx = \frac{1}{8}(x^2+1)^4\Big]_{-1}^{1} = 0$$

40. Let $u = 1 - x^2, \quad du = -2x\,dx$

$$\int_0^1 x\sqrt{1-x^2}\,dx = -\frac{1}{2}\int_0^1 (1-x^2)^{1/2}(-2x)\,dx = -\frac{1}{3}(1-x^2)^{3/2}\Big]_0^1 = 0 + \frac{1}{3} = \frac{1}{3}$$

41. Let $u = 2x + 1, \quad du = 2x\,dx$

$$\int_0^4 \frac{1}{\sqrt{2x+1}}\,dx = \frac{1}{2}\int_0^4 (2x+1)^{-1/2}(2)\,dx = \sqrt{2x+1}\Big]_0^4 = \sqrt{9} - \sqrt{1} = 2$$

42. Let $u = 1 + 2x^2, \quad du = 4x\,dx$

$$\int_0^2 \frac{x}{\sqrt{1+2x^2}}\,dx = \frac{1}{4}\int_0^2 (1+2x^2)^{-1/2}(4x)\,dx = \frac{1}{2}\sqrt{1+2x^2}\Big]_0^2 = \frac{3}{2} - \frac{1}{2} = 1$$

43. Let $u = 1+\sqrt{x}$, $du = \dfrac{1}{2\sqrt{x}}\,dx$

$$\int_1^9 \frac{1}{\sqrt{x}(1+\sqrt{x})^2}\,dx = 2\int_1^9 (1+\sqrt{x})^{-2}\left(\frac{1}{2\sqrt{x}}\right)dx = \left[-\frac{2}{1+\sqrt{x}}\right]_1^9 = -\frac{1}{2}+1 = \frac{1}{2}$$

44. Let $u = 4+x^2$, $du = 2x\,dx$

$$\int_0^2 x\sqrt[3]{4+x^2}\,dx = \frac{1}{2}\int_0^2 (4+x^2)^{1/3}(2x)\,dx = \frac{3}{8}(4+x^2)^{4/3}\Big]_0^2 = \frac{3}{8}(8^{4/3}-4^{4/3}) = 6-\frac{3}{2}\sqrt[3]{4} \approx 3.619$$

45. Let $u = \sqrt{2-x}$, $x = 2-u^2$, $dx = -2u\,du$ when $x = 1 \Rightarrow u = 1$, and when $x = 2 \Rightarrow u = 0$.

$$\int_1^2 (x-1)\sqrt{2-x}\,dx = -2\int_1^0 (1-u^2)u^2\,du = 2\int_0^1 (u^2-u^4)\,du = 2\left[\frac{u^3}{3}-\frac{u^5}{5}\right]_0^1 = \frac{4}{15}$$

46. Let $u = \sqrt{2x+1}$, $x = \dfrac{u^2-1}{2}$, $dx = u\,du$ when $x = 0 \Rightarrow u = 1$, and when $x = 4 \Rightarrow u = 3$.

$$\int_0^4 \frac{x}{\sqrt{2x+1}}\,dx = \frac{1}{2}\int_1^3 \left(\frac{u^2-1}{u}\right)u\,du = \frac{1}{2}\left[\frac{1}{3}u^3-u\right]_1^3 = \frac{10}{3}$$

47. Let $u = \sqrt{x-3}$, $x = u^2+3$, $dx = 2u\,du$ when $x = 3 \Rightarrow u = 0$, and when $x = 7 \Rightarrow u = 2$.

$$\int_3^7 x\sqrt{x-3}\,dx = 2\int_0^2 (u^2+3)u^2\,du = 2\int_0^2 (u^4+3u^2)\,du = 2\left[\frac{1}{5}u^5+u^3\right]_0^2 = \frac{144}{5}$$

48. $$\int_0^1 \frac{1}{\sqrt{x}+\sqrt{x+1}}\,dx = -\int_0^1 (\sqrt{x}-\sqrt{x+1})\,dx \qquad \text{(Rationalize)}$$

$$= -\left[\frac{2}{3}x^{3/2}-\frac{2}{3}(x+1)^{3/2}\right]_0^1 = \frac{4}{3}(\sqrt{2}-1)$$

49. Let $u = \sqrt[3]{x+1}$, $x = u^3-1$, $dx = 3u^2\,du$ when $x = 0 \Rightarrow u = 1$, and when $x = 7 \Rightarrow u = 2$.

$$\int_0^7 x\sqrt[3]{x+1}\,dx = 3\int_1^2 (u^3-1)u^3\,du = 3\int_1^2 (u^6-u^3)\,du = 3\left[\frac{u^7}{7}-\frac{u^4}{4}\right]_0^2 = \frac{1209}{28}$$

50. Let $u = \sqrt[3]{x+2}$, $x = u^3-2$, $dx = 3u^2\,du$ when $x = -2 \Rightarrow u = 0$, and when $x = 6 \Rightarrow u = 2$.

$$\int_{-2}^6 x^2\sqrt[3]{x+2}\,dx = 3\int_0^2 (u^3-2)^2u^3\,du = 3\int_0^2 (u^9-4u^6+4u^3)\,du = 3\left[\frac{u^{10}}{10}-\frac{4u^7}{7}+u^4\right]_0^2 = \frac{4752}{35}$$

51. $$\int_0^2 x^2\,dx = \frac{8}{3}$$

(a) $\displaystyle\int_{-2}^0 x^2\,dx = \int_0^2 x^2\,dx = \frac{8}{3}$

(b) $\displaystyle\int_{-2}^2 x^2\,dx = 2\int_0^2 x^2\,dx = \frac{16}{3}$

(c) $\displaystyle\int_0^2 (-x^2)\,dx = -\int_0^2 x^2\,dx = -\frac{8}{3}$

(d) $\displaystyle\int_{-2}^0 3x^2\,dx = 3\int_0^2 x^2\,dx = 8$

52. $f'(x) = x\sqrt{1-x^2}, \quad \left(0, \frac{7}{3}\right)$

$$f(x) = \int x\sqrt{1-x^2}\,dx = -\frac{1}{2}\int (1-x^2)^{1/2}(-2x)\,dx = -\frac{1}{3}(1-x^2)^{3/2} + C$$

$$f(0) = -\frac{1}{3} + C = \frac{7}{3} \Rightarrow C = \frac{8}{3}$$

$$f(x) = -\frac{1}{3}(1-x^2)^{3/2} + \frac{8}{3} = -\frac{1}{3}[(1-x^2)^{3/2} - 8]$$

53. $\frac{dW}{dt} = \frac{12}{\sqrt{16t+9}}$

(a) $$W = \int \frac{12}{\sqrt{16t+9}}\,dt = \frac{12}{16}\int (16t+9)^{-1/2}16\,dt = \frac{3}{4}\left[\frac{(16t+9)^{1/2}}{1/2}\right] + C = \frac{3}{2}\sqrt{16t+9} + C$$

$$W(0) = \frac{9}{2} + C = 0 \Rightarrow C = -\frac{9}{2}$$

$$W(t) = \frac{3}{2}\sqrt{16t+9} - \frac{9}{2} = \frac{3}{2}[\sqrt{16t+9} - 3]$$

(b) $W(100) = \frac{3}{2}[\sqrt{1609} - 3] \approx 55.67$ lb

54. $\frac{dC}{dx} = \frac{12}{\sqrt[3]{12x+1}}$

(a) $$C(x) = \int \frac{12}{\sqrt[3]{12x+1}}\,dx = \int (12x+1)^{-1/3}12\,dx = \frac{3}{2}(12x+1)^{2/3} + C_1$$

$$C(13) \approx 43.65 + C_1 = 100$$

$$C_1 = 56.35$$

$$C(x) = \frac{3}{2}(12x+1)^{2/3} + 56.35$$

(b)

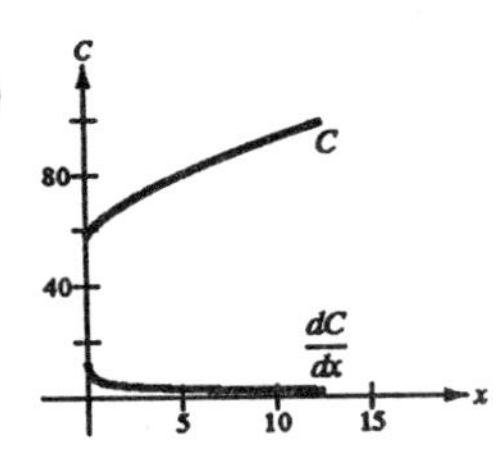

Section 5.6 Numerical Integration

1. Exact: $$\int_0^2 x^2\,dx = \frac{1}{3}x^3\Big]_0^2 = \frac{8}{3} \approx 2.6667$$

Trapezoidal: $$\int_0^2 x^2\,dx \approx \frac{1}{4}\left[0 + 2\left(\frac{1}{2}\right)^2 + 2(1)^2 + 2\left(\frac{3}{2}\right)^2 + (2)^2\right] = \frac{11}{4} = 2.7500$$

Simpson's: $$\int_0^2 x^2\,dx \approx \frac{1}{6}\left[0 + 4\left(\frac{1}{2}\right)^2 + 2(1)^2 + 4\left(\frac{3}{2}\right)^2 + (2)^2\right] = \frac{8}{3} \approx 2.6667$$

2. Exact: $\displaystyle\int_0^1 \left(\frac{x^2}{2}+1\right)dx = \left[\frac{x^3}{6}+x\right]_0^1 = \frac{7}{6} \approx 1.1667$

Trapezoidal: $\displaystyle\int_0^1 \left(\frac{x^2}{2}+1\right)dx \approx \frac{1}{8}\left[1+2\left(\frac{(\frac{1}{4})^2}{2}+1\right)+2\left(\frac{(\frac{1}{2})^2}{2}+1\right)+2\left(\frac{(\frac{3}{4})^2}{2}+1\right)+\left(\frac{1^2}{2}+1\right)\right]$

$\displaystyle= \frac{75}{64} \approx 1.1719$

Simpson's: $\displaystyle\int_0^1 \left(\frac{x^2}{2}+1\right)dx \approx \frac{1}{12}\left[1+4\left(\frac{(\frac{1}{4})^2}{2}+1\right)+2\left(\frac{(\frac{1}{2})^2}{2}+1\right)+4\left(\frac{(\frac{3}{4})^2}{2}+1\right)+\left(\frac{1^2}{2}+1\right)\right]$

$\displaystyle= \frac{7}{6} \approx 1.1667$

3. Exact: $\displaystyle\int_0^2 x^3\,dx = \left.\frac{x^4}{4}\right]_0^2 = 4.000$

Trapezoidal: $\displaystyle\int_0^2 x^3\,dx \approx \frac{1}{4}\left[0+2\left(\frac{1}{2}\right)^3+2(1)^3+2\left(\frac{3}{2}\right)^3+(2)^3\right] = \frac{17}{4} = 4.2500$

Simpson's: $\displaystyle\int_0^2 x^3\,dx \approx \frac{1}{6}\left[0+4\left(\frac{1}{2}\right)^3+2(1)^3+4\left(\frac{3}{2}\right)^3+(2)^3\right] = \frac{24}{6} = 4.0000$

4. Exact: $\displaystyle\int_1^2 \frac{1}{x^2}\,dx = \left.\frac{-1}{x}\right]_1^2 = 0.5000$

Trapezoidal: $\displaystyle\int_1^2 \frac{1}{x^2}\,dx \approx \frac{1}{8}\left[1+2\left(\frac{4}{5}\right)^2+2\left(\frac{4}{6}\right)^2+2\left(\frac{4}{7}\right)^2+\frac{1}{4}\right] \approx 0.5090$

Simpson's: $\displaystyle\int_1^2 \frac{1}{x^2}\,dx \approx \frac{1}{12}\left[1+4\left(\frac{4}{5}\right)^2+2\left(\frac{4}{6}\right)^2+4\left(\frac{4}{7}\right)^2+\frac{1}{4}\right] \approx 0.5004$

5. Exact: $\displaystyle\int_0^2 x^3\,dx = \left.\frac{1}{4}x^4\right]_0^2 = 4.000$

Trapezoidal: $\displaystyle\int_0^2 x^3\,dx \approx \frac{1}{8}\left[0+2\left(\frac{1}{4}\right)^3+2\left(\frac{2}{4}\right)^3+2\left(\frac{3}{4}\right)^3+2(1)^3+2\left(\frac{5}{4}\right)^3+2\left(\frac{6}{4}\right)^3+2\left(\frac{7}{4}\right)^3+8\right]$

$= 4.0625$

Simpson's: $\displaystyle\int_0^2 x^3\,dx \approx \frac{1}{12}\left[0+4\left(\frac{1}{4}\right)^3+2\left(\frac{2}{4}\right)^3+4\left(\frac{3}{4}\right)^3+2(1)^3+4\left(\frac{5}{4}\right)^3+2\left(\frac{6}{4}\right)^3+4\left(\frac{7}{4}\right)^3+8\right]$

$= 4.000$

6. Exact: $\displaystyle\int_0^8 \sqrt[3]{x}\,dx = \left.\frac{3}{4}x^{4/3}\right]_0^8 = 12$

Trapezoidal: $\displaystyle\int_0^8 \sqrt[3]{x}\,dx \approx \frac{1}{2}\left[0+2+2\sqrt[3]{2}+2\sqrt[3]{3}+2\sqrt[3]{4}+2\sqrt[3]{5}+2\sqrt[3]{6}+2\sqrt[3]{7}+2\right] \approx 11.7296$

Simpson's: $\displaystyle\int_0^8 \sqrt[3]{x}\,dx \approx \frac{1}{3}\left[0+4+2\sqrt[3]{2}+4\sqrt[3]{3}+2\sqrt[3]{4}+4\sqrt[3]{5}+2\sqrt[3]{6}+4\sqrt[3]{7}+2\right] \approx 11.8632$

7. Exact: $\displaystyle\int_4^9 \sqrt{x}\,dx = \frac{2}{3}x^{3/2}\Big]_4^9 = 18 - \frac{16}{3} = \frac{38}{3} \approx 12.6667$

Trapezoidal: $\displaystyle\int_4^9 \sqrt{x}\,dx \approx \frac{5}{16}\left[2 + 2\sqrt{\frac{37}{8}} + 2\sqrt{\frac{21}{4}} + 2\sqrt{\frac{47}{8}} + 2\sqrt{\frac{26}{4}} + 2\sqrt{\frac{57}{8}} + 2\sqrt{\frac{31}{4}} + 2\sqrt{\frac{67}{8}} + 3\right]$

≈ 12.6640

Simpson's: $\displaystyle\int_4^9 \sqrt{x}\,dx \approx \frac{5}{24}\left[2 + 4\sqrt{\frac{37}{8}} + \sqrt{21} + 4\sqrt{\frac{47}{8}} + \sqrt{26} + 4\sqrt{\frac{57}{8}} + \sqrt{31} + 4\sqrt{\frac{67}{8}} + 3\right] \approx 12.6667$

8. Exact: $\displaystyle\int_1^3 (4 - x^2)\,dx = \left[4x - \frac{x^3}{3}\right]_1^3 = 3 - \frac{11}{3} = -\frac{2}{3} \approx -0.6667$

Trapezoidal: $\displaystyle\int_1^3 (4 - x^2)\,dx \approx \frac{1}{4}\left\{3 + 2\left[4 - \left(\frac{3}{2}\right)^2\right] + 2(0) + 2\left[4 - \left(\frac{5}{2}\right)^2\right] - 5\right\} = -0.75$

Simpson's: $\displaystyle\int_1^3 (4 - x^2)\,dx \approx \frac{1}{6}\left[3 + 4\left(4 - \frac{9}{4}\right) + 0 + 4\left(4 - \frac{25}{4}\right) - 5\right] \approx -0.6667$

9. Exact: $\displaystyle\int_1^2 \frac{1}{(x+1)^2}\,dx = -\frac{1}{x+1}\Big]_1^2 = -\frac{1}{3} + \frac{1}{2} = \frac{1}{6} \approx 0.1667$

Trapezoidal: $\displaystyle\int_1^2 \frac{1}{(x+1)^2}\,dx \approx \frac{1}{8}\left[\frac{1}{4} + 2\left(\frac{1}{(\frac{5}{4}+1)^2}\right) + 2\left(\frac{1}{(\frac{3}{2}+1)^2}\right) + 2\left(\frac{1}{(\frac{7}{4}+1)^2}\right) + \frac{1}{9}\right]$

$\displaystyle= \frac{1}{8}\left(\frac{1}{4} + \frac{32}{81} + \frac{8}{25} + \frac{32}{121} + \frac{1}{9}\right) \approx 0.1676$

Simpson's: $\displaystyle\int_1^2 \frac{1}{(x+1)^2}\,dx \approx \frac{1}{12}\left[\frac{1}{4} + 4\left(\frac{1}{(\frac{5}{4}+1)^2}\right) + 2\left(\frac{1}{(\frac{3}{2}+1)^2}\right) + 4\left(\frac{1}{(\frac{7}{4}+1)^2}\right) + \frac{1}{9}\right]$

$\displaystyle= \frac{1}{12}\left(\frac{1}{4} + \frac{64}{81} + \frac{8}{25} + \frac{64}{121} + \frac{1}{9}\right) \approx 0.1667$

10. Exact: $\displaystyle\int_0^2 x\sqrt{x^2+1}\,dx = \frac{1}{3}(x^2+1)^{3/2}\Big]_0^2 = \frac{1}{3}(5^{3/2} - 1) \approx 3.3934$

Trapezoidal: $\displaystyle\int_0^2 x\sqrt{x^2+1}\,dx \approx \frac{1}{4}\left[0 + 2\left(\frac{1}{2}\right)\sqrt{\left(\frac{1}{2}\right)^2 + 1} + 2(1)\sqrt{1^2+1} + 2\left(\frac{3}{2}\right)\sqrt{\left(\frac{3}{2}\right)^2 + 1} + 2\sqrt{2^2+1}\right]$

≈ 3.4567

Simpson's: $\displaystyle\int_0^2 x\sqrt{x^2+1}\,dx \approx \frac{1}{6}\left[0 + 4\left(\frac{1}{2}\right)\sqrt{\left(\frac{1}{2}\right)^2 + 1} + 2(1)\sqrt{1^2+1} + 4\left(\frac{3}{2}\right)\sqrt{\left(\frac{3}{2}\right)^2 + 1} + 2\sqrt{2^2+1}\right]$

≈ 3.3922

11. Trapezoidal: $\displaystyle\int_0^4 \frac{1}{x+1}\,dx \approx \frac{1}{2}\left[1 + 2\left(\frac{1}{2}\right) + 2\left(\frac{1}{3}\right) + 2\left(\frac{1}{4}\right) + \frac{1}{5}\right] = \frac{101}{60} \approx 1.6833$

Simpson's: $\displaystyle\int_0^4 \frac{1}{x+1}\,dx \approx \frac{1}{3}\left[1 + 4\left(\frac{1}{2}\right) + 2\left(\frac{1}{3}\right) + 4\left(\frac{1}{4}\right) + \frac{1}{5}\right] = \frac{73}{45} \approx 1.6222$

12. Trapezoidal: $\int_0^4 \sqrt{1+x^2}\,dx \approx \frac{1}{2}\left[1+2\sqrt{2}+2\sqrt{5}+2\sqrt{10}+\sqrt{17}\right] \approx 9.3741$

Simpson's: $\int_0^4 \sqrt{1+x^2}\,dx \approx \frac{1}{3}\left[1+4\sqrt{2}+2\sqrt{5}+4\sqrt{10}+\sqrt{17}\right] \approx 9.3004$

13. Trapezoidal: $\int_0^2 \sqrt{1+x^3}\,dx \approx \frac{1}{2}(1+2\sqrt{2}+3) = 2+\sqrt{2} \approx 3.41$

Simpson's: $\int_0^2 \sqrt{1+x^3}\,dx \approx \frac{1}{3}(1+4\sqrt{2}+3) = \frac{4}{3}(1+\sqrt{2}) \approx 3.22$

14. Trapezoidal: $\int_0^2 \frac{1}{\sqrt{1+x^3}}\,dx \approx \frac{1}{4}\left[1+2\left(\frac{1}{\sqrt{1+\left(\frac{1}{2}\right)^3}}\right)+2\left(\frac{1}{\sqrt{1+1^3}}\right)+2\left(\frac{1}{\sqrt{1+\left(\frac{3}{2}\right)^3}}\right)+\frac{1}{3}\right] \approx 1.3973$

Simpson's: $\int_0^2 \frac{1}{\sqrt{1+x^3}}\,dx \approx \frac{1}{6}\left[1+4\left(\frac{1}{\sqrt{1+\left(\frac{1}{2}\right)^3}}\right)+2\left(\frac{1}{\sqrt{1+1^3}}\right)+4\left(\frac{1}{\sqrt{1+\left(\frac{3}{2}\right)^3}}\right)+\frac{1}{3}\right] \approx 1.4052$

15. $\int_0^1 \sqrt{x}\sqrt{1-x}\,dx = \int_0^1 \sqrt{x(1-x)}\,dx$

Trapezoidal: $\int_0^1 \sqrt{x(1-x)}\,dx \approx \frac{1}{8}\left[0+2\sqrt{\frac{1}{4}\left(1-\frac{1}{4}\right)}+2\sqrt{\frac{1}{2}\left(1-\frac{1}{2}\right)}+2\sqrt{\frac{3}{4}\left(1-\frac{3}{4}\right)}\right] \approx 0.342$

Simpson's: $\int_0^1 \sqrt{x(1-x)}\,dx \approx \frac{1}{12}\left[0+4\sqrt{\frac{1}{4}\left(1-\frac{1}{4}\right)}+2\sqrt{\frac{1}{2}\left(1-\frac{1}{2}\right)}+4\sqrt{\frac{3}{4}\left(1-\frac{3}{4}\right)}\right] \approx 0.372$

16. Trapezoidal: $\int_0^1 \frac{1}{x^2+1}\,dx \approx \frac{1}{4}\left[1+\frac{2}{\left(\frac{1}{2}\right)^2+1}+\frac{1}{2}\right] = 0.775$

Simpson's: $\int_0^1 \frac{1}{x^2+1}\,dx \approx \frac{1}{6}\left[1+\frac{4}{\left(\frac{1}{2}\right)^2+1}+\frac{1}{2}\right] \approx 0.7833$

17. Trapezoidal: $\int_{-2}^2 \frac{1}{x^2+1}\,dx \approx \frac{1}{4}\left[\frac{1}{5}+\frac{2}{\left(-\frac{3}{2}\right)^2+1}+\frac{2}{(-1)^2+1}+\frac{2}{\left(-\frac{1}{2}\right)^2+1}+2+\frac{2}{\left(\frac{1}{2}\right)^2+1}\right.$

$\left.+\frac{2}{(1)^2+1}+\frac{2}{\left(\frac{3}{2}\right)^2+1}+\frac{1}{5}\right] \approx 2.2077$

Simpson's: $\int_{-2}^2 \frac{1}{x^2+1}\,dx \approx \frac{1}{6}\left[\frac{1}{5}+\frac{4}{\left(-\frac{3}{2}\right)^2+1}+1+\frac{4}{\left(-\frac{1}{2}\right)^2+1}+2+\frac{4}{\left(\frac{1}{2}\right)^2+1}+1+\frac{4}{\left(\frac{3}{2}\right)^2+1}+\frac{1}{5}\right]$

≈ 2.2103

18. Trapezoidal: $\int_{-1}^1 x\sqrt{x+1}\,dx \approx \frac{1}{4}\left[0+2\left(-\frac{1}{2\sqrt{2}}\right)+2(0)+2\left(\frac{\sqrt{3}}{2\sqrt{2}}\right)+\sqrt{2}\right] = \frac{1+\sqrt{3}}{4\sqrt{2}} \approx 0.4830$

Simpson's: $\int_{-1}^1 x\sqrt{x+1}\,dx \approx \frac{1}{6}\left[0+4\left(-\frac{1}{2\sqrt{2}}\right)+2(0)+4\left(\frac{\sqrt{3}}{2\sqrt{2}}\right)+\sqrt{2}\right] = \frac{1}{\sqrt{6}} \approx 0.4082$

19. Trapezoidal: $\int_1^7 \frac{\sqrt{x-1}}{x}\,dx \approx \frac{1}{2}\left[0+2\left(\frac{1}{2}\right)+2\left(\frac{\sqrt{2}}{3}\right)+2\left(\frac{\sqrt{3}}{4}\right)+2\left(\frac{2}{5}\right)+2\left(\frac{\sqrt{5}}{6}\right)+\frac{\sqrt{6}}{7}\right] \approx 2.3521$

Simpson's: $\int_1^7 \frac{\sqrt{x-1}}{x}\,dx \approx \frac{1}{3}\left[0+4\left(\frac{1}{2}\right)+2\left(\frac{\sqrt{2}}{3}\right)+4\left(\frac{\sqrt{3}}{4}\right)+2\left(\frac{2}{5}\right)+4\left(\frac{\sqrt{5}}{6}\right)+\frac{\sqrt{6}}{7}\right] \approx 2.4385$

20. Trapezoidal: $\int_2^5 \frac{1}{1+\sqrt{x-1}}\,dx \approx \frac{1}{4}\left[\frac{1}{2}+\frac{2}{1+\sqrt{\left(\frac{5}{2}\right)-1}}+\frac{2}{1+\sqrt{2}}+\frac{2}{1+\sqrt{\left(\frac{7}{2}\right)-1}}\right.$

$\left.+\frac{2}{1+\sqrt{3}}+\frac{2}{1+\sqrt{\left(\frac{9}{2}\right)-1}}+\frac{1}{3}\right] \approx 1.1911$

Simpson's: $\int_2^5 \frac{1}{1+\sqrt{x-1}}\,dx \approx \frac{1}{6}\left[\frac{1}{2}+\frac{4}{1+\sqrt{1.5}}+\frac{2}{1+\sqrt{2}}+\frac{4}{1+\sqrt{2.5}}+\frac{2}{1+\sqrt{3}}+\frac{4}{1+\sqrt{3.5}}+\frac{1}{3}\right]$

≈ 1.1891

21. $f(x) = x^3$

$f'(x) = 3x^2$

$f''(x) = 6x$

$f'''(x) = 6$

$f^{(4)}(x) = 0$

Trapezoidal: Error $\le \frac{(2-0)^3}{12(4^2)}(12) = 0.5$ since $f''(x)$ is maximum in $[0, 2]$ when $x = 2$.

Simpson's: Error $\le \frac{(2-0)^5}{180(4^4)}(0) = 0$ since $f^{(4)}(x) = 0$.

22. $f(x) = x^4$

$f'(x) = 4x^3$

$f''(x) = 12x^2$

$f'''(x) = 24x$

$f^{(4)}(x) = 24$

Trapezoidal: Error $\le \frac{(2-0)^3}{12(4^2)}(48) = 2$ since $f''(x)$ is maximum in $[0, 2]$ when $x = 2$.

Simpson's: Error $\le \frac{(2-0)^5}{180(4^4)}(24) = \frac{1}{60} \approx 0.01667$ since $f^{(4)}(x) = 24$.

23. $f(x) = \frac{1}{x+1}$

$f'(x) = \frac{-1}{(x+1)^2}$

$f''(x) = \frac{2}{(x+1)^3}$

$f'''(x) = \frac{-6}{(x+1)^4}$

$f^{(4)}(x) = \frac{24}{(x+1)^5}$

Trapezoidal: Error $\le \frac{(1-0)^3}{12(4^2)}(2) = \frac{1}{96} \approx 0.01$ since $f''(x)$ is maximum in $[0, 1]$ when $x = 0$.

Simpson's: Error $\le \frac{(1-0)^5}{180(4^4)}(24) = \frac{1}{1920} \approx 0.0005$ since $f^{(4)}(x)$ is maximum in $[0, 1]$ when $x = 0$.

24. $f(x) = \frac{1}{x^2+1}$

$f'(x) = \frac{-2x}{(x^2+1)^2}$

$f''(x) = \frac{-2(1-3x^2)}{(x^2+1)^3}$

$f'''(x) = \frac{-24x(x^2-1)}{(x^2+1)^4}$

$f^{(4)}(x) = (-24)\frac{-5x^4+10x^2-1}{(x^2+1)^5}$

Trapezoidal: Error $\le \frac{(1-0)^3}{12(4^2)}(2) = \frac{1}{96} \approx 0.0104$ since $f''(x)$ is maximum in $[0, 1]$ when $x = 0$.

Simpson's: Error $\le \frac{(1-0)^5}{180(4^5)}(24) = \frac{1}{1920} \approx 0.0005$ since $f^{(4)}(x)$ is maximum in $[0, 1]$ when $x = 0$.

25. $f''(x) = \dfrac{2}{x^3}$ in $[1, 3]$

$|f''(x)|$ is maximum when $x = 1$ and $|f''(1)| = 2$.

Trapezoidal: Error $\leq \dfrac{2^3}{12n^2}(2) < 0.00001$, $n^2 > 133333.33$, $n > 365.15$; let $n = 366$.

$f^{(4)}(x) = \dfrac{24}{x^5}$ in $[1, 3]$

$|f^{(4)}(x)|$ is maximum when $x = 1$ and when $|f^{(4)}(1)| = 24$.

Simpson's: Error $\leq \dfrac{2^5}{180n^4}(24) < 0.00001$, $n^4 > 426666.67$, $n > 25.55$; let $n = 26$.

26. $f''(x) = \dfrac{2}{(1+x)^3}$ in $[0, 1]$

$|f''(x)|$ is maximum when $x = 0$ and $|f''(0)| = 2$.

Trapezoidal: Error $\leq \dfrac{1}{12n^2}(2) < 0.00001$, $n^2 > 16666.67$, $n > 129.099$; let $n = 130$.

$f^{(4)}(x) = \dfrac{24}{(1+x)^4}$ in $[0, 1]$

$|f^{(4)}(x)|$ is maximum when $x = 0$ and $|f^{(4)}(0)| = 24$.

Simpson's: Error $\leq \dfrac{1}{180n^4}(24) < 0.00001$, $n^4 > 13333.33$, $n > 10.75$; let $n = 12$.

(In Simpson's Rule, n must be even.)

27. $f''(x) = -\dfrac{1}{4(1+x)^{3/2}}$ in $[0, 2]$

$|f''(x)|$ is maximum when $x = 0$ and $|f''(0)| = \frac{1}{4}$.

Trapezoidal: Error $\leq \dfrac{8}{12n^2}\left(\dfrac{1}{4}\right) < 0.00001$, $n^2 > 16666.67$, $n > 129.099$; let $n = 130$.

$f^{(4)}(x) = \dfrac{-15}{16(1+x)^{7/2}}$ in $[0, 2]$

$|f^{(4)}(x)|$ is maximum when $x = 0$ and $|f^{(4)}(0)| = \frac{15}{16}$.

Simpson's: Error $\leq \dfrac{32}{180n^4}\left(\dfrac{15}{16}\right) < 0.00001$, $n^4 > 16666.67$, $n > 11.362$; let $n = 12$.

28. $f''(x) = -\dfrac{2}{9(x+1)^{4/3}}$ in $[0, 2]$

$|f''(x)|$ is maximum when $x = 0$ and $|f''(0)| = \frac{2}{9}$.

Trapezoidal: Error $\leq \dfrac{8}{12n^2}\left(\dfrac{2}{9}\right) < 0.00001$, $n^2 > 14814.81$, $n > 121.72$; let $n = 122$.

$f^{(4)}(x) = -\dfrac{56}{81(x+1)^{10/3}}$ in $[0, 2]$

$|f^{(4)}(x)|$ is maximum when $x = 0$ and $|f^{(4)}(0)| = \frac{56}{81}$.

Simpson's: Error $\leq \dfrac{32}{180n^4}\left(\dfrac{56}{81}\right) < 0.00001$, $n^4 > 12290.81$, $n > 10.529$; let $n = 12$.

(In Simpson's Rule, n must be even.)

29. Simpson's Rule: $n = 6$

$$\pi = 4\int_0^1 \frac{1}{1+x^2}\,dx \approx \frac{4}{3(6)}\left[1+\frac{4}{1+(\frac{1}{6})^2}+\frac{2}{1+(\frac{2}{6})^2}+\frac{4}{1+(\frac{3}{6})^2}+\frac{2}{1+(\frac{4}{6})^2}+\frac{4}{1+(\frac{5}{6})^2}+\frac{1}{2}\right]$$

$$\approx 3.14159$$

30. Let $f(x) = ax^3 + bx^2 + cx + d$. Then $f^{(4)}(x) = 0$.

Simpson's: Error $\le \dfrac{(b-a)^5}{180n^4}(0) = 0$

Therefore, Simpson's Rule is exact when approximating the integral of a cubic polynomial.

Example: $\displaystyle\int_0^1 x^3\,dx = \frac{1}{6}\left[0+4\left(\frac{1}{2}\right)^3+1\right] = \frac{1}{4}$

This is the exact value of the integral.

31. Area $\approx \dfrac{1000}{2(10)}[125+2(125)+2(120)+2(112)+2(90)+2(90)+2(95)+2(88)+2(75)+2(35)] = 89{,}250$ sq ft

32. (a) Trapezoidal:

Area $\approx \dfrac{160}{2(8)}[0+2(50)+2(54)+2(82)+2(82)+2(73)+2(75)+2(80)+0] = 9920$ square feet

(b) Simpson's:

Area $\approx \dfrac{160}{3(8)}[0+4(50)+2(54)+4(82)+2(82)+4(73)+2(75)+4(80)+0] = 10{,}413\frac{1}{3}$ square feet

33. Trapezoidal:

$$\int_0^2 f(x)\,dx \approx \frac{2}{2(8)}[4.32+2(4.36)+2(4.58)+2(5.79)+2(6.14)+2(7.25)+2(7.64)+2(8.08)+8.14]$$

$$= 12.5175$$

Simpson's:

$$\int_0^2 f(x)\,dx \approx \frac{2}{3(8)}[4.32+4(4.36)+2(4.58)+4(5.79)+2(6.14)+4(7.25)+2(7.64)+4(8.08)+8.14]$$

$$\approx 12.5917$$

34. The program will vary depending upon the computer or programmable calculator that you use.

35. $f(x) = \sqrt{2+3x^2}$ on $[0, 4]$

n	$L(n)$	$M(n)$	$R(n)$	$T(n)$	$S(n)$
4	12.7771	15.3965	18.4340	15.6055	15.4845
8	14.0868	15.4480	16.9152	15.5010	15.4662
10	14.3569	15.4544	16.6197	15.4883	15.4658
12	14.5386	15.4578	16.4242	15.4814	15.4657
16	14.7674	15.4613	16.1816	15.4745	15.4657
20	14.9056	15.4628	16.0370	15.4713	15.4657

36. $f(x) = \sqrt{1-x^2}$ on [0, 1]

n	$L(n)$	$M(n)$	$R(n)$	$T(n)$	$S(n)$
4	0.8739	0.7960	0.6239	0.7489	0.7709
8	0.8350	0.7892	0.7100	0.7725	0.7803
10	0.8261	0.7881	0.7261	0.7761	0.7818
12	0.8200	0.7875	0.7367	0.7783	0.7826
16	0.8121	0.7867	0.7496	0.7808	0.7836
20	0.8071	0.7864	0.7571	0.7821	0.7841

37. $W = \int_0^5 100x\sqrt{125-x^3}\,dx$

Simpson's Rule: $n = 12$

$$\int_0^5 100x\sqrt{125-x^3}\,dx$$

$$\approx \frac{5}{3(12)}\left[0 + 400\left(\frac{5}{12}\right)\sqrt{125-\left(\frac{5}{12}\right)^3} + 200\left(\frac{10}{12}\right)\sqrt{125-\left(\frac{10}{12}\right)^3} + 400\left(\frac{15}{12}\right)\sqrt{125-\left(\frac{15}{12}\right)^3} + \cdots + 0\right] \approx 10233.58 \text{ ft} \cdot \text{lb}$$

Chapter 5 Review Exercises

1. $\int \frac{2}{3\sqrt[3]{x}}\,dx = \frac{2}{3}\int x^{-1/3}\,dx = x^{2/3} + C$

2. $u = 3x$

$du = 3\,dx$

$$\int \frac{2}{\sqrt[3]{3x}}\,dx = \frac{2}{3}\int (3x)^{-1/3}(3)\,dx = (3x)^{2/3} + C$$

3. $\int (2x^2 + x - 1)\,dx = \frac{2}{3}x^3 + \frac{1}{2}x^2 - x + C$

4. $\int \frac{x^3 - 2x^2 + 1}{x^2}\,dx = \int (x - 2 + x^{-2})\,dx$

$$= \frac{1}{2}x^2 - 2x - \frac{1}{x} + C$$

5. $\int \frac{(1+x)^2}{\sqrt{x}}\,dx = \int (x^{-1/2} + 2x^{1/2} + x^{3/2})\,dx$

$$= 2x^{1/2} + \frac{4}{3}x^{3/2} + \frac{2}{5}x^{5/2} + C$$

$$= \frac{2\sqrt{x}}{15}(15 + 10x + 3x^2) + C$$

6. $u = x^3 + 3$

$du = 3x^2\,dx$

$$\int x^2\sqrt{x^3+3}\,dx = \frac{1}{3}\int (x^3+3)^{1/2}(3x^2)\,dx$$

$$= \frac{2}{9}(x^3+3)^{3/2} + C$$

7. $u = x^3 + 3$

$du = 3x^2\,dx$

$$\int \frac{x^2}{\sqrt{x^3+3}}\,dx = \frac{1}{3}\int (x^3+3)^{-1/2}(3x^2)\,dx$$

$$= \frac{2}{3}\sqrt{x^3+3} + C$$

8. $$\int \frac{x^2+2x}{(x+1)^2}\,dx = \int \frac{(x^2+2x+1)-1}{(x+1)^2}\,dx$$

$$= \int \left[1 - \frac{1}{(x+1)^2}\right]dx$$

$$= \int 1\,dx - \int (x+1)^{-2}\,dx$$

$$= x + (x+1)^{-1} + C$$

$$= x + \frac{1}{x+1} + C$$

9. $$\int (x^2+1)^3\,dx = \int (x^6 + 3x^4 + 3x^2 + 1)\,dx$$

$$= \frac{1}{7}x^7 + \frac{3}{5}x^5 + x^3 + x + C$$

10. $u = 2 - 5x$

$du = -5\,dx$

$$\int \sqrt{2-5x}\,dx = -\frac{1}{5}\int (2-5x)^{1/2}(-5)\,dx$$

$$= -\frac{2}{15}(2-5x)^{3/2} + C$$

11. $$\int x(x^2+1)^3\,dx = \frac{1}{2}\int (x^2+1)^3(2x)\,dx$$

$$= \frac{1}{8}(x^2+1)^4 + C$$

12. $$\int x^2\sqrt{4-x^3}\,dx = -\frac{1}{3}\int (4-x^3)^{1/2}(-3x^2)\,dx$$

$$= -\frac{2}{9}(4-x^3)^{3/2} + C$$

13. $$\int \frac{x}{(x^2+1)^3}\,dx = \frac{1}{2}\int (x^2+1)^{-3}(2x)\,dx$$

$$= \frac{-1}{4(x^2+1)^2} + C$$

14. $$\int \frac{x}{\sqrt{25-9x^2}}\,dx = -\frac{1}{18}\int (25-9x^2)^{-1/2}(-18x)\,dx$$

$$= -\frac{1}{9}\sqrt{25-9x^2} + C$$

15. $u = \sqrt{x+5}, \quad x = u^2 - 5, \quad dx = 2u\,du$

$$\int x^2\sqrt{x+5}\,dx = \int (u^2-5)^2(2u^2)\,du$$

$$= \frac{50u^3}{3} - 4u^5 + \frac{2u^7}{7} + C$$

$$= \frac{2}{21}u^3(175 - 42u^2 + 3u^4) + C$$

$$= \frac{2}{21}(x+5)^{3/2}(3x^2 - 12x + 40) + C$$

16. $u = \sqrt{x+5}, \quad x = u^2 - 5, \quad dx = 2u\,du$

$$\int x\sqrt{x+5}\,dx = \int u(u^2-5)(2u)\,du$$

$$= 2\int (u^4 - 5u^2)\,du$$

$$= 2\left(\frac{u^5}{5} - 5\frac{u^3}{3}\right) + C$$

$$= \frac{2}{5}(x+5)^{5/2} - \frac{10}{3}(x+5)^{3/2} + C$$

$$= \frac{2}{15}(x+5)^{3/2}(3x-10) + C$$

17. $$\int \frac{x^3+1}{x^2}\,dx = \int \left(x + \frac{1}{x^2}\right)dx$$

$$= \frac{1}{2}x^2 - \frac{1}{x} + C$$

18. $$\int \left(x + \frac{1}{x}\right)^2 dx = \int (x^2 + 2 + x^{-2})\,dx$$

$$= \frac{1}{3}x^3 + 2x - \frac{1}{x} + C$$

19. (a) $\displaystyle\sum_{i=1}^{10} (2i-1)$ (b) $\displaystyle\sum_{i=1}^{n} i^3$ (c) $\displaystyle\sum_{i=1}^{10} (4i+2)$

20. $x_1 = 2, \quad x_2 = -1, \quad x_3 = 5, \quad x_4 = 3, \quad x_5 = 7$

(a) $\frac{1}{5}\sum_{i=1}^{5} x_i = \frac{1}{5}(2 - 1 + 5 + 3 + 7) = \frac{16}{5}$

(b) $\sum_{i=1}^{5} \frac{1}{x_i} = \frac{1}{2} - 1 + \frac{1}{5} + \frac{1}{3} + \frac{1}{7} = \frac{37}{210}$

(c) $\sum_{i=1}^{5} (2x_i - x_i^2) = [2(2) - (2)^2] + [2(-1) - (-1)^2] + [2(5) - (5)^2] + [2(3) - (3)^2] + [2(7) - (7)^2] = -56$

(d) $\sum_{i=2}^{5} (x_i - x_{i-1}) = (-1 - 2) + [5 - (-1)] + (3 - 5) + (7 - 3) = 5$

21. $\int_0^4 (2 + x)\,dx = \left[2x + \frac{x^2}{2}\right]_0^4 = 8 + \frac{16}{2} = 16$

22. $\int_{-1}^{1} (t^2 + 2)\,dt = \left[\frac{t^3}{3} + 2t\right]_{-1}^{1} = \frac{14}{3}$

23. $\int_{-1}^{1} (4t^3 - 2t)\,dt = \left[t^4 - t^2\right]_{-1}^{1} = 0$

24. $u = x^2 - 8, \quad du = 2x\,dx$

$$\int_3^6 \frac{x}{3\sqrt{x^2 - 8}}\,dx = \frac{1}{6}\int_3^6 (x^2 - 8)^{-1/2}(2x)\,dx$$

$$= \frac{1}{3}(x^2 - 8)^{1/2}\Big]_3^6$$

$$= \frac{1}{3}(2\sqrt{7} - 1)$$

25. $u = 1 + x$

$du = 1\,dx$

$$\int_0^3 \frac{1}{\sqrt{1 + x}}\,dx = \int_0^3 (1 + x)^{-1/2}\,dx$$

$$= 2(1 + x)^{1/2}\Big]_0^3$$

$$= 4 - 2 = 2$$

26. $u = x^3 + 1$

$du = 3x^2\,dx$

$$\int_0^1 x^2(x^3 + 1)^3\,dx = \frac{1}{3}\int_0^1 (x^3 + 1)^3(3x^2)\,dx$$

$$= \frac{1}{12}(x^3 + 1)^4\Big]_0^1$$

$$= \frac{1}{12}(16 - 1) = \frac{5}{4}$$

27. $\int_4^9 x\sqrt{x}\,dx = \int_4^9 x^{3/2}\,dx = \frac{2}{5}x^{5/2}\Big]_4^9 = \frac{2}{5}\left[(\sqrt{9})^5 - (\sqrt{4})^5\right] = \frac{2}{5}(243 - 32) = \frac{422}{5}$

28. $\int_1^2 \left(\frac{1}{x^2} - \frac{1}{x^3}\right)dx = \int_1^2 (x^{-2} - x^{-3})\,dx = \left[-\frac{1}{x} + \frac{1}{2x^2}\right]_1^2 = \left(-\frac{1}{2} + \frac{1}{8}\right) - \left(-1 + \frac{1}{2}\right) = \frac{1}{8}$

29. $u = \sqrt{1 - y}, \quad y = 1 - u^2, \quad dy = -2u\,du$ when $y = 0 \Rightarrow u = 1$ and when $y = 1 \Rightarrow u = 0$.

$$2\pi \int_0^1 (y + 1)\sqrt{1 - y}\,dy = 2\pi \int_1^0 (2 - u^2)(u)(-2u)\,du$$

$$= -4\pi \int_1^0 (2u^2 - u^4)\,du = -4\pi\left(\frac{2}{3}u^3 - \frac{1}{5}u^5\right)\Big]_1^0 = \frac{28\pi}{15}$$

30. $u = \sqrt{x+1}$, $x = u^2 - 1$, $dx = 2u\,du$ when $x = -1 \Rightarrow u = 0$ and when $x = 0 \Rightarrow u = 1$.

$$2\pi \int_{-1}^{0} x^2\sqrt{x+1}\,dx = 2\pi \int_0^1 (u^2-1)^2(u)(2u)\,du$$

$$= 4\pi \int_0^1 (u^6 - 2u^4 + u^2)\,du = 4\pi\left(\frac{1}{7}u^7 - \frac{2}{5}u^5 + \frac{1}{3}u^3\right)\Big]_0^1 = \frac{32\pi}{105}$$

31. $f'(x) = -2x$, $(-1, 1)$

$$f(x) = \int -2x\,dx = -x^2 + C$$

When $x = -1$:

$y = -1 + C = 1$

$C = 2$

$y = 2 - x^2$

32. $f''(x) = 6(x-1)$

Tangent to $3x - y - 5 = 0$ at $(2, 1)$

$$f'(x) = \int 6(x-1)\,dx = 3(x-1)^2 + C_1$$

$f'(2) = 3 + C_1 = 3$ when $C_1 = 0$.

$f'(x) = 3(x-1)^2$

$$f(x) = \int 3(x-1)^2\,dx = (x-1)^3 + C_2$$

$f(2) = 1 + C_2 = 1$ when $C_2 = 0$.

$f(x) = (x-1)^3$

33. $a(t) = a$

$$v(t) = \int a\,dt = at + C_1$$

$v(0) = 0 + C_1 = 0$ when $C_1 = 0$.

$v(t) = at$

$$s(t) = \int at\,dt = \frac{a}{2}t^2 + C_2$$

$s(0) = 0 + C_2 = 0$ when $C_2 = 0$.

$$s(30) = \frac{a}{2}(30)^2 = 3600 \text{ or}$$

$$a = \frac{2(3600)}{(30)^2} = 8 \text{ ft/sec}^2$$

$v(30) = 8(30) = 240$ ft/sec

34. 45 mph $= 66$ ft/sec

30 mph $= 44$ ft/sec

$a(t) = -a$

$v(t) = -at + 66$ since $v(0) = 66$ ft/sec.

$s(t) = -\frac{a}{2}t^2 + 66t$ since $s(0) = 0$.

Solving the system

$v(t) = -at + 66 = 44$

$s(t) = -\frac{a}{2}t^2 + 66t = 264$

we obtain $t = \frac{24}{5}$ and $a = \frac{55}{12}$.

We now solve $-(\frac{55}{12})t + 66 = 0$ and get $t = \frac{792}{55}$.

Thus,

$$s\left(\frac{792}{55}\right) = -\frac{1}{2}\left(\frac{792}{55}\right)^2 + 66\left(\frac{792}{55}\right) \approx 475.2 \text{ ft.}$$

Stopping distance from 30 mph to rest is $475.2 - 264 = 211.2$ ft.

35. $a(t) = -32$

$v(t) = -32t + 96$

$s(t) = -16t^2 + 96t$

(a) $v(t) = -32t + 96 = 0$ when $t = 3$ sec.

(b) $s(3) = -144 + 288 = 144$ ft

(c) $v(t) = -32t + 96 = \frac{96}{2}$ when $t = \frac{3}{2}$ sec.

(d) $s(\frac{3}{2}) = -16(\frac{9}{4}) + 96(\frac{3}{2}) = 108$ ft

36. $a(t) = -32$

$v(t) = -32t + 128$

$s(t) = -16t^2 + 128t$

(a) $v(t) = -32t + 128 = 0$ when $t = 4$ sec.

(b) $s(4) = 256$ ft

(c) $v(t) = -32t + 128 = \frac{128}{2}$ when $t = 2$ sec.

(d) $s(2) = 192$ ft

37. (a) $S = m\left(\frac{b}{4}\right)\left(\frac{b}{4}\right) + m\left(\frac{2b}{4}\right)\left(\frac{b}{4}\right) + m\left(\frac{3b}{4}\right)\left(\frac{b}{4}\right) + m\left(\frac{4b}{4}\right)\left(\frac{b}{4}\right) = \frac{mb^2}{16}(1+2+3+4) = \frac{5mb^2}{8}$

$s = m(0)\left(\frac{b}{4}\right) + m\left(\frac{b}{4}\right)\left(\frac{b}{4}\right) + m\left(\frac{2b}{4}\right)\left(\frac{b}{4}\right) + m\left(\frac{3b}{4}\right)\left(\frac{b}{4}\right) = \frac{mb^2}{16}(1+2+3) = \frac{3mb^2}{8}$

(b) $S(n) = \sum_{i=1}^{n} f\left(\frac{bi}{n}\right)\left(\frac{b}{n}\right) = \sum_{i=1}^{n}\left(\frac{mbi}{n}\right)\left(\frac{b}{n}\right) = m\left(\frac{b}{n}\right)^2 \sum_{i=1}^{n} i = \frac{mb^2}{n^2}\left(\frac{n(n+1)}{2}\right) = \frac{mb^2(n+1)}{2n}$

$s(n) = \sum_{i=0}^{n-1} f\left(\frac{bi}{n}\right)\left(\frac{b}{n}\right) = \sum_{i=0}^{n-1} m\left(\frac{bi}{n}\right)\left(\frac{b}{n}\right) = m\left(\frac{b}{n}\right)^2 \sum_{i=0}^{n-1} i = \frac{mb^2}{n^2}\left(\frac{(n-1)n}{2}\right) = \frac{mb^2(n-1)}{2n}$

(c) Area $= \lim_{n\to\infty} \frac{mb^2(n+1)}{2n} = \lim_{n\to\infty} \frac{mb^2(n-1)}{2n} = \frac{1}{2}mb^2$

(d) $\int_0^b mx\,dx = \frac{1}{2}mx^2\Big]_0^b = \frac{1}{2}mb^2$

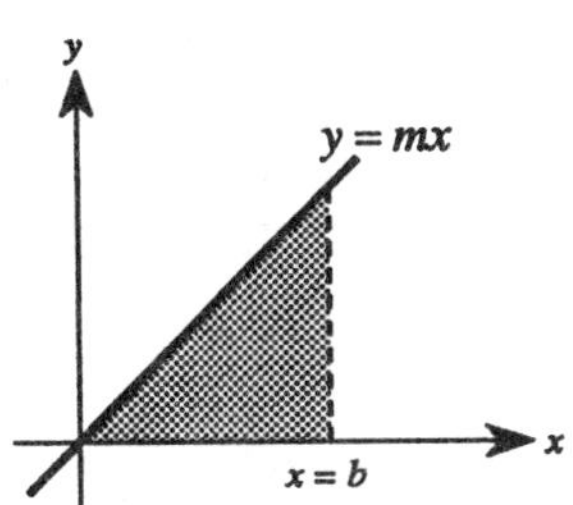

38. (a) $S(n) = \sum_{i=1}^{n} f\left(1 + \frac{2i}{n}\right)\left(\frac{2}{n}\right)$

$= \sum_{i=1}^{n}\left(1 + \frac{2i}{n}\right)^2\left(\frac{2}{n}\right)$

$= \sum_{i=1}^{n}\left(1 + \frac{6i}{n} + \frac{12i^2}{n^2} + \frac{8i^3}{n^3}\right)\left(\frac{2}{n}\right)$

$= 2 + \frac{12}{n^2}\sum_{i=1}^{n} i + \frac{24}{n^3}\sum_{i=1}^{n} i^2 + \frac{16}{n^4}\sum_{i=1}^{n} i^3$

$= 2 + 6\left(1 + \frac{1}{n}\right) + 4\left(2 + \frac{3}{n} + \frac{1}{n^2}\right) + 4\left(1 + \frac{2}{n} + \frac{1}{n^2}\right)$

Area $= \lim_{n\to\infty} S(n) = 2 + 6 + 8 + 4 = 20$

(b) $\int_1^3 x^3\,dx = \frac{x^4}{4}\Big]_1^3$

$= \frac{81}{4} - \frac{1}{4} = 20$

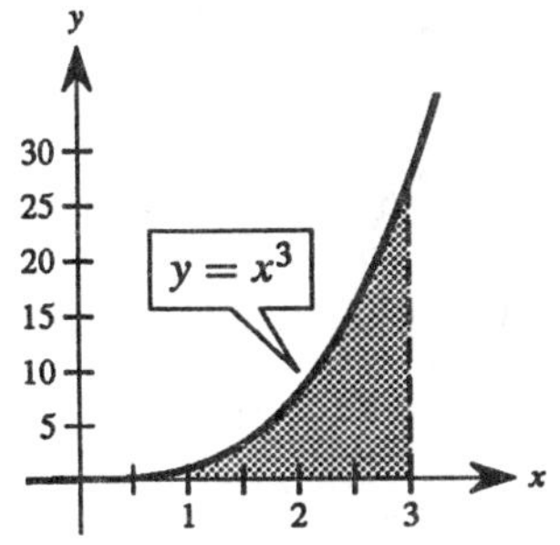

39. $\int_1^3 (2x - 1)\,dx = \left[x^2 - x\right]_1^3 = 6$

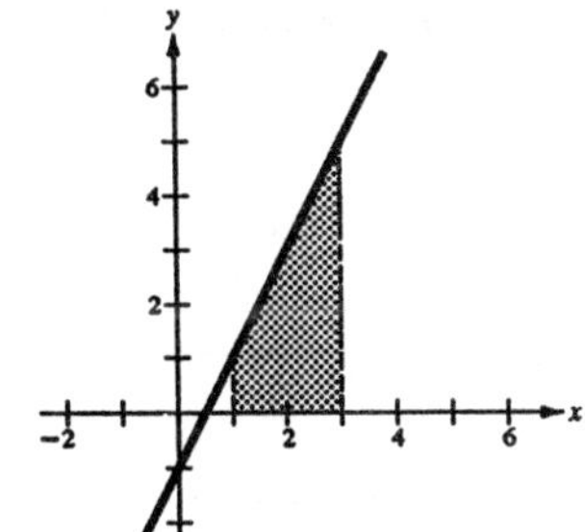

40. $\int_0^2 (x + 4)\,dx = \left[\frac{x^2}{2} + 4x\right]_0^2 = 10$

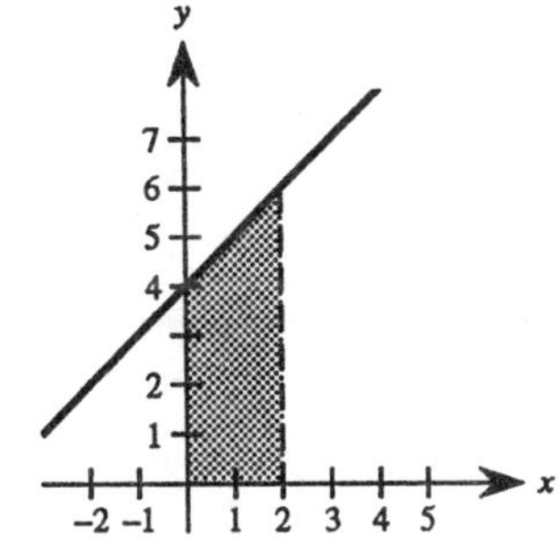

41. $\displaystyle\int_3^4 (x^2 - 9)\,dx = \left[\frac{x^3}{3} - 9x\right]_3^4$

$\displaystyle = \left(\frac{64}{3} - 36\right) - (9 - 27)$

$\displaystyle = \frac{64}{3} - \frac{54}{3} = \frac{10}{3}$

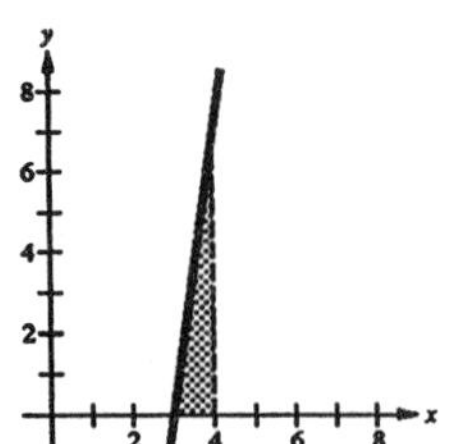

42. $\displaystyle\int_{-1}^2 (-x^2 + x + 2)\,dx = \left[-\frac{x^3}{3} + \frac{x^2}{2} + 2x\right]_{-1}^2$

$\displaystyle = \left(-\frac{8}{3} + 2 + 4\right) - \left(\frac{1}{3} + \frac{1}{2} - 2\right)$

$\displaystyle = \frac{10}{3} + \frac{7}{6} = \frac{9}{2}$

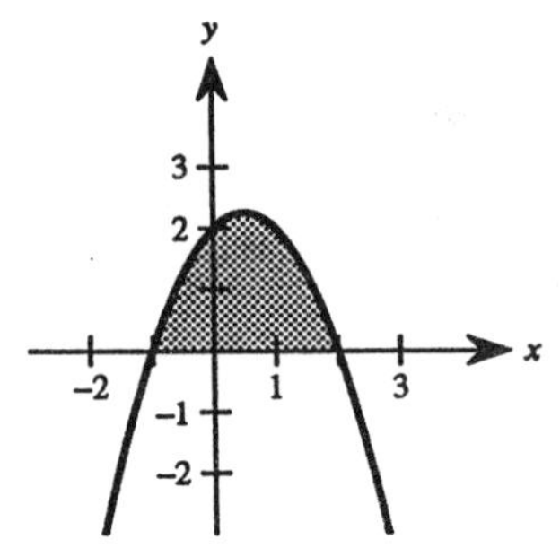

43. $\displaystyle\int_0^1 (x - x^3)\,dx = \left[\frac{x^2}{2} - \frac{x^4}{4}\right]_0^1$

$\displaystyle = \frac{1}{2} - \frac{1}{4} = \frac{1}{4}$

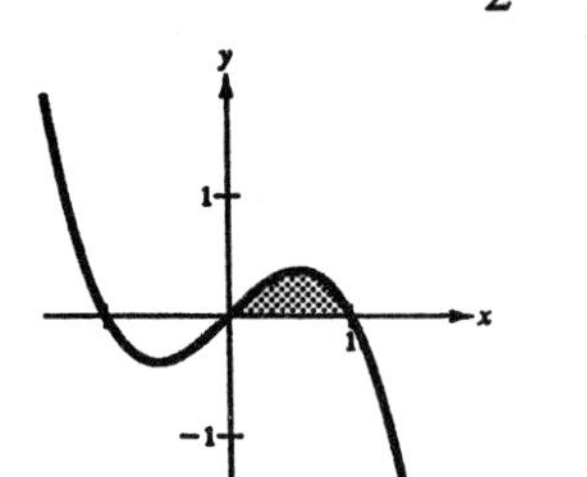

44. $\displaystyle\int_0^1 \sqrt{x}(1 - x)\,dx = \int_0^1 (x^{1/2} - x^{3/2})\,dx$

$\displaystyle = \left[\frac{2}{3}x^{3/2} - \frac{2}{5}x^{5/2}\right]_0^1$

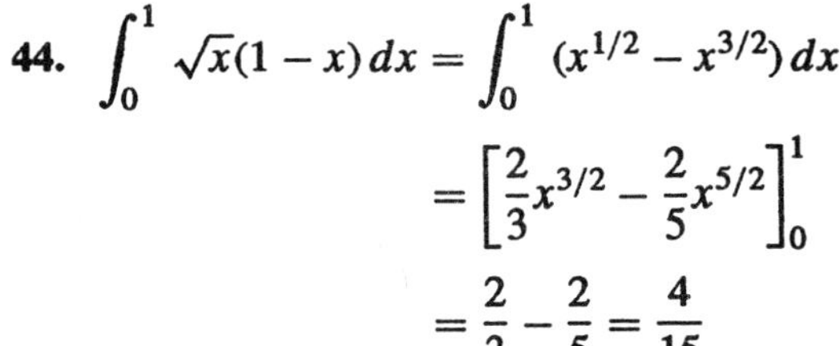

$\displaystyle = \frac{2}{3} - \frac{2}{5} = \frac{4}{15}$

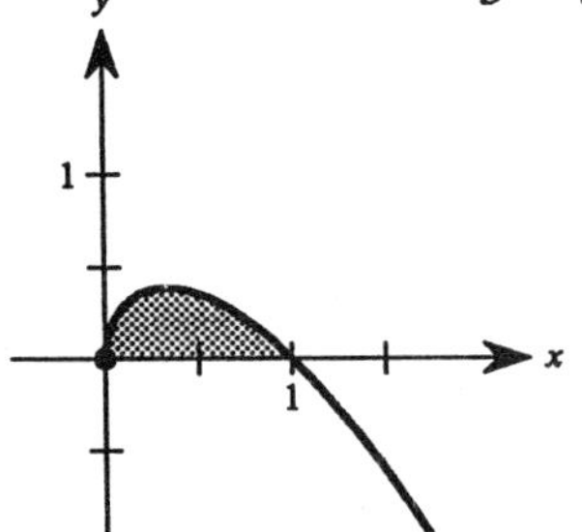

45. $\displaystyle\frac{1}{10 - 5}\int_5^{10} \frac{1}{\sqrt{x - 1}}\,dx = \frac{2}{5}\sqrt{x - 1}\Big]_5^{10} = \frac{2}{5}$

$\displaystyle\frac{1}{\sqrt{x - 1}} = \frac{2}{5}$

$\displaystyle\sqrt{x - 1} = \frac{5}{2}$

$\displaystyle x - 1 = \frac{25}{4}$

$\displaystyle x = \frac{29}{4}$

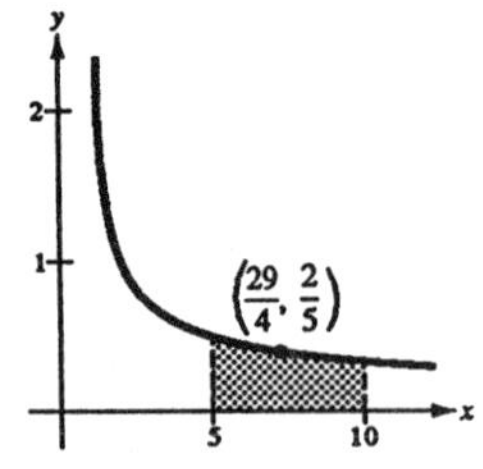

46. $\frac{1}{2-0}\int_0^2 x^3\,dx = \left.\frac{x^4}{8}\right]_0^2 = 2$

$x^3 = 2$

$x = \sqrt[3]{2}$

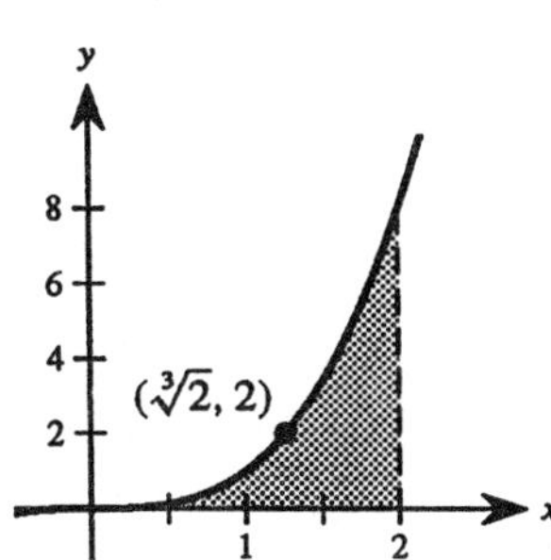

47. $\frac{1}{4-0}\int_0^4 x\,dx = \left.\frac{x^2}{8}\right]_0^4 = 2$

$x = 2$

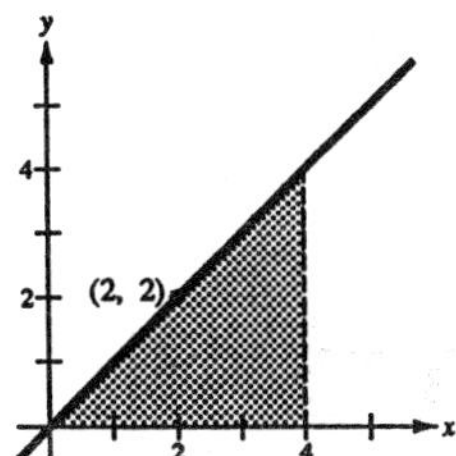

48. $\frac{1}{2-1}\int_1^2 \left(x^2 - \frac{1}{x^2}\right)dx = \left.\left(\frac{x^3}{3} + \frac{1}{x}\right)\right]_1^2 = \left(\frac{8}{3} + \frac{1}{2}\right) - \left(\frac{1}{3} + 1\right) = \frac{11}{6}$

$$x^2 - \frac{1}{x^2} = \frac{11}{6}$$

$$6x^4 - 6 = 11x^2$$

$$6x^4 - 11x^2 - 6 = 0 \text{ when } x^2 = \frac{11 \pm \sqrt{265}}{12}.$$

In the interval [1, 2]: $x = \sqrt{\frac{11 + \sqrt{265}}{12}} \approx 1.508$

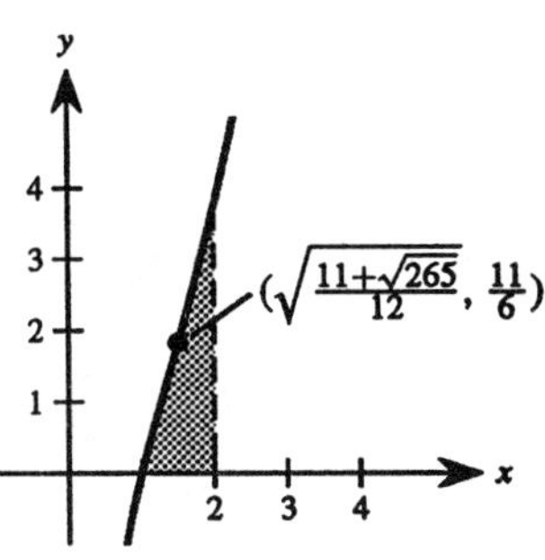

49. Simpson's Rule ($n = 4$):

$$\int_1^2 \frac{1}{1+x^3}\,dx \approx \frac{1}{12}\left[\frac{1}{1+1^3} + \frac{4}{1+(1.25)^3} + \frac{2}{1+(1.5)^3} + \frac{4}{1+(1.75)^3} + \frac{1}{1+2^3}\right] \approx 0.254$$

50. Simpson's Rule ($n = 4$):

$$\int_0^1 \frac{x^{3/2}}{3-x^2}\,dx \approx \frac{1}{12}\left[0 + \frac{4\left(\frac{1}{4}\right)^{3/2}}{3-\left(\frac{1}{4}\right)^2} + \frac{2\left(\frac{1}{2}\right)^{3/2}}{3-\left(\frac{1}{2}\right)^2} + \frac{4\left(\frac{3}{4}\right)^{3/2}}{3-\left(\frac{3}{4}\right)^2} + \frac{1}{2}\right] \approx 0.166$$

51. $p = 1 + 0.1t + 0.02t^2$

$$C = \frac{15{,}000}{M}\int_t^{t+1} (1 + 0.1t + 0.02t^2)\,dt = \frac{15{,}000}{M}\left[t + 0.05t^2 + \frac{0.02}{3}t^3\right]_t^{t+1}$$

(a) Since 1983 is represented by $t = 0$, 1985 is represented by $t = 2$.

$$C = \frac{15{,}000}{M}\left[t + 0.05t^2 + \frac{0.02}{3}t^3\right]_2^3 = \frac{20{,}650}{M}$$

(b) 1990 is represented by $t = 7$.

$$C = \frac{15{,}000}{M}\left[t + 0.05t^2 + \frac{0.02}{3}t^3\right]_7^8 = \frac{43{,}150}{M}$$

52. $u = \sqrt{1-x}, \quad x = 1 - u^2, \quad dx = -2u\,du$

$$P_{a,b} = \int_a^b \frac{15}{4} x\sqrt{1-x}\,dx$$

$$= \frac{15}{4}\int_{\sqrt{1-a}}^{\sqrt{1-b}} (1-u^2)(u)(-2u)\,du = -\frac{15}{2}\int_{\sqrt{1-a}}^{\sqrt{1-b}} (u^2 - u^4)\,du$$

$$= -\frac{15}{2}\left(\frac{u^3}{3} - \frac{u^5}{5}\right)\Bigg]_{\sqrt{1-a}}^{\sqrt{1-b}} = -\frac{u^3}{2}(5 - 3u^2)\Bigg]_{\sqrt{1-a}}^{\sqrt{1-b}} = -\frac{(1-x)^{3/2}}{2}(3x+2)\Bigg]_a^b$$

(a) $P_{0.50,0.75} = -\frac{(1-x)^{3/2}}{2}(3x+2)\Bigg]_{0.50}^{0.75} = 35.3\%$

(b) $P_{0,b} = -\frac{(1-x)^{3/2}}{2}(3x+2)\Bigg]_0^b = -\frac{(1-b)^{3/2}}{2}(3b+2) + 1 = 0.5$

$(1-b)^{3/2}(3b+2) = 1$

$b \approx 58.6\%$

53. $u = \sqrt{1-x}, \quad x = 1 - u^2, \quad dx = -2u\,du$

$$P_{a,b} = \int_a^b \frac{1155}{32} x^3(1-x)^{3/2}\,dx = \frac{1155}{32}\int_{\sqrt{1-a}}^{\sqrt{1-b}} (1-u^2)^3(u^3)(-2u)\,du$$

$$= \frac{1155}{16}\int_{\sqrt{1-a}}^{\sqrt{1-b}} (u^{10} - 3u^8 + 3u^6 - u^4)\,du = \frac{1155}{16}\left(\frac{u^{11}}{11} - \frac{u^9}{3} + \frac{3u^7}{7} - \frac{u^5}{5}\right)\Bigg]_{\sqrt{1-a}}^{\sqrt{1-b}}$$

(a) $P_{0,0.25} = \frac{1155}{16}\left(\frac{u^{11}}{11} - \frac{u^9}{3} + \frac{3u^7}{7} - \frac{u^5}{5}\right)\Bigg]_1^{\sqrt{0.75}} \approx 0.025 = 2.5\%$

(b) $P_{0.50,1} = \frac{1155}{16}\left(\frac{u^{11}}{11} - \frac{u^9}{3} + \frac{3u^7}{7} - \frac{u^5}{5}\right)\Bigg]_{\sqrt{0.50}}^{0} \approx 0.736 = 73.6\%$

CHAPTER 6

Applications of Integration

Section 6.1 Area of a Region Between Two Curves

1. $A = \int_0^6 [0 - (x^2 - 6x)]\, dx = -\left[\frac{x^3}{3} - 3x^2\right]_0^6 = 36$

2. $A = \int_{-2}^{2} [(2x+5) - (x^2 + 2x + 1)]\, dx = \int_{-2}^{2} (-x^2 + 4)\, dx = \left[-\frac{x^3}{3} + 4x\right]_{-2}^{2} = \frac{32}{3}$

3. $A = \int_0^3 [(-x^2 + 2x + 3) - (x^2 - 4x + 3)]\, dx = \int_0^3 (-2x^2 + 6x)\, dx = \left[-\frac{2x^3}{3} + 3x^2\right]_0^3 = 9$

4. $A = \int_0^1 (x^2 - x^3)\, dx = \left[\frac{x^3}{3} - \frac{x^4}{4}\right]_0^1 = \frac{1}{12}$

5. $A = 2\int_{-1}^{0} 3(x^3 - x)\, dx = 6\left[\frac{x^4}{4} - \frac{x^2}{2}\right]_{-1}^{0} = 6\left(\frac{1}{4}\right) = \frac{3}{2}$

6. $A = 2\int_0^1 [(x-1)^3 - (x-1)]\, dx = 2\left[\frac{(x-1)^4}{4} - \left(\frac{x^2}{2} - x\right)\right]_0^1 = \frac{1}{2}$

7. $f(x) = x + 1$
$g(x) = (x-1)^2$
$A \approx 4$
Matches (d)

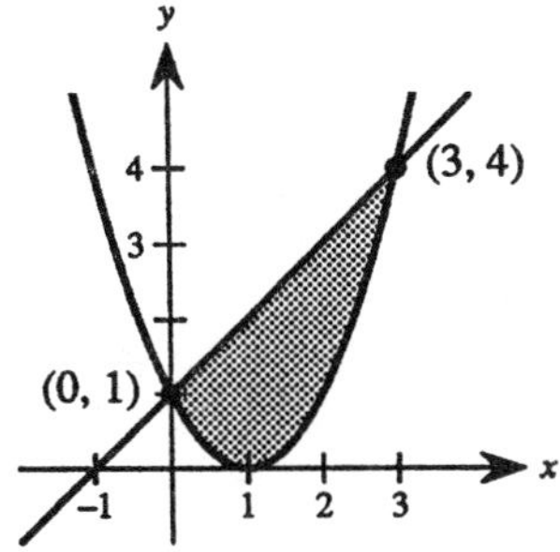

8. $f(x) = 2 - \frac{1}{2}x$
$g(x) = 2 - \sqrt{x}$
$A \approx 1$
Matches (a)

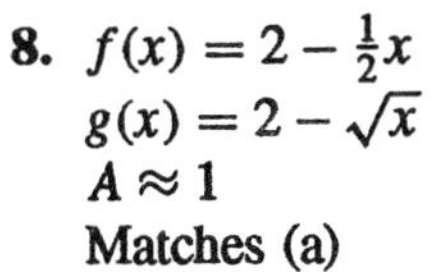

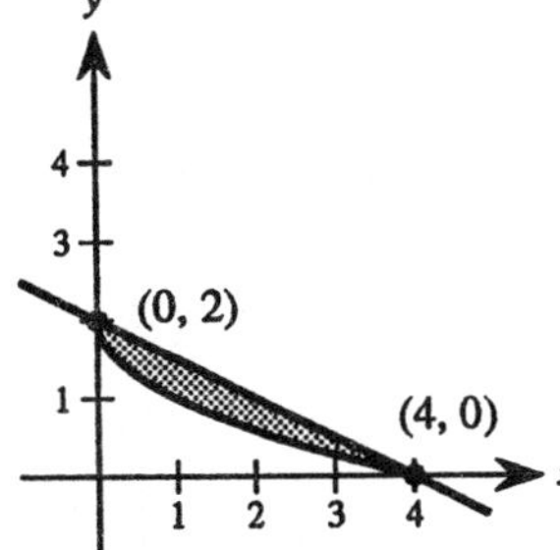

9. The points of intersection are given by:

$x^2 - 4x = 0$

$x(x-4) = 0$ when $x = 0,\ 4$.

$$A = \int_0^4 [g(x) - f(x)]\, dx$$

$$= -\int_0^4 (x^2 - 4x)\, dx$$

$$= -\left[\frac{x^3}{3} - 2x^2\right]_0^4 = \frac{32}{3}$$

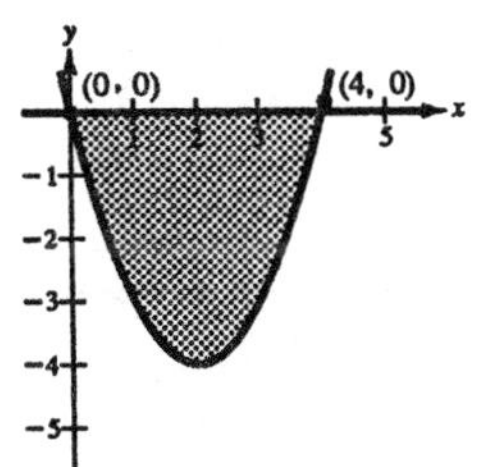

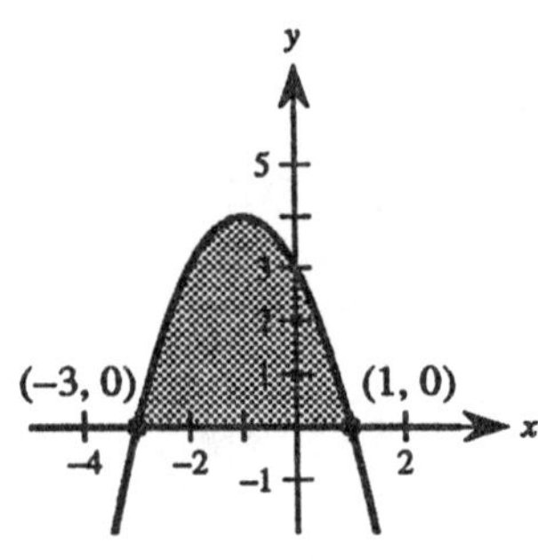

10. The points of intersection are given by:

$$3 - 2x - x^2 = 0$$

$(3 + x)(1 - x) = 0$ when $x = -3,\ 1.$

$$A = \int_{-3}^{1} [f(x) - g(x)]\,dx$$

$$= \int_{-3}^{1} (3 - 2x - x^2)\,dx$$

$$= \left[3x - x^2 - \frac{x^3}{3}\right]_{-3}^{1} = \frac{32}{3}$$

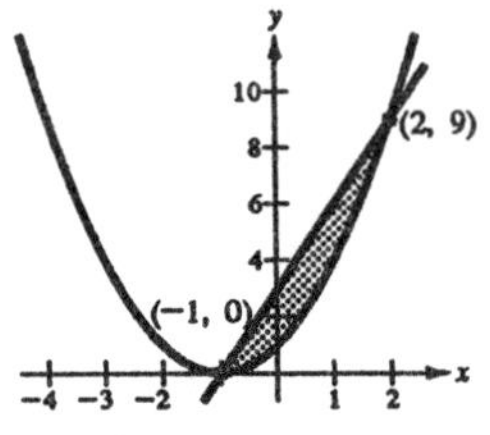

11. The points of intersection are given by:

$$x^2 + 2x + 1 = 3x + 3$$

$(x - 2)(x + 1) = 0$ when $x = -1,\ 2.$

$$A = \int_{-1}^{2} [g(x) - f(x)]\,dx = \int_{-1}^{2} [(3x + 3) - (x^2 + 2x + 1)]\,dx$$

$$= \int_{-1}^{2} (2 + x - x^2)\,dx$$

$$= \left[2x + \frac{x^2}{2} - \frac{x^3}{3}\right]_{-1}^{2} = \frac{9}{2}$$

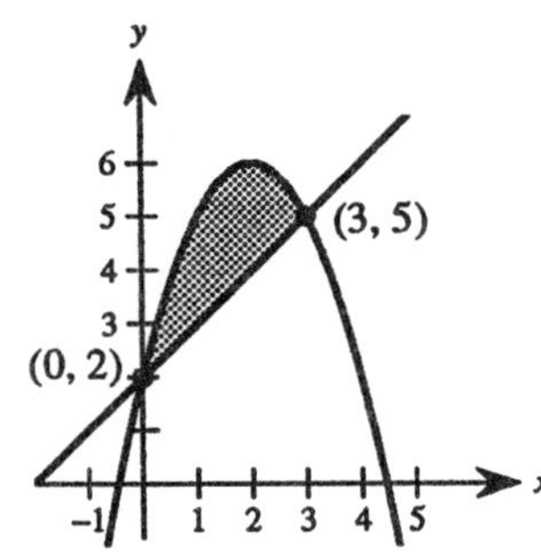

12. The points of intersection are given by:

$$-x^2 + 4x + 2 = x + 2$$

$x(3 - x) = 0$ when $x = 0,\ 3.$

$$A = \int_{0}^{3} [f(x) - g(x)]\,dx = \int_{0}^{3} [(-x^2 + 4x + 2) - (x + 2)]\,dx$$

$$= \int_{0}^{3} (-x^2 + 3x)\,dx$$

$$= \left[\frac{-x^3}{3} + \frac{3}{2}x^2\right]_{0}^{3} = \frac{9}{2}$$

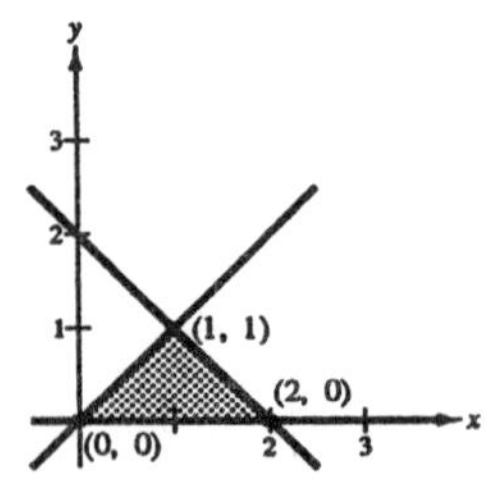

13. The points of intersection are given by:

$x = 2 - x$ and $x = 0$ and $2 - x = 0$

$x = 1$ $\quad x = 0$ $\quad x = 2$

$$A = \int_{0}^{1} [(2 - y) - (y)]\,dy = \left[2y - y^2\right]_{0}^{1} = 1$$

Note that if we integrate with respect to x, we need two integrals.

14. $A = \int_1^5 \left(\frac{1}{x^2} - 0\right) dx = -\frac{1}{x}\Big]_1^5 = \frac{4}{5}$

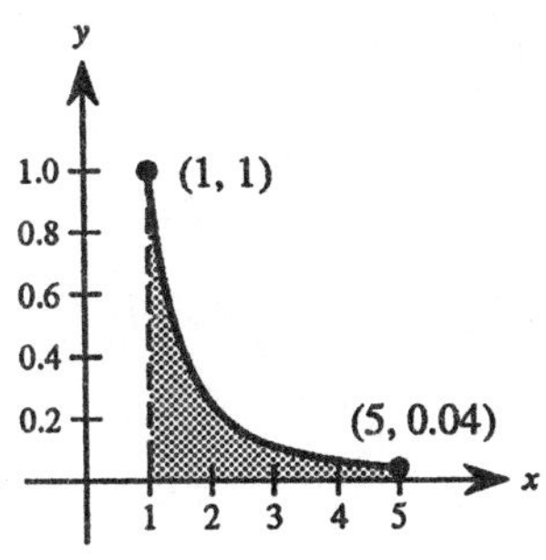

15. The points of intersection are given by:

$$3x^2 + 2x = 8$$

$$(3x - 4)(x + 2) = 0 \text{ when } x = -2, \ \tfrac{4}{3}.$$

$$A = \int_{-2}^{4/3} [g(x) - f(x)]\, dx$$

$$= \int_{-2}^{4/3} (8 - 2x - 3x^2)\, dx = \left[8x - x^2 - x^3\right]_{-2}^{4/3} = \frac{500}{27}$$

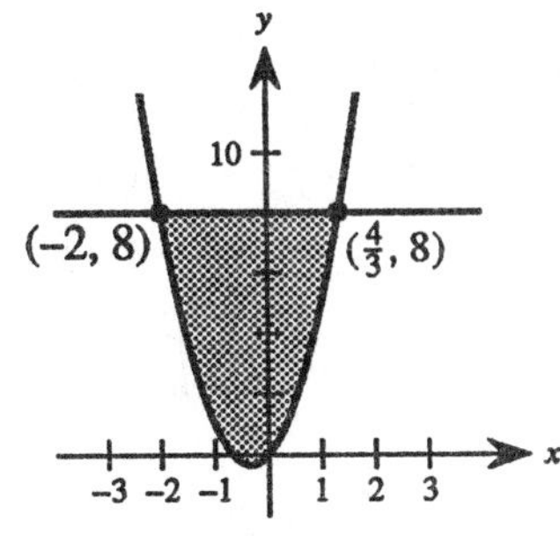

16. The points of intersection are given by:

$$x^3 - 3x^2 + 3x = x^2$$

$$x(x - 1)(x - 3) = 0 \text{ when } x = 0, \ 1, \ 3.$$

$$A = \int_0^1 [f(x) - g(x)]\, dx + \int_1^3 [g(x) - f(x)]\, dx$$

$$= \int_0^1 [(x^3 - 3x^2 + 3x) - x^2]\, dx + \int_1^3 [x^2 - (x^3 - 3x^2 + 3x)]\, dx$$

$$= \int_0^1 (x^3 - 4x^2 + 3x)\, dx + \int_1^3 (-x^3 + 4x^2 - 3x)\, dx$$

$$= \left[\frac{x^4}{4} - \frac{4}{3}x^3 + \frac{3}{2}x^2\right]_0^1 + \left[\frac{-x^4}{4} + \frac{4}{3}x^3 - \frac{3}{2}x^2\right]_1^3 = \frac{37}{12}$$

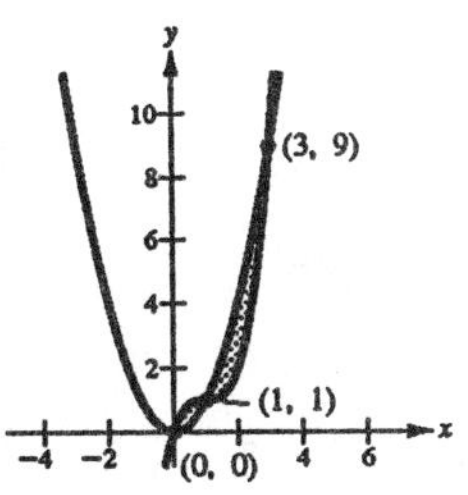

17. The point of intersection is given by:

$$x^3 - 2x + 1 = -2x$$

$$x^3 + 1 = 0 \text{ when } x = -1.$$

$$A = \int_{-1}^1 [f(x) - g(x)]\, dx$$

$$= \int_{-1}^1 [(x^3 - 2x + 1) - (-2x)]\, dx$$

$$= \int_{-1}^1 (x^3 + 1)\, dx$$

$$= \left[\frac{x^4}{4} + x\right]_{-1}^1 = 2$$

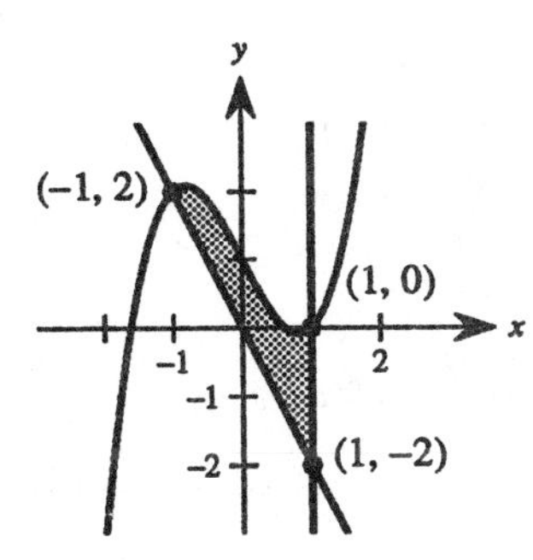

18. The points of intersection are given by:

$$\sqrt[3]{x} = x$$

$$x = -1,\ 0,\ 1.$$

$$A = 2\int_0^1 [f(x) - g(x)]\,dx$$

$$= 2\int_0^1 (\sqrt[3]{x} - x)\,dx = 2\int_0^1 (x^{1/3} - x)\,dx = 2\left[\frac{3}{4}x^{4/3} - \frac{1}{2}x^2\right]_0^1 = \frac{1}{2}$$

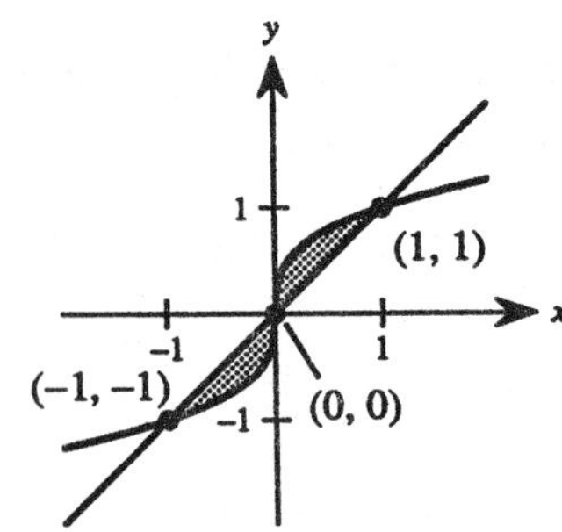

19. The points of intersection are given by:

$$\sqrt{3x} + 1 = x + 1$$

$$\sqrt{3x} = x \text{ when } x = 0,\ 3.$$

$$A = \int_0^3 [f(x) - g(x)]\,dx$$

$$= \int_0^3 [(\sqrt{3x} + 1) - (x + 1)]\,dx$$

$$= \int_0^3 [(3x)^{1/2} - x]\,dx = \left[\frac{2}{9}(3x)^{3/2} - \frac{x^2}{2}\right]_0^3 = \frac{3}{2}$$

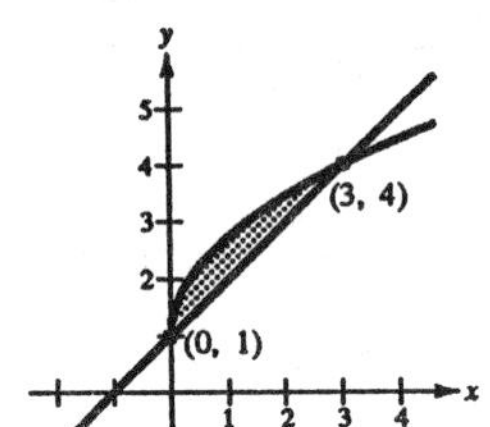

20. The points of intersection are given by:

$$x^2 + 5x - 6 = 6x - 6$$

$$x(x - 1) = 0 \text{ when } x = 0,\ 1.$$

$$A = \int_0^1 [g(x) - f(x)]\,dx$$

$$= \int_0^1 [(6x - 6) - (x^2 + 5x - 6)]\,dx = \int_0^1 (x - x^2)\,dx = \left[\frac{x^2}{2} - \frac{x^3}{3}\right]_0^1 = \frac{1}{6}$$

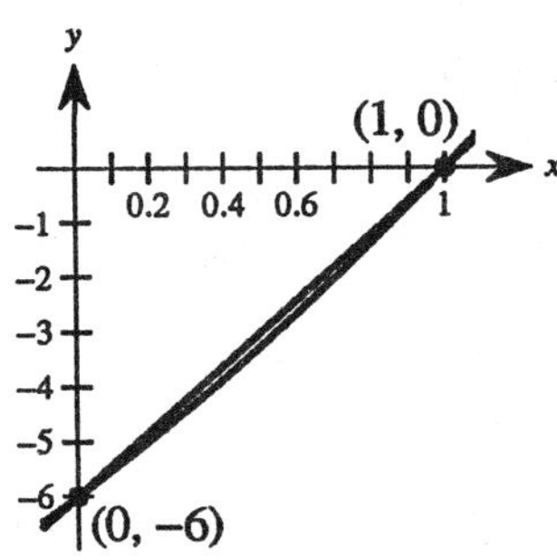

21. The points of intersection are given by:

$$x^2 - 4x + 3 = 3 + 4x - x^2$$

$$2x(x - 4) = 0 \text{ when } x = 0,\ 4.$$

$$A = \int_0^4 [(3 + 4x - x^2) - (x^2 - 4x + 3)]\,dx$$

$$= \int_0^4 (-2x^2 + 8x)\,dx = \left[-\frac{2x^3}{3} + 4x^2\right]_0^4 = \frac{64}{3}$$

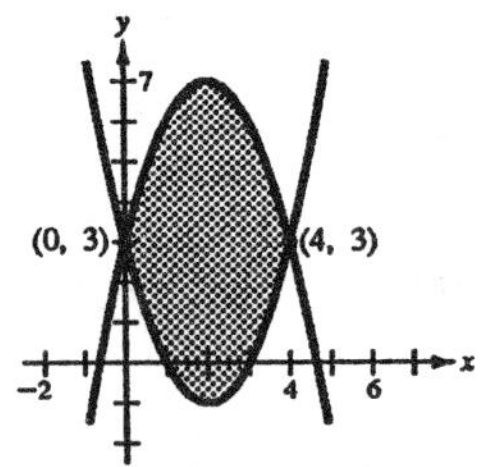

22. The points of intersection are given by:

$$x^4 - 2x^2 = 2x^2$$

$$x^2(x^2 - 4) = 0 \text{ when } x = -2,\ 0,\ 2.$$

$$A = 2\int_0^2 [2x^2 - (x^4 - 2x^2)]\,dx$$

$$= 2\int_0^2 (4x^2 - x^4)\,dx$$

$$= 2\left[\frac{4x^3}{3} - \frac{x^5}{5}\right]_0^2 = \frac{128}{15}$$

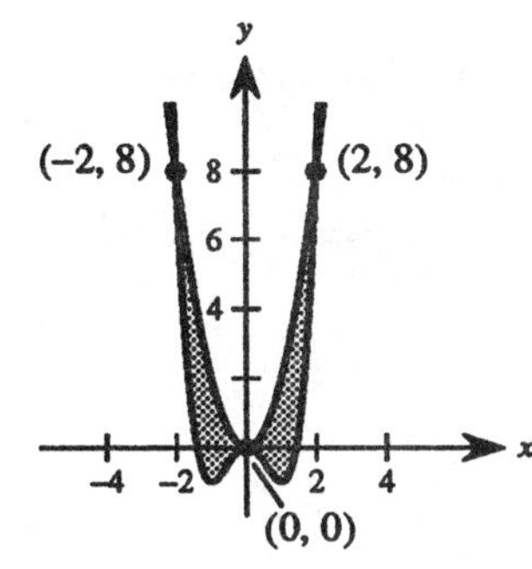

23. The points of intersection are given by:

$$y^2 = y + 2$$

$$(y-2)(y+1) = 0 \text{ when } y = -1,\ 2.$$

$$A = \int_{-1}^{2} [g(y) - f(y)]\,dy$$

$$= \int_{-1}^{2} [(y+2) - y^2]\,dy = \left[2y + \frac{y^2}{2} - \frac{y^3}{3}\right]_{-1}^{2} = \frac{9}{2}$$

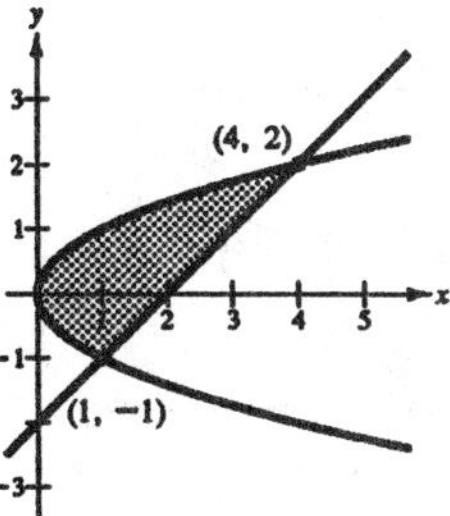

24. The points of intersection are given by:

$$2y - y^2 = -y$$

$$y(y-3) = 0 \text{ when } y = 0,\ 3.$$

$$A = \int_0^3 [f(y) - g(y)]\,dy$$

$$= \int_0^3 [(2y - y^2) - (-y)]\,dy = \int_0^3 (3y - y^2)\,dy = \left[\frac{3}{2}y^2 - \frac{1}{3}y^3\right]_0^3 = \frac{9}{2}$$

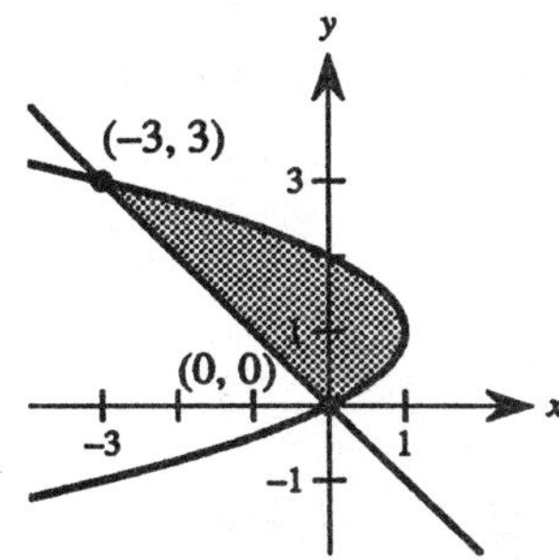

25. $A = \displaystyle\int_{-1}^{2} [f(y) - g(y)]\,dy$

$$= \int_{-1}^{2} [(y^2 + 1) - 0]\,dy$$

$$= \left[\frac{y^3}{3} + y\right]_{-1}^{2} = 6$$

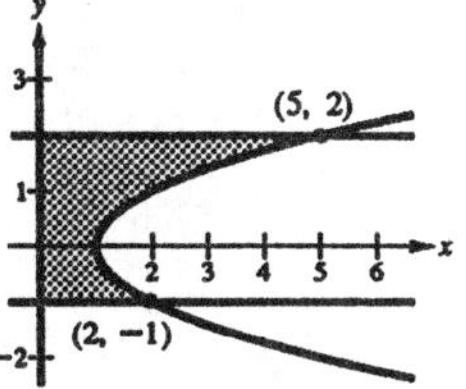

26. $A = \int_0^3 [f(y) - g(y)]\,dy$

$= \int_0^3 \left[\frac{y}{\sqrt{16 - y^2}} - 0\right] dy$

$= -\frac{1}{2}\int_0^3 (16 - y^2)^{-1/2}(-2y)\,dy$

$= -\sqrt{16 - y^2}\Big]_0^3 = 4 - \sqrt{7} \approx 1.354$

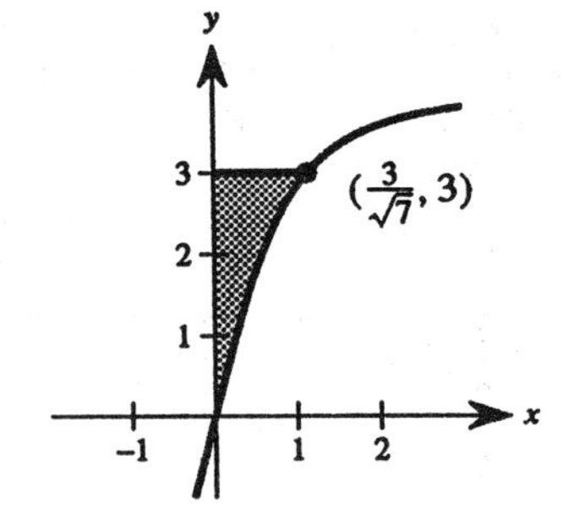

27. $A \approx \int_{-1}^{1.74296} (3x + 4 - x^4)\,dx \approx 10.6115$

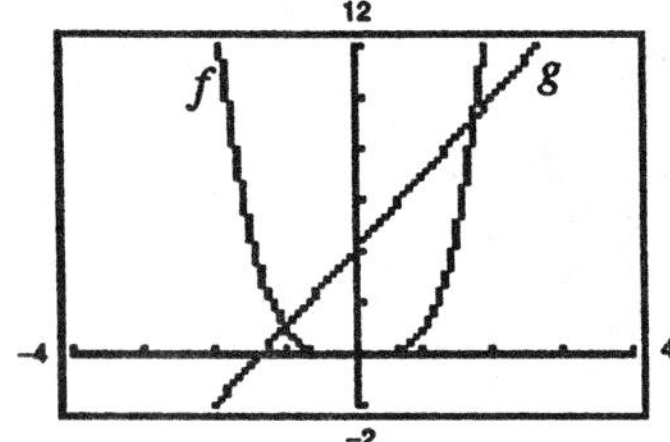

28. $A \approx \int_{-1}^{1.215} (x + 2 - x^6)\,dx \approx 3.967$

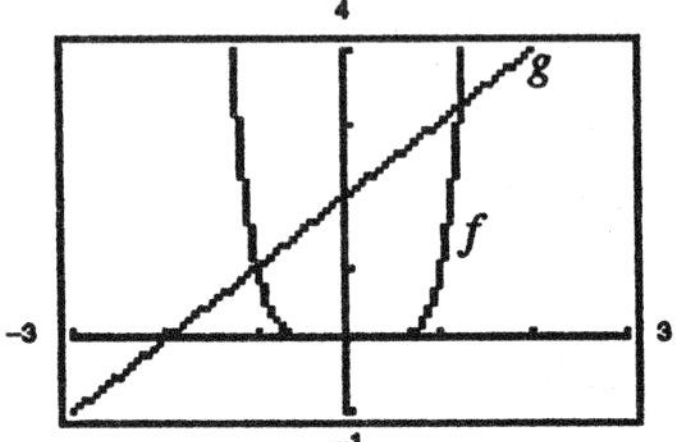

29. $A = \int_0^4 x\,dx = \frac{x^2}{2}\Big]_0^4 = 8$

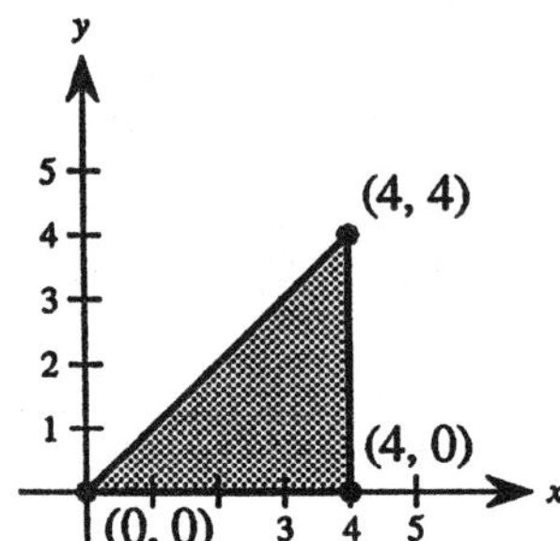

30. $A = \int_0^4 \left[\left(\frac{y}{2} + 4\right) - \frac{3}{2}y\right] dy$

$= \int_0^4 (4 - y)\,dy$

$= \left[4y - \frac{y^2}{2}\right]_0^4 = 8$

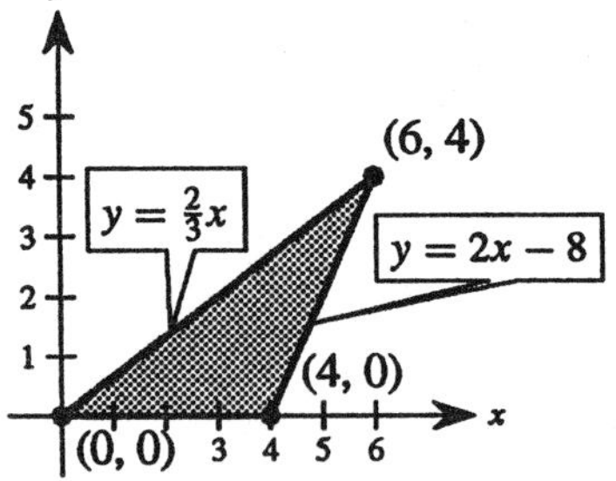

31. $A = \int_0^c \left[\left(\frac{b-a}{c}y + a\right) - \frac{b}{c}y\right] dy$

$= \int_0^c \left(-\frac{a}{c}y + a\right) dy$

$= \left[-\frac{a}{2c}y^2 + ay\right]_0^c$

$= -\frac{ac}{2} + ac$

$= \frac{ac}{2}$

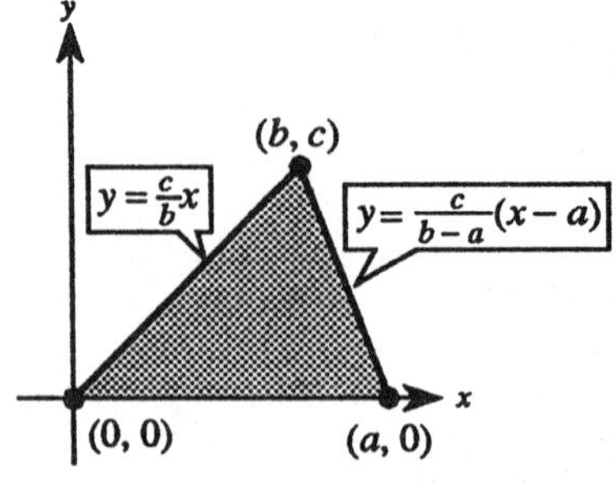

32. $A = \int_2^4 \left[\left(\frac{9}{2}x - 12\right) - (x-5)\right] dx + \int_4^6 \left[\left(-\frac{5}{2}x + 16\right) - (x-5)\right] dx$

$= \int_2^4 \left(\frac{7}{2}x - 7\right) dx + \int_4^6 \left(-\frac{7}{2}x + 21\right) dx$

$= \left[\frac{7}{4}x^2 - 7x\right]_2^4 + \left[-\frac{7}{4}x^2 + 21x\right]_4^6$

$= 14$

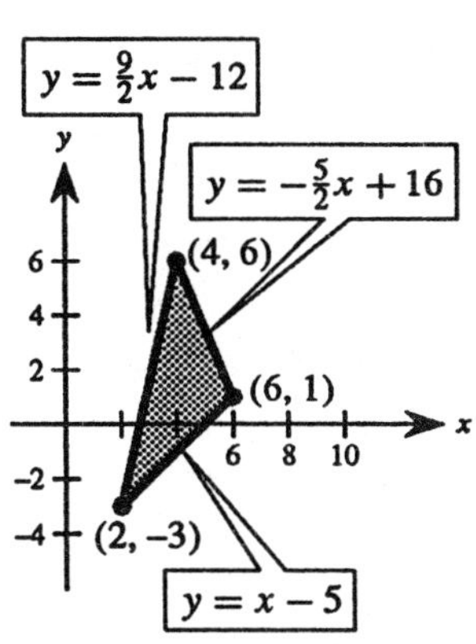

33. $A = \int_{-3}^{3} (9 - x^2)\, dx = 36$

$\int_{-\sqrt{9-b}}^{\sqrt{9-b}} [(9 - x^2) - b]\, dx = 18$

$\int_0^{\sqrt{9-b}} [(9-b) - x^2]\, dx = 9$

$\left[(9-b)x - \frac{x^3}{3}\right]_0^{\sqrt{9-b}} = 9$

$\frac{2}{3}(9-b)^{3/2} = 9$

$(9-b)^{3/2} = \frac{27}{2}$

$9 - b = \frac{9}{\sqrt[3]{4}}$

$b = 9 - \frac{9}{\sqrt[3]{4}} \approx 3.330$

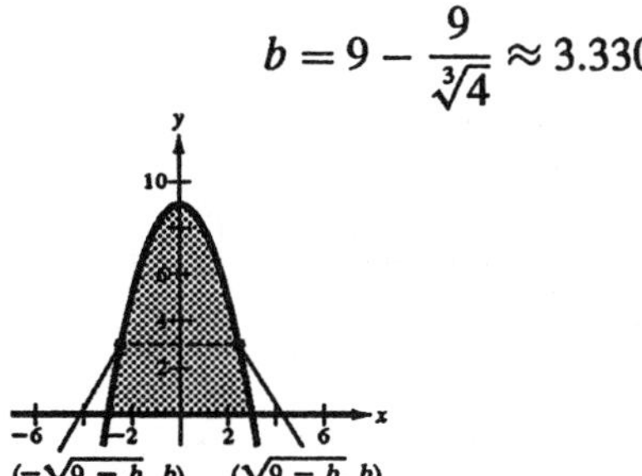

34. $A = 2\int_0^9 (9 - x)\, dx = 2\left[9x - \frac{x^2}{2}\right]_0^9 = 81$

$2\int_0^{9-b} [(9-x) - b]\, dx = \frac{81}{2}$

$2\int_0^{9-b} [(9-b) - x]\, dx = \frac{81}{2}$

$2\left[(9-b)x - \frac{x^2}{2}\right]_0^{9-b} = \frac{81}{2}$

$(9-b)(9-b) = \frac{81}{2}$

$9 - b = \frac{9}{\sqrt{2}}$

$b = 9 - \frac{9}{\sqrt{2}} \approx 2.636$

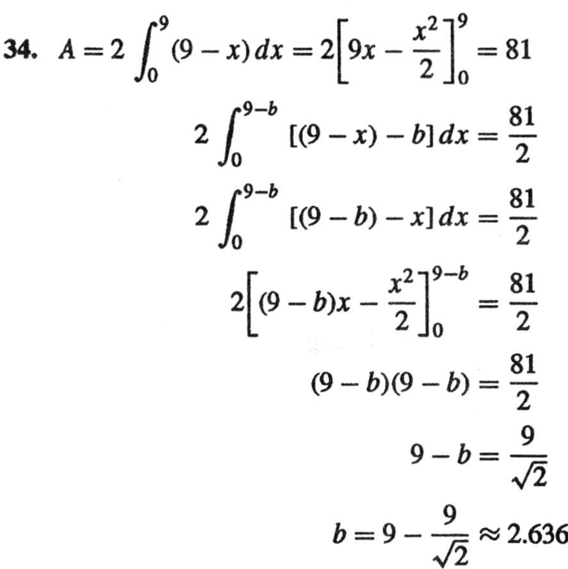

35. $x^4 - 2x^2 + 1 \le 1 - x^2$ on $[-1, 1]$

$A = \int_{-1}^{1} [(1 - x^2) - (x^4 - 2x^2 + 1)]\, dx$

$= \int_{-1}^{1} (x^2 - x^4)\, dx$

$= \left[\frac{x^3}{3} - \frac{x^5}{5}\right]_{-1}^{1} = \frac{4}{15}$

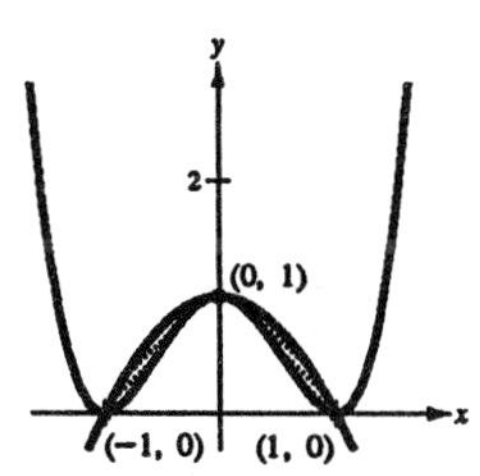

36. $x^3 \geq x$ on $[-1, 0]$

$x^3 \leq x$ on $[0, 1]$

Both functions symmetric to origin

$$\int_{-1}^{0} (x^3 - x)\,dx = -\int_{0}^{1} (x^3 - x)\,dx$$

Thus, $\displaystyle\int_{-1}^{1} (x^3 - x)\,dx = 0.$

$$A = 2\int_{0}^{1} (x - x^3)\,dx = 2\left[\frac{x^2}{2} - \frac{x^4}{4}\right]_0^1 = \frac{1}{2}$$

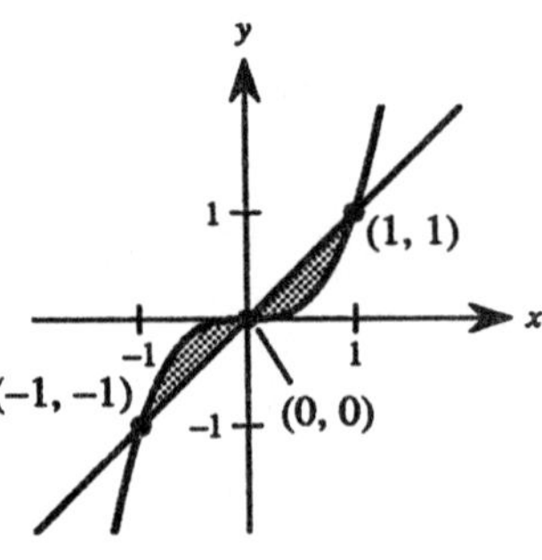

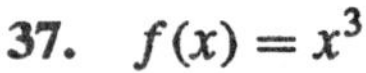

37. $f(x) = x^3$

$f'(x) = 3x^2$

At $(1, 1)$, $f'(1) = 3$.

Tangent line: $y - 1 = 3(x - 1)$ or $y = 3x - 2$

$$A = \int_{-2}^{1} [x^3 - (3x - 2)]\,dx = \left[\frac{x^4}{4} - \frac{3x^2}{2} + 2x\right]_{-2}^{1} = \frac{27}{4}$$

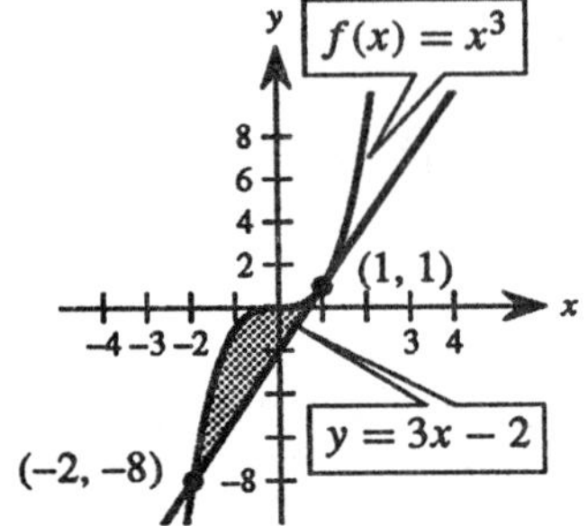
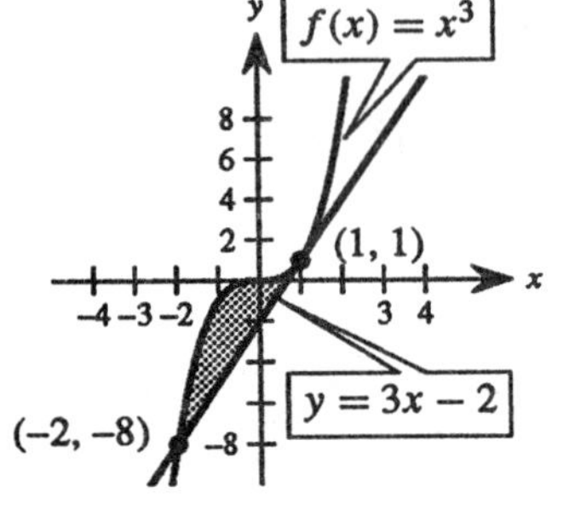

38. $f(x) = \sqrt[3]{x - 1}$

$$f'(x) = \frac{1}{3(x - 1)^{2/3}}$$

At $(2, 1)$, $f'(2) = \frac{1}{3}$.

Tangent line: $y - 1 = \dfrac{1}{3}(x - 2)$

$$y = \frac{1}{3}x + \frac{1}{3}$$

Points of intersection: $\dfrac{1}{3}x + \dfrac{1}{3} = \sqrt[3]{x - 1}$

$$x + 1 = 3\sqrt[3]{x - 1}$$

$$x = 2 \text{ or } x = -7$$

$$A = \int_{-7}^{2} \left[\left(\frac{1}{3}x + \frac{1}{3}\right) - \sqrt[3]{x - 1}\right] dx = \left[\frac{1}{6}x^2 + \frac{1}{3}x - \frac{3}{4}(x - 1)^{4/3}\right]_{-7}^{2} = \left(\frac{2}{3} + \frac{2}{3} - \frac{3}{4}\right) - \left(\frac{49}{6} - \frac{7}{3} - 12\right) = \frac{27}{4}$$

39. $\sqrt{1 + x^3} \leq \dfrac{1}{2}x + 2$ on $[0, 2]$

$$A = \int_{0}^{2} \left[\frac{1}{2}x + 2 - \sqrt{1 + x^3}\right] dx$$

$$\approx \frac{1}{6}\left[1 + 4\left(\frac{9 - 3\sqrt{2}}{4}\right) + 2\left(\frac{5 - 2\sqrt{2}}{2}\right) + 4\left(\frac{11 - \sqrt{70}}{4}\right) + 0\right]$$

$$\approx 1.760$$

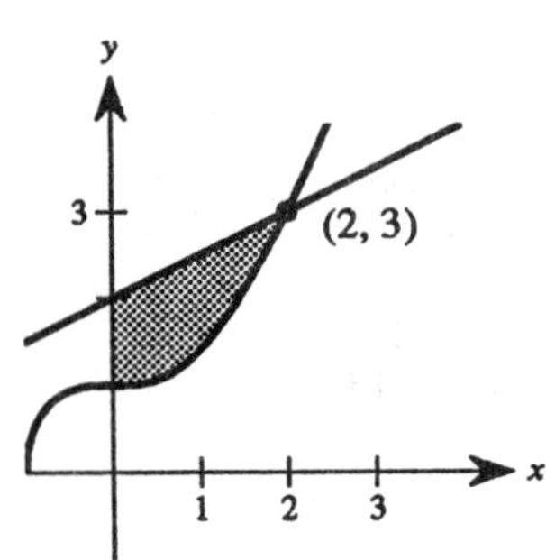

40. $$A = \int_{0}^{1} \sqrt{x + x^2}\,dx = \frac{1}{3(4)}\left[0 + 4\sqrt{\left(\frac{1}{4}\right) + \left(\frac{1}{4}\right)^2} + 2\sqrt{\left(\frac{1}{2}\right) + \left(\frac{1}{2}\right)^2} + 4\sqrt{\left(\frac{3}{4}\right) + \left(\frac{3}{4}\right)^2} + 1\right] \approx 0.8304$$

41. $\lim_{\|\Delta\| \to 0} \sum_{i=1}^{n} (x_i - x_i^2)\Delta x$

where $x_i = i/n$ and $\Delta x = 1/n$ is the same as

$$\int_0^1 (x - x^2)\,dx = \left[\frac{x^2}{2} - \frac{x^3}{3}\right]_0^1 = \frac{1}{6}.$$

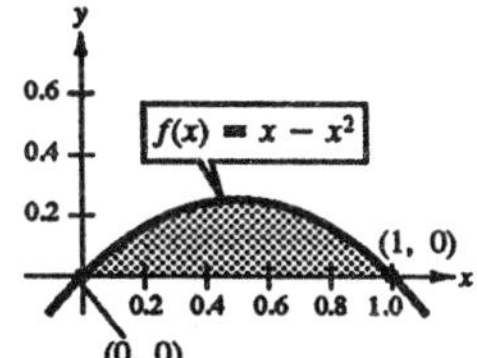

42. $\lim_{\|\Delta\| \to 0} \sum_{i=1}^{n} (4 - x_i^2)\Delta x$

where $x_i = -2 + (4i/n)$ and $\Delta x = 4/n$ is the same as

$$\int_{-2}^{2} (4 - x^2)\,dx = \left[4x - \frac{x^3}{3}\right]_{-2}^{2} = \frac{32}{3}.$$

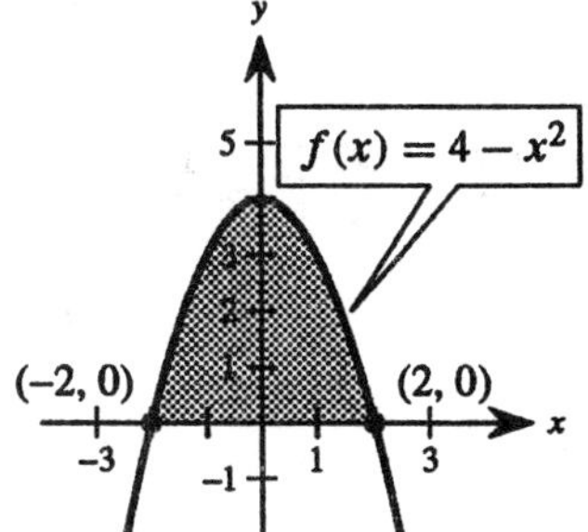

43. $\displaystyle\int_0^5 [(7.21 + 0.58t) - (7.21 + 0.45t)]\,dt = \int_0^5 0.13t\,dt = \left.\frac{0.13t^2}{2}\right]_0^5 = \$1.625 \text{ billion}$

44. $\displaystyle\int_0^5 [(7.21 + 0.26t + 0.02t^2) - (7.21 + 0.1t + 0.01t^2)]\,dt = \int_0^5 (0.01t^2 + 0.16t)\,dt$

$$= \left[\frac{0.01t^3}{3} + \frac{0.16t^2}{2}\right]_0^5 = \frac{29}{12} \text{ billion} \approx \$2.4167 \text{ billion}$$

45. $50 - 0.5x = 0.125x$

$x = 80$

$p(x) = 10$

Point of equilibrium: (80, 10)

$$CS = \int_0^{80} [(50 - 0.5x) - 10]\,dx = \left[-\frac{0.5x^2}{2} + 40x\right]_0^{80} = 1600$$

$$PS = \int_0^{80} [10 - 0.125x]\,dx = \left[10x - \frac{0.125x^2}{2}\right]_0^{80} = 400$$

46. $1000 - 0.4x^2 = 42x$

$x = 20$

$p(x) = 840$

Point of equilibrium: (20, 840)

$$CS = \int_0^{20} [(1000 - 0.4x^2) - 840]\,dx = \left[160x - \frac{0.4x^3}{3}\right]_0^{20} \approx 2133.33$$

$$PS = \int_0^{20} [840 - 42x]\,dx = \left[840x - 21x^2\right]_0^{20} = 8400$$

47. $\dfrac{10{,}000}{\sqrt{x+100}} = 100\sqrt{0.05x+10}$

$$100 = \sqrt{(x+100)(0.05x+10)}$$
$$10{,}000 = 0.05x^2 + 15x + 1000$$
$$0 = x^2 + 300x - 180{,}000$$
$$0 = (x+600)(x-300)$$
$$x = 300$$
$$p(x) = 500$$

Point of equilibrium: (300, 500)

$$CS = \int_0^{300} \left[\frac{10{,}000}{\sqrt{x+100}} - 500\right] dx = \left[20{,}000\sqrt{x+100} - 500x\right]_0^{300} = 250{,}000 - 200{,}000 = 50{,}000$$

$$PS = \int_0^{300} \left(500 - 100\sqrt{0.05x+10}\right) dx = \left[500x - \frac{4000}{3}(0.05x+10)^{3/2}\right]_0^{300}$$
$$= \frac{-50{,}000}{3} + \frac{40{,}000\sqrt{10}}{3} = \frac{10{,}000}{3}(4\sqrt{10}-5) \approx 25{,}497$$

48. $\sqrt{25-0.1x} = \sqrt{9+0.1x} - 2$

$$x^2 - 160x = 0, \quad 0 < x < 250$$
$$x(x-160) = 0$$
$$x = 160$$
$$p(x) = 3$$

Point of equilibrium: (160, 3)

$$CS = \int_0^{160} \left(\sqrt{25-0.1x} - 3\right) dx = \left[-\frac{20}{3}(25-0.1x)^{3/2} - 3x\right]_0^{160} = \frac{520}{3}$$

$$PS = \int_0^{160} \left[3 - \left(\sqrt{0.1x+9} - 2\right)\right] dx = \left[5x - \frac{20}{3}(0.1x+9)^{3/2}\right]_0^{160} = \frac{440}{3}$$

49. $A = \displaystyle\int_0^3 \sqrt{\frac{x^3}{4-x}}\, dx \approx 4.773$

50. $A = \displaystyle\int_0^4 x\sqrt{\frac{4-x}{4+x}}\, dx \approx 3.433$

Section 6.2 Volume: The Disc Method

1. $V = \pi \displaystyle\int_0^1 (-x+1)^2\, dx = \pi \int_0^1 (x^2 - 2x + 1)\, dx = \pi\left[\frac{x^3}{3} - x^2 + x\right]_0^1 = \frac{\pi}{3}$

2. $V = \pi \displaystyle\int_0^2 (4-x^2)^2\, dx = \pi \int_0^2 (x^4 - 8x^2 + 16)\, dx = \pi\left[\frac{x^5}{5} - \frac{8x^3}{3} + 16x\right]_0^2 = \frac{256\pi}{15}$

3. $V = \pi \displaystyle\int_0^2 \left(\sqrt{4-x^2}\right)^2 dx = \pi \int_0^2 (4-x^2)\, dx$

$$= \pi\left[4x - \frac{x^3}{3}\right]_0^2 = \frac{16\pi}{3}$$

4. $V = \pi \displaystyle\int_0^1 (x^2)^2\, dx$

$$= \pi \int_0^1 x^4\, dx = \pi \frac{x^5}{5}\bigg]_0^1 = \frac{\pi}{5}$$

5. $V = \pi \int_1^4 (\sqrt{x})^2\, dx$

$= \pi \int_1^4 x\, dx = \pi \dfrac{x^2}{2}\Big]_1^4 = \dfrac{15\pi}{2}$

6. $V = \pi \int_{-2}^2 \left(\sqrt{4 - x^2}\right)^2 dx$

$= 2\pi \int_0^2 (4 - x^2)\, dx = 2\pi \left[4x - \dfrac{x^3}{3}\right]_0^2 = \dfrac{32\pi}{3}$

7. $V = \pi \int_0^1 [(x^2)^2 - (x^3)^2]\, dx = \pi \int_0^1 (x^4 - x^6)\, dx = \pi \left[\dfrac{x^5}{5} - \dfrac{x^7}{7}\right]_0^1 = \dfrac{2\pi}{35}$

8. $V = \pi \int_{-2}^2 \left[\left(4 - \dfrac{x^2}{2}\right)^2 - (2)^2\right] dx = 2\pi \int_0^2 \left(\dfrac{x^4}{4} - 4x^2 + 12\right) dx = 2\pi \left[\dfrac{x^5}{20} - \dfrac{4x^3}{3} + 12x\right]_0^2 = \dfrac{448\pi}{15}$

9. $y = x^2 \Rightarrow x = \sqrt{y}$

$V = \pi \int_0^4 (\sqrt{y})^2\, dy$

$= \pi \int_0^4 y\, dy = \pi \dfrac{y^2}{2}\Big]_0^4 = 8\pi$

10. $y = \sqrt{16 - x^2} \Rightarrow x = \sqrt{16 - y^2}$

$V = \pi \int_0^4 \left(\sqrt{16 - y^2}\right)^2 dy$

$= \pi \int_0^4 (16 - y^2)\, dy$

$= \pi \left[16y - \dfrac{y^3}{3}\right]_0^4 = \dfrac{128\pi}{3}$

11. $y = x^{2/3} \Rightarrow x = y^{3/2}$

$V = \pi \int_0^1 (y^{3/2})^2\, dy$

$= \pi \int_0^1 y^3\, dy = \pi \dfrac{y^4}{4}\Big]_0^1 = \dfrac{\pi}{4}$

12. $V = \pi \int_1^4 (-y^2 + 4y)^2\, dy$

$= \pi \int_1^4 (y^4 - 8y^3 + 16y^2)\, dy$

$= \pi \left[\dfrac{y^5}{5} - 2y^4 + \dfrac{16y^3}{3}\right]_1^4 = \dfrac{459\pi}{15}$

13. $y = \sqrt{x}, \quad y = 0, \quad x = 4$

(a) $R(x) = \sqrt{x}, \quad r(x) = 0$

$V = \pi \int_0^4 (\sqrt{x})^2\, dx$

$= \pi \int_0^4 x\, dx = \dfrac{\pi}{2} x^2 \Big]_0^4 = 8\pi$

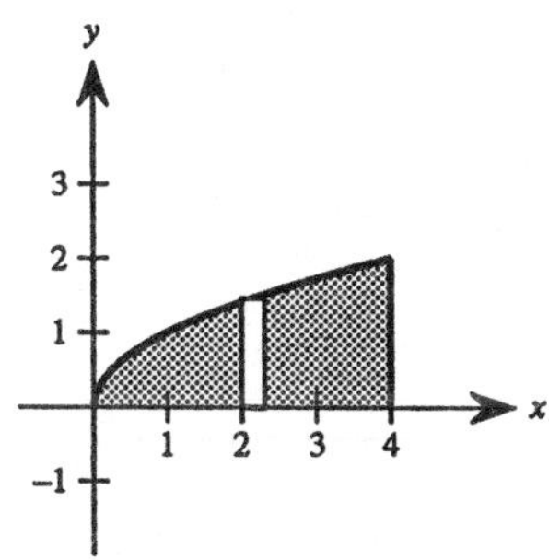

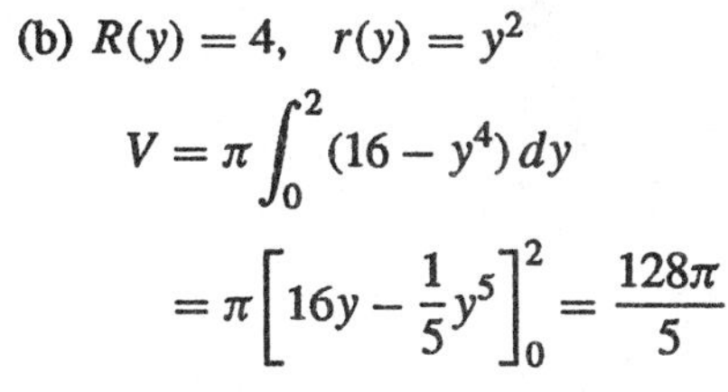

(b) $R(y) = 4, \quad r(y) = y^2$

$V = \pi \int_0^2 (16 - y^4)\, dy$

$= \pi \left[16y - \dfrac{1}{5}y^5\right]_0^2 = \dfrac{128\pi}{5}$

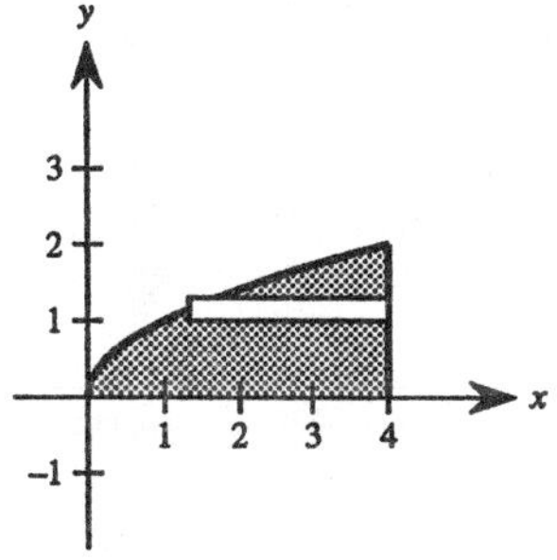

13. —CONTINUED—

(c) $R(y) = 4 - y^2, \quad r(y) = 0$

$$V = \pi \int_0^2 (4 - y^2)^2 \, dy$$

$$= \pi \int_0^2 (16 - 8y^2 + y^4) \, dy$$

$$= \pi \left[16y - \frac{8}{3}y^3 + \frac{1}{5}y^5 \right]_0^2 = \frac{256\pi}{15}$$

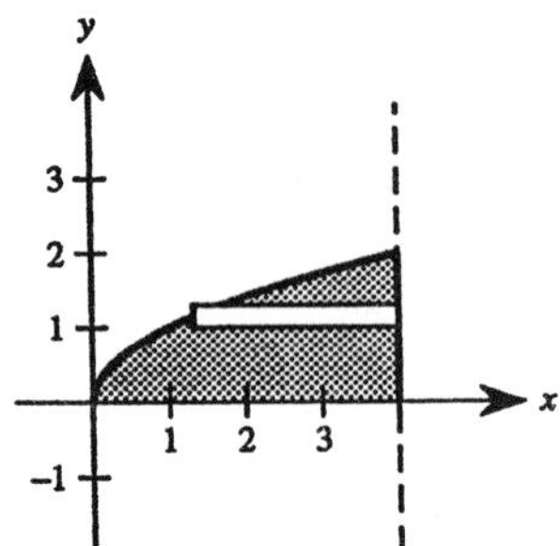

(d) $R(y) = 6 - y^2, \quad r(y) = 2$

$$V = \pi \int_0^2 [(6 - y^2)^2 - 4] \, dy$$

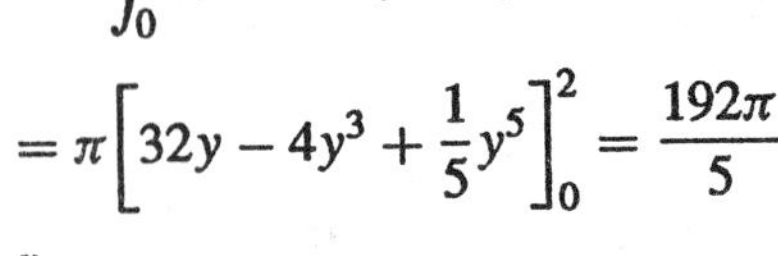

$$= \pi \int_0^2 (32 - 12y^2 + y^4) \, dy$$

$$= \pi \left[32y - 4y^3 + \frac{1}{5}y^5 \right]_0^2 = \frac{192\pi}{5}$$

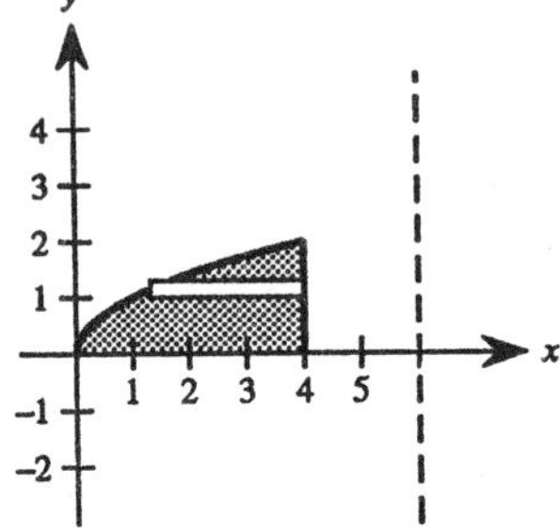

14. $y = 2x^2, \quad y = 0, \quad x = 2$

(a) $R(y) = 2, \quad r(y) = \sqrt{\frac{y}{2}}$

$$V = \pi \int_0^8 \left(4 - \frac{y}{2}\right) dy = \pi \left[4y - \frac{y^2}{4} \right]_0^8 = 16\pi$$

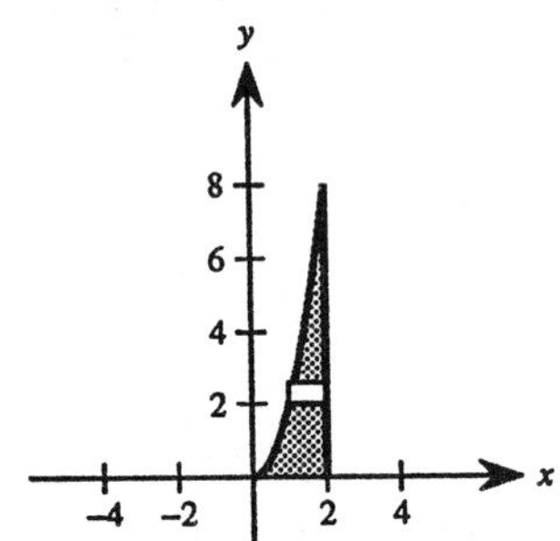

(b) $R(x) = 2x^2, \quad r(x) = 0$

$$V = \pi \int_0^2 4x^4 \, dx = \pi \left[\frac{4x^5}{5} \right]_0^2 = \frac{128\pi}{5}$$

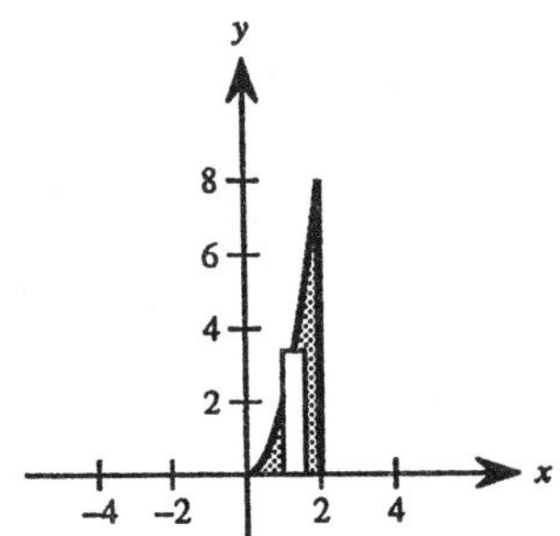

(c) $R(x) = 8, \quad r(x) = 8 - 2x^2$

$$V = \pi \int_0^2 [64 - (64 - 32x^2 + 4x^4)] \, dx$$

$$= \pi \int_0^2 (32x^2 - 4x^4) \, dx = 4\pi \int_0^2 (8x^2 - x^4) \, dx$$

$$= 4\pi \left[\frac{8}{3}x^3 - \frac{1}{5}x^5 \right]_0^2 = \frac{896\pi}{15}$$

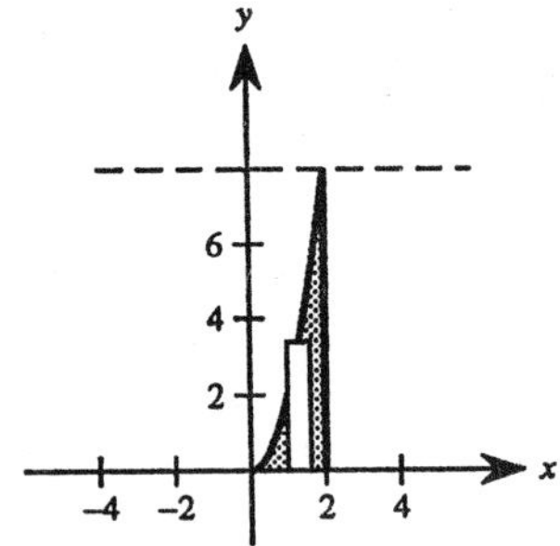

(d) $R(y) = 2 - \sqrt{\frac{y}{2}}, \quad r(y) = 0$

$$V = \pi \int_0^8 \left(2 - \sqrt{\frac{y}{2}}\right)^2 dy$$

$$= \pi \int_0^8 \left(4 - 4\sqrt{\frac{y}{2}} + \frac{y}{2}\right) dy$$

$$= \pi \left[4y - \frac{4\sqrt{2}}{3}y^{3/2} + \frac{y^2}{4} \right]_0^8 = \frac{16\pi}{3}$$

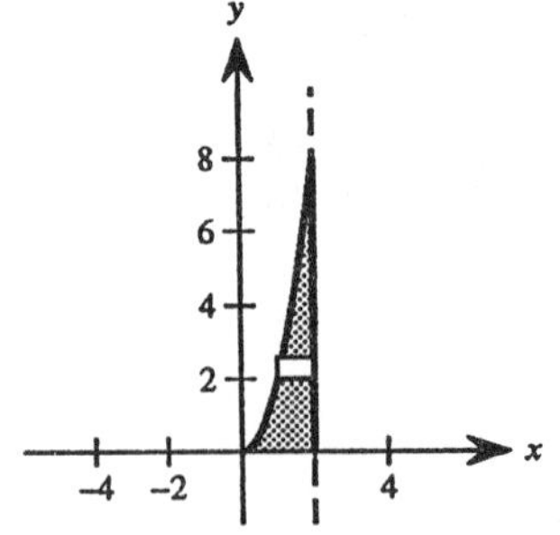

15. $y = x^2, \quad y = 4x - x^2$

(a) $R(x) = 4x - x^2, \quad r(x) = x^2$

$$V = \pi \int_0^2 [(4x - x^2)^2 - x^4]\,dx$$

$$= \pi \int_0^2 (16x^2 - 8x^3)\,dx$$

$$= \pi \left[\frac{16}{3}x^3 - 2x^4\right]_0^2 = \frac{32\pi}{3}$$

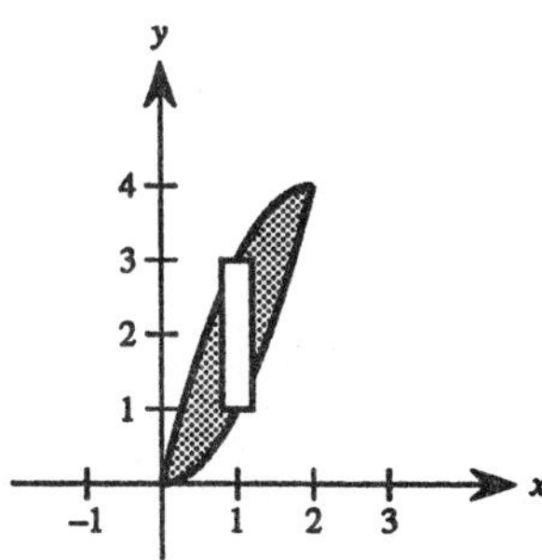

(b) $R(x) = 6 - x^2, \quad r(x) = 6 - (4x - x^2)$

$$V = \pi \int_0^2 [(6 - x^2)^2 - (6 - 4x + x^2)^2]\,dx$$

$$= 8\pi \int_0^2 (x^3 - 5x^2 + 6x)\,dx$$

$$= 8\pi \left[\frac{x^4}{4} - \frac{5}{3}x^3 + 3x^2\right]_0^2 = \frac{64\pi}{3}$$

16. $y = 6 - 2x - x^2, \quad y = x + 6$

(a) $R(x) = 6 - 2x - x^2, \quad r(x) = x + 6$

$$V = \pi \int_{-3}^0 [(6 - 2x - x^2)^2 - (x + 6)^2]\,dx$$

$$= \pi \int_{-3}^0 (x^4 + 4x^3 - 9x^2 - 36x)\,dx$$

$$= \pi \left[\frac{1}{5}x^5 + x^4 - 3x^3 - 18x^2\right]_{-3}^0 = \frac{243\pi}{5}$$

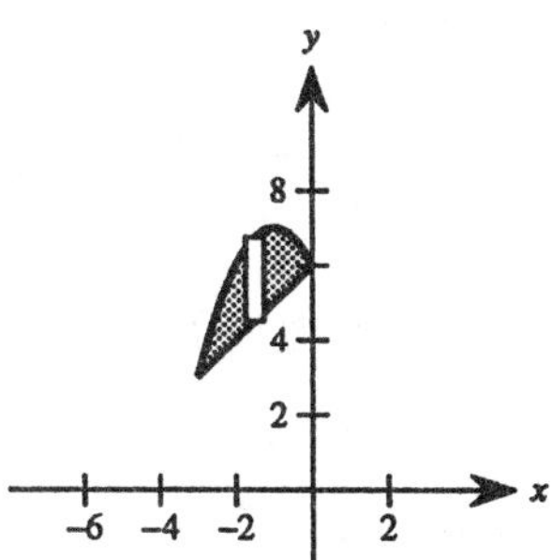

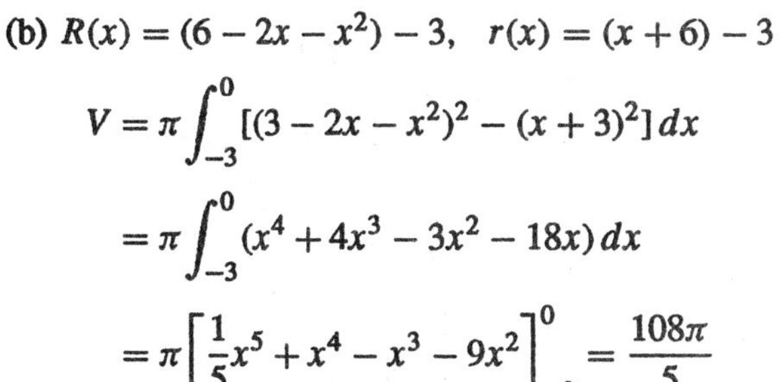

(b) $R(x) = (6 - 2x - x^2) - 3, \quad r(x) = (x + 6) - 3$

$$V = \pi \int_{-3}^0 [(3 - 2x - x^2)^2 - (x + 3)^2]\,dx$$

$$= \pi \int_{-3}^0 (x^4 + 4x^3 - 3x^2 - 18x)\,dx$$

$$= \pi \left[\frac{1}{5}x^5 + x^4 - x^3 - 9x^2\right]_{-3}^0 = \frac{108\pi}{5}$$

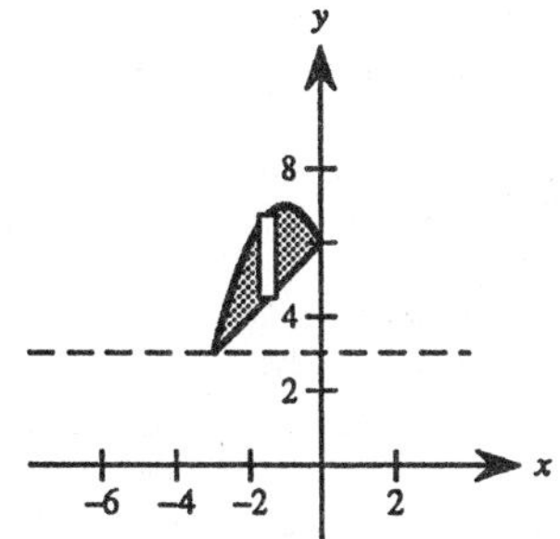

17. $R(x) = x\sqrt{4 - x^2}, \quad r(x) = 0$

$$V = 2\pi \int_0^2 \left[x\sqrt{4 - x^2}\right]^2 dx$$

$$= 2\pi \int_0^2 (4x^2 - x^4)\,dx$$

$$= 2\pi \left[\frac{4x^3}{3} - \frac{x^5}{5}\right]_0^2 = \frac{128\pi}{15}$$

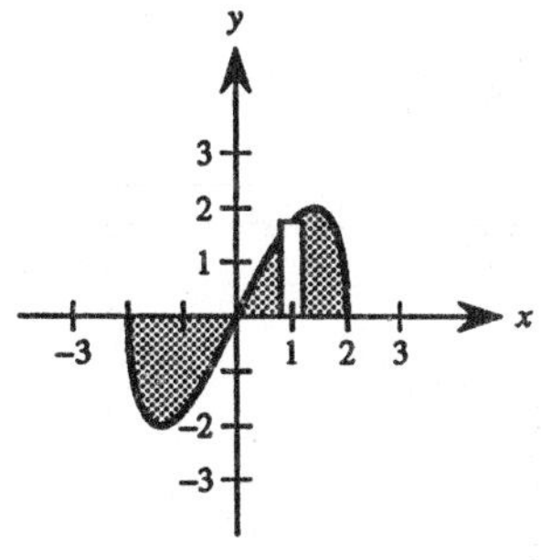

18. $R(x) = 2x^{2/3} - x, \quad r(x) = 0$

$$V = \pi \int_0^8 (2x^{2/3} - x^2)\,dx$$

$$= \pi \int_0^8 (4x^{4/3} - 4x^{5/3} + x^2)\,dx$$

$$= \pi \left[\frac{12}{7}x^{7/3} - \frac{3}{2}x^{8/3} + \frac{1}{3}x^3\right]_0^8 = \frac{128\pi}{21}$$

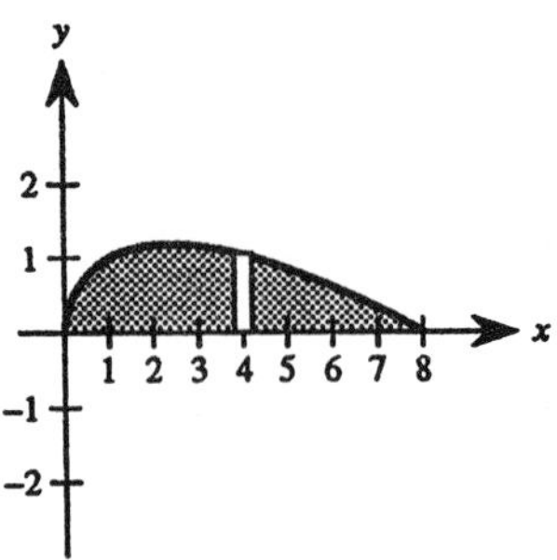

19. $R(x) = \dfrac{1}{x}, \quad r(x) = 0$

$$V = \pi \int_1^4 \left(\frac{1}{x}\right)^2 dx$$

$$= \pi\left[-\frac{1}{x}\right]_1^4 = \frac{3\pi}{4}$$

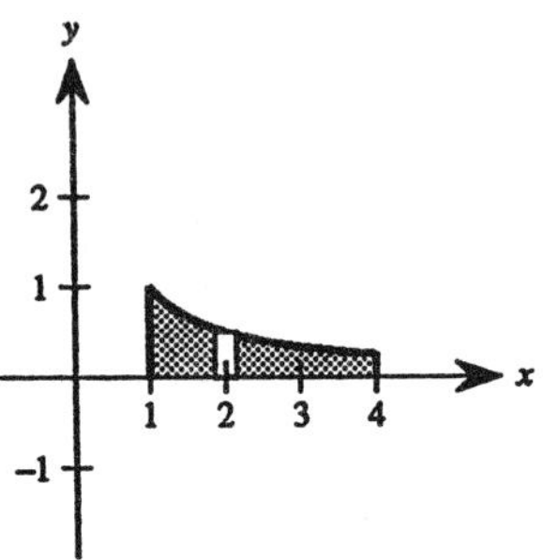

20. $R(x) = \dfrac{3}{x+1}, \quad r(x) = 0$

$$V = \pi \int_0^8 \left(\frac{3}{x+1}\right)^2 dx$$

$$= 9\pi \int_0^8 (x+1)^{-2}\,dx$$

$$= 9\pi\left[-\frac{1}{x+1}\right]_0^8 = 8\pi$$

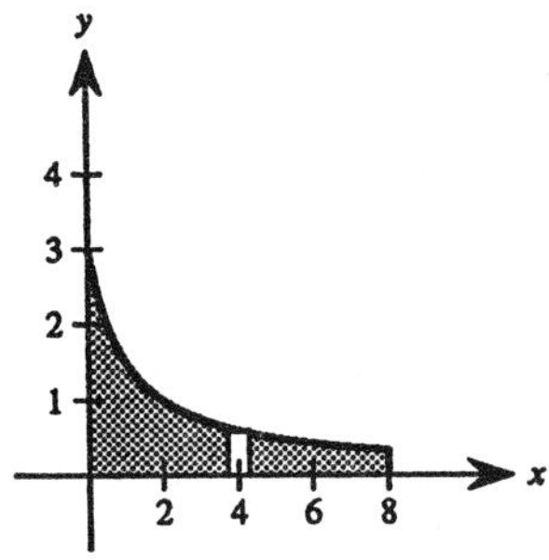

21. $R(x) = 4 - x, \quad r(x) = 1$

$$V = \pi \int_0^3 [(4-x)^2 - (1)^2]\,dx$$

$$= \pi \int_0^3 (x^2 - 8x + 15)\,dx$$

$$= \pi\left[\frac{x^3}{3} - 4x^2 + 15x\right]_0^3 = 18\pi$$

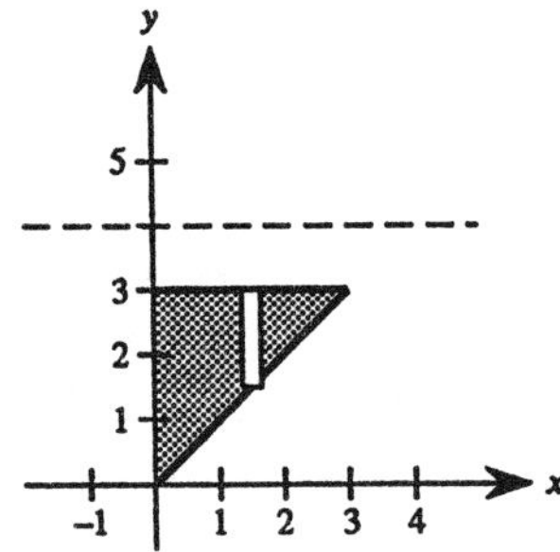

22. $R(x) = 4 - x^2, \quad r(x) = 0$

$$V = 2\pi \int_0^2 (4 - x^2)^2\,dx$$

$$= 2\pi \int_0^2 (x^4 - 8x^2 + 16)\,dx$$

$$= 2\pi\left[\frac{x^5}{5} - \frac{8x^3}{3} + 16x\right]_0^2 = \frac{512\pi}{15}$$

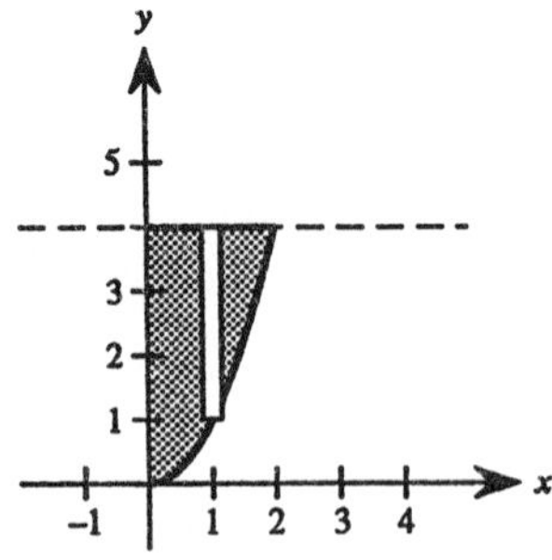

23. $R(x) = 4, \quad r(x) = 4 - \dfrac{1}{x^2}$

$$V = \pi \int_1^4 \left[4^2 - \left(4 - \frac{1}{x^2}\right)^2\right] dx$$

$$= \pi \int_1^4 \left[\frac{8}{x^2} - \frac{1}{x^4}\right] dx$$

$$= \pi \left[-\frac{8}{x} + \frac{1}{3x^3}\right]_1^4$$

$$= \pi \left[-2 + \frac{1}{192} + 8 - \frac{1}{3}\right]$$

$$= \frac{363\pi}{64}$$

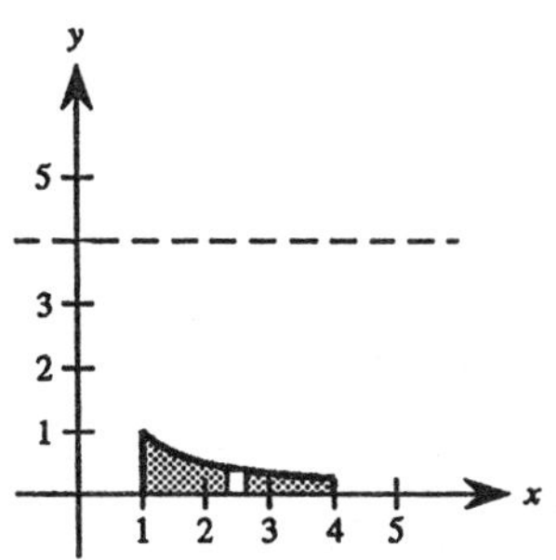

24. $R(x) = 4, \quad r(x) = 4 - \sqrt{x}$

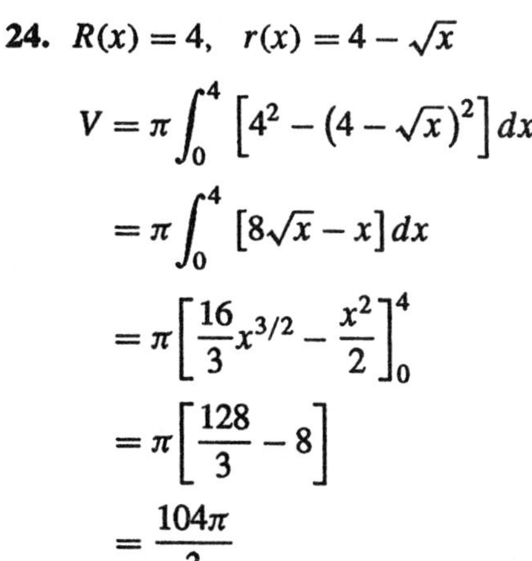

$$V = \pi \int_0^4 \left[4^2 - (4 - \sqrt{x})^2\right] dx$$

$$= \pi \int_0^4 [8\sqrt{x} - x]\, dx$$

$$= \pi \left[\frac{16}{3}x^{3/2} - \frac{x^2}{2}\right]_0^4$$

$$= \pi \left[\frac{128}{3} - 8\right]$$

$$= \frac{104\pi}{3}$$

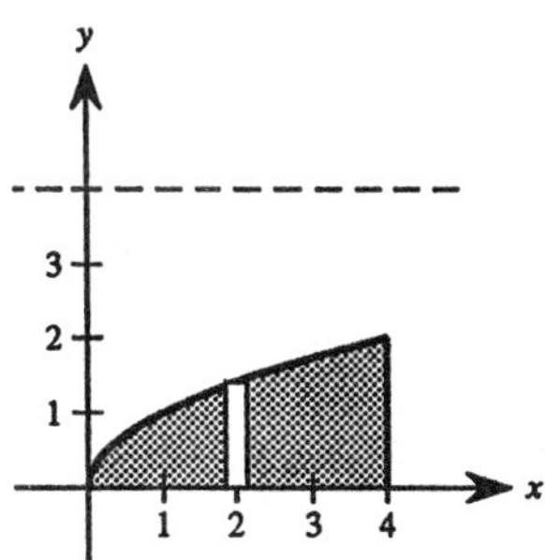

25. $R(y) = 6 - y, \quad r(y) = 0$

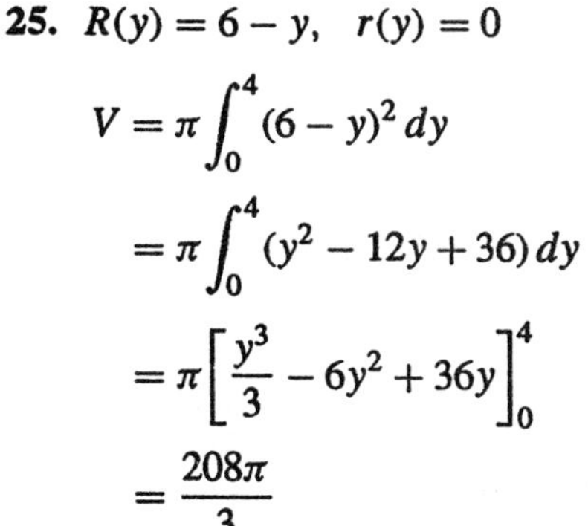

$$V = \pi \int_0^4 (6 - y)^2\, dy$$

$$= \pi \int_0^4 (y^2 - 12y + 36)\, dy$$

$$= \pi \left[\frac{y^3}{3} - 6y^2 + 36y\right]_0^4$$

$$= \frac{208\pi}{3}$$

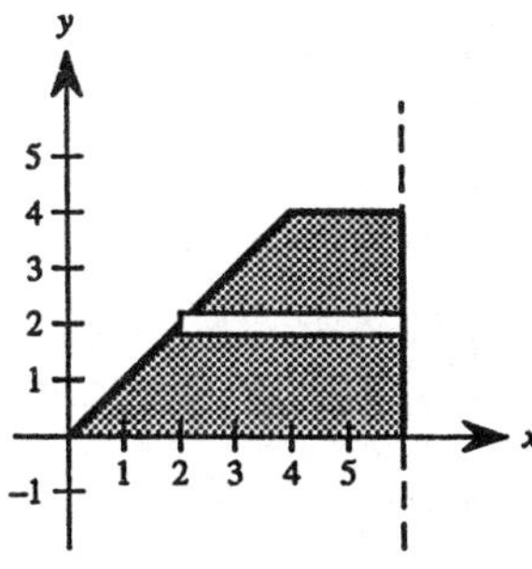

26. $R(y) = 6, \quad r(y) = 6 - (6 - y) = y$

$$V = \pi \int_0^4 [(6)^2 - (y)^2]\, dy$$

$$= \pi \left[36y - \frac{y^3}{3}\right]_0^4$$

$$= \frac{368\pi}{3}$$

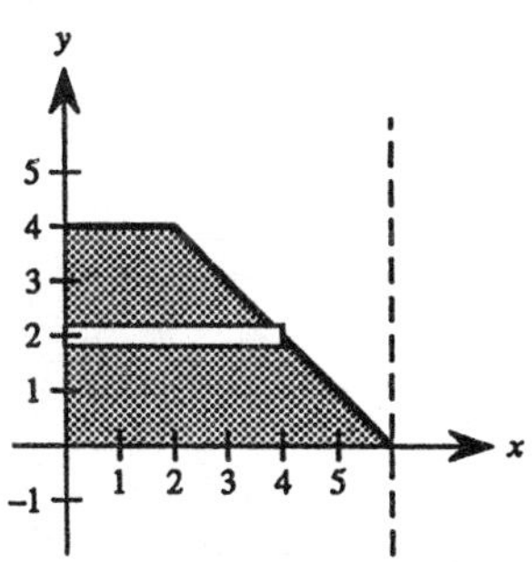

27. $R(y) = 6 - y^2, \quad r(y) = 2$

$$V = \pi \int_{-2}^{2} [(6 - y^2)^2 - (2)^2]\, dy$$

$$= 2\pi \int_{0}^{2} (y^4 - 12y^2 + 32)\, dy$$

$$= 2\pi \left[\frac{y^5}{5} - 4y^3 + 32y \right]_0^2 = \frac{384\pi}{5}$$

28. $R(y) = 6, \quad r(y) = 6 - \dfrac{12}{y^2}$

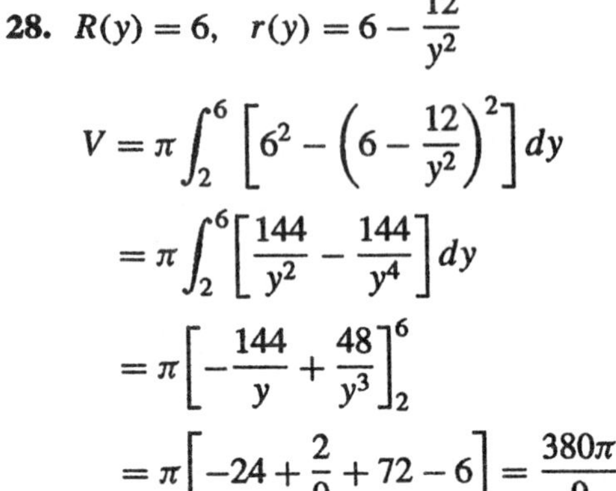

$$V = \pi \int_{2}^{6} \left[6^2 - \left(6 - \frac{12}{y^2} \right)^2 \right] dy$$

$$= \pi \int_{2}^{6} \left[\frac{144}{y^2} - \frac{144}{y^4} \right] dy$$

$$= \pi \left[-\frac{144}{y} + \frac{48}{y^3} \right]_2^6$$

$$= \pi \left[-24 + \frac{2}{9} + 72 - 6 \right] = \frac{380\pi}{9}$$

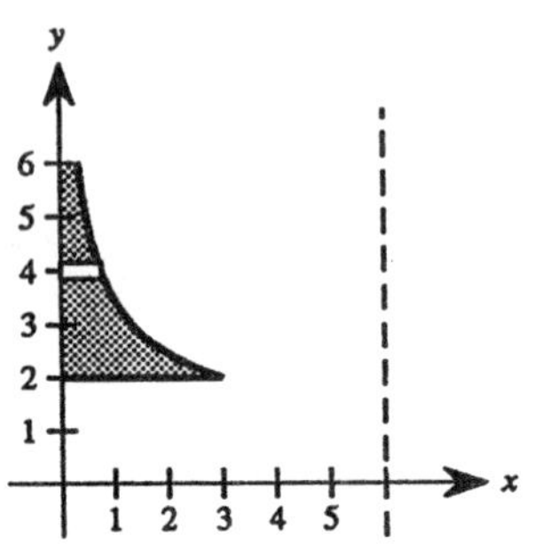

29. $R(x) = 4x - x^2, \quad r(x) = 0$

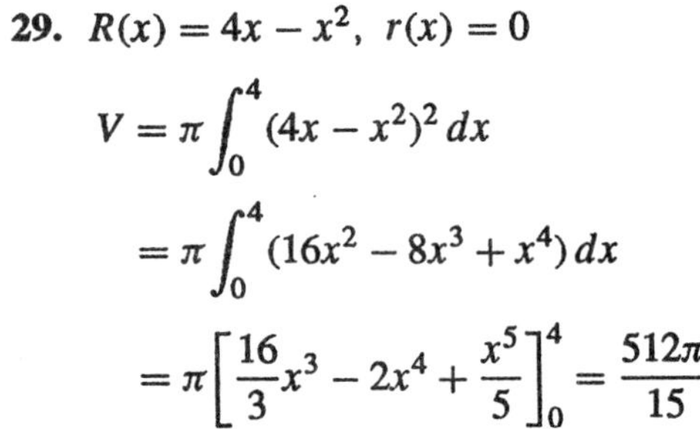

$$V = \pi \int_{0}^{4} (4x - x^2)^2\, dx$$

$$= \pi \int_{0}^{4} (16x^2 - 8x^3 + x^4)\, dx$$

$$= \pi \left[\frac{16}{3}x^3 - 2x^4 + \frac{x^5}{5} \right]_0^4 = \frac{512\pi}{15}$$

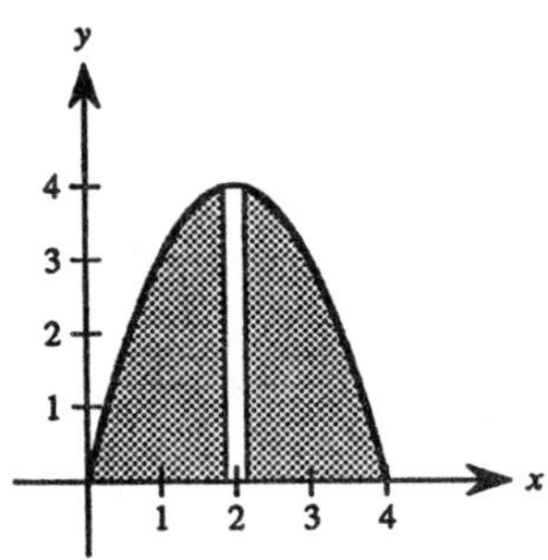

30. Completing the square we have $4x - x^2 = 4 - (x^2 - 4x + 4) = 4 - (x - 2)^2$. Thus, $y = 4 - x^2$ has the same volume as in Exercise 29 since the solid has been translated only horizontally.

31. $R(x) = \dfrac{3}{5}\sqrt{25 - x^2}, \quad r(x) = 0$

$$V = \frac{9\pi}{25} \int_{-5}^{5} (25 - x^2)\, dx$$

$$= \frac{18\pi}{25} \int_{0}^{5} (25 - x^2)\, dx$$

$$= \frac{18\pi}{25} \left[25x - \frac{x^3}{3} \right]_0^5 = 60\pi$$

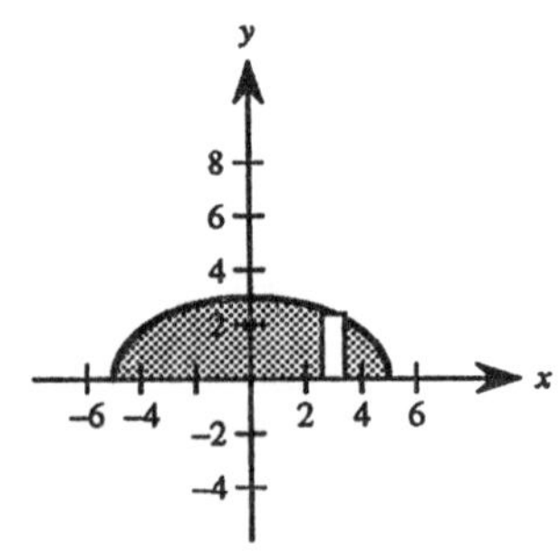

32. $R(y) = \frac{5}{3}\sqrt{9 - y^2}, \quad r(y) = 0$

$$V = \frac{25\pi}{9}\int_{-3}^{3} (9 - y^2)\,dy$$

$$= \frac{50\pi}{9}\int_{0}^{3} (9 - y^2)\,dy$$

$$= \frac{50\pi}{9}\left[9y - \frac{y^3}{3}\right]_0^3 = 100\pi$$

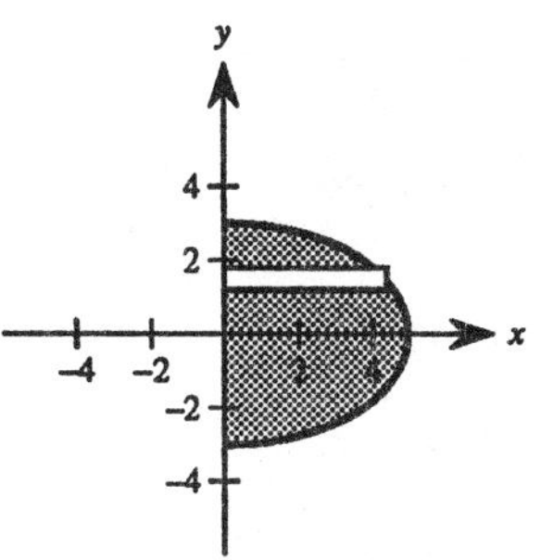

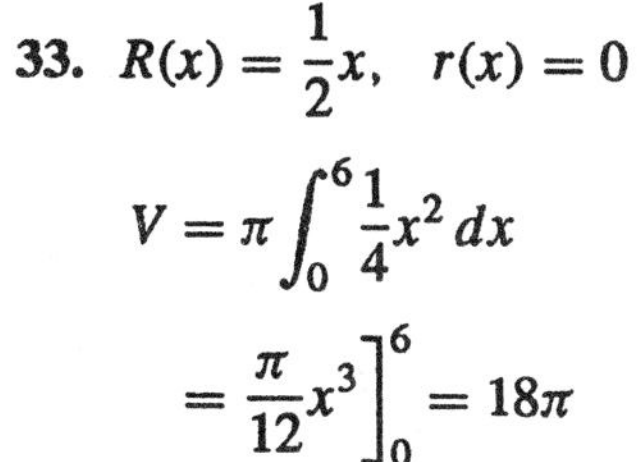

33. $R(x) = \frac{1}{2}x, \quad r(x) = 0$

$$V = \pi\int_0^6 \frac{1}{4}x^2\,dx$$

$$= \frac{\pi}{12}x^3\Big]_0^6 = 18\pi$$

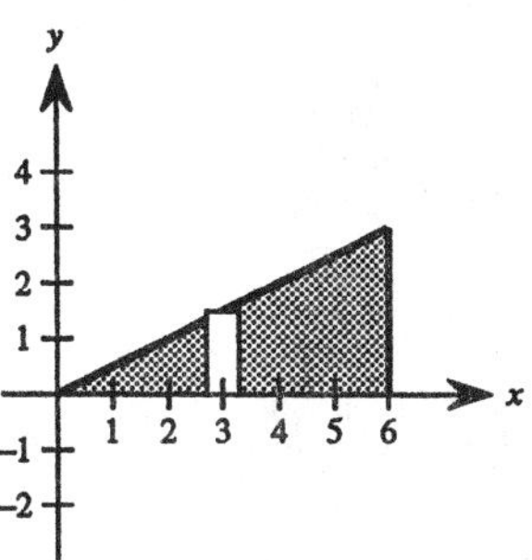

34. $R(x) = \frac{r}{h}x, \quad r(x) = 0$

$$V = \pi\int_0^h \frac{r^2}{h^2}x^2\,dx$$

$$= \frac{r^2\pi}{3h^2}x^3\Big]_0^h$$

$$= \frac{r^2\pi}{3h^2}h^3 = \frac{1}{3}\pi r^2 h$$

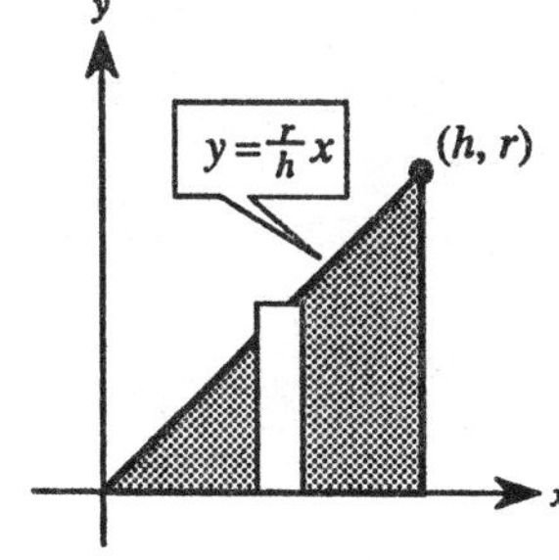

35. $R(x) = \sqrt{r^2 - x^2}, \quad r(x) = 0$

$$V = \pi\int_{-r}^{r} (r^2 - x^2)\,dx$$

$$= 2\pi\int_0^r (r^2 - x^2)\,dx$$

$$= 2\pi\left[r^2x - \frac{1}{3}x^3\right]_0^r$$

$$= 2\pi\left(r^3 - \frac{1}{3}r^3\right) = \frac{4}{3}\pi r^3$$

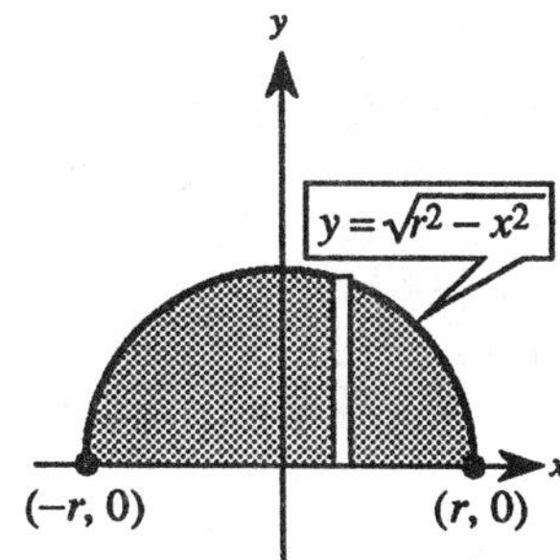

36. $x = \sqrt{r^2 - y^2}, \quad R(y) = \sqrt{r^2 - y^2}, \quad r(y) = 0$

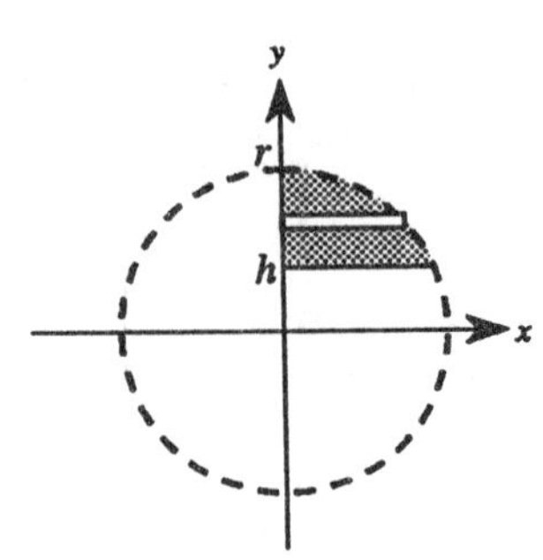

$$\begin{aligned} V &= \pi \int_h^r \left(\sqrt{r^2 - y^2}\right)^2 dy \\ &= \pi \int_h^r (r^2 - y^2)\, dy \\ &= \pi \left[r^2 y - \frac{y^3}{3} \right]_h^r \\ &= \pi \left[\left(r^3 - \frac{r^3}{3} \right) - \left(r^2 h - \frac{h^3}{3} \right) \right] \\ &= \pi \left(\frac{2r^3}{3} - r^2 h + \frac{h^3}{3} \right) \\ &= \frac{\pi}{3} (2r^3 - 3r^2 h + h^3) \end{aligned}$$

37. $x = r - \frac{r}{H} y = r\left(1 - \frac{y}{H}\right), \quad R(y) = r\left(1 - \frac{y}{H}\right), \quad r(y) = 0$

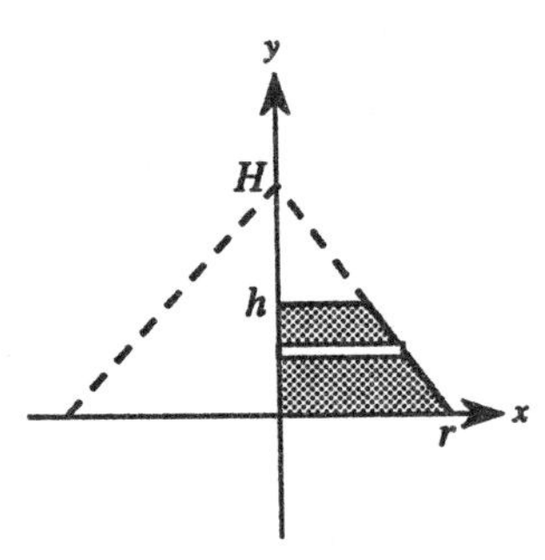

$$\begin{aligned} V = \pi \int_0^h \left[r\left(1 - \frac{y}{H}\right) \right]^2 dy &= \pi r^2 \int_0^h \left(1 - \frac{2}{H} y + \frac{1}{H^2} y^2\right) dy \\ &= \pi r^2 \left[y - \frac{1}{H} y^2 + \frac{1}{3H^2} y^3 \right]_0^h \\ &= \pi r^2 \left(h - \frac{h^2}{H} + \frac{h^3}{3H^2} \right) \\ &= \pi r^2 h \left(1 - \frac{h}{H} + \frac{h^2}{3H^2} \right) \end{aligned}$$

38. (a) $V = \pi \int_0^4 \left(\sqrt{x}\right)^2 dx = \pi \int_0^4 x\, dx = \frac{\pi x^2}{2}\Big]_0^4 = 8\pi$

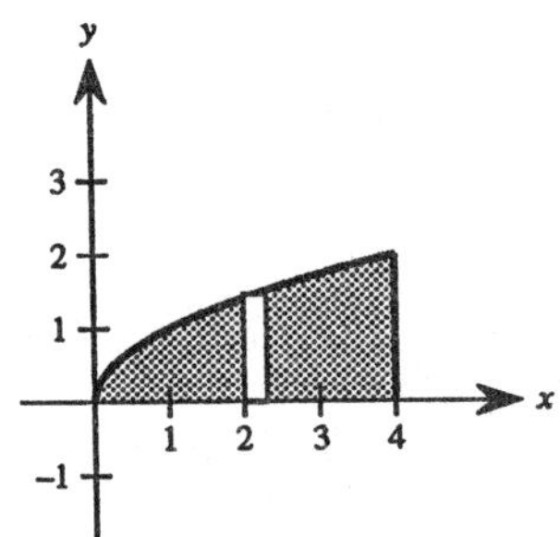

Let $0 < c < 4$ and set $\pi \int_0^c x\, dx = \frac{\pi x^2}{2}\Big]_0^c = \frac{\pi c^2}{2} = 4\pi$

$$c^2 = 8$$

$$c = \sqrt{8} = 2\sqrt{2}.$$

Thus, when $x = 2\sqrt{2}$, the solid is divided into two parts of equal volume.

(b) Set $\pi \int_0^c x\, dx = \frac{8\pi}{3}$ (one third of the volume). Then

$$\frac{\pi c^2}{2} = \frac{8\pi}{3}, \quad c^2 = \frac{16}{3}, \quad c = \frac{4}{\sqrt{3}} = \frac{4\sqrt{3}}{3}.$$

To find the other value, set $\pi \int_0^d x\, dx = (16\pi)/3$ (two thirds of the volume). Then

$$\frac{\pi d^2}{2} = \frac{16\pi}{3}, \quad d^2 = \frac{32}{3}, \quad d = \frac{\sqrt{32}}{\sqrt{3}} = \frac{4\sqrt{6}}{3}.$$

The x values that divide the solid into three parts of equal volume are $x = (4\sqrt{3})/3$ and $x = (4\sqrt{6})/3$.

39. $V = \pi \int_{-5.5}^{5.5} (-0.0944x^2 + 3.4)^2 \, dx = \pi \int_{-5.5}^{5.5} (0.00891136x^4 - 0.64192x^2 + 11.56) \, dx$

$$= \pi \Big[0.001782272x^5 - 0.213973333x^3 + 11.56x \Big]_{-5.5}^{5.5}$$

$$\approx 232.16422 \approx 232 \text{ cubic inches}$$

40. Total volume: $V = \dfrac{4\pi(50)^3}{3} = \dfrac{500{,}000\pi}{3}$ ft^3

Volume of water: $0.216V = 36{,}000\pi = \pi \int_{-50}^{y_0} \left(\sqrt{2500 - y^2}\right)^2 dy$

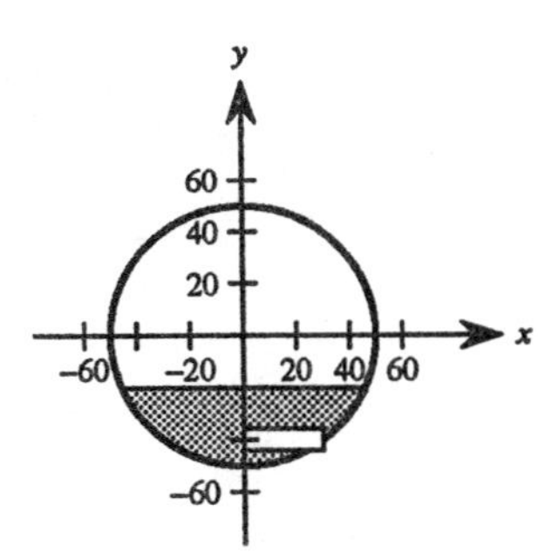

$$= \pi \int_{-50}^{y_0} (2500 - y^2) \, dy = \pi \left[2500y - \frac{y^3}{3} \right]_{-50}^{y_0}$$

$$= \pi \left(2500y_0 - \frac{y_0^3}{3} + \frac{250{,}000}{3} \right)$$

$$108{,}000 = 7500y_0 - y_0^3 + 250{,}000$$

$$y_0^3 - 7500y_0 - 142{,}000 = 0$$

$$(y_0 + 20)(y_0^2 - 20y_0 - 7100) = 0$$

$$y_0 = -20, \quad 10 \pm 60\sqrt{2}$$

Since $-50 < y_0 < 50$, $y_0 = -20$.
Depth: $[-20 - (-50)] = 30$ ft

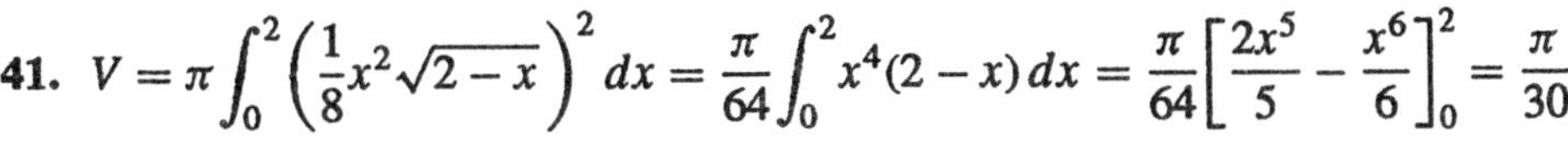

41. $V = \pi \int_0^2 \left(\frac{1}{8} x^2 \sqrt{2 - x} \right)^2 dx = \frac{\pi}{64} \int_0^2 x^4(2 - x) \, dx = \frac{\pi}{64} \left[\frac{2x^5}{5} - \frac{x^6}{6} \right]_0^2 = \frac{\pi}{30}$

42. $V = \pi \int_1^3 \left(\sqrt[3]{x^2 - 1} \right)^2 dx = \pi \int_1^3 (x^2 - 1)^{2/3} \, dx \approx 13.0295$

43. $V = \pi \int_0^2 \left(\sqrt[4]{x^2 + 1} \right)^2 dx = \pi \int_0^2 \sqrt{x^2 + 1} \, dx \approx 9.2927$

44. (a) $\pi \int_0^h r^2 \, dx$ is the volume of a right circular cylinder with radius r and height h.

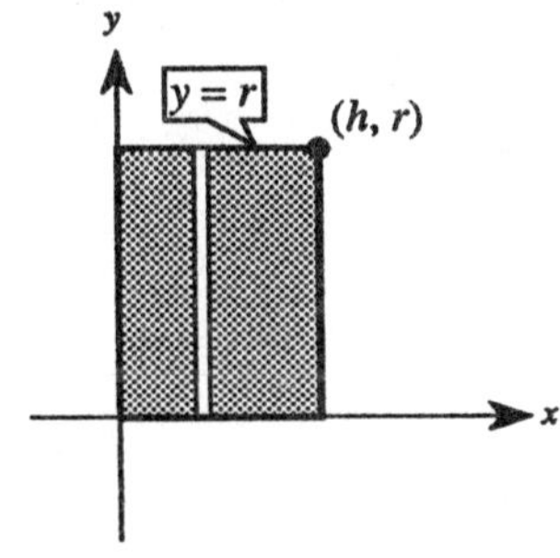

(b) $\pi \int_{-b}^{b} \left(a\sqrt{1 - (x^2/b^2)} \right)^2 dx$ is the volume of an ellipsoid with axes $2a$ and $2b$.

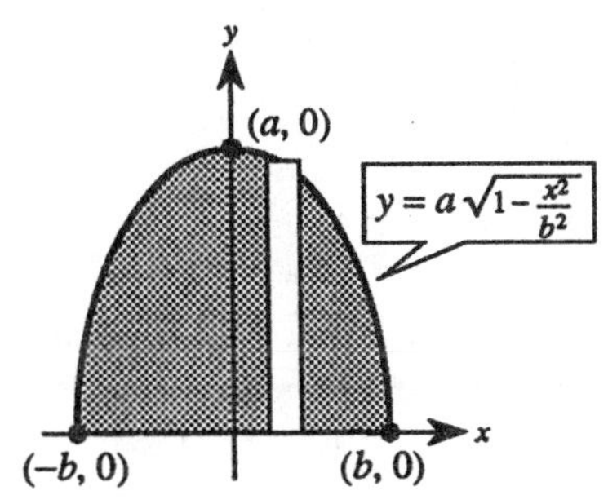

44. —CONTINUED—

(c) $\pi \int_{-r}^{r} \left(\sqrt{r^2 - x^2}\right)^2 dx$ is the volume of a sphere with radius r.

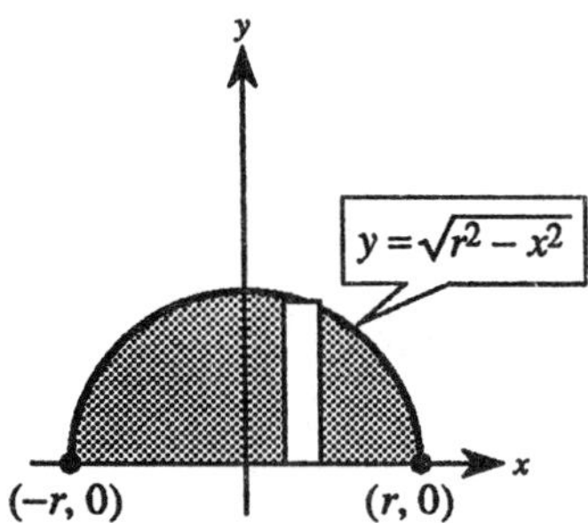

(d) $\pi \int_{0}^{h} (rx/h)^2\, dx$ is the volume of a right circular cone with the radius of the base as r and height h.

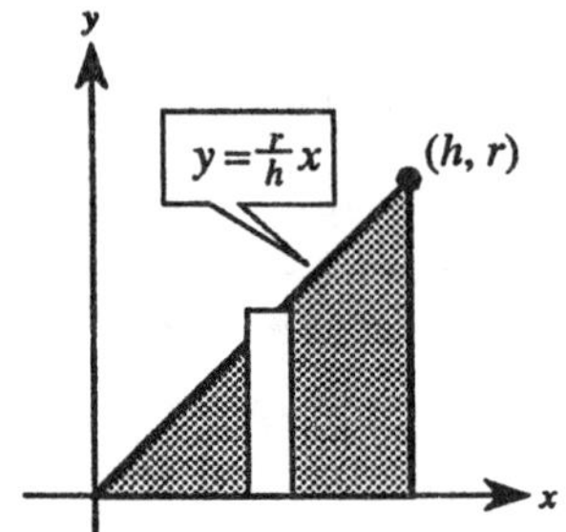

(e) $\pi \int_{-r}^{r} \left[\left(R + \sqrt{r^2 - x^2}\right)^2 - \left(R - \sqrt{r^2 - x^2}\right)^2\right] dx$ is the volume of a torus with the radius of its circular cross section as r and the distance from the axis of the torus to the center of its cross section as R.

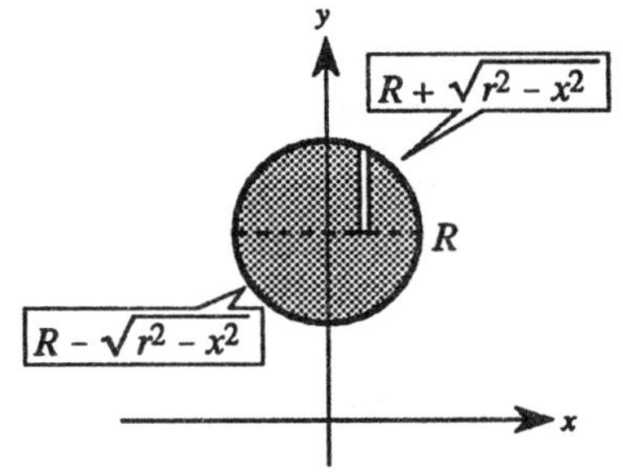

45.

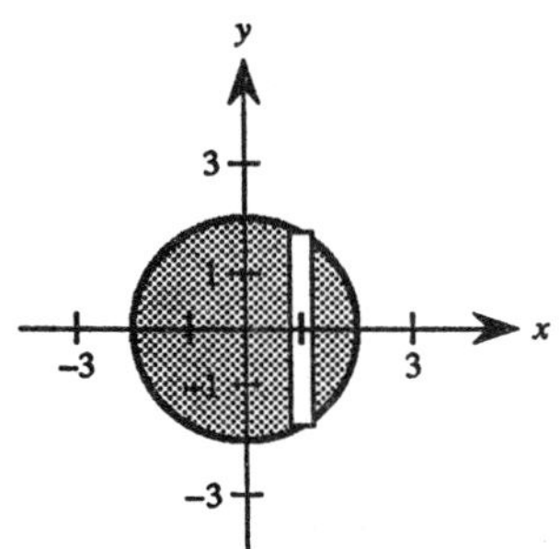

Base of Cross Section $= 2\sqrt{4 - x^2}$

(a) $A(x) = b^2 = \left(2\sqrt{4 - x^2}\right)^2$

$$V = \int_{-2}^{2} 4(4 - x^2)\, dx$$

$$= 4\left[4x - \frac{x^3}{3}\right]_{-2}^{2} = \frac{128}{3}$$

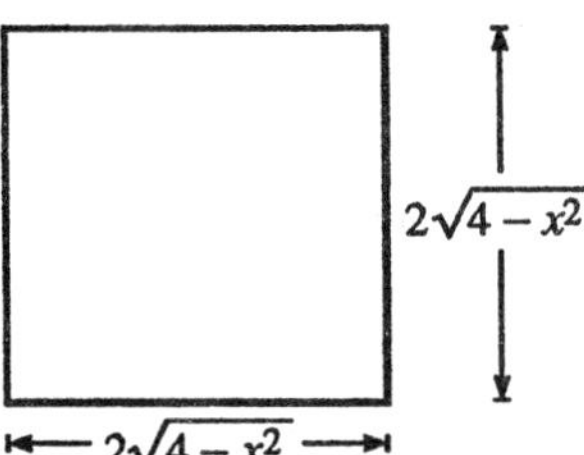

(b) $A(x) = \frac{1}{2}bh = \frac{1}{2}\left(2\sqrt{4 - x^2}\right)\left(\sqrt{3}\sqrt{4 - x^2}\right)$

$$= \sqrt{3}(4 - x^2)$$

$$V = \sqrt{3}\int_{-2}^{2} (4 - x^2)\, dx$$

$$= \sqrt{3}\left[4x - \frac{x^3}{3}\right]_{-2}^{2} = \frac{32\sqrt{3}}{3}$$

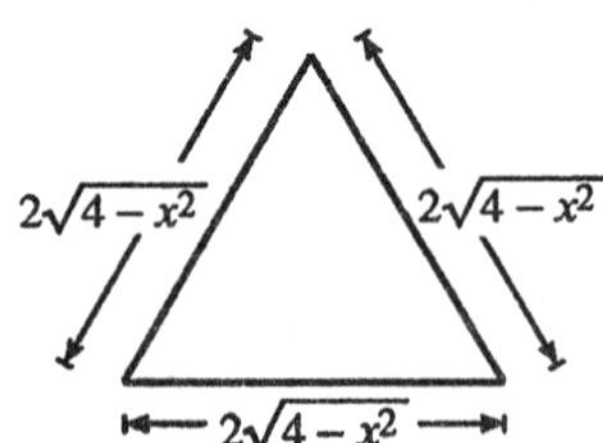

45. —CONTINUED—

(c) $A(x) = \frac{1}{2}\pi r^2 = \frac{\pi}{2}\left(\sqrt{4-x^2}\right)^2 = \frac{\pi}{2}(4-x^2)$

$$V = \frac{\pi}{2}\int_{-2}^{2}(4-x^2)\,dx = \frac{\pi}{2}\left[4x - \frac{x^3}{3}\right]_{-2}^{2} = \frac{16\pi}{3}$$

(d) $A(x) = \frac{1}{2}bh = \frac{1}{2}\left(2\sqrt{4-x^2}\right)\left(\sqrt{4-x^2}\right) = 4 - x^2$

$$V = \int_{-2}^{2}(4-x^2)\,dx = \left[4x - \frac{x^3}{3}\right]_{-2}^{2} = \frac{32}{3}$$

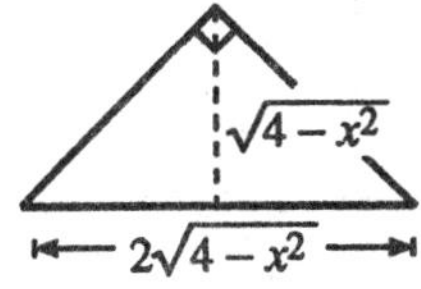

46.

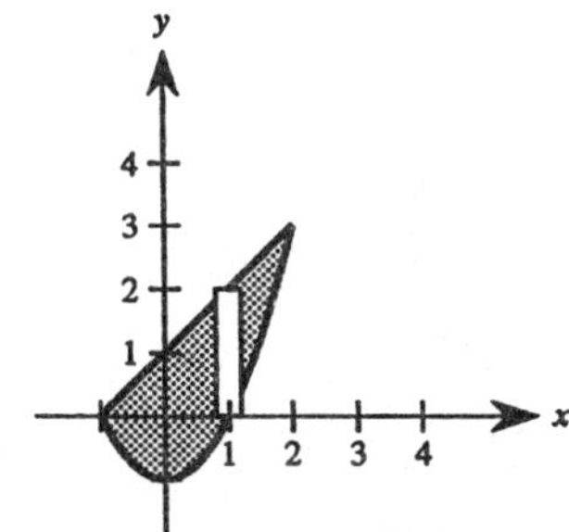

Base of Cross Section
$= (x+1) - (x^2-1) = 2 + x - x^2$

(a) $A(x) = b^2 = (2+x-x^2)^2 = 4 + 4x - 3x^2 - 2x^3 + x^4$

$$V = \int_{-1}^{2}(4 + 4x - 3x^2 - 2x^3 + x^4)\,dx$$

$$= \left[4x + 2x^2 - x^3 - \frac{1}{2}x^4 + \frac{1}{5}x^5\right]_{-1}^{2} = \frac{81}{10}$$

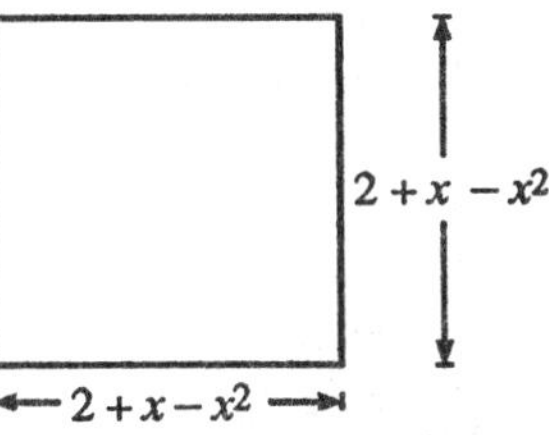

(b) $A(x) = bh = 2 + x - x^2$

$$V = \int_{-1}^{2}(2 + x - x^2)\,dx = \left[2x + \frac{x^2}{2} - \frac{x^3}{3}\right]_{-1}^{2} = \frac{9}{2}$$

(c) $A(x) = \frac{1}{2}\pi ab = \left(\frac{1}{2}\right)\pi(2)\left(\frac{2+x-x^2}{2}\right) = \frac{\pi}{2}(2+x-x^2)$

$$V = \frac{\pi}{2}\int_{-1}^{2}(2+x-x^2)\,dx = \frac{\pi}{2}\left[2x + \frac{x^2}{2} - \frac{x^3}{3}\right]_{-1}^{2} = \frac{\pi}{2}\left(\frac{9}{2}\right) = \frac{9\pi}{4}$$

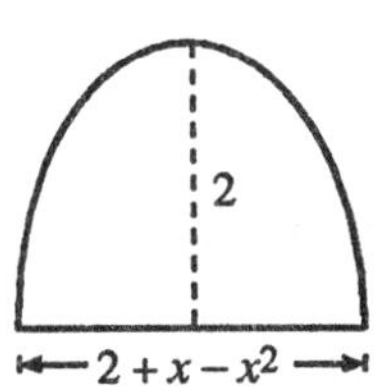

(d) $A(x) = \frac{1}{2}bh = \frac{1}{2}(2+x-x^2)\frac{\sqrt{3}(2+x-x^2)}{2} = \frac{\sqrt{3}}{4}(2+x-x^2)^2$

$$V = \frac{\sqrt{3}}{4}\int_{-1}^{2}(2+x-x^2)^2\,dx$$

$$= \frac{\sqrt{3}}{4}\left[4x + 2x^2 - x^3 - \frac{1}{2}x^4 + \frac{1}{5}x^5\right]_{-1}^{2} = \frac{\sqrt{3}}{4}\left(\frac{81}{10}\right) = \frac{81\sqrt{3}}{40}$$

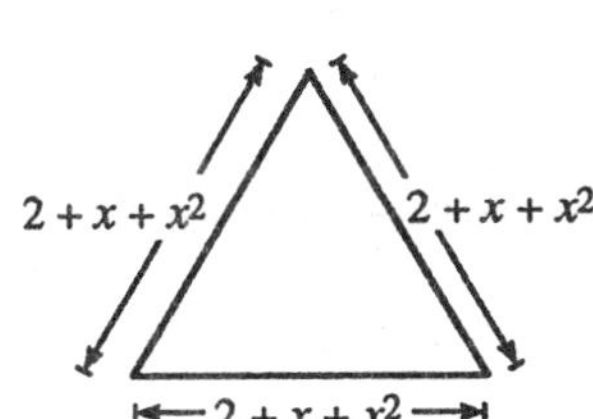

47.

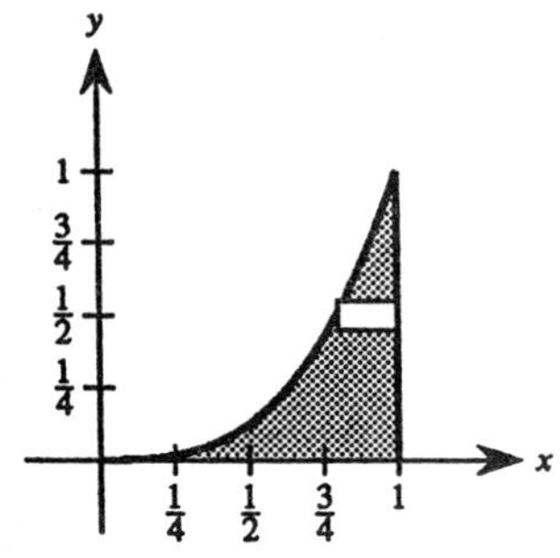

Base of Cross Section $= 1 - \sqrt[3]{y}$

(a) $A(y) = b^2 = \left(1 - \sqrt[3]{y}\right)^2$

$$V = \int_0^1 \left(1 - \sqrt[3]{y}\right)^2 dy$$

$$= \int_0^1 \left(1 - 2y^{1/3} + y^{2/3}\right) dy$$

$$= \left[y - \frac{3}{2}y^{4/3} + \frac{3}{5}y^{5/3}\right]_0^1 = \frac{1}{10}$$

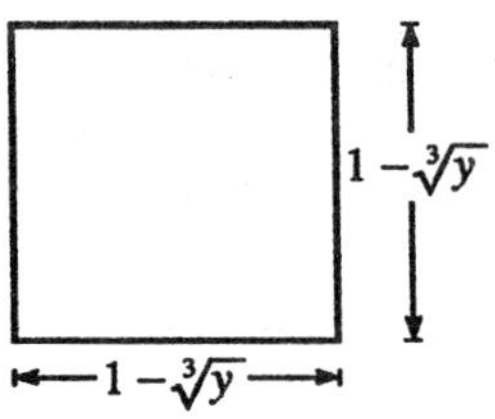

(b) $A(y) = \frac{1}{2}\pi r^2 = \frac{1}{2}\pi\left(\frac{1 - \sqrt[3]{y}}{2}\right)^2 = \frac{1}{8}\pi\left(1 - \sqrt[3]{y}\right)^2$

$$V = \frac{1}{8}\pi \int_0^1 \left(1 - \sqrt[3]{y}\right)^2 dy = \frac{\pi}{8}\left(\frac{1}{10}\right) = \frac{\pi}{80}$$

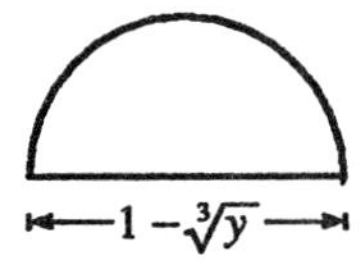

(c) $A(y) = \frac{1}{2}bh = \frac{1}{2}\left(1 - \sqrt[3]{y}\right)\left(\frac{\sqrt{3}}{2}\right)\left(1 - \sqrt[3]{y}\right)$

$$= \frac{\sqrt{3}}{4}\left(1 - \sqrt[3]{y}\right)^2$$

$$V = \frac{\sqrt{3}}{4}\int_0^1 \left(1 - \sqrt[3]{y}\right)^2 dy$$

$$= \frac{\sqrt{3}}{4}\left(\frac{1}{10}\right) = \frac{\sqrt{3}}{40}$$

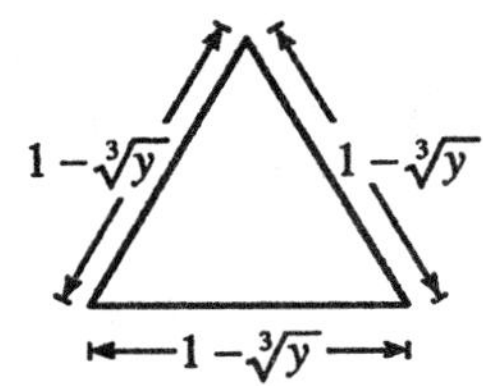

(d) $A(y) = \frac{h}{2}(b_1 + b_2)$

$$= \frac{1}{2}\left[\frac{1 - \sqrt[3]{y}}{2} + \left(1 - \sqrt[3]{y}\right)\right]\left(\frac{1 - \sqrt[3]{y}}{2}\right) = \frac{3}{8}\left(1 - \sqrt[3]{y}\right)^2$$

$$V = \frac{3}{8}\int_0^1 \left(1 - \sqrt[3]{y}\right)^2 dy = \frac{3}{8}\left(\frac{1}{10}\right) = \frac{3}{80}$$

(e) $A(y) = \frac{1}{2}\pi ab = \frac{\pi}{2}(2)\left(1 - \sqrt[3]{y}\right)\frac{1 - \sqrt[3]{y}}{2}$

$$= \frac{\pi}{2}\left(1 - \sqrt[3]{y}\right)^2$$

$$V = \frac{\pi}{2}\int_0^1 \left(1 - \sqrt[3]{y}\right)^2 dy = \frac{\pi}{2}\left(\frac{1}{10}\right) = \frac{\pi}{20}$$

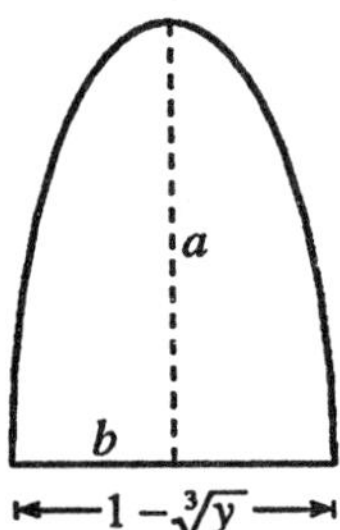

48. Since the cross sections are square:

$$A(y) = b^2 = \left(\sqrt{r^2 - y^2}\right)^2$$

$$V = 8\int_0^r (r^2 - y^2)\,dy$$

$$= 8\left[r^2 y - \frac{1}{3}y^3\right]_0^r$$

$$= \frac{16}{3}r^3$$

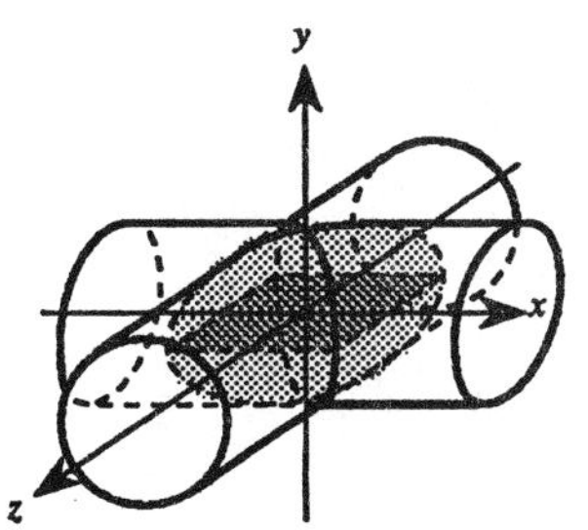

49. Since the cross sections are isosceles right triangles:

$$A(x) = \frac{1}{2}bh = \frac{1}{2}\left(\sqrt{r^2 - y^2}\right)\left(\sqrt{r^2 - y^2}\right) = \frac{1}{2}(r^2 - y^2)$$

$$V = \frac{1}{2}\int_{-r}^{r} (r^2 - y^2)\,dy = \int_0^r (r^2 - y^2)\,dy = \left[r^2 y - \frac{y^3}{3}\right]_0^r = \frac{2}{3}r^3$$

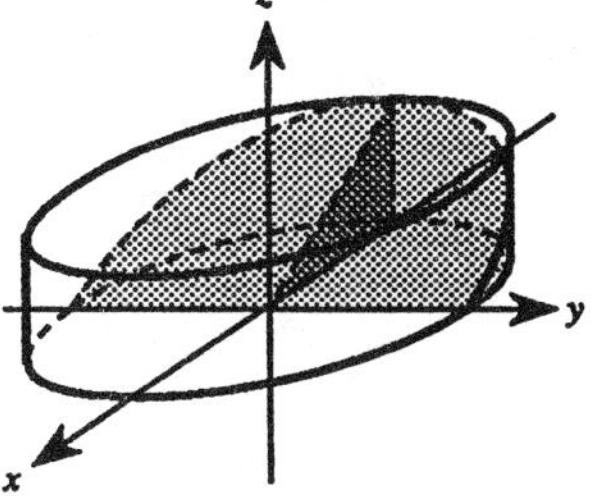

50. Let $A_1(x)$ and $A_2(x)$ equal the areas of the cross sections of the two solids for $a \le x \le b$. Since $A_1(x) = A_2(x)$, we have

$$V_1 = \int_a^b A_1(x)\,dx = \int_a^b A_2(x)\,dx = V_2.$$

Thus, the volumes are the same.

51. $$V = \pi\int_0^4 \left(\frac{10}{\sqrt{x}+2}\right)^2 dx$$

$$= 100\pi\int_0^4 \frac{1}{(\sqrt{x}+2)^2}\,dx$$

$$\approx 122.9220$$

52. $$V = \pi\int_0^5 \left(\frac{8x}{9+x^2}\right)^2 dx$$

$$= \pi\int_0^5 \frac{64x^2}{81 + 18x^2 + x^4}\,dx$$

$$\approx 19.744$$

Section 6.3 Volume: The Shell Method

1. $p(x) = x$

$h(x) = x$

$$V = 2\pi\int_0^2 x(x)\,dx$$

$$= \frac{2\pi x^3}{3}\Bigg]_0^2 = \frac{16\pi}{3}$$

2. $p(x) = x$

$h(x) = 1 - x$

$$V = 2\pi\int_0^1 x(1-x)\,dx$$

$$= 2\pi\int_0^1 (x - x^2)\,dx$$

$$= 2\pi\left[\frac{x^2}{2} - \frac{x^3}{3}\right]_0^1 = \frac{\pi}{3}$$

3. $p(y) = y$

$h(y) = 2 - y$

$$V = 2\pi \int_0^2 y(2 - y)\, dy$$

$$= 2\pi \int_0^2 (2y - y^2)\, dy$$

$$= 2\pi \left[y^2 - \frac{y^3}{3} \right]_0^2 = \frac{8\pi}{3}$$

4. $p(y) = -y$

$h(y) = 4 - (2 - y) = 2 + y$

$$V = 2\pi \int_{-2}^0 (-y)(2 + y)\, dy$$

$$= 2\pi \int_{-2}^0 (-2y - y^2)\, dy$$

$$= 2\pi \left[-y^2 - \frac{y^3}{3} \right]_{-2}^0 = \frac{8\pi}{3}$$

5. $p(x) = x$

$h(x) = \sqrt{x}$

$$V = 2\pi \int_0^4 x\sqrt{x}\, dx$$

$$= 2\pi \int_0^4 x^{3/2}\, dx$$

$$= \frac{4\pi}{5} x^{5/2} \bigg]_0^4 = \frac{128\pi}{5}$$

6. $p(x) = x$

$h(x) = 8 - (x^2 + 4) = 4 - x^2$

$$V = 2\pi \int_0^2 x(4 - x^2)\, dx$$

$$= 2\pi \int_0^2 (4x - x^3)\, dx$$

$$= 2\pi \left[2x^2 - \frac{x^4}{4} \right]_0^2 = 8\pi$$

7. $p(x) = x$

$h(x) = x^2$

$$V = 2\pi \int_0^2 x^3\, dx = \frac{\pi}{2} x^4 \bigg]_0^2 = 8\pi$$

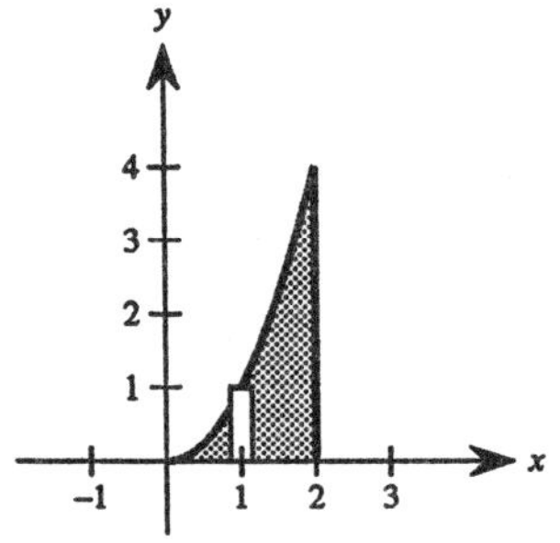

8. $p(x) = x$

$h(x) = x^2$

$$V = 2\pi \int_0^4 x^3\, dx = \frac{\pi}{2} x^4 \bigg]_0^4 = 128\pi$$

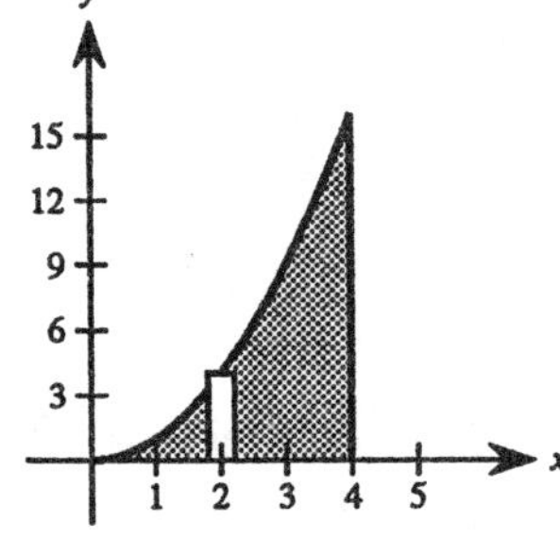

9. $p(x) = x$

$h(x) = (4x - x^2) - x^2 = 4x - 2x^2$

$$V = 2\pi \int_0^2 x(4x - 2x^2)\, dx$$

$$= 4\pi \int_0^2 (2x^2 - x^3)\, dx$$

$$= 4\pi \left[\frac{2}{3}x^3 - \frac{1}{4}x^4 \right]_0^2 = \frac{16\pi}{3}$$

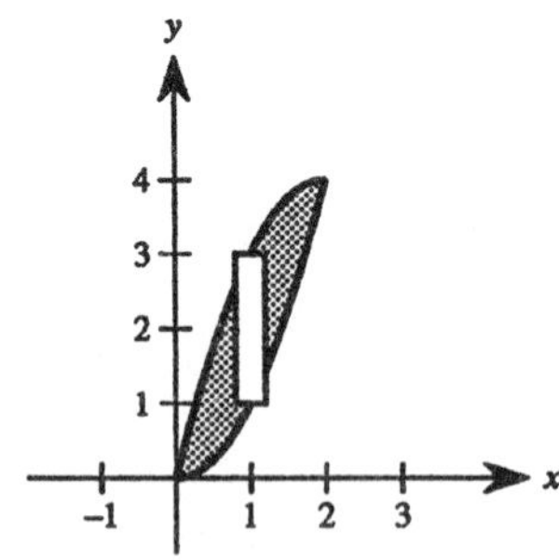

10. $p(x) = 2 - x$

$h(x) = 4x - 2x^2$

$$V = 2\pi \int_0^2 (2 - x)(4x - 2x^2)\,dx$$

$$= 2\pi \int_0^2 (8x - 8x^2 + 2x^3)\,dx$$

$$= 2\pi \left[4x^2 - \frac{8}{3}x^3 + \frac{1}{2}x^4\right]_0^2 = \frac{16\pi}{3}$$

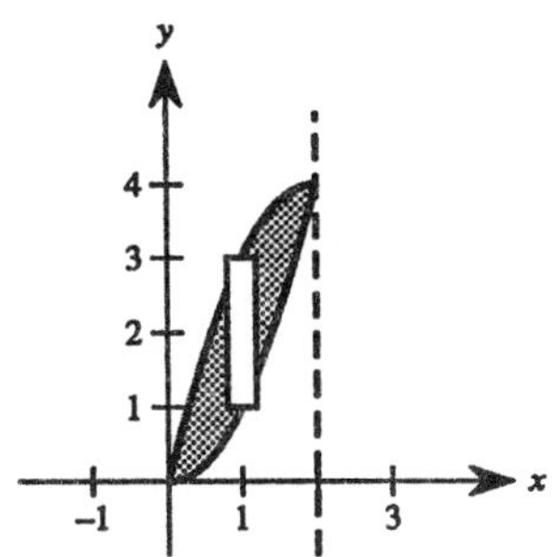

11. $p(x) = 4 - x$

$h(x) = 4x - 2x^2$

$$V = 2\pi \int_0^2 (4 - x)(4x - 2x^2)\,dx$$

$$= 2\pi(2) \int_0^2 (x^3 - 6x^2 + 8x)\,dx$$

$$= 4\pi \left[\frac{x^4}{4} - 2x^3 + 4x^2\right]_0^2 = 16\pi$$

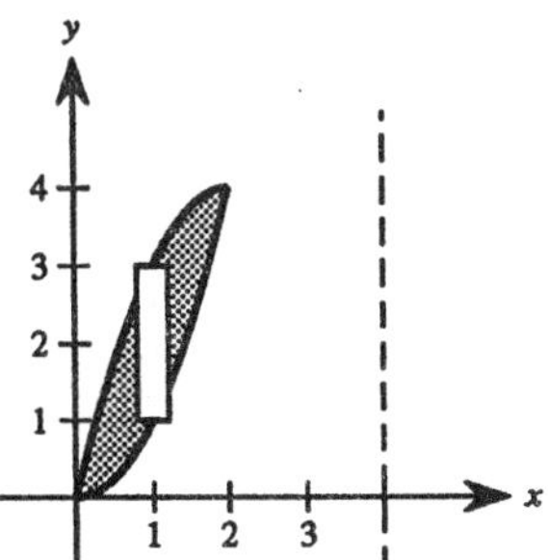

12. $p(y) = y$ and $h(y) = 1$ if $0 \le y < \frac{1}{2}$.

$p(y) = y$ and $h(y) = \frac{1}{y} - 1$ if $\frac{1}{2} \le y \le 1$.

$$V = 2\pi \int_0^{1/2} y\,dy + 2\pi \int_{1/2}^1 (1 - y)\,dy$$

$$= 2\pi \left[\frac{y^2}{2}\right]_0^{1/2} + 2\pi \left[y - \frac{y^2}{2}\right]_{1/2}^1 = \frac{\pi}{2}$$

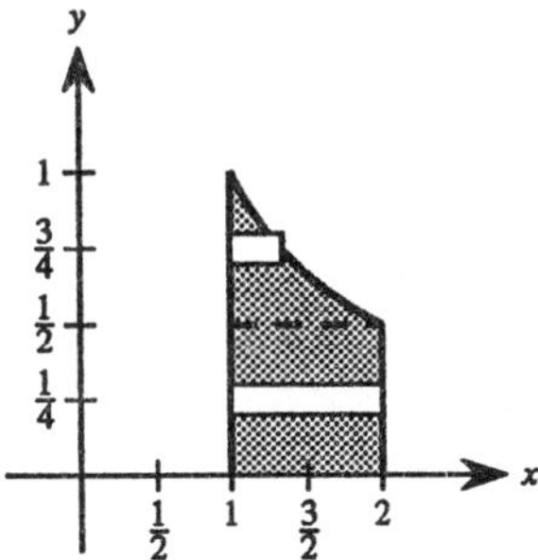

13. $p(x) = 5 - x$

$h(x) = 4x - x^2$

$$V = 2\pi \int_0^4 (5 - x)(4x - x^2)\,dx$$

$$= 2\pi \int_0^4 (x^3 - 9x^2 + 20x)\,dx$$

$$= 2\pi \left[\frac{x^4}{4} - 3x^3 + 10x^2\right]_0^4 = 64\pi$$

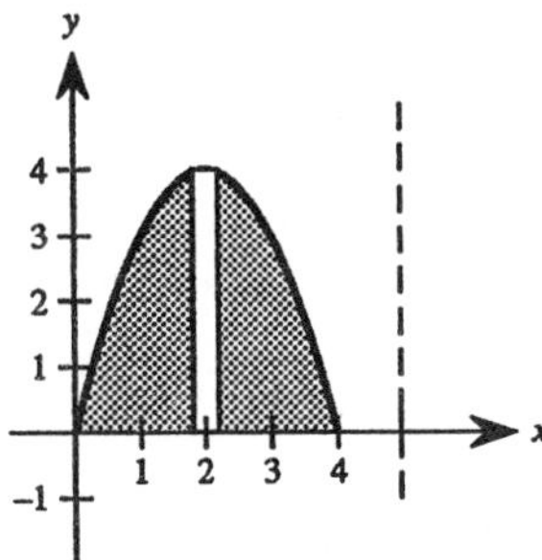

14. $p(y) = y$

$h(y) = 9 - y^2$

$$V = 2\pi \int_0^3 y(9 - y^2)\,dy$$

$$= 2\pi \int_0^3 (9y - y^3)\,dy$$

$$= 2\pi \left[\frac{9}{2}y^2 - \frac{1}{4}y^4\right]_0^3 = \frac{81\pi}{2}$$

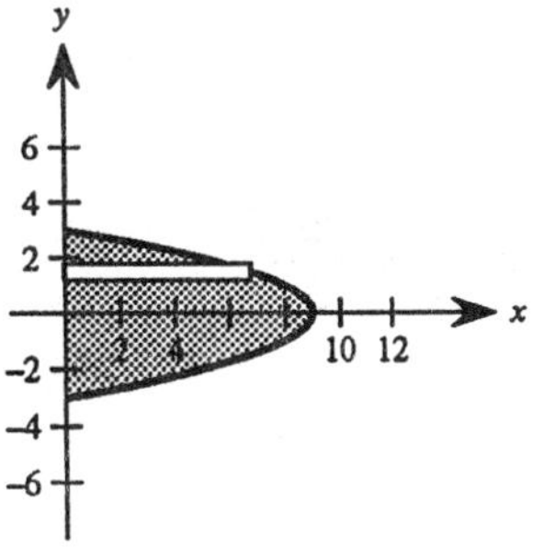

15. $p(x) = x$

$h(x) = 4 - (4x - x^2) = x^2 - 4x + 4$

$$V = 2\pi \int_0^2 (x^3 - 4x^2 + 4x)\, dx$$

$$= 2\pi \left[\frac{x^4}{4} - \frac{4}{3}x^3 + 2x^2\right]_0^2 = \frac{8\pi}{3}$$

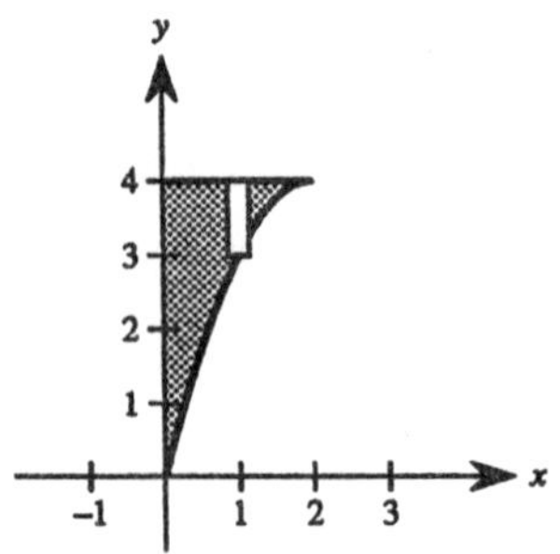

16. $p(x) = x$

$h(x) = 4 - x^2$

$$V = 2\pi \int_0^2 (4x - x^3)\, dx$$

$$= 2\pi \left[2x^2 - \frac{1}{4}x^4\right]_0^2 = 8\pi$$

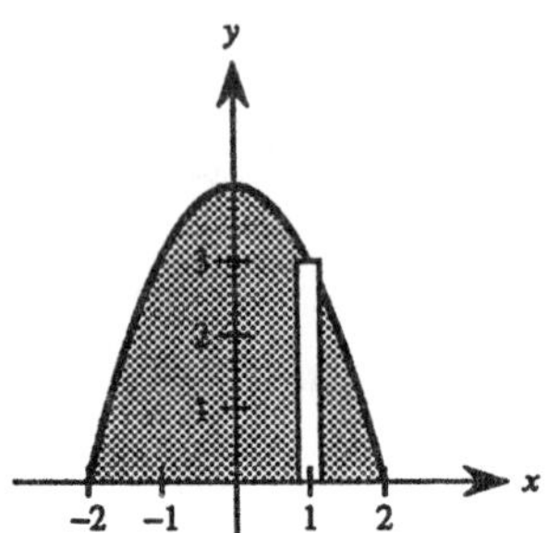

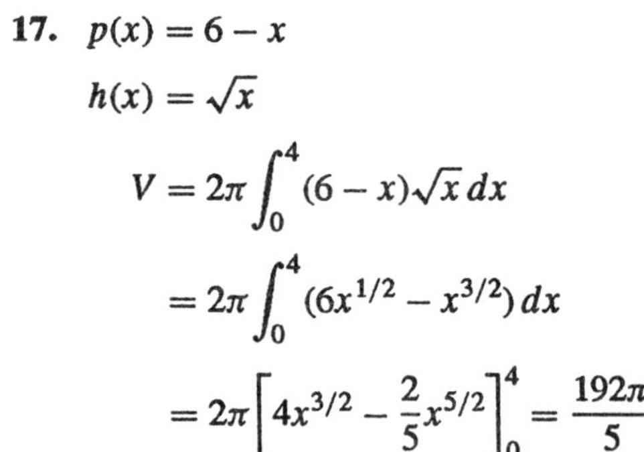

17. $p(x) = 6 - x$

$h(x) = \sqrt{x}$

$$V = 2\pi \int_0^4 (6 - x)\sqrt{x}\, dx$$

$$= 2\pi \int_0^4 (6x^{1/2} - x^{3/2})\, dx$$

$$= 2\pi \left[4x^{3/2} - \frac{2}{5}x^{5/2}\right]_0^4 = \frac{192\pi}{5}$$

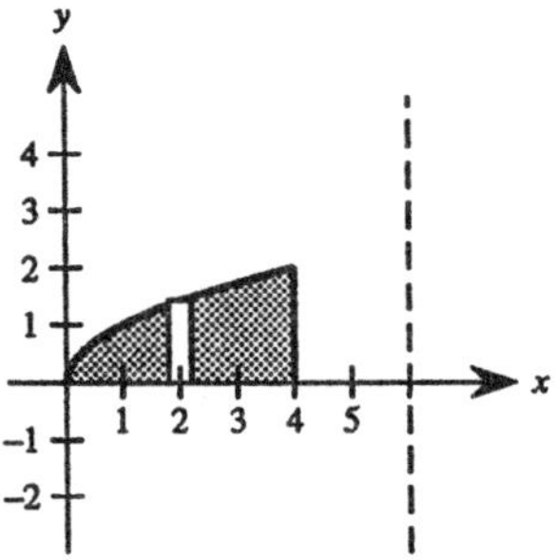

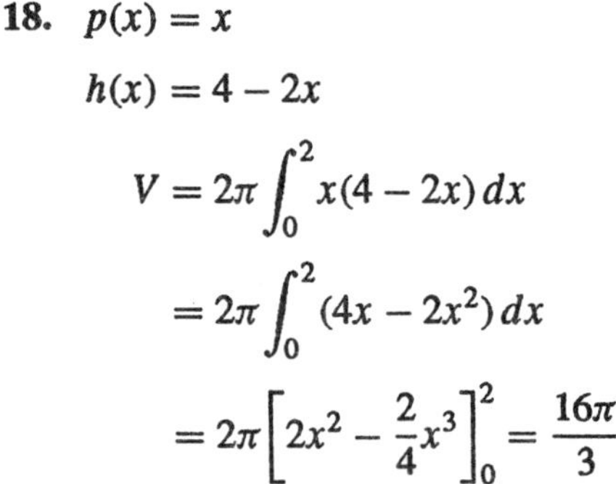

18. $p(x) = x$

$h(x) = 4 - 2x$

$$V = 2\pi \int_0^2 x(4 - 2x)\, dx$$

$$= 2\pi \int_0^2 (4x - 2x^2)\, dx$$

$$= 2\pi \left[2x^2 - \frac{2}{4}x^3\right]_0^2 = \frac{16\pi}{3}$$

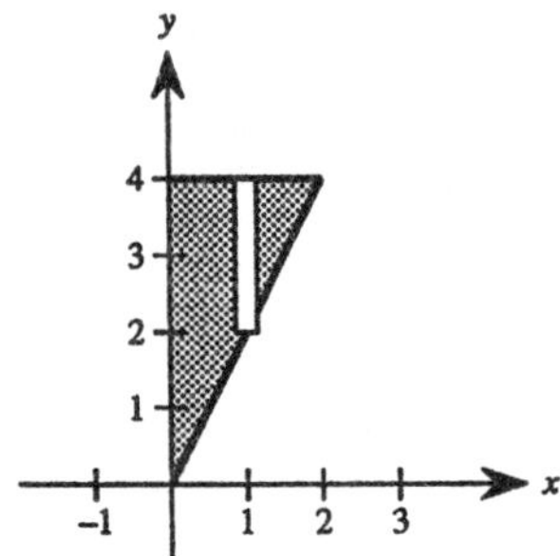

19. $p(y) = 2 - y, \quad h(y) = y - y^2$

$$V = 2\pi \int_0^1 (2 - y)(y - y^2)\, dy$$

$$= 2\pi \int_0^1 (2y - 3y^2 + y^3)\, dy$$

$$= 2\pi \left[y^2 - y^3 + \frac{y^4}{4}\right]_0^1 = 2\pi\left[\frac{1}{4} - 0\right] = \frac{\pi}{2}$$

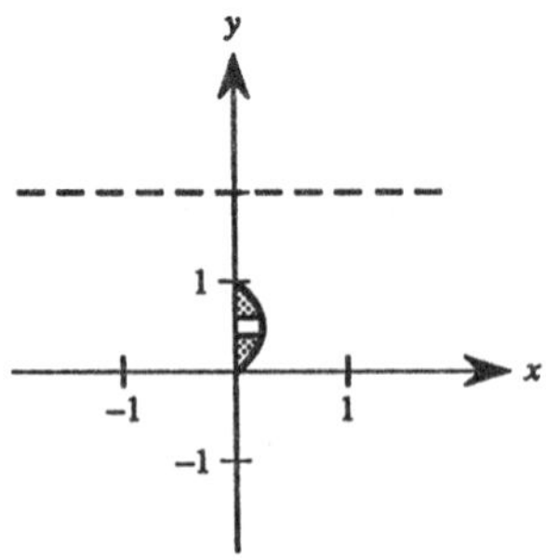

20. $p(y) = 4 + y, \quad h(y) = y - y^2$

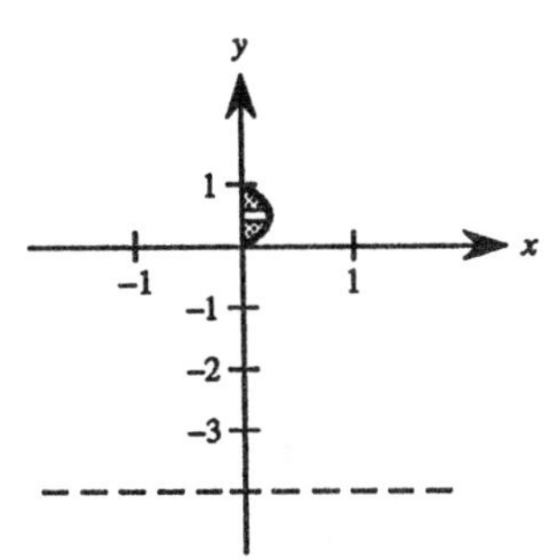

$$V = 2\pi \int_0^1 (4 + y)(y - y^2)\,dy$$

$$= 2\pi \int_0^1 (4y - 3y^2 - y^3)\,dy$$

$$= 2\pi \left[2y^2 - y^3 - \frac{y^4}{4}\right]_0^1$$

$$= 2\pi \left[2 - 1 - \frac{1}{4}\right] = \frac{3\pi}{2}$$

21. (a) **Disc**

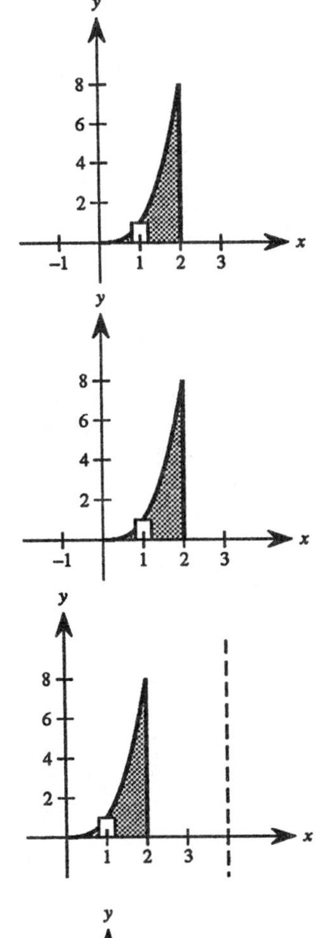

$R(x) = x^3$

$r(x) = 0$

$$V = \pi \int_0^2 x^6\,dx = \pi \frac{x^7}{7}\bigg]_0^2 = \frac{128\pi}{7}$$

(b) **Shell**

$p(x) = x$

$h(x) = x^3$

$$V = 2\pi \int_0^2 x^4\,dx = 2\pi \frac{x^5}{5}\bigg]_0^2 = \frac{64\pi}{5}$$

(c) **Shell**

$p(x) = 4 - x$

$h(x) = x^3$

$$V = 2\pi \int_0^2 (4 - x)x^3\,dx$$

$$= 2\pi \int_0^2 (4x^3 - x^4)\,dx = 2\pi \left[x^4 - \frac{1}{5}x^5\right]_0^2 = \frac{96\pi}{5}$$

(d) **Disc**

$R(x) = 8$

$r(x) = 8 - x^3$

$$V = \pi \int_0^2 [64 - (8 - x^3)^2]\,dx$$

$$= \pi \int_0^2 (16x^3 - x^6)\,dx = \pi \left[4x^4 - \frac{x^7}{7}\right]_0^2 = \frac{320\pi}{7}$$

22. (a) **Disc**

$R(x) = \frac{1}{x^3}, \quad r(x) = 0$

$$V = \pi \int_1^2 \left(\frac{1}{x^3}\right)^2 dx$$

$$= \pi\left(-\frac{1}{5x^5}\right)\Big]_1^2 = \frac{31\pi}{160}$$

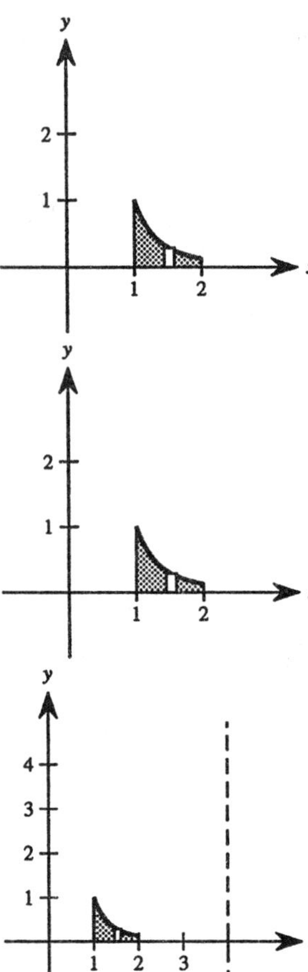

(b) **Shell**

$p(x) = x, \quad h(x) = \frac{1}{x^3}$

$$V = 2\pi \int_1^2 x\left(\frac{1}{x^3}\right) dx$$

$$= 2\pi\left(-\frac{1}{x}\right)\Big]_1^2 = \pi$$

(c) **Shell**

$p(x) = 4 - x, \quad h(x) = \frac{1}{x^3}$

$$V = 2\pi \int_1^2 (4 - x)\left(\frac{1}{x^3}\right) dx$$

$$= 2\pi \int_1^2 \left(\frac{4}{x^3} - \frac{1}{x^2}\right) dx = 2\pi\left(-\frac{2}{x^2} + \frac{1}{x}\right)\Big]_1^2 = 2\pi$$

23. (a) **Shell**

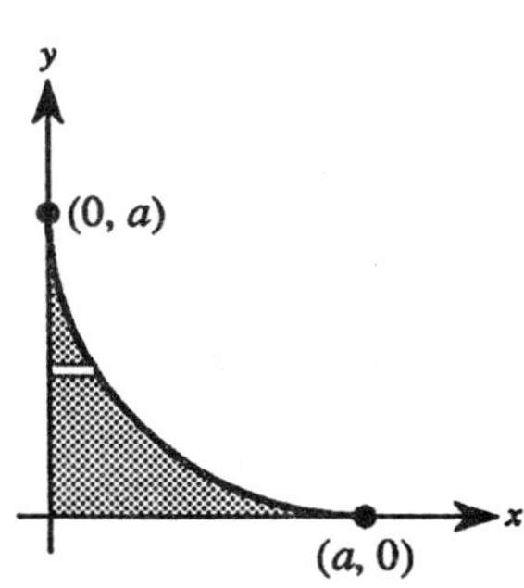

$p(y) = y$

$h(y) = (a^{1/2} - y^{1/2})^2$

$$V = 2\pi \int_0^a y(a - 2a^{1/2}y^{1/2} + y)\, dy = 2\pi \int_0^a (ay - 2a^{1/2}y^{3/2} + y^2)\, dy$$

$$= 2\pi\left[\frac{a}{2}y^2 - \frac{4a^{1/2}}{5}y^{5/2} + \frac{y^3}{3}\right]_0^a = 2\pi\left[\frac{a^3}{2} - \frac{4a^3}{5} + \frac{a^3}{3}\right] = \frac{\pi a^3}{15}$$

(b) Same as part (a) by symmetry

(c) $p(x) = a - x$

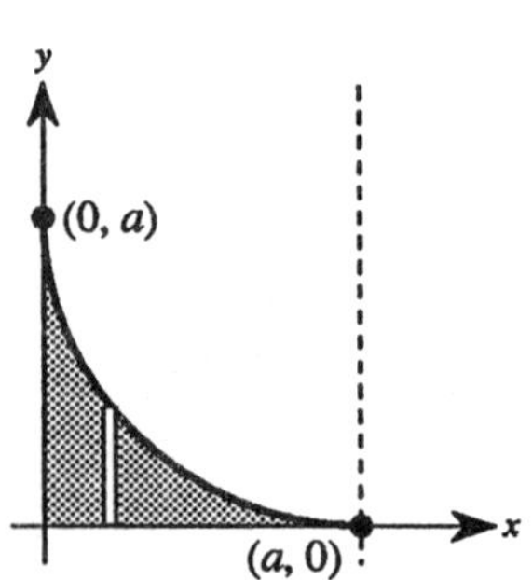

$h(x) = (a^{1/2} - x^{1/2})^2$

$$V = 2\pi \int_0^a (a - x)(a^{1/2} - x^{1/2})^2\, dx$$

$$= 2\pi \int_0^a (a^2 - 2a^{3/2}x^{1/2} + 2a^{1/2}x^{3/2} - x^2)\, dx$$

$$= 2\pi\left[a^2x - \frac{4}{3}a^{3/2}x^{3/2} + \frac{4}{5}a^{1/2}x^{5/2} - \frac{1}{3}x^3\right]_0^a = \frac{4\pi a^3}{15}$$

24. (a) **Disc**

$R(x) = (a^{2/3} - x^{2/3})^{3/2}$

$r(x) = 0$

$$V = \pi \int_{-a}^{a} (a^{2/3} - x^{2/3})^3 \, dx$$

$$= 2\pi \int_{0}^{a} (a^2 - 3a^{4/3}x^{2/3} + 3a^{2/3}x^{4/3} - x^2) \, dx$$

$$= 2\pi \left[a^2 x - \frac{9}{5}a^{4/3}x^{5/3} + \frac{9}{7}a^{2/3}x^{7/3} - \frac{1}{3}x^3 \right]_0^a$$

$$= 2\pi \left(a^3 - \frac{9}{5}a^3 + \frac{9}{7}a^3 - \frac{1}{3}a^3 \right) = \frac{32\pi a^3}{105}$$

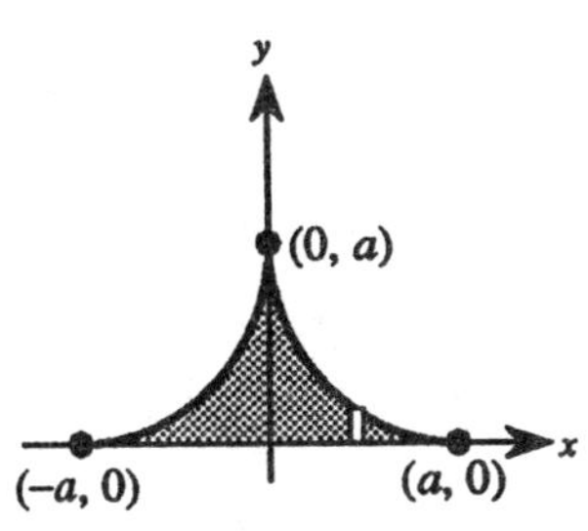

(b) **Disc**

$R(y) = (a^{2/3} - y^{2/3})^{3/2}$

$r(y) = 0$

$$V = \pi \int_{-a}^{a} (a^{2/3} - y^{2/3})^3 \, dy$$

$$= 2\pi \int_{0}^{a} (a^2 - 3a^{4/3}y^{2/3} + 3a^{2/3}y^{4/3} - y^2) \, dy$$

$$= 2\pi \left[a^2 y - \frac{9}{5}a^{4/3}y^{5/3} + \frac{9}{7}a^{2/3}y^{7/3} - \frac{1}{3}y^3 \right]_0^a = \frac{32\pi a^3}{105}$$

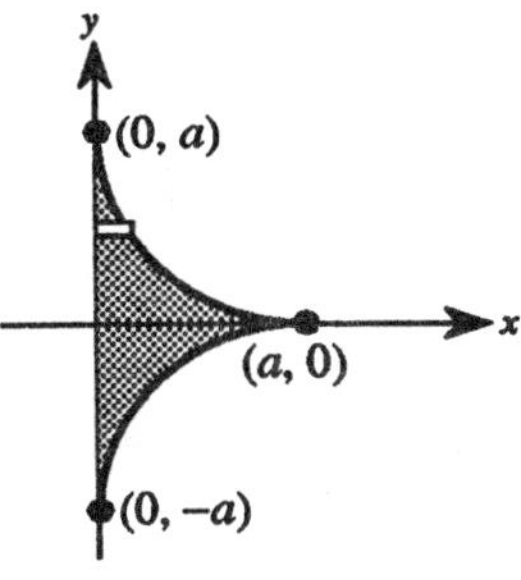

25. $p(x) = x$

$h(x) = 2 - \frac{1}{2}x^2$

$$V = 2\pi \int_0^2 x\left(2 - \frac{1}{2}x^2\right) dx = 2\pi \int_0^2 \left(2x - \frac{1}{2}x^3\right) dx = 2\pi \left[x^2 - \frac{1}{8}x^4 \right]_0^2 = 4\pi \text{ (total volume)}$$

Now find x_0 such that

$$\pi = 2\pi \int_0^{x_0} \left(2x - \frac{1}{2}x^3\right) dx$$

$$1 = 2\left[x^2 - \frac{1}{8}x^4 \right]_0^{x_0}$$

$$1 = 2{x_0}^2 - \frac{1}{4}{x_0}^4$$

$${x_0}^4 - 8{x_0}^2 + 4 = 0$$

$${x_0}^2 = 4 \pm 2\sqrt{3}.$$

Diameter: $2\sqrt{4 - 2\sqrt{3}} \approx 1.464$

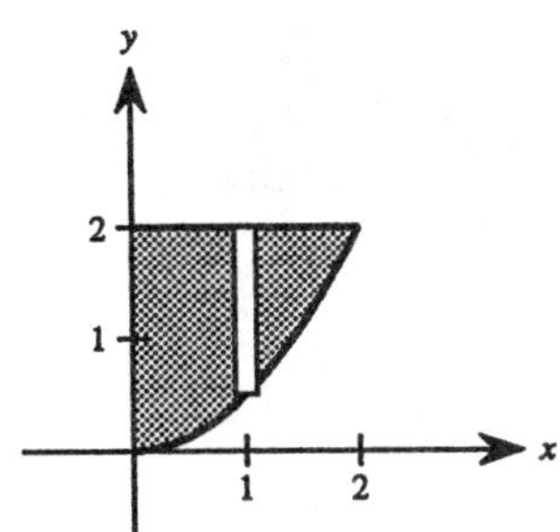

26. Total volume of the hemisphere is $\frac{1}{2}\left(\frac{4}{3}\right)\pi r^3 = \frac{2}{3}\pi(3)^3 = 18\pi$. By the Shell Method, $p(x) = x$, $h(x) = \sqrt{9-x^2}$. Find x_0 such that

$$6\pi = 2\pi\int_0^{x_0} x\sqrt{9-x^2}\,dx$$

$$6 = -\int_0^{x_0}(9-x^2)^{1/2}(-2x)\,dx$$

$$= -\frac{2}{3}(9-x^2)^{3/2}\Big]_0^{x_0} = 18 - \frac{2}{3}(9-x_0^2)^{3/2}$$

$$(9-x_0^2)^{3/2} = 18$$

$$x_0 = \sqrt{9-18^{2/3}}.$$

Diameter: $2\sqrt{9-18^{2/3}} \approx 2.920$

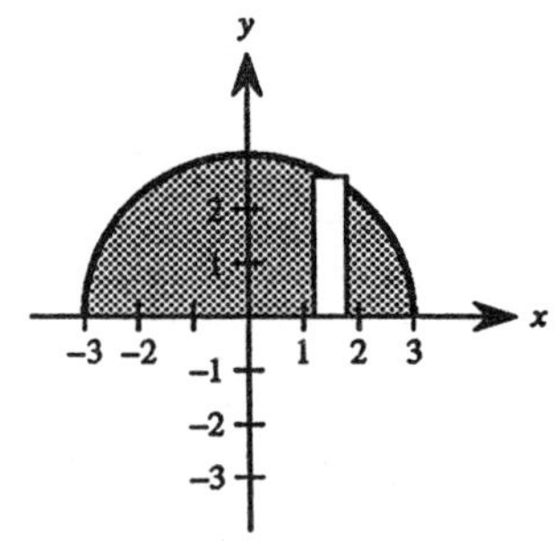

27. Disc

$$R(y) = \sqrt{r^2-y^2}$$

$$r(y) = 0$$

$$V = \pi\int_{r-h}^{r}(r^2-y^2)\,dy$$

$$= \pi\left[r^2y - \frac{y^3}{3}\right]_{r-h}^{r} = \frac{1}{3}\pi h^2(3r-h)$$

28. $V = 2\pi\int_0^1 x\sqrt{1-x^3}\,dx \approx 2.2935$

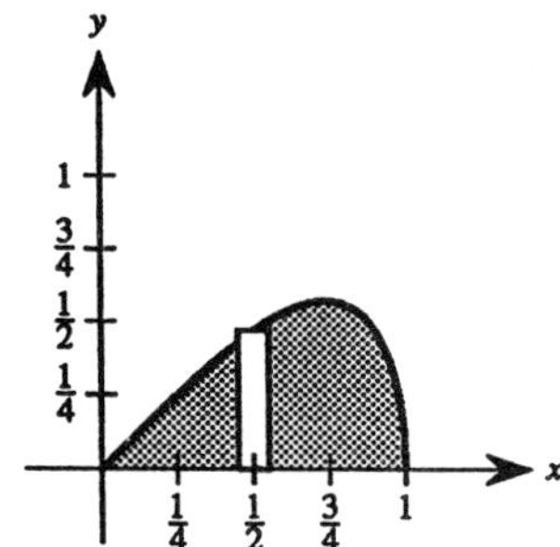

29. $x^{4/3} + y^{4/3} = 1$

$$y = \pm(1-x^{4/3})^{3/4}$$

$$V = 2(2\pi)\int_0^1 x(1-x^{4/3})^{3/4}\,dx \approx 2.970$$

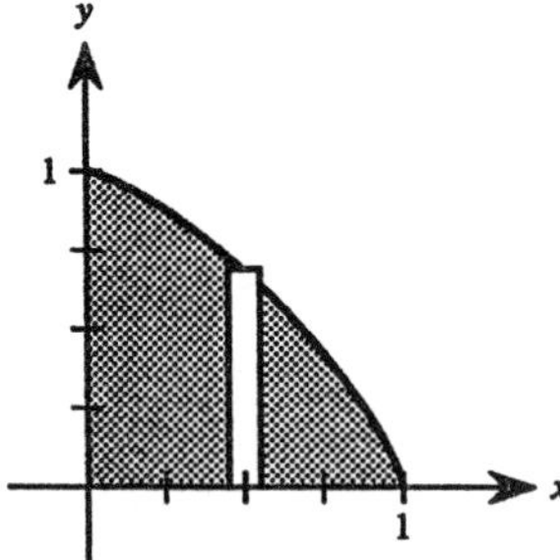

30. $V = 2\pi\int_0^{200} x\,f(x)\,dx$

$$\approx \frac{2\pi(200)}{3(8)}\left[0 + 4(25)(19) + 2(50)(19) + 4(75)(17) + 2(100)(15) + 4(125)(14) + 2(150)(10) + 4(175)(6) + 0\right]$$

$$\approx 1{,}366{,}592.804 \text{ cubic feet}$$

31. $V = 2\pi\int_0^{40} x\,f(x)\,dx \approx \frac{2\pi(40)}{3(4)}\left[0 + 4(10)(46) + 2(20)(40) + 4(30)(20) + 0\right] \approx 122{,}313$ cubic feet

32. (a) $2\pi \int_0^r hx\left(1 - \frac{x}{r}\right) dx$ is the volume of a right circular cone with the radius of the base as r and height h.

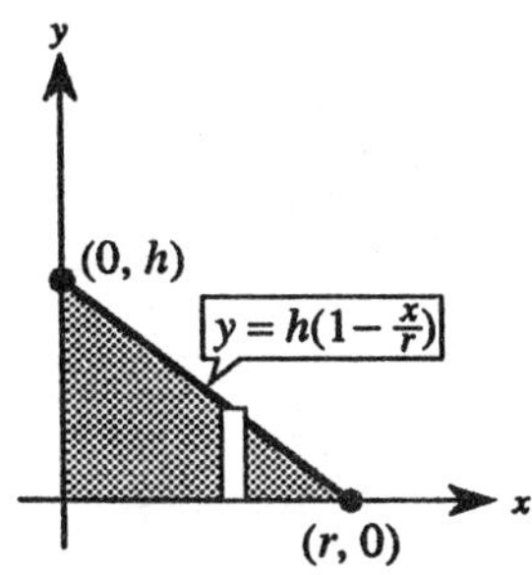

(b) $2\pi \int_{-r}^{r} (R - x)\left(2\sqrt{r^2 - x^2}\right) dx$ is the volume of a torus with the radius of its circular cross section as r and the distance from the axis of the torus to the center of its cross section as R.

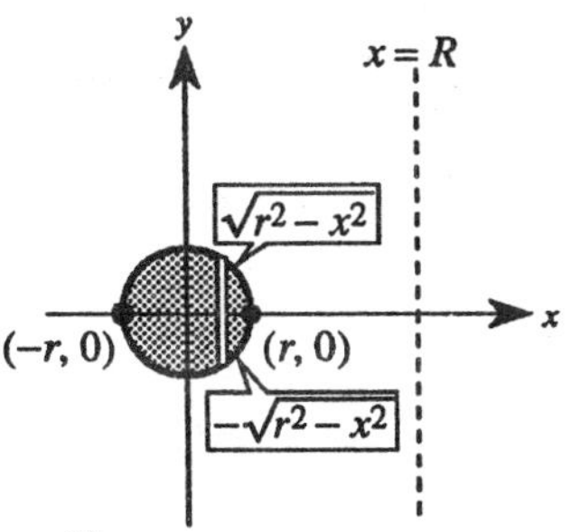

(c) $2\pi \int_0^r 2x\sqrt{r^2 - x^2}\, dx$ is the volume of a sphere with radius r.

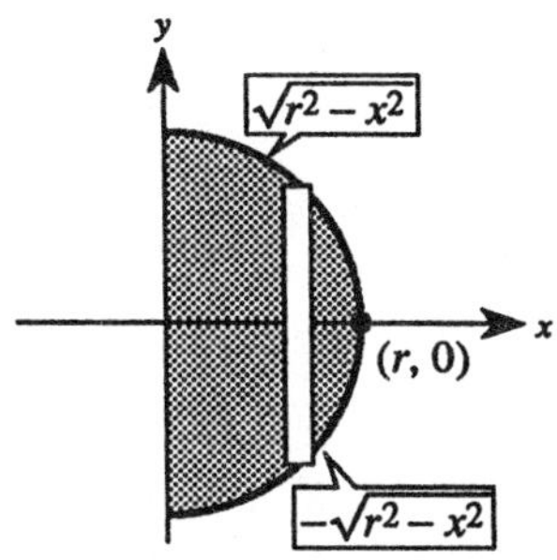

(d) $2\pi \int_0^r hx\, dx$ is the volume of a right circular cylinder with a radius of r and a height of h.

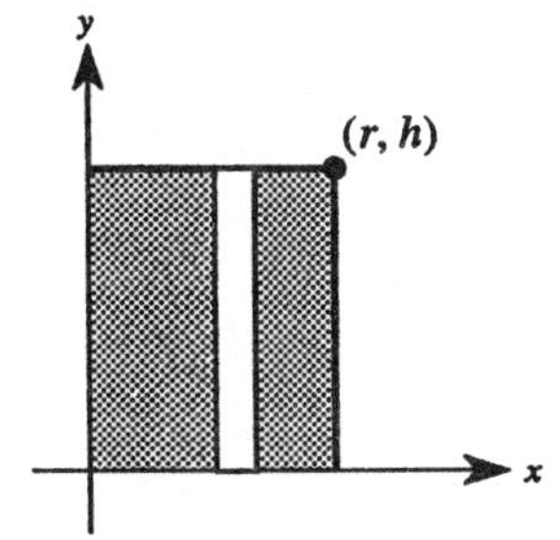

(e) $2\pi \int_0^b 2ax\sqrt{1 - (x^2/b^2)}\, dx$ is the volume of an ellipsoid with axes $2a$ and $2b$.

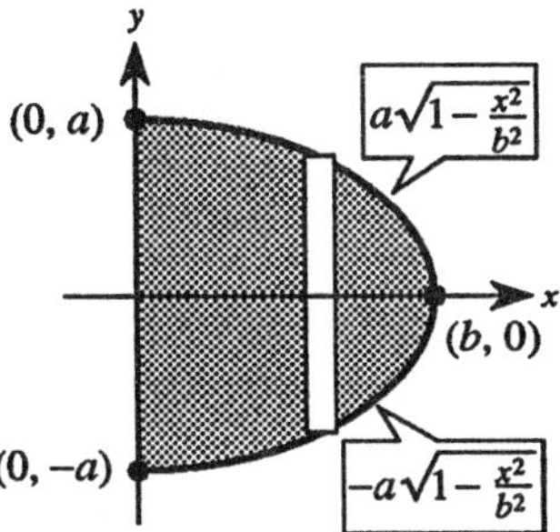

33.
$$\begin{aligned} V &= 4\pi \int_{-1}^{1} (2 - x)\sqrt{1 - x^2}\, dx \\ &= 8\pi \int_{-1}^{1} \sqrt{1 - x^2}\, dx - 4\pi \int_{-1}^{1} x\sqrt{1 - x^2}\, dx \\ &= 8\pi\left(\frac{\pi}{2}\right) + 2\pi \int_{-1}^{1} x(1 - x^2)^{1/2}(-2)\, dx = 4\pi^2 + \left[2\pi\left(\frac{2}{3}\right)(1 - x^2)^{3/2}\right]_{-1}^{1} = 4\pi^2 \end{aligned}$$

34. $x = \sqrt{r^2 - \left(\frac{h}{2}\right)^2} = \frac{\sqrt{4r^2 - h^2}}{2}$

$$V = 4\pi \int_{\sqrt{4r^2-h^2}/2}^{r} x\sqrt{r^2 - x^2}\,dx = -2\pi\left(\frac{2}{3}\right)(r^2 - x^2)^{3/2}\Big]_{\sqrt{4r^2-h^2}/2}^{r}$$

$$= 0 + \frac{4\pi}{3}\left[r^2 - \frac{4r^2 - h^2}{4}\right]^{3/2} = \frac{4\pi}{3}\left(\frac{h^2}{4}\right)^{3/2} = \frac{4\pi}{3}\cdot\frac{h^3}{8} = \frac{\pi h^3}{6}$$

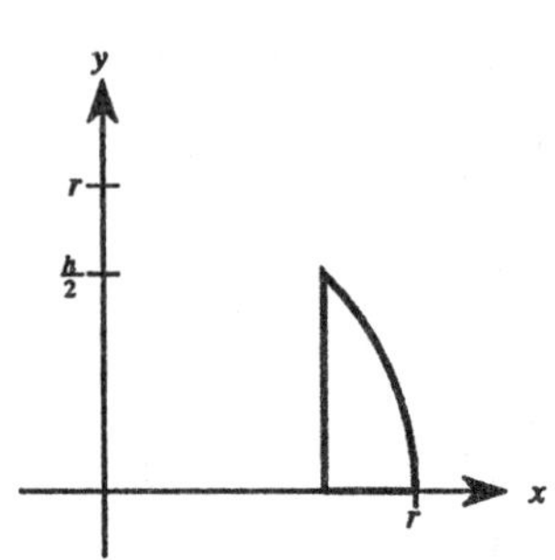

35. $V = 2\pi \int_2^6 x\sqrt[3]{(x-2)^2(x-6)^2}\,dx \approx 186.055$

36. $p(x) = x, \quad h(x) = \frac{2}{1+\sqrt{x}}$

$$V = 2\pi \int_1^4 \frac{2x}{1+\sqrt{x}}\,dx \approx 35.8862$$

Section 6.4 Arc Length and Surfaces of Revolution

1. (0, 0), (5, 12)

(a) $d = \sqrt{(5-0)^2 + (12-0)^2} = 13$

(b) $y = \frac{12}{5}x$

$y' = \frac{12}{5}$

$$s = \int_0^5 \sqrt{1 + \left(\frac{12}{5}\right)^2}\,dx = \frac{13}{5}x\Big]_0^5 = 13$$

2. (1, 2), (7, 10)

(a) $d = \sqrt{(7-1)^2 + (10-2)^2} = 10$

(b) $y = \frac{4}{3}x + \frac{2}{3}$

$y' = \frac{4}{3}$

$$s = \int_1^7 \sqrt{1 + \left(\frac{4}{3}\right)^2}\,dx = \frac{5}{3}x\Big]_1^7 = 10$$

3. $y = \frac{2}{3}x^{3/2} + 1$

$y' = x^{1/2}, \quad [0, 1]$

$$s = \int_0^1 \sqrt{1+x}\,dx = \frac{2}{3}(1+x)^{3/2}\Big]_0^1 = \frac{2}{3}(\sqrt{8} - 1) \approx 1.219$$

4. $y = x^{3/2} - 1$

$y' = \frac{3}{2}x^{1/2}, \quad [0, 4]$

$$s = \int_0^4 \sqrt{1 + \frac{9}{4}x}\,dx = \frac{4}{9}\int_0^4 \left(1 + \frac{9}{4}x\right)^{1/2}\left(\frac{9}{4}\right)dx = \frac{8}{27}\left(1 + \frac{9}{4}x\right)^{3/2}\Big]_0^4 = \frac{8}{27}(10^{3/2} - 1)$$

5. $y = \frac{x^4}{8} + \frac{1}{4x^2}$

$y' = \frac{1}{2}x^3 - \frac{1}{2x^3}$

$1 + (y')^2 = \left(\frac{1}{2}x^3 + \frac{1}{2x^3}\right)^2, \quad [1, 2]$

$$s = \int_a^b \sqrt{1 + (y')^2}\,dx = \int_1^2 \left(\frac{1}{2}x^3 + \frac{1}{2x^3}\right)dx = \left[\frac{1}{8}x^4 - \frac{1}{4x^2}\right]_1^2 = \frac{33}{16}$$

6. $y = \frac{3}{2}x^{2/3}$

$y' = x^{-1/3}, \quad [1, 8]$

$$s = \int_1^8 \sqrt{1 + \frac{1}{x^{2/3}}}\,dx = \int_1^8 \frac{\sqrt{x^{2/3}+1}}{x^{1/3}}\,dx$$

$$= \frac{3}{2}\int_1^8 (x^{2/3}+1)^{1/2}\left(\frac{2}{3x^{1/3}}\right)dx = \left[(x^{2/3}+1)^{3/2}\right]_1^8 = 5^{3/2} - 2^{3/2} = 5\sqrt{5} - 2\sqrt{2}$$

7. $y = \frac{x^5}{10} + \frac{1}{6x^3}$

$y' = \frac{1}{2}x^4 - \frac{1}{2x^4}$

$1 + (y')^2 = \left(\frac{1}{2}x^4 + \frac{1}{2x^4}\right)^2, \quad [1, 2]$

$$s = \int_a^b \sqrt{1 + (y')^2}\,dx = \int_1^2 \sqrt{\left(\frac{1}{2}x^4 + \frac{1}{2x^4}\right)^2}\,dx = \int_1^2 \left(\frac{1}{2}x^4 + \frac{1}{2x^4}\right)dx = \left[\frac{1}{10}x^5 - \frac{1}{6x^3}\right]_1^2 = \frac{779}{240}$$

8. $y = 2 - \frac{2}{3}x$

$y' = -\frac{2}{3}$

$1 + (y')^2 = \frac{13}{9}, \quad [0, 3]$

$$s = \int_a^b \sqrt{1 + (y')^2}\,dx = \int_0^3 \sqrt{\frac{13}{9}}\,dx = \left[\frac{\sqrt{13}}{3}x\right]_0^3 = \sqrt{13}$$

9. $y = x^2 + x - 2$

$y' = 2x + 1, \quad [-2, 1]$

$$s = \int_{-2}^1 \sqrt{1 + (2x+1)^2}\,dx$$

$$= \int_{-2}^1 \sqrt{2 + 4x + 4x^2}\,dx$$

10. $y = \frac{1}{(x+1)}$

$y' = -\frac{1}{(x+1)^2}, \quad [0, 1]$

$$s = \int_0^1 \sqrt{1 + \frac{1}{(x+1)^4}}\,dx$$

$$= \int_0^1 \frac{\sqrt{(x+1)^4 + 1}}{(x+1)^2}\,dx$$

11. $y = 4 - x^2$

$y' = -2x, \quad [0, 2]$

$$s = \int_0^2 \sqrt{1 + (-2x)^2}\,dx$$

$$= \int_0^2 \sqrt{1 + 4x^2}\,dx$$

12. $y = \sqrt[3]{x}$

$y' = \frac{1}{3x^{2/3}}$

$1 + (y')^2 = 1 + \frac{1}{9x^{4/3}}$

$$s = \int_{-8}^8 \sqrt{1 + \frac{1}{9x^{4/3}}}\,dx$$

$$= \int_{-8}^8 \frac{\sqrt{9x^{4/3}+1}}{3x^{2/3}}\,dx$$

13.
$$x = \frac{1}{y^2}$$
$$x' = -\frac{2}{y^3}$$
$$1 + (x')^2 = 1 + \frac{4}{y^6}$$
$$s = \int_1^2 \sqrt{1 + \frac{4}{y^6}}\,dy = \int_1^2 \frac{\sqrt{y^6 + 4}}{y^3}\,dy$$

14. $x = \sqrt{a^2 - y^2}$
$$x' = \frac{-y}{\sqrt{a^2 - y^2}}, \quad \left[0, \frac{a}{2}\right]$$
$$s = \int_0^{a/2} \sqrt{1 + \frac{y^2}{a^2 - y^2}}\,dy$$
$$= \int_0^{a/2} \frac{a}{\sqrt{a^2 - y^2}}\,dy$$

15. $y = \frac{1}{x}$
$$y' = -\frac{1}{x^2}, \quad [1, 3]$$
$$s = \int_1^3 \sqrt{1 + \frac{1}{x^4}}\,dx$$
$$= \int_1^3 \frac{\sqrt{x^4 + 1}}{x^2}\,dx \approx \frac{3-1}{3(4)}\left[\sqrt{2} + \frac{4\sqrt{(1.5)^4 + 1}}{(1.5)^2} + \frac{2\sqrt{17}}{4} + \frac{4\sqrt{(2.5)^4 + 1}}{(2.5)^2} + \frac{\sqrt{82}}{9}\right] \approx 2.1517$$

16. $y = x^2$
$$y' = 2x, \quad [0, 1]$$
$$s = \int_0^1 \sqrt{1 + 4x^2}\,dx \approx \frac{1}{3(4)}\left[1 + 4\sqrt{1 + \frac{1}{4}} + 2\sqrt{1 + 1} + 4\sqrt{1 + \frac{9}{4}} + \sqrt{5}\right] \approx 1.4790$$

17.
$$y = x^3$$
$$y' = 3x^2$$
$$1 + (y')^2 = 1 + 9x^4$$
$$s = \int_0^2 \sqrt{1 + 9x^4}\,dx \approx 8.6101$$

18.
$$y = \sqrt{x}$$
$$y' = \frac{1}{2\sqrt{x}}$$
$$1 + (y')^2 = 1 + \frac{1}{4x} = \frac{4x + 1}{4x}$$
$$s = \int_1^3 \sqrt{\frac{4x + 1}{4x}}\,dx \approx 2.1327$$

19. $y = \frac{1}{3}[x^{3/2} - 3x^{1/2} + 2]$

When $x = 0$, $y = \frac{2}{3}$. Thus, the fleeing object has traveled $\frac{2}{3}$ unit when it is caught.

20.
$$y' = \frac{1}{3}\left[\frac{3}{2}x^{1/2} - \frac{3}{2}x^{-1/2}\right] = \left(\frac{1}{2}\right)\frac{x - 1}{x^{1/2}}$$
$$1 + (y')^2 = 1 + \frac{(x-1)^2}{4x} = \frac{(x+1)^2}{4x}$$
$$s = \int_0^1 \frac{x + 1}{2x^{1/2}}\,dx = \frac{1}{2}\int_0^1 (x^{1/2} + x^{-1/2})\,dx = \frac{1}{2}\left[\frac{2}{3}x^{3/2} + 2x^{1/2}\right]_0^1 = \frac{4}{3} = 2\left(\frac{2}{3}\right)$$

The pursuer has traveled twice the distance that the fleeing object has traveled when it is caught.

21.
$$y = \sqrt{9 - x^2}$$
$$y' = \frac{-x}{\sqrt{9 - x^2}}$$
$$1 + (y')^2 = \frac{9}{9 - x^2}$$
$$s = \int_0^2 \sqrt{\frac{9}{9 - x^2}}\,dx = \int_0^2 \frac{3}{\sqrt{9 - x^2}}\,dx = 3\arcsin\frac{x}{3}\Big]_0^2 = 3\left(\arcsin\frac{2}{3} - \arcsin 0\right) = 3\arcsin\frac{2}{3} \approx 2.1892$$

22.
$$y = \sqrt{25 - x^2}$$
$$y' = \frac{-x}{\sqrt{25 - x^2}}$$
$$1 + (y')^2 = \frac{25}{25 - x^2}$$
$$s = \int_{-3}^4 \sqrt{\frac{25}{25 - x^2}}\,dx = \int_{-3}^4 \frac{5}{\sqrt{25 - x^2}}\,dx = 5\arcsin\frac{x}{5}\Big]_{-3}^4 = 5\left[\arcsin\frac{4}{5} - \arcsin\left(-\frac{3}{5}\right)\right] \approx 7.854$$
$$\frac{1}{4}[2\pi(5)] \approx 7.854 = s$$

23. $y = \dfrac{x^3}{3}$

$y' = x^2$, $[0, 3]$
$$S = 2\pi\int_0^3 \frac{x^3}{3}\sqrt{1 + x^4}\,dx = \frac{\pi}{6}\int_0^3 (1 + x^4)^{1/2}(4x^3)\,dx = \frac{\pi}{9}(1 + x^4)^{3/2}\Big]_0^3 = \frac{\pi}{9}(82\sqrt{82} - 1) \approx 258.85$$

24. $y = \sqrt{x}$

$y' = \dfrac{1}{2\sqrt{x}}$, $[1, 4]$
$$S = 2\pi\int_1^4 \sqrt{x}\sqrt{1 + \frac{1}{4x}}\,dx = 2\pi\int_1^4 \frac{\sqrt{x}}{2\sqrt{x}}\sqrt{4x + 1}\,dx$$
$$= \pi\int_1^4 \sqrt{4x + 1}\,dx = \frac{\pi}{4}\int_1^4 (4x + 1)^{1/2}(4)\,dx = \frac{\pi}{6}(4x + 1)^{3/2}\Big]_1^4 = \frac{\pi}{6}(17\sqrt{17} - 5\sqrt{5})$$

25.
$$y = \frac{x^3}{6} + \frac{1}{2x}$$
$$y' = \frac{x^2}{2} - \frac{1}{2x^2}$$
$$1 + (y')^2 = \left(\frac{x^2}{2} + \frac{1}{2x^2}\right)^2, \quad [1, 2]$$
$$S = 2\pi\int_1^2 \left(\frac{x^3}{6} + \frac{1}{2x}\right)\left(\frac{x^2}{2} + \frac{1}{2x^2}\right)dx$$
$$= 2\pi\int_1^2 \left(\frac{x^5}{12} + \frac{x}{3} + \frac{1}{4x^3}\right)dx = 2\pi\left[\frac{x^6}{72} + \frac{x^2}{6} - \frac{1}{8x^2}\right]_1^2 = \frac{47\pi}{16}$$

26.
$$y = \frac{x}{2}$$
$$y' = \frac{1}{2}$$
$$1 + (y')^2 = \frac{5}{4}, \quad [0, 6]$$
$$S = 2\pi \int_0^6 \frac{x}{2}\sqrt{\frac{5}{4}}\,dx = \frac{2\pi\sqrt{5}}{8}x^2\Big]_0^6 = 9\sqrt{5}\,\pi$$

27. $y = \sqrt[3]{x} + 2$
$$y' = \frac{1}{3x^{2/3}}, \quad [1, 8]$$
$$S = 2\pi \int_1^8 x\sqrt{1 + \frac{1}{9x^{4/3}}}\,dx = \frac{2\pi}{3}\int_1^8 x^{1/3}\sqrt{9x^{4/3}+1}\,dx$$
$$= \frac{\pi}{18}\int_1^8 (9x^{4/3}+1)^{1/2}(12x^{1/3})\,dx = \frac{\pi}{27}(9x^{4/3}+1)^{3/2}\Big]_1^8 = \frac{\pi}{27}(145\sqrt{145} - 10\sqrt{10}) \approx 199.48$$

28. $y = 4 - x^2$
$$y' = -2x, \quad [0, 2]$$
$$S = 2\pi \int_0^2 x\sqrt{1+4x^2}\,dx = \frac{\pi}{4}\int_0^2 (1+4x^2)^{1/2}(8x)\,dx = \frac{\pi}{6}(1+4x^2)^{3/2}\Big]_0^2 = \frac{\pi}{6}(17\sqrt{17} - 1)$$

29.
$$y = \frac{hx}{r}$$
$$y' = \frac{h}{r}$$
$$1 + (y')^2 = \frac{r^2 + h^2}{r^2}$$
$$S = 2\pi \int_0^r x\sqrt{\frac{r^2+h^2}{r^2}}\,dx = \left[\frac{2\pi\sqrt{r^2+h^2}}{r}\left(\frac{x^2}{2}\right)\right]_0^r = \pi r\sqrt{r^2+h^2}$$

30.
$$y = \sqrt{r^2 - x^2}$$
$$y' = \frac{-x}{\sqrt{r^2 - x^2}}$$
$$1 + (y')^2 = \frac{r^2}{r^2 - x^2}$$
$$S = 2\pi \int_{-r}^r \sqrt{r^2 - x^2}\sqrt{\frac{r^2}{r^2 - x^2}}\,dx = 2\pi \int_{-r}^r r\,dx = 2\pi r x\Big]_{-r}^r = 4\pi r^2$$

31.

$$y = \sqrt{9 - x^2}$$

$$y' = \frac{-x}{\sqrt{9 - x^2}}$$

$$\sqrt{1 + (y')^2} = \frac{3}{\sqrt{9 - x^2}}$$

$$S = 2\pi \int_0^2 \frac{3x}{\sqrt{9 - x^2}}\,dx = -3\pi \int_0^2 \frac{-2x}{\sqrt{9 - x^2}}\,dx = -6\pi\sqrt{9 - x^2}\Big]_0^2 = 6\pi(3 - \sqrt{5}) \approx 14.40$$

See figure in Exercise 32.

32. From Exercise 31 we have:

$$S = 2\pi \int_0^a \frac{rx}{\sqrt{r^2 - x^2}}\,dx$$

$$= -r\pi \int_0^a \frac{-2x\,dx}{\sqrt{r^2 - x^2}}$$

$$= -2r\pi\sqrt{r^2 - x^2}\Big]_0^a$$

$$= 2r^2\pi - 2r\pi\sqrt{r^2 - a^2}$$

$$= 2r\pi\left(r - \sqrt{r^2 - a^2}\right)$$

$= 2\pi rh$ (where h is height of zone)

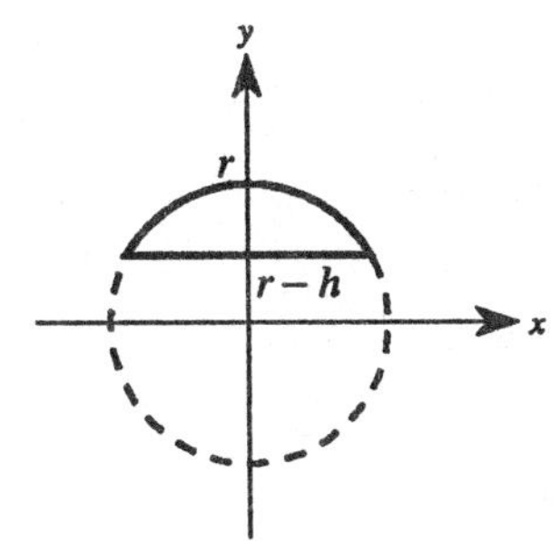

33.

$$y = \frac{1}{3}x^{1/2} - x^{3/2}$$

$$y' = \frac{1}{6}x^{-1/2} - \frac{3}{2}x^{1/2} = \frac{1}{6}(x^{-1/2} - 9x^{1/2})$$

$$1 + (y')^2 = 1 + \frac{1}{36}(x^{-1} - 18 + 81x) = \frac{1}{36}(x^{-1/2} + 9x^{1/2})^2$$

$$S = 2\pi \int_0^{1/3} \left(\frac{1}{3}x^{1/2} - x^{3/2}\right)\sqrt{\frac{1}{36}(x^{-1/2} + 9x^{1/2})^2}\,dx$$

$$= \frac{2\pi}{6} \int_0^{1/3} \left(\frac{1}{3}x^{1/2} - x^{3/2}\right)(x^{-1/2} + 9x^{1/2})\,dx$$

$$= \frac{\pi}{3} \int_0^{1/3} \left(\frac{1}{3} + 2x - 9x^2\right)dx = \frac{\pi}{3}\left[\frac{1}{3}x + x^2 - 3x^3\right]_0^{1/3} = \frac{\pi}{27}$$

Amount of glass needed: $V = \frac{\pi}{27}\left(\frac{0.015}{12}\right) \approx 0.00015 \text{ ft}^3 \approx 0.25 \text{ in}^3$

34. Area of circle with radius L: $A_0 = \pi L^2$

Area of sector with central angle θ (in radians): $A = \frac{\theta}{2\pi}A_0 = \frac{\theta}{2\pi}(\pi L^2) = \frac{1}{2}L^2\theta$

35. $S = \frac{1}{2}L^2\theta$ (from Exercise 34)

$= \frac{1}{2}L^2\left(\frac{S}{L}\right)$ where S is the arc length of the sector

$= \frac{1}{2}LS = \frac{1}{2}L(2\pi r) = \pi rL$

36. The lateral surface area of the frustrum is the difference of the large cone and the small cone.

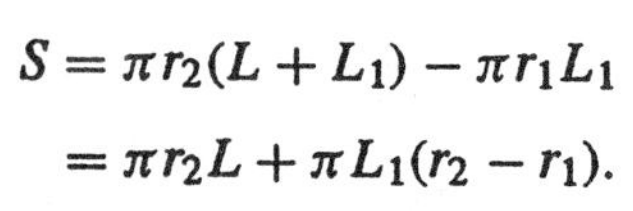

$$S = \pi r_2(L + L_1) - \pi r_1 L_1$$
$$= \pi r_2 L + \pi L_1(r_2 - r_1).$$

By similar triangles,

$$\frac{L + L_1}{r_2} = \frac{L_1}{r_1} \Rightarrow L r_1 = L_1(r_2 - r_1).$$

Hence, $S = \pi r_2 L + \pi(L r_1) = \pi L(r_1 + r_2)$.

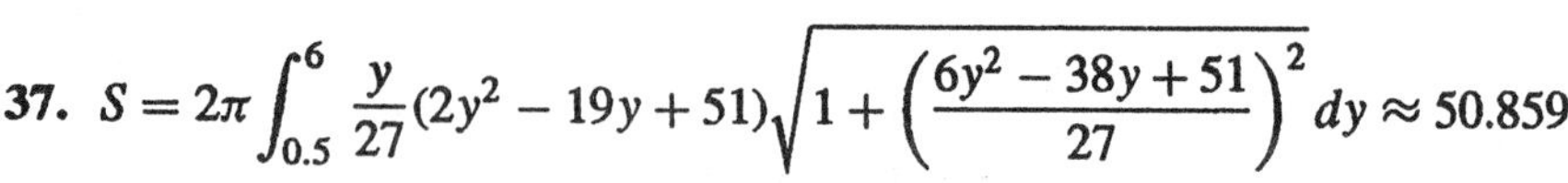

37. $S = 2\pi \int_{0.5}^{6} \frac{y}{27}(2y^2 - 19y + 51)\sqrt{1 + \left(\frac{6y^2 - 38y + 51}{27}\right)^2}\, dy \approx 50.859$

38. $S = 2\pi \int_{0}^{6} \frac{18}{y^2 + 9}\sqrt{1 + \left[\frac{-36y}{(y^2 + 9)^2}\right]^2}\, dy \approx 43.713$

Section 6.5 Work

1. $W = Fd = (100)(10) = 1000$ ft · lb

2. $W = Fd = (2400)(6) = 14{,}400$ ft · lb

3. $W = Fd = (25)(12) = 300$ ft · lb

4. $W = Fd = [9(2000)]\left[\frac{1}{2}(5280)\right] = 47{,}520{,}000$ ft · lb

5. $F(x) = kx$

$5 = k(4)$

$k = \frac{5}{4}$

$$W = \int_0^7 \frac{5}{4}x\, dx$$
$$= \frac{5}{8}x^2\Big]_0^7 = \frac{245}{8} \text{ in} \cdot \text{lb} \approx 2.55 \text{ ft} \cdot \text{lb}$$

6. $W = \int_5^9 \frac{5}{4}x\, dx = \frac{5}{8}x^2\Big]_5^9 = 35$ in · lb ≈ 2.92 ft · lb

7. $F(x) = kx$

$60 = 12k$

$k = 5$

$$W = \int_9^{15} 5x\, dx$$
$$= \frac{5}{2}x^2\Big]_9^{15} = 360 \text{ in} \cdot \text{lb} = 30 \text{ ft} \cdot \text{lb}$$

8. $F(x) = kx$

$200 = 2k$

$k = 100$

$$W = \int_0^2 100x\, dx = 50x^2\Big]_0^2 = 200 \text{ ft} \cdot \text{lb}$$

9. $F(x) = kx$

$15 = 6k$

$k = \frac{5}{2}$

$W = \int_0^{12} \frac{5}{2}x\,dx$

$= \frac{5}{4}x^2\Big]_0^{12} = 180 \text{ in} \cdot \text{lb} = 15 \text{ ft} \cdot \text{lb}$

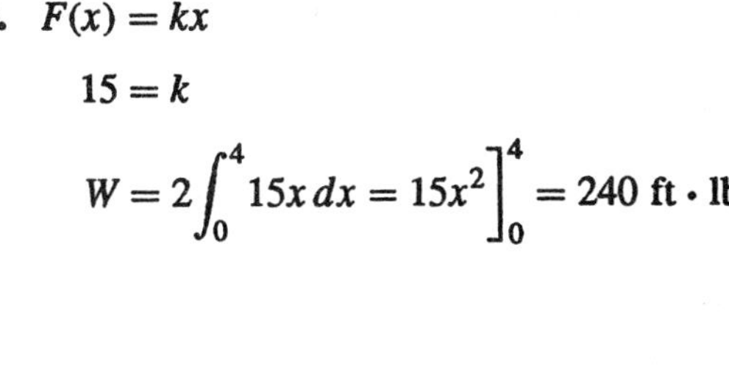

10. $F(x) = kx$

$15 = k$

$W = 2\int_0^4 15x\,dx = 15x^2\Big]_0^4 = 240 \text{ ft} \cdot \text{lb}$

11. Weight of each layer: $62.4(20)\,\Delta y$
Distance: $4 - y$

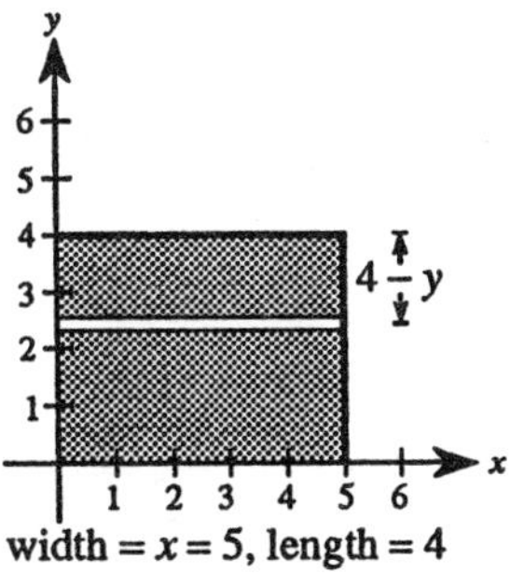

width = x = 5, length = 4

(a) $W = \int_2^4 62.4(20)(4 - y)\,dy$

$= \Big[4992y - 624y^2\Big]_2^4 = 2496 \text{ ft} \cdot \text{lb}$

(b) $W = \int_0^4 62.4(20)(4 - y)\,dy$

$= \Big[4992y - 624y^2\Big]_0^4 = 9984 \text{ ft} \cdot \text{lb}$

12. (a) $W = \int_2^4 42(20)(4 - y)\,dy$

$= \Big[3360y - 420y^2\Big]_2^4$

$= 1680 \text{ ft} \cdot \text{lb}$

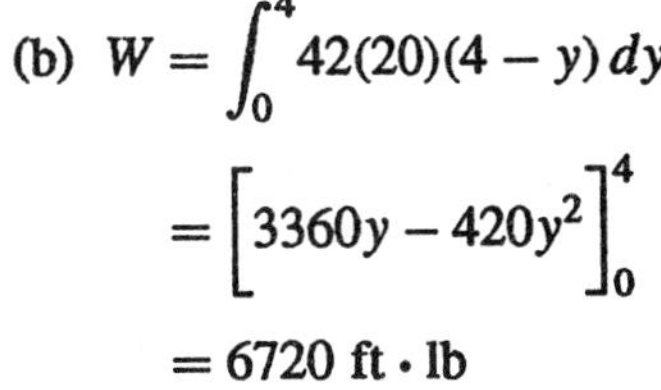

(b) $W = \int_0^4 42(20)(4 - y)\,dy$

$= \Big[3360y - 420y^2\Big]_0^4$

$= 6720 \text{ ft} \cdot \text{lb}$

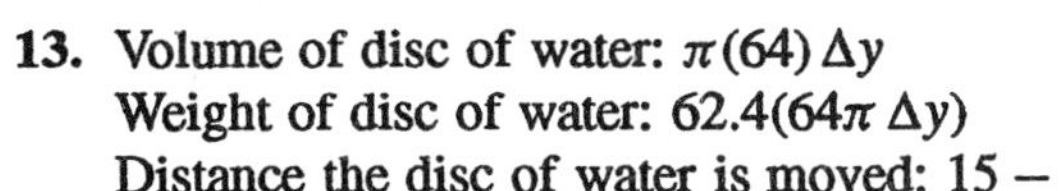

13. Volume of disc of water: $\pi(64)\,\Delta y$
Weight of disc of water: $62.4(64\pi\,\Delta y)$
Distance the disc of water is moved: $15 - y$

$$W = \int_0^{12} (15 - y)(62.4)(64\pi)\,dy = 3993.6\pi \int_0^{12} (15 - y)\,dy = 3993.6\pi\Big[15y - \frac{y^2}{2}\Big]_0^{12} = 431{,}308.8\pi \text{ ft} \cdot \text{lb}$$

14. $$W = \int_{20}^{26} y(62.4)\pi(64)\,dy = 3993.6\pi \int_{20}^{26} y\,dy = 1996.8\pi y^2\Big]_{20}^{26} = 551{,}116.8\pi \text{ ft} \cdot \text{lb}$$

15. Volume of disc: $\pi\left(\sqrt{36 - y^2}\right)^2 \Delta y$

Weight of disc: $62.4\pi(36 - y^2)\,\Delta y$

Distance: y

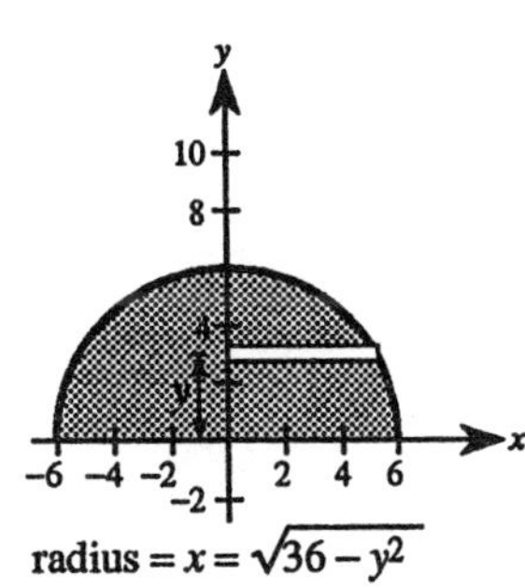

radius = $x = \sqrt{36 - y^2}$

$W = 62.4\pi \int_0^6 y(36 - y^2)\,dy$

$= 62.4\pi \int_0^6 (36y - y^3)\,dy$

$= 62.4\pi\Big[18y^2 - \frac{1}{4}y^4\Big]_0^6$

$= 20{,}217.6\pi \text{ ft} \cdot \text{lb}$

16. Volume of disc: $\pi(36 - y^2)\,\Delta y$

Weight of disc: $62.4\pi(36 - y^2)\,\Delta y$

Distance: $-y$

$$W = \int_{-2}^{0} 62.4\pi(36 - y^2)(-y)\,dy$$

$$= -62.4\pi\left[18y^2 - \frac{1}{4}y^4\right]_{-2}^{0} = 4243.2\pi \text{ ft} \cdot \text{lb}$$

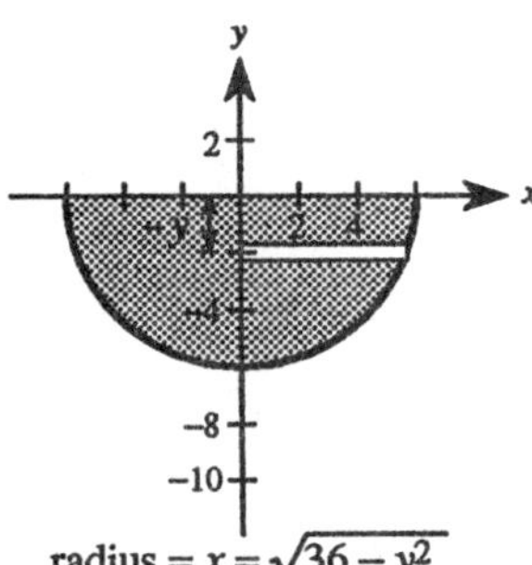

17. Volume of disc: $\pi\left(\frac{2}{3}y\right)^2 \Delta y$

Weight of disc: $62.4\pi\left(\frac{2}{3}y\right)^2 \Delta y$

Distance: $6 - y$

$$W = \frac{4(62.4)\pi}{9}\int_{0}^{6} (6 - y)y^2\,dy$$

$$= \frac{4}{9}(62.4)\pi\left[2y^3 - \frac{1}{4}y^4\right]_{0}^{6} = 2995.2\pi \text{ ft} \cdot \text{lb}$$

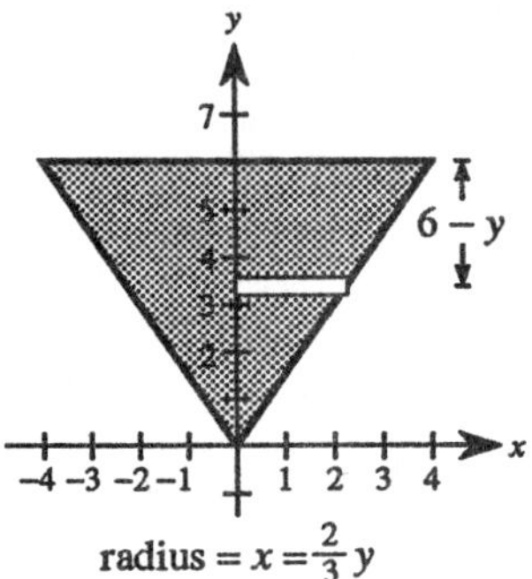

18. Volume of disc: $\pi\left(\frac{2}{3}y\right)^2 \Delta y$

Weight of disc: $62.4\pi\left(\frac{2}{3}y\right)^2 \Delta y$

Distance: y

(a) $W = \frac{4}{9}(62.4)\pi\int_{0}^{2} y^3\,dy = \frac{4}{9}(62.4)\pi\left(\frac{1}{4}y^4\right)\Big]_{0}^{2} = 110.9\pi \text{ ft} \cdot \text{lb}$

(b) $W = \frac{4}{9}(62.4)\pi\int_{4}^{6} y^3\,dy = \frac{4}{9}(62.4)\pi\left(\frac{1}{4}y^4\right)\Big]_{4}^{6} = 7210.7\pi \text{ ft} \cdot \text{lb}$

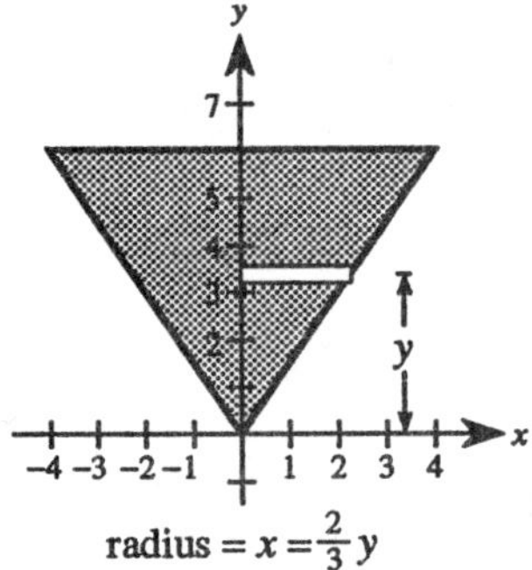

19. Volume of layer: $V = lwh = 4(2)\sqrt{\left(\frac{9}{4}\right) - y^2}\,\Delta y$

Weight of layer: $W = 42(8)\sqrt{\left(\frac{9}{4}\right) - y^2}\,\Delta y$

Distance: $\frac{13}{2} - y$

Work $W = \int_{-1.5}^{1.5} 42(8)\sqrt{\left(\frac{9}{4}\right) - y^2}\left(\frac{13}{2} - y\right)dy$

$$= 336\left[\frac{13}{2}\int_{-1.5}^{1.5}\sqrt{\left(\frac{9}{4}\right) - y^2}\,dy - \int_{-1.5}^{1.5}\sqrt{\left(\frac{9}{4}\right) - y^2}\,y\,dy\right]$$

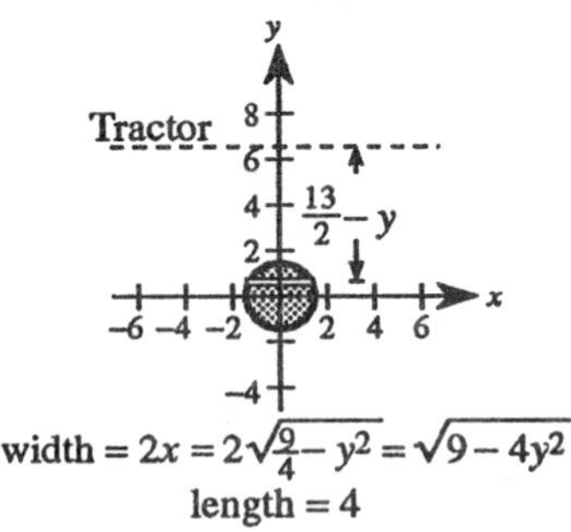

The second integral is zero since the integrand is odd and the limits of integration are symmetric to the origin. The first integral represents the area of a semicircle of radius $\frac{3}{2}$. Thus, the work is

$$W = 336\left(\frac{13}{2}\right)\pi\left(\frac{3}{2}\right)^2\left(\frac{1}{2}\right) = 2457\pi \text{ ft} \cdot \text{lb}.$$

20. Volume of layer: $V = 12(2)\sqrt{\left(\frac{25}{4}\right) - y^2}\,\Delta y$

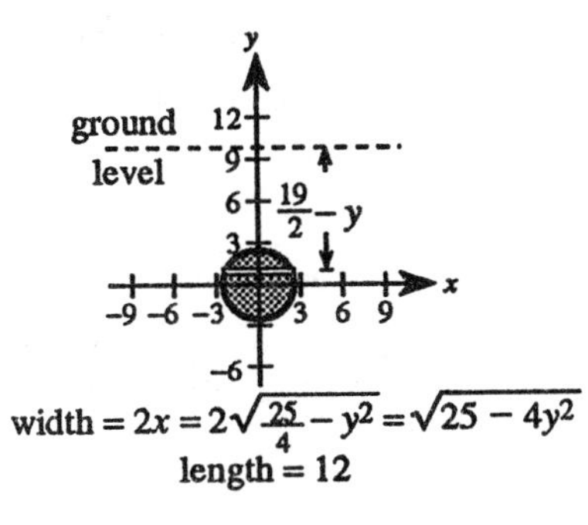

Weight of layer: $W = 42(24)\sqrt{\left(\frac{25}{4}\right) - y^2}\,\Delta y$

Distance: $\frac{19}{2} - y$

Work $W = \int_{-2.5}^{2.5} 42(24)\sqrt{\left(\frac{25}{4}\right) - y^2}\left(\frac{19}{2} - y\right) dy$

$$= 1008\left[\frac{19}{2}\int_{-2.5}^{2.5}\sqrt{\left(\frac{25}{4}\right) - y^2}\,dy + \int_{-2.5}^{2.5}\sqrt{\left(\frac{25}{4}\right) - y^2}(-y)\,dy\right]$$

The second integral is zero since the integrand is odd and the limits of integration are symmetric to the origin. The first integral represents the area of a semicircle of radius $\frac{5}{2}$. Thus, the work is

$$W = 1008\left(\frac{19}{2}\right)\pi\left(\frac{5}{2}\right)^2\left(\frac{1}{2}\right) = 29{,}925\pi \text{ ft} \cdot \text{lb} \approx 94{,}012.16 \text{ ft} \cdot \text{lb}.$$

21. Assume that the earth has a radius of 4000 miles.

$$F(x) = \frac{k}{x^2}$$

$$4 = \frac{k}{(4000)^2}$$

$$k = 64{,}000{,}000$$

$$F(x) = \frac{64{,}000{,}000}{x^2}$$

(a) $W = \int_{4000}^{4200} \frac{64{,}000{,}000}{x^2}\,dx = -\frac{64{,}000{,}000}{x}\Big]_{4000}^{4200} \approx -15{,}238.095 + 16{,}000 = 761.905 \text{ mi} \cdot \text{tons}$

$\approx 8.046 \times 10^9 \text{ ft} \cdot \text{lb}$

(b) $W = \int_{4000}^{4400} \frac{64{,}000{,}000}{x^2}\,dx = -\frac{64{,}000{,}000}{x}\Big]_{4000}^{4400} \approx -14{,}545.455 + 16{,}000 = 1454.545 \text{ mi} \cdot \text{tons}$

$\approx 1.536 \times 10^{10} \text{ ft} \cdot \text{lb}$

22. $W = \int_{4000}^{h} \frac{64{,}000{,}000}{x^2}\,dx = -\frac{64{,}000{,}000}{x}\Big]_{4000}^{h} = -\frac{64{,}000{,}000}{h} + 16{,}000$

$\lim_{h\to\infty} = 16{,}000 \text{ mi} \cdot \text{tons} \approx 1.6896 \times 10^{11} \text{ ft} \cdot \text{lb}$

23. Assume that the earth has a radius of 4000 miles.

$$F(x) = \frac{k}{x^2}$$

$$10 = \frac{k}{(4000)^2}$$

$$k = 160{,}000{,}000$$

$$F(x) = \frac{160{,}000{,}000}{x^2}$$

(a) $W = \int_{4000}^{15{,}000} \frac{160{,}000{,}000}{x^2}\,dx = -\frac{160{,}000{,}000}{x}\Big]_{4000}^{15{,}000} \approx -10{,}666.667 + 40{,}000 = 29{,}333.333 \text{ mi} \cdot \text{tons}$

$\approx 3.098 \times 10^{11} \text{ ft} \cdot \text{lb}$

(b) $W = \int_{4000}^{26{,}000} \frac{160{,}000{,}000}{x^2}\,dx = -\frac{160{,}000{,}000}{x}\Big]_{4000}^{26{,}000} \approx -6153.846 + 40{,}000 = 33{,}846.154 \text{ mi} \cdot \text{tons}$

$\approx 3.574 \times 10^{11} \text{ ft} \cdot \text{lb}$

24. Weight on surface of moon: $\frac{1}{6}(12) = 2$ tons

Weight varies inversely as the square of distance from the center of the moon. Therefore,

$$F(x) = \frac{k}{x^2}$$

$$2 = \frac{k}{(1100)^2}$$

$$k = 2.42 \times 10^6$$

$$W = \int_{1100}^{1150} \frac{2.42 \times 10^6}{x^2}\,dx = \frac{-2.42 \times 10^6}{x}\Big]_{1100}^{1150} = 2.42 \times 10^6\left(\frac{1}{1100} - \frac{1}{1150}\right)$$

$$= 95.652 \text{ mi} \cdot \text{tons} = 1.010 \times 10^9 \text{ ft} \cdot \text{lb}.$$

25. $F(x) = \frac{k}{(2-x)^2}$

$$W = \int_{-2}^{1} \frac{k}{(2-x)^2}\,dx = \frac{k}{2-x}\Big]_{-2}^{1}$$

$$= k\left(1 - \frac{1}{4}\right) = \frac{3k}{4} \text{ (units of work)}$$

26. $W = \int_0^4 10{,}000\sqrt{1+x^5}\,dx$

$\approx 376{,}417.3423 \text{ ft} \cdot \text{lb}$

27. Weight of section of chain: $3\,\Delta y$

Distance: $15 - y$

$$W = 3\int_0^{15} (15 - y)\,dy$$

$$= -\frac{3}{2}(15 - y)^2\Big]_0^{15} = 337.5 \text{ ft} \cdot \text{lb}$$

28. The lower ten feet of chain are raised five feet with a constant force.

$$W_1 = 3(10)5 = 150 \text{ ft} \cdot \text{lb}$$

The top five feet will be raised with variable force.

Weight of section: $3\,\Delta y$

Distance: $5 - y$

$$W_2 = 3\int_0^5 (5 - y)\,dy$$

$$= -\frac{3}{2}(5 - y)^2\Big]_0^5 = \frac{75}{2} \text{ ft} \cdot \text{lb}$$

$$W = W_1 + W_2 = 150 + \frac{75}{2} = \frac{375}{2} \text{ ft} \cdot \text{lb}$$

29. The lower five feet of chain are raised ten feet with a constant force.

$W_1 = 3(5)(10) = 150$ ft • lb

The top ten feet of chain are raised with a variable force.
Weight per section: $3\,\Delta y$
Distance: $10 - y$

$$W_2 = 3\int_0^{10} (10 - y)\,dy = -\frac{3}{2}(10 - y)^2\Big]_0^{10} = 150 \text{ ft} \cdot \text{lb}$$

$$W = W_1 + W_2 = 300 \text{ ft} \cdot \text{lb}$$

30. The work required to lift the chain is 337.5 ft • lb (from Exercise 27). The work required to lift the 100-pound load is $W = (100)(15) = 1500$. The work required to lift the chain with a 100-pound load attached is $W = 337.5 + 1500 = 1837.5$ ft • lb.

31. Weight of section of chain: $3\Delta y$
Distance: $15 - 2y$

$$W = 3\int_0^{7.5} (15 - 2y)\,dy$$

$$= -\frac{3}{4}(15 - 2y)^2\Big]_0^{7.5} = \frac{3}{4}(15)^2 = 168.75 \text{ ft} \cdot \text{lb}$$

32. $W = 3\int_0^{6} (12 - 2y)\,dy$

$$= -\frac{3}{4}(12 - 2y)^2\Big]_0^{6} = \frac{3}{4}(12)^2 = 108 \text{ ft} \cdot \text{lb}$$

33. Work to pull up the ball: $W_1 = 500(15) = 7500$ ft • lb
Work to wind up the top 15 feet of cable: force is variable
Weight per section: $1\Delta x$
Distance: $15 - x$

$$W_2 = \int_0^{15} (15 - x)\,dx$$

$$= -\frac{1}{2}(15 - x)^2\Big]_0^{15} = 112.5 \text{ ft} \cdot \text{lb}$$

Work to lift the lower 25 feet of cable with a constant force:

$$W_3 = (1)(25)(15) = 375 \text{ ft} \cdot \text{lb}$$

$$W = W_1 + W_2 + W_3$$

$$= 7500 + 112.5 + 375 = 7987.5 \text{ ft} \cdot \text{lb}$$

34. Work to pull up the ball:
$W_1 = 500(40) = 20{,}000$ ft • lb
Work to pull up the cable: force is variable
Weight per section: $1\Delta x$
Distance: $40 - x$

$$W_2 = \int_0^{40} (40 - x)\,dx$$

$$= -\frac{1}{2}(40 - x)^2\Big]_0^{40} = 800 \text{ ft} \cdot \text{lb}$$

$$W = W_1 + W_2 = 20{,}000 + 800$$

$$= 20{,}800 \text{ ft} \cdot \text{lb}$$

35. $p = \dfrac{k}{V}$

$$1000 = \frac{k}{2}$$

$$k = 2000$$

$$W = \int_2^3 \frac{2000}{V}\,dV = 2000 \ln|V|\Big]_2^3$$

$$= 2000 \ln\left(\frac{3}{2}\right) \approx 810.93 \text{ ft} \cdot \text{lb}$$

36. $p = \dfrac{k}{V}$

$$2000 = \frac{k}{1}$$

$$k = 2000$$

$$W = \int_1^4 \frac{2000}{V}\,dV = 2000 \ln|V|\Big]_1^4$$

$$= 2000 \ln 4 \approx 2772.59 \text{ ft} \cdot \text{lb}$$

37. $W = \displaystyle\int_0^5 100x\sqrt{125 - x^3}\,dx \approx 10{,}203$ ft • lb

Section 6.6 Fluid Pressure and Fluid Force

1. $F = PA = [62.4(5)](3) = 936 \text{ lb}$

2. $F = PA = [62.4(5)](18) = 5616 \text{ lb}$

3. $F = 62.4(h+2)(6) - (62.4)(h)(6)$

$= 62.4(2)(6) = 748.8 \text{ lb}$

4. $F = 62.4(h+4)(48) - (62.4)(h)(48)$

$= 62.4(4)(48) = 11{,}980.8 \text{ lb}$

5. $h(y) = 3 - y$

$L(y) = 4$

$$F = 62.4\int_0^3 (3-y)(4)\,dy$$

$$= 249.6\int_0^3 (3-y)\,dy = 249.6\left[3y - \frac{y^2}{2}\right]_0^3 = 1123.2 \text{ lb}$$

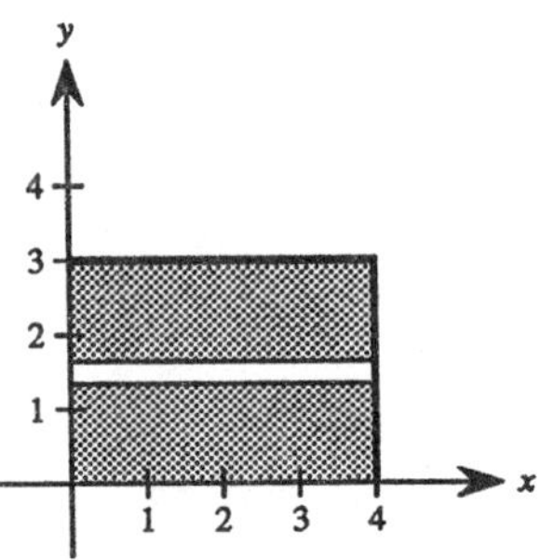

6. $h(y) = 3 - y$

$L(y) = \frac{4}{3}y$

$$F = 62.4\int_0^3 (3-y)\left(\frac{4}{3}y\right)dy$$

$$= \frac{4}{3}(62.4)\int_0^3 (3y - y^2)\,dy = \frac{4}{3}(62.4)\left[\frac{3y^2}{2} - \frac{y^3}{3}\right]_0^3 = 374.4 \text{ lb}$$

Force is one-third that of Exercise 5.

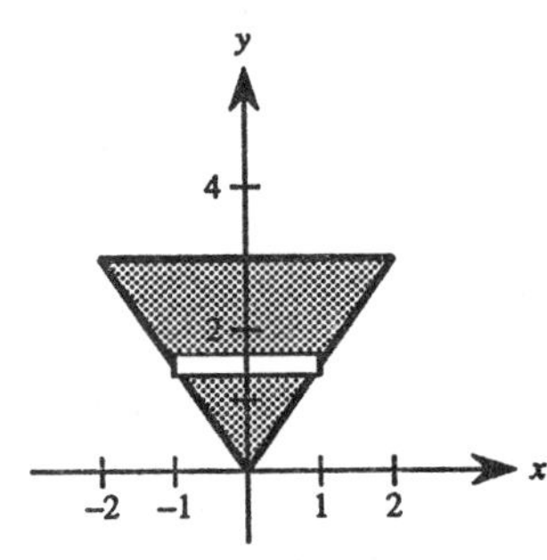

7. $h(y) = 3 - y$

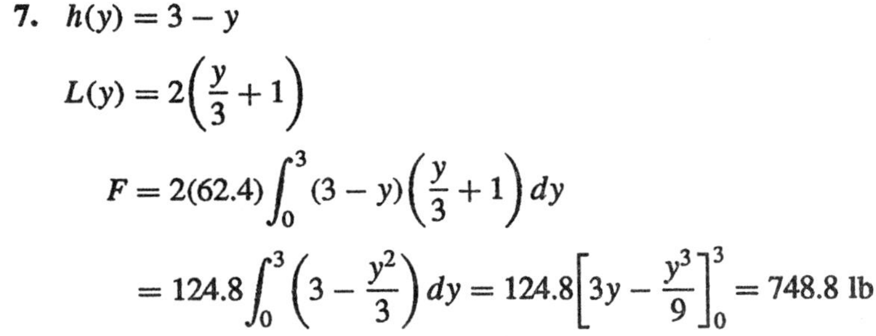

$L(y) = 2\left(\frac{y}{3} + 1\right)$

$$F = 2(62.4)\int_0^3 (3-y)\left(\frac{y}{3} + 1\right)dy$$

$$= 124.8\int_0^3 \left(3 - \frac{y^2}{3}\right)dy = 124.8\left[3y - \frac{y^3}{9}\right]_0^3 = 748.8 \text{ lb}$$

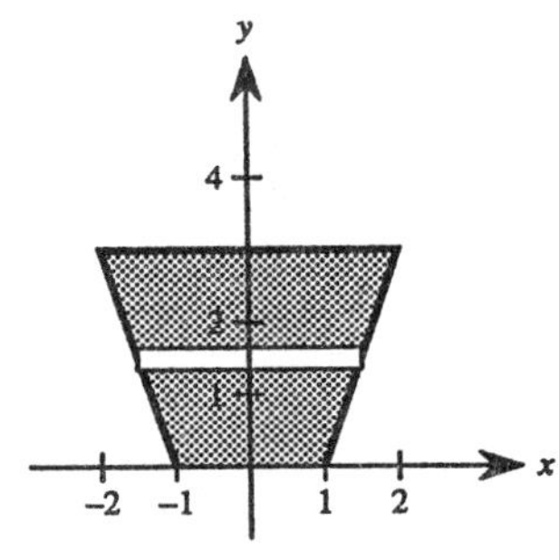

8. $h(y) = -y$

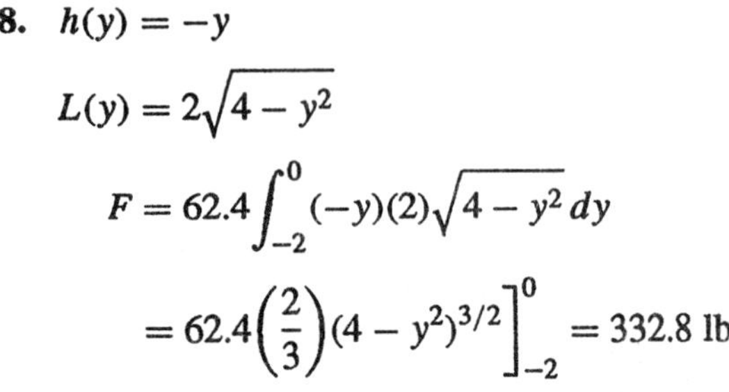

$L(y) = 2\sqrt{4 - y^2}$

$$F = 62.4\int_{-2}^0 (-y)(2)\sqrt{4-y^2}\,dy$$

$$= 62.4\left(\frac{2}{3}\right)(4 - y^2)^{3/2}\Big]_{-2}^0 = 332.8 \text{ lb}$$

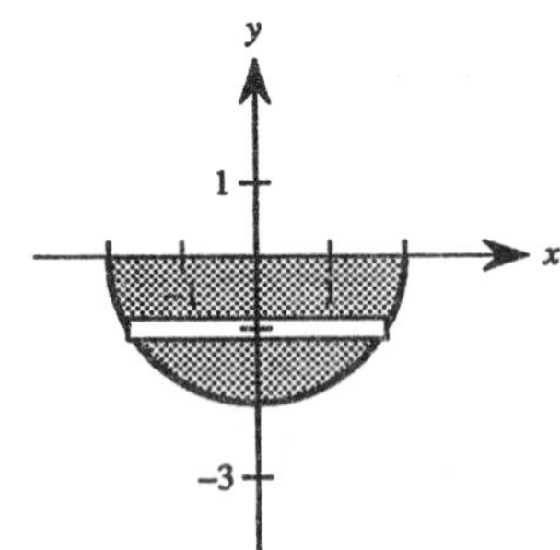

9. $h(y) = 4 - y$

$L(y) = 2\sqrt{y}$

$$F = 2(62.4)\int_0^4 (4 - y)\sqrt{y}\,dy$$

$$= 124.8\int_0^4 (4y^{1/2} - y^{3/2})\,dy = 124.8\left[\frac{8y^{3/2}}{3} - \frac{2y^{5/2}}{5}\right]_0^4 = 1064.96 \text{ lb}$$

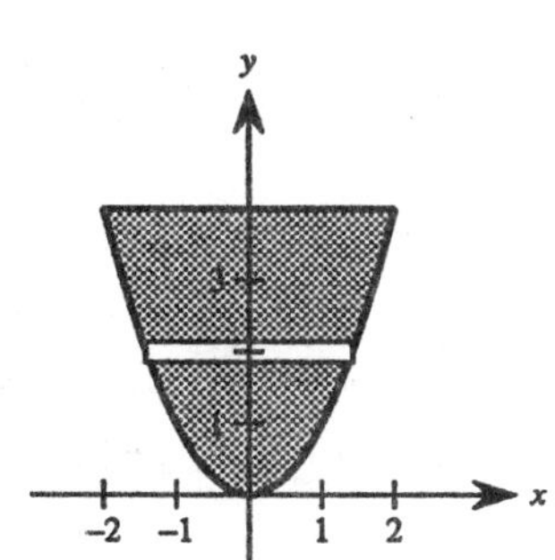

10. $h(y) = -y$

$L(y) = \frac{4}{3}\sqrt{9 - y^2}$

$$F = 62.4\int_{-3}^0 (-y)\frac{4}{3}\sqrt{9 - y^2}\,dy$$

$$= 62.4\left(\frac{2}{3}\right)\int_{-3}^0 (9 - y^2)^{1/2}(-2y)\,dy = 62.4\left(\frac{4}{9}\right)(9 - y^2)^{3/2}\Big]_{-3}^0 = 748.8 \text{ lb}$$

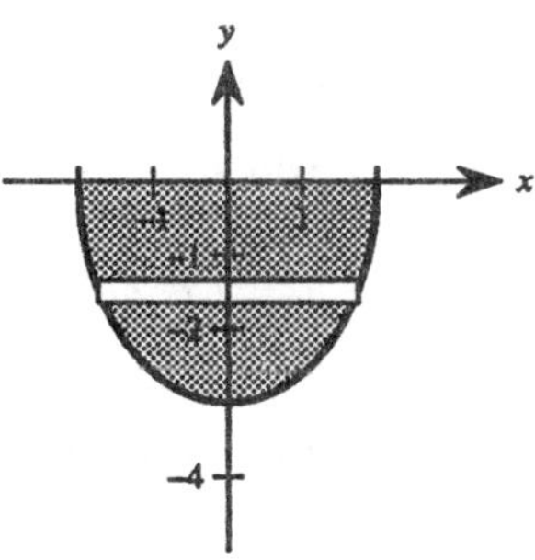

11. $h(y) = 4 - y$

$L(y) = 2$

$$F = 62.4\int_0^2 2(4 - y)\,dy$$

$$= 62.4\Big[8y - y^2\Big]_0^2 = 748.8 \text{ lb}$$

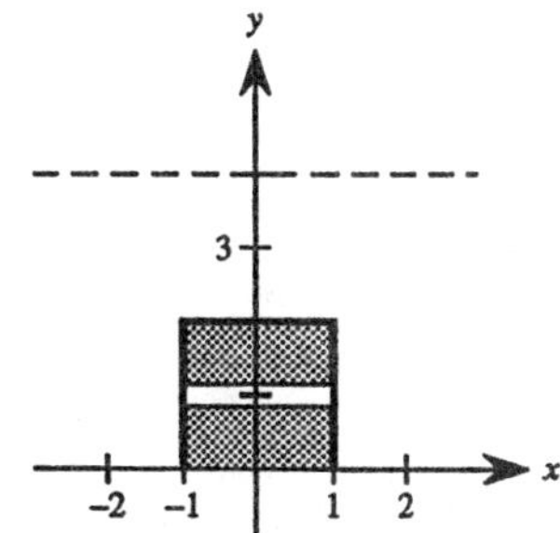

12. $h(y) = (1 + 2\sqrt{2}) - y$

$L_1(y) = 2y$ [lower part]

$L_2(y) = 2(2\sqrt{2} - y)$ [upper part]

$$F = 124.8\left[\int_0^{\sqrt{2}} (1 + 2\sqrt{2} - y)y\,dy + \int_{\sqrt{2}}^{2\sqrt{2}} (1 + 2\sqrt{2} - y)(2\sqrt{2} - y)\,dy\right]$$

$$= 124.8\left(\left[\frac{y^2}{2} + \sqrt{2}y^2 - \frac{y^3}{3}\right]_0^{\sqrt{2}} + \left[2\sqrt{2}y + 8y - 2\sqrt{2}y^2 - \frac{y^2}{2} + \frac{y^3}{3}\right]_{\sqrt{2}}^{2\sqrt{2}}\right)$$

$$= 249.6(1 + \sqrt{2}) \approx 602.6 \text{ lb}$$

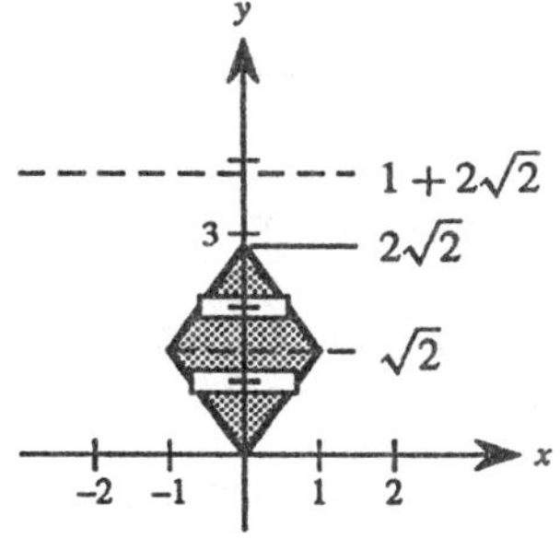

13. $h(y) = 12 - y$

$L(y) = 6 - \frac{2y}{3}$

$$F = 62.4\int_0^9 (12 - y)\left(6 - \frac{2y}{3}\right)dy$$

$$= 62.4\left[72y - 7y^2 + \frac{2y^3}{9}\right]_0^9 = 15{,}163.2 \text{ lb}$$

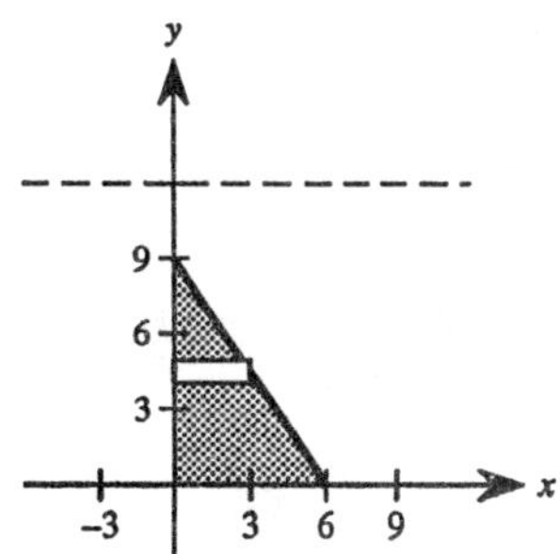

14. $h(y) = 6 - y$

$L(y) = 1$

$$F = 62.4\int_0^5 1(6-y)\,dy$$

$$= 62.4\left[6y - \frac{y^2}{2}\right]_0^5 = 1092.0 \text{ lb}$$

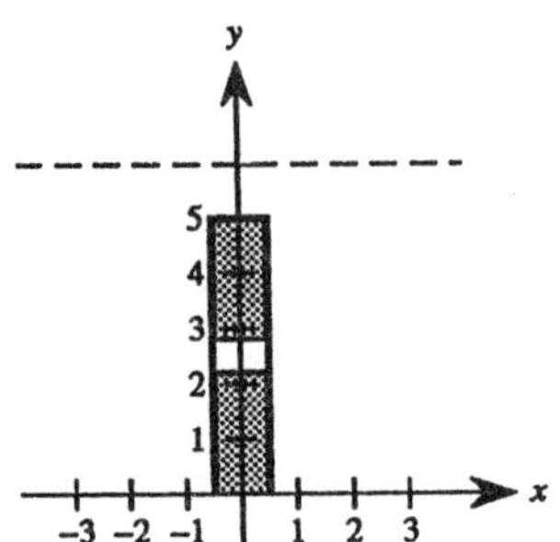

15. $h(y) = 2 - y$

$L(y) = 10$

$$F = 140.7\int_0^2 (2-y)(10)\,dy$$

$$= 1407\int_0^2 (2-y)\,dy$$

$$= 1407\left[2y - \frac{y^2}{2}\right]_0^2 = 2814 \text{ lb}$$

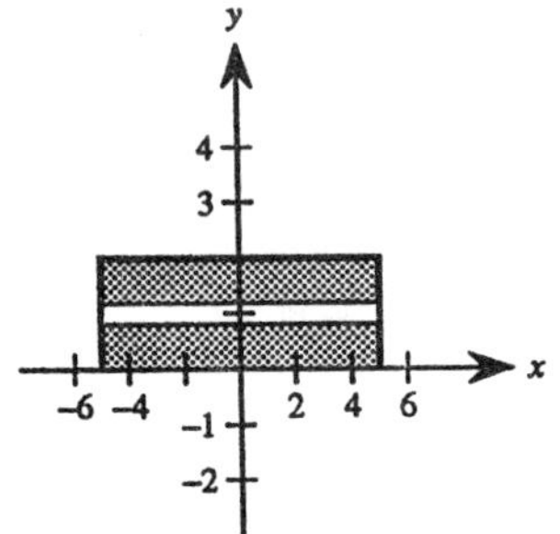

16. $h(y) = 4 - y$

$L(y) = 6$

$$F = 140.7\int_0^4 (4-y)(6)\,dy$$

$$= 844.2\int_0^4 (4-y)\,dy$$

$$= 844.2\left[4y - \frac{y^2}{2}\right]_0^4 = 6753.6 \text{ lb}$$

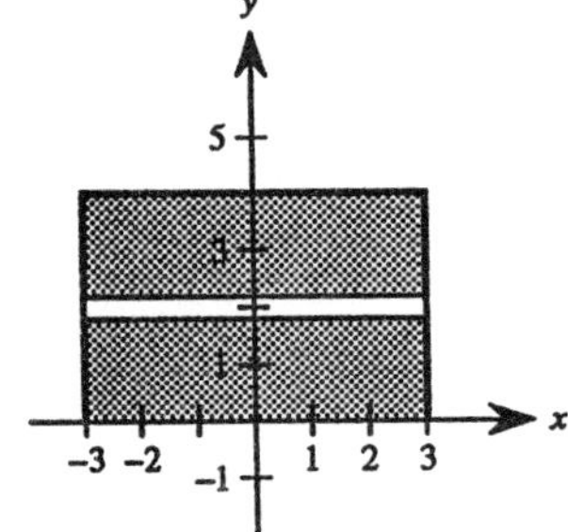

17. $h(y) = -y$

$$L(y) = 2\left(\frac{4}{3}\sqrt{9-y^2}\right)$$

$$F = 140.7\int_{-3}^0 (-y)(2)\left(\frac{4}{3}\sqrt{9-y^2}\right)dy$$

$$= \frac{(140.7)(4)}{3}\int_{-3}^0 \sqrt{9-y^2}(-2y)\,dy$$

$$= \frac{(140.7)(4)}{3}\left(\frac{2}{3}\right)(9-y^2)^{3/2}\Bigg]_{-3}^0 = 3376.8 \text{ lb}$$

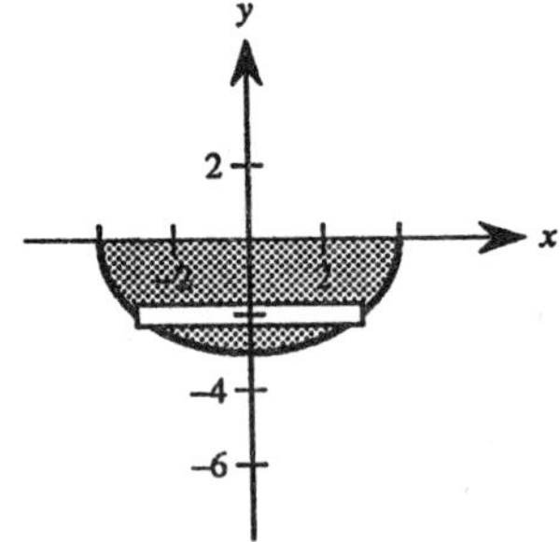

18. $h(y) = -y$

$$L(y) = 6 + \frac{3}{2}y$$

$$F = 140.7\int_{-4}^0 (-y)\left(6 + \frac{3}{2}y\right)dy$$

$$= -140.7\left(3y^2 + \frac{y^3}{2}\right)\Bigg]_{-4}^0 = 2251.2 \text{ lb}$$

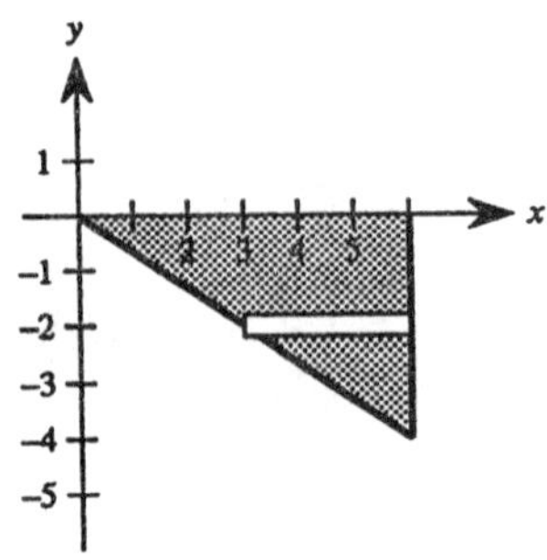

19. $h(y) = -y$

$L(y) = 2\left(\frac{1}{2}\right)\sqrt{9-4y^2}$

$$F = 42\int_{-3/2}^{0} (-y)\sqrt{9-4y^2}\,dy$$

$$= \frac{42}{8}\int_{-3/2}^{0} (9-4y^2)^{1/2}(-8y)\,dy$$

$$= \left(\frac{21}{4}\right)\left(\frac{2}{3}\right)(9-4y^2)^{3/2}\Big]_{-3/2}^{0} = 94.5 \text{ lb}$$

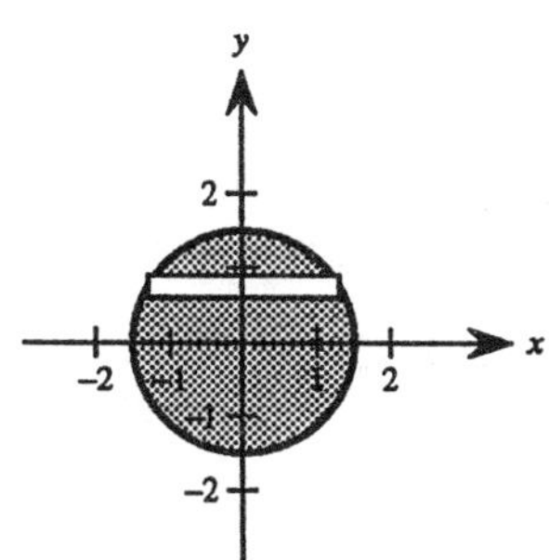

20. $h(y) = \frac{3}{2} - y$

$L(y) = 2\left(\frac{1}{2}\right)\sqrt{9-4y^2}$

$$F = 42\int_{-3/2}^{3/2}\left(\frac{3}{2}-y\right)\sqrt{9-4y^2}\,dy = 63\int_{-3/2}^{3/2}\sqrt{9-4y^2}\,dy + \frac{21}{4}\int_{-3/2}^{3/2}\sqrt{9-4y^2}(-8y)\,dy$$

The second integral is zero since it is an odd function and the limits of integration are symmetric to the origin. The first integral is twice the area of a semicircle of radius $\frac{3}{2}$.

$$\left(\sqrt{9-4y^2} = 2\sqrt{\left(\frac{9}{4}\right) - y^2}\right)$$

Thus, the force is $63\left(\frac{9}{4}\pi\right) = 141.75\pi \approx 445.32$ lb.

21. $h(y) = k - y$

$L(y) = 2\sqrt{r^2 - y^2}$

$$F = w\int_{-r}^{r}(k-y)\sqrt{r^2-y^2}(2)\,dy$$

$$= w\left[2k\int_{-r}^{r}\sqrt{r^2-y^2}\,dy + \int_{-r}^{r}\sqrt{r^2-y^2}(-2y)\,dy\right]$$

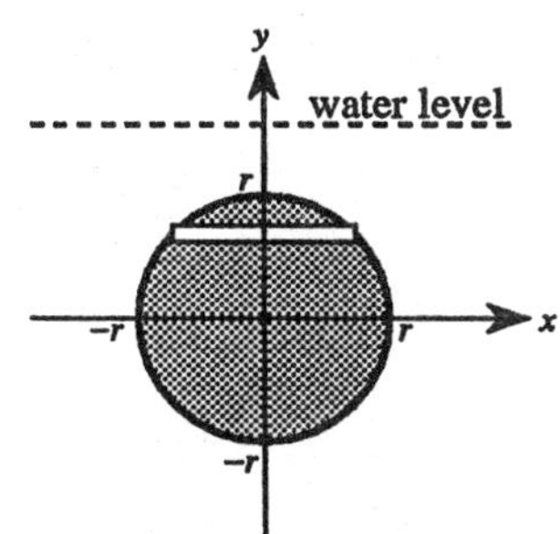

The second integral is zero since its integrand is odd and the limits of integration are symmetric to the origin. The first integral is the area of a semicircle with radius r.

$$F = w\left[(2k)\frac{\pi r^2}{2} + 0\right] = wk\pi r^2$$

22. $h(y) = k - y$

$L(y) = b$

$$F = w\int_{-h/2}^{h/2}(k-y)b\,dy$$

$$= wb\left[ky - \frac{y^2}{2}\right]_{-h/2}^{h/2} = wb(hk) = wkhb$$

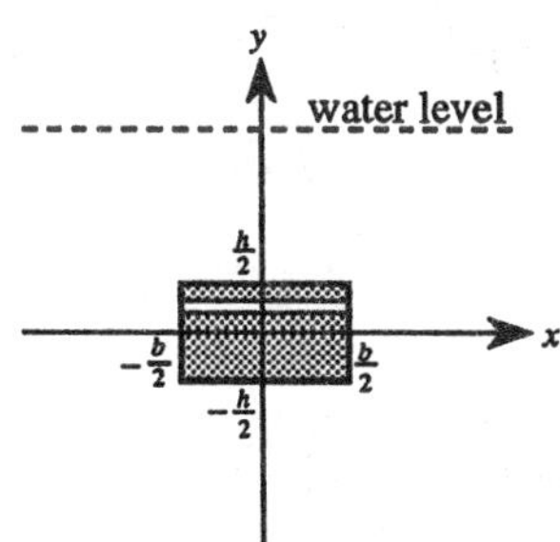

23. From Exercise 22:

$F = 64(15)(1)(1) = 960$ lb

24. From Exercise 21:

$F = 64(15)\pi\left(\frac{1}{2}\right)^2 = 753.98$ lb

25. (a) Wall at shallow end

From Exercise 22: $F = 62.4(2)(4)(20) = 9984$ lb

(b) Wall at deep end

From Exercise 22: $F = 62.4(4)(8)(20) = 39{,}936$ lb

(c) Side wall

From Exercise 22: $F_1 = 62.4(2)(4)(40) = 19{,}968$ lb

$$F_2 = 62.4\int_0^4 (8-y)(10y)\,dy$$

$$= 624\int_0^4 (8y - y^2)\,dy = 624\left[4y^2 - \frac{y^3}{3}\right]_0^4 = 26{,}624 \text{ lb}$$

Total force: $F_1 + F_2 = 46{,}592$ lb

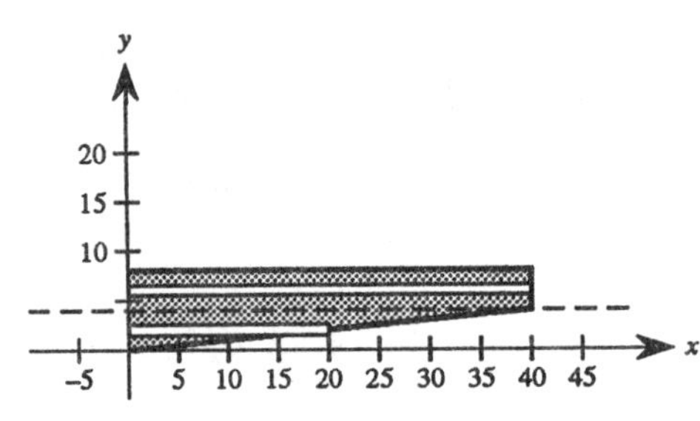

26. $h(y) = 3 - y$

$$L(y) = \sqrt{\frac{4y}{5-y}}$$

$$F = 62.4\int_0^3 (3-y)\sqrt{\frac{4y}{5-y}}\,dy$$

$$= 2(62.4)\int_0^3 (3-y)\sqrt{\frac{y}{5-y}}\,dy$$

≈ 268.2 lb

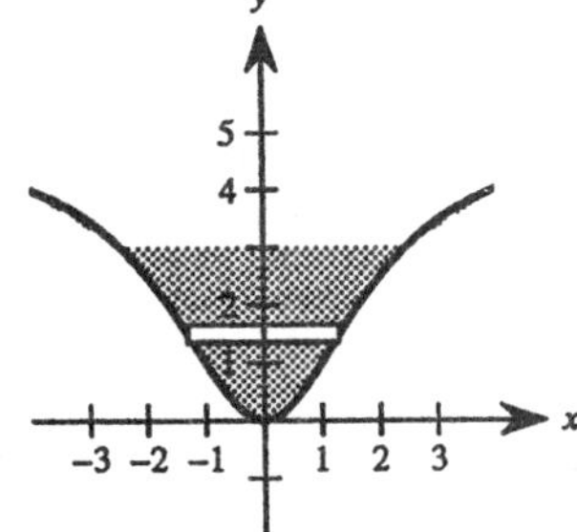

27. $h(y) = 12 - y$

$$L(y) = (4^{2/3} - y^{2/3})^{3/2}$$

$$F = 62.4\int_0^4 2(12-y)(4^{2/3} - y^{2/3})^{3/2}\,dy$$

≈ 6483 lb

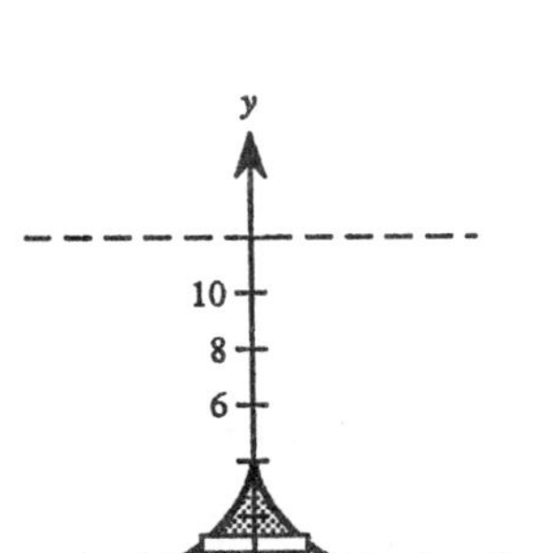

28. $h(y) = 12 - y$

$$L(y) = \frac{\sqrt{7(16-y^2)}}{2}$$

$$F = 62.4\int_0^4 (12-y)\frac{\sqrt{7(16-y^2)}}{2}\,dy$$

$$= 31.2\sqrt{7}\int_0^4 (12-y)\sqrt{16-y^2}\,dy$$

$\approx 10{,}686.847$ lb

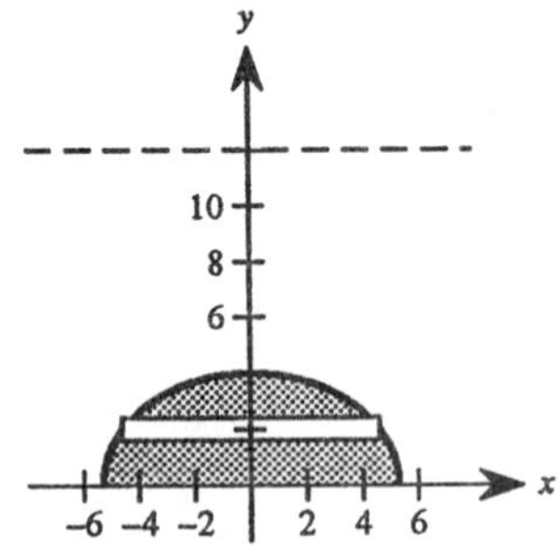

Section 6.7 Moments, Centers of Mass, and Centroids

1. $\bar{x} = \dfrac{6(-5)+3(1)+5(3)}{6+3+5} = -\dfrac{6}{7}$

2. $\bar{x} = \dfrac{7(-3)+4(-2)+3(5)+8(6)}{7+4+3+8} = \dfrac{17}{11}$

3. $\bar{x} = \dfrac{1(7)+1(8)+1(12)+1(15)+1(18)}{1+1+1+1+1} = 12$

4. $\bar{x} = \dfrac{12(-3)+1(-2)+6(-1)+3(0)+11(4)}{12+1+6+3+11} = 0$

5. $\bar{x} = \dfrac{(7+5)+(8+5)+(12+5)+(15+5)+(18+5)}{5} = 17 = 12+5$

6. $\bar{x} = \dfrac{12(-3-3)+1(-2-3)+6(-1-3)+3(0-3)+11(4-3)}{12+1+6+3+11} = -3 = 0-3$

7. $\bar{x} = \dfrac{3(-2)+4(-1)+2(7)+1(0)+6(-3)}{3+4+2+1+6} = -\dfrac{7}{8}$

$\bar{y} = \dfrac{3(-3)+4(0)+2(1)+1(0)+6(0)}{3+4+2+1+6} = -\dfrac{7}{16}$

$(\bar{x}, \bar{y}) = \left(-\dfrac{7}{8}, -\dfrac{7}{16}\right)$

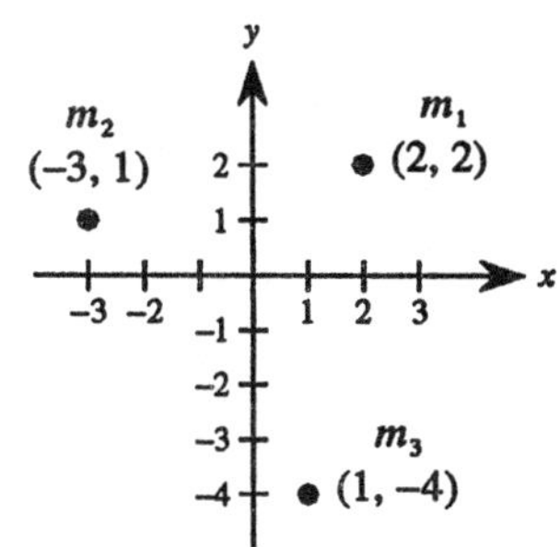

8. $\bar{x} = \dfrac{4(2)+2(-1)+2.5(6)+5(2)}{4+2+2.5+5} = \dfrac{31}{13.5} = \dfrac{62}{27}$

$\bar{y} = \dfrac{4(3)+2(5)+2.5(8)+5(-2)}{4+2+2.5+5} = \dfrac{32}{13.5} = \dfrac{64}{27}$

$(\bar{x}, \bar{y}) = \left(\dfrac{62}{27}, \dfrac{64}{27}\right)$

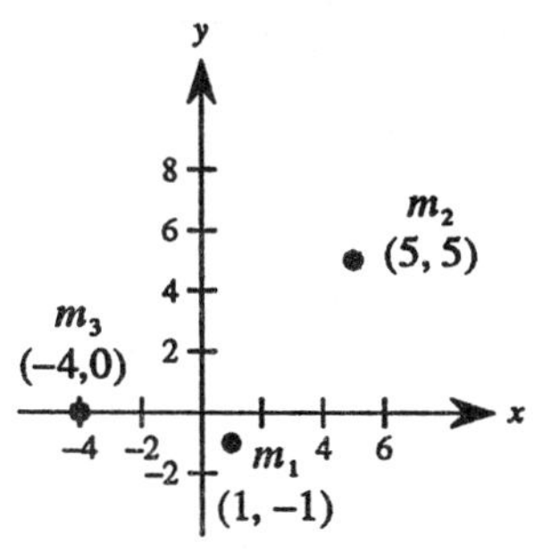

9. For simplicity, we assume the density to be 1 unit of mass/unit of area.

$m_1 = 4, \quad P_1 = (1, 0)$

$m_2 = \pi, \quad P_2 = (3, 0)$

$\bar{x} = \dfrac{4(1)+\pi(3)}{4+\pi} = \dfrac{4+3\pi}{4+\pi}$

$\bar{y} = \dfrac{4(0)+\pi(0)}{4+\pi} = 0$

$(\bar{x}, \bar{y}) = \left(\dfrac{4+3\pi}{4+\pi}, 0\right)$

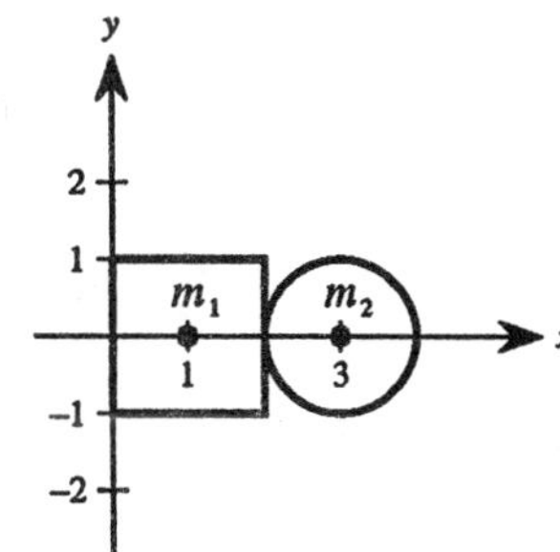

10. For simplicity, we assume the density to be 1 unit of mass/unit of area.

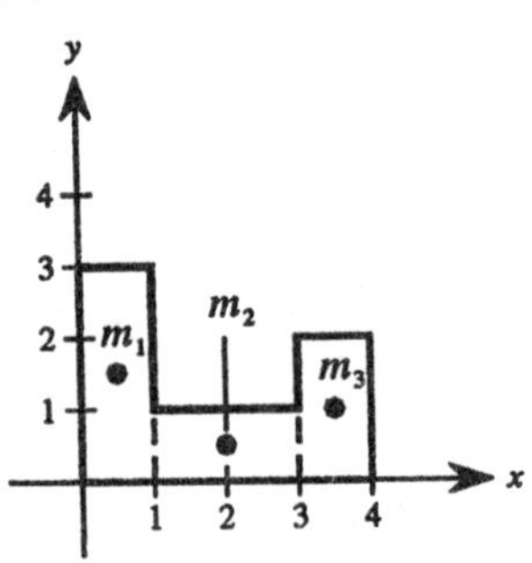

$$m_1 = 3, \quad P_1 = \left(\frac{1}{2}, \frac{3}{2}\right)$$

$$m_2 = 2, \quad P_2 = \left(2, \frac{1}{2}\right)$$

$$m_3 = 2, \quad P_3 = \left(\frac{7}{2}, 1\right)$$

$$\bar{x} = \frac{3(1/2) + 2(2) + 2(7/2)}{3+2+2} = \frac{25/2}{7} = \frac{25}{14}$$

$$\bar{y} = \frac{3(3/2) + 2(1/2) + 2(1)}{3+2+2} = \frac{15/2}{7} = \frac{15}{14}$$

$$(\bar{x}, \bar{y}) = \left(\frac{25}{14}, \frac{15}{14}\right)$$

11.

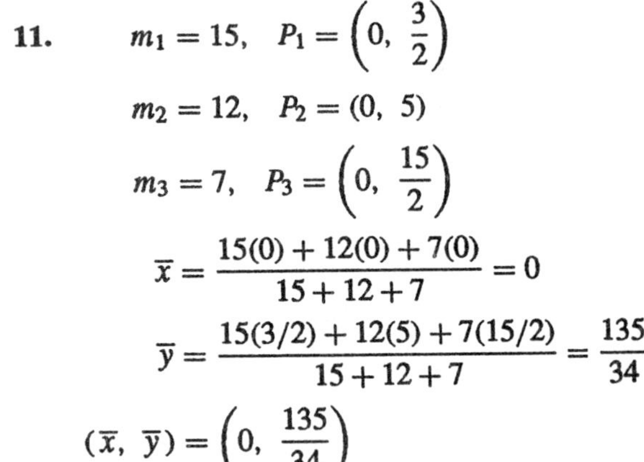

$$m_1 = 15, \quad P_1 = \left(0, \frac{3}{2}\right)$$

$$m_2 = 12, \quad P_2 = (0, 5)$$

$$m_3 = 7, \quad P_3 = \left(0, \frac{15}{2}\right)$$

$$\bar{x} = \frac{15(0) + 12(0) + 7(0)}{15+12+7} = 0$$

$$\bar{y} = \frac{15(3/2) + 12(5) + 7(15/2)}{15+12+7} = \frac{135}{34}$$

$$(\bar{x}, \bar{y}) = \left(0, \frac{135}{34}\right)$$

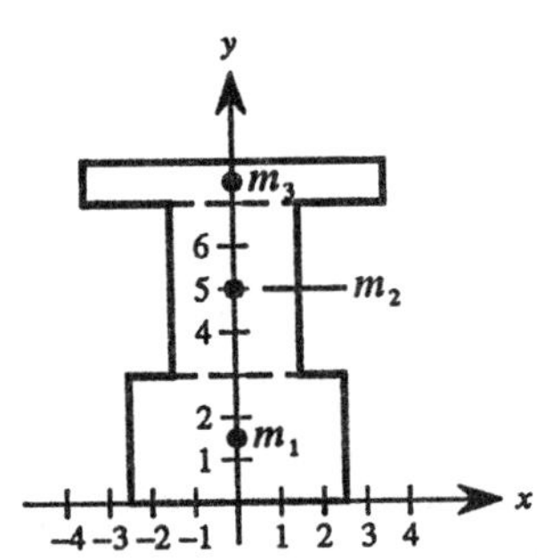

12.

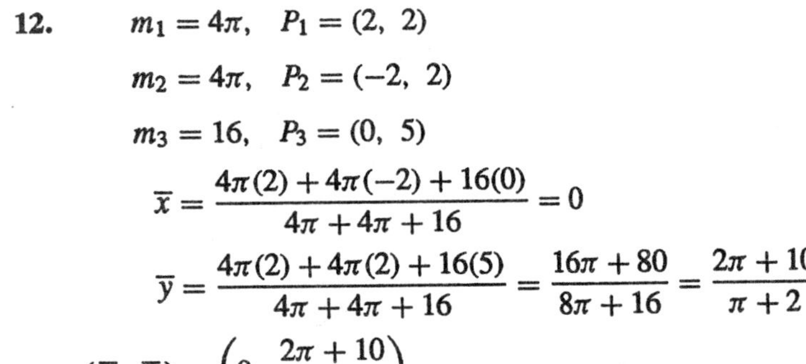

$$m_1 = 4\pi, \quad P_1 = (2, 2)$$

$$m_2 = 4\pi, \quad P_2 = (-2, 2)$$

$$m_3 = 16, \quad P_3 = (0, 5)$$

$$\bar{x} = \frac{4\pi(2) + 4\pi(-2) + 16(0)}{4\pi + 4\pi + 16} = 0$$

$$\bar{y} = \frac{4\pi(2) + 4\pi(2) + 16(5)}{4\pi + 4\pi + 16} = \frac{16\pi + 80}{8\pi + 16} = \frac{2\pi + 10}{\pi + 2}$$

$$(\bar{x}, \bar{y}) = \left(0, \frac{2\pi + 10}{\pi + 2}\right)$$

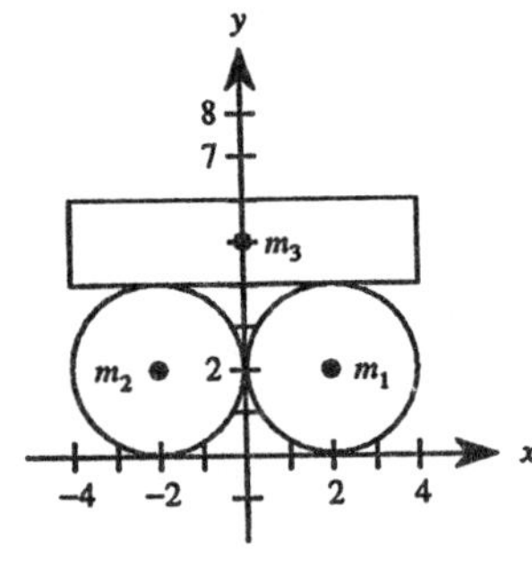

13.

$$m_1 = 4, \quad P_1 = (0, 1)$$

$$m_2 = 2\pi, \quad P_2 = (0, 3)$$

$$\bar{x} = 0$$

$$\bar{y} = \frac{4(1) + 2\pi(3)}{4 + 2\pi} = \frac{2 + 3\pi}{2 + \pi}$$

$$(\bar{x}, \bar{y}) = \left(0, \frac{2 + 3\pi}{2 + \pi}\right)$$

14.

$$m_1 = 8, \quad P_1 = (0, 1)$$

$$m_2 = \pi, \quad P_2 = (0, 3)$$

$$\bar{x} = 0$$

$$\bar{y} = \frac{8(1) + \pi(3)}{8 + \pi} = \frac{8 + 3\pi}{8 + \pi}$$

$$(\bar{x}, \bar{y}) = \left(0, \frac{8 + 3\pi}{8 + \pi}\right)$$

15. $$m = \rho\int_0^4 \sqrt{x}\,dx = \frac{2\rho}{3}x^{3/2}\Big]_0^4 = \frac{16\rho}{3}$$

$$M_x = \rho\int_0^4 \frac{\sqrt{x}}{2}(\sqrt{x})\,dx = \rho\frac{x^2}{4}\Big]_0^4 = 4\rho$$

$$\overline{y} = 4\rho\left(\frac{3}{16\rho}\right) = \frac{3}{4}$$

$$M_y = \rho\int_0^4 x\sqrt{x}\,dx = \rho\frac{2}{5}x^{5/2}\Big]_0^4 = \frac{64\rho}{5}$$

$$\overline{x} = \frac{64\rho}{5}\left(\frac{3}{16\rho}\right) = \frac{12}{5}$$

$$(\overline{x}, \overline{y}) = \left(\frac{12}{5}, \frac{3}{4}\right)$$

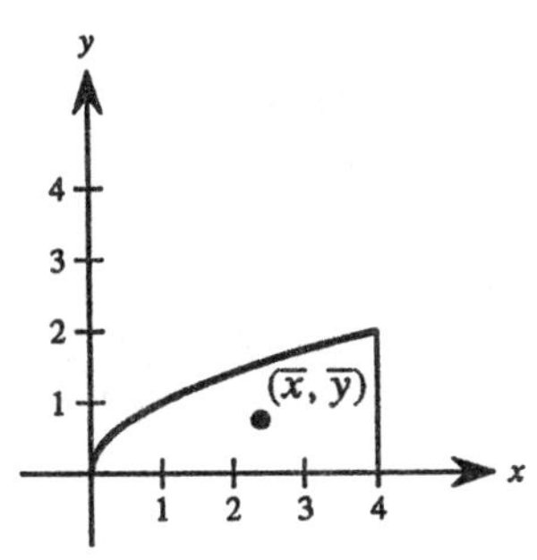

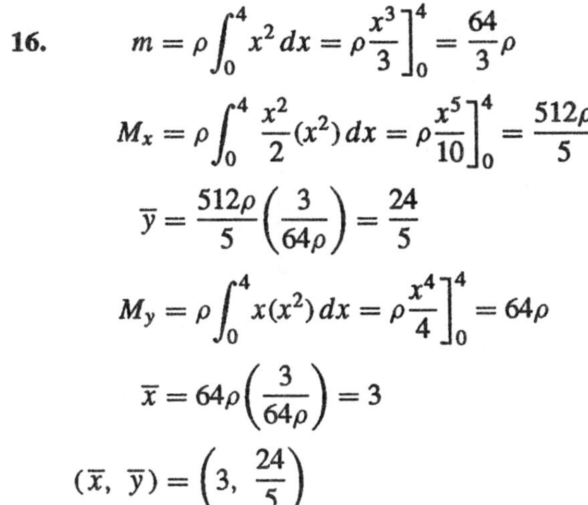

16. $$m = \rho\int_0^4 x^2\,dx = \rho\frac{x^3}{3}\Big]_0^4 = \frac{64}{3}\rho$$

$$M_x = \rho\int_0^4 \frac{x^2}{2}(x^2)\,dx = \rho\frac{x^5}{10}\Big]_0^4 = \frac{512\rho}{5}$$

$$\overline{y} = \frac{512\rho}{5}\left(\frac{3}{64\rho}\right) = \frac{24}{5}$$

$$M_y = \rho\int_0^4 x(x^2)\,dx = \rho\frac{x^4}{4}\Big]_0^4 = 64\rho$$

$$\overline{x} = 64\rho\left(\frac{3}{64\rho}\right) = 3$$

$$(\overline{x}, \overline{y}) = \left(3, \frac{24}{5}\right)$$

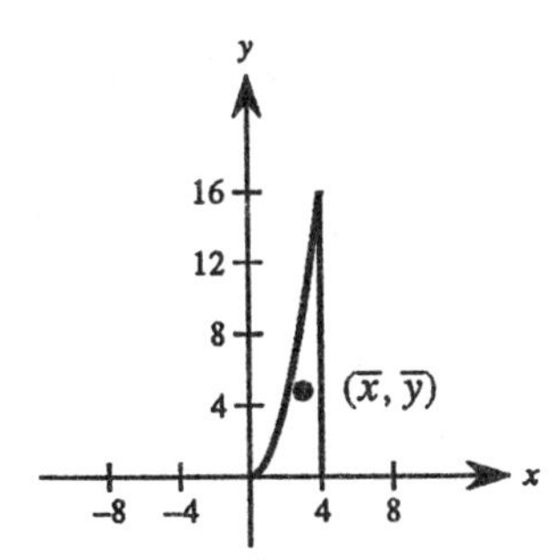

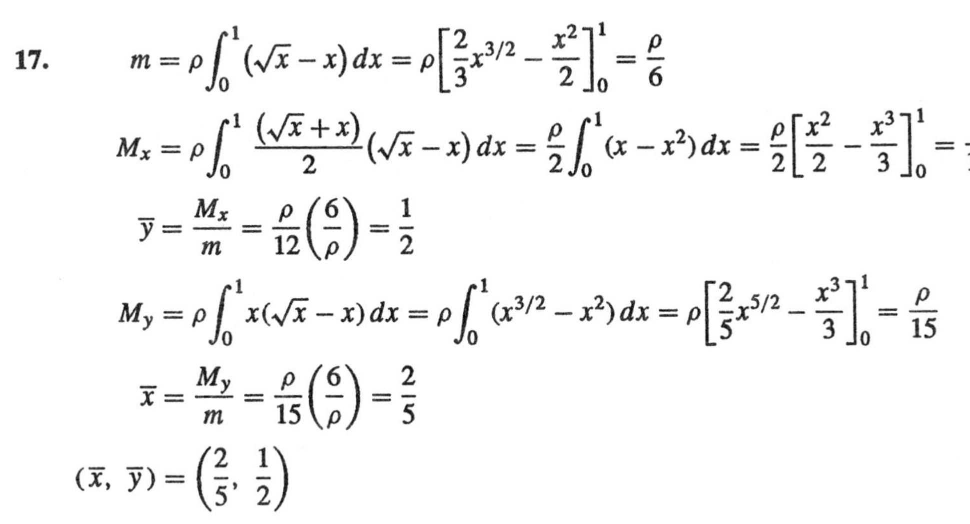

17. $$m = \rho\int_0^1 (\sqrt{x} - x)\,dx = \rho\left[\frac{2}{3}x^{3/2} - \frac{x^2}{2}\right]_0^1 = \frac{\rho}{6}$$

$$M_x = \rho\int_0^1 \frac{(\sqrt{x}+x)}{2}(\sqrt{x} - x)\,dx = \frac{\rho}{2}\int_0^1 (x - x^2)\,dx = \frac{\rho}{2}\left[\frac{x^2}{2} - \frac{x^3}{3}\right]_0^1 = \frac{\rho}{12}$$

$$\overline{y} = \frac{M_x}{m} = \frac{\rho}{12}\left(\frac{6}{\rho}\right) = \frac{1}{2}$$

$$M_y = \rho\int_0^1 x(\sqrt{x} - x)\,dx = \rho\int_0^1 (x^{3/2} - x^2)\,dx = \rho\left[\frac{2}{5}x^{5/2} - \frac{x^3}{3}\right]_0^1 = \frac{\rho}{15}$$

$$\overline{x} = \frac{M_y}{m} = \frac{\rho}{15}\left(\frac{6}{\rho}\right) = \frac{2}{5}$$

$$(\overline{x}, \overline{y}) = \left(\frac{2}{5}, \frac{1}{2}\right)$$

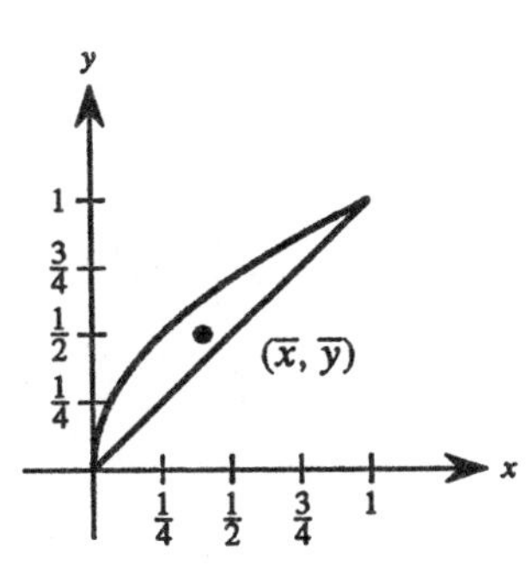

18. $$m = \rho \int_0^1 (x^2 - x^3)\,dx = \rho\left[\frac{x^3}{3} - \frac{x^4}{4}\right]_0^1 = \frac{\rho}{12}$$

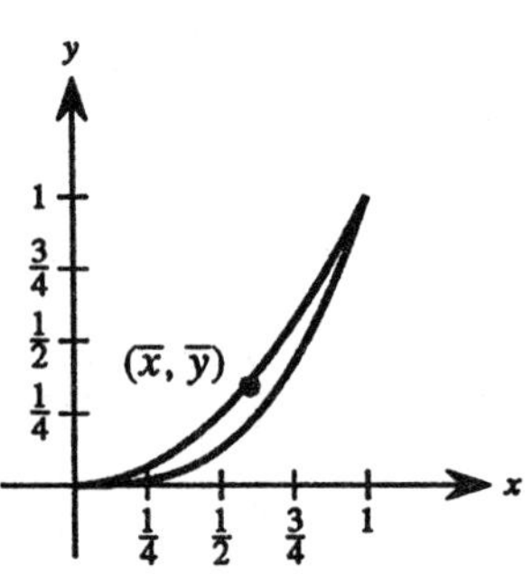

$$M_x = \rho \int_0^1 \frac{(x^2 + x^3)}{2}(x^2 - x^3)\,dx$$

$$= \frac{\rho}{2}\int_0^1 (x^4 - x^6)\,dx = \frac{\rho}{2}\left[\frac{x^5}{5} - \frac{x^7}{7}\right]_0^1 = \frac{\rho}{35}$$

$$\bar{y} = \frac{M_x}{m} = \frac{\rho}{35}\left(\frac{12}{\rho}\right) = \frac{12}{35}$$

$$M_y = \rho \int_0^1 x(x^2 - x^3)\,dx = \rho\int_0^1 (x^3 - x^4)\,dx = \rho\left[\frac{x^4}{4} - \frac{x^5}{5}\right]_0^1 = \frac{\rho}{20}$$

$$\bar{x} = \frac{M_y}{m} = \frac{\rho}{20}\left(\frac{12}{\rho}\right) = \frac{3}{5}$$

$$(\bar{x}, \bar{y}) = \left(\frac{3}{5}, \frac{12}{35}\right)$$

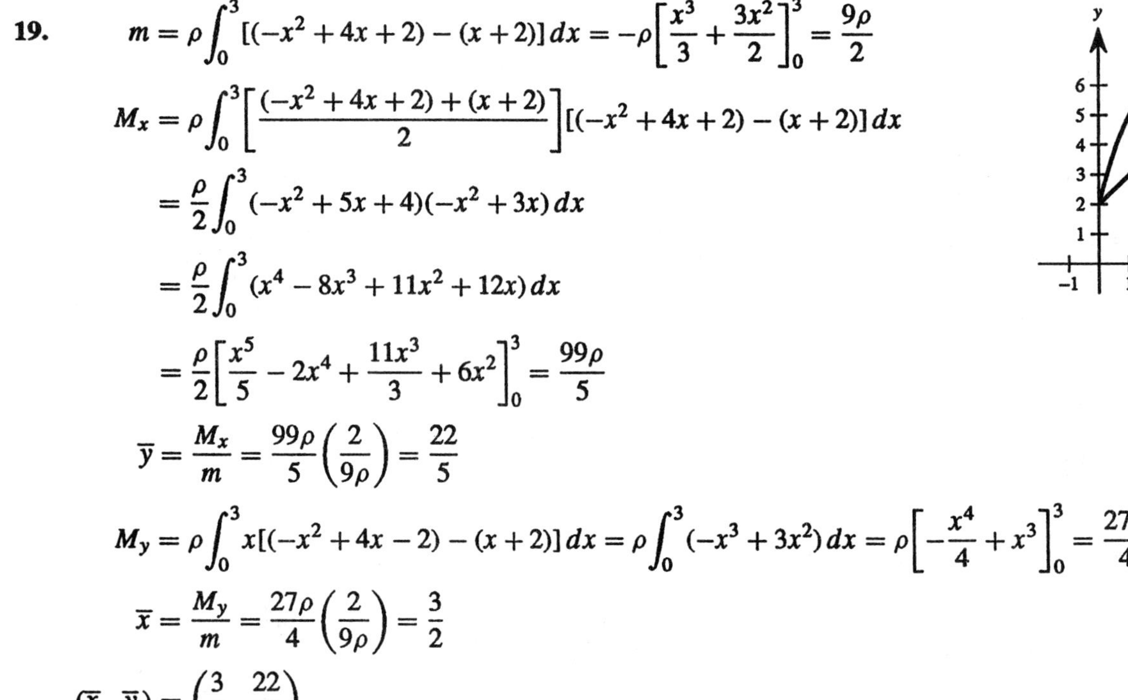

19. $$m = \rho \int_0^3 [(-x^2 + 4x + 2) - (x + 2)]\,dx = -\rho\left[\frac{x^3}{3} + \frac{3x^2}{2}\right]_0^3 = \frac{9\rho}{2}$$

$$M_x = \rho \int_0^3 \left[\frac{(-x^2 + 4x + 2) + (x + 2)}{2}\right][(-x^2 + 4x + 2) - (x + 2)]\,dx$$

$$= \frac{\rho}{2}\int_0^3 (-x^2 + 5x + 4)(-x^2 + 3x)\,dx$$

$$= \frac{\rho}{2}\int_0^3 (x^4 - 8x^3 + 11x^2 + 12x)\,dx$$

$$= \frac{\rho}{2}\left[\frac{x^5}{5} - 2x^4 + \frac{11x^3}{3} + 6x^2\right]_0^3 = \frac{99\rho}{5}$$

$$\bar{y} = \frac{M_x}{m} = \frac{99\rho}{5}\left(\frac{2}{9\rho}\right) = \frac{22}{5}$$

$$M_y = \rho \int_0^3 x[(-x^2 + 4x - 2) - (x + 2)]\,dx = \rho\int_0^3 (-x^3 + 3x^2)\,dx = \rho\left[-\frac{x^4}{4} + x^3\right]_0^3 = \frac{27\rho}{4}$$

$$\bar{x} = \frac{M_y}{m} = \frac{27\rho}{4}\left(\frac{2}{9\rho}\right) = \frac{3}{2}$$

$$(\bar{x}, \bar{y}) = \left(\frac{3}{2}, \frac{22}{5}\right)$$

20. $m = \rho\int_0^3 [(\sqrt{3x}+1)-(x+1)]\,dx = \rho\left[\frac{2}{9}(3x)^{3/2} - \frac{x^2}{2}\right]_0^3 = \frac{3}{2}\rho$

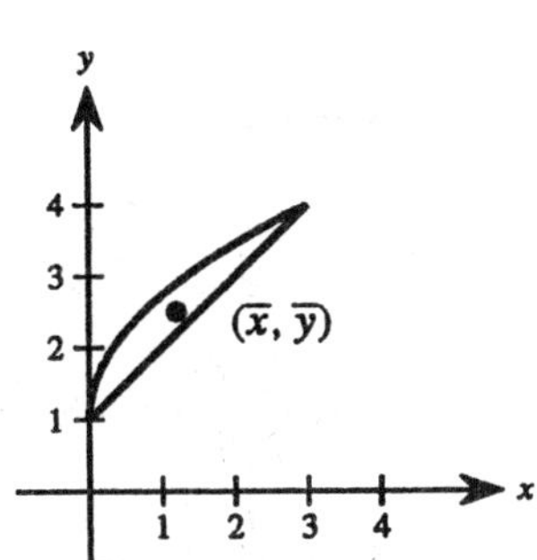

$$M_x = \rho\int_0^3 \frac{[(\sqrt{3x}+1)+(x+1)]}{2}[(\sqrt{3x}+1)-(x+1)]\,dx$$

$$= \frac{\rho}{2}\int_0^3 (\sqrt{3x}+x+2)(\sqrt{3x}-x)\,dx$$

$$= \frac{\rho}{2}\int_0^3 \left(2\sqrt{3x}+x-x^2\right)dx$$

$$= \frac{\rho}{2}\left[\frac{4}{9}(3x)^{3/2}+\frac{x^2}{2}-\frac{x^3}{3}\right]_0^3 = \frac{15\rho}{4}$$

$$\bar{y} = \frac{M_x}{m} = \frac{15\rho}{4}\left(\frac{2}{3\rho}\right) = \frac{5}{2}$$

$$M_y = \rho\int_0^3 x[(\sqrt{3x}+1)-(x+1)]\,dx = \rho\int_0^3 (x\sqrt{3x}-x^2)\,dx = \rho\left[\frac{2\sqrt{3}}{5}x^{5/2}-\frac{x^3}{3}\right]_0^3 = \frac{9\rho}{5}$$

$$\bar{x} = \frac{M_y}{m} = \frac{9\rho}{5}\left(\frac{2}{3\rho}\right) = \frac{6}{5}$$

$$(\bar{x}, \bar{y}) = \left(\frac{6}{5}, \frac{5}{2}\right)$$

21. 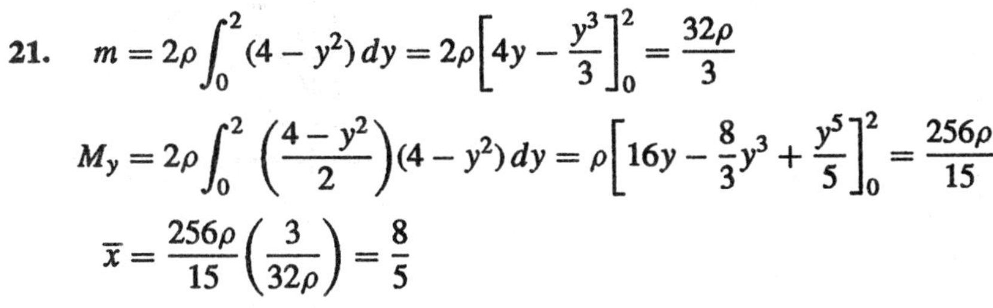

$m = 2\rho\int_0^2 (4-y^2)\,dy = 2\rho\left[4y - \frac{y^3}{3}\right]_0^2 = \frac{32\rho}{3}$

$$M_y = 2\rho\int_0^2 \left(\frac{4-y^2}{2}\right)(4-y^2)\,dy = \rho\left[16y - \frac{8}{3}y^3 + \frac{y^5}{5}\right]_0^2 = \frac{256\rho}{15}$$

$$\bar{x} = \frac{256\rho}{15}\left(\frac{3}{32\rho}\right) = \frac{8}{5}$$

By symmetry, M_x and $\bar{y} = 0$.

$$(\bar{x}, \bar{y}) = \left(\frac{8}{5}, 0\right)$$

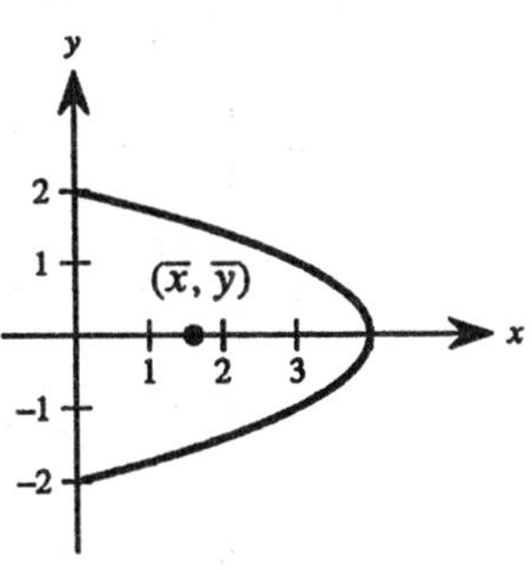

22. 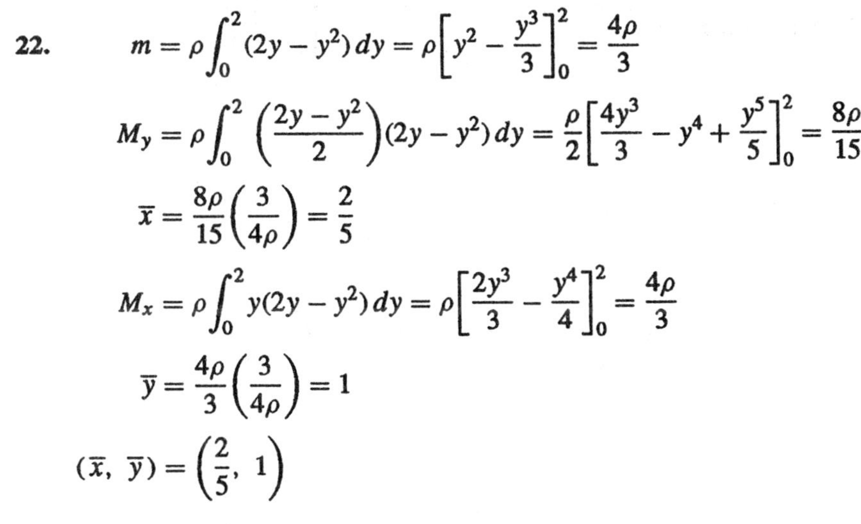

$m = \rho\int_0^2 (2y-y^2)\,dy = \rho\left[y^2 - \frac{y^3}{3}\right]_0^2 = \frac{4\rho}{3}$

$$M_y = \rho\int_0^2 \left(\frac{2y-y^2}{2}\right)(2y-y^2)\,dy = \frac{\rho}{2}\left[\frac{4y^3}{3} - y^4 + \frac{y^5}{5}\right]_0^2 = \frac{8\rho}{15}$$

$$\bar{x} = \frac{8\rho}{15}\left(\frac{3}{4\rho}\right) = \frac{2}{5}$$

$$M_x = \rho\int_0^2 y(2y-y^2)\,dy = \rho\left[\frac{2y^3}{3} - \frac{y^4}{4}\right]_0^2 = \frac{4\rho}{3}$$

$$\bar{y} = \frac{4\rho}{3}\left(\frac{3}{4\rho}\right) = 1$$

$$(\bar{x}, \bar{y}) = \left(\frac{2}{5}, 1\right)$$

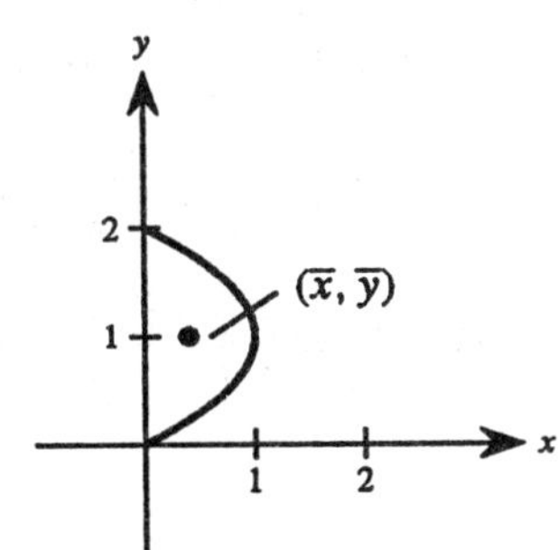

23. $$m = \rho \int_0^3 [(2y - y^2) - (-y)]\,dy = \rho\left[\frac{3y^2}{2} - \frac{y^3}{3}\right]_0^3 = \frac{9\rho}{2}$$

$$M_y = \rho \int_0^3 \frac{[(2y - y^2) + (-y)]}{2}[(2y - y^2) - (-y)]\,dy$$

$$= \frac{\rho}{2}\int_0^3 (y - y^2)(3y - y^2)\,dy = \frac{\rho}{2}\int_0^3 (y^4 - 4y^3 + 3y^2)\,dy$$

$$= \frac{\rho}{2}\left[\frac{y^5}{5} - y^4 + y^3\right]_0^3 = -\frac{27\rho}{10}$$

$$\bar{x} = \frac{M_y}{m} = -\frac{27\rho}{10}\left(\frac{2}{9\rho}\right) = -\frac{3}{5}$$

$$M_x = \rho \int_0^3 y[(2y - y^2) - (-y)]\,dy = \rho \int_0^3 (3y^2 - y^3)\,dy = \rho\left[y^3 - \frac{y^4}{4}\right]_0^3 = \frac{27\rho}{4}$$

$$\bar{y} = \frac{M_x}{m} = \frac{27\rho}{4}\left(\frac{2}{9\rho}\right) = \frac{3}{2}$$

$$(\bar{x}, \bar{y}) = \left(-\frac{3}{5}, \frac{3}{2}\right)$$

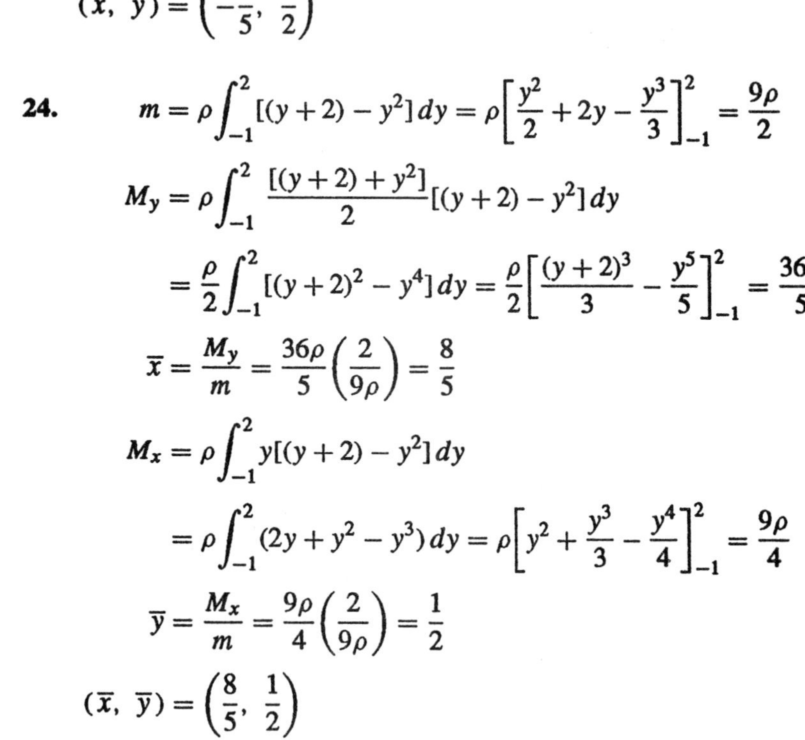

24. $$m = \rho \int_{-1}^2 [(y + 2) - y^2]\,dy = \rho\left[\frac{y^2}{2} + 2y - \frac{y^3}{3}\right]_{-1}^2 = \frac{9\rho}{2}$$

$$M_y = \rho \int_{-1}^2 \frac{[(y + 2) + y^2]}{2}[(y + 2) - y^2]\,dy$$

$$= \frac{\rho}{2}\int_{-1}^2 [(y + 2)^2 - y^4]\,dy = \frac{\rho}{2}\left[\frac{(y + 2)^3}{3} - \frac{y^5}{5}\right]_{-1}^2 = \frac{36\rho}{5}$$

$$\bar{x} = \frac{M_y}{m} = \frac{36\rho}{5}\left(\frac{2}{9\rho}\right) = \frac{8}{5}$$

$$M_x = \rho \int_{-1}^2 y[(y + 2) - y^2]\,dy$$

$$= \rho \int_{-1}^2 (2y + y^2 - y^3)\,dy = \rho\left[y^2 + \frac{y^3}{3} - \frac{y^4}{4}\right]_{-1}^2 = \frac{9\rho}{4}$$

$$\bar{y} = \frac{M_x}{m} = \frac{9\rho}{4}\left(\frac{2}{9\rho}\right) = \frac{1}{2}$$

$$(\bar{x}, \bar{y}) = \left(\frac{8}{5}, \frac{1}{2}\right)$$

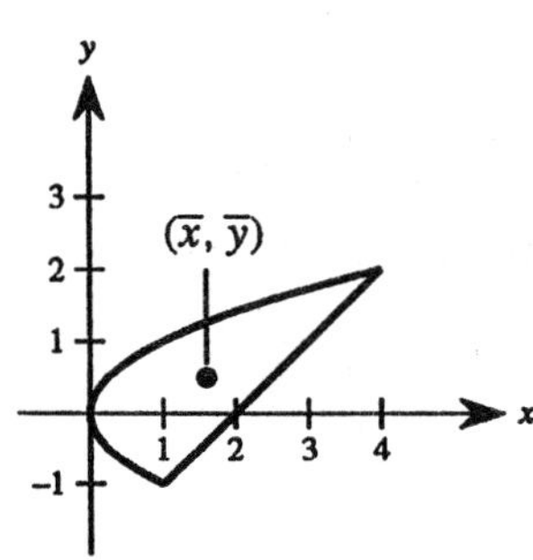

25. $m = \rho\int_0^8 x^{2/3}\,dx = \rho\left[\frac{3}{5}x^{5/3}\right]_0^8 = \frac{96\rho}{5}$

$M_x = \rho\int_0^8 \frac{x^{2/3}}{2}(x^{2/3})\,dx = \frac{\rho}{2}\left[\frac{3}{7}x^{7/3}\right]_0^8 = \frac{192\rho}{7}$

$\bar{y} = \frac{192\rho}{7}\left(\frac{5}{96\rho}\right) = \frac{10}{7}$

$M_y = \rho\int_0^8 x(x^{2/3})\,dx = \rho\left[\frac{3}{8}x^{8/3}\right]_0^8 = 96\rho$

$\bar{x} = 96\rho\left(\frac{5}{96\rho}\right) = 5$

$(\bar{x},\ \bar{y}) = \left(5,\ \frac{10}{7}\right)$

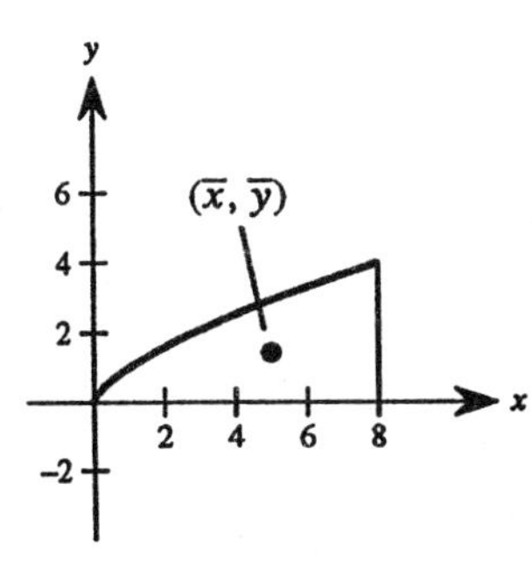

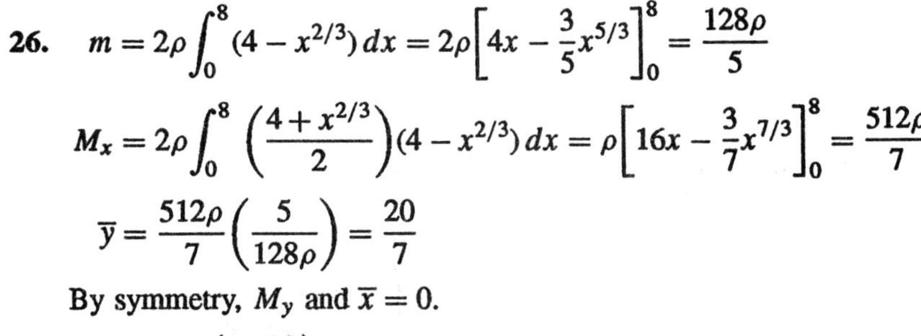

26. $m = 2\rho\int_0^8 (4 - x^{2/3})\,dx = 2\rho\left[4x - \frac{3}{5}x^{5/3}\right]_0^8 = \frac{128\rho}{5}$

$M_x = 2\rho\int_0^8 \left(\frac{4 + x^{2/3}}{2}\right)(4 - x^{2/3})\,dx = \rho\left[16x - \frac{3}{7}x^{7/3}\right]_0^8 = \frac{512\rho}{7}$

$\bar{y} = \frac{512\rho}{7}\left(\frac{5}{128\rho}\right) = \frac{20}{7}$

By symmetry, M_y and $\bar{x} = 0$.

$(\bar{x},\ \bar{y}) = \left(0,\ \frac{20}{7}\right)$

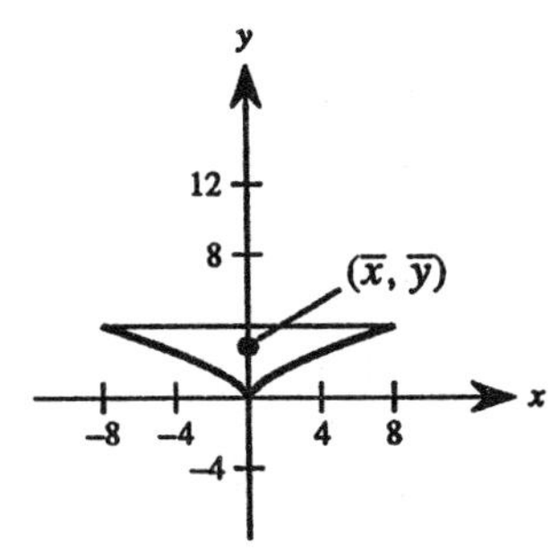

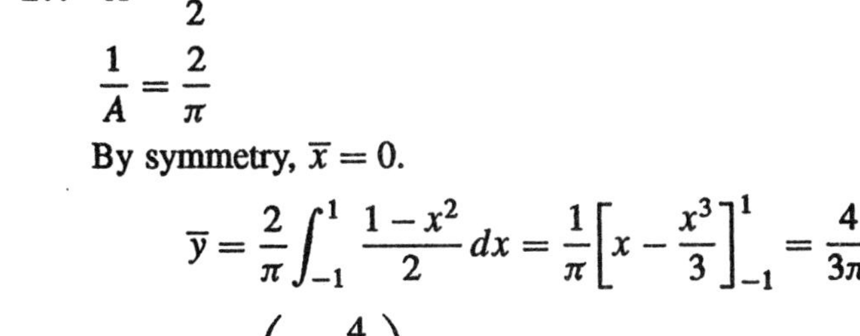

27. $A = \frac{\pi}{2}$

$\frac{1}{A} = \frac{2}{\pi}$

By symmetry, $\bar{x} = 0$.

$\bar{y} = \frac{2}{\pi}\int_{-1}^1 \frac{1 - x^2}{2}\,dx = \frac{1}{\pi}\left[x - \frac{x^3}{3}\right]_{-1}^1 = \frac{4}{3\pi}$

$(\bar{x},\ \bar{y}) = \left(0,\ \frac{4}{3\pi}\right)$

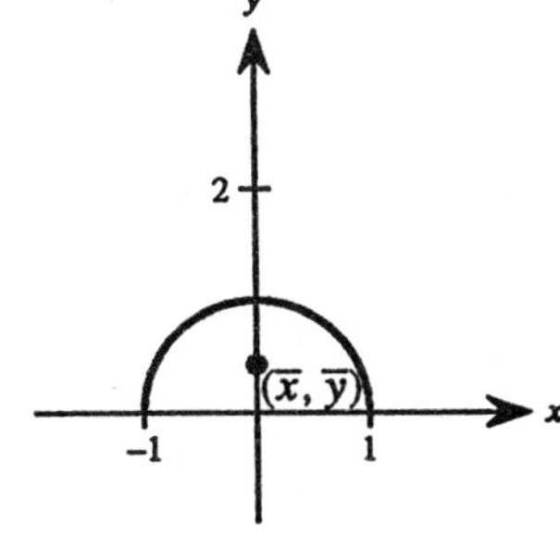

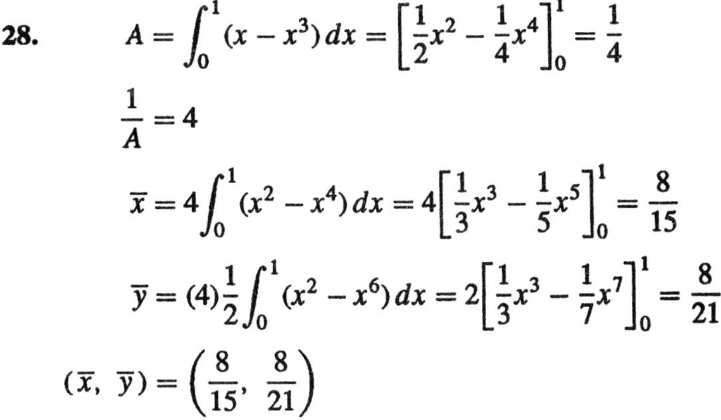

28. $A = \int_0^1 (x - x^3)\,dx = \left[\frac{1}{2}x^2 - \frac{1}{4}x^4\right]_0^1 = \frac{1}{4}$

$\frac{1}{A} = 4$

$\bar{x} = 4\int_0^1 (x^2 - x^4)\,dx = 4\left[\frac{1}{3}x^3 - \frac{1}{5}x^5\right]_0^1 = \frac{8}{15}$

$\bar{y} = (4)\frac{1}{2}\int_0^1 (x^2 - x^6)\,dx = 2\left[\frac{1}{3}x^3 - \frac{1}{7}x^7\right]_0^1 = \frac{8}{21}$

$(\bar{x},\ \bar{y}) = \left(\frac{8}{15},\ \frac{8}{21}\right)$

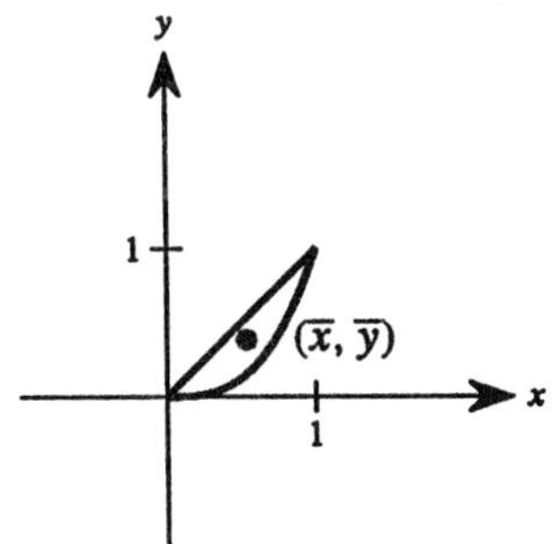

29. $$A = \int_0^1 (x - x^2)\,dx = \left[\frac{1}{2}x^2 - \frac{1}{3}x^3\right]_0^1 = \frac{1}{6}$$

$$\frac{1}{A} = 6$$

$$\bar{x} = 6\int_0^1 (x^2 - x^3)\,dx = 6\left[\frac{1}{3}x^3 - \frac{1}{4}x^4\right]_0^1 = \frac{1}{2}$$

$$\bar{y} = (6)\frac{1}{2}\int_0^1 (x^2 - x^4)\,dx = 3\left[\frac{1}{3}x^3 - \frac{1}{5}x^5\right]_0^1 = \frac{2}{5}$$

$$(\bar{x}, \bar{y}) = \left(\frac{1}{2}, \frac{2}{5}\right)$$

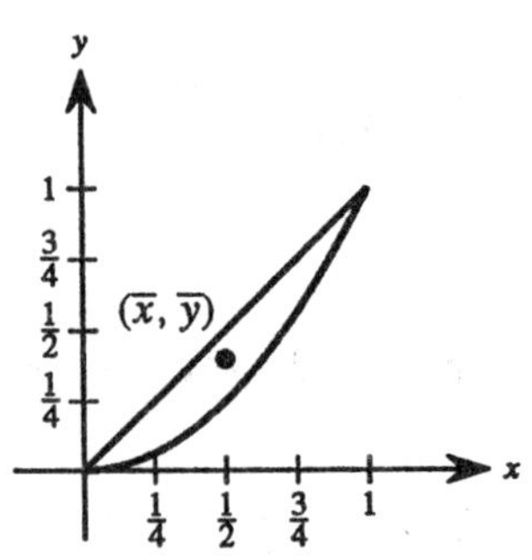

30.

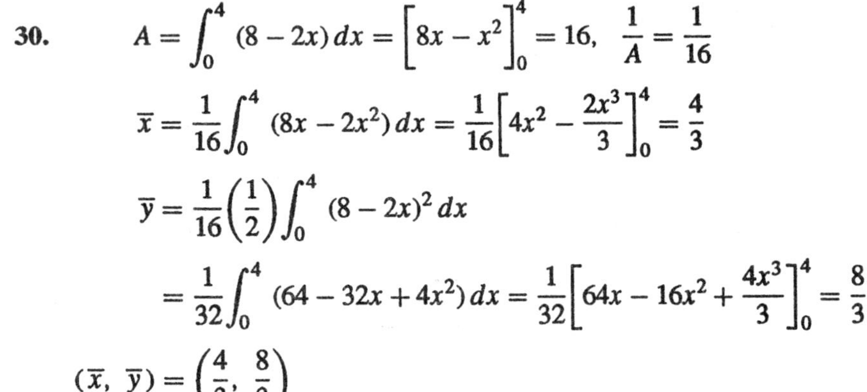

$$A = \int_0^4 (8 - 2x)\,dx = \left[8x - x^2\right]_0^4 = 16, \quad \frac{1}{A} = \frac{1}{16}$$

$$\bar{x} = \frac{1}{16}\int_0^4 (8x - 2x^2)\,dx = \frac{1}{16}\left[4x^2 - \frac{2x^3}{3}\right]_0^4 = \frac{4}{3}$$

$$\bar{y} = \frac{1}{16}\left(\frac{1}{2}\right)\int_0^4 (8 - 2x)^2\,dx$$

$$= \frac{1}{32}\int_0^4 (64 - 32x + 4x^2)\,dx = \frac{1}{32}\left[64x - 16x^2 + \frac{4x^3}{3}\right]_0^4 = \frac{8}{3}$$

$$(\bar{x}, \bar{y}) = \left(\frac{4}{3}, \frac{8}{3}\right)$$

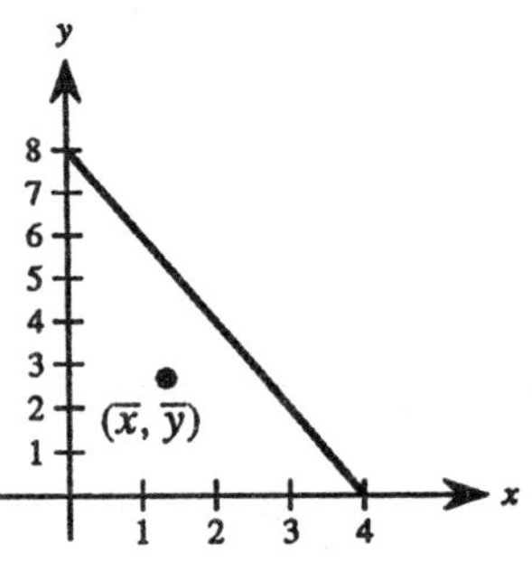

31.

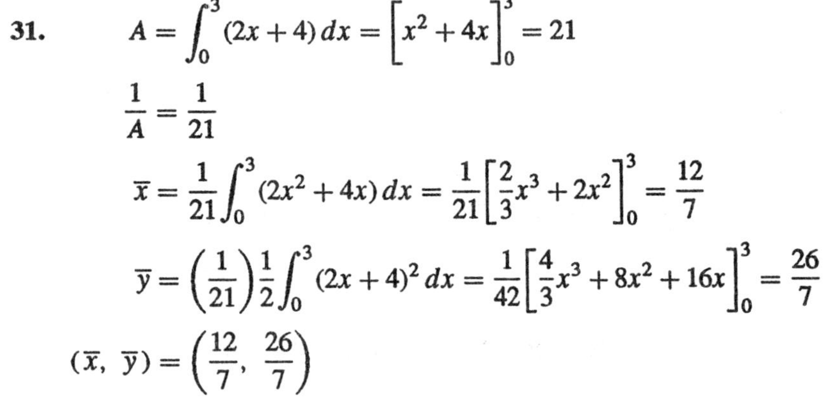

$$A = \int_0^3 (2x + 4)\,dx = \left[x^2 + 4x\right]_0^3 = 21$$

$$\frac{1}{A} = \frac{1}{21}$$

$$\bar{x} = \frac{1}{21}\int_0^3 (2x^2 + 4x)\,dx = \frac{1}{21}\left[\frac{2}{3}x^3 + 2x^2\right]_0^3 = \frac{12}{7}$$

$$\bar{y} = \left(\frac{1}{21}\right)\frac{1}{2}\int_0^3 (2x + 4)^2\,dx = \frac{1}{42}\left[\frac{4}{3}x^3 + 8x^2 + 16x\right]_0^3 = \frac{26}{7}$$

$$(\bar{x}, \bar{y}) = \left(\frac{12}{7}, \frac{26}{7}\right)$$

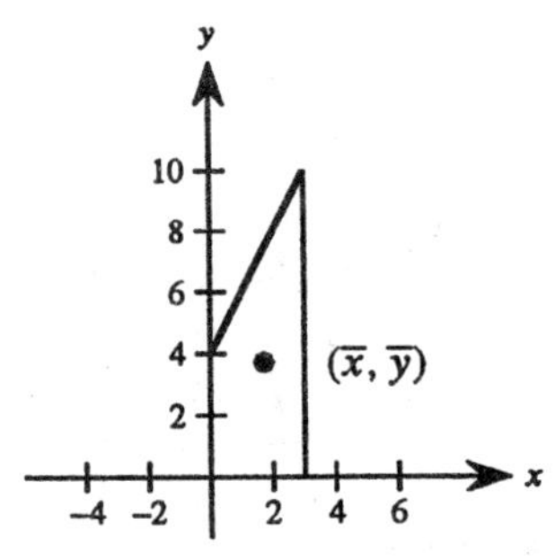

32. $A = \int_{-2}^{2} (4 - x^2)\,dx = \left[4x - \frac{x^3}{3}\right]_{-2}^{2} = \frac{32}{3}$

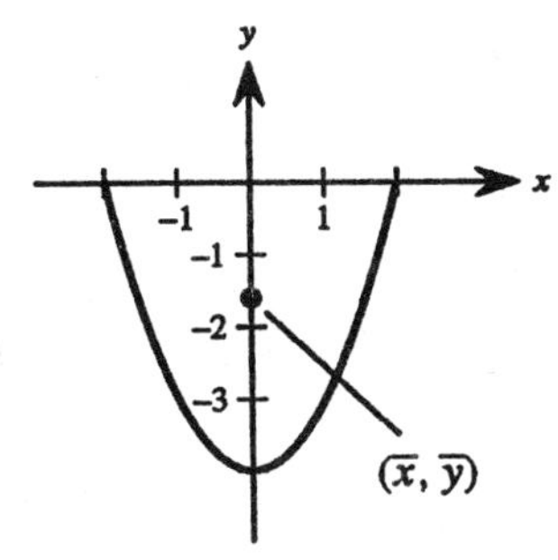

$\frac{1}{A} = \frac{3}{32}$

$$\bar{x} = \frac{3}{32}\int_{-2}^{2} x(4 - x^2)\,dx = \frac{3}{32}(0) = 0$$

$$\bar{y} = \left(\frac{3}{32}\right)\frac{1}{2}\int_{-2}^{2} (x^2 - 4)(4 - x^2)\,dx$$

$$= -\frac{3}{64}\int_{-2}^{2} (x^4 - 8x^2 + 16)\,dx$$

$$= -\frac{3}{64}\left[\frac{x^5}{5} - \frac{8}{3}x^3 + 16x\right]_{-2}^{2} = -\frac{3}{64}\left(\frac{512}{15}\right) = -\frac{8}{5}$$

$$(\bar{x}, \bar{y}) = \left(0, -\frac{8}{5}\right)$$

33. $A = \frac{1}{2}(2a)c = ac$

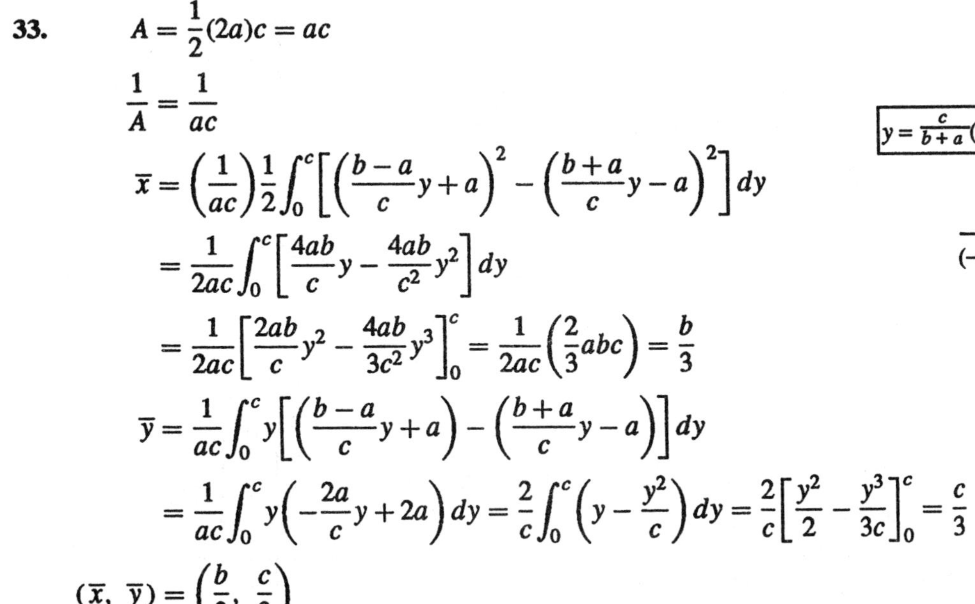

$\frac{1}{A} = \frac{1}{ac}$

$$\bar{x} = \left(\frac{1}{ac}\right)\frac{1}{2}\int_0^c \left[\left(\frac{b-a}{c}y + a\right)^2 - \left(\frac{b+a}{c}y - a\right)^2\right]dy$$

$$= \frac{1}{2ac}\int_0^c \left[\frac{4ab}{c}y - \frac{4ab}{c^2}y^2\right]dy$$

$$= \frac{1}{2ac}\left[\frac{2ab}{c}y^2 - \frac{4ab}{3c^2}y^3\right]_0^c = \frac{1}{2ac}\left(\frac{2}{3}abc\right) = \frac{b}{3}$$

$$\bar{y} = \frac{1}{ac}\int_0^c y\left[\left(\frac{b-a}{c}y + a\right) - \left(\frac{b+a}{c}y - a\right)\right]dy$$

$$= \frac{1}{ac}\int_0^c y\left(-\frac{2a}{c}y + 2a\right)dy = \frac{2}{c}\int_0^c \left(y - \frac{y^2}{c}\right)dy = \frac{2}{c}\left[\frac{y^2}{2} - \frac{y^3}{3c}\right]_0^c = \frac{c}{3}$$

$$(\bar{x}, \bar{y}) = \left(\frac{b}{3}, \frac{c}{3}\right)$$

In Exercise 58 of Section 1.4, we showed that $(b/3,\ c/3)$ is the point of intersection of the medians.

34. $A = \int_0^1 [1 - (2x - x^2)]\,dx = \frac{1}{3}$

$\frac{1}{A} = 3$

$$\bar{x} = 3\int_0^1 x[1 - (2x - x^2)]\,dx = 3\int_0^1 [x - 2x^2 + x^3]\,dx = 3\left[\frac{x^2}{2} - \frac{2}{3}x^3 + \frac{x^4}{4}\right]_0^1 = \frac{1}{4}$$

$$\bar{y} = 3\int_0^1 \frac{[1 + (2x - x^2)]}{2}[1 - (2x - x^2)]\,dx$$

$$= \frac{3}{2}\int_0^1 [1 - (2x - x^2)^2]\,dx = \frac{3}{2}\int_0^1 [1 - 4x^2 + 4x^3 - x^4]\,dx = \frac{3}{2}\left[x - \frac{4}{3}x^3 + x^4 - \frac{x^5}{5}\right]_0^1 = \frac{7}{10}$$

$$(\bar{x}, \bar{y}) = \left(\frac{1}{4}, \frac{7}{10}\right)$$

35. $A = \frac{c}{2}(a+b)$

$$\frac{1}{A} = \frac{2}{c(a+b)}$$

$$\bar{x} = \frac{2}{c(a+b)}\int_0^c x\left(\frac{b-a}{c}x+a\right)dx = \frac{2}{c(a+b)}\int_0^c \left(\frac{b-a}{c}x^2+ax\right)dx = \frac{2}{c(a+b)}\left[\frac{b-a}{c}\frac{x^3}{3}+\frac{ax^2}{2}\right]_0^c$$

$$= \frac{2}{c(a+b)}\left[\frac{(b-a)c^2}{3}+\frac{ac^2}{2}\right] = \frac{2}{c(a+b)}\left[\frac{2bc^2-2ac^2+3ac^2}{6}\right] = \frac{c(2b+a)}{3(a+b)} = \frac{(a+2b)c}{3(a+b)}$$

$$\bar{y} = \frac{2}{c(a+b)}\frac{1}{2}\int_0^c \left(\frac{b-a}{c}x+a\right)^2 dx = \frac{1}{c(a+b)}\int_0^c \left[\left(\frac{b-a}{c}\right)^2x^2+\frac{2a(b-a)}{c}x+a^2\right]dx$$

$$= \frac{1}{c(a+b)}\left[\left(\frac{b-a}{c}\right)^2\frac{x^3}{3}+\frac{2a(b-a)}{c}\frac{x^2}{2}+a^2x\right]_0^c = \frac{1}{c(a+b)}\left[\frac{(b-a)^2c}{3}+ac(b-a)+a^2c\right]$$

$$= \frac{1}{3c(a+b)}[(b^2-2ab+a^2)c+3ac(b-a)+3a^2c]$$

$$= \frac{1}{3(a+b)}[b^2-2ab+a^2+3ab-3a^2+3a^2] = \frac{a^2+ab+b^2}{3(a+b)}$$

Thus,

$$(\bar{x}, \bar{y}) = \left(\frac{(a+2b)c}{3(a+b)}, \frac{a^2+ab+b^2}{3(a+b)}\right).$$

The one line passes through $(0, a/2)$ and $(c, b/2)$. It's equation is

$$y = \frac{b-a}{2c}x+\frac{a}{2}.$$

The other line passes through $(0, -b)$ and $(c, a+b)$. It's equation is

$$y = \frac{a+2b}{c}x-b.$$

$(\bar{x}, \bar{y})$ is the point of intersection of these two lines.

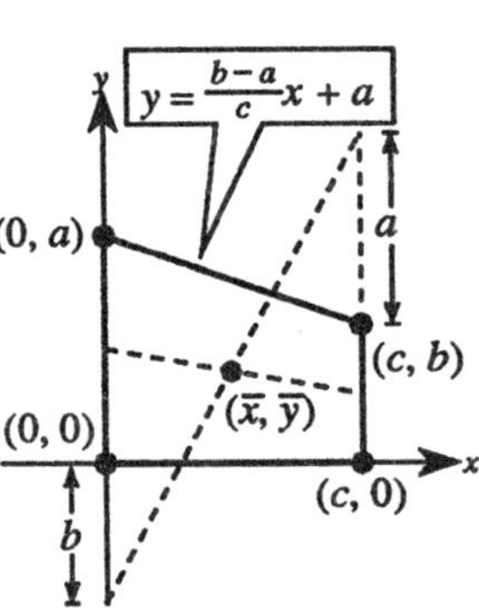

36. $\bar{x} = 0$ by symmetry. Using Simpson's Rule to approximate the area and M_x, we have:

$$A \approx \frac{40-(-40)}{3(8)}(2)[30+4(29)+2(26)+4(20)+0] = \frac{5560}{3}$$

$$m = \rho A = \frac{5560\rho}{3}$$

$$M_x = \rho\int_{-40}^{40}\frac{1}{2}[f(x)]^2$$

$$\approx \rho\left(\frac{40-(-40)}{3(8)}\right)(2)\left(\frac{1}{2}\right)[(30)^2+4(29)^2+2(26)^2+4(20)^2+0] = \frac{72{,}160\rho}{3}$$

$$\bar{y} = \frac{M_x}{m} = \frac{72{,}160\rho}{3}\cdot\frac{3}{5560\rho} \approx 12.98$$

Center of mass: $(\bar{x}, \bar{y}) = (0, 12.98)$

37. $m = \rho \int_{-20}^{20} 5\sqrt[3]{400 - x^2}\,dx \approx 1210.0\rho$

$$M_x = \rho \int_{-20}^{20} \frac{5\sqrt[3]{400 - x^2}}{2}\left(5\sqrt[3]{400 - x^2}\right)dx$$

$$= \frac{25\rho}{2}\int_{-20}^{20} (400 - x^2)^{2/3}\,dx$$

$$\approx 19936\rho$$

$$\bar{y} = \frac{M_x}{m} \approx 16.5 \text{ ft}$$

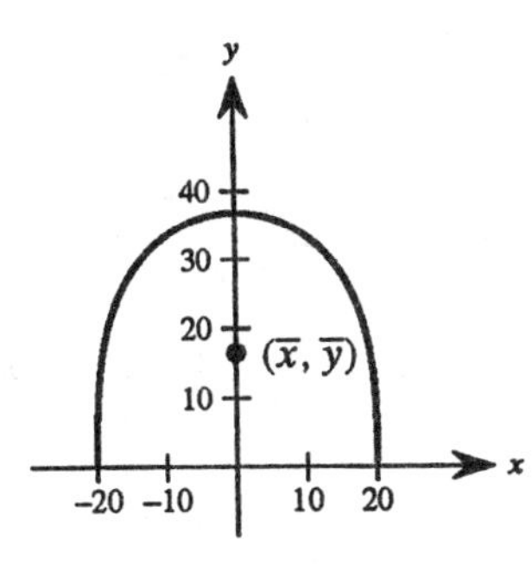

38. $m = \rho \int_{-2}^{2} \frac{8}{x^2 + 4}\,dx \approx 6.2831\rho$

$$M_x = \rho \int_{-2}^{2} \frac{1}{2}\left(\frac{8}{x^2 + 4}\right)\left(\frac{8}{x^2 + 4}\right)dx$$

$$= 32\rho \int_{-2}^{2} \frac{1}{(x^2 + 4)^2}\,dx$$

$$\approx 5.1411\rho$$

$$\bar{y} = \frac{M_x}{m} \approx 0.818$$

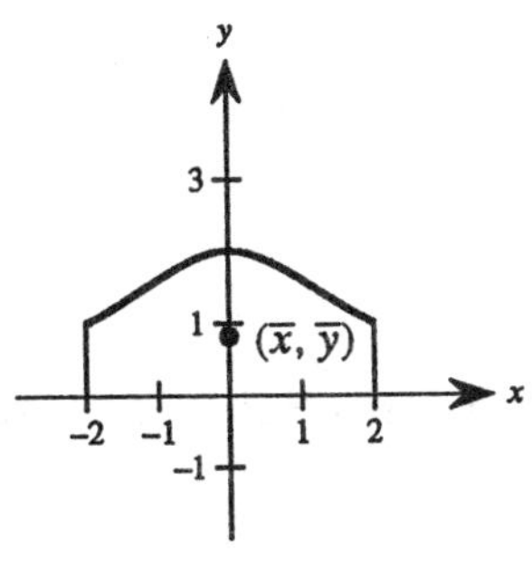

39. $V = 2\pi r A = 2\pi(5)(16\pi) = 160\pi^2 \approx 1579.14$

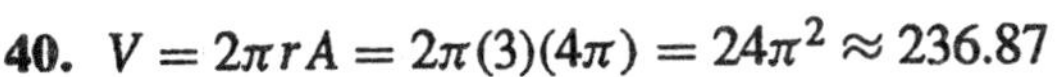
40. $V = 2\pi r A = 2\pi(3)(4\pi) = 24\pi^2 \approx 236.87$

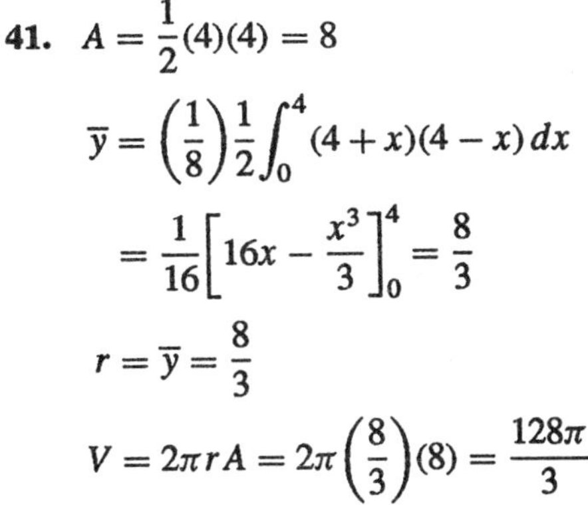
41. $A = \frac{1}{2}(4)(4) = 8$

$$\bar{y} = \left(\frac{1}{8}\right)\frac{1}{2}\int_0^4 (4 + x)(4 - x)\,dx$$

$$= \frac{1}{16}\left[16x - \frac{x^3}{3}\right]_0^4 = \frac{8}{3}$$

$$r = \bar{y} = \frac{8}{3}$$

$$V = 2\pi r A = 2\pi\left(\frac{8}{3}\right)(8) = \frac{128\pi}{3}$$

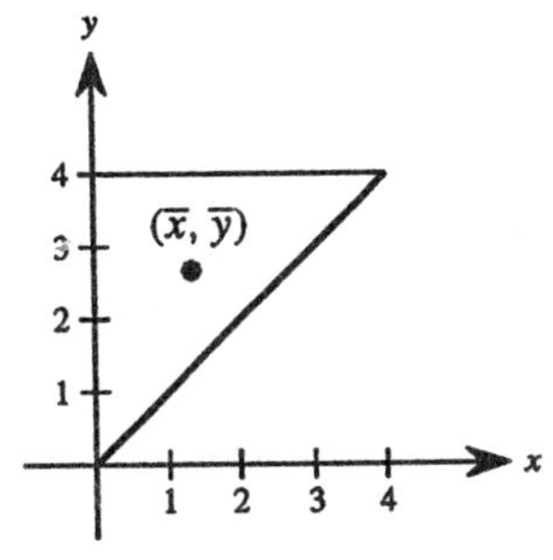

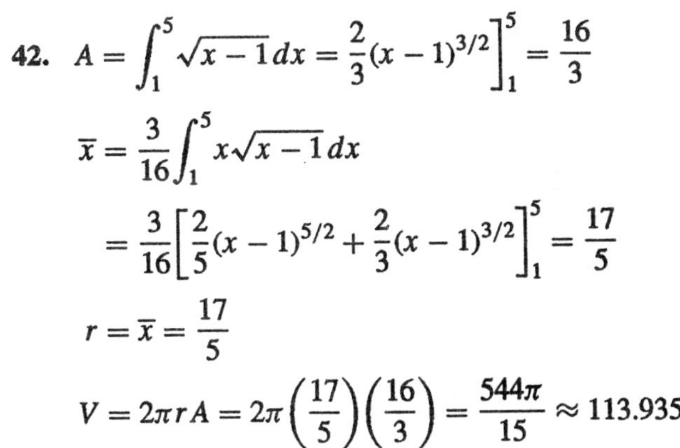
42. $A = \int_1^5 \sqrt{x - 1}\,dx = \frac{2}{3}(x - 1)^{3/2}\Big]_1^5 = \frac{16}{3}$

$$\bar{x} = \frac{3}{16}\int_1^5 x\sqrt{x - 1}\,dx$$

$$= \frac{3}{16}\left[\frac{2}{5}(x - 1)^{5/2} + \frac{2}{3}(x - 1)^{3/2}\right]_1^5 = \frac{17}{5}$$

$$r = \bar{x} = \frac{17}{5}$$

$$V = 2\pi r A = 2\pi\left(\frac{17}{5}\right)\left(\frac{16}{3}\right) = \frac{544\pi}{15} \approx 113.9351$$

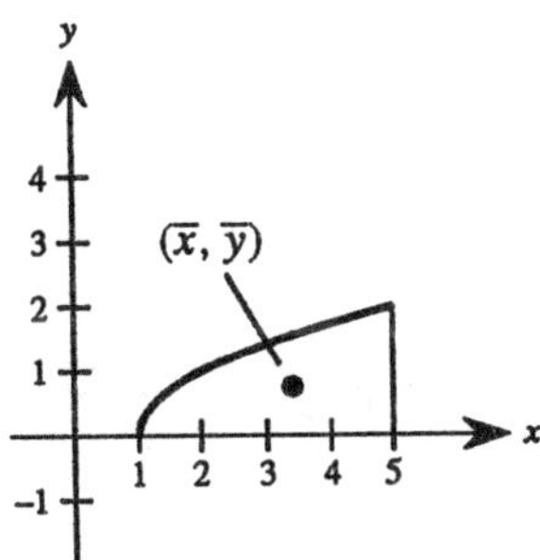

Chapter 6 Review Exercises

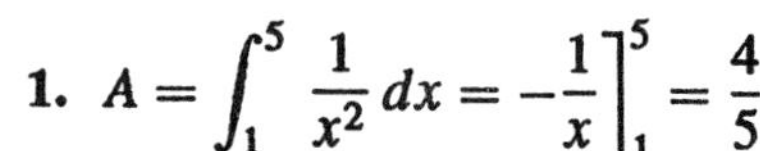

1. $A = \int_1^5 \frac{1}{x^2}\,dx = -\frac{1}{x}\Big]_1^5 = \frac{4}{5}$

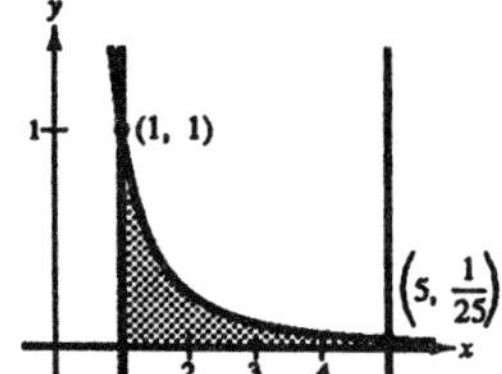

2. $A = \int_{1/2}^5 \left(4 - \frac{1}{x^2}\right) dx = \left[4x + \frac{1}{x}\right]_{1/2}^5 = \frac{81}{5}$

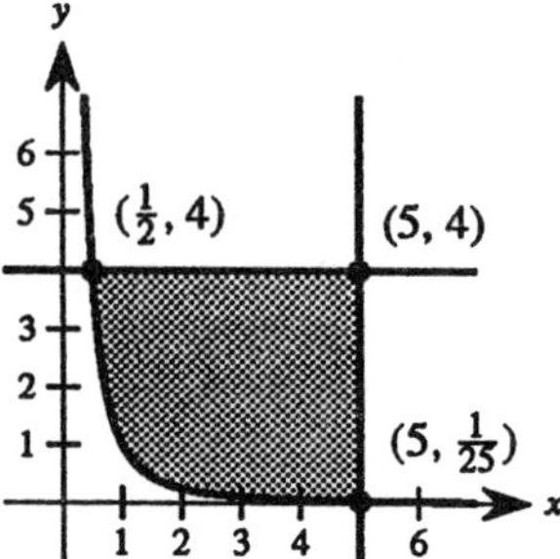

3. $A = \int_0^1 \frac{x}{(x^2+1)^2}\,dx$

$= -\frac{1}{2(x^2+1)}\Big]_0^1$

$= -\frac{1}{4} + \frac{1}{2}$

$= \frac{1}{4}$

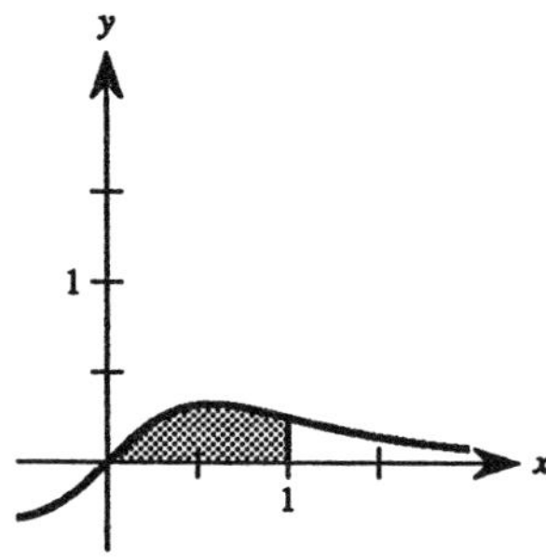

4. $y = 1 - \frac{x}{2} \Rightarrow x = 2 - 2y$

$y = x - 2 \Rightarrow x = y + 2, \quad y = 1$

$A = \int_0^1 [(y+2) - (2-2y)]\,dy$

$= \int_0^1 3y\,dy$

$= \frac{3}{2}y^2\Big]_0^1 = \frac{3}{2}$

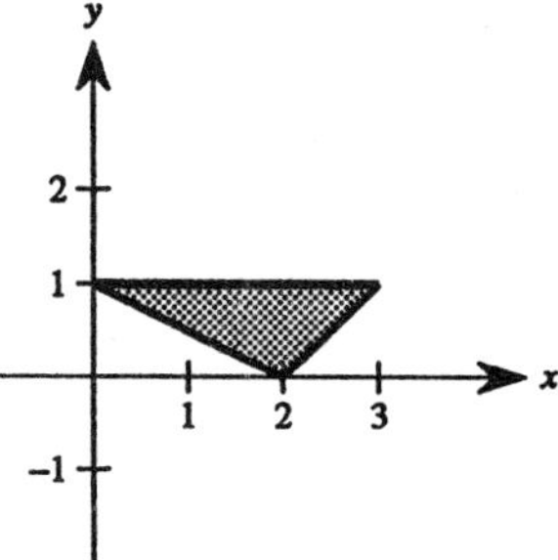

5. $A = \int_0^2 [0 - (y^2 - 2y)]\,dy$

$= \int_0^2 (2y - y^2)\,dy$

$= \left[y^2 - \frac{1}{3}y^3\right]_0^2$

$= \frac{4}{3}$

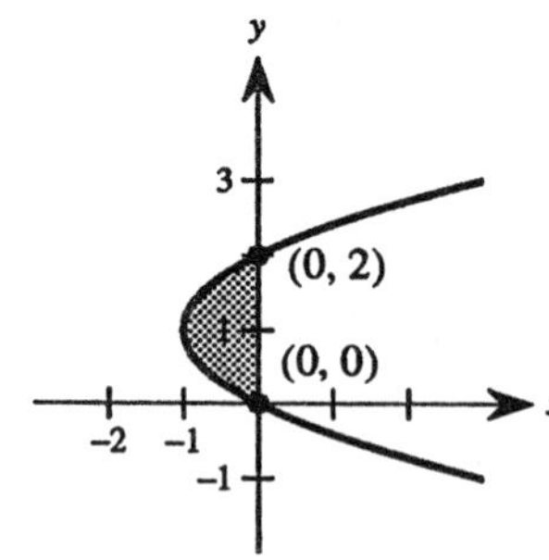

6. $A = \int_0^1 [(y^2 - 2y) - (-1)]\,dy$

$= \int_0^1 (y^2 - 2y + 1)\,dy$

$= \int_0^1 (y-1)^2\,dy$

$= \left.\frac{(y-1)^3}{3}\right]_0^1 = \frac{1}{3}$

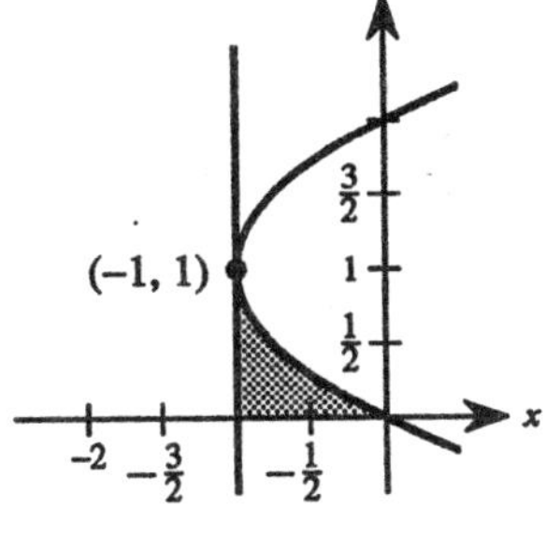

7. $A = 2\int_0^1 (x - x^3)\,dx$

$= 2\left[\frac{1}{2}x^2 - \frac{1}{4}x^4\right]_0^1$

$= \frac{1}{2}$

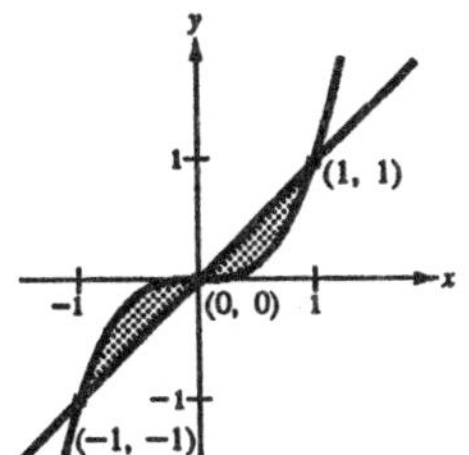

8. $A = \int_{-1}^{2} [(y+3) - (y^2+1)]\,dy$

$= \int_{-1}^{2} (2 + y - y^2)\,dy$

$= \left[2y + \frac{1}{2}y^2 - \frac{1}{3}y^3\right]_{-1}^{2}$

$= \frac{9}{2}$

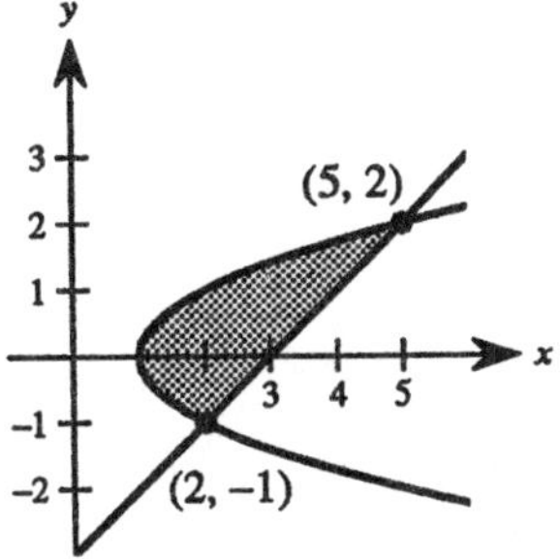

9. $A = \int_0^8 [(3 + 8x - x^2) - (x^2 - 8x + 3)]\,dx$

$= \int_0^8 (16x - 2x^2)\,dx$

$= \left[8x^2 - \frac{2}{3}x^3\right]_0^8$

$= \frac{512}{3}$

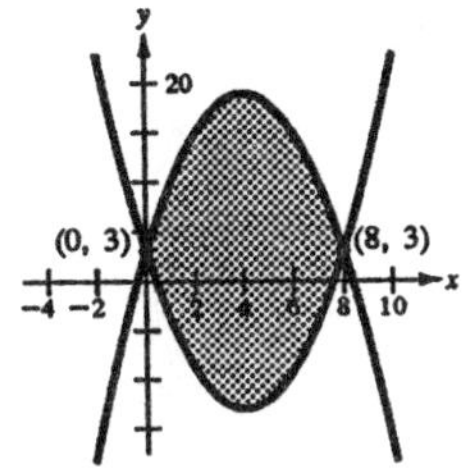

10. Point of intersection is given by:
$x^3 - x^2 + 4x - 3 = 0 \Rightarrow x \approx 0.783.$

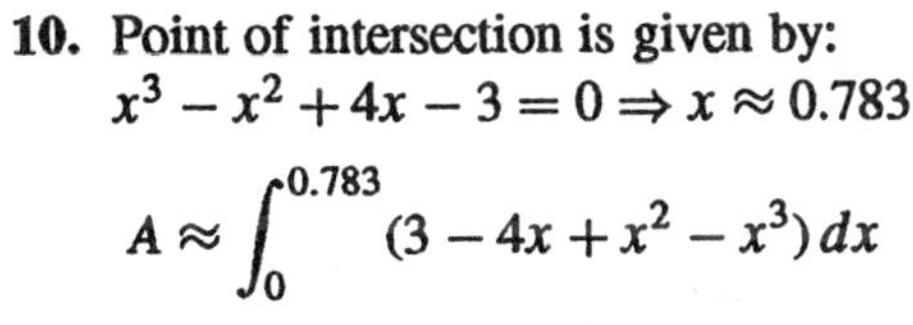

$A \approx \int_0^{0.783} (3 - 4x + x^2 - x^3)\,dx$

$= \left[3x - 2x^2 + \frac{1}{3}x^3 - \frac{1}{4}x^4\right]_0^{0.783}$

≈ 1.189

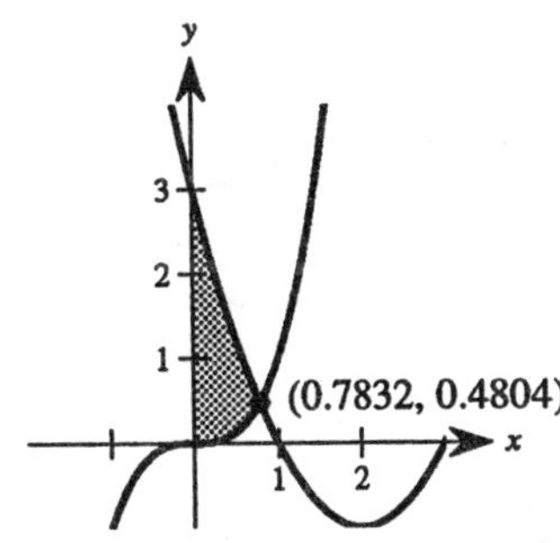

11. $x = y^2 + 1$

$$A = \int_0^2 (y^2 + 1)\,dy = \left[\frac{1}{3}y^3 + y\right]_0^2 = \frac{14}{3}$$

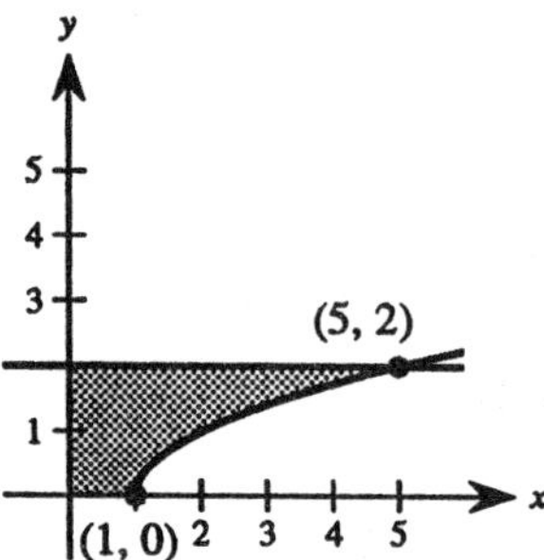

12. $A = \int_1^5 \left[\sqrt{x-1} - \frac{x-1}{2}\right] dx = \left[\frac{2}{3}(x-1)^{3/2} - \frac{1}{4}(x-1)^2\right]_1^5 = \frac{4}{3}$

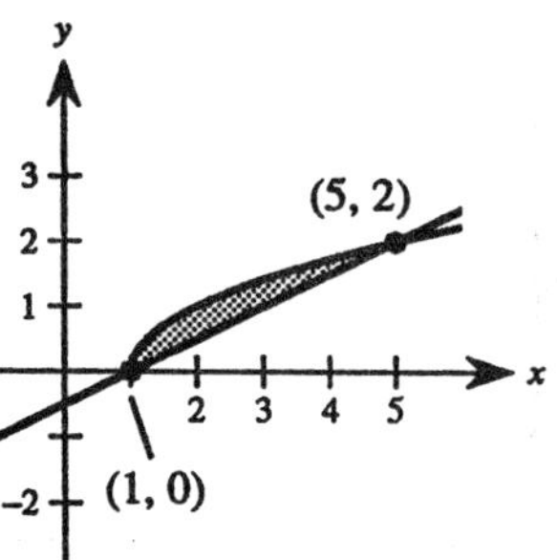

13. $y = (1 - \sqrt{x})^2$

$$A = \int_0^1 (1 - \sqrt{x})^2\,dx = \int_0^1 (1 - 2x^{1/2} + x)\,dx = \left[x - \frac{4}{3}x^{3/2} + \frac{1}{2}x^2\right]_0^1 = \frac{1}{6}$$

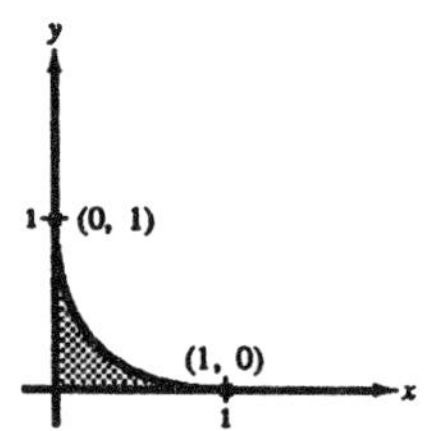

14. $A = 2\int_0^2 [2x^2 - (x^4 - 2x^2)]\,dx = 2\int_0^2 (4x^2 - x^4)\,dx = 2\left[\frac{4}{3}x^3 - \frac{1}{5}x^5\right]_0^2 = \frac{128}{15}$

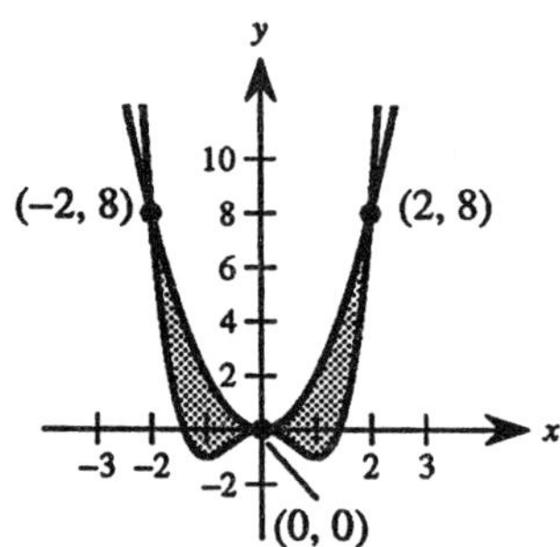

15. **(a) Disc**

$$V = \pi\int_0^4 x^2\,dx = \frac{\pi x^3}{3}\Big]_0^4 = \frac{64\pi}{3}$$

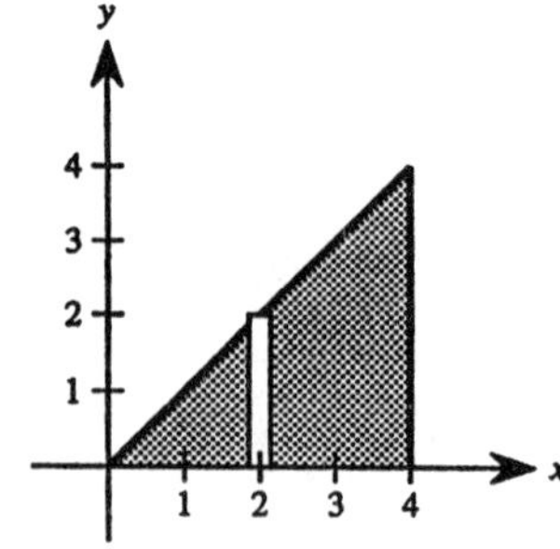

(b) Shell

$$V = 2\pi\int_0^4 x^2\,dx = \frac{2\pi}{3}x^3\Big]_0^4 = \frac{128\pi}{3}$$

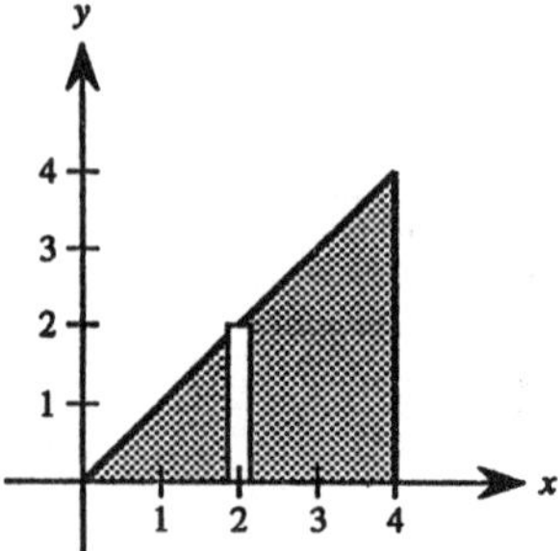

15. —CONTINUED—

(c) **Shell**

$$V = 2\pi \int_0^4 (4 - x)x\,dx$$

$$= 2\pi \int_0^4 (4x - x^2)\,dx$$

$$= 2\pi \left[2x^2 - \frac{x^3}{3}\right]_0^4 = \frac{64\pi}{3}$$

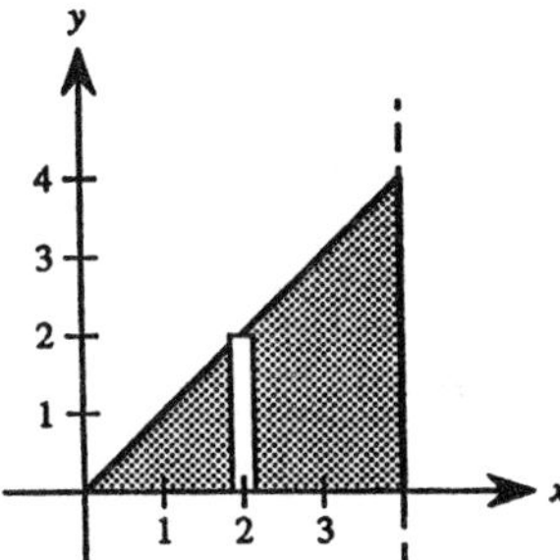

(d) **Shell**

$$V = 2\pi \int_0^4 (6 - x)x\,dx$$

$$= 2\pi \int_0^4 (6x - x^2)\,dx$$

$$= 2\pi \left[3x^2 - \frac{1}{3}x^3\right]_0^4 = \frac{160\pi}{3}$$

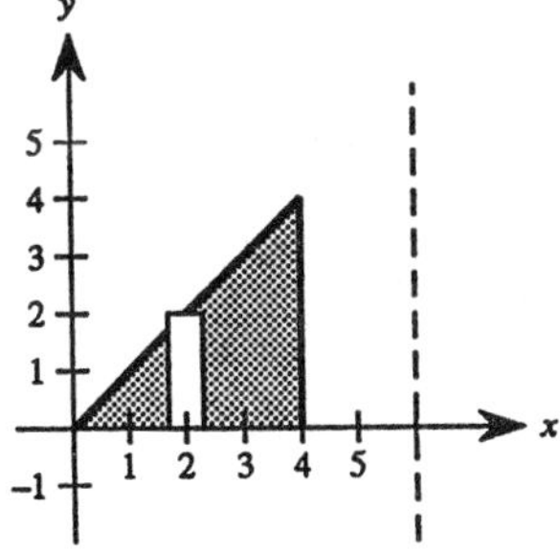

16. (a) **Shell**

$$V = 2\pi \int_0^2 y^3\,dy = \frac{\pi}{2}y^4\Big]_0^2 = 8\pi$$

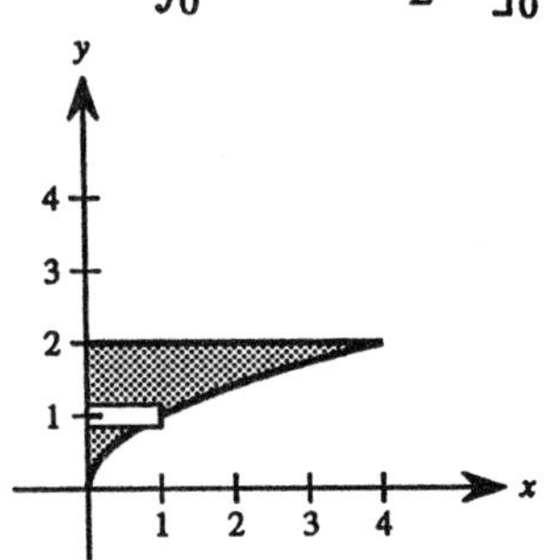

(b) **Shell**

$$V = 2\pi \int_0^2 (2 - y)y^2\,dy$$

$$= 2\pi \int_0^2 (2y^2 - y^3)\,dy$$

$$= 2\pi \left[\frac{2}{3}y^3 - \frac{1}{4}y^4\right]_0^2 = \frac{8\pi}{3}$$

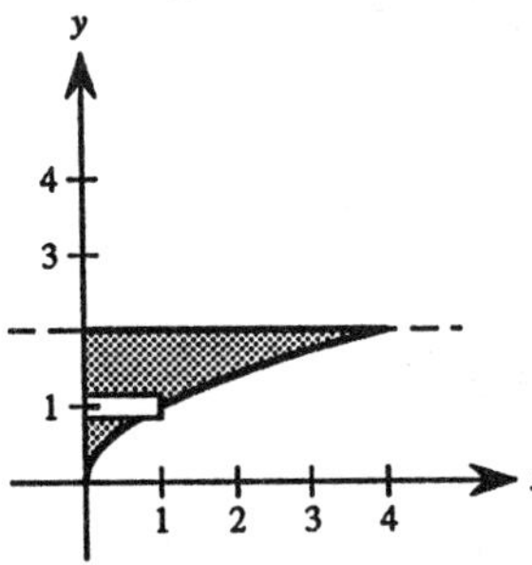

(c) **Disc**

$$V = \pi \int_0^2 y^4\,dy$$

$$= \frac{\pi}{5}y^5\Big]_0^2 = \frac{32\pi}{5}$$

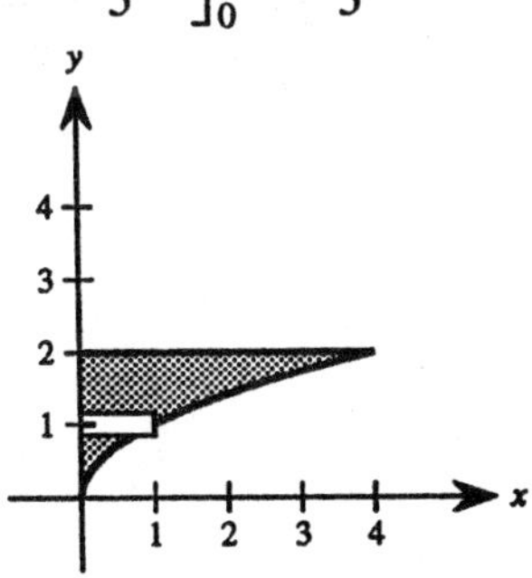

(d) **Disc**

$$V = \pi \int_0^2 [(y^2 + 1)^2 - 1^2]\,dy$$

$$= \pi \int_0^2 (y^4 + 2y^2)\,dy$$

$$= \pi \left[\frac{1}{5}y^5 + \frac{2}{3}y^3\right]_0^2 = \frac{176\pi}{15}$$

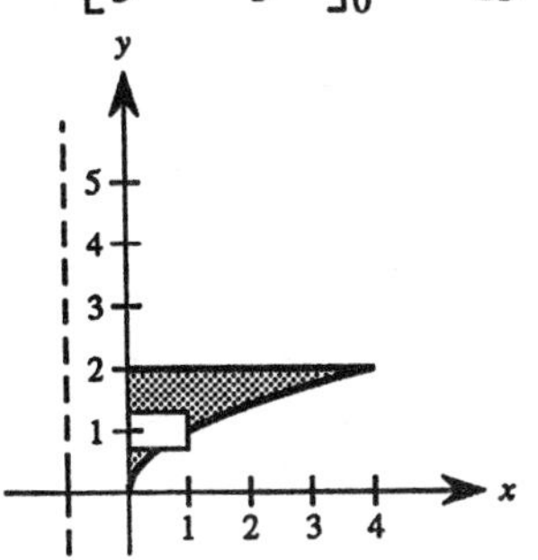

17. From Exercise 18, letting $a = 4$ and $b = 3$ we have:

(a) $V = \frac{4}{3}\pi(4^2)(3) = 64\pi$

(b) $V = \frac{4}{3}\pi(4)(3)^2 = 48\pi$

18. (a) **Shell**

$$V = 4\pi \int_0^a (x)\frac{b}{a}\sqrt{a^2 - x^2}\,dx$$
$$= \frac{-2\pi b}{a}\int_0^a (a^2 - x^2)^{1/2}(-2x)\,dx$$
$$= \frac{-4\pi b}{3a}(a^2 - x^2)^{3/2}\Big]_0^a = \frac{4}{3}\pi a^2 b$$

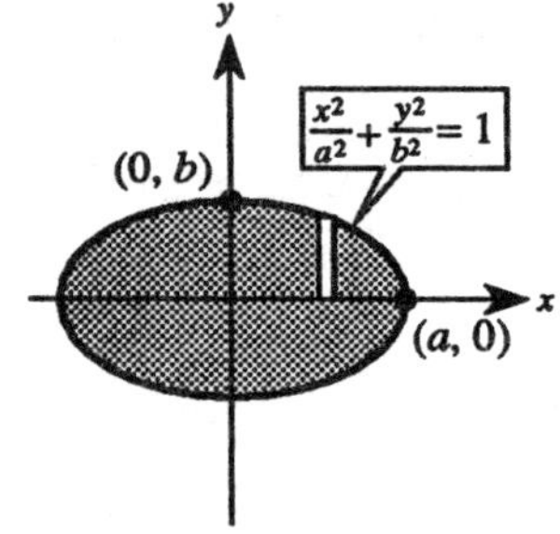

(b) **Disc**

$$V = 2\pi \int_0^a \frac{b^2}{a^2}(a^2 - x^2)\,dx$$
$$= \frac{2\pi b^2}{a^2}\left[a^2x - \frac{1}{3}x^3\right]_0^a$$
$$= \frac{4}{3}\pi a b^2$$

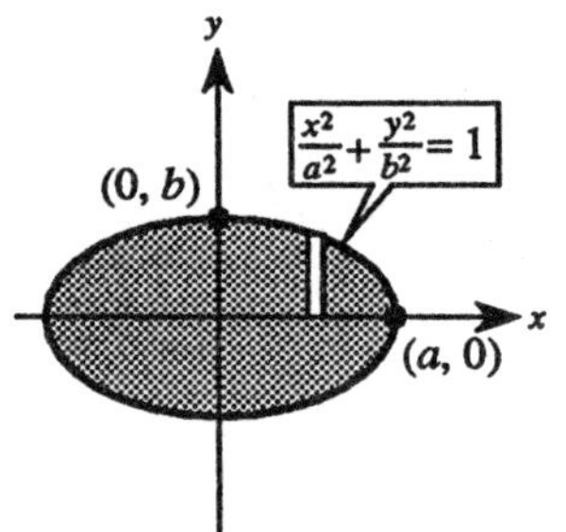

19. $V = 2\pi \int_0^1 \frac{x}{(x^2+1)^2}\,dx$

$$= \pi \int_0^1 (x^2+1)^{-2}2x\,dx$$
$$= -\frac{\pi}{x^2+1}\Big]_0^1$$
$$= \frac{\pi}{2}$$

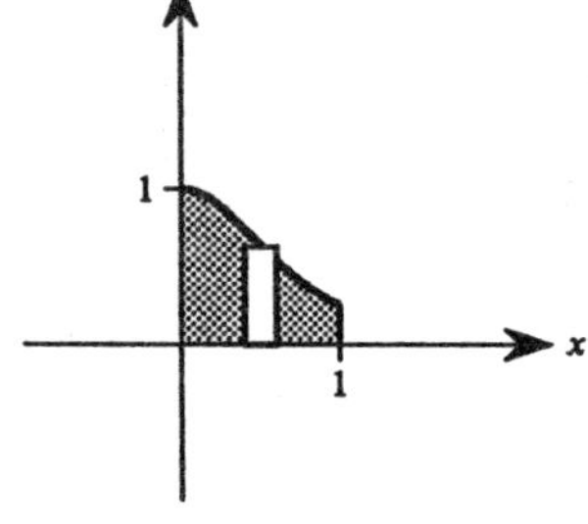

20. $V = \pi \int_{-1}^1 (x+1)^3\,dx$

$$= \frac{\pi(x+1)^4}{4}\Big]_{-1}^1$$
$$= 4\pi$$

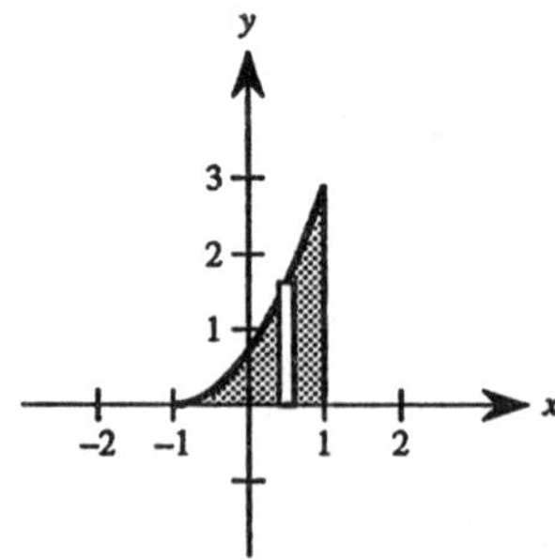

21. (a) **Disc**

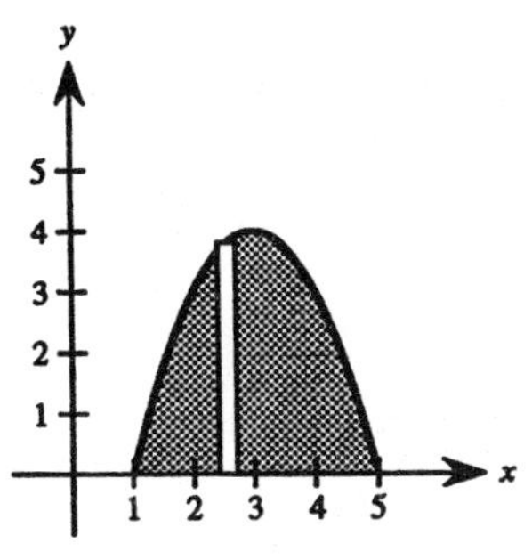

$$V = \pi \int_1^5 (-x^2 + 6x - 5)^2\,dx$$
$$= \pi \int_1^5 (x^4 - 12x^3 + 46x^2 - 60x + 25)\,dx$$
$$= \pi \left[\frac{x^5}{5} - 3x^4 + \frac{46}{3}x^3 - 30x^2 + 25x\right]_1^5$$
$$= \frac{512\pi}{15}$$

(b) **Shell**

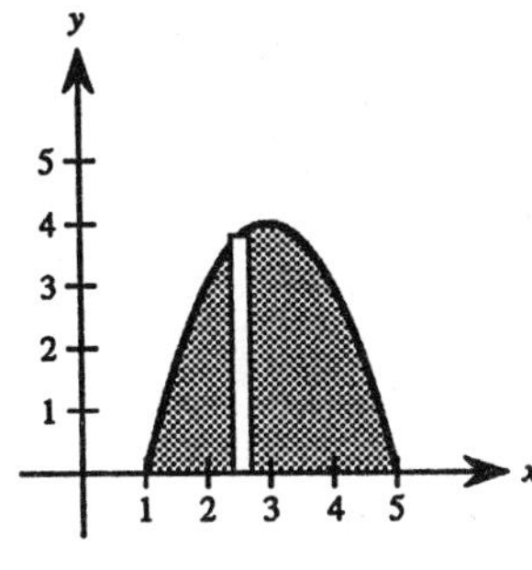

$$V = 2\pi \int_1^5 x(-x^2 + 6x - 5)\,dx$$
$$= 2\pi \int_1^5 (-x^3 + 6x^2 - 5x)\,dx$$
$$= 2\pi \left[-\frac{x^4}{4} + 2x^3 - \frac{5}{2}x^2\right]_1^5$$
$$= 64\pi$$

22. Disc

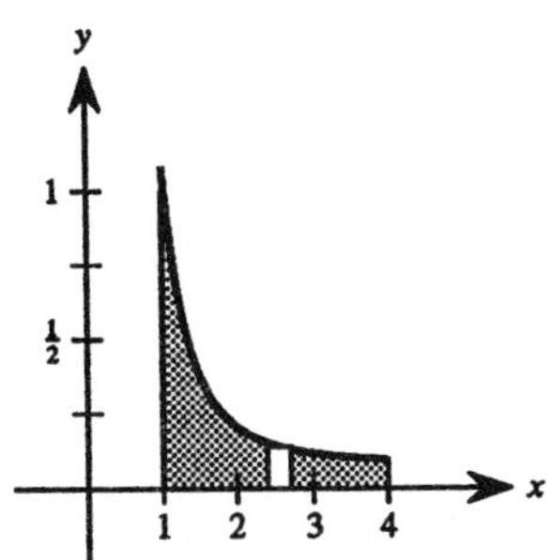

$$V = \pi \int_0^1 (x^3 + 1)^2\,dx$$
$$= \pi \int_0^1 (x^6 + 2x^3 + 1)\,dx$$
$$= \pi \left[\frac{x^7}{7} + \frac{x^4}{2} + x\right]_0^1 = \pi\left[\frac{1}{7} + \frac{1}{2} + 1\right] = \frac{23\pi}{14}$$

23. $F = kx$

$4 = k(1)$

$F = 4x$

$$W = \int_0^5 4x\,dx = 2x^2\Big]_0^5 = 50 \text{ in} \cdot \text{lb}$$

24. $9k = 50$

$$k = \frac{50}{9}$$
$$F = \frac{50}{9}x$$
$$W = \int_0^9 \frac{50}{9}x\,dx = \frac{25}{9}x^2\Big]_0^9 = 225 \text{ in} \cdot \text{lb}$$

25. Volume of disc: $\pi\left(\frac{1}{3}\right)^2 \Delta y$

Weight of disc: $62.4\pi\left(\frac{1}{3}\right)^2 \Delta y$

Distance: $175 - y$

$$W = \frac{62.4\pi}{9}\int_0^{150} (175 - y)\,dy = \frac{62.4\pi}{9}\left[175y - \frac{y^2}{2}\right]_0^{150} = 104{,}000\pi \text{ ft} \cdot \text{lb} = 52\pi \text{ ft} \cdot \text{ton}$$

26. We know that:

$$\frac{dV}{dt} = \frac{4 \text{ gal/min} - 12 \text{ gal/min}}{7.481 \text{ gal/ft}^3} = -\frac{8}{7.481} \text{ ft}^3\text{/min}$$

$$V = \pi r^2 h = \pi\left(\frac{1}{9}\right)h$$

$$\frac{dV}{dt} = \frac{\pi}{9}\left(\frac{dh}{dt}\right)$$

$$\frac{dh}{dt} = \frac{9}{\pi}\left(\frac{dV}{dt}\right) = \frac{9}{\pi}\left(-\frac{8}{7.481}\right) \approx -3.064 \text{ ft/min}$$

Depth of water: $-3.064t + 150$

Time to drain well: $t = \dfrac{150}{3.064} \approx 49$ min

$(49)(12) = 588$ gallons pumped

Volume of water pumped in Exercise 25: 391.7 gallons.

$$\frac{391.7}{52\pi} = \frac{588}{x\pi}$$

$$x = \frac{588(52)}{391.7} \approx 78$$

Work $\approx 78\pi$ ft · ton

27. Weight of section of chain: $5\,\Delta x$
Distance moved: $10 - x$

$$W = 5\int_0^{10} (10 - x)\,dx = -\frac{5}{2}(10 - x)^2\Big]_0^{10} = 250 \text{ ft} \cdot \text{lb}$$

28. (a) Weight of section of cable: $4\Delta x$
Distance: $200 - x$

$$W = 4\int_0^{200} (200 - x)\,dx = \Big[-2(200 - x)^2\Big]_0^{200} = 80{,}000 \text{ ft} \cdot \text{lb} = 40 \text{ ft} \cdot \text{ton}$$

(b) Work to move 300 pounds 200 feet vertically: $200(300) = 60{,}000$ ft · lb $= 30$ ft · ton

Total work = work for drawing up the cable + work of lifting the load

= 40 ft · ton + 30 ft · ton = 70 ft · ton

29. Wall at shallow end:

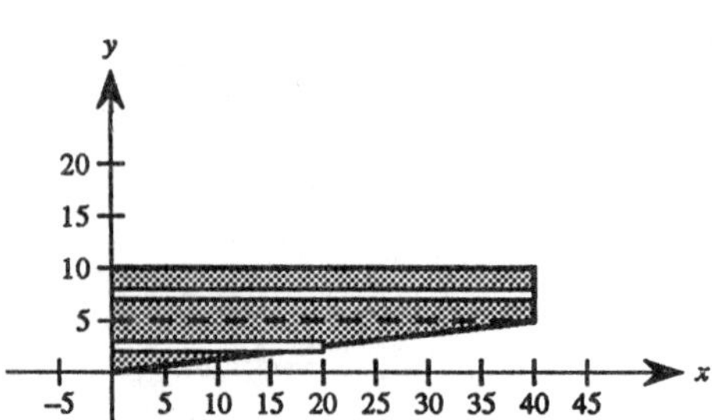

$$F = 62.4\int_0^5 y(20)\,dy = (1248)\frac{y^2}{2}\Big]_0^5 = 15{,}600 \text{ lb}$$

Wall at deep end:

$$F = 62.4\int_0^{10} y(20)\,dy = (624)y^2\Big]_0^{10} = 62{,}400 \text{ lb}$$

Side wall:

$$F_1 = 62.4\int_0^5 y(40)\,dy = (1248)y^2\Big]_0^5 = 31{,}200 \text{ lb}$$

$$F_2 = 62.4\int_0^5 (10 - y)8y\,dy = 62.4\int_0^5 (80y - 8y^2)\,dy = 62.4\left[40y^2 - \frac{8}{3}y^3\right]_0^5 = 41{,}600 \text{ lb}$$

$$F = F_1 + F_2 = 72{,}800 \text{ lb}$$

30. Let D = surface of liquid; ρ = weight per cubic volume.

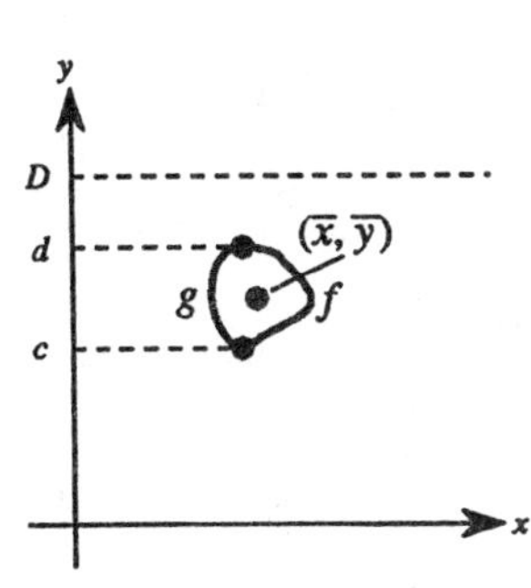

$$F = \rho\int_c^d (D - y)[f(y) - g(y)]\,dy$$

$$= \rho\left[\int_c^d D[f(y) - g(y)]\,dy - \int_c^d y[f(y) - g(y)]\,dy\right]$$

$$= \rho\left[\int_c^d [f(y) - g(y)]\,dy\right]\left[D - \frac{\int_c^d y[f(y) - g(y)]\,dy}{\int_c^d [f(y) - g(y)]\,dy}\right]$$

$$= \rho(\text{Area})(D - \overline{y}) = \rho(\text{Area})(\text{depth of centroid})$$

31. $F = 62.4(16\pi)5 = 4992\pi$ lb

32. Raise water level five more feet

33. $$A = \int_0^a (\sqrt{a} - \sqrt{x})^2\,dx = \int_0^a \left(a - 2\sqrt{a}\,x^{1/2} + x\right)dx = \left[ax - \frac{4}{3}\sqrt{a}\,x^{3/2} + \frac{1}{2}x^2\right]_0^a = \frac{a^2}{6}$$

$$\frac{1}{A} = \frac{6}{a^2}$$

$$\overline{x} = \frac{6}{a^2}\int_0^a x(\sqrt{a} - \sqrt{x})^2\,dx = \frac{6}{a^2}\int_0^a \left(ax - 2\sqrt{a}\,x^{3/2} + x^2\right)dx = \frac{6}{a^2}\left[\frac{ax^2}{2} - \frac{4}{5}\sqrt{a}\,x^{5/2} + \frac{1}{3}x^3\right]_0^a = \frac{a}{5}$$

$$\overline{y} = \left(\frac{6}{a^2}\right)\frac{1}{2}\int_0^a (\sqrt{a} - \sqrt{x})^4\,dx$$

$$= \frac{3}{a^2}\int_0^a (a^2 - 4a^{3/2}x^{1/2} + 6ax - 4a^{1/2}x^{3/2} + x^2)\,dx$$

$$= \frac{3}{a^2}\left[a^2x - \frac{8}{3}a^{3/2}x^{3/2} + 3ax^2 - \frac{8}{5}a^{1/2}x^{5/2} + \frac{1}{3}x^3\right]_0^a = \frac{a}{5}$$

$$(\overline{x},\ \overline{y}) = \left(\frac{a}{5},\ \frac{a}{5}\right)$$

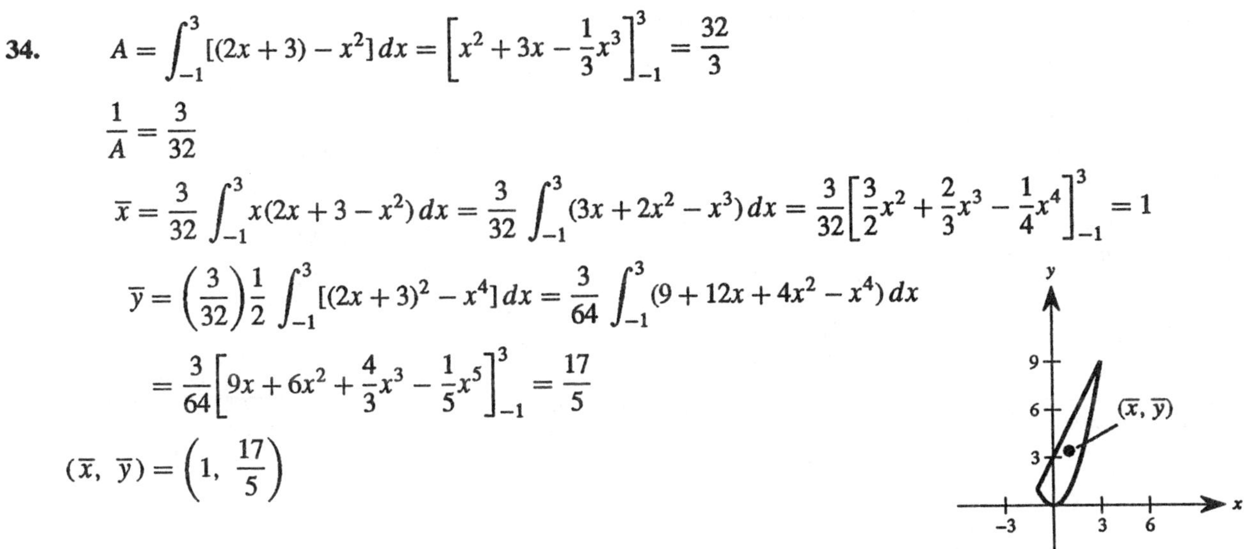

34. $$A = \int_{-1}^3 [(2x + 3) - x^2]\,dx = \left[x^2 + 3x - \frac{1}{3}x^3\right]_{-1}^3 = \frac{32}{3}$$

$$\frac{1}{A} = \frac{3}{32}$$

$$\overline{x} = \frac{3}{32}\int_{-1}^3 x(2x + 3 - x^2)\,dx = \frac{3}{32}\int_{-1}^3 (3x + 2x^2 - x^3)\,dx = \frac{3}{32}\left[\frac{3}{2}x^2 + \frac{2}{3}x^3 - \frac{1}{4}x^4\right]_{-1}^3 = 1$$

$$\overline{y} = \left(\frac{3}{32}\right)\frac{1}{2}\int_{-1}^3 [(2x + 3)^2 - x^4]\,dx = \frac{3}{64}\int_{-1}^3 (9 + 12x + 4x^2 - x^4)\,dx$$

$$= \frac{3}{64}\left[9x + 6x^2 + \frac{4}{3}x^3 - \frac{1}{5}x^5\right]_{-1}^3 = \frac{17}{5}$$

$$(\overline{x},\ \overline{y}) = \left(1,\ \frac{17}{5}\right)$$

35. By symmetry, $x = 0$.

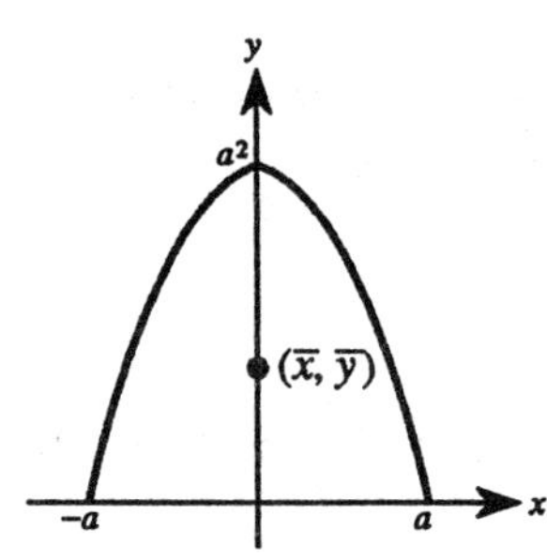

$$A = 2\int_0^1 (a^2 - x^2)\,dx = 2\left[a^2x - \frac{x^3}{3}\right]_0^a = \frac{4a^3}{3}$$

$$\frac{1}{A} = \frac{3}{4a^3}$$

$$\bar{y} = \left(\frac{3}{4a^3}\right)\frac{1}{2}\int_{-a}^{a}(a^2 - x^2)^2\,dx = \frac{6}{8a^3}\int_0^a (a^4 - 2a^2x^2 + x^4)\,dx$$

$$= \frac{6}{8a^3}\left[a^4x - \frac{2a^2}{3}x^3 + \frac{1}{5}x^5\right]_0^a = \frac{6}{8a^3}\left(a^5 - \frac{2}{3}a^5 + \frac{1}{5}a^5\right) = \frac{2a^2}{5}$$

$$(\bar{x},\ \bar{y}) = \left(0,\ \frac{2a^2}{5}\right)$$

36.

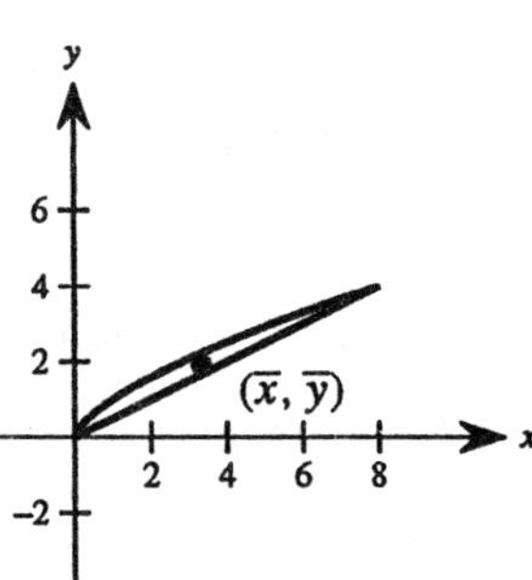

$$A = \int_0^8 \left(x^{2/3} - \frac{1}{2}x\right)dx = \left[\frac{3}{5}x^{5/3} - \frac{1}{4}x^2\right]_0^8 = \frac{16}{5}$$

$$\frac{1}{A} = \frac{5}{16}$$

$$\bar{x} = \frac{5}{16}\int_0^8 x\left(x^{2/3} - \frac{1}{2}x\right)dx = \frac{5}{16}\left[\frac{3}{8}x^{8/3} - \frac{1}{6}x^3\right]_0^8 = \frac{10}{3}$$

$$\bar{y} = \left(\frac{5}{16}\right)\frac{1}{2}\int_0^8 \left(x^{4/3} - \frac{1}{4}x^2\right)dx = \frac{1}{2}\left(\frac{5}{16}\right)\left[\frac{3}{7}x^{7/3} - \frac{1}{12}x^3\right]_0^8 = \frac{40}{21}$$

$$(\bar{x},\ \bar{y}) = \left(\frac{10}{3},\ \frac{40}{21}\right)$$

37.

$$y = \sqrt{4 - x^2}$$

$$y' = \frac{-x}{\sqrt{4 - x^2}}$$

$$1 + (y')^2 = \frac{4}{4 - x^2}$$

$$s = \int_{-\sqrt{3}}^{\sqrt{3}} \sqrt{\frac{4}{4 - x^2}}\,dx$$

$$= 2\int_0^{\sqrt{3}} \frac{2}{\sqrt{4 - x^2}}\,dx$$

38.

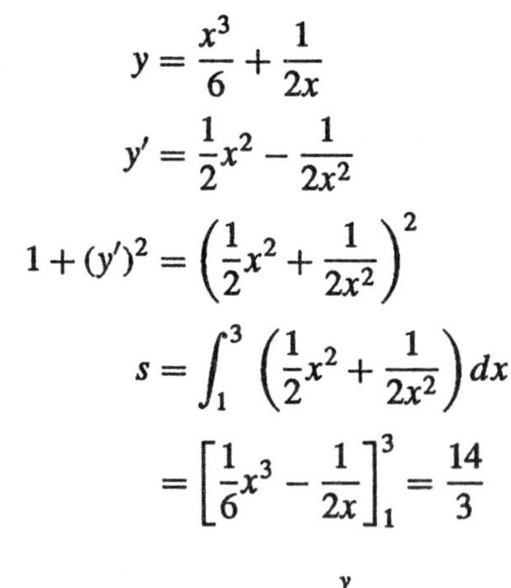

$$y = \frac{x^3}{6} + \frac{1}{2x}$$

$$y' = \frac{1}{2}x^2 - \frac{1}{2x^2}$$

$$1 + (y')^2 = \left(\frac{1}{2}x^2 + \frac{1}{2x^2}\right)^2$$

$$s = \int_1^3 \left(\frac{1}{2}x^2 + \frac{1}{2x^2}\right)dx$$

$$= \left[\frac{1}{6}x^3 - \frac{1}{2x}\right]_1^3 = \frac{14}{3}$$

39.

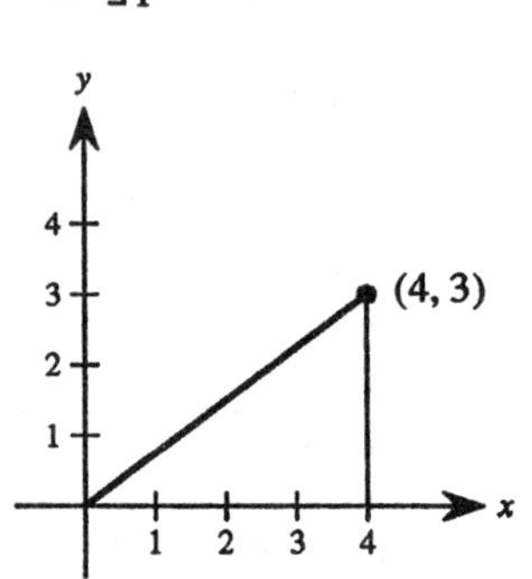

$$y = \frac{3}{4}x$$

$$y' = \frac{3}{4}$$

$$1 + (y')^2 = \frac{25}{16}$$

$$S = 2\pi\int_0^4 \left(\frac{3}{4}x\right)\sqrt{\frac{25}{16}}\,dx = \left(\frac{15\pi}{8}\right)\left[\frac{x^2}{2}\right]_0^4 = 15\pi$$

40. From Exercise 17(a) we have

$$V = 64\pi \text{ ft}^3$$

$$\frac{1}{4}V = 16\pi.$$

Disc

$$\pi\int_{-3}^{y_0} \frac{16}{9}(9 - y^2)\,dy = 16\pi$$

$$\frac{1}{9}\int_{-3}^{y_0}(9 - y^2)\,dy = 1$$

$$\left[9y - \frac{1}{3}y^3\right]_{-3}^{y_0} = 9$$

$$\left(9y_0 - \frac{1}{3}y_0^3\right) - (-27 + 9) = 9$$

$$y_0^3 - 27y_0 - 27 = 0$$

By Newton's Method, $y_0 \approx -1.042$ and the depth of the gasoline is $3 - 1.042 = 1.958$ ft.

41. Since $y \le 0$, $A = -\int_{-1}^{0} x\sqrt{x+1}\,dx.$

By the substitution method we obtain

$$A = -\int_{-1}^{0}\left[(x+1)^{3/2} - (x+1)^{1/2}\right]dx$$

$$= \left[-\frac{2}{5}(x+1)^{5/2} + \frac{2}{3}(x+1)^{3/2}\right]_{-1}^{0} = \frac{4}{15}.$$

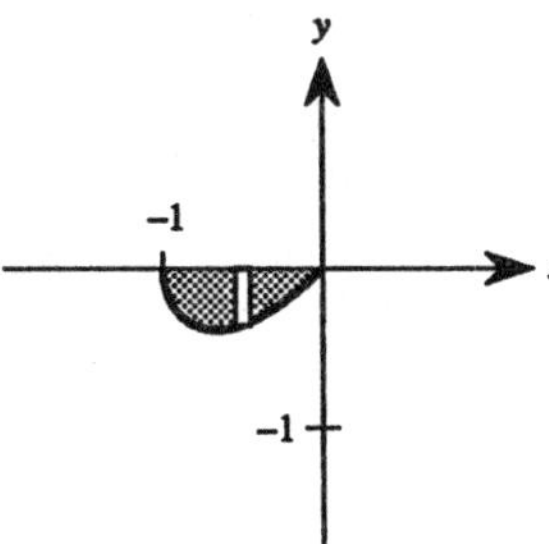

42. Disc

$$V = \pi\int_{-1}^{0} x^2(x+1)\,dx$$

$$= \pi\int_{-1}^{0}(x^3 + x^2)\,dx$$

$$= \pi\left[\frac{x^4}{4} + \frac{x^3}{3}\right]_{-1}^{0}$$

$$= \frac{\pi}{12}$$

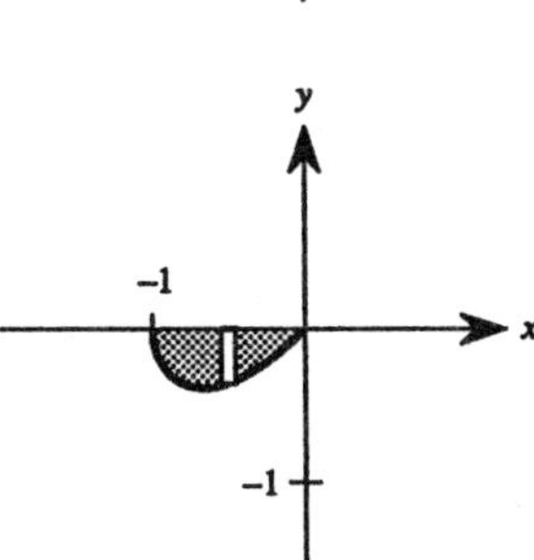

43. Shell

$$u = \sqrt{x+1}$$

$$x = u^2 - 1$$

$$dx = 2u\,du$$

$$V = 2\pi\int_{-1}^{0} x^2\sqrt{x+1}\,dx = 4\pi\int_{0}^{1}(u^2-1)^2u^2\,du$$

$$= 4\pi\int_{0}^{1}(u^6 - 2u^4 + u^2)\,du = 4\pi\left[\frac{1}{7}u^7 - \frac{2}{5}u^5 + \frac{1}{3}u^3\right]_{0}^{1} = \frac{32\pi}{105}$$

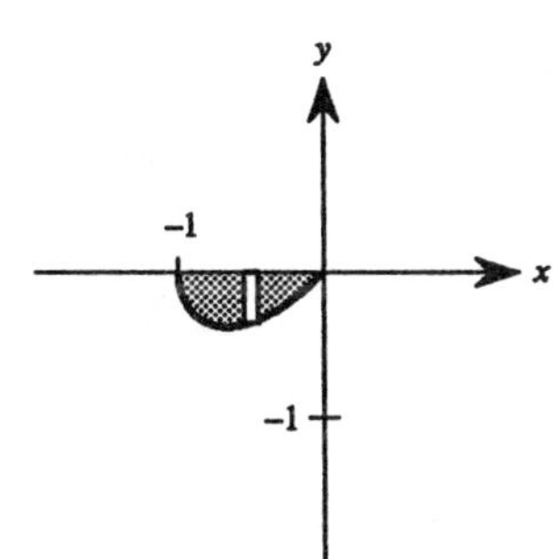

44.

$$y = 2\sqrt{x}$$

$$y' = \frac{1}{\sqrt{x}}$$

$$1 + (y')^2 = 1 + \frac{1}{x} = \frac{x+1}{x}$$

$$S = 2\pi \int_0^3 2\sqrt{x}\sqrt{\frac{x+1}{x}}\,dx = 4\pi \int_0^3 \sqrt{x+1}\,dx = 4\pi\left(\frac{2}{3}\right)(x+1)^{3/2}\Big]_0^3 = \frac{56\pi}{3}$$

45.

$$f(x) = \frac{4}{5}x^{5/4}$$

$$f'(x) = x^{1/4}$$

$$1 + [f'(x)]^2 = 1 + \sqrt{x}$$

$$u = 1 + \sqrt{x}$$

$$x = (u-1)^2$$

$$dx = 2(u-1)\,du$$

$$s = \int_0^4 \sqrt{1+\sqrt{x}}\,dx = 2\int_1^3 \sqrt{u}(u-1)\,du = 2\int_1^3 (u^{3/2} - u^{1/2})\,du$$

$$= 2\left[\frac{2}{5}u^{5/2} - \frac{2}{3}u^{3/2}\right]_1^3 = \frac{4}{15}u^{3/2}(3u-5)\Big]_1^3 = \frac{8}{15}(1 + 6\sqrt{3}) \approx 6.076$$

CHAPTER 7
Exponential and Logarithmic Functions

Section 7.1 Exponential Functions

1. (a) $25^{3/2} = (\sqrt{25})^3 = 125$

(b) $81^{1/2} = \sqrt{81} = 9$

(c) $3^{-2} = \frac{1}{3^2} = \frac{1}{9}$

(d) $27^{-1/3} = \frac{1}{\sqrt[3]{27}} = \frac{1}{3}$

2. (a) $64^{1/3} = \sqrt[3]{64} = 4$

(b) $5^{-4} = \frac{1}{5^4} = \frac{1}{625}$

(c) $\left(\frac{1}{8}\right)^{1/3} = \sqrt[3]{\frac{1}{8}} = \frac{1}{2}$

(d) $\left(\frac{1}{4}\right)^3 = \frac{1}{64}$

3. (a) $(5^2)(5^3) = 5^{2+3} = 5^5 = 3125$

(b) $(5^2)(5^{-3}) = 5^{2+(-3)} = 5^{-1} = \frac{1}{5}$

(c) $\frac{5^3}{25^2} = \frac{5^3}{5^4} = \frac{1}{5}$

(d) $\left(\frac{1}{4}\right)^2(2^6) = \left(\frac{1}{2}\right)^4(2^6) = 2^2 = 4$

4. (a) $(4^2)^3 = 4^{2 \cdot 3} = 4^6$

(b) $(6^4)^{1/2} = 6^{4(1/2)} = 6^2 = 36$

(c) $[(8^{-1})(8^{2/3})]^3 = [8^{-1/3}]^3 = 8^{-1} = \frac{1}{8}$

(d) $(32^{3/2})(4^2) = (2^5)^{3/2}(2^2)^2$

$= (2^{15/2})(2^4) = 2^{23/2}$

5. (a) $e^2(e^4) = e^{2+4} = e^6$

(b) $(e^3)^4 = e^{3 \cdot 4} = e^{12}$

(c) $(e^3)^{-2} = e^{-6} = \frac{1}{e^6}$

(d) $\frac{e^5}{e^3} = e^{5-3} = e^2$

6. (a) $\left(\frac{1}{e}\right)^{-2} = \left(\frac{e}{1}\right)^2 = e^2$

(b) $\left(\frac{e^5}{e^2}\right)^{-1} = (e^3)^{-1} = \frac{1}{e^3}$

(c) $e^0 = 1$

(d) $\frac{1}{e^{-3}} = e^3$

7. $3^x = 81$

$3^x = 3^4$

$x = 4$

8. $5^{x+1} = 125$

$5^{x+1} = 5^3$

$x + 1 = 3$

$x = 2$

9. $\left(\frac{1}{3}\right)^{x-1} = 27$

$3^{1-x} = 3^3$

$1 - x = 3$

$x = -2$

10. $\left(\frac{1}{5}\right)^{2x} = 625$

$5^{-2x} = 5^4$

$-2x = 4$

$x = -2$

11. $4^3 = (x + 2)^3$

$4 = x + 2$

$x = 2$

12. $18^2 = (5x - 7)^2$

$18 = 5x - 7$

$25 = 5x$

$x = 5$

13. $x^{3/4} = 8$

$x = 8^{4/3}$

$x = (\sqrt[3]{8})^4$

$x = 16$

14. $(x+3)^{4/3} = 16$

$x+3 = 16^{3/4}$

$x+3 = (\sqrt[4]{16})^3$

$x+3 = 8$

$x = 5$

15. $e^{-2x} = e^5$

$-2x = 5$

$x = -\frac{5}{2}$

16. $e^x = 1$

$x = 0$

17. $\left(1+\dfrac{1}{1{,}000{,}000}\right)^{1{,}000{,}000} \approx 2.718280469$

$e \approx 2.718281828$

$e > \left(1+\dfrac{1}{1{,}000{,}000}\right)^{1{,}000{,}000}$

18. $1+1+\frac{1}{2}+\frac{1}{6}+\frac{1}{24}+\frac{1}{120}+\frac{1}{720}+\frac{1}{5040} = 2.71\overline{825396}$

$e \approx 2.718281828$

$e > 1+1+\frac{1}{2}+\frac{1}{6}+\frac{1}{24}+\frac{1}{120}+\frac{1}{720}+\frac{1}{5040}$

19. $y = 3^x$

x	-2	-1	0	1	2
y	$\frac{1}{9}$	$\frac{1}{3}$	1	3	9

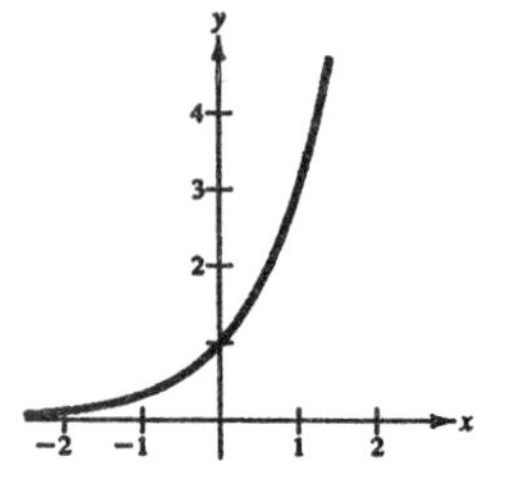

20. $y = 3^{x-1}$

x	-1	0	1	2	3
y	$\frac{1}{9}$	$\frac{1}{3}$	1	3	9

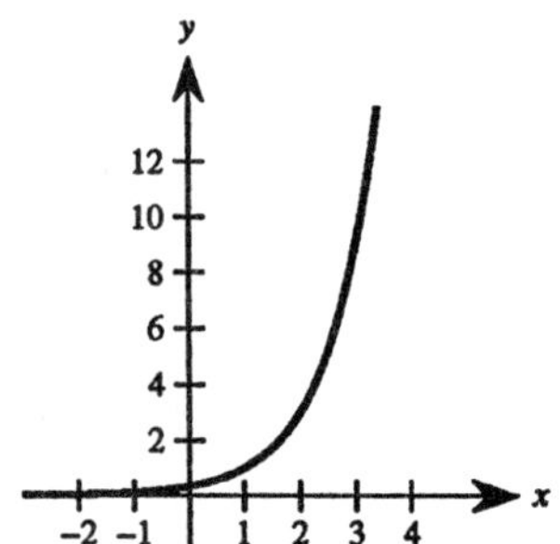

21. $y = \left(\frac{1}{3}\right)^x = 3^{-x}$

x	-2	-1	0	1	2
y	9	3	1	$\frac{1}{3}$	$\frac{1}{9}$

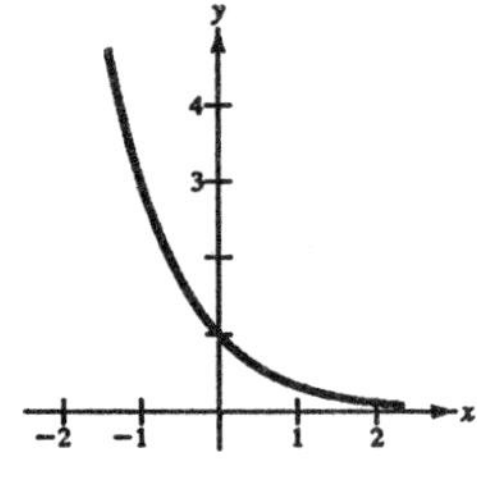

22. $y = 2^{x^2}$

x	-2	-1	0	1	2
y	16	2	1	2	16

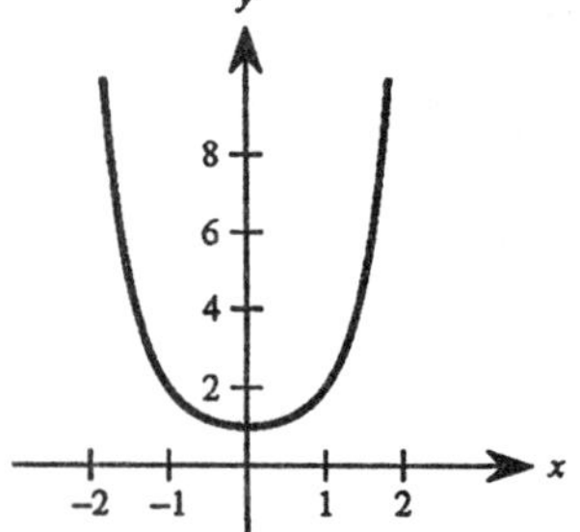

23. $f(x) = 3^{-x^2}$

x	-2	-1	0	1	2
$f(x)$	$\frac{1}{81}$	$\frac{1}{3}$	1	$\frac{1}{3}$	$\frac{1}{81}$

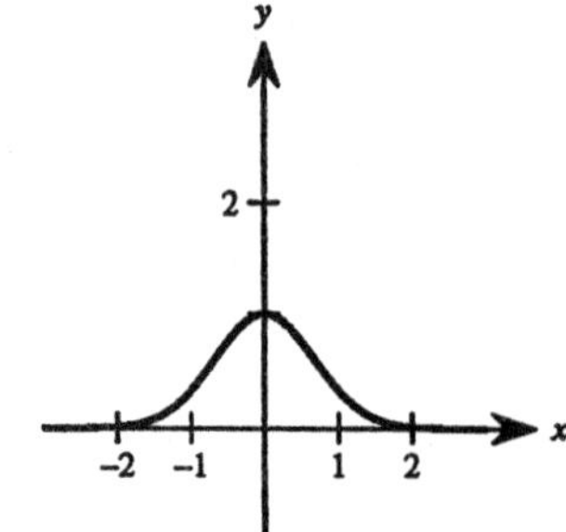

24. $f(x) = 3^{\lvert x\rvert}$

x	-2	-1	0	1	2
$f(x)$	9	3	1	3	9

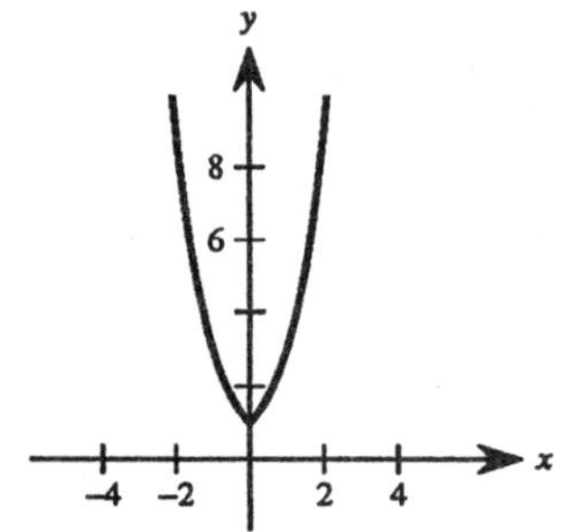

25. $h(x) = e^{x-2}$

x	-1	0	1	2	3
$h(x)$	0.05	0.14	0.37	1	2.72

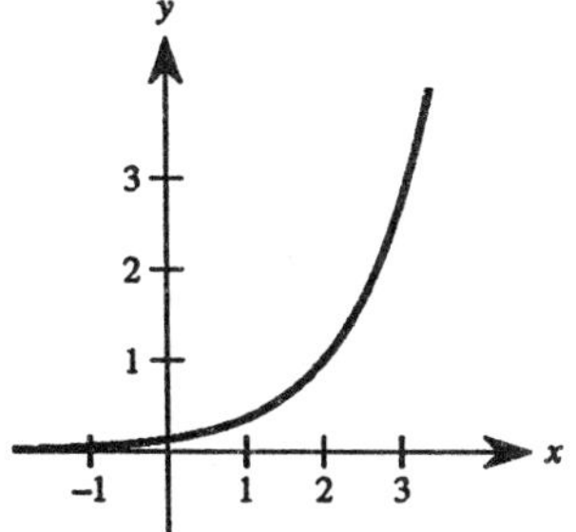

26. $g(x) = -e^{x/2}$

x	-2	-1	0	1	2
$g(x)$	-0.37	-0.61	-1	-1.65	-2.72

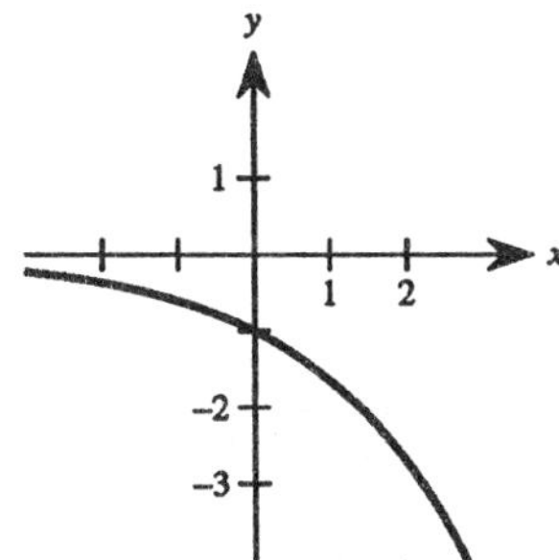

27. $y = e^{-x^2}$
Symmetric with respect to the y-axis
$y = 0$
Horizontal asymptote

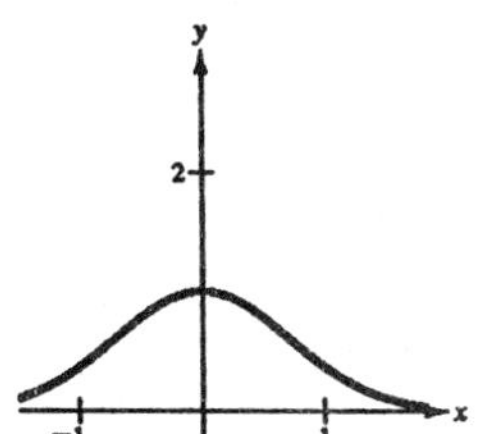

28. $y = e^{-x/2}$

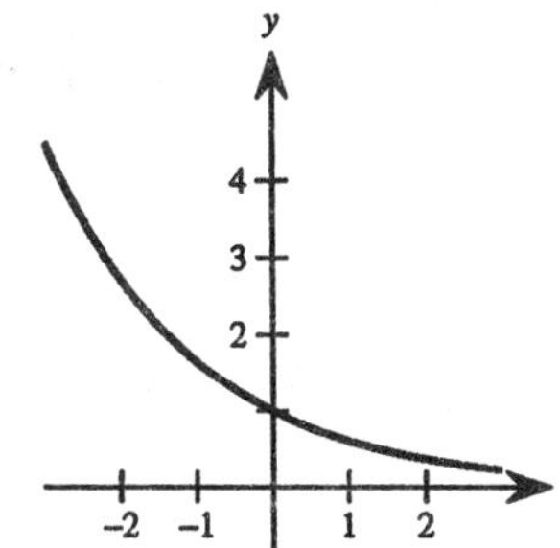

29. (a)

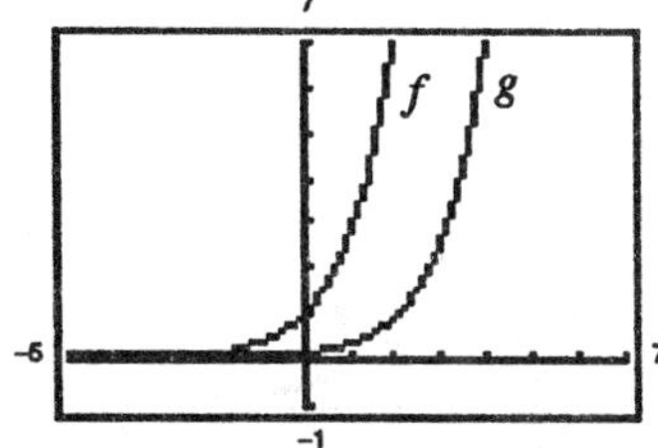

Horizontal shift 2 units to the right

(b) A reflection in the x-axis and a vertical shrink

(c)

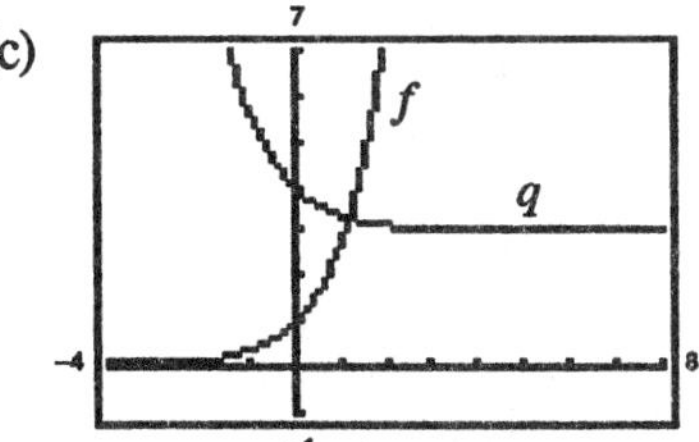

Vertical shift 3 units upward and a reflection in the y-axis

30. (a)

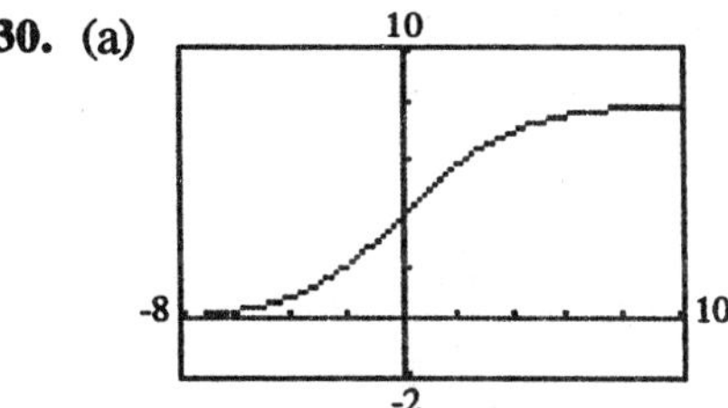

Horizontal asymptotes: $y = 0$ and $y = 8$

(b)

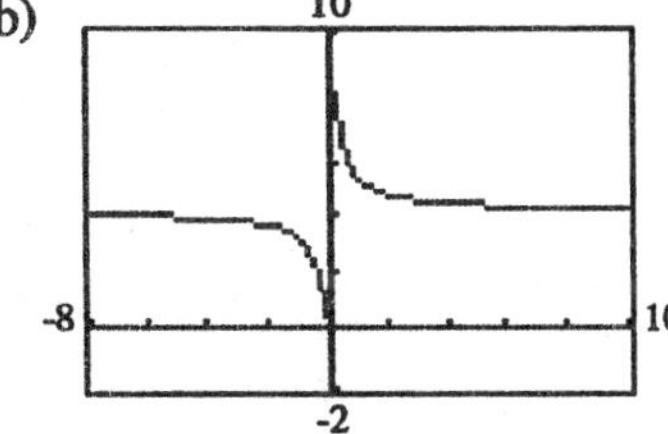

Horizontal asymptote: $y = 4$

31. $y = Ce^{ax}$
Horizontal asymptote: $y = 0$
Matches (c)

32. $y = Ce^{-ax}$
Horizontal asymptote: $y = 0$
Reflection in the y-axis
Matches (d)

33. $y = C(1 - e^{-ax})$
Vertical shift C units
Reflection in both the x- and y-axes
Matches (a)

34. $y = \dfrac{C}{1 + e^{-ax}}$

$$\lim_{x \to \infty} \frac{C}{1 + e^{-ax}} = C$$

$$\lim_{x \to -\infty} \frac{C}{1 + e^{-ax}} = 0$$

Horizontal asymptotes: $y = C$ and $y = 0$
Matches (b)

35. $P = \$1000, \quad r = 10\%, \quad t = 10$ years

(a) $A = 1000\left[1 + \dfrac{0.10}{1}\right]^{1(10)} \approx \$2593.74$

(b) $A = 1000\left[1 + \dfrac{0.10}{2}\right]^{2(10)} \approx \$2653.30$

(c) $A = 1000\left[1 + \dfrac{0.10}{12}\right]^{12(10)} \approx \$2707.04$

(d) $A = 1000\left[1 + \dfrac{0.10}{365}\right]^{365(10)} \approx \$2717.91$

(e) $A = 1000e^{0.10(10)} \approx \$2718.28$

36. $P = \$2500, \quad r = 12\%, \quad t = 20$ years

(a) $A = 2500\left[1 + \dfrac{0.12}{1}\right]^{1(20)} \approx \$24{,}115.73$

(b) $A = 2500\left[1 + \dfrac{0.12}{2}\right]^{2(20)} \approx \$25{,}714.29$

(c) $A = 2500\left[1 + \dfrac{0.12}{12}\right]^{12(20)} \approx \$27{,}231.38$

(d) $A = 2500\left[1 + \dfrac{0.12}{365}\right]^{365(20)} \approx \$27{,}547.07$

(e) $A = 2500e^{0.12(20)} \approx \$27{,}557.94$

37. $A = 100{,}000, \quad r = 0.12, \quad P = \dfrac{A}{e^{rt}}$

(a) $P = \dfrac{100{,}000}{e^{0.12(1)}} = \$88{,}692.04$

(b) $P = \dfrac{100{,}000}{e^{0.12(10)}} = \$30{,}119.42$

(c) $P = \dfrac{100{,}000}{e^{0.12(20)}} = \$9071.80$

(d) $P = \dfrac{100{,}000}{e^{0.12(50)}} = \$247.88$

38. $A = 100{,}000, \quad r = 0.09, \quad P = \dfrac{A}{e^{rt}}$

(a) $P = \dfrac{100{,}000}{e^{0.09(1)}} \approx \$91{,}393.12$

(b) $P = \dfrac{100{,}000}{e^{0.09(10)}} \approx \$40{,}656.97$

(c) $P = \dfrac{100{,}000}{e^{0.09(20)}} \approx \$16{,}529.89$

(d) $P = \dfrac{100{,}000}{e^{0.09(50)}} \approx \$1110.90$

39. (a) $P\left(\frac{1}{2}\right) = 1 - e^{-1/6} \approx 0.1535 = 15.35\%$

(b) $P(2) = 1 - e^{-2/3} \approx 0.4866 = 48.66\%$

(c) $P(5) = 1 - e^{-5/3} \approx 0.8111 = 81.11\%$

40. $y = 28e^{0.6-0.012s}, \quad s \geq 50$

Speed (s)	50	55	60	65	70
Miles per gallon (y)	28	26.3694	24.8338	23.3876	22.0256

41. (a) $\lim_{t\to\infty} \frac{850}{1+e^{-0.2t}} = \frac{850}{1+0} = 850$

(b)

t	0	5	10	15
y	425	621.4	748.7	809.7

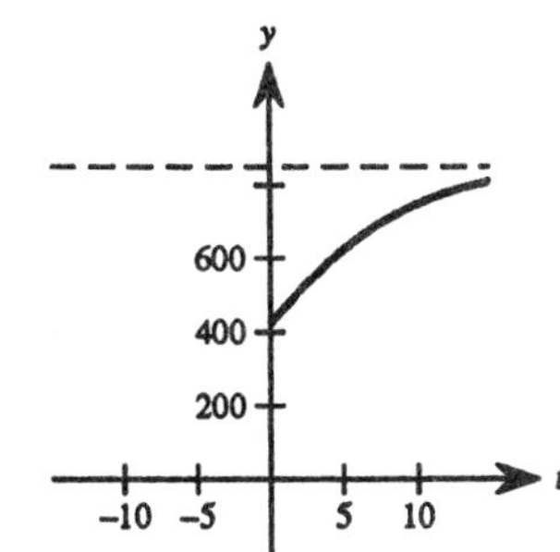

42. (a) $V(20) = 6.7e^{-48/20} \approx 0.6078$

$V(50) = 6.7e^{-48/50} \approx 2.5654$

(b) $\lim_{t\to\infty} 6.7e^{-48/t} = 6.7(1) = 6.7$

(c)

t	10	20	30	40	50
V	0.0551	0.6078	1.3527	2.0180	2.5654

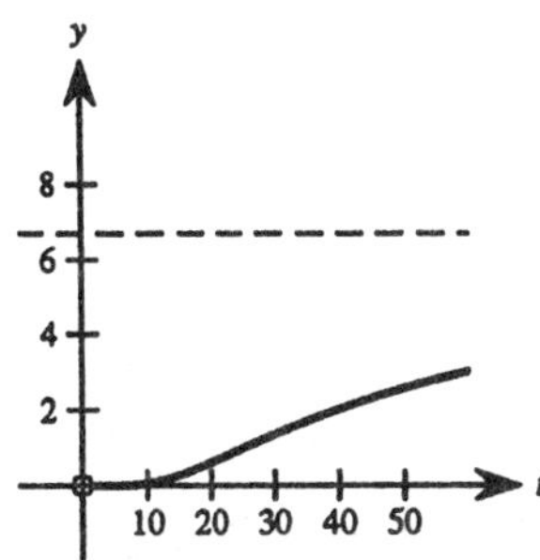

43. (a) $P(10) = \frac{0.83}{1+e^{-0.2(10)}} \approx 0.731$

(b) $\lim_{n\to\infty} \frac{0.83}{1+e^{-0.2n}} = \frac{0.83}{1+0} = 0.83$

44. (a) $N(10) = \frac{157}{1+5.4e^{-0.12(10)}}$

≈ 59.7765 words/minute

(b) $\lim_{t\to\infty} \frac{157}{1+5.4e^{-0.12t}} = \frac{157}{1+0}$

$= 157$ words/minute

45. $y(0) = 757.71 - 127.71\left(\frac{1+1}{2}\right) = 630$ feet which is the same as the distance between its two legs.

46. (a)

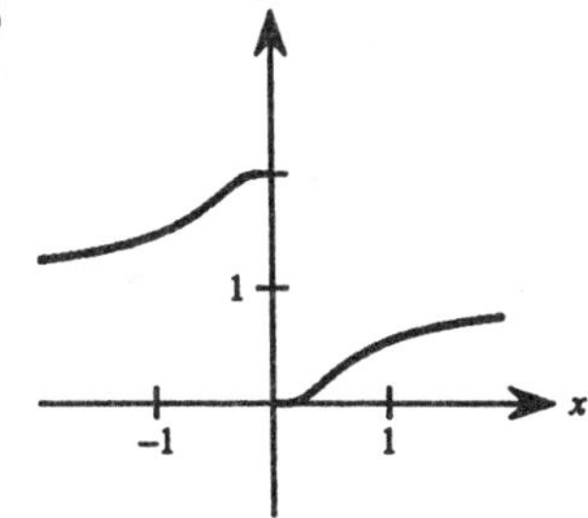

(b) $\lim_{x\to\infty} \frac{2}{1+e^{1/x}} = \lim_{x\to-\infty} \frac{2}{1+e^{1/x}} = \frac{2}{1+1} = 1$

Horizontal asymptote: $y = 1$

(c) $\lim_{x\to 0} \frac{2}{1+e^{1/x}} = \frac{2}{\infty} = 0$

Section 7.2 Differentiation and Integration of Exponential Functions

1. $y = e^{3x}$

$y' = 3e^{3x}$

At $(0, 1)$, $y' = 3$.

2. $y = e^{2x}$

$y' = 2e^{2x}$

At $(0, 1)$, $y' = 2$.

3. $y = e^{x}$

$y' = e^{x}$

At $(0, 1)$, $y' = 1$.

4. $y = e^{-3x}$

$y' = -3e^{-3x}$

At $(0, 1)$, $y' = -3$.

5. $y = e^{-2x}$

$y' = -2e^{-2x}$

At $(0, 1)$, $y' = -2$.

6. $y = e^{-x}$

$y' = -e^{-x}$

At $(0, 1)$, $y' = -1$.

7. $y = e^{2x}$

$\frac{dy}{dx} = 2e^{2x}$

8. $y = e^{1-x}$

$\frac{dy}{dx} = -e^{1-x}$

9. $y = e^{-2x+x^2}$

$\dfrac{dy}{dx} = 2(x-1)e^{-2x+x^2}$

10. $y = e^{-x^2}$

$\dfrac{dy}{dx} = -2xe^{-x^2}$

11. $f(x) = e^{1/x} = e^{x^{-1}}$

$f'(x) = (-x^{-2})e^{x^{-1}} = -\dfrac{e^{1/x}}{x^2}$

12. $y = e^{-1/x^2}$

$\dfrac{dy}{dx} = \dfrac{2e^{-1/x^2}}{x^3}$

13. $y = e^{\sqrt{x}}$

$\dfrac{dy}{dx} = \dfrac{e^{\sqrt{x}}}{2\sqrt{x}}$

14. $g(x) = e^{x^3}$

$g'(x) = 3x^2e^{x^3}$

15. $f(x) = (x+1)e^{3x}$

$$f'(x) = (x+1)(3e^{3x}) + e^{3x}(1)$$
$$= e^{3x}(3x+4)$$

16. $y = x^2e^{-x}$

$\dfrac{dy}{dx} = -x^2e^{-x} + 2xe^{-x} = xe^{-x}(2-x)$

17. $f(x) = \dfrac{e^{x^2}}{x}$

$$f'(x) = \frac{x(2xe^{x^2}) - e^{x^2}}{x^2} = \frac{e^{x^2}(2x^2-1)}{x^2}$$

18. $f(x) = \dfrac{e^{x/2}}{\sqrt{x}}$

$$f'(x) = \frac{\sqrt{x}[(1/2)e^{x/2}] - e^{x/2}[1/(2\sqrt{x})]}{x} \cdot \frac{2\sqrt{x}}{2\sqrt{x}}$$
$$= \frac{xe^{x/2} - e^{x/2}}{2x\sqrt{x}} = \frac{e^{x/2}(x-1)}{2x\sqrt{x}}$$

19. $y = (e^{-x} + e^{x})^3$

$\dfrac{dy}{dx} = 3(e^{-x} + e^{x})^2(e^{x} - e^{-x})$

20. $y = (1 - e^{-x})^2 = 1 - 2e^{-x} + e^{-2x}$

$$y' = 2e^{-x} - 2e^{-2x}$$
$$= 2e^{-x}(1 - e^{-x}) = \frac{2(e^x - 1)}{e^{2x}}$$

21. $y = \dfrac{2}{e^x + e^{-x}} = 2(e^x + e^{-x})^{-1}$

$$\frac{dy}{dx} = -2(e^x + e^{-x})^{-2}(e^x - e^{-x})$$
$$= \frac{-2(e^x - e^{-x})}{(e^x + e^{-x})^2}$$

22. $f(x) = \dfrac{e^x - e^{-x}}{2} = \dfrac{1}{2}(e^x - e^{-x})$

$f'(x) = \dfrac{1}{2}(e^x + e^{-x}) = \dfrac{e^x + e^{-x}}{2}$

23. $y = xe^x - e^x = e^x(x-1)$

$\dfrac{dy}{dx} = e^x + e^x(x-1) = xe^x$

24. $y = x^2e^x - 2xe^x + 2e^x = e^x(x^2 - 2x + 2)$

$\dfrac{dy}{dx} = e^x(2x-2) + e^x(x^2 - 2x + 2) = x^2e^x$

25.

$$xe^y - 10x + 3y = 0$$
$$xe^y\frac{dy}{dx} + e^y - 10 + 3\frac{dy}{dx} = 0$$
$$\frac{dy}{dx}(xe^y + 3) = 10 - e^y$$
$$\frac{dy}{dx} = \frac{10 - e^y}{xe^y + 3}$$

26.

$$e^{xy} + x^2 - y^2 = 10$$
$$\left(x\frac{dy}{dx} + y\right)e^{xy} + 2x - 2y\frac{dy}{dx} = 0$$
$$\frac{dy}{dx}(xe^{xy} - 2y) = -ye^{xy} - 2x$$
$$\frac{dy}{dx} = -\frac{ye^{xy} + 2x}{xe^{xy} - 2y}$$

27. $f(x) = 2e^{3x} + 3e^{-2x}$

$$f'(x) = 6e^{3x} - 6e^{-2x}$$
$$f''(x) = 18e^{3x} + 12e^{-2x}$$

28. $f(x) = 5e^{-x} - 2e^{-5x}$

$$f'(x) = -5e^{-x} + 10e^{-5x}$$
$$f''(x) = 5e^{-x} - 50e^{-5x}$$

29. $g(x) = (1 + 2x)e^{4x}$

$$g'(x) = (1 + 2x)(4e^{4x}) + e^{4x}(2)$$
$$= 2e^{4x}(4x + 3)$$
$$g''(x) = 2e^{4x}(4) + (4x + 3)(8e^{4x})$$
$$= 8e^{4x}(4x + 4)$$
$$= 32e^{4x}(x + 1)$$

30. $g(x) = (3 + 2x)e^{-3x}$

$$g'(x) = (3 + 2x)(-3e^{-3x}) + e^{-3x}(2)$$
$$= e^{-3x}(-6x - 7)$$
$$g''(x) = e^{-3x}(-6) + (-6x - 7)(-3e^{-3x})$$
$$= 3e^{-3x}(6x + 5)$$

31. $f(x) = \dfrac{2}{1 + e^{-x}} = 2(1 + e^{-x})^{-1}$

$$f'(x) = -2(1 + e^{-x})^{-2}(-e^{-x}) = \frac{2e^{-x}}{(1 + e^{-x})^2} > 0 \text{ for all } x.$$

Always increasing
No relative extrema

$$f''(x) = 2e^{-x}[-2(1 + e^{-x})^{-3}(-e^{-x})] + (1 + e^{-x})^{-2}(-2e^{-x})$$
$$= 2e^{-x}(1 + e^{-x})^{-3}[2e^{-x} - (1 + e^{-x})]$$
$$= \frac{2e^{-x}(e^{-x} - 1)}{(1 + e^{-x})^3} = 0 \Rightarrow e^{-x} = 1 \Rightarrow x = 0$$

Inflection point: (0, 1)

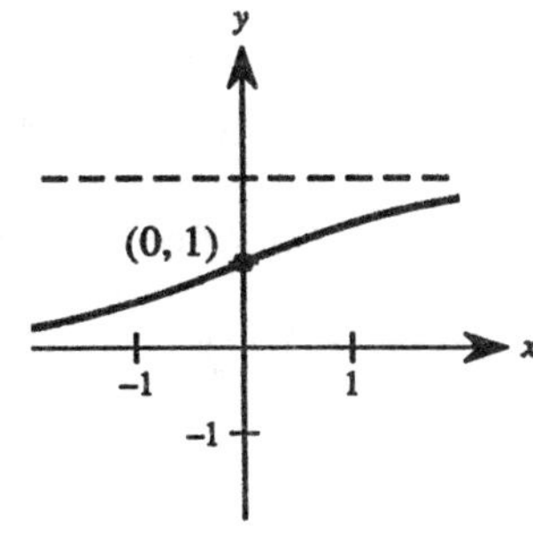

32.

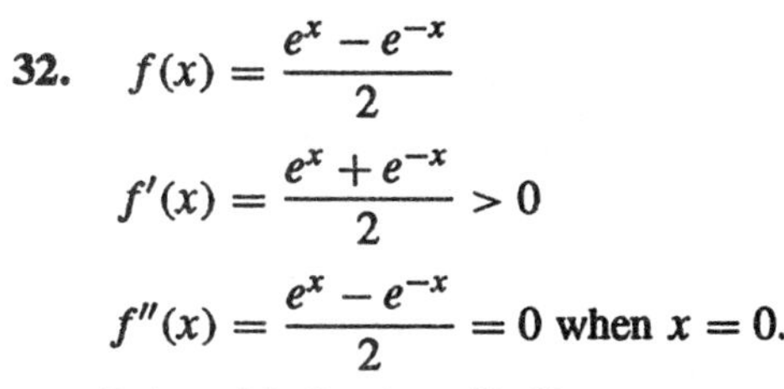

$$f(x) = \frac{e^x - e^{-x}}{2}$$
$$f'(x) = \frac{e^x + e^{-x}}{2} > 0$$
$$f''(x) = \frac{e^x - e^{-x}}{2} = 0 \text{ when } x = 0.$$

Point of inflection: (0, 0)

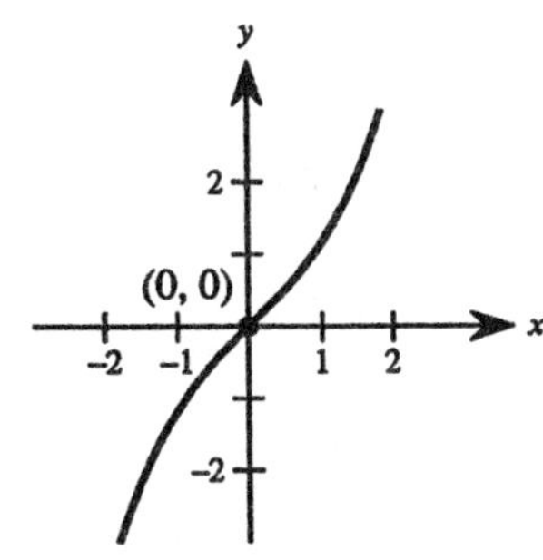

33. $f(x) = x^2e^{-x}$

$$f'(x) = -x^2e^{-x} + 2xe^{-x} = xe^{-x}(2 - x) = 0 \text{ when } x = 0, \ 2.$$
$$f''(x) = -e^{-x}(2x - x^2) + e^{-x}(2 - 2x)$$
$$= e^{-x}(x^2 - 4x + 2) = 0 \text{ when } x = 2 \pm \sqrt{2}.$$

Relative minimum: (0, 0)
Relative maximum: $(2, \ 4e^{-2})$

$$x = 2 \pm \sqrt{2}$$
$$y = (2 \pm \sqrt{2})^2 e^{-(2 \pm \sqrt{2})}$$

Point of inflection: (3.414, 0.384), (0.586, 0.191)

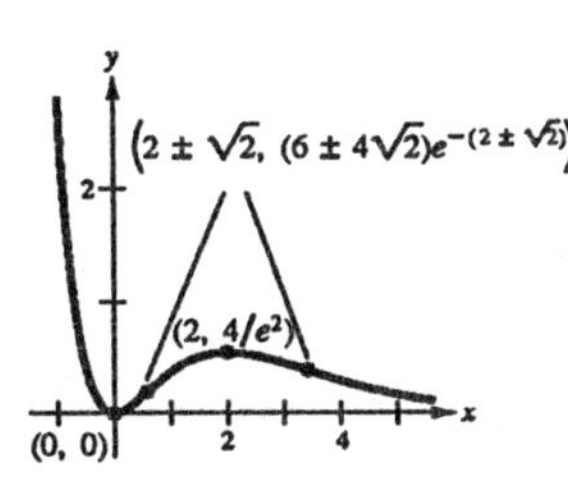

34. $f(x) = xe^{-x}$

$f'(x) = -xe^{-x} + e^{-x} = e^{-x}(1 - x) = 0$ when $x = 1$.

$f''(x) = -e^{-x} + (-e^{-x})(1 - x) = e^{-x}(x - 2) = 0$ when $x = 2$.

Relative maximum: $(1,\ e^{-1})$
Point of inflection: $(2,\ 2e^{-2})$

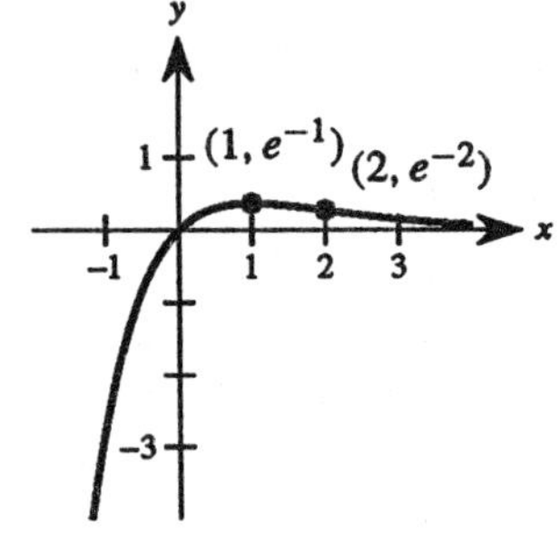

35. $y = e^{-x}$ Point: (0, 1)

$y' = -e^{-x}$ (Slope of tangent line)

$-\dfrac{1}{y'} = e^x$ (Slope of normal line)

At (0, 1), $e^x = 1$, $y = x + 1$.

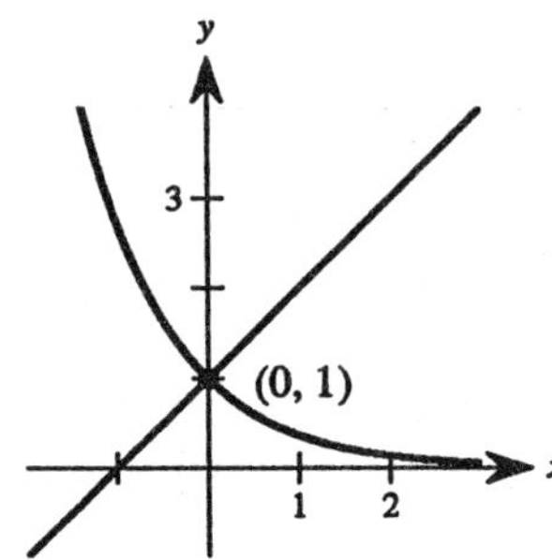

36.

$y = e^{-x}$

$y' = -e^{-x}$ (Slope of tangent line)

$-\dfrac{1}{y'} = e^x$ (Slope of normal line)

$y - e^{-x_0} = e^{x_0}(x - x_0)$

We want (0, 0) to satisfy the equation:

$-e^{-x_0} = -x_0 e^{x_0}$

$1 = x_0 e^{2x_0}$

$x_0 e^{2x_0} - 1 = 0$ when $x_0 \approx 0.4263$.

$(0.4263,\ e^{-0.4263})$

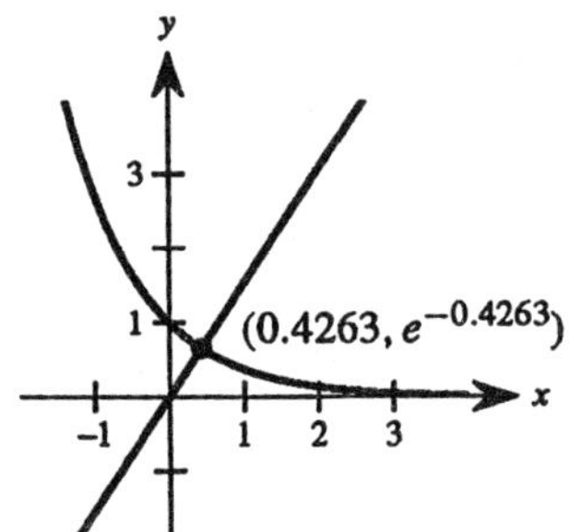

37. $A = 2xe^{-x^2}$

$$\frac{dA}{dx} = -4x^2e^{-x^2} + 2e^{-x^2}$$

$$= 2e^{-x^2}(1 - 2x^2) = 0 \text{ when } x = \frac{\sqrt{2}}{2}.$$

$A = \sqrt{2}e^{-1/2}$

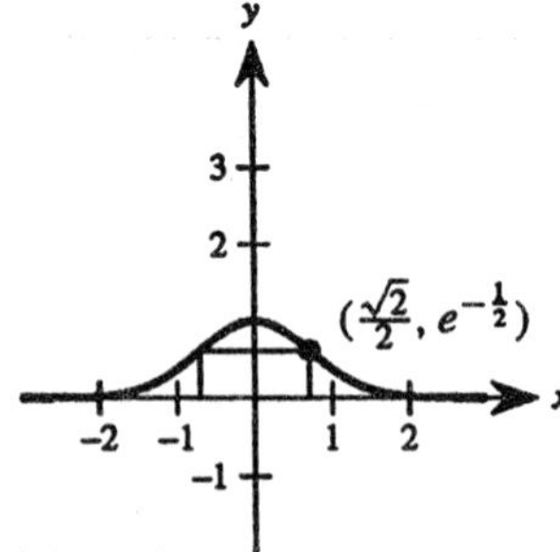

38. $e^{-x} = x \Rightarrow f(x) = x - e^{-x}$

$f'(x) = 1 + e^{-x}$

$$x_{n+1} = x_n - \frac{f(x_n)}{f'(x_n)} = x_n - \frac{x_n - e^{-x_n}}{1 + e^{-x_n}}$$

$x_1 = 1$

$$x_2 = x_1 - \frac{f(x_1)}{f'(x_1)} \approx 0.5379$$

$$x_3 = x_2 - \frac{f(x_2)}{f'(x_2)} \approx 0.5670$$

$$x_4 = x_3 - \frac{f(x_3)}{f'(x_3)} \approx 0.5671$$

We approximate the root of f to be $x = 0.567$.

39. $f(x) = e^{x/2}, \quad f(0) = 1$

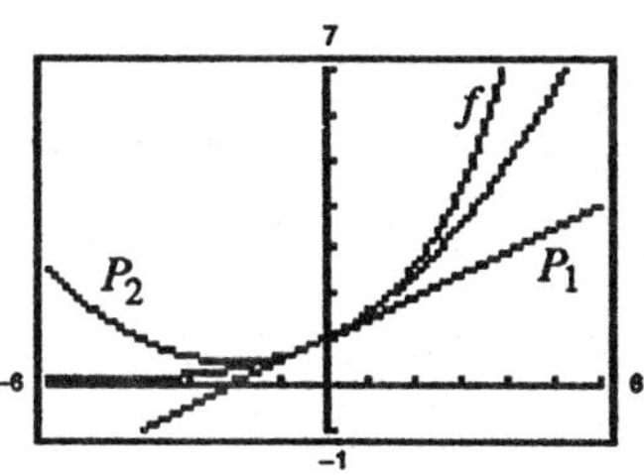

$f'(x) = \frac{1}{2}e^{x/2}, \quad f'(0) = \frac{1}{2}$

$f''(x) = \frac{1}{4}e^{x/2}, \quad f''(0) = \frac{1}{4}$

$P_1(x) = 1 + \frac{1}{2}(x - 0) = \frac{x}{2} + 1, \quad P_1(0) = 1$

$P_1'(x) = \frac{1}{2}, \quad P_1'(0) = \frac{1}{2}$

$P_2(x) = 1 + \frac{1}{2}(x - 0) + \frac{1}{8}(x - 0)^2 = \frac{x^2}{8} + \frac{x}{2} + 1, \quad P_2(0) = 1$

$P_2'(x) = \frac{1}{4}x + \frac{1}{2}, \quad P_2'(0) = \frac{1}{2}$

$P_2''(x) = \frac{1}{4}, \quad P_2''(0) = \frac{1}{4}$

The values of f, P_1, P_2 and their first derivatives agree at $x = 0$. The values of the second derivatives of f and P_2 agree at $x = 0$.

40. $f(x) = e^{-x^2/2}, \quad f(0) = 1$

$f'(x) = -xe^{-x^2/2}, \quad f'(0) = 0$

$f''(x) = x^2e^{-x^2/2} - e^{-x^2/2} = e^{-x^2/2}(x^2 - 1), \quad f''(0) = -1$

$P_1(x) = 1 + 0(x - 0) = 1, \quad P_1(0) = 1$

$P_1'(x) = 0, \quad P_1'(0) = 0$

$P_2(x) = 1 + 0(x - 0) - \frac{1}{2}(x - 0)^2 = 1 - \frac{x^2}{2}, \quad P_2(0) = 1$

$P_2'(x) = -x, \quad P_2'(0) = 0$

$P_2''(x) = -1, \quad P_2''(0) = -1$

The values of f, P_1, P_2 and their first derivatives agree at $x = 0$. The values of the second derivatives of f and P_2 agree at $x = 0$.

41. 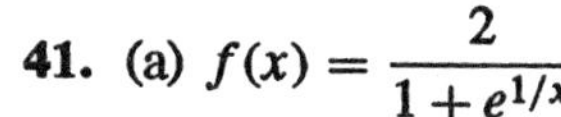
(a) $f(x) = \dfrac{2}{1 + e^{1/x}}$

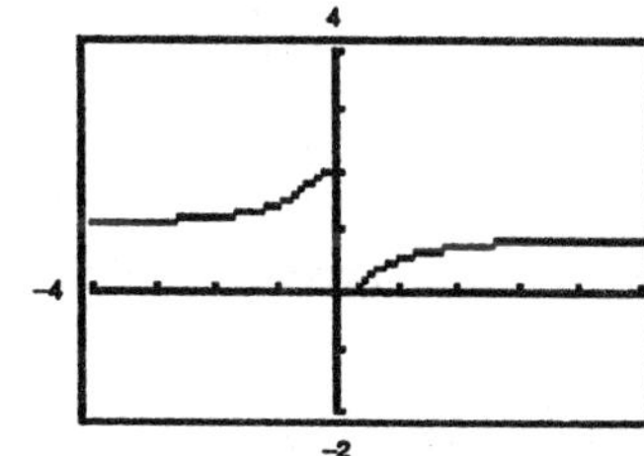

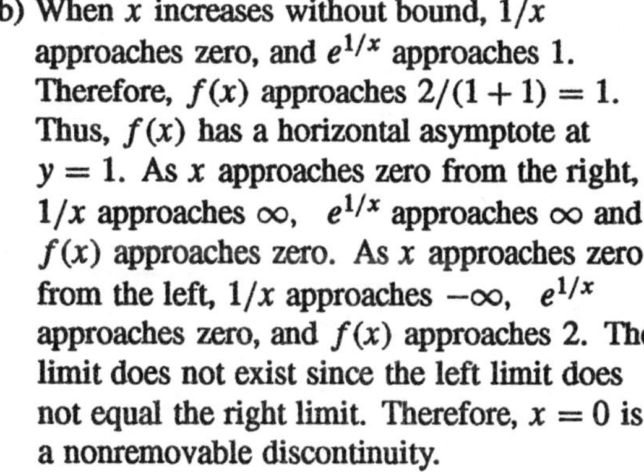
(b) When x increases without bound, $1/x$ approaches zero, and $e^{1/x}$ approaches 1. Therefore, $f(x)$ approaches $2/(1 + 1) = 1$. Thus, $f(x)$ has a horizontal asymptote at $y = 1$. As x approaches zero from the right, $1/x$ approaches ∞, $e^{1/x}$ approaches ∞ and $f(x)$ approaches zero. As x approaches zero from the left, $1/x$ approaches $-\infty$, $e^{1/x}$ approaches zero, and $f(x)$ approaches 2. The limit does not exist since the left limit does not equal the right limit. Therefore, $x = 0$ is a nonremovable discontinuity.

42. $V = 15{,}000e^{-0.6286t}, \quad 0 \le t \le 10$

(a)

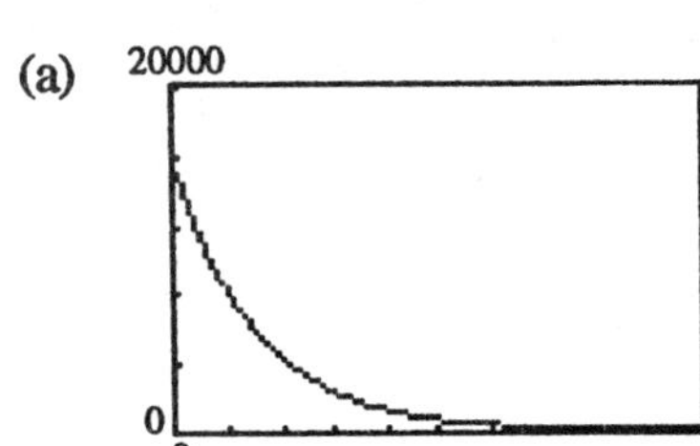

(b)

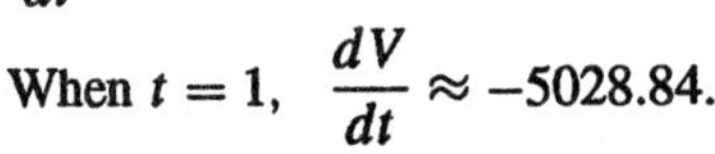

$$\frac{dV}{dt} = -9429e^{-0.6286t}$$

When $t = 1$, $\dfrac{dV}{dt} \approx -5028.84$.

When $t = 5$, $\dfrac{dV}{dt} \approx -406.89$.

(c)

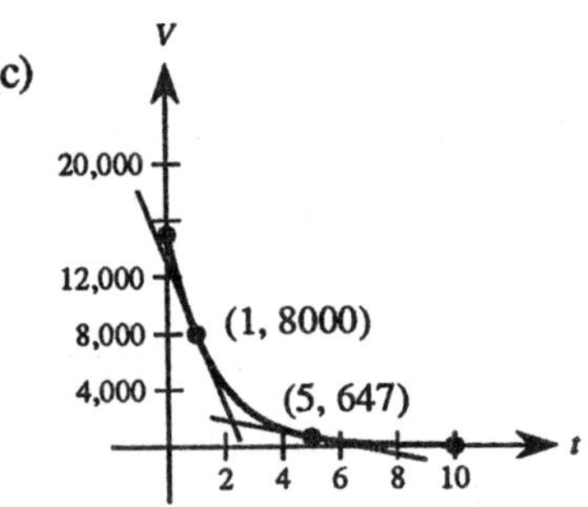

43. Let $u = -2x$, $du = -2\,dx$.

$$\int_0^1 e^{-2x}\,dx = -\frac{1}{2}\int_0^1 e^{-2x}(-2)\,dx = -\frac{1}{2}e^{-2x}\Big]_0^1 = \frac{1}{2}(1 - e^{-2}) = \frac{e^2 - 1}{2e^2}$$

44. Let $u = 1 - x$, $du = -dx$.

$$\int_1^2 e^{1-x}\,dx = -\int_1^2 e^{1-x}(-1)\,dx = -e^{1-x}\Big]_1^2 = 1 - e^{-1} = \frac{e-1}{e}$$

45. Let $u = x^3 - 3x + 1$, $du = 3(x^2 - 1)\,dx$.

$$\int_0^2 (x^2 - 1)e^{x^3-3x+1}\,dx = \frac{1}{3}\int_0^2 e^{x^3-3x+1}3(x^2 - 1)\,dx = \frac{1}{3}e^{x^3-3x+1}\Big]_0^2 = \frac{e}{3}(e^2 - 1)$$

46. Let $u = x^3$, $du = 3x^2\,dx$.

$$\int x^2e^{x^3}\,dx = \frac{1}{3}\int e^{x^3}(3x^2)\,dx = \frac{1}{3}e^{x^3} + C$$

47. $$\int \frac{e^{-x}}{(1+e^{-x})^2}\,dx = -\int (1+e^{-x})^{-2}(-e^{-x})\,dx = (1+e^{-x})^{-1} + C = \frac{1}{1+e^{-x}} + C = \frac{e^x}{e^x+1} + C$$

48. $$\int \frac{e^{2x}}{(1+e^{2x})^2}\,dx = \frac{1}{2}\int (1+e^{2x})^{-2}(2e^{2x})\,dx = \frac{1}{2}\left[\frac{(1+e^{2x})^{-1}}{-1}\right] + C = -\frac{1}{2(1+e^{2x})} + C$$

49. Let $u = ax^2$, $du = 2ax\,dx$.

$$\int xe^{ax^2}\,dx = \frac{1}{2a}\int e^{ax^2}(2ax)\,dx = \frac{1}{2a}e^{ax^2} + C$$

50. Let $u = \dfrac{-x^2}{2}$, $du = -x\,dx$.

$$\int_0^{\sqrt{2}} xe^{-x^2/2}\,dx = -\int_0^{\sqrt{2}} e^{-x^2/2}(-x)\,dx = -e^{-x^2/2}\Big]_0^{\sqrt{2}} = 1 - e^{-1} = \frac{e-1}{e}$$

51. Let $u = \dfrac{3}{x}$, $du = -\dfrac{3}{x^2}\,dx$.

$$\int_1^3 \frac{e^{3/x}}{x^2}\,dx = -\frac{1}{3}\int_1^3 e^{3/x}\left(-\frac{3}{x^2}\right)dx = -\frac{1}{3}e^{3/x}\Big]_1^3 = \frac{e}{3}(e^2 - 1)$$

52. $$\int (e^x - e^{-x})^2\,dx = \int (e^{2x} - 2 + e^{-2x})\,dx = \frac{1}{2}e^{2x} - 2x + \left(-\frac{1}{2}\right)e^{-2x} + C$$

53. $$\int e^{-x}(1+e^{-x})^2\,dx = -\int (1+e^{-x})^2(-e^{-x})\,dx = -\frac{1}{3}(1+e^{-x})^3 + C$$

54. $\int e^{2x}(1-3e^{2x})^{-2}\,dx = -\frac{1}{6}\int (1-3e^{2x})^{-2}(-6e^{2x})\,dx = \frac{1}{6}(1-3e^{2x})^{-1} + C = \frac{1}{6(1-3e^{2x})} + C$

55. Let $u = 1 - e^x$, $du = -e^x\,dx$.

$$\int e^x\sqrt{1-e^x}\,dx = -\int (1-e^x)^{1/2}(-e^x)\,dx = -\frac{2}{3}(1-e^x)^{3/2} + C$$

56. $\int e^x(e^x - e^{-x})\,dx = \int (e^{2x} - 1)\,dx = \frac{1}{2}e^{2x} - x + C$

57. $\int \frac{e^x + e^{-x}}{\sqrt{e^x - e^{-x}}}\,dx = \int (e^x - e^{-x})^{-1/2}(e^x + e^{-x})\,dx = 2(e^x - e^{-x})^{1/2} + C = 2\sqrt{e^x - e^{-x}} + C$

58. Let $u = e^x + e^{-x}$, $du = (e^x - e^{-x})\,dx$.

$$\int \frac{2e^x - 2e^{-x}}{(e^x + e^{-x})^2}\,dx = 2\int (e^x + e^{-x})^{-2}(e^x - e^{-x})\,dx = \frac{-2}{e^x + e^{-x}} + C$$

59. $\int \frac{5 - e^x}{e^{2x}}\,dx = \int 5e^{-2x}\,dx - \int e^{-x}\,dx = -\frac{5}{2}e^{-2x} + e^{-x} + C$

60. $\int \frac{e^{2x} + 2e^x + 1}{e^x}\,dx = \int (e^x + 2 + e^{-x})\,dx = e^x + 2x - e^{-x} + C$

61. $\int_{-2}^{0} (3^3 - 5^2)\,dx = \int_{-2}^{0} 2\,dx = 2x\Big]_{-2}^{0} = 4$

62. $\int (3-x)e^{(3-x)^2}\,dx = -\frac{1}{2}\int e^{(3-x)^2}(-2)(3-x)\,dx = -\frac{1}{2}e^{(3-x)^2} + C$

63. $f'(x) = \int \frac{1}{2}(e^x + e^{-x})\,dx = \frac{1}{2}(e^x - e^{-x}) + C_1$

$f'(0) = C_1 = 0$

$f(x) = \int \frac{1}{2}(e^x - e^{-x})\,dx = \frac{1}{2}(e^x + e^{-x}) + C_2$

$f(0) = 1 + C_2 = 1 \Rightarrow C_2 = 0$

$f(x) = \frac{1}{2}(e^x + e^{-x})$

64. $f'(x) = \int (x + e^{-2x})\,dx = \frac{x^2}{2} - \frac{e^{-2x}}{2} + C_1$

$f'(0) = 0 - \frac{1}{2} + C_1 = -\frac{1}{2} \Rightarrow C_1 = 0$

$f'(x) = \frac{1}{2}(x^2 - e^{-2x})$

$f(x) = \int \frac{1}{2}(x^2 - e^{-2x})\,dx = \frac{1}{2}\left(\frac{x^3}{3} + \frac{e^{-2x}}{2}\right) + C_2$

$f(0) = \frac{1}{2}\left(0 + \frac{1}{2}\right) + C_2 = 3 \Rightarrow C_2 = \frac{11}{4}$

$f(x) = \frac{x^3}{6} + \frac{e^{-2x}}{4} + \frac{11}{4} = \frac{1}{12}(3e^{-2x} + 2x^3 + 33)$

65. $\int_0^5 e^x\,dx = e^x\Big]_0^5 = e^5 - 1 \approx 147.413$

66. $\int_a^b e^{-x}\,dx = -e^{-x}\Big]_a^b = e^{-a} - e^{-b}$

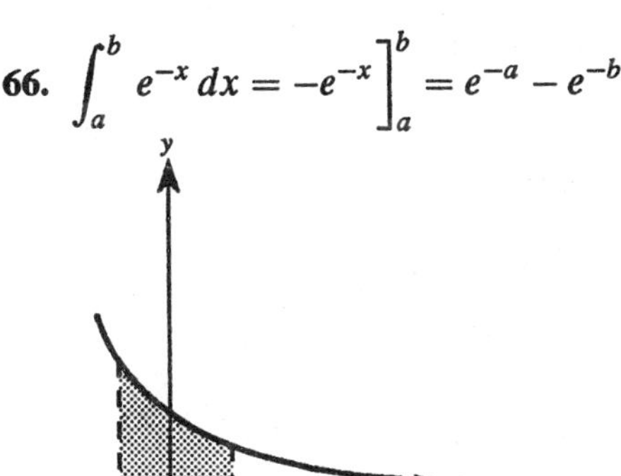

67. $\displaystyle\int_0^{\sqrt{2}} xe^{-(x^2/2)}\,dx = -e^{-(x^2/2)}\Big]_0^{\sqrt{2}}$

$= -e^{-1} + 1 \approx 0.632$

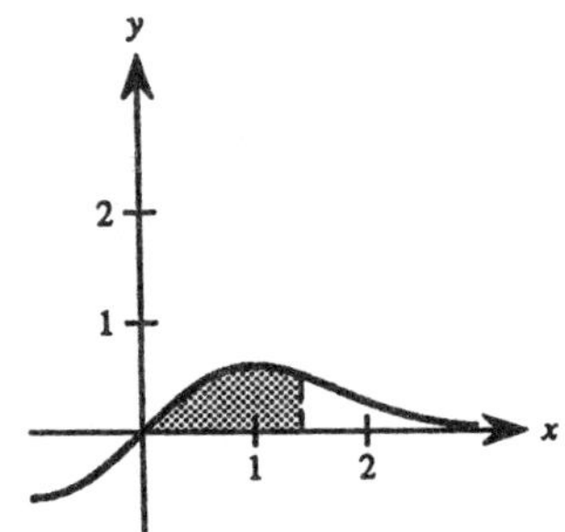

68. $\displaystyle\int_0^2 (e^{-2x} + 2)\,dx = \left[-\frac{1}{2}e^{-2x} + 2x\right]_0^2$

$= -\frac{1}{2}e^{-4} + 4 + \frac{1}{2} \approx 4.4908$

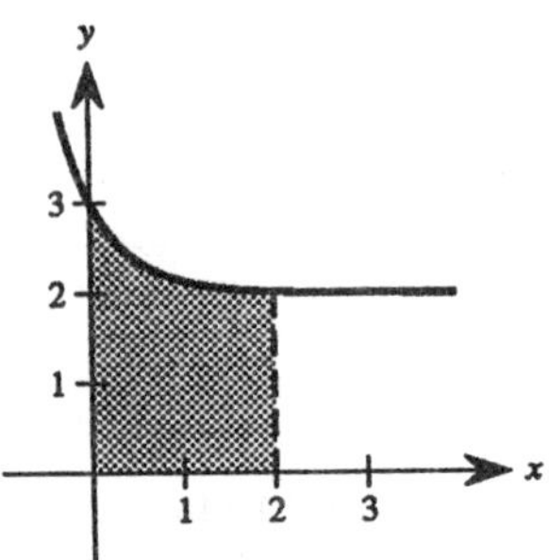

69. $V = \pi \displaystyle\int_0^1 (e^x)^2\,dx$

$= \dfrac{\pi e^{2x}}{2}\Big]_0^1$

$= \dfrac{\pi}{2}(e^2 - 1)$

70. $V = \pi \displaystyle\int_0^4 (e^{-x/2})^2\,dx$

$= -\pi e^{-x}\Big]_0^4$

$= \pi(1 - e^{-4})$

71. $\displaystyle\int_0^x e^t\,dt \ge \int_0^x 1\,dt$

$e^t\Big]_0^x \ge t\Big]_0^x$

$e^x - 1 \ge x \Rightarrow e^x \ge 1 + x$

$x \ge 0$

72. (a) $e^x \ge 1 + x$

$\displaystyle\int_0^x e^t\,dt \ge \int_0^x (1 + t)\,dt$

$e^t\Big]_0^x \ge \left[t + \left(\frac{t^2}{2}\right)\right]_0^x$

$e^x - 1 \ge x + \dfrac{x^2}{2} \Rightarrow e^x \ge 1 + x + \dfrac{x^2}{2}$

$x \ge 0$

When $x = 1$, $2.718281828 > 2.5$.

(b) $e^x \ge 1 + x + \dfrac{x^2}{2}$

$\displaystyle\int_0^x e^t\,dt \ge \int_0^x \left(1 + t + \frac{t^2}{2}\right) dt$

$e^t\Big]_0^x \ge t + \dfrac{t^2}{2} + \dfrac{t^3}{6}\Big]_0^x$

$e^x - 1 \ge x + \dfrac{x^2}{2} + \dfrac{x^3}{6} \Rightarrow e^x \ge 1 + x + \dfrac{x^2}{2} + \dfrac{x^3}{6}$

$x \ge 0$

When $x = 1$, $2.718281828 > 2.666\overline{6}$.

(c) $e^x \ge 1 + x + \dfrac{x^2}{2} + \dfrac{x^3}{6}$

$\displaystyle\int_0^x e^t\,dt > \int_0^x \left(1 + t + \frac{t^2}{2} + \frac{t^3}{6}\right) dt$

$e^x - 1 \ge x + \dfrac{x^2}{2} + \dfrac{x^3}{6} + \dfrac{x^4}{24} \Rightarrow$

$e^x \ge 1 + x + \dfrac{x^2}{2} + \dfrac{x^3}{6} + \dfrac{x^4}{24}$

$x \ge 0$

When $x = 1$, $2.718281828 > 2.7083\overline{3}$.

(d) $e^x \ge 1 + x + \dfrac{x^2}{2} + \dfrac{x^3}{6} + \dfrac{x^4}{24}$

$\displaystyle\int_0^x e^t\,dt \ge \int_0^x \left(1 + t + \frac{t^2}{2} + \frac{t^3}{6} + \frac{t^4}{24}\right) dt$

$e^x - 1 \ge x + \dfrac{x^2}{2} + \dfrac{x^3}{6} + \dfrac{x^4}{24} + \dfrac{x^5}{120}$

$e^x \ge 1 + x + \dfrac{x^2}{2} + \dfrac{x^3}{6} + \dfrac{x^4}{24} + \dfrac{x^5}{120}$

$x \ge 0$

When $x = 1$, $2.718281828 > 2.716\overline{6}$.

73. $f(x) = e^{x^3}$

$f''(x) = 3(3x^4 + 2x)e^{x^3}$ which in [0, 1] is maximum when $x = 1$ and $f''(1) = 15e$.

$f^{(4)}(x) = 3(27x^8 + 108x^5 + 60x^2)e^{x^3}$ which in [0, 1] is maximum when $x = 1$ and $f^{(4)}(1) = 585e$.

Trapezoidal: $\text{Error} \le \dfrac{(1-0)^3}{12(4^2)}(15e) = \dfrac{5e}{64} \approx 0.212$

Simpson's: $\text{Error} \le \dfrac{(1-0)^5}{180(4^4)}(585e) = \dfrac{13e}{1024} \approx 0.035$

74. $f(x) = e^{-x^2}$

$f''(x) = -2(-2x^2 + 1)e^{-x^2}$ and in [0, 1], $|f''(x)|$ is maximum when $x = 0$ and $|f''(0)| = 2$.

$f^{(4)}(x) = -2(-8x^4 + 24x^2 - 6)e^{-x^2}$ and in [0, 1], $|f^{(4)}(x)|$ is maximum when $x = 0$ and $|f^{(4)}(0)| = 12$.

Trapezoidal: $\text{Error} \le \dfrac{(1-0)^3}{12(4^2)}(2) = \dfrac{1}{96} \approx 0.01$

Simpson's: $\text{Error} \le \dfrac{(1-0)^5}{180(4^4)}(12) = \dfrac{1}{3840} \approx 0.00026$

75. $f(x) = \dfrac{1}{\sqrt{2\pi}}e^{-x^2/2}$

$$P(a \le x \le b) = \int_a^b f(x)\,dx$$

$$P(0 \le x \le 1) = \int_0^1 \frac{1}{\sqrt{2\pi}}e^{-x^2/2}\,dx$$

$$\approx \frac{1}{\sqrt{2\pi}}\frac{1}{3(6)}\left[1 + 4e^{-1/72} + 2e^{-1/18} + 4e^{-1/8} + 2e^{-2/9} + 4e^{-25/72} + e^{-1/2}\right] \approx 0.3413 \approx 34.13\%$$

76. $$P(0 \le x \le 2) = \int_0^2 \frac{1}{\sqrt{2\pi}}e^{-x^2/2}\,dx$$

$$\approx \frac{1}{\sqrt{2\pi}}\frac{2}{3(6)}\left[1 + 4e^{-1/18} + 2e^{-2/9} + 4e^{-1/2} + 2e^{-8/9} + 4e^{-25/18} + e^{-2}\right] \approx 0.4772 \approx 47.72\%$$

77. (a) $\displaystyle\int_0^4 \sqrt{x}\,e^x\,dx, \quad n = 12$

Midpoint Rule: 92.1898
Trapezoidal Rule: 93.8371
Simpson's Rule: 92.7385

(b) $\displaystyle\int_0^2 2xe^{-x}\,dx, \quad n = 12$

Midpoint Rule: 1.1906
Trapezoidal Rule: 1.1827
Simpson's Rule: 1.1880

78. $\displaystyle 0.0665\int_{48}^{60} e^{-0.0139(t-48)^2}\,dt$

$n = 24$

Simpson's Rule:
$0.4772 = 47.72\%$

Section 7.3 Inverse Functions

1. $f(x) = x^3$

$g(x) = \sqrt[3]{x}$

$f(g(x)) = f(\sqrt[3]{x})$

$= (\sqrt[3]{x})^3 = x$

$g(f(x)) = g(x^3)$

$= \sqrt[3]{x^3} = x$

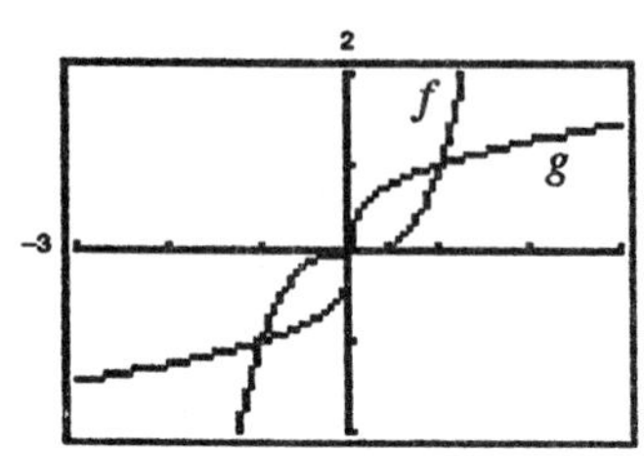

2. $f(x) = \dfrac{1}{x}$

$g(x) = \dfrac{1}{x}$

$f(g(x)) = g(f(x))$

$= \dfrac{1}{1/x} = x$

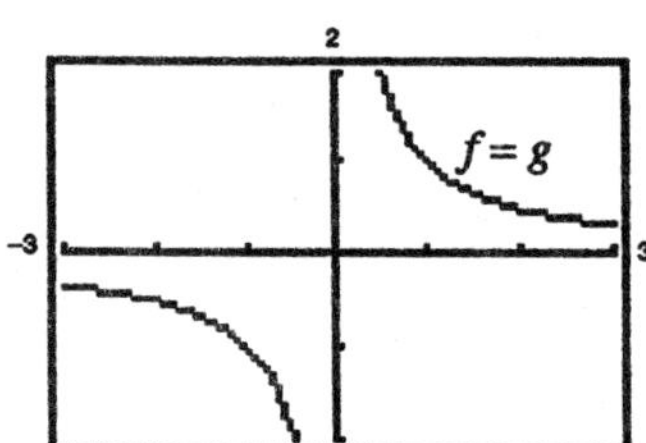

3. $f(x) = 5x + 1$

$g(x) = \dfrac{x-1}{5}$

$f(g(x)) = f\left(\dfrac{x-1}{5}\right)$

$= 5\left(\dfrac{x-1}{5}\right) + 1 = x$

$g(f(x)) = g(5x+1)$

$= \dfrac{(5x+1)-1}{5} = x$

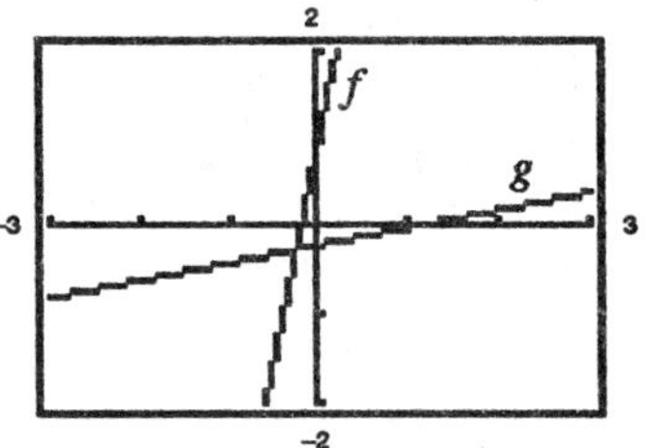

4. $f(x) = 3 - 4x$

$g(x) = \dfrac{3-x}{4}$

$f(g(x)) = f\left(\dfrac{3-x}{4}\right)$

$= 3 - 4\left(\dfrac{3-x}{4}\right) = x$

$g(f(x)) = g(3-4x)$

$= \dfrac{3-(3-4x)}{4} = x$

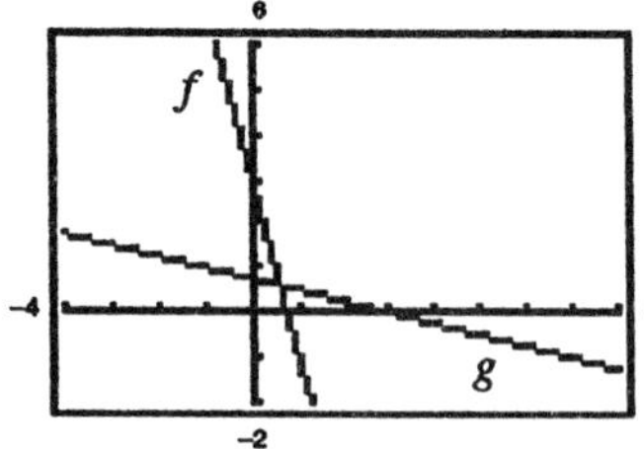

5. $f(x) = \sqrt{x-4}$

$g(x) = x^2 + 4, \quad x \geq 0$

$f(g(x)) = f(x^2+4)$

$= \sqrt{(x^2+4)-4}$

$= \sqrt{x^2} = x$

$g(f(x)) = g(\sqrt{x-4})$

$= (\sqrt{x-4})^2 + 4$

$= x - 4 + 4 = x$

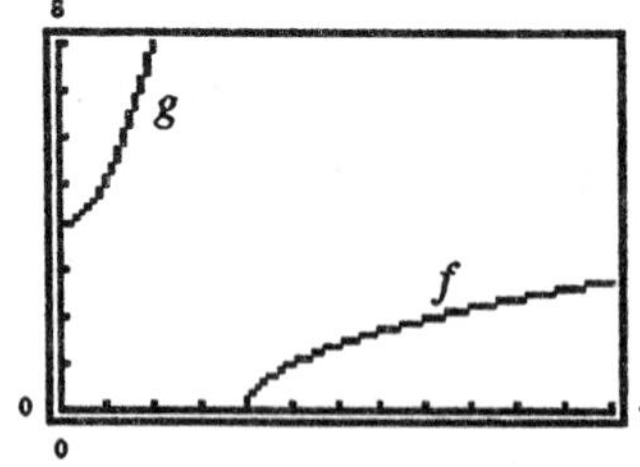

6. $f(x) = 9 - x^2, \quad x \geq 0$

$g(x) = \sqrt{9-x}, \quad x \leq 9$

$f(g(x)) = f(\sqrt{9-x})$

$= 9 - (\sqrt{9-x})^2$

$= 9 - (9-x) = x$

$g(f(x)) = g(9-x^2)$

$= \sqrt{9-(9-x^2)}$

$= \sqrt{x^2} = x$

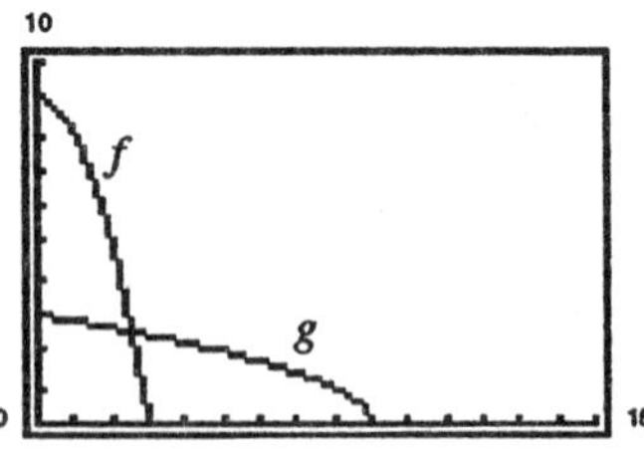

7. $f(x) = 1 - x^3$

$g(x) = \sqrt[3]{1-x}$

$f(g(x)) = f(\sqrt[3]{1-x})$

$= 1 - (\sqrt[3]{1-x})^3$

$= 1 - (1-x) = x$

$g(f(x)) = g(1-x^3)$

$= \sqrt[3]{1-(1-x^3)}$

$= \sqrt[3]{x^3} = x$

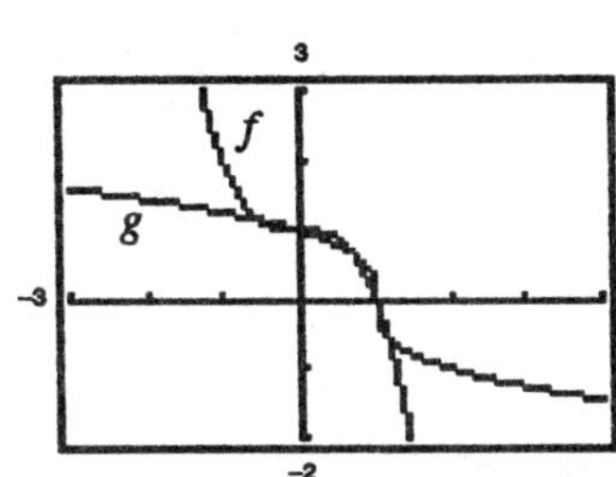

8. $f(x) = x^{-2},\ x > 0$

$g(x) = x^{-1/2},\ x > 0$

$f(g(x)) = f(x^{-1/2})$

$= (x^{-1/2})^{-2} = x,\ x > 0$

$g(f(x)) = g(x^{-2})$

$= (x^{-2})^{-1/2} = x,\ x > 0$

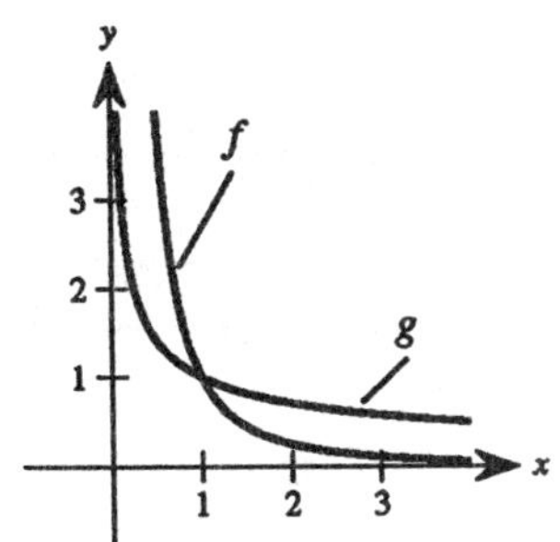

9. $f(x) = 2x - 3 = y$

$x = \dfrac{y+3}{2}$

$f^{-1}(x) = \dfrac{x+3}{2}$

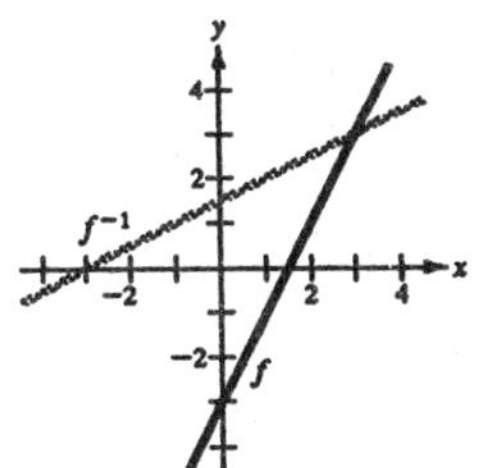

10. $f(x) = 3x = y$

$x = \dfrac{y}{3}$

$f^{-1}(x) = \dfrac{x}{3}$

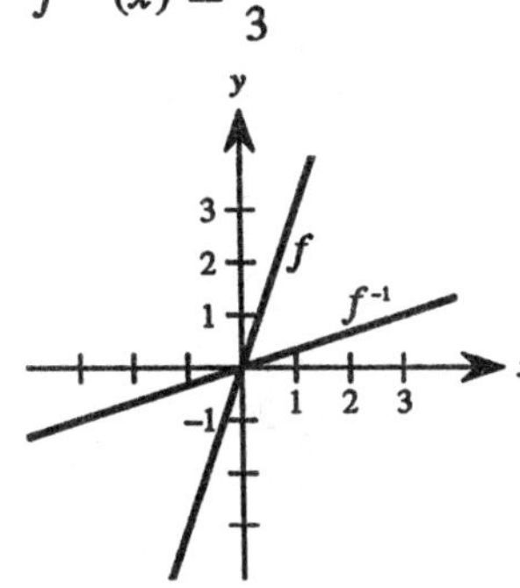

11. $f(x) = x^5 = y$

$x = \sqrt[5]{y}$

$f^{-1}(x) = \sqrt[5]{x}$

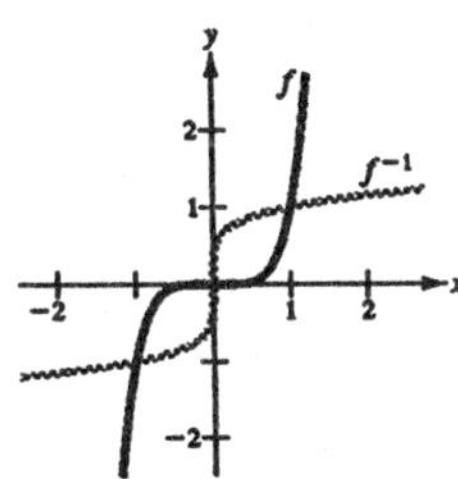

12. $f(x) = x^3 + 1 = y$

$x = \sqrt[3]{y-1}$

$f^{-1}(x) = \sqrt[3]{x-1}$

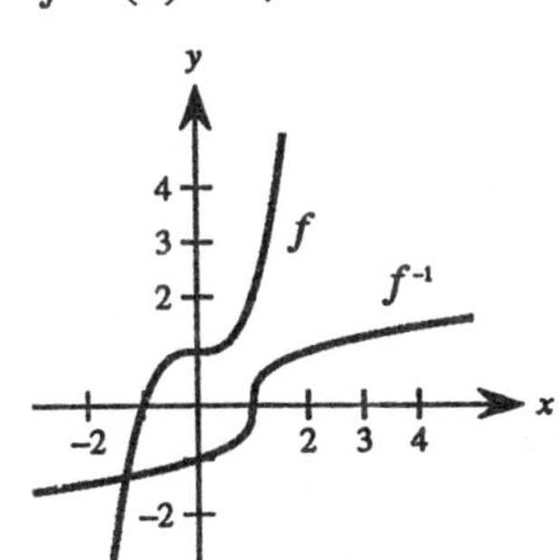

13. $f(x) = \sqrt{x} = y$

$x = y^2$

$f^{-1}(x) = x^2,\quad x \geq 0$

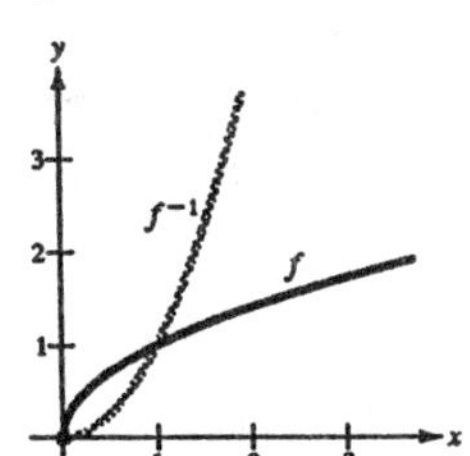

14. $f(x) = x^2 = y,\quad 0 \leq x$

$x = \sqrt{y}$

$f^{-1}(x) = \sqrt{x}$

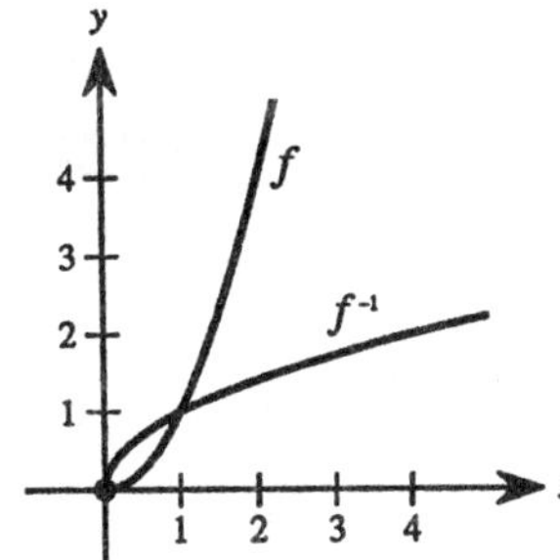

15. $f(x) = \sqrt{4 - x^2} = y, \quad 0 \le x \le 2$

$x = \sqrt{4 - y^2}$

$f^{-1}(x) = \sqrt{4 - x^2}, \quad 0 \le x \le 2$

16. $f(x) = \sqrt{x^2 - 4} = y, \quad x \ge 2$

$x = \sqrt{y^2 + 4}$

$f^{-1}(x) = \sqrt{x^2 + 4}, \quad x \ge 0$

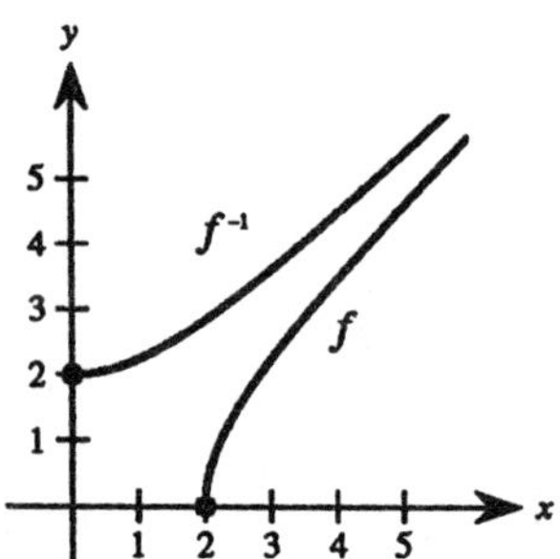

17. $f(x) = \sqrt[3]{x - 1} = y$

$x = y^3 + 1$

$f^{-1}(x) = x^3 + 1$

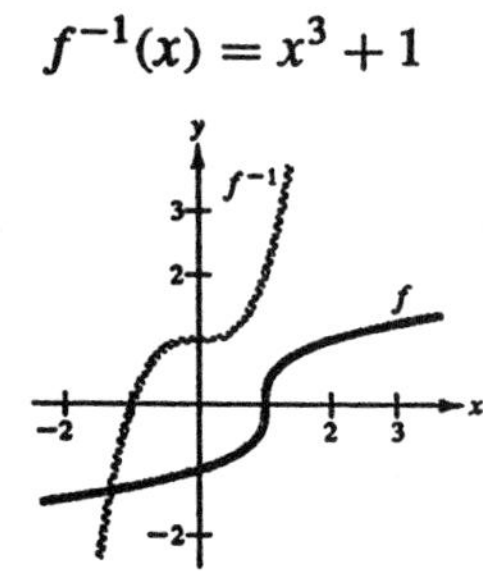

18. $f(x) = 3\sqrt[5]{2x - 1} = y$

$x = \dfrac{y^5 + 243}{486}$

$f^{-1}(x) = \dfrac{x^5 + 243}{486}$

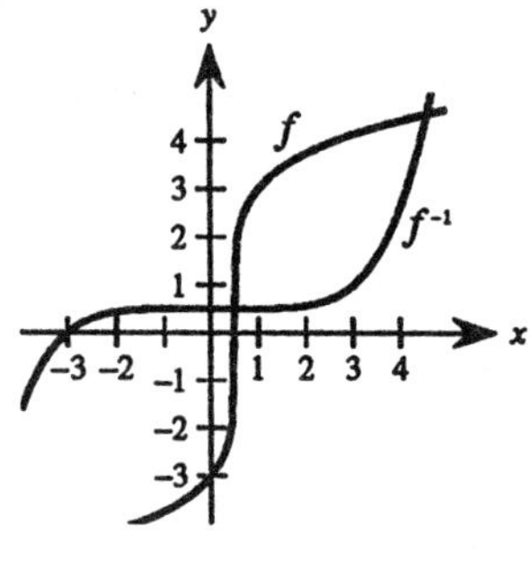

19. $f(x) = x^{2/3} = y, \quad x \ge 0$

$x = y^{3/2}$

$f^{-1}(x) = x^{3/2}, \quad x \ge 0$

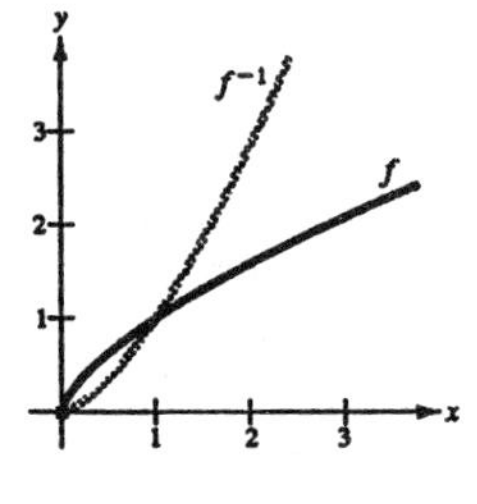

20. $f(x) = x^{3/5}$

$x = y^{5/3}$

$f^{-1}(x) = x^{5/3}$

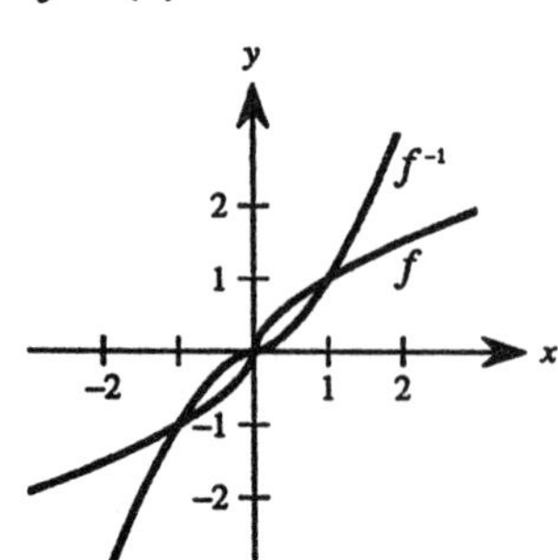

21. $f(x) = \dfrac{x}{\sqrt{x^2 + 7}} = y$

$x = \dfrac{\sqrt{7}\,y}{\sqrt{1 - y^2}}$

$f^{-1}(x) = \dfrac{\sqrt{7}\,x}{\sqrt{1 - x^2}}, \quad -1 < x < 1$

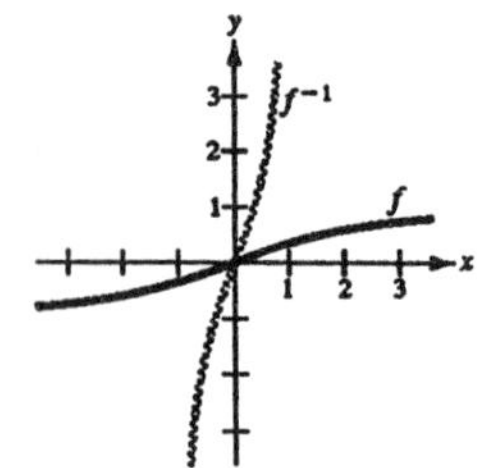

22. $f(x) = \dfrac{x + 2}{x} = y$

$x = \dfrac{2}{y - 1}$

$f^{-1}(x) = \dfrac{2}{x - 1}$

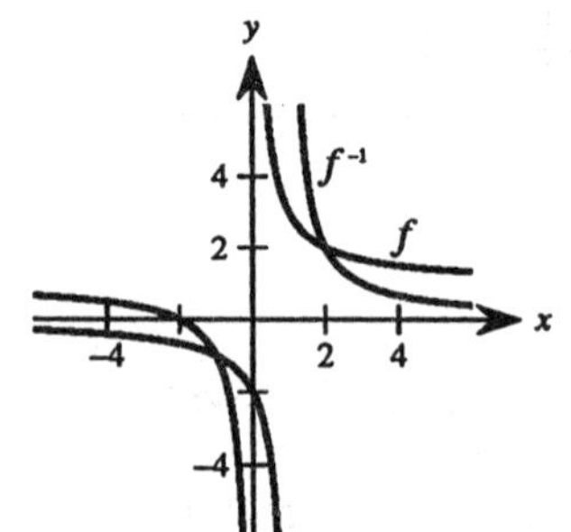

23. $f(x) = \dfrac{x}{x^2-4} = y$ on $(-2,\ 2)$

$$x^2y - 4y = x$$

$$x^2y - x - 4y = 0$$

$$a = y,\quad b = -1,\quad c = -4y$$

$$x = \frac{1 \pm \sqrt{1 - 4(y)(-4y)}}{2y} = \frac{1 \pm \sqrt{1 + 16y^2}}{2y}$$

On $(-2,\ 2)$: $$y = \begin{cases} \dfrac{1 - \sqrt{1 + 16x^2}}{2x}, & \text{if } x \neq 0 \\ 0, & \text{if } x = 0 \end{cases}$$

$$f^{-1}(x) = \begin{cases} \dfrac{1 - \sqrt{1 + 16x^2}}{2x}, & \text{if } x \neq 0 \\ 0, & \text{if } x = 0 \end{cases}$$

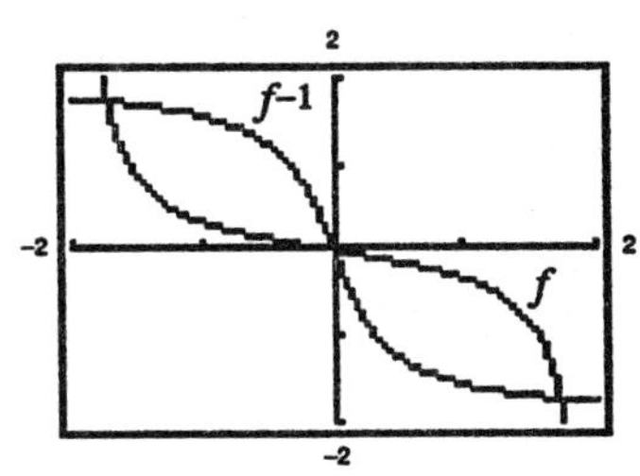

24. $f(x) = 2 - \dfrac{3}{x^2} = y$ on $(0,\ 10)$

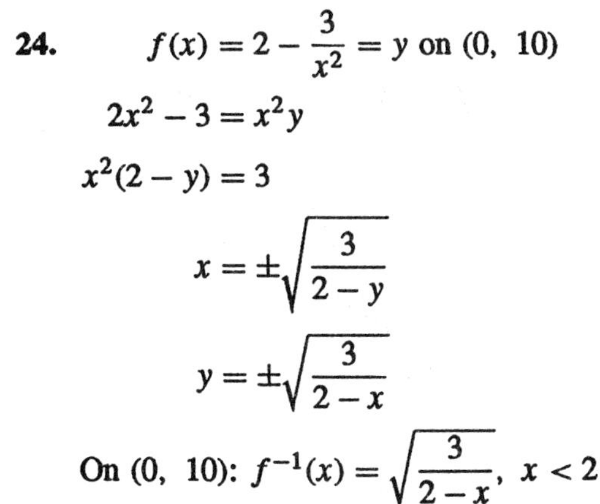

$$2x^2 - 3 = x^2y$$

$$x^2(2 - y) = 3$$

$$x = \pm\sqrt{\frac{3}{2-y}}$$

$$y = \pm\sqrt{\frac{3}{2-x}}$$

On $(0,\ 10)$: $f^{-1}(x) = \sqrt{\dfrac{3}{2-x}},\ x < 2$

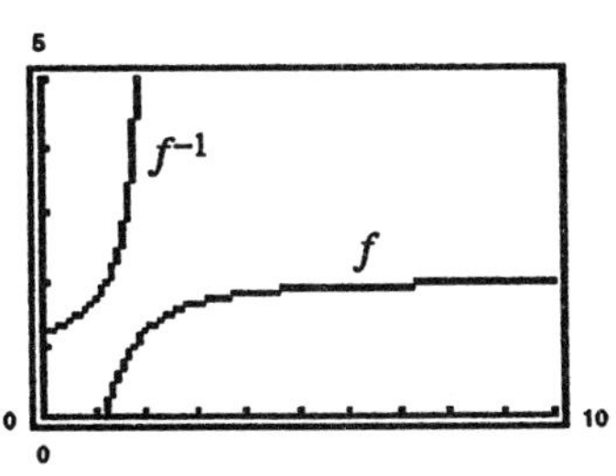

25.

x	1	2	3	4
$f^{-1}(x)$	0	1	2	4

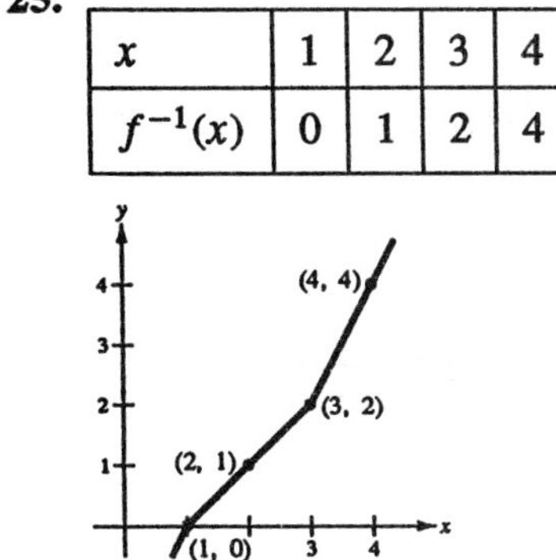

26.

x	0	2	4
$f^{-1}(x)$	6	2	0

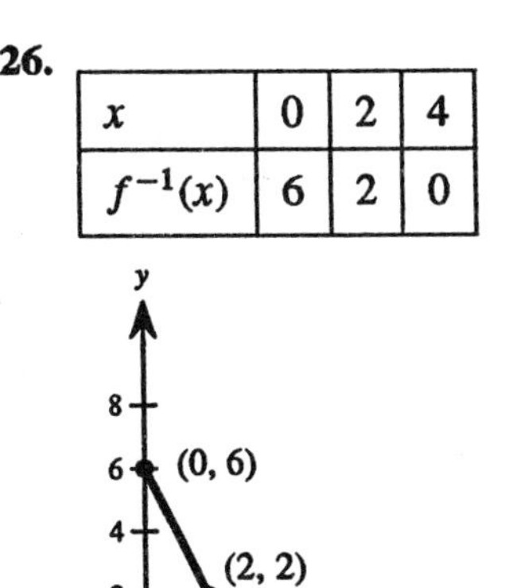

In Exercises 27–30, use the following.

$f(x) = \frac{1}{8}x - 3$ and $g(x) = x^3$

$f^{-1}(x) = 8(x + 3)$ and $g^{-1}(x) = \sqrt[3]{x}$

27. $(f^{-1} \circ g^{-1})(1) = f^{-1}(g^{-1}(1)) = f^{-1}(1) = 32$

28. $(g^{-1} \circ f^{-1})(-3) = g^{-1}(f^{-1}(-3)) = g^{-1}(0) = 0$

29. $(f^{-1} \circ f^{-1})(6) = f^{-1}(f^{-1}(6)) = f^{-1}(72) = 600$

30. $(g^{-1} \circ g^{-1})(-4) = g^{-1}(g^{-1}(-4)) = g^{-1}(\sqrt[3]{-4})$
$= \sqrt[3]{\sqrt[3]{-4}} = -\sqrt[9]{4}$

31. $f(x) = \frac{3}{4}x + 6$
One-to-one; has an inverse

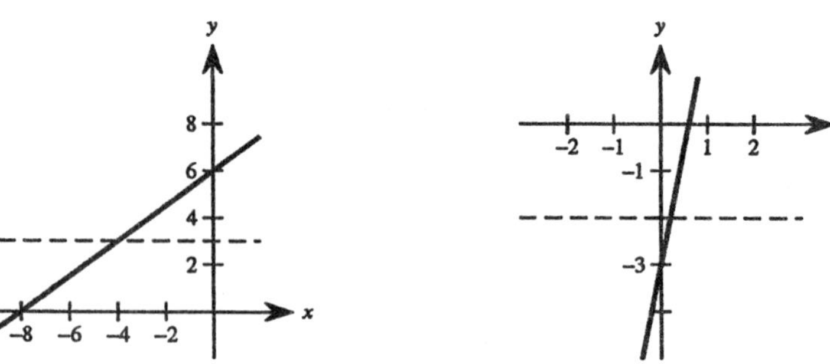
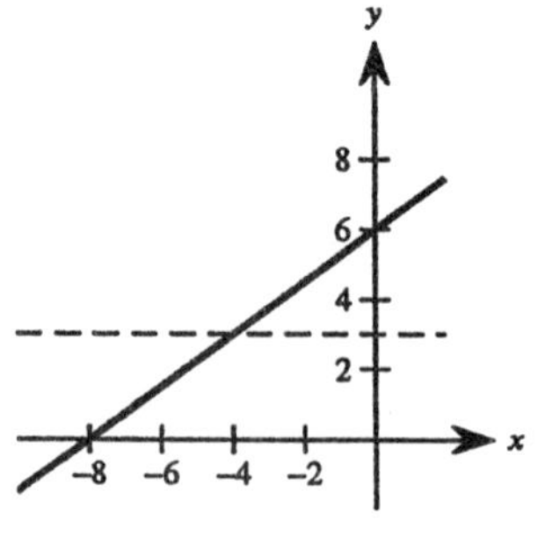

32. $f(x) = 5x - 3$
One-to-one; has an inverse

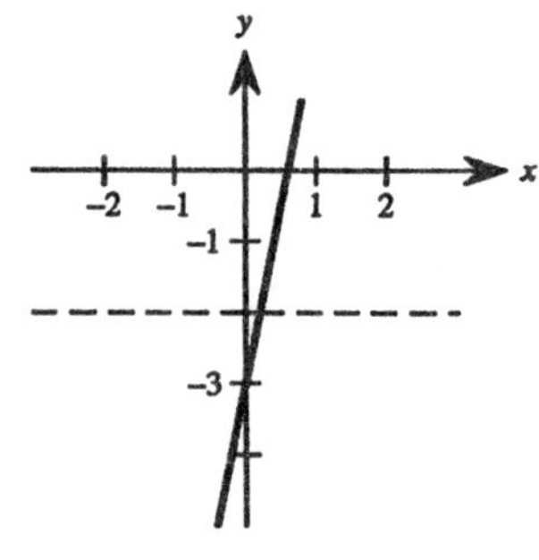

33. $F(x) = \dfrac{x^2}{x^2 + 4}$
Not one-to-one; does not have an inverse

34. $h(s) = \dfrac{1}{s - 2} - 3$
One-to-one; has an inverse

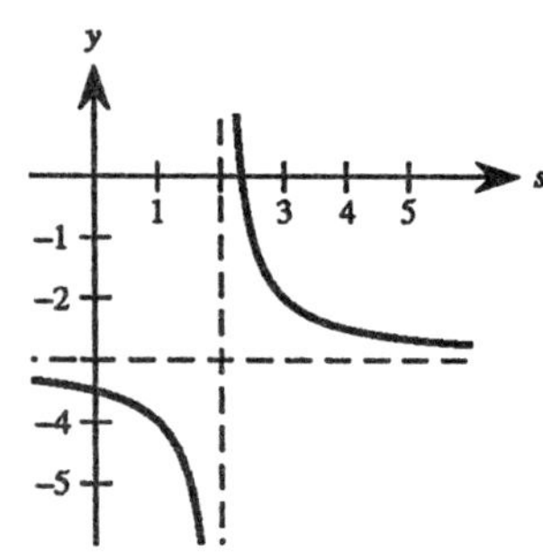

35. $g(t) = \dfrac{t}{\sqrt{t^2 + 1}}$
One-to-one; has an inverse

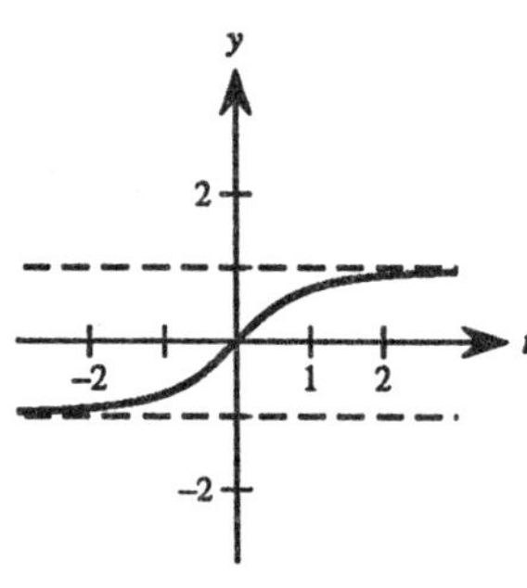

36. $f(x) = 3x\sqrt{x + 1}$
Not one-to-one; does not have an inverse

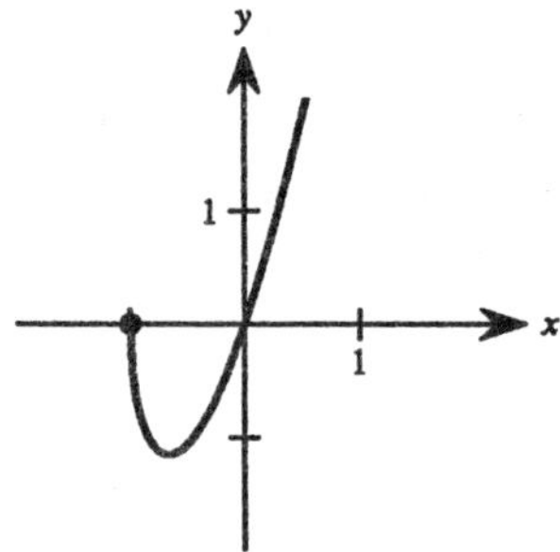

37. $f(x) = (x + a)^3 + b$

$f'(x) = 3(x + a)^2 \geq 0$ for all x

f is increasing on $(-\infty, \infty)$. Therefore, f is strictly monotonic and has an inverse.

38. $f(x) = \dfrac{x^4}{4} - 2x^2$

$f'(x) = x^3 - 4x = 0$ when $x = 0, 2, -2$

f is not strictly monotonic on $(-\infty, \infty)$. Therefore, f does not have an inverse.

39. $f(x) = x^3 - 6x^2 + 12x$

$f'(x) = 3x^2 - 12x + 12 = 3(x - 2)^2 \geq 0$ for all x

f is increasing on $(-\infty, \infty)$. Therefore, f is strictly monotonic and has an inverse.

40. $f(x) = 2 - x - x^3$

$f'(x) = -1 - 3x^2 < 0$ for all x

f is decreasing on $(-\infty, \infty)$. Therefore, f is strictly monotonic and has an inverse.

41. $h(s) = \dfrac{1}{s-2} - 3$

One-to-one; has an inverse

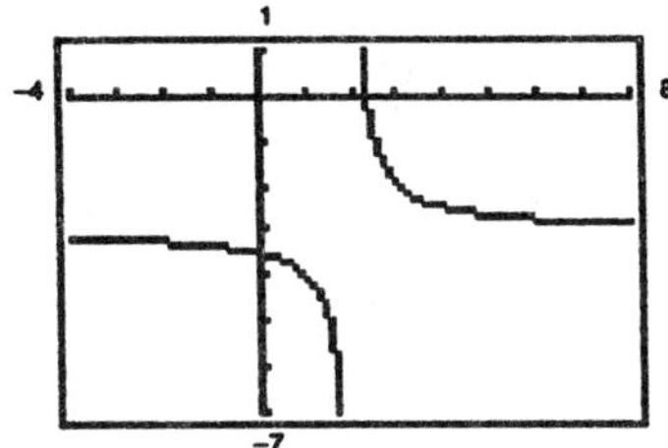

42. $g(t) = \dfrac{1}{\sqrt{t^2+1}}$

Not one-to-one; does not have an inverse

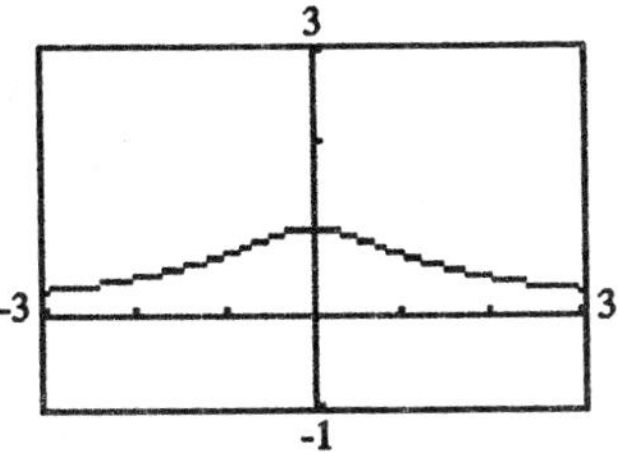

43. $h(x) = |x+4| - |x-4|$

Not one-to-one; does not have an inverse

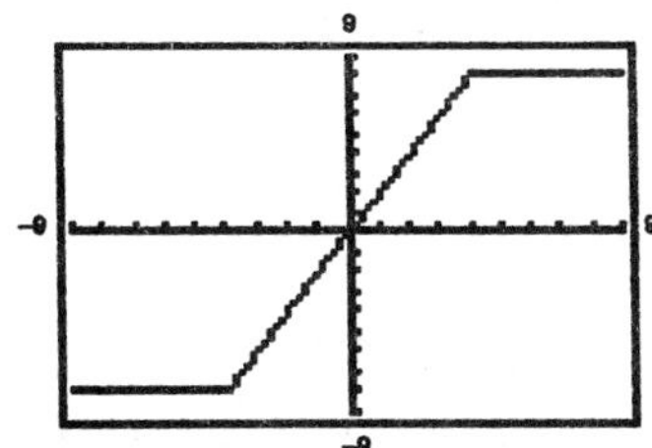

44. $g(x) = (x+5)^3$

One-to-one; has an inverse

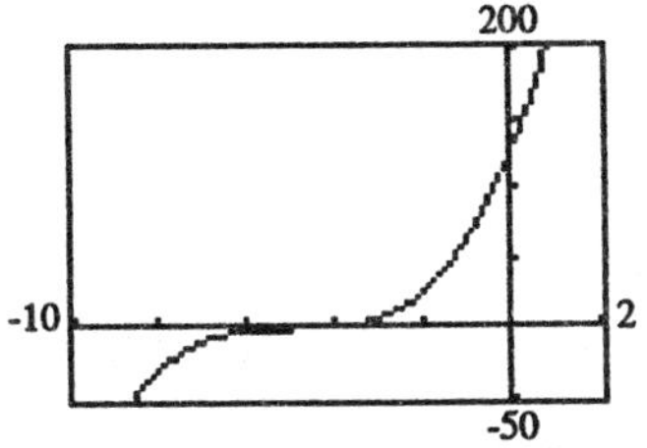

45. $f(x) = (x-4)^2$ on $[4, \infty)$

$f'(x) = 2(x-4) > 0$ on $(4, \infty)$

f is increasing on $[4, \infty)$. Therefore, f is strictly monotonic and has an inverse.

46. $f(x) = |x+2|$ on $[-2, \infty)$

$f'(x) = \dfrac{|x+2|}{x+2}(1) = 1 > 0$ on $(-2, \infty)$

f is increasing on $[-2, \infty)$. Therefore, f is strictly monotonic and has an inverse.

47. $f(x) = \dfrac{4}{x^2}$ on $(0, \infty)$

$f'(x) = -\dfrac{8}{x^3} < 0$ on $(0, \infty)$

f is decreasing on $(0, \infty)$. Therefore, f is strictly monotonic and has an inverse.

48. $f(x) = x^3 - x$

$f'(x) = 3x^2 - 1 > 0$ on $[1, \infty)$

f is increasing on $[1, \infty)$. Therefore, f is strictly monotonic and has an inverse.

49.
$$f(x) = x^3, \quad \left(\frac{1}{2}, \frac{1}{8}\right)$$
$$f'(x) = 3x^2$$
$$f'\left(\frac{1}{2}\right) = \frac{3}{4}$$
$$f^{-1}(x) = \sqrt[3]{x}, \quad \left(\frac{1}{8}, \frac{1}{2}\right)$$
$$(f^{-1})'(x) = \frac{1}{3\sqrt[3]{x^2}}$$
$$(f^{-1})'(x)\left(\frac{1}{8}\right) = \frac{4}{3}$$

50.
$$f(x) = 3 - 4x, \quad (1, -1)$$
$$f'(x) = -4$$
$$f'(1) = -4$$
$$f^{-1}(x) = \frac{3-x}{4}, \quad (-1, 1)$$
$$(f^{-1})'(x) = -\frac{1}{4}$$
$$(f^{-1})'(-1) = -\frac{1}{4}$$

51. $f(x) = \sqrt{x-4}, \quad (5, 1)$

$f'(x) = \dfrac{1}{2\sqrt{x-4}}$

$f'(5) = \dfrac{1}{2}$

$f^{-1}(x) = x^2 + 4, \quad (1, 5)$

$(f^{-1})'(x) = 2x$

$(f^{-1})'(x) = 2$

52. $f(x) = \dfrac{1}{1+x^2}, \quad \left(1, \dfrac{1}{2}\right)$

$f'(x) = \dfrac{-2x}{(1+x^2)^2}$

$f'(1) = -\dfrac{1}{2}$

$f^{-1}(x) = \sqrt{\dfrac{1-x}{x}}, \quad \left(\dfrac{1}{2}, 1\right)$

$(f^{-1})'(x) = -\dfrac{1}{2x^2}\sqrt{\dfrac{x}{1-x}}$

$(f^{-1})'\left(\dfrac{1}{2}\right) = -2$

53. $f(x) = \dfrac{1}{x}$

$f'(x) = -\dfrac{1}{x^2} < 0$

f is not strictly monotonic since f is not continuous at $x = 0$.

54. $f(x) = \dfrac{x}{x^2 - 4}$

$f'(x) = -\dfrac{x^2+4}{(x^2-4)^2} < 0$

f is not one-to-one since f is not continuous at $x = \pm 2$.

Example: $f\left(\dfrac{1+\sqrt{17}}{2}\right) = f\left(\dfrac{1-\sqrt{17}}{2}\right) = 1$

55. Suppose $g(x)$ and $h(x)$ are both inverses of $f(x)$. Then the graph of $f(x)$ contains the point (a, b) if and only if the graphs of $g(x)$ and $h(x)$ contain the point (b, a). Since the graphs of $g(x)$ and $h(x)$ are the same, $g(x) = h(x)$. Therefore, the inverse of $f(x)$ is unique.

56. If f has an inverse, then f and f^{-1} are both one-to-one. Let $(f^{-1})^{-1}(x) = y$ then $x = f^{-1}(y)$ and $f(x) = y$. Thus, $(f^{-1})^{-1} = f$.

57. If f has an inverse and $f(x_1) = f(x_2)$, then $f^{-1}(f(x_1)) = f^{-1}(f(x_2)) \Rightarrow x_1 = x_2$. Therefore, f is one-to-one. If $f(x)$ is one-to-one, then for every value b in the range, there corresponds exactly one value a in the domain. Define $g(x)$ such that the domain of g equals the range of f and $g(b) = a$. By the reflexive property of inverses, $g = f^{-1}$.

58. Let $(f \circ g)(x) = y$ then $x = (f \circ g)^{-1}(y)$. Also,

$$(f \circ g)(x) = y$$
$$f(g(x)) = y$$
$$g(x) = f^{-1}(y)$$
$$x = g^{-1}(f^{-1}(y)) = (g^{-1} \circ f^{-1})(y)$$

Since f and g are one-to-one functions, $(f \circ g)^{-1} = g^{-1} \circ f^{-1}$.

Section 7.4 Logarithmic Functions

1. (a) $2^3 = 8$

$\log_2 8 = 3$

(b) $3^{-1} = \frac{1}{3}$

$\log_3 \frac{1}{3} = -1$

2. (a) $27^{2/3} = 9$

$\log_{27} 9 = \frac{2}{3}$

(b) $16^{3/4} = 8$

$\log_{16} 8 = \frac{3}{4}$

3. (a) $\log_{10} 0.01 = -2$

$10^{-2} = 0.01$

(b) $\log_{0.5} 8 = -3$

$0.5^{-3} = 8$

$\left(\frac{1}{2}\right)^{-3} = 8$

4. (a) $e^{\circ} = 1$

$\ln 1 = 0$

(b) $e^2 = 7.389\ldots$

$\ln 7.389\ldots = 2$

5. (a) $\ln 2 = 0.6931\ldots$

$e^{0.6931\ldots} = 2$

(b) $\ln 8.4 = 2.128\ldots$

$e^{2.128\ldots} = 8.4$

6. (a) $\ln 0.5 = -0.6931\ldots$

$e^{-0.6931\ldots} = 0.5$

(b) $49^{1/2} = 7$

$\log_{49} 7 = \frac{1}{2}$

7. (a) $\log_{10} 1000 = x$

$10^x = 1000$

$x = 3$

(b) $\log_{10} 0.1 = x$

$10^x = 0.1$

$x = -1$

8. (a) $\log_4 \frac{1}{64} = x$

$4^x = \frac{1}{64}$

$x = -3$

(b) $\log_5 25 = x$

$5^x = 25$

$x = 2$

9. (a) $\log_3 x = -1$

$3^{-1} = x$

$x = \frac{1}{3}$

(b) $\log_2 x = -4$

$2^{-4} = x$

$x = \frac{1}{16}$

10. (a) $\log_b 27 = 3$

$b^3 = 27$

$b = 3$

(b) $\log_b 125 = 3$

$b^3 = 125$

$b = 5$

11. (a) $\log_{27} x = -\frac{2}{3}$

$27^{-2/3} = x$

$\frac{1}{9} = x$

(b) $\ln e^x = 3$

$e^x = e^3$

$x = 3$

12. (a) $e^{\ln x} = 4$

$x = 4$

(b) $\ln x = 2$

$x = e^2$

13. (a) $x^2 - x = \log_5 25$

$5^{x^2 - x} = 25$

$5^{x^2 - x} = 5^2$

$x^2 - x = 2$

$x^2 - x - 2 = 0$

$(x + 1)(x - 2) = 0$

$x = -1$ or $x = 2$

(b) $3x + 5 = \log_2 64$

$2^{3x+5} = 64$

$2^{3x+5} = 2^6$

$3x + 5 = 6$

$3x = 1$

$x = \frac{1}{3}$

14. (a) $\log_3 x + \log_3(x-2) = 1$

$$\log_3[x(x-2)] = 1$$
$$x(x-2) = 3^1$$
$$x^2 - 2x = 3$$
$$x^2 - 2x - 3 = 0$$
$$(x+1)(x-3) = 0$$
$$x = -1 \text{ or } x = 3$$

$x = 3$ is the only solution since the domain of the logarithmic function is the set of all *positive* real numbers.

(b) $\log_{10}(x+3) - \log_{10} x = 1$

$$\log_{10} \frac{x+3}{x} = 1$$
$$\frac{x+3}{x} = 10^1$$
$$x + 3 = 10x$$
$$3 = 9x$$
$$x = \frac{1}{3}$$

15. $f(x) = 3 \ln x$

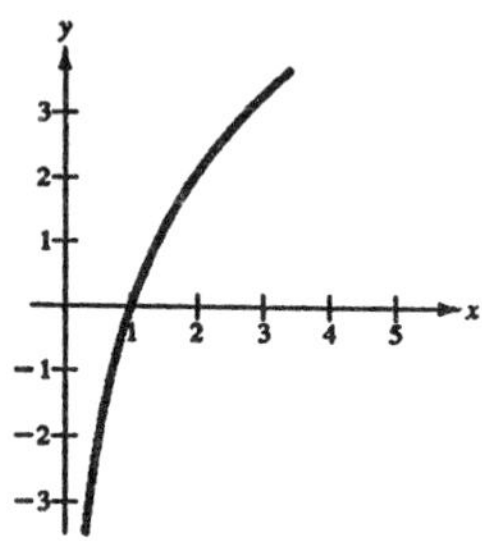

16. $f(x) = -2 \ln x$

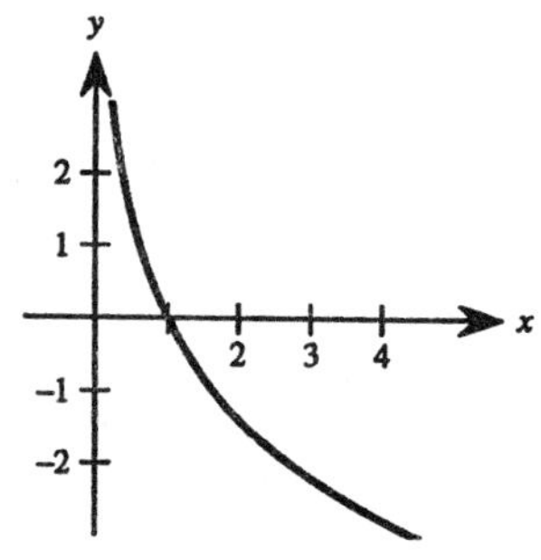

17. $f(x) = \ln 2x$

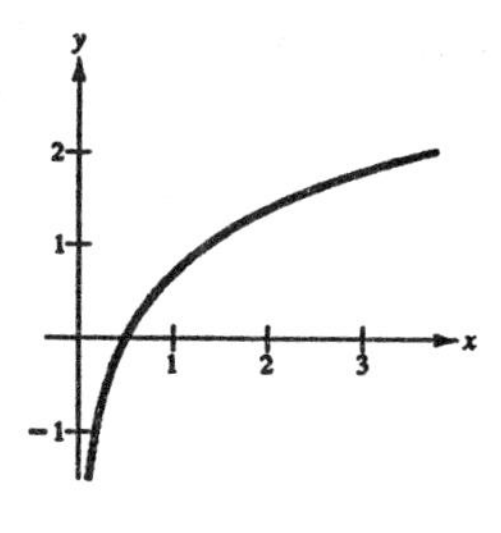

18. $f(x) = \ln |x|$

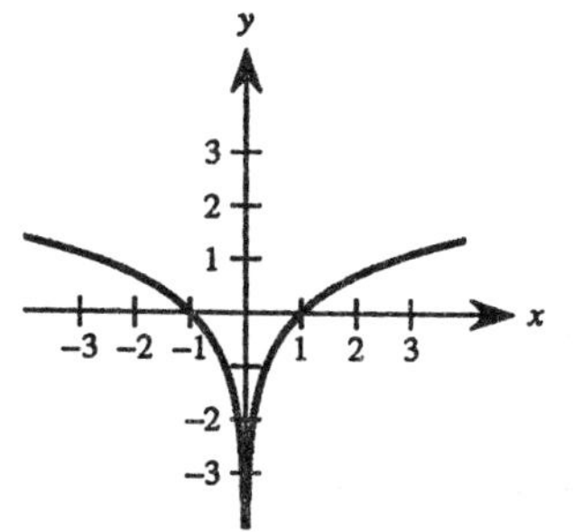

19. $f(x) = \ln(x-1)$

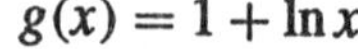

20. $f(x) = 2 + \ln x$

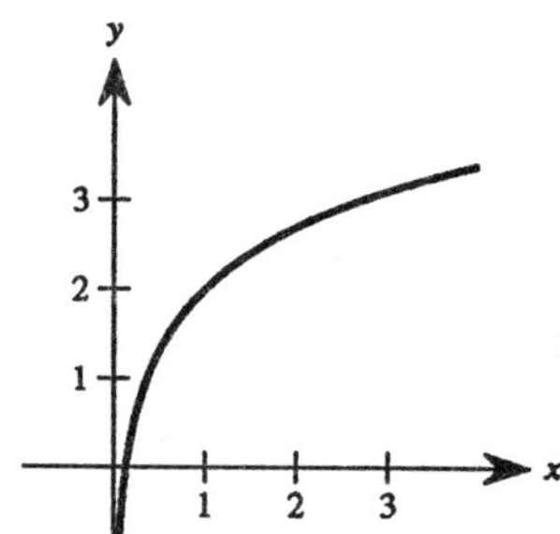

21. $f(x) = e^{2x}$

$g(x) = \ln \sqrt{x} = \frac{1}{2} \ln x$

22. $f(x) = e^x - 1$

$g(x) = \ln(x+1)$

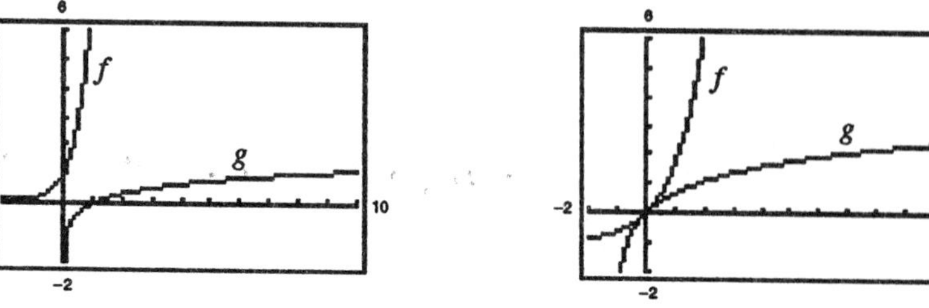

23. $f(x) = e^{x-1}$

$g(x) = 1 + \ln x$

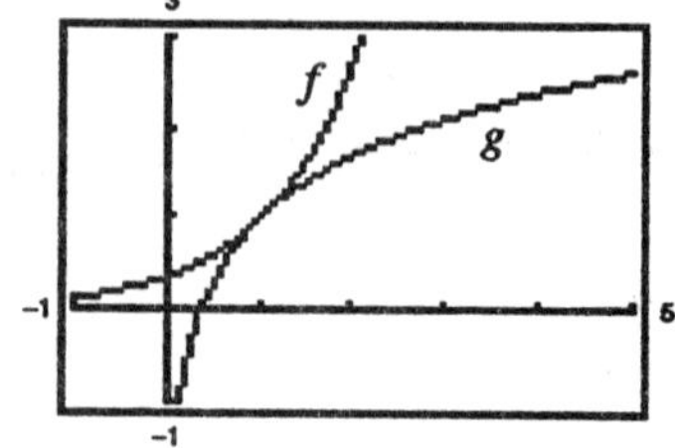

24. $f(x) = e^{x/3}$

$g(x) = \ln x^3 = 3 \ln x$

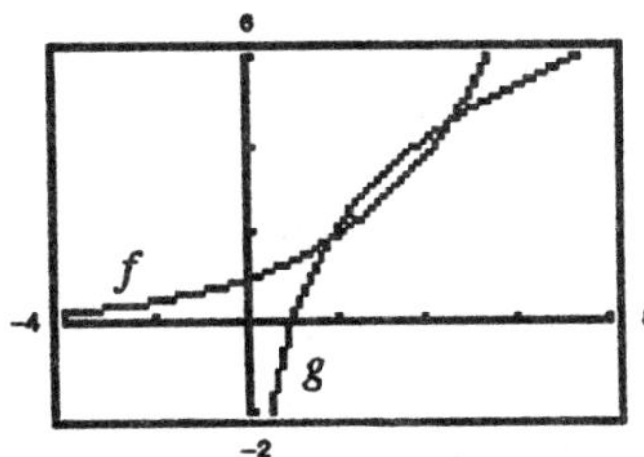

25. $\ln e^{x^2} = x^2$

26. $\ln e^{2x-1} = 2x - 1$

27. $e^{\ln(5x+2)} = 5x + 2$

28. $-1 + \ln e^{2x} = -1 + 2x$

29. $e^{\ln\sqrt{x}} = \sqrt{x}$

30. $-8 + e^{\ln x^3} = -8 + x^3$

31. (a) $\ln 6 = \ln 2 + \ln 3 = 1.7917$
(b) $\ln \frac{2}{3} = \ln 2 - \ln 3 = -0.4055$
(c) $\ln 81 = 4 \ln 3 = 4.3944$
(d) $\ln \sqrt{3} = \frac{1}{2} \ln 3 = 0.5493$

32. (a) $\ln 0.25 = \ln \frac{1}{4} = \ln 1 - 2 \ln 2 = -1.3862$
(b) $\ln 24 = 3 \ln 2 + \ln 3 = 3.1779$
(c) $\ln \sqrt[3]{12} = \frac{1}{3}(2 \ln 2 + \ln 3) = 0.8283$
(d) $\ln \frac{1}{72} = \ln 1 - (3 \ln 2 + 2 \ln 3) = -4.2765$

33. $\ln \frac{2}{3} = \ln 2 - \ln 3$

34. $\ln(xyz) = \ln x + \ln y + \ln z$

35. $\ln \dfrac{xy}{z} = \ln x + \ln y - \ln z$

36. $\ln \sqrt{a-1} = \ln(a-1)^{1/2} = \left(\frac{1}{2}\right) \ln(a-1)$

37. $\ln \sqrt{2^3} = \ln 2^{3/2} = \frac{3}{2} \ln 2$

38. $\ln \frac{1}{5} = \ln 1 - \ln 5 = -\ln 5$

39.
$$\begin{aligned}\ln\left(\frac{x^2-1}{x^3}\right)^3 &= 3[\ln(x^2-1) - \ln x^3] \\ &= 3[\ln(x+1) + \ln(x-1) - 3 \ln x]\end{aligned}$$

40. $\ln 3e^2 = \ln 3 + 2 \ln e = 2 + \ln 3$

41. $\ln z(z-1)^2 = \ln z + \ln(z-1)^2 = \ln z + 2 \ln(z-1)$

42. $\ln \dfrac{1}{e} = \ln 1 - \ln e = -1$

43. $\ln(x-2) - \ln(x+2) = \ln \dfrac{x-2}{x+2}$

44.
$$\begin{aligned}3 \ln x + 2 \ln y - 4 \ln z &= \ln x^3 + \ln y^2 - \ln z^4 \\ &= \ln \frac{x^3 y^2}{z^4}\end{aligned}$$

45. $\dfrac{1}{3}[2 \ln(x+3) + \ln x - \ln(x^2-1)] = \ln \sqrt[3]{\dfrac{x(x+3)^2}{x^2-1}}$

46. $2[\ln x - \ln(x+1) - \ln(x-1)] = \ln\left(\dfrac{x}{x^2-1}\right)^2$

47. $2 \ln 3 - \dfrac{1}{2} \ln(x^2+1) = \ln \dfrac{9}{\sqrt{x^2+1}}$

48. $\dfrac{3}{2}[\ln(x^2+1) - \ln(x+1) - \ln(x-1)] = \dfrac{3}{2} \ln \dfrac{x^2+1}{(x+1)(x-1)} = \ln \sqrt{\left(\dfrac{x^2+1}{x^2-1}\right)^3}$

49.
$$\begin{aligned}e^{\ln x} &= 4 \\ x &= 4\end{aligned}$$

50.
$$\begin{aligned}e^{\ln x^2} - 9 &= 0 \\ x^2 - 9 &= 0 \\ x^2 &= 9 \\ x &= \pm 3\end{aligned}$$

51.
$$\begin{aligned}\ln x &= 0 \\ x &= e^0 \\ x &= 1\end{aligned}$$

52.
$$\begin{aligned}2 \ln x &= 4 \\ \ln x &= 2 \\ x &= e^2 \approx 7.3891\end{aligned}$$

53. $e^{x+1} = 4$

$x + 1 = \ln 4$

$x = -1 + \ln 4 \approx 0.3863$

54. $e^{-0.5x} = 0.075$

$-0.5x = \ln 0.075$

$x = -2 \ln 0.075 \approx 5.1805$

55. $500e^{-0.11t} = 600$

$$e^{-0.11t} = \frac{6}{5}$$

$$-0.11t = \ln \frac{6}{5}$$

$$t = \frac{\ln(6/5)}{-0.11}$$

$$\approx -1.6575$$

56. $e^{-0.0174t} = 0.5$

$-0.0174t = \ln 0.5$

$$t = \frac{\ln 0.5}{-0.0174}$$

$$\approx 39.8360$$

57. $5^{2x} = 15$

$\ln 5^{2x} = \ln 15$

$2x \ln 5 = \ln 15$

$$x = \frac{\ln 15}{2 \ln 5} \approx 0.8413$$

58. $2^{1-x} = 6$

$\ln 2^{1-x} = \ln 6$

$$1 - x = \frac{\ln 6}{\ln 2}$$

$$x = 1 - \frac{\ln 6}{\ln 2} \approx -1.5850$$

59. $500(1.07)^t = 1000$

$(1.07)^t = 2$

$\ln 1.07^t = \ln 2$

$t \ln 1.07 = \ln 2$

$$t = \frac{\ln 2}{\ln 1.07} \approx 10.2448$$

60.

$$1000\left(1 + \frac{0.07}{12}\right)^{12t} = 3000$$

$$\left(1 + \frac{0.07}{12}\right)^{12t} = 3$$

$$12t \ln\left(1 + \frac{0.07}{12}\right) = \ln 3$$

$$t = \frac{\ln 3}{12 \ln[1 + (0.07/12)]} \approx 15.7402$$

61. $P = \$1000,\quad r = 0.11$

(a) $$1000\left(1 + \frac{0.11}{1}\right)^t = 2000$$

$1.11^t = 2$

$t \ln 1.11 = \ln 2$

$$t = \frac{\ln 2}{\ln 1.11} \approx 6.6419$$

(b) $$1000\left(1 + \frac{0.11}{12}\right)^{12t} = 2000$$

$$12t \ln\left(1 + \frac{0.11}{12}\right) = \ln 2$$

$$t = \frac{\ln 2}{12 \ln[1 + (0.11/12)]} \approx 6.3302$$

(c) $$1000\left(1 + \frac{0.11}{365}\right)^{365t} = 2000$$

$$365t \ln\left(1 + \frac{0.11}{365}\right) = \ln 2$$

$$t = \frac{\ln 2}{365 \ln[1 + (0.11/365)]} \approx 6.3023$$

(d) $1000e^{0.11t} = 2000$

$0.11t = \ln 2$

$$t = \frac{\ln 2}{0.11} \approx 6.3013$$

62. $P = \$1000, \quad r = 0.105$

(a) $1000\left(1 + \frac{0.105}{1}\right)^t = 3000$

$$t \ln 1.105 = \ln 3$$

$$t = \frac{\ln 3}{\ln 1.105} \approx 11.0031$$

(b) $1000\left(1 + \frac{0.105}{12}\right)^{12t} = 3000$

$$12t \ln\left(1 + \frac{0.105}{12}\right) = \ln 3$$

$$t = \frac{\ln 3}{12 \ln[1 + (0.105/12)]} \approx 10.5087$$

(c) $1000\left(1 + \frac{0.105}{365}\right)^{365t} = 3000$

$$365t \ln\left(1 + \frac{0.105}{365}\right) = \ln 3$$

$$t = \frac{\ln 3}{365 \ln[1 + (0.105/365)]} \approx 10.4645$$

(d) $1000e^{0.105t} = 3000$

$$0.105t = \ln 3$$

$$t = \frac{\ln 3}{0.105} \approx 10.4630$$

63. $Pe^{rt} = 3P$

$$e^{rt} = 3$$

$$rt = \ln 3$$

$$t = \frac{\ln 3}{r}$$

r	2%	4%	6%	8%	10%	12%
t	54.9306	27.4653	18.3102	13.7327	10.9861	9.1551

64. (a) $350 = 500 - 0.5e^{0.004x}$

$$300 = e^{0.004x}$$

$$x = \frac{\ln 300}{0.004} \approx 1425.95 \approx 1426$$

(b) $300 = 500 - 0.5e^{0.004x}$

$$400 = e^{0.004x}$$

$$x = \frac{\ln 400}{0.004} \approx 1497.87 \approx 1498$$

65.

x	y	$\frac{\ln x}{\ln y}$	$\ln \frac{x}{y}$	$\ln x - \ln y$
1	2	0	−0.6931	−0.6931
3	4	0.7925	−0.2877	−0.2877
10	5	1.4307	0.6931	0.6931
4	0.5	−2	2.0794	2.0794

66. When $x = 1000$, $\frac{x}{\ln x} \approx 144.7648$.

When $x = 1{,}000{,}000$, $\frac{x}{\ln x} \approx 72{,}382.4137$.

When $x = 1{,}000{,}000{,}000$, $\frac{x}{\ln x} \approx 48{,}254{,}942.43$.

$$\frac{p(1000)}{144.7648} \approx 1.1605$$

$$\frac{p(10^6)}{72{,}382.4137} \approx 1.0844$$

$$\frac{p(10^9)}{48{,}254{,}942.43} \approx 1.0537$$

67.

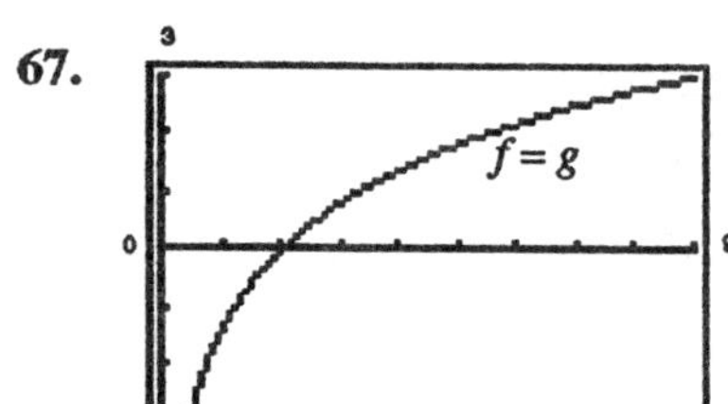

68.

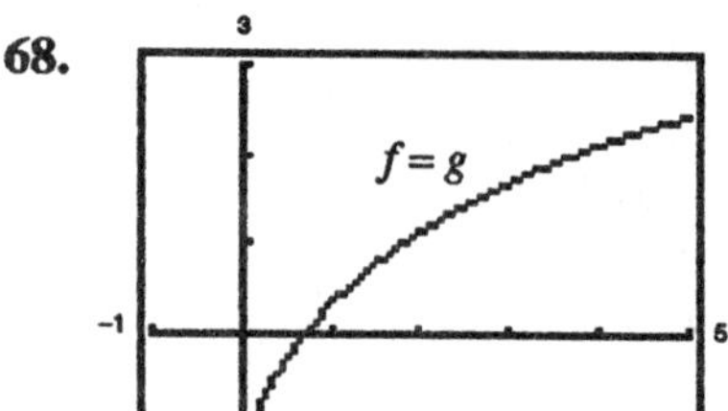

69. (a) $\log_3 7 = \dfrac{\log 7}{\log 3} = \dfrac{\ln 7}{\ln 3} \approx 1.771$

(b) $\log_7 4 = \dfrac{\log 4}{\log 7} = \dfrac{\ln 4}{\ln 7} \approx 0.712$

(c) $\log_{1/2} 10 = \dfrac{1}{\log 1/2} = \dfrac{\ln 10}{\ln 1/2} \approx -3.322$

(d) $\log_4 0.55 = \dfrac{\log 0.55}{\log 4} = \dfrac{\ln 0.55}{\ln 4} \approx -0.431$

70. (a) $\log_9 0.4 = \dfrac{\log 0.4}{\log 9} = \dfrac{\ln 0.4}{\ln 9} \approx -0.417$

(b) $\log_2 0.125 = \dfrac{\log 0.125}{\log 2} = \dfrac{\ln 0.125}{\ln 2} \approx -3$

(c) $\log_{15} 1250 = \dfrac{\log 1250}{\log 15} = \dfrac{\ln 1250}{\ln 15} \approx 2.633$

(d) $\log_{1/3} 0.015 = \dfrac{\log 0.015}{\log(1/3)} = \dfrac{\ln 0.015}{\ln(1/3)} \approx 3.823$

71. Let $a = \ln \dfrac{x}{y}$, $b = \ln x$, $c = \ln y$, then

$$e^a = \frac{x}{y}, \quad e^b = x, \quad e^c = y$$

$$\frac{x}{y} = \frac{e^b}{e^c} = e^{b-c} = e^a \Rightarrow b - c = a.$$

Thus, $\ln x - \ln y = \ln \dfrac{x}{y}$.

72. Let $a = \ln x^y$ and $b = y \ln x$.

$$e^a = x^y \qquad e^{b/y} = x$$

$$e^{a/y} = x = e^{b/y} \quad \Rightarrow \quad \frac{a}{y} = \frac{b}{y} \Rightarrow a = b$$

Thus, $\ln x^y = y \ln x$.

Section 7.5 Logarithmic Functions and Differentiation

1. $y = \ln x^3 = 3 \ln x$

$y' = \dfrac{3}{x}$

At (1, 0), $y' = 3$.

2. $y = \ln x^{3/2} = \dfrac{3}{2} \ln x$

$y' = \dfrac{3}{2x}$

At (1, 0), $y' = \frac{3}{2}$.

3. $y = \ln x^2 = 2 \ln x$

$y' = \dfrac{2}{x}$

At (1, 0), $y' = 2$.

4. $y = \ln x^{1/2} = \dfrac{1}{2} \ln x$

$y' = \dfrac{1}{2x}$

At (1, 0), $y' = \frac{1}{2}$.

5. $y = \ln x^2 = 2 \ln x$

$\dfrac{dy}{dx} = \dfrac{2}{x}$

6. $y = \ln(x^2 + 3)$

$\dfrac{dy}{dx} = \dfrac{2x}{x^2 + 3}$

7. $y = \ln \sqrt{x^4 - 4x} = \dfrac{1}{2} \ln(x^4 - 4x)$

$$\frac{dy}{dx} = \frac{(4x^3 - 4)}{2(x^4 - 4x)} = \frac{2(x^3 - 1)}{x(x^3 - 4)}$$

8. $y = \ln(1 - x)^{3/2} = \dfrac{3}{2} \ln(1 - x)$

$$\frac{dy}{dx} = \frac{3}{2}\left(\frac{-1}{1 - x}\right) = \frac{3}{2(x - 1)}$$

9. $y = (\ln x)^4$

$$\frac{dy}{dx} = 4(\ln x)^3 \left(\frac{1}{x}\right) = \frac{4(\ln x)^3}{x}$$

10. $y = x \ln x$

$$\frac{dy}{dx} = x\left(\frac{1}{x}\right) + \ln x = 1 + \ln x$$

11. $y = \ln x\sqrt{x^2-1} = \ln x + \frac{1}{2}\ln(x^2-1)$

$\frac{dy}{dx} = \frac{1}{x} + \frac{1}{2}\left(\frac{2x}{x^2-1}\right) = \frac{2x^2-1}{x(x^2-1)}$

12. $y = \ln\frac{x}{x+1} = \ln x - \ln(x+1)$

$\frac{dy}{dx} = \frac{1}{x} - \frac{1}{x+1} = \frac{1}{x^2+x}$

13. $y = \ln\frac{x}{x^2+1} = \ln x - \ln(x^2+1)$

$\frac{dy}{dx} = \frac{1}{x} - \frac{2x}{x^2+1} = \frac{1-x^2}{x(x^2+1)}$

14. $y = \frac{\ln x}{x}$

$\frac{dy}{dx} = \frac{x(1/x) - \ln x}{x^2} = \frac{1-\ln x}{x^2}$

15. $y = \frac{\ln x}{x^2}$

$\frac{dy}{dx} = \frac{x^2(1/x) - 2x\ln x}{x^4} = \frac{1-2\ln x}{x^3}$

16. $y = \ln(\ln x)$

$\frac{dy}{dx} = \frac{1/x}{\ln x} = \frac{1}{x\ln x}$

17. $y = \ln(\ln x^2)$

$\frac{dy}{dx} = \frac{(2x/x^2)}{\ln x^2} = \frac{2}{x\ln x^2} = \frac{1}{x\ln x}$

18. $y = \ln\sqrt{\frac{x-1}{x+1}} = \frac{1}{2}[\ln(x-1) - \ln(x+1)]$

$\frac{dy}{dx} = \frac{1}{2}\left[\frac{1}{x-1} - \frac{1}{x+1}\right] = \frac{1}{x^2-1}$

19. $y = \ln\sqrt{\frac{x+1}{x-1}} = \frac{1}{2}[\ln(x+1) - \ln(x-1)]$

$\frac{dy}{dx} = \frac{1}{2}\left[\frac{1}{x+1} - \frac{1}{x-1}\right] = \frac{1}{1-x^2}$

20. $y = \ln\sqrt{x^2-4} = \frac{1}{2}\ln(x^2-4)$

$\frac{dy}{dx} = \frac{1}{2}\left(\frac{2x}{x^2-4}\right)(2x) = \frac{x}{x^2-4}$

21. $y = \ln\frac{\sqrt{4+x^2}}{x} = \frac{1}{2}\ln(4+x^2) - \ln x$

$\frac{dy}{dx} = \frac{x}{4+x^2} - \frac{1}{x} = \frac{-4}{x(x^2+4)}$

22. $y = \ln\left(x+\sqrt{4+x^2}\right)$

$\frac{dy}{dx} = \frac{1}{x+\sqrt{4+x^2}}\left(1+\frac{x}{\sqrt{4+x^2}}\right) = \frac{1}{\sqrt{4+x^2}}$

23. $y = \frac{-\sqrt{x^2+1}}{x} + \ln\left(x+\sqrt{x^2+1}\right)$

$$\frac{dy}{dx} = \frac{-x\left(x/\sqrt{x^2+1}\right)+\sqrt{x^2+1}}{x^2} + \left(\frac{1}{x+\sqrt{x^2+1}}\right)\left(1+\frac{x}{\sqrt{x^2+1}}\right)$$

$$= \frac{1}{x^2\sqrt{x^2+1}} + \left(\frac{1\left(x-\sqrt{x^2+1}\right)}{-1}\right)\left(\frac{\sqrt{x^2+1}+x}{\sqrt{x^2+1}}\right) = \frac{\sqrt{x^2+1}}{x^2}$$

24. $y = \frac{-\sqrt{x^2+4}}{2x^2} - \frac{1}{4}\ln\left(\frac{2+\sqrt{x^2+4}}{x}\right) = \frac{-\sqrt{x^2+4}}{2x^2} - \frac{1}{4}\ln\left(2+\sqrt{x^2+4}\right) + \frac{1}{4}\ln x$

$$\frac{dy}{dx} = \frac{-2x^2\left(x/\sqrt{x^2+4}\right)+4x\sqrt{x^2+4}}{4x^4} - \frac{1}{4}\left(\frac{1}{2+\sqrt{x^2+4}}\right)\left(\frac{x}{\sqrt{x^2+4}}\right) + \frac{1}{4x} = \frac{\sqrt{x^2+4}}{x^3}$$

25. $y = 4^x$

$\frac{dy}{dx} = (\ln 4)4^x$

26. $y = 2^{-x}$

$\frac{dy}{dx} = -(\ln 2)2^{-x}$

27. $y = 5^{x-2}$

$$\frac{dy}{dx} = (\ln 5)5^{x-2}$$

28. $y = x(7^{-3x})$

$$\frac{dy}{dx} = x\left[-3(\ln 7)7^{-3x}\right] + 7^{-3x}$$
$$= 7^{-3x}[-3x(\ln 7) + 1]$$
$$= 7^{-3x}(1 - 3x \ln 7)$$

29. $y = x^2 2^x$

$$\frac{dy}{dx} = x^2(\ln 2)2^x + (2x)2^x$$
$$= x2^x(x \ln 2 + 2) = 2^x x(2 + x \ln 2)$$

30. $y = 2^{x^2}3^{-x}$

$$\frac{dy}{dx} = 2^{x^2}[-(\ln 3)3^{-x}] + 3^{-x}\left[2x(\ln 2)2^{x^2}\right]$$
$$= 2^{x^2}3^{-x}[-(\ln 3) + 2x(\ln 2)]$$
$$= 2^{x^2}3^{-x}(2x \ln 2 - \ln 3)$$

31. $y = \log_3 x$

$$\frac{dy}{dx} = \frac{1}{x \ln 3}$$

32. $y = \log_{10}(2x) = \log_{10} 2 + \log_{10} x$

$$\frac{dy}{dx} = 0 + \frac{1}{x \ln 10} = \frac{1}{x \ln 10}$$

33. $y = \log_2 \dfrac{x^2}{x-1} = 2\log_2 x - \log_2(x-1) = \dfrac{2}{x \ln 2} - \dfrac{1}{(x-1)\ln 2} = \dfrac{x-2}{(\ln 2)x(x-1)}$

34. $y = \log_3 \dfrac{x\sqrt{x-1}}{2} = \log_3 x + \dfrac{1}{2}\log_3(x-1) - \log_3 2$

$$\frac{dy}{dx} = \frac{1}{x \ln 3} + \frac{1}{2} \cdot \frac{1}{(x-1)\ln 3} - 0 = \frac{1}{\ln 3}\left[\frac{1}{x} + \frac{1}{2(x-1)}\right] = \frac{1}{\ln 3}\left[\frac{3x-2}{2x(x-1)}\right]$$

35. $y = \log_5 \sqrt{x^2-1} = \dfrac{1}{2}\log_5(x^2-1)$

$$\frac{dy}{dx} = \frac{1}{2} \cdot \frac{2x}{(x^2-1)\ln 5}$$
$$= \frac{x}{(x^2-1)\ln 5}$$

36. $y = \log_{10} \dfrac{x^2-1}{x} = \log_{10}(x^2-1) - \log_{10} x$

$$\frac{dy}{dx} = \frac{2x}{(x^2-1)\ln 10} - \frac{1}{x \ln 10}$$
$$= \frac{1}{\ln 10}\left[\frac{2x}{x^2-1} - \frac{1}{x}\right]$$
$$= \frac{1}{\ln 10}\left[\frac{x^2+1}{x(x^2-1)}\right]$$

37. $y = \ln|x^2 - 1|$

$$y' = \frac{[(x^2-1)/|x^2-1|](2x)}{|x^2-1|}$$
$$= \frac{2x(x^2-1)}{|x^2-1|^2} = \frac{2x}{x^2-1}$$

38. $y = \ln\left|\dfrac{x+5}{x}\right| = \ln|x+5| - \ln|x|$

$$y' = \frac{[(x+5)/|x+5|](1)}{|x+5|} - \frac{(x/|x|)(1)}{|x|}$$
$$= \frac{x+5}{|x+5|^2} - \frac{x}{|x|^2} = \frac{1}{x+5} - \frac{1}{x}$$

39.
$$y = x\sqrt{x^2 - 1}$$
$$\ln y = \ln x + \frac{1}{2}\ln(x^2 - 1)$$
$$\frac{1}{y}\left(\frac{dy}{dx}\right) = \frac{1}{x} + \frac{x}{x^2 - 1}$$
$$\frac{dy}{dx} = y\left[\frac{2x^2 - 1}{x(x^2 - 1)}\right] = \frac{2x^2 - 1}{\sqrt{x^2 - 1}}$$

40.
$$y = \sqrt{(x - 1)(x - 2)(x - 3)}$$
$$\ln y = \frac{1}{2}[\ln(x - 1) + \ln(x - 2) + \ln(x - 3)]$$
$$\frac{1}{y}\left(\frac{dy}{dx}\right) = \frac{1}{2}\left[\frac{1}{x - 1} + \frac{1}{x - 2} + \frac{1}{x - 3}\right]$$
$$= \frac{1}{2}\left[\frac{3x^2 - 12x + 11}{(x - 1)(x - 2)(x - 3)}\right]$$
$$\frac{dy}{dx} = \frac{3x^2 - 12x + 11}{2y}$$

41.
$$y = \frac{x^2\sqrt{3x - 2}}{(x - 1)^2}$$
$$\ln y = 2\ln x + \frac{1}{2}\ln(3x - 2) - 2\ln(x - 1)$$
$$\frac{1}{y}\left(\frac{dy}{dx}\right) = \frac{2}{x} + \frac{3}{2(3x - 2)} - \frac{2}{x - 1}$$
$$\frac{dy}{dx} = y\left[\frac{3x^2 - 15x + 8}{2x(3x - 2)(x - 1)}\right]$$
$$= \frac{3x^3 - 15x^2 + 8x}{2(x - 1)^3\sqrt{3x - 2}}$$

42.
$$y = \sqrt[3]{\frac{x^2 + 1}{x^2 - 1}}$$
$$\ln y = \frac{1}{3}[\ln(x^2 + 1) - \ln(x + 1) - \ln(x - 1)]$$
$$\frac{1}{y}\left(\frac{dy}{dx}\right) = \frac{1}{3}\left[\frac{2x}{x^2 + 1} - \frac{1}{x + 1} - \frac{1}{x - 1}\right]$$
$$\frac{dy}{dx} = \frac{-4xy}{3(x^4 - 1)}$$

43.
$$y = \frac{x(x - 1)^{3/2}}{\sqrt{x + 1}}$$
$$\ln y = \ln x + \frac{3}{2}\ln(x - 1) - \frac{1}{2}\ln(x + 1)$$
$$\frac{1}{y}\left(\frac{dy}{dx}\right) = \frac{1}{x} + \frac{3}{2}\left(\frac{1}{x - 1}\right) - \frac{1}{2}\left(\frac{1}{x + 1}\right)$$
$$\frac{dy}{dx} = \frac{y}{2}\left[\frac{2}{x} + \frac{3}{x - 1} - \frac{1}{x + 1}\right]$$
$$= \frac{y}{2}\left[\frac{4x^2 + 4x - 2}{x(x^2 - 1)}\right]$$
$$= \frac{(2x^2 + 2x - 1)\sqrt{x - 1}}{(x + 1)^{3/2}}$$

44.
$$y = \frac{(x + 1)(x + 2)}{(x - 1)(x - 2)}$$
$$\ln y = \ln(x + 1) + \ln(x + 2) - \ln(x - 1) - \ln(x - 2)$$
$$\frac{1}{y}\left(\frac{dy}{dx}\right) = \frac{1}{x + 1} + \frac{1}{x + 2} - \frac{1}{x - 1} - \frac{1}{x - 2}$$
$$\frac{dy}{dx} = y\left[\frac{-2}{x^2 - 1} + \frac{-4}{x^2 - 4}\right]$$
$$= y\left[\frac{-6x^2 + 12}{(x^2 - 1)(x^2 - 4)}\right]$$
$$= \frac{6y(2 - x^2)}{(x^2 - 1)(x^2 - 4)}$$

45.
$$y = x^{2/x}$$
$$\ln y = \frac{2}{x}\ln x$$
$$\frac{1}{y}\left(\frac{dy}{dx}\right) = \frac{2}{x}\left(\frac{1}{x}\right) + \ln x\left(-\frac{2}{x^2}\right)$$
$$\frac{dy}{dx} = \frac{2y}{x^2}(1 - \ln x) = 2x^{(2/x)-2}(1 - \ln x)$$

46.
$$y = x^{x-1}$$
$$\ln y = (x - 1)(\ln x)$$
$$\frac{1}{y}\left(\frac{dy}{dx}\right) = (x - 1)\left(\frac{1}{x}\right) + \ln x$$
$$\frac{dy}{dx} = y\left[\frac{x - 1}{x} + \ln x\right] = x^{x-2}(x - 1 + x\ln x)$$

47.
$$y = (x-2)^{x+1}$$
$$\ln y + (x+1)\ln(x-2)$$
$$\frac{1}{y}\left(\frac{dy}{dx}\right) = (x+1)\left(\frac{1}{x-2}\right) + \ln(x-2)$$
$$\frac{dy}{dx} = y\left[\frac{x+1}{x-2} + \ln(x-2)\right]$$
$$= (x-2)^{x+1}\left[\frac{x+1}{x-2} + \ln(x-2)\right]$$

48.
$$y = (1+x)^{1/x}$$
$$\ln y = \frac{1}{x}\ln(1+x)$$
$$\frac{1}{y}\left(\frac{dy}{dx}\right) = \frac{1}{x}\left(\frac{1}{1+x}\right) + \ln(1+x)\left(-\frac{1}{x^2}\right)$$
$$\frac{dy}{dx} = \frac{y}{x}\left[\frac{1}{x+1} - \frac{\ln(x+1)}{x}\right]$$

49.
$$y = 2(\ln x) + 3$$
$$y' = \frac{2}{x}$$
$$y'' = -\frac{2}{x^2}$$
$$xy'' + y' = x\left(-\frac{2}{x^2}\right) + \frac{2}{x} = 0$$

50.
$$y = x(\ln x) - 4x$$
$$y' = x\left(\frac{1}{x}\right) + \ln x - 4 = -3 + \ln x$$
$$(x+y) - xy' = x + x\ln x - 4x - x(-3 + \ln x) = 0$$

51.
$$x^2 - 3\ln y + y^2 = 10$$
$$2x - \frac{3}{y}\frac{dy}{dx} + 2y\frac{dy}{dx} = 0$$
$$2x = \frac{dy}{dx}\left(\frac{3}{y} - 2y\right)$$
$$\frac{dy}{dx} = \frac{2x}{(3/y) - 2y} = \frac{2xy}{3 - 2y^2}$$

52.
$$\ln(xy) + 5x = 30$$
$$\ln x + \ln y + 5x = 30$$
$$\frac{1}{x} + \frac{1}{y}\frac{dy}{dx} + 5 = 0$$
$$\frac{1}{y}\frac{dy}{dx} = -\frac{1}{x} - 5$$
$$\frac{dy}{dx} = -\frac{y}{x} - 5y = -\left(\frac{y + 5xy}{x}\right)$$

53. $y = 3x^2 - \ln x, \quad (1,\ 3)$
$$\frac{dy}{dx} = 6x - \frac{1}{x}$$
When $x = 1$, $dy/dx = 5$.

Tangent line: $y - 3 = 5(x-1)$
$$y = 5x - 2$$

54. $x^2 + \ln(x+1) + y^2 = 4, \quad (0,\ 2)$
$$2x + \frac{1}{x+1} + 2y\frac{dy}{dx} = 0$$
$$\frac{dy}{dx} = -\frac{2x^2 + 2x + 1}{2y(x+1)}$$
At $(0, 2)$, $dy/dx = -1/4$.

Tangent line: $y = -\frac{1}{4}x + 2$

55. $y = \frac{x^2}{2} - \ln x, \quad$ Domain: $(0,\ \infty)$
$$y' = x - \frac{1}{x}$$
$$= \frac{(x+1)(x-1)}{x} = 0 \text{ when } x = \pm 1.$$
$$y'' = 1 + \frac{1}{x^2} > 0$$
Relative minimum: $\left(1,\ \frac{1}{2}\right)$

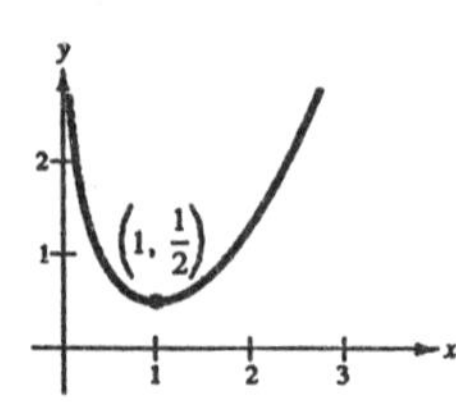

56. $y = x - \ln x$, Domain: $(0, \infty)$

$y' = 1 - \dfrac{1}{x} = 0$ when $x = 1$.

$y'' = \dfrac{1}{x^2} > 0$

Relative minimum: $(1, 1)$

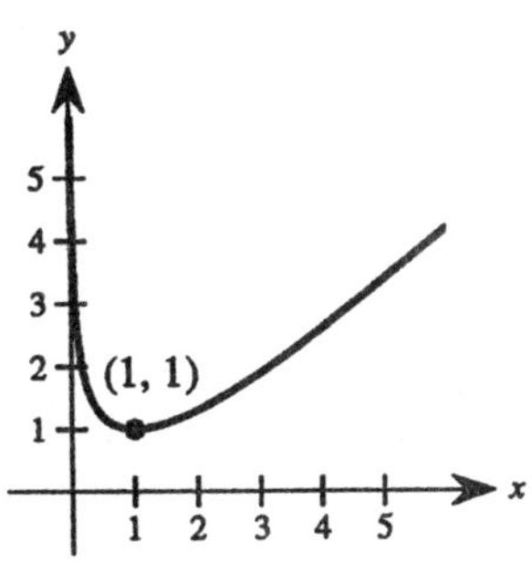

57. $y = x \ln x$, Domain: $(0, \infty)$

$y' = x\left(\dfrac{1}{x}\right) + \ln x = 1 + \ln x = 0$ when $x = e^{-1}$.

$y'' = \dfrac{1}{x} > 0$

Relative minimum: $(e^{-1}, -e^{-1})$

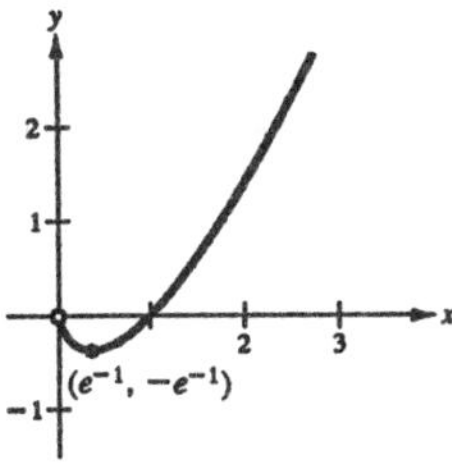

58. $y = \dfrac{\ln x}{x}$, Domain: $(0, \infty)$

$y' = \dfrac{x(1/x) - \ln x}{x^2} = \dfrac{1 - \ln x}{x^2} = 0$ when $x = e$.

$y'' = \dfrac{x^2(-1/x) - (1 - \ln x)(2x)}{x^4} = \dfrac{2(\ln x) - 3}{x^3} = 0$ when $x = e^{3/2}$.

Relative maximum: (e, e^{-1})

Point of inflection: $\left(e^{3/2}, \frac{3}{2}e^{-3/2}\right)$

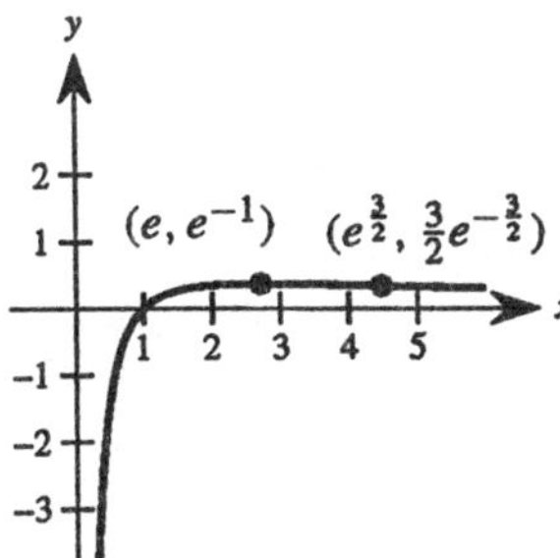

59. $y = \dfrac{x}{\ln x}$, Domain: $(0, 1) \cup (1, \infty)$

$y' = \dfrac{(\ln x)(1) - (x)(1/x)}{(\ln x)^2} = \dfrac{\ln x - 1}{(\ln x)^2} = 0$ when $x = e$.

$y'' = \dfrac{2 - \ln x}{x(\ln x)^3} = 0$ when $x = e^2$.

Relative minimum: (e, e)

Point of inflection: $(e^2, e^2/2)$

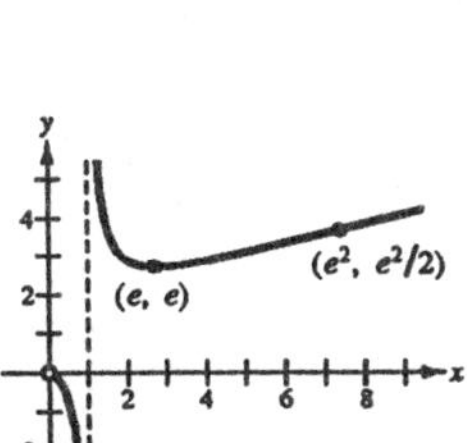

60. $y = x^2 \ln x$, Domain: $(0, \infty)$

$y' = x^2\left(\dfrac{1}{x}\right) + 2x(\ln x) = x(1 + 2\ln x) = 0$ when $x = e^{-1/2}$.

$y'' = x\left(\dfrac{2}{x}\right) + (1 + 2\ln x) = 3 + 2(\ln x) = 0$ when $x = e^{-3/2}$.

Relative minimum: $\left(e^{-1/2}, -\frac{1}{2}e^{-1}\right)$

Point of inflection: $\left(e^{-3/2}, -\frac{3}{2}e^{-3}\right)$

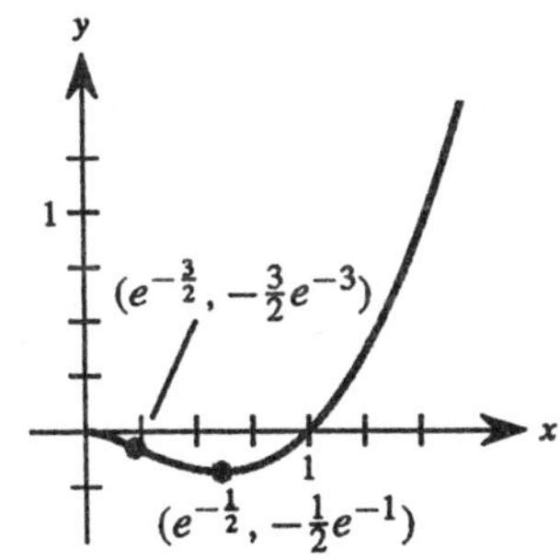

61. Find x such that $\ln x = -x$.

$f(x) = (\ln x) + x = 0$

$f'(x) = \frac{1}{x} + 1$

$x_{n+1} = x_n - \frac{f(x_n)}{f'(x_n)} = x_n\left[\frac{1 - \ln x_n}{1 + x_n}\right]$

n	1	2	3
x_n	0.5	0.5644	0.5671
$f(x_n)$	−0.1932	−0.0076	−0.0000

Approximate root: $x = 0.567$

62. Find x such that $\ln x = 3 - x$.

$f(x) = x + (\ln x) - 3 = 0$

$f'(x) = 1 + \frac{1}{x}$

$x_{n+1} = x_n - \frac{f(x_n)}{f'(x_n)} = x_n\left[\frac{4 - \ln x_n}{1 + x_n}\right]$

n	1	2	3
x_n	2	2.2046	2.2079
$f(x_n)$	−0.3069	−0.0049	−0.0000

Approximate root: $x = 2.208$

63. $L(x) = \int_1^x \frac{1}{t}\,dt$

(a) $L(1) = \int_1^1 \frac{1}{t}\,dt = 0$

(b) $L'(x) = \frac{1}{x}$

$L'(1) = 1$

(c) By trial and error, $x \approx 2.718$.

(d) $L'(x) = \frac{1}{x} > 0$ on $(0, \infty)$. Thus, L is increasing on $(0, \infty)$.

(e) $L(x_1x_2) = \int_1^{x_1x_2} \frac{1}{t}\,dt = \int_1^{x_1} \frac{1}{t}\,dt + \int_{x_1}^{x_1x_2} \frac{1}{t}\,dt$

But, this second integral is equal to $\int_1^{x_2} (1/t)\,dt$, by substitution $u = t/x_1$, $du = dt/x_1$.

64. $f(x) = \frac{\ln x^n}{x} = \frac{n \ln x}{x}$

$f'(x) = n\left[\frac{x(1/x) - (\ln x)(1)}{x^2}\right] = n\left[\frac{1 - \ln x}{x^2}\right]$

If $n > 0$ and $x > e$, then $f'(x) < 0$. Therefore, $f(x)$ is decreasing.

65. $y = 10\ln\left(\frac{10 + \sqrt{100 - x^2}}{x}\right) - \sqrt{100 - x^2} = 10\left[\ln\left(10 + \sqrt{100 - x^2}\right) - \ln x\right] - \sqrt{100 - x^2}$

$$\frac{dy}{dx} = 10\left[\frac{-x}{\sqrt{100 - x^2}\left(10 + \sqrt{100 - x^2}\right)} - \frac{1}{x}\right] + \frac{x}{\sqrt{100 - x^2}} = -\frac{\sqrt{100 - x^2}}{x}$$

(a) When $x = 10$, $dy/dx = 0$.

(b) When $x = 5$, $dy/dx = -\sqrt{3}$.

Section 7.6 Logarithmic Functions and Integration

1. $u = x+1,\quad du = dx$

$$\int \frac{1}{x+1}\,dx = \ln|x+1| + C$$

2. $u = x-5,\quad du = dx$

$$\int \frac{1}{x-5}\,dx = \ln|x-5| + C$$

3. $u = 3-2x,\quad du = -2\,dx$

$$\int \frac{1}{3-2x}\,dx = -\frac{1}{2}\int \frac{1}{3-2x}(-2)\,dx = -\frac{1}{2}\ln|3-2x| + C$$

4. $u = 6x+1,\quad du = 6\,dx$

$$\int \frac{1}{6x+1}\,dx = \frac{1}{6}\int \frac{1}{6x+1}(6)\,dx = \frac{1}{6}\ln|6x+1| + C$$

5. $u = x^2+1,\quad du = 2x\,dx$

$$\int \frac{x}{x^2+1}\,dx = \frac{1}{2}\int \frac{1}{x^2+1}(2x)\,dx = \frac{1}{2}\ln(x^2+1) + C = \ln\sqrt{x^2+1} + C$$

6. $u = 3-x^3,\quad du = -3x^2\,dx$

$$\int \frac{x^2}{3-x^3}\,dx = -\frac{1}{3}\int \frac{1}{3-x^3}(-3x^2)\,dx = -\frac{1}{3}\ln|3-x^3| + C$$

7. $\displaystyle\int \frac{x^2-4}{x}\,dx = \int\left(x - \frac{4}{x}\right)dx = \frac{x^2}{2} - 4\ln|x| + C$

8. $\displaystyle\int \frac{x+5}{x}\,dx = \int\left(1 + \frac{5}{x}\right)dx = x + 5\ln|x| + C$

9. $u = \ln x,\quad du = \dfrac{1}{x}\,dx$

$$\int_1^e \frac{\ln x}{2x}\,dx = \frac{1}{2}\int_1^e \frac{\ln x}{x}\,dx = \left[\frac{1}{4}\ln^2|x|\right]_1^e = \frac{1}{4}$$

10. $u = \ln x,\quad du = \dfrac{1}{x}\,dx$

$$\int_e^{e^2} \frac{1}{x\ln x}\,dx = \int_e^{e^2}\left(\frac{1}{\ln x}\right)\frac{1}{x}\,dx = \Big[\ln|\ln|x||\Big]_e^{e^2} = \ln 2$$

11. $u = 1+\ln x,\quad du = \dfrac{1}{x}\,dx$

$$\int_1^e \frac{(1+\ln x)^2}{x}\,dx = \left[\frac{1}{3}(1+\ln|x|)^3\right]_1^e = \frac{7}{3}$$

12. $\displaystyle\int_0^1 \frac{x-1}{x+1}\,dx = \int_0^1 1\,dx + \int_0^1 \frac{-2}{x+1}\,dx = \Big[x - 2\ln|x+1|\Big]_0^1 = 1 - 2\ln 2$

13. $\displaystyle\int_0^2 \frac{x^2-2}{x+1}\,dx = \int_0^2\left(x - 1 - \frac{1}{x+1}\right)dx = \left[\frac{1}{2}x^2 - x - \ln|x+1|\right]_0^2 = -\ln 3$

14. $u = x+1,\quad du = dx$

$$\int \frac{1}{(x+1)^2}\,dx = \int (x+1)^{-2}\,dx = -(x+1)^{-1} + C = \frac{-1}{x+1} + C$$

15. $u = x+1,\quad du = dx$

$$\int \frac{1}{\sqrt{x+1}}\,dx = \int (x+1)^{-1/2}\,dx = 2(x+1)^{1/2} + C = 2\sqrt{x+1} + C$$

16. $u = x^2+6x+7,\quad du = 2(x+3)\,dx$

$$\int \frac{x+3}{x^2+6x+7}\,dx = \frac{1}{2}\int \frac{2x+6}{x^2+6x+7}\,dx = \frac{1}{2}\ln|x^2+6x+7| + C$$

17. $u = x^3 + 3x^2 + 9x, \quad du = 3(x^2 + 2x + 3)\,dx$

$$\int \frac{x^2 + 2x + 3}{x^3 + 3x^2 + 9x}\,dx = \frac{1}{3}\int \frac{3(x^2 + 2x + 3)}{x^3 + 3x^2 + 9x}\,dx$$
$$= \frac{1}{3}\ln|x^3 + 3x^2 + 9x| + C$$

18. $u = \ln x, \quad du = \frac{1}{x}\,dx$

$$\int \frac{(\ln x)^2}{x}\,dx = \frac{1}{3}(\ln x)^3 + C$$

19. $u = 1 + x^{1/3}, \quad du = \frac{1}{3x^{2/3}}\,dx$

$$\int \frac{1}{x^{2/3}(1 + x^{1/3})}\,dx = 3\int \frac{1}{1 + x^{1/3}}\left(\frac{1}{3x^{2/3}}\right)dx$$
$$= 3\ln|1 + x^{1/3}| + C$$

20. $u = \ln x, \quad du = \frac{1}{x}\,dx$

$$\int \frac{1}{x\ln x^2}\,dx = \frac{1}{2}\int \frac{1}{x\ln x}\,dx$$
$$= \frac{1}{2}\ln|\ln|x|| + C$$

21. $u = 1 + \sqrt{x}, \quad du = \frac{dx}{2\sqrt{x}} \Rightarrow 2(u - 1)\,du = dx$

$$\int \frac{1}{1 + \sqrt{x}}\,dx = 2\int \frac{u - 1}{u}\,du = 2\int 1\,du - 2\int \frac{1}{u}\,du$$
$$= 2u - 2\ln u + C_1 = 2(1 + \sqrt{x}) - 2\ln(1 + \sqrt{x}) + C_1 = 2[\sqrt{x} - \ln(1 + \sqrt{x})] + C$$

22. $u = 1 + \sqrt{x}, \quad du = \frac{1}{2\sqrt{x}}\,dx \Rightarrow 2(u - 1)\,du = dx$

$$\int \frac{1 - \sqrt{x}}{1 + \sqrt{x}}\,dx = 2\int \frac{(2 - u)(u - 1)}{u}\,du = 2\int \frac{-u^2 + 3u - 2}{u}\,du$$
$$= 2\int\left(-u + 3 - \frac{2}{u}\right)du = 2\left[-\frac{u^2}{2} + 3u - 2\ln|u|\right] + C_1 = -u^2 + 6u - 4\ln|u| + C_1$$
$$= -(1 + \sqrt{x})^2 + 6(1 + \sqrt{x}) - 4\ln(1 + \sqrt{x}) + C_1 = -x + 4\sqrt{x} - 4\ln(1 + \sqrt{x}) + C$$

23. $u = \sqrt{x} - 3, \quad du = \frac{1}{2\sqrt{x}}\,dx \Rightarrow 2(u + 3)\,du = dx$

$$\int \frac{\sqrt{x}}{\sqrt{x} - 3}\,dx = 2\int \frac{(u + 3)^2}{u}\,du = 2\int \frac{u^2 + 6u + 9}{u}\,du$$
$$= 2\int\left(u + 6 + \frac{9}{u}\right)du = 2\left[\frac{u^2}{2} + 6u + 9\ln|u|\right] + C_1 = u^2 + 12u + 18\ln|u| + C_1$$
$$= (\sqrt{x} - 3)^2 + 12(\sqrt{x} - 3) + 18\ln|\sqrt{x} - 3| + C_1 = x + 6\sqrt{x} + 18\ln|\sqrt{x} - 3| + C$$

24. $u = 1 + \sqrt{2x}, \quad du = \frac{1}{\sqrt{2x}}\,dx \Rightarrow (u - 1)\,du = dx$ when $x = 0 \Rightarrow u = 1$ and when $x = 2 \Rightarrow u = 3$.

$$\int_0^2 \frac{1}{1 + \sqrt{2x}}\,dx = \int_1^3 \frac{(u - 1)}{u}\,du = \int_1^3 \left(1 - \frac{1}{u}\right)du = \Big[u - \ln u\Big]_1^3 = (3 - \ln 3) - (1 - \ln 1)$$
$$= 2 - \ln 3 \approx 0.901$$

25. $u = 1 - x\sqrt{x} = 1 - x^{3/2}, \quad du = -\frac{3}{2}x^{1/2}\,dx$

$$\int \frac{\sqrt{x}}{1 - x\sqrt{x}}\,dx = -\frac{2}{3}\int \frac{1}{1 - x\sqrt{x}}\left(-\frac{3}{2}\sqrt{x}\right)dx = -\frac{2}{3}\ln|1 - x\sqrt{x}| + C$$

26. $\displaystyle\int \frac{2x}{(x-1)^2}\,dx = \int \frac{2x-2+2}{(x-1)^2}\,dx = \int \frac{2(x-1)}{(x-1)^2}\,dx + 2\int \frac{1}{(x-1)^2}\,dx = 2\int \frac{1}{x-1}\,dx + 2\int \frac{1}{(x-1)^2}\,dx$

$$= 2\ln|x-1| - \frac{2}{(x-1)} + C$$

27. $\displaystyle\int \frac{x(x-2)}{(x-1)^3}\,dx = \int \frac{x^2-2x+1-1}{(x-1)^3}\,dx = \int \frac{(x-1)^2}{(x-1)^3}\,dx - \int \frac{1}{(x-1)^3}\,dx = \int \frac{1}{x-1}\,dx - \int \frac{1}{(x-1)^3}\,dx$

$$= \ln|x-1| + \frac{1}{2(x-1)^2} + C$$

28. $\displaystyle\int \frac{x\sqrt{x}}{1+x^2\sqrt{x}}\,dx = \frac{2}{5}\int \frac{(5/2)x\sqrt{x}}{1+x^2\sqrt{x}}\,dx = \frac{2}{5}\ln\left|1+x^2\sqrt{x}\right| + C$

29. $\displaystyle\int 3^x\,dx = \frac{3^x}{\ln 3} + C$

30. $\displaystyle\int 4^{-x}\,dx = -\frac{4^{-x}}{\ln 4} + C$

31. $\displaystyle\int_{-1}^{2} 2^x\,dx = \left[\frac{2^x}{\ln 2}\right]_{-1}^{2}$

$$= \frac{1}{\ln 2}\left[4 - \frac{1}{2}\right] = \frac{7}{2\ln 2} = \frac{7}{\ln 4}$$

32. $\displaystyle\int 2^3\,dx = \int 8\,dx = 8x + C$

33. $\displaystyle\int x5^{x^2}\,dx = \frac{1}{2}\int 5^{x^2}(2x)\,dx = \left(\frac{1}{2}\right)\frac{5^{x^2}}{\ln 5} + C = \frac{1}{2\ln 5}(5^{x^2}) + C$

34. $\displaystyle\int (3-x)7^{(3-x)^2}\,dx = -\frac{1}{2}\int -2(3-x)7^{(3-x)^2}\,dx = -\frac{1}{2\ln 7}\left[7^{(3-x)^2}\right] + C$

35. Let $u = 1 + e^{-x}$, $du = -e^{-x}\,dx$.

$$\int \frac{e^{-x}}{1+e^{-x}}\,dx = -\int \frac{-e^{-x}}{1+e^{-x}}\,dx = -\ln(1+e^{-x}) + C = \ln\left(\frac{e^x}{e^x+1}\right) + C = x - \ln(e^x+1) + C$$

36. Let $u = e^x + e^{-x}$, $du = (e^x - e^{-x})\,dx$.

$$\int \frac{e^x - e^{-x}}{e^x + e^{-x}}\,dx = \ln(e^x + e^{-x}) + C$$

37. $\displaystyle A = \int_1^4 \frac{x^2+4}{x}\,dx$

$$= \int_1^4 \left(x + \frac{4}{x}\right)dx = \left[\frac{x^2}{2} + 4\ln x\right]_1^4 = (8 + 4\ln 4) - \frac{1}{2} = \frac{15}{2} + 8\ln 2 \approx 13.045 \text{ square units}$$

38. $\displaystyle A = \int_1^5 \frac{x+5}{x}\,dx = \int_1^5 \left(1 + \frac{5}{x}\right)dx = \Big[x + 5\ln x\Big]_1^5 = 4 + 5\ln 5 \approx 12.047 \text{ square units}$

39. Disc

$R(x) = \dfrac{1}{\sqrt{x+1}}$

$r(x) = 0$

$$V = \pi \int_0^3 \left(\frac{1}{\sqrt{x+1}}\right)^2 dx$$

$$= \pi \int_0^3 \frac{1}{x+1}\,dx = \pi \ln|x+1|\Big]_0^3 = \pi \ln 4$$

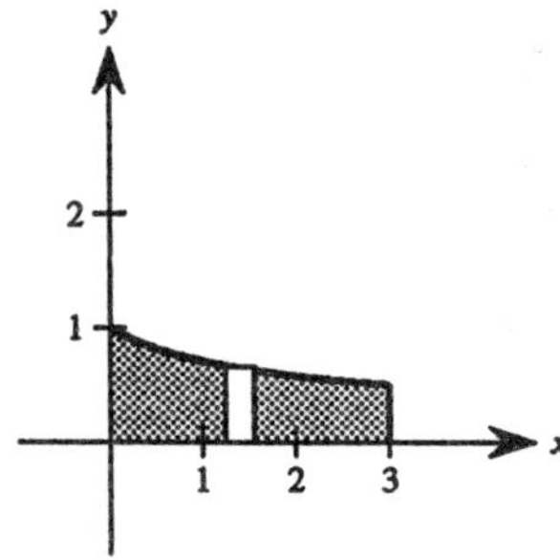

40. Shell

$p(x) = x$

$h(x) = \dfrac{1}{x(x+1)}$

$$V = 2\pi \int_1^3 x\left[\frac{1}{x(x+1)}\right] dx$$

$$= 2\pi\Big[\ln(x+1)\Big]_1^3 = 2\pi[\ln 4 - \ln 2] = 2\pi \ln 2$$

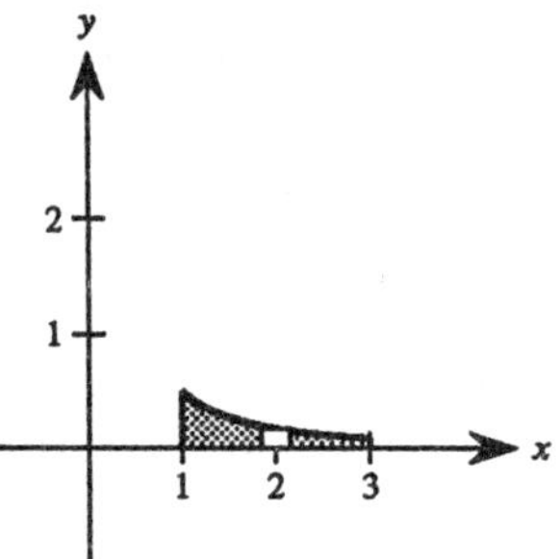

41. Disc

$R(x) = 4$

$r(x) = 4 - \dfrac{1}{x}$

$$V = \pi \int_1^4 \left[4^2 - \left(4 - \frac{1}{x}\right)^2\right] dx$$

$$= \pi\left[8\ln x + \frac{1}{x}\right]_1^4 = \frac{1}{4}(32\ln 4 - 3)\pi$$

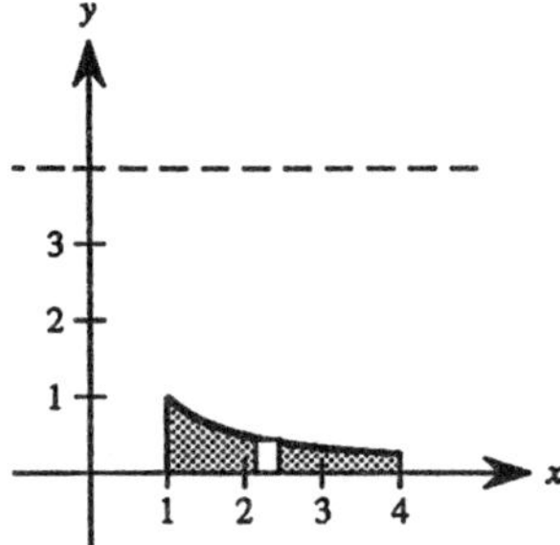

42. Shell

$u = \sqrt{x-2}, \quad x = u^2 + 2, \quad dx = 2u\,du$

$$V = 2\pi \int_2^6 \frac{x}{1+\sqrt{x-2}}\,dx$$

$$= 4\pi \int_0^2 \frac{(u^2+2)u}{1+u}\,du = 4\pi \int_0^2 \frac{u^3+2u}{1+u}\,du$$

$$= 4\pi \int_0^2 \left(u^2 - u + 3 - \frac{3}{1+u}\right) du$$

$$= 4\pi\left[\frac{1}{3}u^3 - \frac{1}{2}u^2 + 3u - 3\ln(1+u)\right]_0^2 = \frac{4\pi}{3}(20 - 9\ln 3)$$

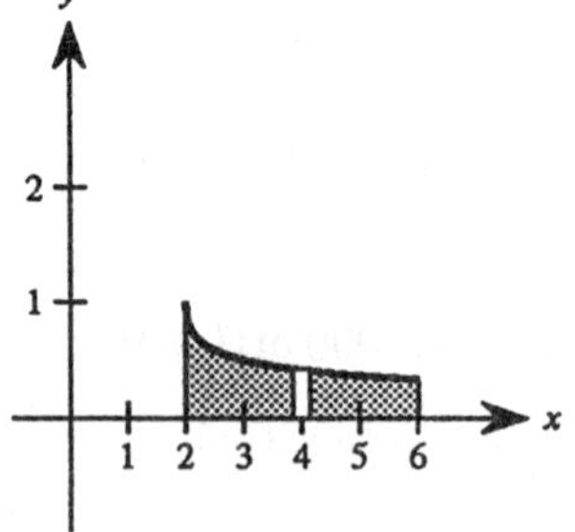

43. $A = \int_0^3 3^x\,dx$

$= \left.\frac{3^x}{\ln 3}\right]_0^3$

$= \frac{26}{\ln 3}$

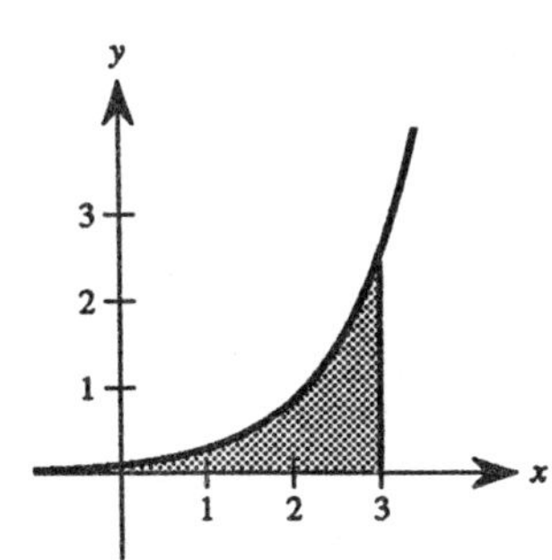

44. $A = \int_1^4 \frac{1}{x}\,dx = \ln|x|\Big]_1^4 = \ln 4$

$\bar{x} = \frac{1}{\ln 4}\int_1^4 \frac{x}{x}\,dx = \left.\frac{x}{\ln 4}\right]_1^4 = \frac{3}{\ln 4}$

$\bar{y} = \frac{1}{2\ln 4}\int_1^4 \frac{1}{x^2}\,dx = -\left.\frac{1}{2x\ln 4}\right]_1^4 = \frac{3}{8\ln 4}$

$(\bar{x},\ \bar{y}) = \left(\frac{3}{\ln 4},\ \frac{3}{8\ln 4}\right) \approx (2.164,\ 0.271)$

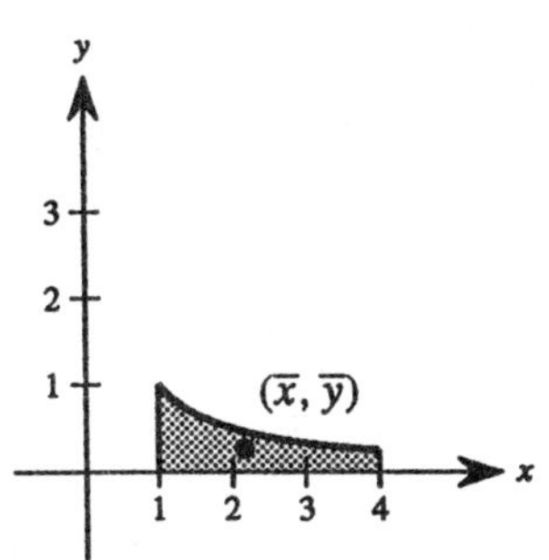

45. $p = \frac{k}{V}$

$1000 = \frac{k}{2}$

$k = 2000$

$W = \int_2^3 \frac{2000}{V}\,dV = 2000\ln|V|\Big]_2^3$

$= 2000\ln\left(\frac{3}{2}\right) \approx 810.93$ ft • lb

46. $p = \frac{k}{V}$

$2000 = \frac{k}{1}$

$k = 2000$

$W = \int_1^4 \frac{2000}{V}\,dV = 2000\ln|V|\Big]_1^4$

$= 2000\ln 4 \approx 2772.59$ ft • lb

47. $P = \int \frac{3000}{1+0.25t}\,dt = (3000)(4)\int \frac{0.25}{1+0.25t}\,dt = 12{,}000\ln|1+0.25t| + C$

$P(0) = 12{,}000\ln|1+0.25(0)| + C = 1000$

$C = 1000$

$P = 12{,}000\ln|1+0.25t| + 1000 = 1000[12\ln|1+0.25t| + 1]$

$P(3) = 1000[12(\ln 1.75) + 1] \approx 7715$

48. $\frac{1}{50-40}\int_{40}^{50} \frac{90{,}000}{400+3x}\,dx = \Big[3000\ln|400+3x|\Big]_{40}^{50} \approx \$168.27$

49. $t = \frac{10}{\ln 2}\int_{250}^{300} \frac{1}{T-100}\,dt$

$= \frac{10}{\ln 2}\Big[\ln(T-100)\Big]_{250}^{300} = \frac{10}{\ln 2}[\ln 200 - \ln 150] = \frac{10}{\ln 2}\left[\ln\left(\frac{4}{3}\right)\right] \approx 4.1504$ units of time

50. $\frac{1}{5-1}\int_1^5 \frac{1}{x}\,dx = \frac{1}{4}\ln x\Big]_1^5 = \frac{1}{4}\ln 5 - \frac{1}{4}\ln 1 = \frac{1}{4}\ln 5 \approx 0.4024$

51. $\int_1^{2.7} \frac{1}{t}\,dt \approx \frac{2.7-1}{3(4)}\left[\frac{1}{1}+\frac{4}{1.425}+\frac{2}{1.85}+\frac{4}{2.275}+\frac{1}{2.7}\right] \approx 0.9940$

$\int_1^{2.8} \frac{1}{t}\,dt \approx \frac{2.8-1}{3(4)}\left[\frac{1}{1}+\frac{4}{1.45}+\frac{2}{1.9}+\frac{4}{2.35}+\frac{1}{2.8}\right] \approx 1.0306$

Therefore, $\int_1^{2.7} \frac{1}{t}\,dt < 1 < \int_1^{2.8} \frac{1}{t}\,dt.$

52. (a) $\int_1^{3} \frac{1}{t}\,dt \approx \frac{3-1}{3(10)}\left[\frac{1}{1}+4\left(\frac{5}{6}\right)+2\left(\frac{5}{7}\right)+4\left(\frac{5}{8}\right)+2\left(\frac{5}{9}\right)+4\left(\frac{1}{2}\right)+2\left(\frac{5}{11}\right)\right.$

$\left.+4\left(\frac{5}{12}\right)+2\left(\frac{5}{13}\right)+4\left(\frac{5}{14}\right)+\frac{1}{3}\right] \approx 1.0987$

(b) $\int_1^{8.7} \frac{1}{t}\,dt \approx \frac{8.7-1}{3(10)}\left[\frac{1}{1}+4\left(\frac{1}{1.77}\right)+2\left(\frac{1}{2.54}\right)+4\left(\frac{1}{3.31}\right)+2\left(\frac{1}{4.08}\right)+4\left(\frac{1}{4.85}\right)+2\left(\frac{1}{5.62}\right)\right.$

$\left.+4\left(\frac{1}{6.39}\right)+2\left(\frac{1}{7.16}\right)+4\left(\frac{1}{7.93}\right)+\frac{1}{8.7}\right] \approx 2.1691$

Section 7.7 Growth and Decay

1. $y = Ce^{kt}$, $\left(0, \frac{1}{2}\right)$, $(5, 5)$

$C = \frac{1}{2}$

$y = \frac{1}{2}e^{kt}$

$5 = \frac{1}{2}e^{5k}$

$k = \frac{\ln 10}{5} \approx 0.4605$

$y = \frac{1}{2}e^{0.4605t}$

2. $y = Ce^{kt}$, $(0, 4)$, $\left(5, \frac{1}{2}\right)$

$C = 4$

$y = 4e^{kt}$

$\frac{1}{2} = 4e^{5k}$

$k = \frac{\ln(1/8)}{5} \approx -0.4159$

$y = 4e^{-0.4159t}$

3. $y = Ce^{kt}$, $(1, 1)$, $(5, 5)$

$1 = Ce^{k}$

$5 = Ce^{5k}$

$5Ce^{k} = Ce^{5k}$

$5e^{k} = e^{5k}$

$5 = e^{4k}$

$k = \frac{\ln 5}{4} \approx 0.4024$

$y = Ce^{0.4024t}$

$1 = Ce^{0.4024}$

$C \approx 0.6687$

$y = 0.6687e^{0.4024t}$

4. $y = Ce^{kt}$, $\left(3, \frac{1}{2}\right)$, $(4, 5)$

$\frac{1}{2} = Ce^{3k}$

$5 = Ce^{4k}$

$2Ce^{3k} = \frac{1}{5}Ce^{4k}$

$10e^{3k} = e^{4k}$

$10 = e^{k}$

$k = \ln 10 \approx 2.3026$

$y = Ce^{2.3026t}$

$5 = Ce^{2.3026(4)}$

$C \approx 0.0005$

$y = 0.0005e^{2.3026t}$

5. Since $A = 1000e^{0.12t}$, the time to double is given by $2000 = 1000e^{0.12t}$ and we have

$$t = \frac{\ln 2}{0.12} \approx 5.78 \text{ years.}$$

Amount after 10 years:

$$A = 1000e^{1.2} \approx \$3320.12$$

Amount after 25 years:

$$A = 1000e^{3} \approx \$20,085.54$$

6. Since $A = 20,000e^{0.105t}$, the time to double is given by the following.

$$40,000 = 20,000e^{0.105t}$$

$$\ln 2 = 0.105t$$

$$t = \frac{\ln 2}{0.105} \approx 6.6 = 6\tfrac{3}{5} \text{ years}$$

Amount after 10 years:

$$A = 20,000e^{0.105(10)} \approx \$57,153.02$$

Amount after 25 years:

$$A = 20,000e^{0.105(25)} \approx \$276,091.48$$

7. Since $A = 750e^{rt}$ and $A = 1500$ when $t = 7.75$, we have the following.

$$1500 = 750e^{7.75r}$$

$$r = \frac{\ln 2}{7.75} \approx 0.0894 = 8.94\%$$

Amount after 10 years:

$$A = 750e^{0.0894(10)} \approx \$1833.67$$

Amount after 25 years:

$$A = 750e^{0.0894(25)} \approx \$7009.86$$

8. Since $A = 10,000e^{rt}$ and $A = 20,000$ when $t = 5$, we have the following.

$$20,000 = 10,000e^{5r}$$

$$r = \frac{\ln 2}{5} \approx 0.1386 = 13.86\%$$

Amount after 10 years:

$$A = 10,000e^{[(\ln 2)/5](10)} = \$40,000$$

Amount after 25 years:

$$a = 10,000e^{[(\ln 2)/5](25)} = \$320,000$$

9. Since $A = 500e^{rt}$ and $A = 1292.85$ when $t = 10$, we have the following.

$$1292.85 = 500e^{10r}$$

$$r = \frac{\ln(1292.85/500)}{10} \approx 0.0950 = 9.50\%$$

The time to double is given by:

$$1000 = 500e^{0.0950t}$$

$$t = \frac{\ln 2}{0.095} \approx 7.30 \text{ years}$$

Amount after 25 years:

$$a = 500e^{0.095(25)} \approx \$5375.51$$

10. Since $A = 2000e^{rt}$ and $A = \$6008.33$ when $t = 25$, we have the following.

$$6008.33 = 2000e^{25r}$$

$$r = \frac{\ln(6008.33/2000)}{25} \approx 0.044 = 4.4\%$$

The time to double is given by:

$$4000 = 2000e^{0.044t}$$

$$t = \frac{\ln 2}{0.044} \approx 15.75 \text{ years}$$

Amount after 10 years:

$$A = 2000e^{0.044(10)} \approx \$3105.41$$

11. (a) $$19 = 30(1 - e^{20k})$$

$$30e^{20k} = 11$$

$$k = \frac{\ln(11/30)}{20} \approx -0.0502$$

$$N = 30(1 - e^{-0.0502t})$$

(b) $$25 = 30(1 - e^{-0.0502t})$$

$$e^{-0.0502t} = \frac{1}{6}$$

$$t = \frac{-\ln 6}{-0.0502} \approx 36 \text{ days}$$

12. (a) $$20 = 30(1 - e^{30k})$$

$$30e^{30k} = 10$$

$$k = \frac{\ln(1/3)}{30} = \frac{-\ln 3}{30} \approx -0.0366$$

$$N = 30(1 - e^{-0.0366t})$$

(b) $$25 = 30(1 - e^{-0.0366t})$$

$$e^{-0.0366t} = \frac{1}{6}$$

$$t = \frac{-\ln 6}{-0.0366} \approx 49 \text{ days}$$

13. $S = Ce^{k/t}$

(a) $S = 5$ when $t = 1$

$5 = Ce^k$

$\lim_{t\to\infty} Ce^{k/t} = C = 30$

$5 = 30e^k$

$k = \ln \frac{1}{6} \approx -1.7918$

$S = 30e^{-1.7918/t}$

(b) $S(5) = 30e^{-1.7918/5} \approx 20.9646$
(rounded to 20,965 units)

(c)

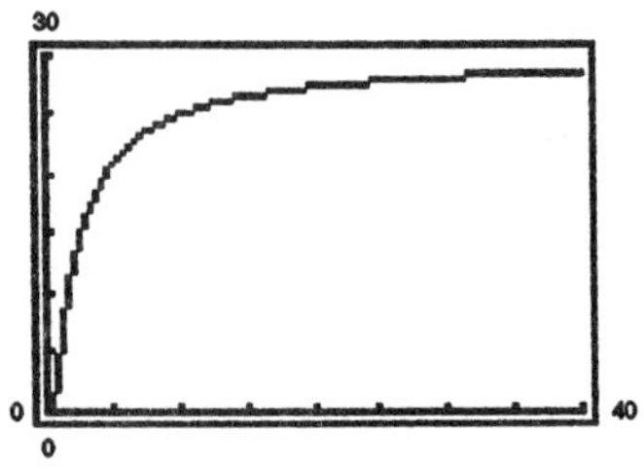

30, 0, 0, 40

14. $S = 30(1 - e^{kt})$

(a) $5 = 30(1 - e^{k(1)}) \Rightarrow k = \ln \frac{5}{6}$

$S = 30(1 - e^{\ln(5/6)t}) \approx 30(1 - e^{-0.1823t})$

(b) 30,000 units

(c) 17,944 units

(d)

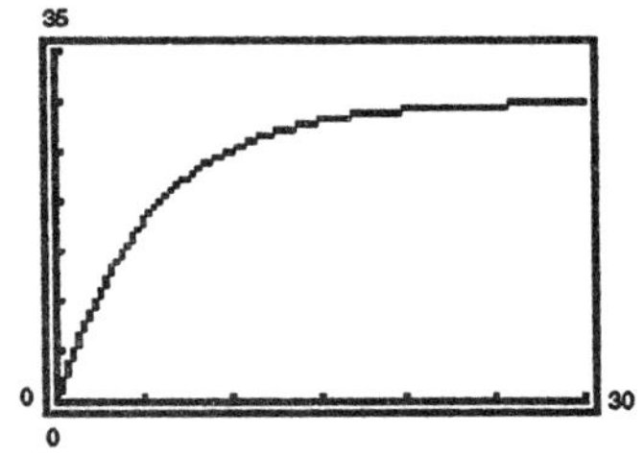

35, 0, 0, 30

15. $y = Ce^{kt}$, (0, 100), (5, 300)

$C = 100$

$300 = 100e^{5k}$

$k = \dfrac{\ln 3}{5} \approx 0.2197$

$y = 100e^{0.2197t}$

$y(10) \approx 900$

16. $200 = 100e^{0.2197t}$

$t = \dfrac{\ln 2}{0.2197} \approx 3.15$ hours

17. Since $\dfrac{dy}{dt} = ky$, $y = Ce^{kt}$.

When $t = 0$ (1970), $y = 2500$. Thus, $C = 2500$.

When $t = 10$ (1980), $y = 3350$.

Therefore, $3350 = 2500e^{10k}$.

$k = \dfrac{\ln(3350/2500)}{10} = \dfrac{\ln 67 - \ln 50}{10}$

$y = 2500e^{t(\ln 67 - \ln 50)/10}$

When $t = 30$ (2000),

$y = 2500e^{(\ln 67 - \ln 50)(30)/10} \approx 6015$.

18. $5000 = 2500e^{t(\ln 67 - \ln 50)/10}$

$\ln 2 = \dfrac{\ln 67 - \ln 50}{10} t$

$t = \dfrac{10 \ln 2}{\ln 67 - \ln 50} \approx 23.68$ years

19. Since the initial quantity is 10 grams, $y = 10e^{[\ln(1/2)/1620]t}$. When $t = 1000$, $y = 10e^{[\ln(1/2)/1620](1000)} \approx 6.52$ grams. When $t = 10{,}000$, $y = 10e^{[\ln(1/2)/1620](10{,}000)} \approx 0.14$ gram.

20. Since $y = Ce^{[\ln(1/2)/1620]t}$, we have $1.5 = Ce^{[\ln(1/2)/1620](1000)} \Rightarrow C \approx 2.30$ which implies that the initial quantity is 2.30 grams. When $t = 10{,}000$, we have $y = 2.30e^{[\ln(1/2)/1620](10{,}000)} \approx 0.03$ gram.

21. Since $y = Ce^{[\ln(1/2)/5730]t}$, we have $2.0 = Ce^{[\ln(1/2)/5730](10{,}000)} \Rightarrow C \approx 6.70$ which implies that the initial quantity is 6.70 grams. When $t = 1000$, we have $y = 6.70e^{[\ln(1/2)/5730](1000)} \approx 5.94$ grams.

22. Since the initial quantity is 3.0 grams, we have $y = 3.0e^{[\ln(1/2)/5730]t}$. When $t = 1000$, $y = 3.0e^{[\ln(1/2)/5730](1000)} \approx 2.66$ grams. When $t = 10{,}000$, $y = 3.0e^{[\ln(1/2)/5730](10{,}000)} \approx 0.89$ gram.

23. Since $y = Ce^{[\ln(1/2)/24{,}360]t}$, we have $2.1 = Ce^{[\ln(1/2)/24{,}360](1000)} \Rightarrow C \approx 2.16$. Thus, the initial quantity is 2.16 grams. When $t = 10{,}000$, $y = 2.16e^{[\ln(1/2)/24{,}360](10{,}000)} \approx 1.63$ grams.

24. Since $y = Ce^{[\ln(1/2)/24{,}360]t}$, we have $0.4 = Ce^{[\ln(1/2)/24{,}360](10{,}000)} \Rightarrow C \approx 0.53$ which implies that the initial quantity is 0.53 gram. When $t = 1000$, we have $y = 0.53e^{[\ln(1/2)/24{,}360](1000)} \approx 0.52$ gram.

25. Since $\dfrac{dy}{dx} = ky$, $y = Ce^{kt}$ or $y = y_0e^{kt}$.

$$\frac{1}{2}y_0 = y_0e^{1620k}$$

$$k = \frac{-\ln 2}{1620}$$

$$y = y_0e^{-(\ln 2)t/1620}.$$

When $t = 100$, $y = y_0e^{-(\ln 2)/16.2} = y_0(0.9581)$.
Therefore, 95.81% of the present amount still exists.

26. Since $\dfrac{dy}{dt} = ky$, $y = Ce^{kt}$ or $y = y_0e^{kt}$.

When $t = 1$,

$$0.9957y_0 = y_0e^{k}$$

$$k = \ln(0.9957)$$

$$y = y_0e^{[\ln(0.9957)]t}.$$

Solving $\frac{1}{2}y_0 = y_0e^{[\ln(0.9957)]t}$, $-\ln 2 = [\ln(0.9957)]t$.

Therefore, $t = \dfrac{-\ln 2}{\ln(0.9957)} \approx 160.85$ years.

27. Since $\dfrac{dy}{dx} = ky$,

$$y = Ce^{kt}$$

$$y = y_0e^{kt}$$

$$\frac{1}{2}y_0 = y_0e^{5730k}$$

$$k = -\frac{\ln 2}{5730}$$

$$0.15y_0 = y_0e^{(-\ln 2/5730)t}$$

$$\ln 0.15 = -\frac{(\ln 2)t}{5730}$$

$$t = -\frac{5730\ln 0.15}{\ln 2} \approx 15{,}682.813 \text{ years.}$$

28. $R = \dfrac{\ln I - 0}{\ln 10}$, $I = e^{R\ln 10} = 10^R$

(a) $8.3 = \dfrac{\ln I - 0}{\ln 10}$

$$I = 10^{8.3} = 199{,}526{,}231.5$$

(b) $2R = \dfrac{\ln I - 0}{\ln 10}$

$$I = e^{2R\ln 10} = e^{2R\ln 10} = (e^{R\ln 10})^2 = (10^R)^2$$

Increases by a factor of $e^{2R\ln 10}$ or 10^R.

(c) $\dfrac{dR}{dI} = \dfrac{1}{I\ln 10}$

29. Since $\dfrac{dy}{dt} = k(y - 20)$,

$$\int \frac{1}{y-20}\,dy = \int k\,dt$$

$$\ln(y - 20) = kt + C$$

$$y = Ce^{kt} + 20.$$

When $t = 0$, $y = 72$. Therefore, $C = 52$.

When $t = 1$, $y = 48$. Therefore, $48 = 52e^k + 20$,

$e^k = \frac{28}{52} = \frac{7}{13}$, and $k = \ln\frac{7}{13}$. Thus,

$y = 52e^{[\ln(7/13)]t} + 20$.

When $t = 5$, $y = 52e^{5\ln(7/13)} + 20 \approx 22.35°$.

30. Since $\dfrac{dy}{dt} = k(y - 70)$,

$$\int \frac{1}{y-70}\,dy = \int k\,dt$$

$$\ln(y - 70) = kt + C.$$

When $t = 0$, $y = 350$. Thus,

$$C = \ln 280$$

$$kt = \ln(y - 70) - \ln 280$$

$$kt = \ln\left(\frac{y-70}{280}\right).$$

When $t = 45$, $y = 150$. Thus,

$$k = \frac{1}{45}\ln\left(\frac{150-70}{280}\right) = \frac{1}{45}\ln\frac{2}{7}.$$

When $y = 80°$,

$$t = \frac{\ln[(80-70)/280]}{(1/45)\ln(2/7)}$$

$$= \frac{-45\ln 28}{(\ln 2 - \ln 7)} \approx 119.7 \text{ minutes.}$$

31. Since $\dfrac{dy}{dt} = k(y - T)$,

$$\int \frac{1}{y-T}\,dy = \int k\,dt$$

$$\ln(y - T) = kt + C.$$

When $t = 0$ minute, $y = 68°$. Thus, $\ln(68 - T) = k(0) + C$ and $C = \ln(68 - T)$.

When $t = 0.5$ minute, $y = 53°$. Thus, $\ln(53 - T) = k(0.5) + \ln(68 - T)$ and $k = \ln\left(\dfrac{53-T}{68-T}\right)^2$.

When $t = 1$ minute, $y = 42°$. Thus

$$\ln(42 - T) = k(1) + \ln(68 - T)$$

$$\ln(42 - T) = \ln\left(\frac{53-T}{68-T}\right)^2 + \ln(68 - T)$$

$$\ln(42 - T) = \ln\frac{(53-T)^2}{68-T}$$

$$42 - T = \frac{(53-T)^2}{68-T}$$

$$2856 - 110T + T^2 = 2809 - 106T + T^2$$

$$-4T = -47$$

$$T = \frac{47}{4} = 11.75°.$$

32. Since $\dfrac{dy}{dt} = k(y - 90)$,

$$\int \frac{1}{y-90}\,dy = \int k\,dt$$

$$\ln(y - 90) = kt + C.$$

When $t = 0$, $y = 1500$. Thus, $C = \ln 1410$.

When $t = 1$,

$$y = 1120$$

$$k(1) = \ln 1030 - \ln 1410$$

$$k = \ln \frac{103}{141}.$$

Thus, $y = e^{[\ln(103/141)t + \ln 1410]} + 90 = 1410e^{[\ln(103/141)]t} + 90$.

When $t = 5$, $y = 1410e^{5\ln(103/141)} + 90 \approx 383.298°$.

33. $y = Ce^{kt}$, $(0, 760)$, $(1000, 672.71)$

$$C = 760$$

$$672.71 = 760e^{1000k}$$

$$k = \frac{\ln(672.71/760)}{1000} \approx -0.000122$$

$$y = 760e^{-0.000122t}$$

$y(3000) \approx 527.06$ mm Hg

34. $y = Ce^{kt}$, $(0, 742{,}000)$, $(2, 632{,}000)$

$$C = 742{,}000$$

$$632{,}000 = 742{,}000e^{2k}$$

$$k = \frac{\ln(632/742)}{2}$$

$$y = 742{,}000e^{(3/2)\ln(632/742)}$$

$$= 742{,}000\left(\frac{632}{742}\right)^{3/2}$$

$y(3) \approx \$275.41$

Section 7.8 Indeterminate Forms and L'Hôpital's Rule

1. (a) $\displaystyle\lim_{x\to 3} \frac{2(x-3)}{x^2-9} = \lim_{x\to 3} \frac{2(x-3)}{(x+3)(x-3)} = \lim_{x\to 3} \frac{2}{x+3} = \frac{1}{3}$

(b) $\displaystyle\lim_{x\to 3} \frac{2(x-3)}{x^2-9} = \lim_{x\to 3} \frac{(d/dx)[2(x-3)]}{(d/dx)[x^2-9]} = \lim_{x\to 3} \frac{2}{2x} = \frac{1}{3}$

2. (a) $\displaystyle\lim_{x\to -1} \frac{2x^2-x-3}{x+1} = \lim_{x\to -1} \frac{(2x-3)(x+1)}{x+1} = \lim_{x\to -1} (2x-3) = -5$

(b) $\displaystyle\lim_{x\to -1} \frac{2x^2-x-3}{x+1} = \lim_{x\to -1} \frac{(d/dx)[2x^2-x-3]}{(d/dx)[x+1]} = \lim_{x\to -1} \frac{4x-1}{1} = -5$

3. (a) $\displaystyle\lim_{x\to \infty} \frac{5x^2-3x+1}{3x^2-5} = \lim_{x\to \infty} \frac{5-(3/x)+(1/x^2)}{3-(5/x^2)} = \frac{5}{3}$

(b) $\displaystyle\lim_{x\to \infty} \frac{5x^2-3x+1}{3x^2-5} = \lim_{x\to \infty} \frac{(d/dx)[5x^2-3x+1]}{(d/dx)[3x^2-5]}$

$$= \lim_{x\to \infty} \frac{10x-3}{6x} = \lim_{x\to \infty} \frac{(d/dx)[10x-3]}{(d/dx)[6x]} = \lim_{x\to \infty} \frac{10}{6} = \frac{5}{3}$$

4. (a) $\lim_{x\to\infty} \frac{2x+1}{4x^2+x} = \lim_{x\to\infty} \frac{(2/x)+(1/x^2)}{4+(1/x)} = \frac{0}{4} = 0$

(b) $\lim_{x\to\infty} \frac{2x+1}{4x^2+x} = \lim_{x\to\infty} \frac{(d/dx)[2x+1]}{(d/dx)[4x^2+x]} = \lim_{x\to\infty} \frac{2}{8x+1} = 0$

5. $\lim_{x\to 2} \frac{x^2-x-2}{x-2} = \lim_{x\to 2} \frac{2x-1}{1} = 3$

6. $\lim_{x\to -1} \frac{x^2-x-2}{x+1} = \lim_{x\to -1} \frac{2x-1}{1} = -3$

7. $\lim_{x\to 0} \frac{\sqrt{4-x^2}-2}{x} = \lim_{x\to 0} \frac{-x/\sqrt{4-x^2}}{1} = 0$

8. $\lim_{x\to 2^-} \frac{\sqrt{4-x^2}}{x-2} = \lim_{x\to 2^-} \frac{-x/\sqrt{4-x^2}}{1}$
$= \lim_{x\to 2^-} \frac{-x}{\sqrt{4-x^2}} = -\infty$

9. $\lim_{x\to 0} \frac{e^x-(1-x)}{x} = \lim_{x\to 0} \frac{e^x+1}{1} = 2$

10. $\lim_{x\to 0^+} \frac{e^x-(1+x)}{x^3} = \lim_{x\to 0^+} \frac{e^x-1}{3x^2} = \lim_{x\to 0^+} \frac{e^x}{6x} = \infty$

11. *Case 1:* $n = 1$

$$\lim_{x\to 0^+} \frac{e^x-(1+x)}{x} = \lim_{x\to 0^+} \frac{e^x-1}{1} = 0$$

Case 2: $n = 2$

$$\lim_{x\to 0^+} \frac{e^x-(1+x)}{x^2} = \lim_{x\to 0^+} \frac{e^x-1}{2x} = \lim_{x\to 0^+} \frac{e^x}{2} = \frac{1}{2}$$

Case 3: $n \ge 3$

$$\lim_{x\to 0^+} \frac{e^x-(1+x)}{x^n} = \lim_{x\to 0^+} \frac{e^x-1}{nx^{n-1}} = \lim_{x\to 0^+} \frac{e^x}{n(n-1)x^{n-2}} = \infty$$

12. $\lim_{x\to 1} \frac{\ln x}{x^2-1} = \lim_{x\to 1} \frac{(1/x)}{2x} = \lim_{x\to 1} \frac{1}{2x^2} = \frac{1}{2}$

13. $\lim_{x\to\infty} \frac{\ln x}{x} = \lim_{x\to\infty} \frac{(1/x)}{1} = 0$

14. $\lim_{x\to\infty} \frac{e^x}{x} = \lim_{x\to\infty} \frac{e^x}{1} = \infty$

15. $\lim_{x\to\infty} \frac{3x^2-2x+1}{2x^2+3} = \lim_{x\to\infty} \frac{6x-2}{4x}$
$= \lim_{x\to\infty} \frac{6}{4} = \frac{3}{2}$

16. $\lim_{x\to\infty} \frac{x-1}{x^2+2x+3} = \lim_{x\to\infty} \frac{1}{2x+2} = 0$

17. $\lim_{x\to\infty} \frac{x^2+2x+3}{x-1} = \lim_{x\to\infty} \frac{2x+2}{1} = \infty$

18. $\lim_{x\to\infty} \frac{x^2}{e^x} = \lim_{x\to\infty} \frac{2x}{e^x} = \lim_{x\to\infty} \frac{2}{e^x} = 0$

19. $\lim_{x\to 0^+} x^2 \ln x = \lim_{x\to 0^+} \frac{\ln x}{(1/x^2)}$
$= \lim_{x\to 0^+} \frac{(1/x)}{(-2/x^3)} = \lim_{x\to 0^+} \frac{-x^2}{2} = 0$

20. $\lim_{x\to\infty} \left(\frac{1}{x} - \frac{1}{x^2}\right) = \lim_{x\to 0} \frac{x-1}{x^2} = -\infty$

21. $\lim_{x\to 2} \left(\frac{8}{x^2-4} - \frac{x}{x-2}\right) = \lim_{x\to 2} \frac{8-x(x+2)}{x^2-4} = \lim_{x\to 2} \frac{(2-x)(4+x)}{(x+2)(x-2)} = \lim_{x\to 2} \frac{-(x+4)}{x+2} = \frac{-3}{2}$

22. $\lim_{x\to 2} \left(\frac{1}{x^2-4} - \frac{\sqrt{x-1}}{x^2-4}\right) = \lim_{x\to 2} \frac{1-\sqrt{x-1}}{x^2-4} = \lim_{x\to 2} \frac{-1/(2\sqrt{x-1})}{2x} = \lim_{x\to 2} \frac{-1}{4x\sqrt{x-1}} = \frac{-1}{8}$

23. $\lim_{x\to\infty} \frac{x}{\sqrt{x^2+1}} = \lim_{x\to\infty} \frac{1}{\sqrt{1+(1/x^2)}} = 1$

24. $\lim_{x\to 1^+}\left(\frac{3}{\ln x}-\frac{2}{x-1}\right)=\lim_{x\to 1^+}\frac{3x-3-2\ln x}{(x-1)\ln x}=\lim_{x\to 1^+}\frac{3-(2/x)}{[(x-1)/x]+\ln x}=\infty$

25. $\lim_{x\to 0^+} x^{1/x}=0^\infty=0$

26. $\lim_{x\to 0}(e^x+x)^{1/x}$

$$y=(e^x+x)^{1/x}$$

$$\ln y=\frac{1}{x}\ln(e^x+1)$$

$$\lim_{x\to 0}\frac{1}{x}\ln(e^x+x)=\lim_{x\to 0}\frac{\ln(e^x+x)}{x}=\lim_{x\to 0}\frac{e^x+1}{e^x+x}=2$$

As $x\to 0$, $\ln y=2$. Therefore, $y=e^2$, $\lim_{x\to 0}(e^x+x)^{1/x}=e^2$.

27. $\lim_{x\to\infty} x^{1/x}$

$$y=x^{1/x}$$

$$\ln y=\frac{1}{x}\ln x$$

$$\lim_{x\to\infty}\frac{\ln x}{x}=\lim_{x\to\infty}\frac{(1/x)}{1}=0$$

As $x\to\infty$, $\ln y=0$, $y=1$, and $\lim_{x\to\infty}x^{1/x}=1$.

28. $\lim_{x\to\infty}\left(1+\frac{1}{x}\right)^x$

$$y=\lim_{x\to\infty}\left(1+\frac{1}{x}\right)^x$$

$$\ln y=\lim_{x\to\infty}\left[x\ln\left(1+\frac{1}{x}\right)\right]$$

$$\ln y=\lim_{x\to\infty}\frac{\ln[1+(1/x)]}{1/x}$$

$$=\lim_{x\to\infty}\frac{\left[\frac{(-1/x^2)}{1+(1/x)}\right]}{(-1/x^2)}=\lim_{x\to\infty}\frac{1}{1+(1/x)}=1$$

As $x\to\infty$,

$\ln y=1$, $y=e$, and $\lim_{x\to\infty}\left(1+\frac{1}{x}\right)^x=e$.

29. $\lim_{x\to\infty}(1+x)^{1/x}$

$$y=(1+x)^{1/x}$$

$$\ln y=\frac{1}{x}\ln(1+x)$$

$$\lim_{x\to\infty}\frac{1}{x}\ln(1+x)=\lim_{x\to\infty}\frac{1/(1+x)}{1}=0$$

As $x\to\infty$, $\ln y=0$, $y=1$, and

$\lim_{x\to\infty}(1+x)^{1/x}=1$.

30. $\lim_{x\to\infty}\frac{\sqrt{x}}{\sqrt{x-1}}=\lim_{x\to\infty}\frac{1}{\sqrt{1-(1/x)}}=1$

31. (a) Let $f(x)=x^2-25$ and $g(x)=x-5$.

(b) Let $f(x)=(x-5)^2$ and $g(x)=x^2-25$.

(c) Let $f(x)=x^2-25$ and $g(x)=(x-5)^3$.

32. Let $f(x)=x+25$ and $g(x)=x$.

33. $\lim_{x\to\infty}\frac{x^2}{e^{5x}}=\lim_{x\to\infty}\frac{2x}{5e^{5x}}=\lim_{x\to\infty}\frac{2}{25e^{5x}}=0$

34. $\lim_{x\to\infty}\frac{x^3}{e^{2x}}=\lim_{x\to\infty}\frac{3x^2}{2e^{2x}}$

$$=\lim_{x\to\infty}\frac{6x}{4e^{2x}}=\lim_{x\to\infty}\frac{6}{8e^{2x}}=0$$

35. $\lim_{x\to\infty} \frac{(\ln x)^3}{x} = \lim_{x\to\infty} \frac{3(\ln x)^2(1/x)}{1} = \lim_{x\to\infty} \frac{3(\ln x)^2}{x} = \lim_{x\to\infty} \frac{6(\ln x)(1/x)}{1} = \lim_{x\to\infty} \frac{6(\ln x)}{x} = \lim_{x\to\infty} \frac{6}{x} = 0$

36. $\lim_{x\to\infty} \frac{(\ln x)^2}{x^3} = \lim_{x\to\infty} \frac{(2\ln x)/x}{3x^2} = \lim_{x\to\infty} \frac{2\ln x}{3x^3} = \lim_{x\to\infty} \frac{2/x}{9x^2} = \lim_{x\to\infty} \frac{2}{9x^3} = 0$

37. $\lim_{x\to\infty} \frac{(\ln x)^n}{x^m} = \lim_{x\to\infty} \frac{n(\ln x)^{n-1}/x}{mx^{m-1}} = \lim_{x\to\infty} \frac{n(\ln x)^{n-1}}{mx^m} = \lim_{x\to\infty} \frac{n(n-1)(\ln x)^{n-2}}{m^2x^m} = \cdots = \lim_{x\to\infty} \frac{n!}{m^nx^m} = 0$

38. $\lim_{x\to\infty} \frac{x^m}{e^{nx}} = \lim_{x\to\infty} \frac{mx^{m-1}}{ne^{nx}} = \lim_{x\to\infty} \frac{m(m-1)x^{m-2}}{n^2e^{nx}} = \cdots = \lim_{x\to\infty} \frac{m!}{n^me^{nx}} = 0$

39. $\lim_{x\to 0} \frac{e^{2x}-1}{e^x} = \frac{0}{1} = 0$

L'Hôpital's Rule does not apply.

40. $\lim_{x\to\infty} \frac{e^{-x}}{1+e^{-x}} = \frac{0}{1+0} = 0$

L'Hôpital's Rule does not apply.

41.

x	10	10^2	10^4	10^6	10^8	10^{10}
$\frac{(\ln x)^4}{x}$	2.811	4.498	0.720	0.036	0.001	0.000

42.

x	1	5	10	20	30	40	50	100
$\frac{e^x}{x^5}$	2.718	0.047	0.220	151.614	4.39×10^5	2.30×10^9	1.66×10^{13}	2.69×10^{23}

43. $\lim_{x\to a} f(x)^{g(x)}$

$y = f(x)^{g(x)}$

$\ln y = g(x) \ln f(x)$

$\lim_{x\to a} g(x) \ln f(x) = (\infty)(-\infty) = -\infty$

As $x \to a$, $\ln y = -\infty$, $y = 0$, and

$\lim_{x\to a} f(x)^{g(x)} = 0.$

44. $\lim_{x\to a} f(x)^{g(x)}$

$y = f(x)^{g(x)}$

$\ln y = g(x) \ln f(x)$

$\lim_{x\to a} g(x) \ln f(x) = (-\infty)(-\infty) = \infty$

As $x \to a$, $\ln y = \infty$, $y = \infty$, and

$\lim_{x\to a} f(x)^{g(x)} = \infty.$

45. (a) $\lim_{x\to 0} (e^x - 1)^{1/x^2} = 0$ since $\lim_{x\to 0} (e^x - 1) = 0$ and $\lim_{x\to 0} \frac{1}{x^2} = \infty$. (See Exercise 43.)

(b) $\lim_{x\to 0} (e^x - 1)^{-1/x^2} = \infty$ since $\lim_{x\to 0} (e^x - 1) = 0$ and $\lim_{x\to 0} -\frac{1}{x^2} = -\infty$. (See Exercise 44.)

46. Let N be a fixed value for n. Then

$$\lim_{x\to\infty} \frac{x^{N-1}}{e^x} = \lim_{x\to\infty} \frac{(N-1)x^{N-2}}{e^x} = \lim_{x\to\infty} \frac{(N-1)(N-2)x^{N-3}}{e^x} = \cdots = \lim_{x\to\infty} \left[\frac{(N-1)!}{e^x}\right] = 0.$$

47. $$\lim_{k\to 0} \frac{32\left(1 - e^{-kt} + \frac{v_0ke^{-kt}}{32}\right)}{k} = \lim_{k\to 0} \frac{32(1-e^{-kt})}{k} + \lim_{k\to 0} (v_0e^{-kt})$$

$$= \lim_{k\to 0} \frac{32(0 + te^{-kt})}{1} + \lim_{k\to 0} \left(\frac{v_0}{e^{kt}}\right) = 32t + v_0$$

48. $A = P\left(1 + \frac{r}{n}\right)^{nt}$

$$\ln A = \ln P + nt\ln\left(1 + \frac{r}{n}\right) = \ln P + \frac{\ln\left(1 + \frac{r}{n}\right)}{\frac{1}{nt}}$$

$$\lim_{n\to\infty}\left[\frac{\ln\left(1 + \frac{r}{n}\right)}{\frac{1}{nt}}\right] = \lim_{n\to\infty}\left[\frac{-\frac{r}{n^2}\left(\frac{1}{1 + (r/n)}\right)}{-\left(\frac{1}{n^2 t}\right)}\right] = \lim_{n\to\infty}\left[rt\left(\frac{1}{1 + \frac{r}{n}}\right)\right] = rt$$

Since $\lim_{n\to\infty} \ln A = \ln P + rt$, we have $\lim_{n\to\infty} A = e^{(\ln P + rt)} = e^{\ln P}e^{rt} = Pe^{rt}$.

49. $y = x^{1/x}, \quad x > 0$

Horizontal asymptote: $y = 1$ (See Exercise 27.)

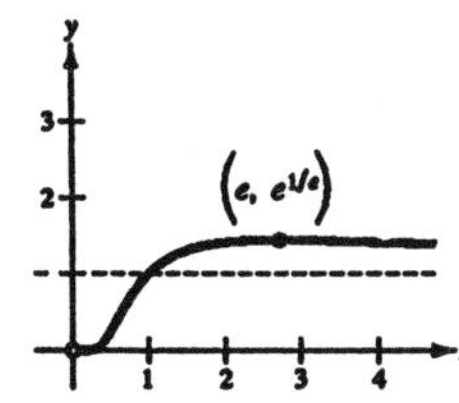

$$\ln y = \frac{1}{x}\ln x$$

$$\frac{1}{y}\frac{dy}{dx} = \frac{1}{x}\left(\frac{1}{x}\right) + (\ln x)\left(-\frac{1}{x^2}\right)$$

$$\frac{dy}{dx} = x^{1/x}\left(\frac{1}{x^2}\right)(1 - \ln x)$$

$$= x^{(1/x)-2}(1 - \ln x) = 0$$

Critical number:	$x = e$	
Intervals:	$(0, e)$	(e, ∞)
Sign of dy/dx:	$+$	$-$
$y = f(x)$:	Increasing	Decreasing

Relative maximum: $(e, e^{1/e})$

50. $y = x^x, \quad x > 0$

$\lim_{x\to\infty} x^x \to \infty$

No horizontal asymptotes

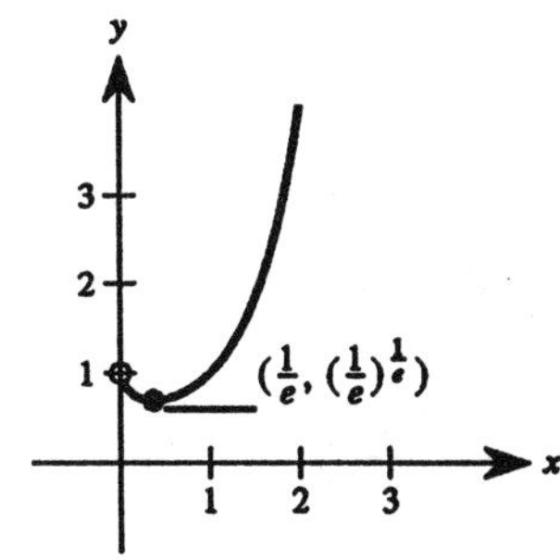

$$\ln y = x\ln x$$

$$\frac{1}{y}\frac{dy}{dx} = x\left(\frac{1}{x}\right) + \ln x$$

$$\frac{dy}{dx} = x^x(1 + \ln x) = 0$$

Critical number:	$x = e^{-1}$	
Intervals:	$(0, e^{-1})$	$(e^{-1}, 0)$
Sign of dy/dx:	$-$	$+$
$y = f(x)$:	Decreasing	Increasing

Relative minimum: $\left(e^{-1}, (e^{-1})^{e^{-1}}\right) = (1/e, (1/e)^{1/e})$

51. $y = 2xe^{-x}$

$\lim_{x\to\infty} \frac{2x}{e^x} = \lim_{x\to\infty} \frac{2}{e^x} = 0$

Horizontal asymptote: $y = 0$

$\frac{dy}{dx} = 2x(-e^{-x}) + 2e^{-x} = 2e^{-x}(1 - x) = 0$

Critical number:	$x = 1$	
Intervals:	$(-\infty, 1)$	$(1, \infty)$
Sign of dy/dx:	$+$	$-$
$y = f(x)$:	Increasing	Decreasing

Relative maximum: $(1, 2/e)$

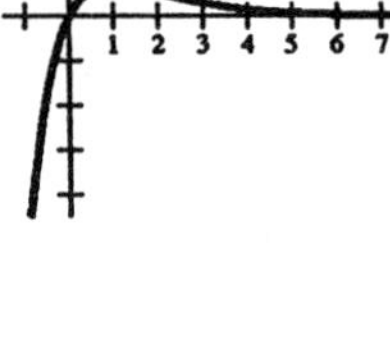

52. $y = \frac{\ln x}{x}$

Horizontal asymptote: $y = 0$ (See Exercise 13.)

$\frac{dy}{dx} = \frac{x(1/x) - (\ln x)(1)}{x^2} = \frac{1 - \ln x}{x^2} = 0$

Critical number:	$x = e$	
Intervals:	$(0, e)$	(e, ∞)
Sign of dy/dx:	$+$	$-$
$y = f(x)$:	Increasing	Decreasing

Relative maximum: $(e, 1/e)$

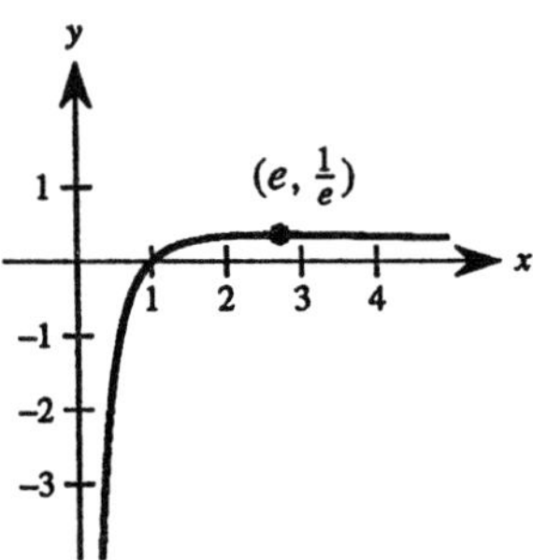

53. (a)

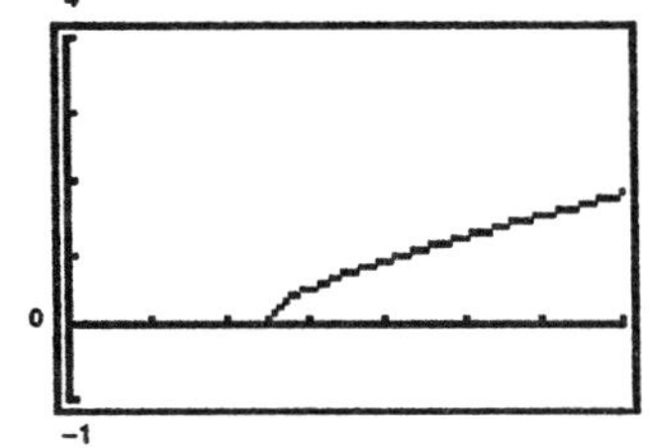

(b) $\lim_{x\to3} \frac{x-3}{\ln(2x-5)} = \lim_{x\to3} \frac{1}{2/(2x-5)}$

$= \lim_{x\to3} \frac{2x-5}{2} = \frac{1}{2}$

54. (a)

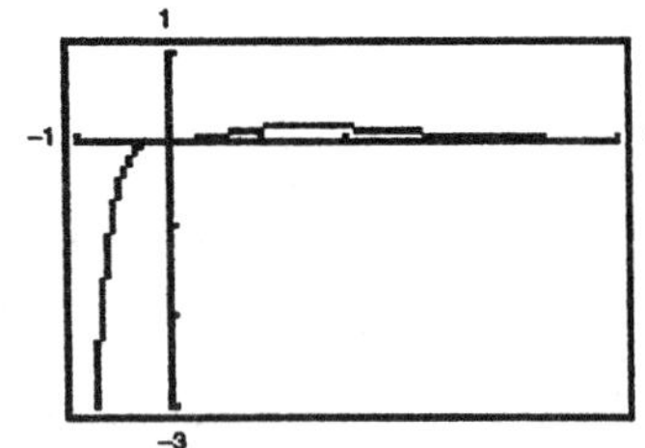

(b) $\lim_{x\to\infty} \frac{x^3}{e^{2x}} = \lim_{x\to\infty} \frac{3x^2}{2e^{2x}}$

$= \lim_{x\to\infty} \frac{6x}{4e^{2x}}$

$= \lim_{x\to\infty} \frac{6}{8e^{2x}} = 0$

55. $f(x) = \frac{x^k - 1}{k}$

$k = 1, \quad f(x) = x - 1$

$k = 0.1, \quad f(x) = \frac{x^{0.1} - 1}{0.1} = 10(x^{0.1} - 1)$

$k = 0.01, \quad f(x) = \frac{x^{0.01} - 1}{0.01} = 100(x^{0.01} - 1)$

$\lim_{k\to0^+} \frac{x^k - 1}{k} = \lim_{k\to0^+} \frac{x^k(\ln x)}{1} = \ln x$

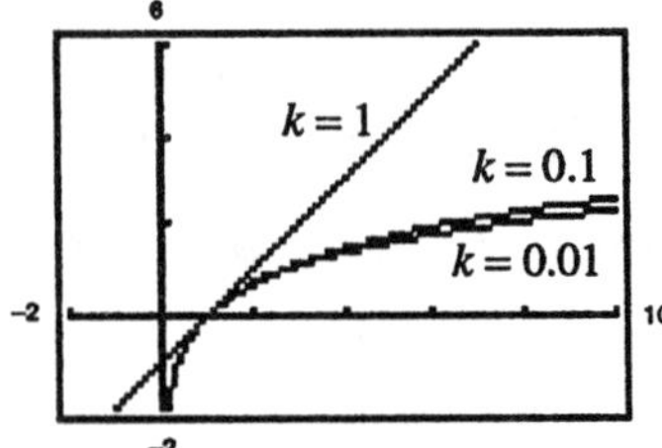

Chapter 7 Review Exercises

1. (a)

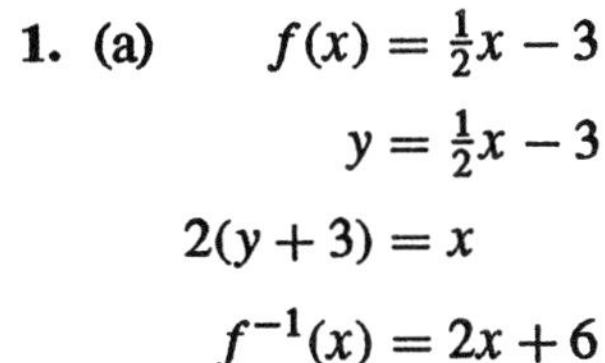

$$f(x) = \tfrac{1}{2}x - 3$$
$$y = \tfrac{1}{2}x - 3$$
$$2(y + 3) = x$$
$$f^{-1}(x) = 2x + 6$$

(b)

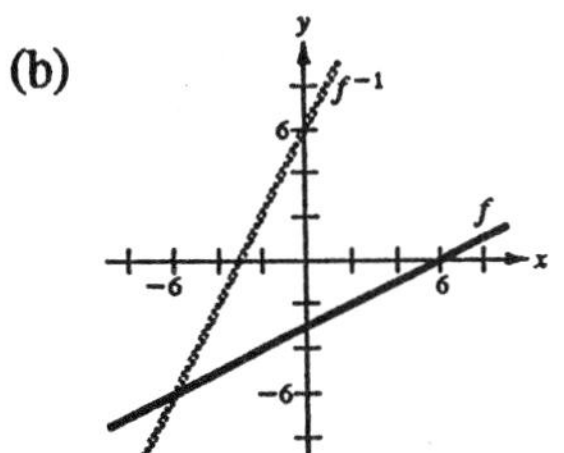

2. (a)

$$f(x) = 5x - 7$$
$$y = 5x - 7$$
$$\frac{y + 7}{5} = x$$
$$f^{-1}(x) = \frac{x + 7}{5}$$

(b)

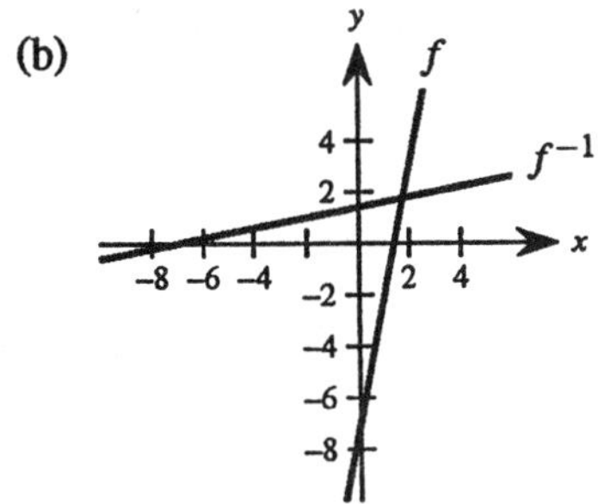

3. (a)

$$f(x) = \sqrt{x + 1}$$
$$y = \sqrt{x + 1}$$
$$y^2 - 1 = x$$
$$f^{-1}(x) = x^2 - 1, \quad x \geq 0$$

(b)

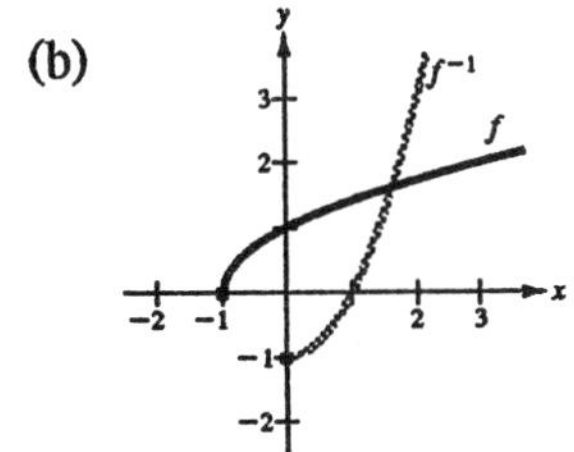

4. (a)

$$f(x) = x^3 + 2$$
$$y = x^3 + 2$$
$$\sqrt[3]{y - 2} = x$$
$$f^{-1}(x) = \sqrt[3]{x - 2}$$

(b)

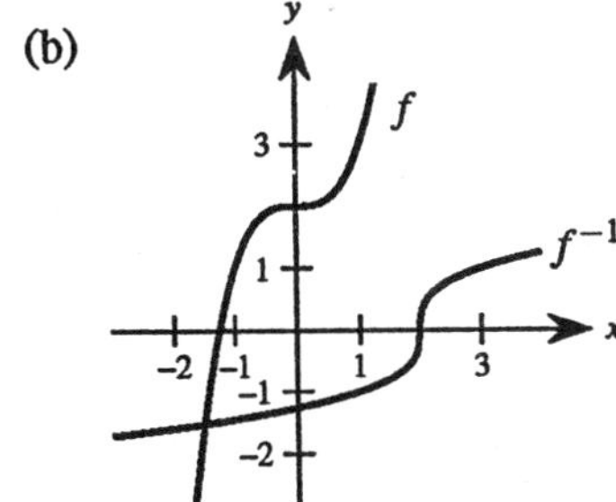

5. (a)

$$f(x) = x^2 - 5, \quad x \geq 0$$
$$y = x^2 - 5$$
$$\sqrt{y + 5} = x$$
$$f^{-1}(x) = \sqrt{x + 5}$$

(b)

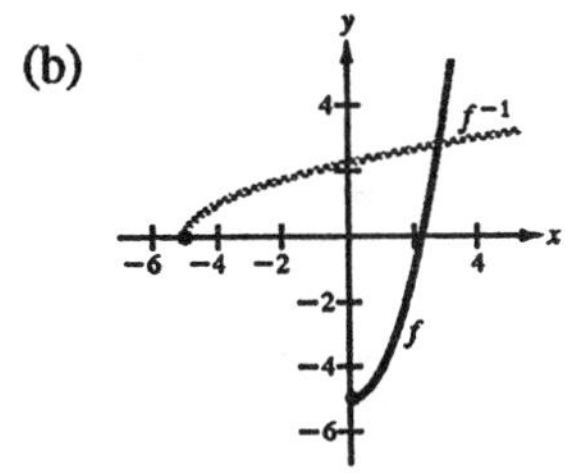

6. (a)

$$f(x) = \sqrt[3]{x + 1}$$
$$y = \sqrt[3]{x + 1}$$
$$y^3 - 1 = x$$
$$f^{-1}(x) = x^3 - 1$$

(b)

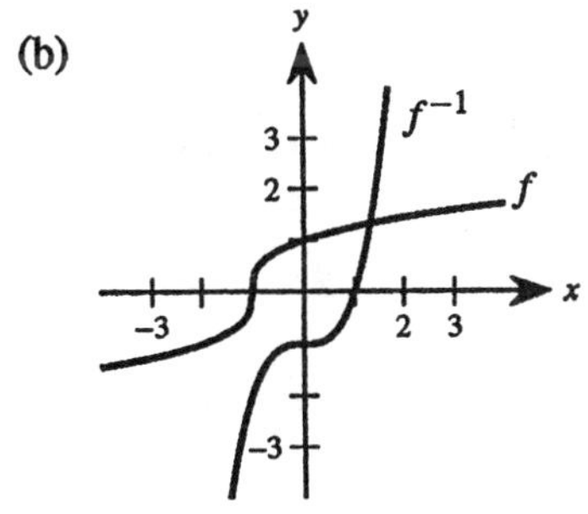

7. $f(x) = 2(x-4)^2, \quad x \geq 4$

$$y = 2(x-4)^2$$

$$\sqrt{y/2} + 4 = x$$

$$f^{-1}(x) = \sqrt{x/2} + 4$$

8. $f(x) = |x-2| = x-2, \quad x \geq 2$

$$y = x - 2$$

$$y + 2 = x$$

$$f^{-1}(x) = x + 2, \quad x \geq 0$$

9. $e^{\ln x} = 3$

$$x = 3$$

10. $\ln x + \ln(x-3) = 0$

$$\ln[x(x-3)] = 0$$

$$x^2 - 3x = e^0$$

$$x^2 - 3x - 1 = 0$$

$$x = \frac{3 \pm \sqrt{13}}{2}$$

$$x = \frac{3 + \sqrt{13}}{2}$$

11. $\log_3 x + \log_3(x-1) - \log_3(x-2) = 2$

$$\log_3 \frac{x(x-1)}{x-2} = 2$$

$$\frac{x(x-1)}{x-2} = 3^2$$

$$x^2 - x = 9(x-2)$$

$$x^2 - 10x + 18 = 0$$

$$x = \frac{10 \pm \sqrt{28}}{2} = 5 \pm \sqrt{7}$$

12. $\log_x 125 = 3$

$$125 = x^3$$

$$x = 5$$

13. $y = \ln\sqrt{x} = \frac{1}{2}\ln x$

$$\frac{dy}{dx} = \frac{1}{2x}$$

14. $y = \ln\frac{x(x-1)}{x-2}$

$$= \ln x + \ln(x-1) - \ln(x-2)$$

$$\frac{dy}{dx} = \frac{1}{x} + \frac{1}{x-1} - \frac{1}{x-2} = \frac{x^2 - 4x + 2}{x^3 - 3x^2 + 2x}$$

15. $y = x\sqrt{\ln x}$

$$\frac{dy}{dx} = \left(\frac{x}{2}\right)(\ln x)^{-1/2}\left(\frac{1}{x}\right) + \sqrt{\ln x}$$

$$= \frac{1}{2\sqrt{\ln x}} + \sqrt{\ln x} = \frac{1 + 2\ln x}{2\sqrt{\ln x}}$$

16. $y = \ln[x(x^2-2)^{2/3}] = \ln x + \frac{2}{3}\ln(x^2-2)$

$$\frac{dy}{dx} = \frac{1}{x} + \frac{2}{3}\left(\frac{2x}{x^2-2}\right) = \frac{7x^2-6}{3x^3-6x}$$

17. $y(\ln x) + y^2 = 0$

$$y\left(\frac{1}{x}\right) + (\ln x)\left(\frac{dy}{dx}\right) + 2y\left(\frac{dy}{dx}\right) = 0$$

$$(2y + \ln x)\frac{dy}{dx} = \frac{-y}{x}$$

$$\frac{dy}{dx} = \frac{-y}{x(2y + \ln x)}$$

18. $\ln(x+y) = x$

$$\frac{1}{x+y}\left(1 + \frac{dy}{dx}\right) = 1$$

$$\left(\frac{1}{x+y}\right)\frac{dy}{dx} = 1 - \frac{1}{x+y}$$

$$\frac{dy}{dx} = x + y - 1$$

19. $\ln y = x \ln x$

$$\frac{1}{y}\frac{dy}{dx} = x\left(\frac{1}{x}\right) + \ln x$$

$$\frac{dy}{dx} = y(1 + \ln x)$$

20. $\ln y = x \ln \sqrt{x^2+1} = \frac{1}{2}x \ln(x^2+1)$

$$\frac{1}{y}\frac{dy}{dx} = \frac{1}{2}x\left(\frac{2x}{x^2+1}\right) + \frac{1}{2}\ln(x^2+1)$$

$$\frac{dy}{dx} = y\left[\frac{x^2}{x^2+1} + \ln\sqrt{x^2+1}\right]$$

21. $y = \frac{1}{b^2}\left[\ln(a+bx) + \frac{a}{a+bx}\right]$

$$\frac{dy}{dx} = \frac{1}{b^2}\left[\frac{b}{a+bx} - \frac{ab}{(a+bx)^2}\right] = \frac{x}{(a+bx)^2}$$

22. $y = \frac{1}{b^2}[a + bx - a\ln(a+bx)]$

$$\frac{dy}{dx} = \frac{1}{b^2}\left(b - \frac{ab}{a+bx}\right) = \frac{x}{a+bx}$$

23. $y = -\frac{1}{a}\ln\left(\frac{a+bx}{x}\right)$

$$= -\frac{1}{a}[\ln(a+bx) - \ln x]$$

$$\frac{dy}{dx} = -\frac{1}{a}\left(\frac{b}{a+bx} - \frac{1}{x}\right) = \frac{1}{x(a+bx)}$$

24. $y = -\frac{1}{ax} + \frac{b}{a^2}\ln\frac{a+bx}{x}$

$$= -\frac{1}{ax} + \frac{b}{a^2}[\ln(a+bx) - \ln x]$$

$$\frac{dy}{dx} = \frac{1}{ax^2} + \left(\frac{b}{a^2}\right)\frac{-a}{x(a+bx)} = \frac{1}{x^2(a+bx)}$$

25. $y = \ln(e^{-x^2}) = -x^2$

$$y' = -2x$$

26. $y = \ln\left(\frac{e^x}{1+e^x}\right)$

$$= \ln e^x - \ln(1+e^x) = x - \ln(1+e^x)$$

$$y' = 1 - \frac{e^x}{1+e^x} = \frac{1}{1+e^x}$$

27. $y = x^2e^x$

$$y' = x^2e^x + 2xe^x = xe^x(x+2)$$

28. $y = e^{-x^2/2}$

$$y' = -xe^{-x^2/2}$$

29. $y = \sqrt{e^{2x} + e^{-2x}}$

$$y' = \frac{1}{2}(e^{2x} + e^{-2x})^{-1/2}(2e^{2x} - 2e^{-2x})$$

$$= \frac{e^{2x} - e^{-2x}}{\sqrt{e^{2x} + e^{-2x}}}$$

30. $y = x^{2x+1}$

$$\ln y = (2x+1)\ln x$$

$$\frac{y'}{y} = \frac{2x+1}{x} + 2\ln x$$

$$y' = y\left(\frac{2x+1}{x} + 2\ln x\right)$$

$$= x^{2x+1}\left(\frac{2x+1}{x} + 2\ln x\right)$$

31. $y = 3^{x-1}$

$$y' = 3^{x-1}\ln 3$$

32. $y = 4^xe^x$

$$y' = 4^xe^x + (\ln 4)4^xe^x = 4^xe^x(1 + \ln 4)$$

33. $ye^x + xe^y = xy$

$$ye^x + y'e^x + xy'e^y + e^y = y + xy'$$

$$y'(e^x + xe^y - x) = y - ye^x - e^y$$

$$y' = \frac{y - ye^x - e^y}{e^x + xe^y - x}$$

34. $y = \frac{x^2}{e^x}$

$$y' = \frac{e^x(2x) - x^2e^x}{e^{2x}} = \frac{x(2-x)}{e^x}$$

35. $y = x^a$
$y' = ax^{a-1}$

36. $y = a^x$
$y' = (\ln a)a^x$

37. $y = x^x$
$y' = x^x(1 + \ln x)$

38. $y = a^a$
$y' = 0$

39. $u = 7x - 2, \quad du = 7\,dx$

$$\int \frac{1}{7x-2}\,dx = \frac{1}{7}\int \frac{1}{7x-2}(7)\,dx = \frac{1}{7}\ln|7x-2| + C$$

40. $u = x^2 - 1, \quad du = 2x\,dx$

$$\int \frac{x}{x^2-1}\,dx = \frac{1}{2}\int \frac{2x}{x^2-1}\,dx = \frac{1}{2}\ln|x^2-1| + C$$

41. $\displaystyle\int \frac{1}{x\ln(3x)}\,dx = \int \frac{1/x}{\ln(3x)}\,dx = \ln|\ln(3x)| + C$

42. $u = \ln x, \quad du = \dfrac{1}{x}\,dx$

$$\int \frac{\ln\sqrt{x}}{x}\,dx = \frac{1}{2}\int (\ln x)\left(\frac{1}{x}\right)dx = \frac{1}{4}(\ln x)^2 + C$$

43. $\displaystyle\int \frac{x^2+3}{x}\,dx = \int \left(x + \frac{3}{x}\right)dx = \frac{x^2}{2} + 3\ln|x| + C$

44. $\displaystyle\int \frac{x^3+1}{x^2}\,dx = \int \left(x + \frac{1}{x^2}\right)dx = \frac{1}{2}x^2 - \frac{1}{x} + C$

45. $u = x^3 - 1, \quad du = 3x^2\,dx$

$$\int \frac{x^2}{x^3-1}\,dx = \frac{1}{3}\int \frac{3x^2}{x^3-1}\,dx = \frac{1}{3}\ln|x^3-1| + C$$

46. $$\int \left(x + \frac{1}{x}\right)^2 dx = \int (x^2 + 2 + x^{-2})\,dx = \frac{1}{3}x^3 + 2x - \frac{1}{x} + C$$

47. $u = \ln x, \quad du = \dfrac{1}{x}\,dx$

$$\int \frac{1}{x\sqrt{\ln x}}\,dx = \int (\ln x)^{-1/2}\left(\frac{1}{x}\right)dx = 2\sqrt{\ln x} + C$$

48. $$\int \frac{x+2}{2x+3}\,dx = \int \left(\frac{1}{2} + \frac{1/2}{2x+3}\right)dx$$
$$= \frac{1}{2}\left(\int dx + \int \frac{1}{2x+3}\,dx\right) = \frac{1}{2}\left(x + \frac{1}{2}\ln|2x+3|\right) + C = \frac{1}{4}(2x + \ln|2x+3|) + C$$

49. $$\int xe^{-3x^2}\,dx = -\frac{1}{6}\int e^{-3x^2}(-6x)\,dx = -\frac{1}{6}e^{-3x^2} + C$$

50. $\displaystyle\int \frac{e^{1/x}}{x^2}\,dx = -\int e^{1/x}\left(-\frac{1}{x^2}\right)dx = -e^{1/x} + C$

51. $$\int \frac{e^{4x} - e^{2x} + 1}{e^x}\,dx = \int (e^{3x} - e^x + e^{-x})\,dx = \frac{1}{3}e^{3x} - e^x - e^{-x} + C = \frac{e^{4x} - 3e^{2x} - 3}{3e^x} + C$$

52. Let $u = e^{2x} + e^{-2x}, \quad du = (2e^{2x} - 2e^{-2x})\,dx$.

$$\int \frac{e^{2x} - e^{-2x}}{e^{2x} + e^{-2x}}\,dx = \frac{1}{2}\int \frac{2e^{2x} - 2e^{-2x}}{e^{2x} + e^{-2x}}\,dx = \frac{1}{2}\ln(e^{2x} + e^{-2x}) + C$$

53. $\displaystyle\int \frac{e^x}{e^x - 1}\,dx = \ln|e^x - 1| + C$

54. $\displaystyle\int x^2 e^{x^3+1}\,dx = \frac{1}{3}\int e^{x^3+1}(3x^2)\,dx = \frac{1}{3}e^{x^3+1} + C$

55. $\displaystyle\int xe^{-x^2/2}\,dx = -\int e^{-x^2/2}(-x)\,dx = -e^{-x^2/2} + C$

56. $$\int \frac{x-1}{3x^2-6x-1}\,dx = \frac{1}{6}\int \frac{6x-6}{3x^2-6x-1}\,dx = \frac{1}{6}\ln|3x^2-6x-1| + C$$

57. $\displaystyle\int_1^4 \frac{x+1}{x}\,dx = \int_1^4 \left(1+\frac{1}{x}\right)dx$

$\displaystyle= \Big[x + \ln|x|\Big]_1^4 = 3 + \ln 4$

58. $\displaystyle\int_1^4 \frac{\ln x}{x}\,dx = \left[\frac{(\ln x)^2}{2}\right]_1^4 = \frac{(\ln 4)^2}{2}$

59. $\displaystyle\int_3^4 \frac{1}{x-2}\,dx - \int_3^4 \frac{1}{x+2}\,dx = \ln|x-2|\Big]_3^4 - \ln|x+2|\Big]_3^4 = \ln 2 - \ln 6 + \ln 5 = \ln\frac{5}{3}$

60. $\displaystyle\int_0^2 xe^{-x^2}\,dx = -\frac{1}{2}\int_0^2 e^{-x^2}(-2x)\,dx = -\frac{1}{2}e^{-x^2}\Big]_0^2 = -\frac{1}{2}(e^{-4}-1) = \frac{1}{2}(1-e^{-4})$

61. Area $\displaystyle= \int_0^2 4e^{-2x}\,dx$

$\displaystyle= -2e^{-2x}\Big]_0^2$

$= -2[e^{-4}-1]$

$= 2[1-e^{4}] \approx 1.963$

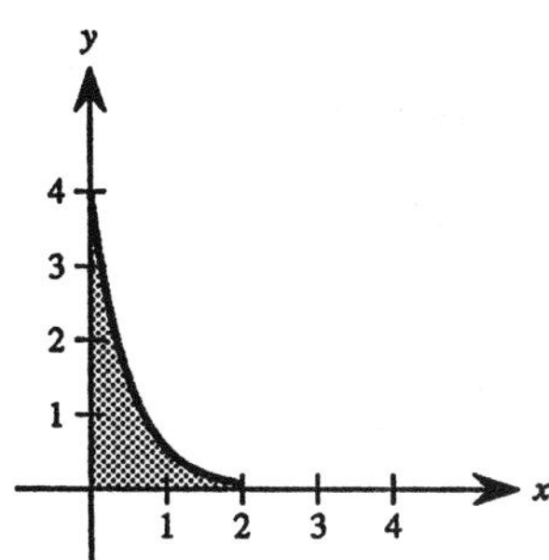

62. $\displaystyle\int_0^1 \frac{1}{x+1}\,dx = \ln|x+1|\Big]_0^1 = \ln 2$

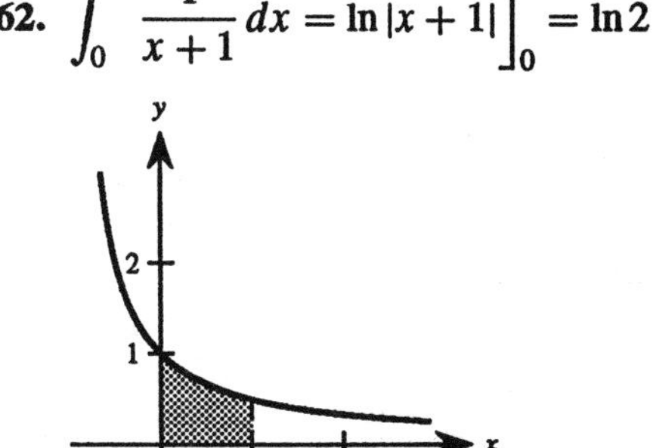

63. $\displaystyle\frac{1}{5}\int_5^{10} \frac{1}{x-1}\,dx = \frac{1}{5}\ln\frac{9}{4} = \frac{2}{5}\ln\frac{3}{2}$

$\approx 0.162 = \text{Average}$

$\displaystyle x = 1 + \frac{5}{\ln(9/4)}$

$\displaystyle= 1 + \frac{5}{2\ln(3/2)} \approx 7.166$

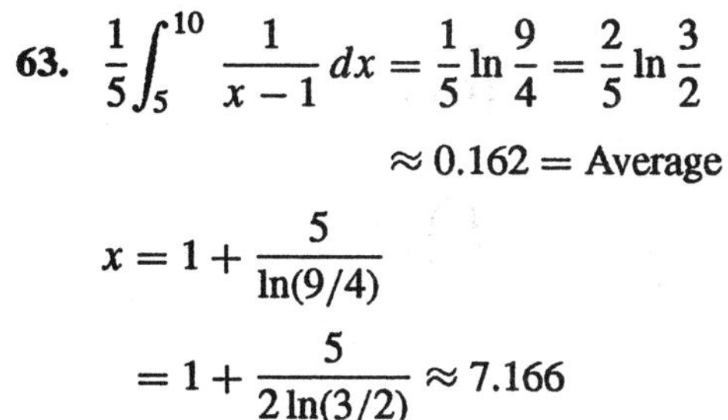

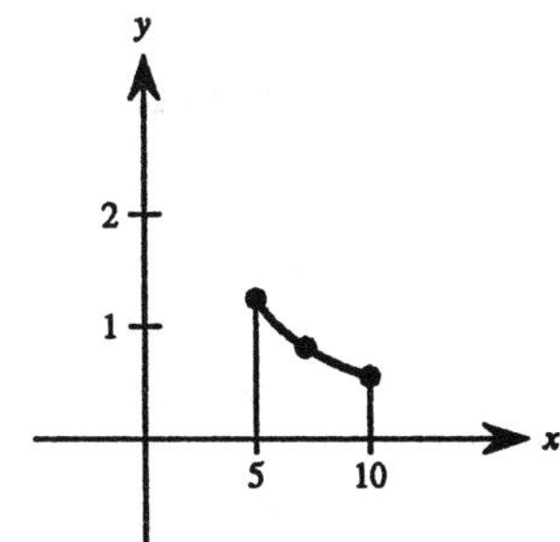

64. $\displaystyle\frac{1}{2}\int_0^2 e^{-x}\,dx = -\frac{1}{2}e^{-x}\Big]_0^2$

$\displaystyle= \frac{1}{2}(1-e^{-2}) \approx 0.432 = \text{Average}$

$\displaystyle e^{-x} = \frac{1}{2}(1-e^{-2})$

$\displaystyle x = -\ln\frac{1-e^{-2}}{2}$

$= \ln 2 - \ln(1-e^{-2}) \approx 0.839$

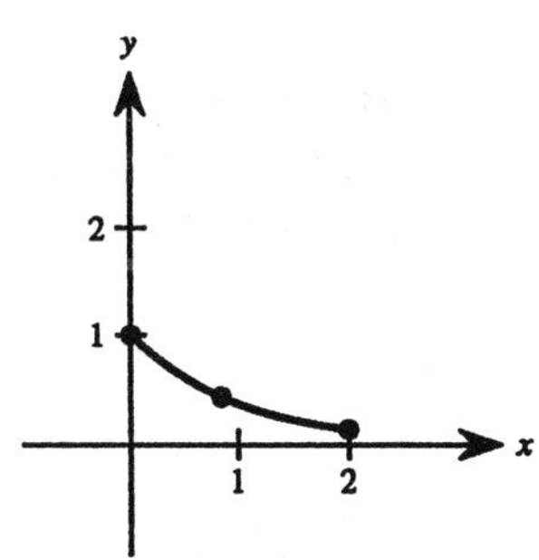

65. Area $= \int_0^4 xe^{-x^2}\,dx$

$$= -\frac{1}{2}e^{-x^2}\Big]_0^4 = -\frac{1}{2}(e^{-16} - 1) \approx 0.500$$

66. Area $= \int_0^4 3e^{-x/2}\,dx$

$$= -6e^{-x/2}\Big]_0^4 = -6(e^{-2} - 1) \approx 5.188$$

67. (a) $A = 500e^{(0.05)(1)} \approx \$525.64$

(b) $A = 500e^{(0.05)(10)} \approx \$824.36$

(c) $A = 500e^{(0.05)(100)} \approx \$74,206.58$

68. $2P = Pe^{10r}$

$$2 = e^{10r}$$

$$\ln 2 = 10r$$

$$r = \frac{\ln 2}{10} \approx 6.93\%$$

69. $10,000 = Pe^{(0.07)(15)}$

$$P = \frac{10,000}{e^{1.05}} \approx \$3499.38$$

70. $\frac{1}{5}\int 2500e^{0.12t}\,dt = \frac{12,500}{3}e^{0.12t}\Big]_0^5 \approx \$3425.50$

71. (a) $2P_0 = P_0e^{0.025t}$

$$2 = e^{0.025t}$$

$$t = \frac{\ln 2}{0.025} \approx 27.73 \text{ yrs}$$

(b) $3P_0 = P_0e^{0.025t}$

$$3 = e^{0.025t}$$

$$t = \frac{\ln 3}{0.025} \approx 43.94 \text{ yrs}$$

72. $P(h) = 30e^{kh}$

$$P(18,000) = 30e^{18,000k} = 15$$

$$k = \frac{\ln(1/2)}{18,000} \approx -0.00004$$

$$P(h) = 30e^{-(h\ln 2)/18,000}$$

$$P(35,000) = 30e^{-(35,000\ln 2)/18,000} \approx 7.79$$

73. $p(t) = 80e^{-0.5t} + 20$

$$p'(t) = -40e^{-0.5t}$$

$$p'(1) = -40e^{-0.5(1)} \approx -24.26\%$$

$$p'(2) = -40e^{-0.5(2)} \approx -14.72\%$$

74. (a) $\frac{1}{2}\int_0^2 (80e^{-0.5t} + 20)\,dt = \frac{1}{2}(-160e^{-0.5t} + 20t)\Big]_0^2$

$$= 10(-8e^{-0.5t} + t)\Big]_0^2$$

$$\approx 70.57\%$$

(b) $\frac{1}{2}\int_2^4 (80e^{-0.5t} + 20)\,dt = 10(-8e^{-0.5t} + t)\Big]_2^4$

$$\approx 38.60\%$$

75. $\int_{t_1}^{t_2} \frac{1}{20}e^{-t/20}\,dt = -e^{-t/20}\Big]_{t_1}^{t_2}$

(a) $\int_0^{10} \frac{1}{20}e^{-t/20}\,dt = -e^{-t/20}\Big]_0^{10} \approx 0.3935$

(b) $\int_0^{30} \frac{1}{20}e^{-t/20}\,dt = -e^{-t/20}\Big]_0^{30} \approx 0.7769$

(c) $\int_{15}^{30} \frac{1}{20}e^{-t/20}\,dt = -e^{-t/20}\Big]_{15}^{30} \approx 0.2492$

(d) $\int_0^{60} \frac{1}{20}e^{-t/20}\,dt = -e^{-t/20}\Big]_0^{60} \approx 0.9502$

76. $\int_{t_1}^{t_2} \frac{1}{3}e^{-t/3}\,dt = -e^{-t/3}\Big]_{t_1}^{t_2}$

0 – 2	2 – 4	4 – 6	6 – 8	8 – 10
0.4866	0.2498	0.1283	0.0658	0.0338

77. (a) $P = \frac{1}{100}\left(25 + \int_{2.5}^{10} \frac{25}{x}\,dx\right)$

$= \frac{1}{100}\left(25 + \left[25\ln x\right]_{2.5}^{10}\right)$

$= \frac{1}{4}(1 + \ln 4) \approx 0.60$

(b) $P = \frac{1}{100}\left(50 + \int_{5}^{10} \frac{50}{x}\,dx\right)$

$= \frac{1}{100}\left(50 + \left[50\ln x\right]_{5}^{10}\right)$

$= \frac{1}{2}(1 + \ln 2) \approx 0.85$

78. $A = A_0 e^{kt}$

$A_0 = 500$

When $t = 40$: $A = 300$

$300 = 500e^{40k}$

$\ln \frac{3}{5} = 40k$

$k \approx -0.0128$

$A = 500e^{-0.0128t}$

79. $\frac{dv}{dt} = -kv$

$v = v_0 e^{-kt}$

$s = \int v_0 e^{-kt}\,dt = \frac{-v_0}{k}(e^{-kt}) + C$

$s(0) = 0 = \frac{-v_0}{k} + C$

$C = \frac{v_0}{k}$

$s = \frac{-v_0}{k}(e^{-kt}) + \frac{v_0}{k} = \frac{v_0}{k}(1 - e^{-kt})$

80. $\int \frac{1}{kv - 32}\,dv = \int dt$

$\frac{1}{k}\ln|kv - 32| = t + C_1$

$kv - 32 = Ce^{kt}$

$kv = 32 + Ce^{kt}$

$v(t) = \frac{1}{k}(32 + Ce^{kt})$

$v(0) = \frac{1}{k}(32 + C) = 0 \Rightarrow C = -32$

$v(t) = \frac{32}{k}(1 - e^{kt})$

Note that $k < 0$ since the object is moving downward.

81. From Exercise 80, we have $v(t) = (1/k)(32 + Ce^{kt})$.

$v(0) = \frac{1}{k}(32 + C) = -20 \Rightarrow C = -20k - 32$

$v(t) = \frac{1}{k}[32 - (20k + 32)e^{kt}]$

82. $v(t) = \frac{32}{k}(1 - e^{kt}),\quad k < 0$

$\lim_{t\to\infty} \frac{32}{k}(1 - e^{kt}) = \frac{32}{k}(1 - 0) = \frac{32}{k}$

The velocity does not increase indefinitely but approaches a limiting velocity of $(32/k)$ ft/sec.

83. $s(t) = \frac{32}{k}\int (1 - e^{kt})\,dt = \frac{32}{k}\left(t - \frac{1}{k}e^{kt}\right) + C$

$s(0) = \frac{32}{k}\left(-\frac{1}{k}\right) + C = s_0 \Rightarrow C = s_0 + \frac{32}{k^2}$

$s(t) = \frac{32}{k}\left(t - \frac{1}{k}e^{kt}\right) + s_0 + \frac{32}{k^2}$

$= \frac{32t}{k} + \frac{32}{k^2}(1 - e^{kt}) + s_0$

84. $y = Ce^{-0.012(s)}$, $28 = Ce^{-0.6}$, $C = 28e^{0.6} \approx 51.02$

$y = 28e^{0.6 - 0.012s}$

85. $\lim_{x\to 1}\left[\frac{(\ln x)^2}{x - 1}\right] = \lim_{x\to 1}\left[\frac{2(1/x)\ln x}{1}\right] = 0$

86. $\lim_{x\to k}\left(\frac{x^{1/3} - k^{1/3}}{x - k}\right) = \lim_{x\to k}\left(\frac{1/3x^{-2/3}}{1}\right) = \frac{1}{3\sqrt[3]{k^2}}$

87. $\lim_{x\to\infty} \frac{e^{2x}}{x^2} = \lim_{x\to\infty} \frac{2e^{2x}}{2x} = \lim_{x\to\infty} \frac{4e^{2x}}{2} = \infty$

88. $\lim_{x\to 1^+} \left(\frac{2}{\ln x} - \frac{3}{x-1}\right) = \lim_{x\to 1^+} \left[\frac{2x - 2 - 3\ln x}{(\ln x)(x-1)}\right]$

$= \lim_{x\to 1^+} \left[\frac{2 - (3/x)}{(x-1)(1/x) + \ln x}\right]$

$= -\infty$

89. $y = \lim_{x\to\infty} (\ln x)^{2/x}$

$\ln y = \lim_{x\to\infty} \frac{2\ln(\ln x)}{x} = \lim_{x\to\infty} \left[\frac{\frac{2}{x\ln x}}{1}\right] = 0$

Since $\ln y = 0$, $y = 1$.

90. $y = \lim_{x\to 1} (x-1)^{\ln x}$

$\ln y = \lim_{x\to 1} \left[(\ln x)\ln(x-1)\right]$

$= \lim_{x\to 1} \left[\frac{\ln(x-1)}{\frac{1}{\ln x}}\right] = \lim_{x\to 1} \left[\frac{\frac{1}{x-1}}{\left(\frac{1}{x}\right)\frac{-1}{\ln^2 x}}\right] = \lim_{x\to 1} \left[\frac{-\ln^2 x}{\frac{x-1}{x}}\right] = \lim_{x\to 1} \left[\frac{-2\left(\frac{1}{x}\right)(\ln x)}{\frac{1}{x^2}}\right] = \lim_{x\to 1} 2x(\ln x) = 0$

Since $\ln y = 0$, $y = 1$.

91. $\lim_{n\to\infty} 1000\left(1 + \frac{0.09}{n}\right)^n = 1000e^{0.09} \approx 1094.17$

92. $\lim_{x\to\infty} xe^{-x^2} = \lim_{x\to\infty} \frac{x}{e^{x^2}} = \lim_{x\to\infty} \frac{1}{2xe^{x^2}} = 0$